Global Environment Outlook 2000

Global Environment Outlook 2000

Earthscan Publications Ltd, London

First published in the United Kingdom in 1999 by Earthscan Publications Ltd
for and on behalf of the United Nations Environment Programme

ISBN: 1 85383 588 9 (paperback)
1 85383 587 0 (hardback)
ISSN: 0 1366 8080

United Nations Environment Programme
PO Box 30552, Nairobi, Kenya
Tel: +254 2 621234
Fax: +254 2 623943/44
E-mail: geo@unep.org
http://www.unep.org

Printed and bound by Polestar, Exeter

Earthscan Publications Ltd
120 Pentonville Road
London N1 9JN, United Kingdom
Tel: +44 (0)171 278 0433
Fax: +44 (0)171 278 1142
E-mail: earthinfo@earthscan.co.uk
http://www.earthscan.co.uk

Earthscan is an editorially independent subsidiary of Kogan Page Ltd and publishes in association with WWF-UK and the International Institute for Environment and Development.

This book is printed on 100 per cent recycled, chlorine free paper.

GEO-2000

UNEP's Millennium Report on the Environment

Acknowledgements

UNEP acknowledges the contributions made by the many individuals and institutions that have contributed to *Global Environment Outlook 2000*. A full list of names is included on page 380. Special thanks are extended to:

GEO-2000 collaborating centres

Arab Centre for the Studies of Arid Zones and Drylands (ACSAD), Republic of Syria

Arabian Gulf University (AGU), Bahrain

Asian Institute of Technology (AIT), Thailand

Bangladesh Centre for Advanced Studies (BCAS), Bangladesh

Central European University (CEU), Hungary

Centre for Environment and Development for the Arab Region and Europe (CEDARE), Egypt

European Environment Agency (EEA), Denmark

Instituto Brasileiro do Meio Ambiente e dos Recursos Naturais Renováveis (IBAMA), Brazil

International Institute for Sustainable Development (IISD), Canada

Moscow State University (MSU), Russian Federation

National Institute for Environmental Studies (NIES), Japan

National Institute of Public Health and the Environment (RIVM), Netherlands

Network for Environment and Sustainable Development in Africa (NESDA), Côte d'Ivoire

Regional Environmental Centre for Central and Eastern Europe (REC), Hungary

Southern African Research and Documentation Centre (SARDC), Zimbabwe

State Environmental Protection Administration (SEPA), China

Stockholm Environment Institute (SEI), Sweden, United Kingdom and United States

Tata Energy Research Institute (TERI), India

Thailand Environment Institute (TEI), Thailand

University of Chile, Sustainable Development Programme, Chile

University of Costa Rica, Development Observatory, Costa Rica

World Resources Institute (WRI), United States

GEO-2000 associated centres

African Centre for Technology Studies (ACTS), Kenya

Asociación Latinoamericana de Derecho Ambiental (ALDA), Mexico

Centro Internacional de Agricultura Tropical (CIAT), Colombia

Commission for Environmental Cooperation (CEC) of the North American Agreement on Environmental Cooperation (NAAEC), Canada

Earth Council, Costa Rica

UNEP's Global Resource Information Database Centre in Arendal, Norway; Christchurch, New Zealand; Geneva, Switzerland; and Sioux Falls, United States

National Environment Management Authority (NEMA), Uganda

Indian Ocean Commission (IOC), Mauritius

Scientific Committee on Problems of the Environment (SCOPE) of the International Council for Science (ICSU), France

South Pacific Regional Environment Programme (SPREP), Samoa

University of the West Indies, Centre for Environment and Development, Jamaica

World Conservation Monitoring Centre (WCMC), United Kingdom

Funding

The Government of the Netherlands (Ministry for Development Cooperation) provided funding to support the involvement of collaborating centres in developing countries.

Global Environment Outlook 2000: the production team

GEO Coordinating Team
Marion Cheatle
Miriam Schomaker
Megumi Seki
Veerle Vandeweerd
Kaveh Zahedi

GEO Support Team
Berna Bayinder, Ulf Carlsson, Choudhury Rudra Charan Mohanty, Dan Claasen, Garth Edward, Arthur Dahl, Norberto Fernandez, David Henry, Bob Kakuyo, Danielle Mitchell, Surendra Shrestha, Ashbindu Singh, Anna Stabrawa and Ron Witt

Editor
Robin Clarke

Graphics
Philippe Rekacewicz

Cover and page design
Paul Sands

Editorial assistants
Isabelle Fleuraud, Peter Saunders

Contents

List of illustrations

List of boxes

List of tables

Foreword

UNEP's *Global Environment Outlook 2000* is a unique product of a unique process. Prepared with the participation of more than 850 individuals around the world, and in collaboration with more than 30 environmental institutes as well as other United Nations agencies, the resulting report presents a comprehensive integrated assessment of the global environment at the turn of the millennium. It is a summing up of where we have reached to date as users and custodians of the environmental goods and services provided by our planet. It is also a forward-looking document, providing a vision into the 21st century

GEO-2000 shows that even as we struggle with the traditional environmental problems, new ones continue to emerge. We are still grappling with environmental issues such as pollution of freshwater resources, atmospheric pollution, extinction of biodiversity and urbanization. And today we are being asked to sort out, with an unprecedented urgency, the long-term effects on our climate of the atmospheric build-up of greenhouse gases. We are being asked to consider the potential impacts of genetically-modified organisms. And we are trying to understand and deal with the rush of exposure to synthetic chemicals.

In charting a new course for global environmental policy, we must look for the causes of these environmental problems. *GEO-2000* identifies many underlying causes including consumption patterns that continue to be unsustainable in many parts of the world, high population densities that place impossible demands on the environmental resources available and armed conflicts causing environmental stress and degradation, locally and regionally. *GEO-2000* further acknowledges the efforts being made to halt environmental deterioration but recognizes that many of these are too few and too late; signs of improvements are few and far between. This is further exacerbated by the low priority that continues to be afforded to the environment in national and regional planning, and the sparse funding the environment receives in relation to other areas.

There are, however, positive signs in all this: cleaner production and promotion of more sustainable ways of producing energy, including increased efficiency; raised environmental awareness among the public, leading to new action and initiatives at all levels of society; and innovative local solutions to environmental problems in almost all countries of the world.

The number of policy responses is growing, and they are increasing in effectiveness. *GEO-2000* documents many policy successes that have been recorded in all continents in the recent past.

GEO-2000 stresses the need for more comprehensive, integrated policy-making. In itself this call is not new. But it gains urgency in view of the increasingly cross-cutting nature of environmental issues. Thus, rather than trying to tackle issues such as deforestation and land degradation on a piecemeal basis, these must be integrated and in turn be connected with the needs and aspirations of the people.

It is usually impossible to determine which policy contributes to what change in the state of the environment, and furthermore there are few mechanisms, concepts, methodologies or criteria for making these policy assessments. Linkages between human actions and environmental outcomes are still poorly understood. A more complete and precise analysis will require the development of more comprehensive and long-term mechanisms for monitoring and assessing the effects of environmental policies on environmental quality.

These current limitations pose a new challenge to the GEO assessment process. *GEO-2000* is the culmination of a participatory process involving the work of experts from more than 100 countries. 'Our goal', we wrote in *GEO-1*, 'is that by the turn of the century a truly global participatory assessment process may be in operation to effectively keep under review the state of the world's environment, as well as to guide international policy setting'. I am pleased to be able to report that we have achieved this goal with a few months to spare.

While commending UNEP's millennium report on the state of the environment to you, I emphasize also that *GEO-2000* is but one output of a process designed to provide a continual assessment of the global state of the environment. Even as this book is launched, we are strengthening assessment capabilities, preparing a range of associated products and beginning work on the next issue in the GEO series.

Klaus Töpfer
United Nations Under-Secretary General and Executive Director,
United Nations Environment Programme

The GEO Project

In response to the need for comprehensive, integrated, policy-relevant assessments of the global environment, UNEP launched the Global Environment Outlook (GEO) Project in 1995. The GEO Project has two main components:

- A global environmental assessment process, the GEO Process, that is cross-sectoral and participatory. It incorporates regional views and perceptions, and builds consensus on priority issues and actions through dialogue among policy-makers and scientists at regional and global levels.
- GEO outputs, in printed and electronic formats, including the GEO Report series. This series makes periodic reviews of the state of the world's environment, identifying major environmental concerns, trends and emerging issues together with their causes, and their social and economic impacts. It also provides guidance for decision-making processes such as the formulation of environmental policies, action planning and resource allocation. Other outputs include technical reports, a Web site and a publication for young people.

The GEO process

The GEO Process is a collaborative effort involving and supported by a range of partners around the world. A coordinated network of Collaborating Centres forms the core of the process. GEO Collaborating Centres are multidisciplinary institutes with a regional outlook that work at the interface between science and policy. They undertake studies with the dual aim of keeping the state of the region and world environment under review and providing guidance to regional and international policy setting and planning for sustainable development. The Centres work with other institutions in their region to bring together the required expertise to cover all the environmental sectors relevant to sustainable development.

The Collaborating Centres have played an increasingly important role in preparing GEO reports as the GEO process has progressed. They are now responsible for almost all the regional inputs, thus combining top-down integrated assessment with bottom-up environmental reporting. To promote these activities and improve regional capacities, a training component on integrated environmental assessment and reporting has been developed within the GEO process.

A number of Associated Centres also participate in the GEO process. According to their specialized areas of expertise, they contribute to the assessment and analysis activities as well as providing specific input to GEO reports and related products.

Four working groups – on modelling, scenarios, policy and data – provide advice and support to the GEO process. Composed of experts from around the world, they help coordinate the work of the Collaborating Centres and advise on the use of methodologies to make outputs from the Centres as

GEO project organization and outputs

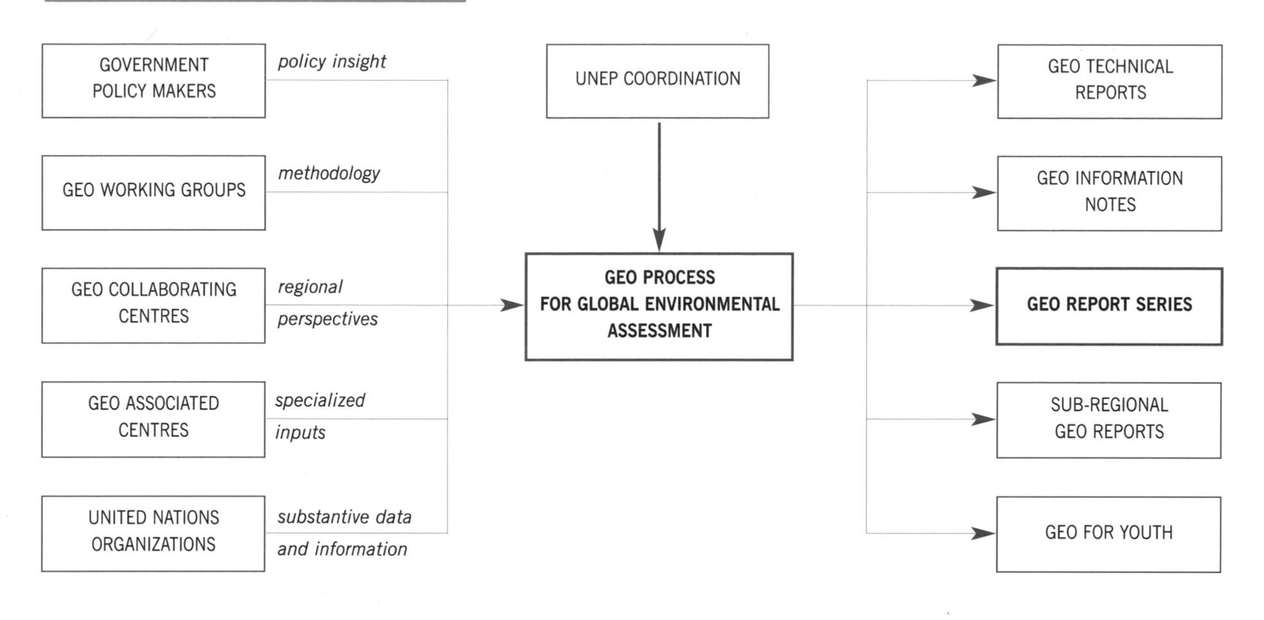

comparable as possible.

Other United Nations agencies contribute to the GEO Process through the United Nations System-wide Earthwatch, coordinated by UNEP. In particular, they provide substantive data and information on the many environmentally-related issues that fall under their individual mandates; they also help review drafts.

An essential component of the GEO Process is the set of regional consultations and other consultative mechanisms designed to promote and contribute to a regular dialogue between scientists and policy makers. These consultations help to guide the GEO Process and, in the framework of the GEO report, are used to review draft material and ensure that the report is geared towards policy formulation and action planning.

Many of the activities that take place within the GEO process are funded from sources other than UNEP. It is this external support that makes worldwide participation in GEO a reality. More than 800 people have contributed to the production of this report (see page 380).

GEO outputs

The GEO Report Series is the main output of the GEO Process. *GEO-1*, the first report in the series, was published in January 1997. It reviewed major environmental issues from regional and global perspectives, and made an initial evaluation of some of the existing policy responses that address priority environmental concerns. The second in the series, *GEO-2000,* addresses three main areas: the state of the environment; trends and progress in policy development, including multilateral environmental agreements; and the future, with a focus on emerging environmental issues and region-specific alternative policies.

A range of other outputs is prepared within the GEO Process and in association with the GEO Reports. These include technical and meeting reports, booklets, extracts and information notes. Some of these are available on the Internet (http://www.unep.org/geo2000) as well as in hard copy, and some are available in the other official languages of the United Nations (Arabic, Chinese, French, Russian and Spanish). Some of the associated outputs of particular relevance to *GEO-2000* are:

- technical reports on region-specific alternative policy studies, global datasets and emerging environmental issues for the 21st century;
- Environment Outlooks for Small Island Developing States in the Caribbean, Indian Ocean and the Pacific; and
- *Pachamama: Our Earth, Our Future* – a book for young people based on the GEO reports.

The Data Issue

Integrated environmental assessment poses many questions:

- what are the main environmental trends?
- what are the environmental impacts of policies?
- are environmental policies meeting their targets and achieving their objectives?

Answering these types of questions requires accurate, harmonized, time-series datasets on a wide range of issues, with appropriate resolution and geographical coverage, which can be turned into policy-relevant information.

Data are produced by ground- and space-based monitoring networks and statistical surveys carried out by national, regional and international organizations. These core datasets are often further processed to derive indicators for easier communication, increased policy relevance, and further use in analytical work and reporting. Addressing data issues is not simply a technical and methodological question; it also involves monitoring capacity and reporting, and broad international coordination and consensus.

Although core environmental datasets are improving, expanding and becoming more easily accessible, numerous inconsistencies and shortcomings still exist. Even the straightforward mapping of such basic indicators as current GDP, water consumption and fertilizer use is difficult. In particular, little information is yet available on environmental impacts on human health and natural ecosystems, social response and policy effectiveness. The conversion, integration and moulding of data to information is a complex process. In part, the difficulties persist because there is not yet enough feedback and validation during the process that connects collectors, collators, disseminators and end-users.

Since *GEO-1*, steps were taken within the GEO Process to identify and analyse existing global datasets in order to improve the data underlying GEO reports. The main finding of this exercise was that missing data and data of uncertain quality are seriously hindering integrated environmental assessment at global and regional levels. As future GEO assessment requires more detail on selected issues, the demands on input data and constraints (coverage, reliability, resolution, frequency, timeliness) are expected to become more pronounced. More specific findings of the GEO data exercise are given below.

Data quality and availability

The lack of relevant data is a common experience. Within the environmental domain, there are still serious data gaps related to, for example, pesticide application, the state of fish stocks, forest quality, groundwater and biological diversity. The quality of existing data is of equal concern. Causes of data gaps and poor data quality are complex and diverse.

There are inherent challenges in working with datasets on a global scale. From the perspective of GEO as a high-level global assessment, linkages of data across scale are particularly important. Given that in general only data with the same definition, standards and date of measurement can be safely aggregated to a regional or global level, even small discrepancies or gaps can make datasets incomplete or otherwise deficient. On the other hand, even with good quality data, aggregation and averaging may mask important spatial or temporal diversity. In large-scale aggregations, issues unique to smaller regions disappear. Therefore the scale of aggregation and reporting of averages should be carefully matched with the scale of environmental issues or policies and the purpose of assessment.

Most of the available data apply to quantitative attributes of the environment. While measuring qualitative variables is usually more difficult, it is often through qualitative change that major trends can be detected. Monitoring ecosystem quality – for example, for forests or fisheries – needs to be improved.

Some new global or regional compendia of environment-related data have considerably improved the global stock of data resources. Notable examples are the Dobříš data compilations in Europe and the World Bank's World Development Indicators. In addition, a small but steadily growing number of countries has set up systematic compilations of environmental data, in part following the guidelines of the United Nations Statistical Office. This is resulting in national environmental reports being issued by more countries, and in the gradual improvement and harmonization of reporting to the Commission on Sustainable Development (CSD) and within the frameworks of multilateral environmental agreements. The relatively widespread testing by countries of the CSD indicator methodology may well see the demand for input data developing and becoming more concrete.

Major institutional and technical constraints currently affecting data issues are listed in the tables on this and the following page. These tables are based on the experience of GEO Collaborating Centres in both developed and developing countries.

Geo-referenced data and space-based observations

There is a gradual recognition of the need to use geo-referenced data in environmental assessment. The same holds true for the need to have some information broken down by spatial units other than administrative units. Some important global, geo-referenced data sets, such as population and landcover, have been produced in the past few years. However, this should be regarded only as a beginning and few if any of these new datasets seem to be routinely updated.

Institutional constraints affecting data issues

General institutional constraints	The monitoring and data collection infrastructure of most developing countries is severely handicapped or non-existent due to limitations in resources, personnel and equipment. Constraints are also faced by international organizations. Keeping well-trained personnel in publicly-funded institutions is difficult. In some cases, there is no organization mandated to collect and report time-series data internationally on specific issues on a regular basis.
Data reporting units	Data are reported for different geographical areas by different agencies and organizations. As a result, it may be impossible to use and compare otherwise valuable aggregated datasets in global and regional assessments.
Data management	The data management infrastructure of many countries is weak and data reporting is fragmented. Without a central compiling system, environmental data may remain scattered across many sectoral organizations and departments.
Relevance	Many issues are not universally relevant. In such cases, not all countries will collect associated data and global datasets will therefore be incomplete.

Technical constraints affecting data issues

Definition differences	In some cases the definition of what is being measured is vague and open to misinterpretation. In other cases national reporting is simply incompatible with international standards. 'Wetlands', for instance, include different categories in different countries.
Coverage of monitoring networks	Collection of time-series data requires permanent monitoring networks with adequate geographic coverage and sufficient resources. Although the availability of remotely-sensed data has led to improvements in the cost, quality and availability of environmental data, remote sensing cannot entirely substitute for measurements on the ground.
Different reporting periods	Time-series data rarely match between countries or across a whole region. Essentially, the problem is that 1990 data, for instance, from one country cannot be compared with similar data from another country in 1995. Similarly, if data for different indicators exist for different time periods, their comparison is problematic.
Gap filling	Various statistical methods are used to fill data gaps and smooth curves. In addition, gaps are often filled with estimates provided by experts. Although in the absence of real data these methods are necessary, the risks of using them should be understood. Furthermore, they are clearly not substitutes for monitoring, measurement and the verification of data obtained through remote sensing and on the ground.
Conceptual and technical difficulties of measurement	Some variables are inherently difficult and/or costly to measure for large geographic areas. Two examples are the measurement of particulate matter in air and the measurement of biological diversity. Measuring the effectiveness of policy implementation may be equally challenging given that outcomes are often the results of several parallel policy actions. This makes the separation of the impact of one single policy from the others difficult.
Differences in measurement method	Frequently, there are underlying differences in data collection methods for data with the same label from different sources. Without going into a detailed analysis of data collection and measurement methods and standards, there is a risk that incompatible data may end up in aggregated datasets.

Use of satellite data for environment reporting has increased but the full potential remains untapped. The common belief that space observations will make ground-based measurements redundant is seldom justified; while space observations may reduce the need for conventional *in situ* measurements, they do not remove the need for direct reporting and ground truthing. More importantly, many of the data categories that are needed to draw up policy-oriented assessments (for example on resource efficiency, finance and impacts on human well-being) cannot be detected from space.

Data access

Data may be inaccessible because of copyright issues, high cost, professional jealousy or organizational competition. Although some parameters are accurately and routinely measured, the information may be classified or otherwise publicly unavailable. Difficulty of access to data on shared aquifers and surface water is an example which occurs in many parts of the world.

However, public and institutional attitudes towards access to data have changed noticeably during this decade. With Internet access becoming widespread, mass data processing cheaper and easier, and Cold-War style security no longer neccessary, the public has become more demanding and institutions more pro-active and open. This is true for a wide range of issues and organizations while the most symbolic event is the partial de-classification of military satellite imagery.

This opening-up of data holdings and data exchange brings two potential problems for their use in broad assessments such as GEO. First, access to essential data which is currently taken for granted may become more commercialized and therefore more difficult for multilateral organizations and other compilers of environment assessments. In particular, this applies to satellite data and to large integrated databases. Secondly, as data become more widely distributed and recycled, critical validation will become even more important than it is today, making good scientific links essential for GEO-type assessments.

Synthesis

Two over-riding trends characterize the beginning of the third millennium. First, the global human ecosystem is threatened by grave imbalances in productivity and in the distribution of goods and services. A significant proportion of humanity still lives in dire poverty, and projected trends are for an increasing divergence between those that benefit from economic and technological development, and those that do not. This unsustainable progression of extremes of wealth and poverty threatens the stability of the whole human system, and with it the global environment.

Secondly, the world is undergoing accelerating change, with internationally-coordinated environmental stewardship lagging behind economic and social development. Environmental gains from new technology and policies are being overtaken by the pace and scale of population growth and economic development. The processes of globalization that are so strongly influencing social evolution need to be directed towards resolving rather than aggravating the serious imbalances that divide the world today. All the partners involved – governments, intergovernmental organizations, the private sector, the scientific community, NGOs and other major groups – need to work together to resolve this complex and interacting set of economic, social and environmental challenges in the interests of a more sustainable future for the planet and human society.

While each part of the Earth's surface is endowed with its own combination of environmental attributes, each area must also contend with a unique, but interlinked, set of current and emerging problems. *GEO-2000* provides an overview of this range of issues. This synthesis provides a summary of the main conclusions of *GEO-2000*.

The state of the environment: a global overview

Climate change

In the late 1990s, annual emissions of carbon dioxide were almost four times the 1950 level and atmospheric concentrations of carbon dioxide had reached their highest level in 160 000 years. According to the Intergovernmental Panel on Climate Change, 'the balance of evidence suggests that there is a discernible human influence on global climate'. Expected results include a shifting of climatic zones, changes in species composition and the productivity of ecosystems, an increase in extreme weather events and impacts on human health.

Through the United Nations Framework Convention on Climate Change and the Kyoto Protocol, efforts are under way to start controlling and reducing greenhouse gas emissions. During the Third Conference of the Parties in Buenos Aires in 1998, a plan of action was developed on how to use the new international policy instruments such as emission trading and the Clean Development Mechanism. However, the Kyoto Protocol alone will be insufficient to stabilize carbon dioxide levels in the atmosphere.

Stratospheric ozone depletion

Major reductions in the production, consumption and release of ozone-depleting substances (ODS) have been, and continue to be, achieved by the Montreal Protocol and its related amendments. The abundance of ODS in the lower atmosphere peaked in about 1994 and is now slowly declining. This is expected to bring about a recovery of the ozone layer to pre-1980 levels by around 2050.

Illegal trading, still a problem, is being addressed by national governments but substantial quantities of ODS are still being smuggled across national borders. The Multilateral Fund and the Global Environment Facility are helping developing countries and countries in transition to phase out ODS. Since 1 July 1999, these countries have, for the first time, had to start meeting obligations under the Montreal Protocol.

Nitrogen loading

We are fertilizing the Earth on a global scale through intensive agriculture, fossil fuel combustion and widespread cultivation of leguminous crops. Evidence is growing that the huge additional quantities of nitrogen being used are exacerbating acidification, causing changes in the species composition of ecosystems, raising nitrate levels in freshwater supplies above acceptable limits for human consumption and causing eutrophication in many freshwater habitats. In addition, river discharges laden with nitrogen-rich sewage and fertilizer run-off tend to stimulate algal blooms in coastal waters, which can lead to oxygen starvation and subsequent fish kills at lower depths, and reduce marine biodiversity through competition. Nitrogen emissions to the atmosphere contribute to global warming. Consensus among researchers is growing that the scale of disruption to the nitrogen cycle may have global implications comparable to those caused by disruption of the carbon cycle.

Chemical risks

With the massive expansion in the availability and use of chemicals throughout the world, exposure to pesticides, heavy metals, small particulates and other substances poses an increasing threat to the health of humans and their environment. Pesticide use causes 3.5 to 5 million acute poisonings a year. Worldwide, 400 million tonnes of hazardous waste are generated each year. About 75 per cent of pesticide use and hazardous waste generation occurs in developed countries. Despite restrictions on toxic and persistent chemicals such as DDT, PCBs and dioxin in many developed countries, they are still manufactured for export and remain widely used in developing countries. Efforts are under way to promote cleaner production, to limit the emissions and phase out the use of some persistent organic pollutants, to control waste production and trade, and improve waste management.

Disasters

The frequency and effects of natural disasters such as earthquakes, volcanic eruptions, hurricanes, fires and floods are increasing. This not only affects the lives of millions of people directly, through death, injury and economic losses, but adds to environmental problems. As just one example, in 1996-98 uncontrolled wildfires swept through forests in Brazil, Canada, China's north-eastern Inner Mongolia Autonomous Region, France, Greece, Indonesia, Italy, Mexico, Turkey, the Russian Federation and the United States. The health impacts of forest fires can be serious. Experts consider a pollution index of 100 $\mu g/m^3$ unhealthy; in Malaysia, the index reached 800 $\mu g/m^3$. The estimated health cost of forest fires to the people of Southeast Asia was US$1 400 million. Fires are also a serious threat to biodiversity, especially when protected areas are burnt. Early warning and response systems are still weak, particularly in developing countries; there is an urgent need for improved information infrastructures and increased technical response capabilities.

El Niño

Unusual weather conditions over the past two years are also attributed to the *El Niño* Southern Oscillation (ENSO). The 1997/98 *El Niño* developed more quickly and resulted in higher temperatures in the Pacific Ocean than ever recorded before. The presence of this mass of warm water dominated world climate patterns up to mid-1998, causing substantial disruption and damage in many areas, including temperate zones. Extreme rainfall and flooding, droughts and forest fires were among the major impacts. Forecasting and early warning systems, together with human, agricultural and infrastructural

Sea surface temperature anomalies in January 1998, at the height of the 1997/98 *El Niño* – see page 33

protection, have been substantially improved as a result of the most recent *El Niño*.

Land, forests and biodiversity

Forests, woodlands and grasslands are still being degraded or destroyed, marginal lands turned into deserts, and natural ecosystems reduced or fragmented, further threatening biodiversity. New evidence confirms that climate change may further aggravate soil erosion in many regions in the coming decades, and threaten food production. Deforestation continues at high rates in developing countries, mainly driven by the demand for wood products and the need for land for agriculture and other purposes. Some 65 million hectares of forest were lost between 1990 and 1995, out of a total of 3500 million hectares. An increase of 9 million hectares in the developed world only slightly offset this loss. The quality of the remaining forest is threatened by a range of pressures including acidification, fuelwood and water abstraction, and fire. Reduced or degraded habitats threaten biodiversity at gene, species and ecosystems level, hampering the provision of key products and services. The widespread introduction of exotic species is a further major cause of biodiversity loss. Most of the threatened species are land-based, with more than half occurring in forests. Freshwater and marine habitats, especially coral reefs, are also very vulnerable.

Freshwater

Rapid population growth combined with industrialization, urbanization, agricultural intensification and water-intensive lifestyles is resulting in a global water crisis. About 20 per cent of the population currently lacks access to safe drinking water, while 50 per cent lacks access to a safe sanitation system. Falling water tables are widespread and cause serious problems, both because they lead to water shortages and, in coastal areas, to salt intrusion. Contamination of drinking water is mostly felt in megacities, while nitrate pollution and increasing loads of heavy metals affect water quality nearly everywhere. The world supply of freshwater cannot be increased; more and more people depend on this fixed supply; and more and more of it is polluted. Water security, like food security, will become a major national and regional priority in many areas of the world in the decades to come.

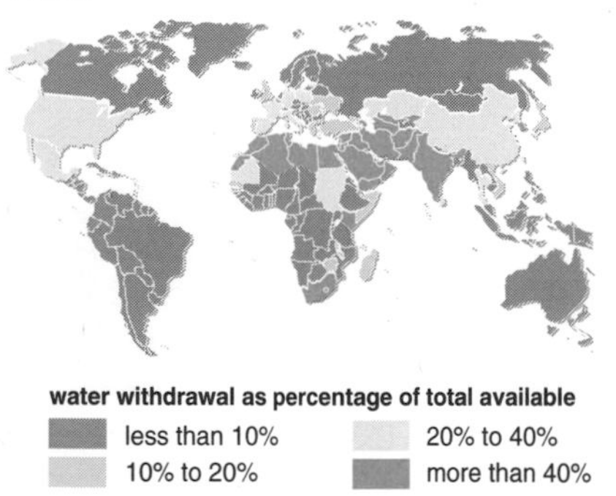

By the year 2025, as much as two-thirds of the world population may be subject to moderate to high water stress – see page 42

Marine and coastal areas

Urban and industrial development, tourism, aquaculture, waste dumping and discharges into marine areas are degrading coastal areas around the world and destroying ecosystems such as wetlands, mangroves and coral reefs. Climatic changes also affect the quality of ocean water as well as sea levels. Low-lying areas, including many small islands, risk inundation. The global marine fish catch almost doubled between 1975 and 1995, and the state of the world's fisheries has now reached crisis point. About 60 per cent of the world's fisheries are at or near the point at which yields decline.

Atmosphere

There is a major difference between air pollution trends in developed and developing countries. Strenuous efforts have begun to abate atmospheric pollution in many industrialized countries but urban air pollution is reaching crisis dimensions in most large cities of the developing world. Road traffic, the burning of coal and high-sulphur fuels, and forest fires are the major causes of air pollution. People in developing countries are also exposed to high levels of indoor pollutants from open fires. Some 50 per cent of chronic respiratory illness is now thought to be associated with air pollution. Large areas of forest and farmland are also being degraded by acid rain.

Urban impacts

Many environmental problems reinforce one another in small, densely-populated areas. Air pollution, garbage, hazardous wastes, noise and water contamination turn these areas into environmental hot spots. Children are the most vulnerable to the inevitable health risks. Some 30–60 per cent of the urban population in low-income countries still lack adequate housing with sanitary facilities, drainage systems and piping for clean water. Continuing urbanization and industrialization, combined with a lack of resources and expertise, are increasing the severity of the problem. However, many local authorities are now joining forces to promote the concept of the sustainable city.

Policy responses: a global overview

As awareness of environmental issues and their causes develops, the focus of policy questions shifts towards the policy response itself: what is being done, is it adequate and what are the alternatives? *GEO-2000* includes a unique assessment of environmental policies worldwide.

Environmental laws and institutions have been strongly developed over the past few years in almost all countries. Command and control policy via direct regulation is the most prominent policy instrument but its effectiveness depends on the manpower available, methods of implementation and control, and level of institutional coordination and policy integration. In most regions, such policies are still organized by sector but environmental planning and environmental impact assessment are becoming increasingly common everywhere.

While most regions are now trying to strengthen their institutions and regulations, some are shifting towards deregulation, increased use of economic instruments and subsidy reform, reliance on voluntary action by the private sector, and more public and NGO participation. This development is fed by the increasing complexity of environmental regulation and high control costs as well as demands from the private sector for more flexibility, self-regulation and cost-effectiveness.

GEO-2000 confirms the overall assessment of *GEO-1*: the global system of environmental management is moving in the right direction but much too slowly. Yet effective and well tried policy instruments do exist that could lead much more quickly to sustainability. If the new millennium is not to be marred by major environmental disasters, alternative policies will have to be swiftly implemented.

One of the major conclusions of the policy review concerns the implementation and effectiveness of existing policy instruments. The assessment of implementation, compliance and effectiveness of policy initiatives is complicated and plagued by gaps in data, conceptual difficulties and methodological problems.

Multilateral environmental agreements (MEAs) have proven to be powerful tools for attacking environmental problems. Each region has its own regional and sub-regional agreements, mostly relating to the common management or protection of natural resources such as water supply in river basins and transboundary air pollution. There are also many global-level agreements, including those on climate change and biodiversity that resulted from the United Nations Conference on Environment and Development, held in Rio de Janeiro, Brazil, in 1992.

In addition to the binding MEAs, there are non-binding agreements (such as *Agenda 21*) and environmental clauses or principles in wider agreements (such as regional trade treaties). A major trend in MEAs over the years has been a widening focus from issue-specific approaches (such as provisions for shared rivers) to trans-sectoral approaches (such as the Basel Convention), to globalization and to the general recognition of the linkage between environment and development. Another trend is still unfolding: the step-by-step establishment of common principles (such as the Forest Principles) in different sectors.

The *GEO-2000* review of MEAs highlights two issues:

- the effectiveness of MEAs depends strongly on the institutional arrangements, the financial and compliance mechanisms, and the enforcement systems that have been set up for them;
- it is still difficult to assess accurately the effectiveness of MEAs and non-binding instruments because of the lack of accepted indicators.

Regional trends

Africa

Poverty is a major cause and consequence of the environmental degradation and resource depletion that threaten the region. Major environmental challenges include deforestation, soil degradation and desertification, declining biodiversity and marine resources, water scarcity, and deteriorating water and air quality. Urbanization is an emerging issue, bringing with it the range of human health and environmental problems well known in urban areas throughout the world. Growing ‘environmental debts’ in many countries are a major concern because the cost of remedial action will be far greater than preventive action.

Although many African countries are implementing new national and multilateral environmental policies, their effectiveness is often low due to lack of adequate staff, expertise, funds and equipment for implementation and enforcement. Current environmental policies are mainly based on regulatory instruments but some countries have begun to consider a broader range, including economic incentives implemented through different tax systems. Although cleaner production centres have been created in a few countries, most industries have made little effort to adopt cleaner production approaches. However, some multinational corporations, large-scale mining companies and even local enterprises have recently voluntarily adopted precautionary environmental standards.

There is growing recognition that national environmental policies are more likely to be effectively implemented if they are supported by an informed and involved public. Environmental awareness and education programmes are expanding almost everywhere, while indigenous knowledge receives greater recognition and is increasingly used. Environmental information systems are still weak.

There is fairly high interest in many of the global MEAs, and several regional MEAs have been developed to support the global ones. The compliance and implementation rate is, however, quite low, mainly due to lack of funds.

Asia and the Pacific

Asia and the Pacific is the largest region and it is facing serious environmental challenges. High population densities are putting enormous stress on the environment. Continued rapid economic growth and industrialization is likely to cause further environmental damage, with the region becoming more degraded, less forested, more polluted and less ecologically diverse in the future.

The region, which has only 30 per cent of the world's land area, supports 60 per cent of the world population. This is leading to land degradation, especially in marginal areas, and habitat fragmentation. Increasing habitat fragmentation has depleted the wide variety of forest products that used to be an important source of food, medicine and income for indigenous people. Forest fires caused extensive damage in 1997–98.

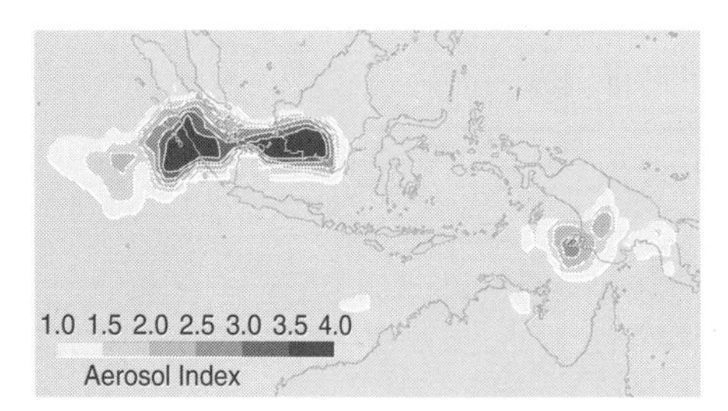

Smoke haze over Indonesia on 19 October 1997, caused by forest fires – see page 90

Water supply is a serious problem. Already at least one in three Asians has no access to safe drinking water and freshwater will be the major limiting factor to producing more food in the future, especially in populous and arid areas. Energy demand is rising faster than in any other part of the world. The proportion of people living in urban centres is rising rapidly, and is focused on a few urban centres. Asia's particular style of urbanization – towards megacities – is likely to increase environmental and social stresses.

Widespread concern over pollution and natural resources has led to legislation to curb emissions and conserve natural resources. Governments have been particularly active in promoting environmental compliance and enforcement although the latter is still a problem in parts of the region. Economic incentives and disincentives are beginning to be used for environmental protection and the promotion of resource efficiency. Pollution fines are common and deposit-refund schemes are being promoted to encourage reuse and recycling. Industry groups in both low- and high-income countries are becoming increasingly sensitive to environmental concerns over industrial production. There is keen interest in ISO14 000 standards for manufacturing and in eco-labelling.

In most countries, domestic investment in environmental issues is increasing. A major thrust, particularly among developing countries, is on water supply, waste reduction and waste recycling. Environment funds have also been established in many countries and have contributed to the prominent role that NGOs now play in environmental action. Many countries are in favour of public participation, and in some this is now required by law. However, education and awareness levels amongst the public are often low, and the environmental information base in the region is weak.

Whilst there is uneven commitment to global MEAs, regional MEAs are important. They include a number of important environmental policy initiatives developed by sub-regional cooperative mechanisms.

One of the greatest challenges is to promote liberal trade yet maintain and strengthen the protection of the environment and natural resources. Some governments are now taking action to reconcile trade and environmental interests through special policies, agreements on product standards, enforcement of the Polluter Pays Principle, and the enforcement of health and sanitary standards for food exports.

Europe and Central Asia

Environmental trends reflect the political and socio-economic legacy of the region. In Western Europe, overall consumption levels have remained high but measures to curb environmental degradation have led to considerable improvements in some, though not all, environmental parameters. Sulphur dioxide emissions, for example, were reduced by more than one-half between 1980 and 1995. In the other sub-regions, recent political change has resulted in sharp though probably temporary reductions in industrial activity, reducing many environmental pressures.

A number of environmental characteristics are common to much of the region. Large areas of forests are damaged by acidification, pollution, drought and forest fires. In many European countries, as much as half the known vertebrate species are under threat and most stocks of commercially-exploited fish in the North Sea have been seriously over-fished. More than half of the large cities in Europe are overexploiting their groundwater resources. Marine and coastal areas are susceptible to damage from a variety of sources. Road transport is now the main source of urban air pollution, and overall emissions are high – Western Europe produces nearly 15 per cent of global CO_2 emissions and eight of the ten countries with the highest per capita SO_2 emissions are in Central and Eastern Europe.

Regional action plans have been effective in forging policies consistent with the principles of sustainable development and in catalysing national and local action. However, some targets have yet to be met and plans in Eastern Europe and Central Asia are less advanced than elsewhere because of weak institutional capacities and the slower pace of economic restructuring and political reform.

Public participation in environmental issues is considered satisfactory in Western Europe, and there are some positive trends in Central and Eastern Europe. Many countries, however, still lack a proper legislative framework for public participation although the Convention on Access to Environmental Information and Public Participation in Environmental Decision Making signed by most of the ECE countries in 1998 should improve the situation. Access to environmental information has significantly increased with the formation of the European Environment Agency and other information resource centres in Europe. The level of support for global and regional MEAs, in terms of both ratification and compliance, is high.

There has been significant success, particularly in Western Europe, in implementing cleaner production programmes and eco-labelling. Within the European Union, green taxation and mitigating the adverse effects of subsidies are important priorities. Legislation is being adopted on entirely new subjects. Examples include the Nitrates Directive, the Habitat Directive and the Natura 2000 plan for a European Ecological Network. Implementation is, however, proving difficult.

The transition countries need to strengthen their institutional capacities, improve the enforcement of fees and fines, and build up the capacity of enterprises to introduce environmental management systems. The major challenge for the region as a whole is to integrate environmental, economic and social policies.

Latin America and the Caribbean

Two major environmental issues stand out in the region. The first is to find solutions to the problems of the urban environment – nearly three-quarters of the population are already urbanized, many in mega-cities. The air quality in most major cities threatens human health and water shortages are common. The second major issue is the depletion and destruction of forest resources, especially in the Amazon basin.

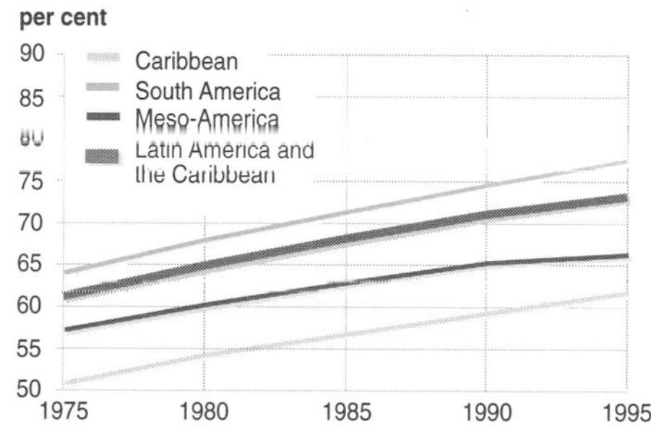

In Latin America and the Caribbean nearly 75 per cent of the population is already urbanized – see page 133

Natural forest cover continues to decrease in all countries. A total of 5.8 million hectares a year was lost during 1990–95, resulting in a 3 per cent total loss for the period. This is a major threat to biodiversity. More than 1 000 vertebrate species are now threatened with extinction.

The region has the largest reserves of cultivable land in the world but soil degradation is threatening much cultivated land. In addition, the environmental costs of improved farm technologies have been high. During the 1980s, Central America increased production by 32 per cent but doubled its consumption of pesticides. On the plus side, many countries have substantial potential for curbing their contributions to the build-up of greenhouse gases, given the region's renewable energy sources and the potential of forest conservation and reforestation programmes to provide valuable carbon sinks.

During the past decade, concern for environmental

issues has greatly increased, and many new institutions and policies have been put in place. However, these changes have apparently not yet greatly improved environmental management which continues to concentrate on sectoral issues, without integration with economic and social strategies. The lack of financing, technology, personnel and training and, in some cases, large and complex legal frameworks are the most common problems.

Most Latin American economies still rely on the growth of the export sector and on foreign capital inflows, regardless of the consequences to the environment. One feature of such policies is their failure to include environmental costs. Economic development efforts and programmes aimed at fighting poverty continue to be unrelated to environmental policy, due to poor inter-agency coordination and the lack of focus on a broader picture. On the industrial side, some producers have adopted ISO 14 000 standards as a means of demonstrating compliance with international rules.

An encouraging aspect is the trend towards regional collaboration, particularly on transboundary issues. For example, a Regional Response Mechanism for natural disasters has been established with telecommunication networks that link key agencies so that they can make quick assessments of damage, establish needs and mobilize resources to provide initial relief to affected communities. There is considerable interest in global and regional MEAs, and a high level of ratification. However, the level of implementing new policies to comply with these MEAs is generally low.

North America

North Americans use more energy and resources per capita than people in any other region. This causes acute problems for the environment and human health. The region has succeeded, however, in reducing many environmental impacts through stricter legislation and improved management. Whilst emissions of many air pollutants have been markedly reduced over the past 20 years, the region is the largest per capita contributor to greenhouse gases, mainly due to high energy consumption. Fuel use is high – in 1995 the average North American used more than 1600 litres of fuel a year (compared to about 330 litres in Europe). There is continuing concern about the effects of exposure to pesticides, organic pollutants and other toxic compounds. Changes to ecosystems caused by the introduction of non-indigenous species are threatening biodiversity and, in the longer term, global warming could move the ideal range for many North American forest species some 300 km to the north, undermining the utility of forest reserves established to protect particular plant and animal species. Locally, coastal and marine resources are close to depletion or are being seriously threatened.

The environmental policy scene is changing in North America. In Canada, most emphasis is on regulatory reform, federal/provincial policy harmonization and voluntary initiatives. In the United States, the impetus for introducing new types of environmental policies has increased and the country is developing market-based policies such as the use of tradeable emissions permits and agricultural subsidy reform. Voluntary policies and private sector initiatives, often in combination with civil society, are also gaining in importance. These include voluntary pollution reduction initiatives and programmes to ensure responsible management of chemical products. The region is generally active in supporting and complying with regional and global MEAs.

Public participation has been at the heart of many local resource management initiatives. Environmental policy instruments are increasingly developed in consultation with the public and the business community. Participation by NGOs and community residents is increasingly viewed as a valuable part of any environmental protection programme.

Increasing accountability and capacity to measure the performance of environmental policies is an overarching trend. Target setting, monitoring, scientific analysis and the public reporting of environmental policy performance are used to keep stakeholders involved and policies under control. Access to information has been an important incentive for industries to improve their environmental performance.

Despite the many areas where policies have made a major difference, environmental problems have not been eliminated. Economic growth has negated many of the improvements made so far and new problems – such as climate change and biodiversity loss – have emerged.

West Asia

The region is facing a number of major environmental issues, of which degradation of water and land resources is the most pressing. Groundwater resources are in a critical condition because the volumes withdrawn far exceed natural recharge rates. Unless improved water management plans are put in place, major environmental problems are likely to occur in the future.

Land degradation is a serious problem, and the region's rangelands – important for food security – are deteriorating, mainly as a result of overstocking what are essentially fragile ecosystems. Drought, mismanagement of land resources, intensification of agriculture, poor irrigation practices and uncontrolled urbanization have also contributed. Marine and coastal environments have been degraded by overfishing, pollution and habitat destruction. Industrial pollution and management of hazardous wastes also threaten socio-economic development in the region with the oil-producing countries generating two to eight times more hazardous waste per capita than the United States. Over the next decade, urbanization, industrialization, population growth, abuse of agrochemicals, and uncontrolled fishing and hunting are expected to increase pressures on the region's fragile ecosystems and their endemic species.

The command and control approach, through legislation, is still the main environmental management tool in almost all states. However, several new initiatives, such as public awareness campaigns, have been taken to protect environmental resources and control pollution. In addition, many enterprises such as refineries, petrochemical complexes and metal smelters have begun procedures for obtaining certification under the ISO 14 000 series. Another important approach to resource conservation has been a growing interest in recycling scarce resources, particularly water. In many states on the Arabian Peninsula, municipal wastewater is subjected at least to secondary treatment, and is widely used to irrigate trees planted to green the landscape.

Success in implementing global and regional MEAs in the region is mixed and commitment to such policy tools quite weak. At a national level there has, however, been a significant increase in commitment to sustainable development, and environmental institutions have been given a higher priority and status.

Polar regions

The Arctic and Antarctic play a significant role in the dynamics of the global environment and act as barometers of global change. Both areas are mainly affected by events occurring outside the polar regions. Stratospheric ozone depletion has resulted in high levels of ultraviolet radiation, and polar ice caps, shelves and glaciers are melting as a result of global warming. Both areas act as sinks for persistent organic pollutants, heavy metals and radioactivity, mostly originating from other parts of the world. The contaminants accumulate in food chains and pose a health hazard to polar inhabitants. Wild flora and fauna are also affected by human activities. For example, capelin stocks have collapsed twice in the Arctic since the peak catch of 3 million tonnes in 1977. In the Southern Ocean, the Patagonian toothfish is being over-fished and there is a large accidental mortality of seabirds caught up in fishing equipment. On land, wild communities have been modified by introductions of exotic species and, particularly in northern Europe, by overgrazing of domestic reindeer.

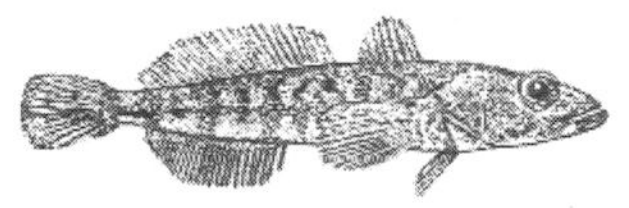

The Patagonian toothfish, *Dissostichus eleginoides*, is being severely overfished – see page 193

In the Arctic, the end of Cold War tensions has led to new environmental cooperation. The eight Arctic countries have adopted the Arctic Environmental Protection Strategy which includes monitoring and assessment, environmental emergencies, conservation of flora and fauna, and protection of the marine environment. Cooperation amongst groups of indigenous peoples has also been organized. The Antarctic environment benefits from the continuing commitment of Parties to the Antarctic Treaty aimed at reducing the chance of the region becoming a source of discord between states. The Treaty originally focused on mineral and living resources but this focus has now shifted towards broader environmental issues. A similar shift is expected in the Arctic, within the broader context of European environmental policies. In both polar areas, limited financial resources and political attention still constrain the development and implementation of effective policies.

Prospects for the future

Issues for the 21st century

Environmental issues that may become priorities in the 21st century can be clustered in three groups – unforeseen events and scientific discoveries; sudden, unexpected transformations of old issues; and already well-known issues to which the present response is inadequate.

The Scientific Committee on Problems of the

Environment of the International Council for Science conducted a special survey for *GEO-2000* on environmental issues that may require attention in the 21st century. The survey was conducted among 200 scientists in 50 countries. Most of the responding scientists expect that the major environmental problems of the next century will stem from the continuation and aggravation of existing problems that currently do not receive enough policy attention.

The issues cited most frequently are climate change, and the quantity and quality of water resources. These are followed by deforestation and desertification, and problems arising from poor governance at national and international levels. Two social issues, population growth and changing social values, also received considerable attention. Many scientists emphasized that the interlinkages between climate change and other environmental problems could be important. This includes the emerging scientific understanding of complex interactions in the atmosphere-biosphere-cryosphere-ocean system – which could lead to irreversible changes such as shifts in ocean currents and changes in biodiversity.

The emphasis on interlinkages is not surprising. It has been repeatedly shown that sectoral policies taken in isolation do not always yield the desired results. One reason is that sectoral policies can solve one problem while aggravating others, particularly over a long time frame. Although the existence of interlinkages between environmental problems is now better known, we still lack understanding of exactly how the issues are linked, to what degree they interact and what the most effective measures are likely to be. One such issue that is identified throughout *GEO-2000* is the need to integrate land- and water-use planning to provide food and water security.

Alternative policies

Since current policies will not lead to a sustainable future, at either the regional or the global level, region-specific studies were undertaken for *GEO-2000* to investigate possible alternative policies. Each regional study focused on one or two specific issues selected on the basis of regional challenges identified in *GEO-1* (see table opposite).

In each study, several alternative policy responses were identified to address the issues at hand. Each of the selected responses has been implemented elsewhere with success. The results confirm that, in principle, the knowledge and technological base to solve environmental issues are available, and that if these alternative policies were implemented immediately and pursued with vigour they could indeed set the world on a more sustainable course.

A number of key conclusions emerge from the alternative policy studies.

- There is a clear need for integrated policies. For example, in Latin America a broad intersectoral approach is advocated to achieve sustainable forest development. In Europe and Central Asia, combined strategies to deal with acidification, urban air pollution and climate change could lead to an optimal use of opportunities for energy efficiency and fuel switching.
- Market-based incentives, particularly subsidy reforms, have a role to play in all regions. Reform of unnecessary subsidies can encourage the more efficient use of resources such as energy, and thus help reduce pollution and degradation.
- Effective institutional mechanisms are essential. Too many institutions are weak and plagued with limited mandates and power, small financial resources and few human resources.
- A main obstacle to successful policy implementation is lack of money. Attention is drawn to the crucial point that environmental management usually needs financing.

The regional studies highlight major gaps in our knowledge and experience when it comes to analysing and directing macro-economical processes relating to the environment. A number of issues, including trade and financial flows, were not addressed because of a lack of relevant information and knowledge. There is an urgent need to improve understanding of the effects of economic and social developments on the environment, and vice versa.

Environmental focus of region-specific alternative policy studies

Asia and the Pacific	Air pollution
Africa	Land and water resource management
Europe and Central Asia	Energy-related issues
Latin America	Use and conservation of forests
North America	Resource use, greenhouse gas emissions
West Asia	Land and water resource management

Outlook and recommendations

GEO-2000 confirms the overall assessment of *GEO-1*: the global system of environmental policy and management is moving in the right direction but much too slowly. On balance, gains by better management and technology are still being outpaced by the environmental impacts associated with the speed and scale of population and economic growth. Substantial improvement in the environment is rarely achieved.

The continued poverty of the majority of the planet's inhabitants and excessive consumption by the minority are the two major causes of environmental degradation. The present course is unsustainable and postponing action is no longer an option. Inspired political leadership and intense cooperation across all regions and sectors will be needed to put both existing and new policy instruments to work.

One of GEO's tasks is to recommend measures and actions that could reverse unwelcome trends and reduce threats to the environment. This publication therefore concludes with recommendations made by UNEP after consideration of the findings of the *GEO-2000* assessment. These recommendations are focused on four areas.

Filling the knowledge gaps

GEO-2000 shows that we still lack a comprehensive view of the interactions and impacts of global and inter-regional processes. Information on the current state of the environment is riddled with weakness. There are few tools to assess how developments in one region affect other regions, and whether the dreams and aspirations of one region are compatible with the sustainability of the global commons.

Another serious omission is the lack of effort to find out whether new environmental policies and expenditures have the desired results. These knowledge gaps act as a collective blindfold that hides both the road to environmental sustainability and the direction in which we are travelling. However, whilst it is imperative to address these gaps, they should not be used as an excuse for delaying action on environmental issues that are known to be a problem.

Tackling root causes

Means must be found to tackle the root causes of environmental problems, many of which are unaffected by strictly environmental policies. Resource consumption, for example, is a key driver of environmental degradation. Policy measures to attack this issue must reduce population growth, reorient consumption patterns, increase resource use efficiency and make structural changes to the economy. Ideally, such measures must simultaneously maintain the living standards of the wealthy, upgrade the living standards of the disadvantaged, and increase sustainability. This will require a shift in values away from material consumption. Without such a shift, environmental policies can effect only marginal improvements.

Taking an integrated approach

Changes are needed in the ways we think about the environment and in the ways in which we manage it. First, environmental issues need to be integrated into mainstream thinking. Options for add-on environmental policies have been exhausted in many sub-regions. Better integration of environmental thinking into decision-making about agriculture, trade, investment, research and development, infrastructure and finance is now the best chance for effective action.

Secondly, environmental policies that move away from strictly sectoral issues to encompass broad social considerations are the most likely to make a lasting impact. This holds good across the gamut of environmental issues – for example, water, land and other forms of natural resource management, forest conservation, air pollution and coastal area management.

Thirdly, there is a need for better integration of international action to improve the environment – particularly in relation to regional and multilateral environment agreements.

Mobilizing action

Solutions to environmental issues must come from cooperative action between all those involved – individuals, NGOs, industry, local and national governments, and international organizations. The need to involve all the parties concerned is emphasized throughout *GEO-2000*. Specific examples include the increasing role of NGOs in multilateral agreements, the involvement of stakeholders in property rights issues, and the leading role played by some manufacturing and resource industries in setting ambitious but voluntary environmental targets.

The GEO-2000 Regions

There are seven GEO-2000 regions, each divided into sub-regions:
Africa, Asia and the Pacific, Europe and Central Asia, Latin America and the Caribbean, North America, West Asia and the Polar Regions.

Polar Regions
The Arctic: the eight Arctic countries are: Canada, Greenland (Denmark), Finland, Iceland, Norway, Russia, Sweden, Alaska (United States)
The Antarctic

AFRICA

Northern Africa:
Algeria, Egypt, Libyan Arab Jamahiriya, Mauritania, Morocco, Tunisia

Western and Central Africa:
Benin, Burkina Faso, Burundi, Cameroon, Cape Verde, Central African Republic, Chad, Congo, Democratic Republic of the Congo, Côte d'Ivoire, Equatorial Guinea, Gabon, Gambia, Ghana, Guinea, Guinea-Bissau, Liberia, Mali, Niger, Nigeria, Rwanda, Sao Tomé and Principe, Senegal, Sierra Leone, Togo

Eastern Africa and Indian Ocean Islands:
Comoros, Djibouti, Eritrea, Ethiopia, Kenya, Madagascar, Mauritius, Réunion (France), Seychelles, Somalia, The Sudan, Uganda

Southern Africa:
Angola, Botswana, Lesotho, Malawi, Mozambique, Namibia, South Africa, Swaziland, United Republic of Tanzania, Zambia, Zimbabwe

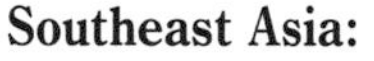

ASIA AND THE PACIFIC

South Asia:
Afghanistan, Bangladesh, Bhutan, India, Islamic Republic of Iran, Maldives, Nepal, Pakistan, Sri Lanka

Southeast Asia:
Brunei Darussalam, Indonesia, Malaysia, Philippines, Singapore

Greater Mekong Region:
Cambodia, China-Yunnan, Lao People's Democratic Republic, Myanmar, Thailand, Viet Nam

Northwest Pacific and East Asia:
China, Democratic People's Republic of Korea, Japan, Republic of Korea, Mongolia

Australasia and the Pacific:
American Samoa (United States), Australia, Cook Islands, Fiji, French Polynesia (France), Guam (United States), Kiribati, Micronesia, Marshall Islands, Nauru, New Caledonia (France), New Zealand, Northern Mariana Islands (United States), Niue, Palau, Papua New Guinea, Pitcairn Island (United Kingdom), Samoa, Solomon Islands, Tokelau (New Zealand), Tonga, Tuvalu, Vanuatu, Wallis and Futuna (France)

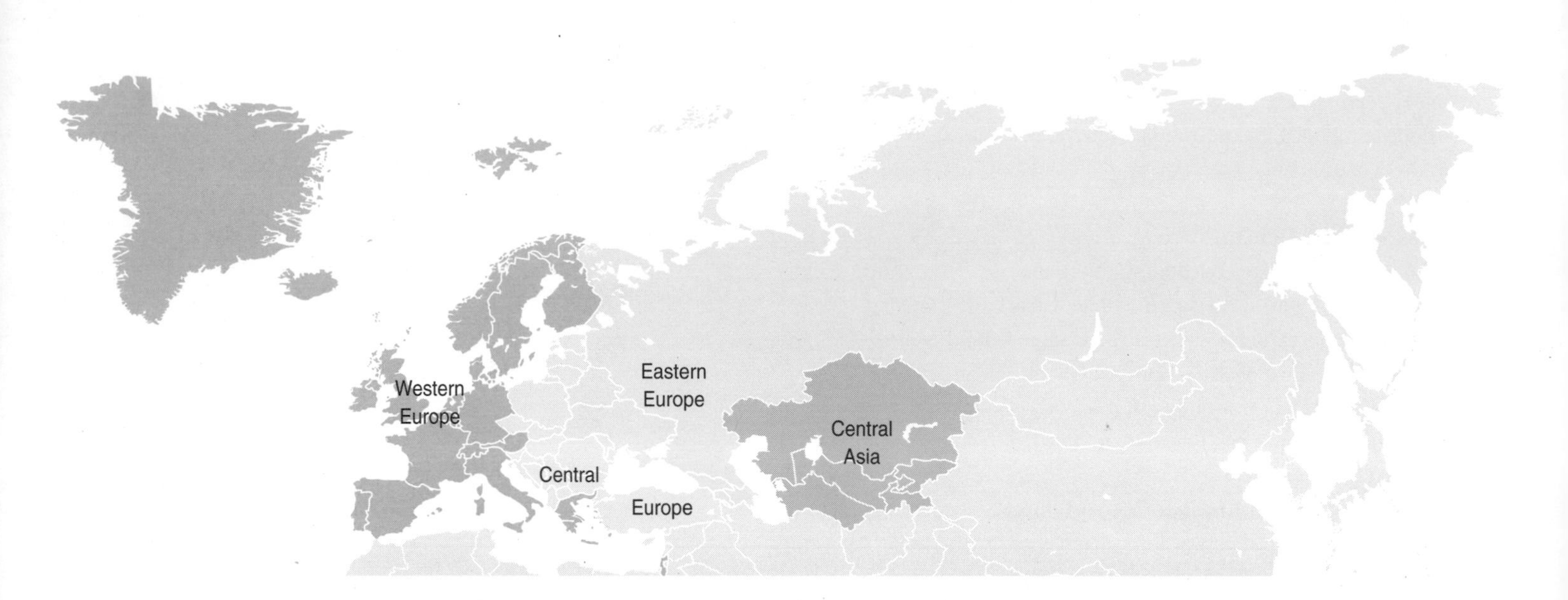

EUROPE AND CENTRAL ASIA

Western Europe:
Andorra, Austria, Belgium, Denmark, Finland, France, Germany, Greece, Holy See, Iceland, Ireland, Israel, Italy, Liechtenstein, Luxembourg, Malta, Monaco, Netherlands, Norway, Portugal, San Marino, Spain, Sweden, Switzerland, United Kingdom

Central Europe:
Albania, Bosnia and Herzegovina, Bulgaria, Croatia, Cyprus, Czech Republic, Estonia, Hungary, Latvia, Lithuania, Poland, Romania, Slovakia, Slovenia, The Former Yugoslav Republic of Macedonia, Turkey, Yugoslavia

Eastern Europe
Armenia, Azerbaijan, Belarus, Russian Federation, Georgia, Republic of Moldova, Ukraine

Central Asia:
Kazakhstan, Kyrgyzstan, Tajikistan, Turkmenistan, Uzbekistan

WEST ASIA

Arabian Peninsula:
Bahrain, Kuwait, Oman, Qatar, Saudi Arabia, United Arab Emirates, Yemen

Mashriq:
Iraq, Jordan, Lebanon, Syrian Arab Republic, West Bank and Gaza

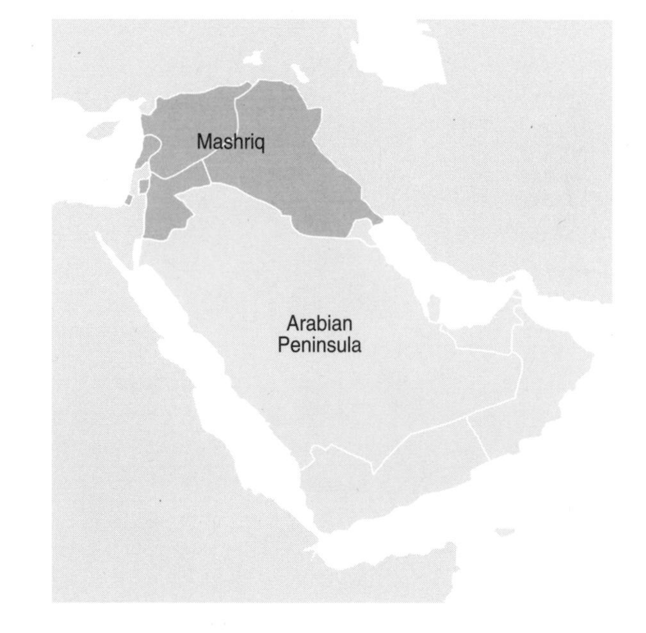

NORTH AMERICA

Canada
United States
Mexico
(for selected sections only)

LATIN AMERICA AND THE CARIBBEAN

Caribbean:
Anguilla (United Kingdom), Antigua and Barbuda, Aruba (The Netherlands), Bahamas, Barbados, British Virgin Islands (United Kingdom), Cayman Islands (United Kingdom), Cuba, Dominica, Dominican Republic, Grenada, Guadeloupe (France), Haiti, Jamaica, Martinique (France), Montserrat (United Kingdom), Netherlands Antilles (The Netherlands), Puerto Rico (United States), St Kitts and Nevis, St Lucia, St Vincent and the Grenadines, Trinidad and Tobago, Turks and Caicos (United Kingdom), Virgin Islands (United States)

Meso-America:
Belize, Costa Rica, El Salvador, Guatemala, Honduras, Mexico, Nicaragua, Panama

South America:
Argentina, Bolivia, Brazil, Chile, Colombia, Ecuador, French Guyana (France), Guyana, Paraguay, Peru, Suriname, Uruguay, Venezuela

Chapter One

Global Perspectives

Global Perspectives

KEY FACTS

- Average global per capita income has now passed US$5 000 a year – 2.6 times that of 1950 – but more than 1 300 million people still live on less than US$1 per day.
- The high-income countries, home to 20 per cent of the world's population, account for about 60 per cent of commercial energy use.
- Total carbon emissions from China now exceed those of the European Union although China's per capita emissions are much lower.
- World military expenditure fell by an average of 4.5 per cent a year during the decade 1988–97.
- In 1996, private investment, concentrated in a limited number of developing countries, was about US$250 000 million, compared to overseas development assistance of less than US$50 000 million.
- A tenfold reduction in resource consumption in the industrialized countries is a necessary long-term target if adequate resources are to be released for the needs of developing countries.
- There are encouraging signs of real interest among consumers in more environmentally-sustainable products and services. A number of cooperative organizations have sprung up to promote the 'Fair Trade' movement, which aims to achieve fair prices for small farmers who use environmentally-friendly methods. Such products are beginning to move from niche markets to the mainstream.
- The processes of globalization that are so strongly influencing the evolution of society need to be directed towards resolving rather than aggravating the serious imbalances that divide the world today.

The last millennium change on this planet took place under very different conditions from those found today. In China, the Sung dynasty, with its giant metropolitan centres, delicate paintings and moving poetry, had by the year AD 1000 been established for 40 years. Islamic culture had welded disparate peoples over an area stretching from Spain to central Asia and northern India into a single cultural unit. In Mexico, the lowland Mayan civilization had collapsed and the Toltecs were building the first great Meso-American civilization. In Africa, Arab culture flourished in the north, the kingdoms of Kanem and Ghana, with their substantial stone-built houses, held sway in the west, and in the east the influence of the Ethiopian empire was waning. In Europe, the Cluny Abbey had just been rebuilt for the first time. Waterpower was being more effectively harnessed than in Roman times and innovative credit instruments were being developed. After centuries of exporting unskilled labour and raw material, the region was now becoming an exporter of industrial products – while importing chemicals for cloth manufacture in the cities of northern Italy and Flanders (Gies 1994, Lacey and Danziger 1999).

One thousand years later, the planet is also poised on the threshold of a new era – one in which the disparate divisions that have always separated human beings in one area from those in another are finally disappearing. Globalization and electronic communications are effecting a profound revolution. The Industrial Revolution of the 19th century is being replaced by the Communications Revolution.

This chapter provides a background perspective to the environmental changes covered in the rest of the report. It describes the main drivers of environmental change – the economy, population growth, political organization, conflict, peace and security, and regionalization. It then assesses the main dangers and opportunities presented by the beginning of the third millennium: globalization, trade, international debt, demography, the consumer culture, technology and transport. Finally, it examines responses to the situation, covering environmental policies, the changing concept of development, science and research, business and industry, employment and consumer awareness.

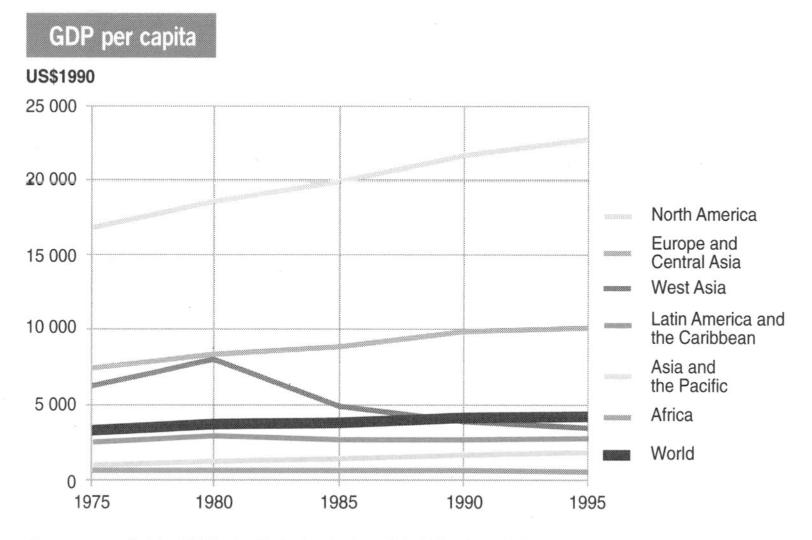

Source: compiled by RIVM, the Netherlands, from World Bank and UN data

Steady growth in global GDP/ capita hides large differences both between and within regions

Social and economic background

Since 1950, the global economy has more than quintupled in size. Despite the financial turmoil in East Asia starting in late 1997, the world's economy continues to expand, growing by 4.1 per cent in 1997. In terms of income, the global per capita average has now passed US$5 000 a year – 2.6 times that of 1950 (in real terms).

Average figures for income hide great discrepancies between regions (see graph), between countries, and between population groups within countries. Despite remarkable improvements in many places, one-quarter of the world's population remains in severe poverty. In 1993, more than 1 300 million people were living on less than US$1 per day. Of these, the largest number, nearly 1 000 million people, are in the Asia and Pacific region; the highest proportion and the fastest growth are in sub-Saharan Africa, where half the population is expected to be poor by 2000; a growing number, 110 million in 1993, are in Latin America; the number below the poverty level in Eastern Europe and the former Soviet Union had risen to 120 million people by 1993/94; and, in industrialized countries, 80 million people are still below the poverty line (UNDP 1997).

Nearly half of all people now live in cities; an increasing number of them travel enormous distances every year by private car and in aircraft. In the developed world, technology has transformed patterns of work and family life, communications, leisure activities, diet and health. Similar transformations are well under way in the more prosperous parts of the developing world.

The impacts of these changes on the natural environment are complex. The modern industrial economies of North America, Europe and parts of East Asia consume immense quantities of energy and raw materials, and produce high volumes of wastes and polluting emissions. The magnitude of this economic activity is causing environmental damage on a global scale (notably climate change) and widespread pollution and disruption of ecosystems, often in countries far removed from the site of consumption. Considerable progress has been made in controlling pollution at local and transboundary levels in the wealthier industrialized countries but the wider-scale impacts (apart from ozone depletion) have yet to be tackled effectively.

In other regions, particularly in many parts of the developing world, poverty combined with rapid population growth is leading to widespread

Annual average growth of per capita GDP (1975–95)

Africa	-0.20%
Asia and the Pacific	3.09%
Europe and Central Asia	1.54%
Latin America and the Caribbean	0.66%
North America	1.53%
West Asia	-2.93%
WORLD	**1.17%**

Source: compiled by RIVM, the Netherlands, from World Bank and UN data degradation of renewable resources – primarily forests, soils and water. People living in subsistence economies are faced with few alternatives to depleting their natural resources. Renewable resources still sustain the livelihood of nearly one-third of the world's population; environmental deterioration therefore directly reduces living standards and prospects for economic improvement among rural peoples. At the same time, rapid urbanization and industrialization in many developing countries are creating high levels of air and water pollution, which often hit the poor hardest. Worldwide, the urban poor tend to live in neglected neighbourhoods, enduring pollution, waste dumping and ill health, but lacking the political influence to effect improvements. Consumption and waste generation among the newly industrialized nations are rising very steeply – approaching, and in some cases even overtaking, per capita consumption levels in industrialized countries. In 1995, per capita energy consumption in the Republic of Korea, for example, equalled that of Italy (UNSTAT 1997). The same is true of many consumers in prosperous enclaves of the developing countries and those in economies in transition.

So what does the future hold in store? *GEO-1* included an account of a 'business-as-usual' scenario in which the world population nearly doubled between 1990 and 2050, and GDP per capita, expressed in constant prices, grew 2.4 times. Simultaneously, food requirements doubled, energy consumption rose by a factor of 2.6 and water consumption by a factor of nearly 1.5. The world economy continued its rapid growth with a projected rise in GDP of 4.5 times. Under this scenario, sufficient food would be available globally to feed all the growing population but inequalities of access would mean that hunger would remain.

From what follows in *GEO-2000*, it is clear that if present trends in population growth, economic growth and consumption patterns were continued, the natural environment would be increasingly stressed. Distinct environmental gains and improvements would probably be offset by the pace and scale of global economic growth, increased global environmental pollution and accelerated degradation of the Earth's renewable resource base.

The negative impacts of environmental degradation would fall most heavily (as they do now) on the poorer developing regions. The income gap between rich and poor countries, and between the rich and poor within countries, would increase for several decades. The ratio of income between the richest and poorest 20 per cent of the world population doubled from 30:1 to more than 80:1 between 1960 and 1995 (UNDP 1998). Under a business-as-usual scenario, current inequities in the distribution of the environmental costs and benefits of consumption seem likely to grow worse. This could be expected to have a destabilizing influence on the physical, social and political environment.

However, trends towards environmental degradation can be slowed, and economic activity can be shifted to a more sustainable pattern. Choices for development, and levels and patterns of consumption, are shaped by human aspirations and values, and these choices can be influenced by policy intervention. Many promising policy responses are being developed and tested, as described in this report.

Some environmental trends over the past half-century demonstrate the potential of regulation, information and, above all, prices to encourage both more efficient and less polluting uses of energy and materials. Technology has already delivered astonishing improvements in product performance but innovation to improve resource productivity – the utility that can be squeezed out of any given amount of resource input – has so far lagged behind. Better public understanding and awareness of the environmental and social consequences of the consumer society have begun to catalyse profound shifts in purchasing behaviour and lifestyle choices. The challenge for policy-makers in the next century will be to devise approaches that encourage a more efficient, fair and responsible use of natural resources by the production sectors of the economy, that encourage consumers to support and demand such changes, and that will lead to a more equitable use of resources by the entire world population. In this context, policy-makers are not necessarily government officials. Business leaders also make policy, and in some of the major industrialized countries business leaders are already leading the way to improved systems of resource use (Rabobank 1998).

The key drivers

The forces that are driving global change are a complex mix of economic and political factors magnified by a high rate of population growth. These interact in ways that are not always predictable. While it is possible to identify overall trends in each of these factors, we are often less successful in identifying feedback loops and interrelationships between them that may be critical to the ultimate outcome. In this report, many trends are described and projections made, each of which is based on sound reasoning. However, we are still far from being able to understand, model and forecast all the complex interactions in the global human and natural system. Just as engineers allow for a considerable margin of safety, so we should not rely only on the most optimistic assumptions in each sector as the basis for decisions on our future well-being and survival.

Economy

The industrialized countries still dominate economic activity; absolute and per capita levels of consumption of most – if not all – natural resources remain far higher in the OECD economies than in the developing countries. A recent, detailed study of four industrialized countries indicates that the total quantity of natural resources, or materials flow, required to support their economies ranges from 45 to 85 tonnes per person per year. A significant proportion of those resources is imported from developing countries (Adriaanse and others 1997). In 1995, the high-income countries, home to 20 per cent of the world's population, accounted for about 60 per cent of commercial energy use (UNSTAT 1997). The bar chart above right shows total and per capita energy consumption by region.

The United States, Japan and the European Union produce more than 40 per cent of global carbon dioxide emissions (CDIAC 1998). However, there have been unprecedented rates of economic growth in many developing countries, particularly the populous economies of east and south Asia, over the past 25 years. The highest consumption growth rates are now found in the developing world and, because of the large populations in these regions, their total consumption is catching up with the industrialized world. Total carbon emissions from China now exceed those of the European Union, although China's per capita emissions are much lower (CDIAC 1998).

The pattern of industrial activity has undergone important shifts in recent decades. Heavy industry is expanding rapidly in the developing Asian and South American economies, while expansion of the industrial base in Europe, the United States and Japan is directed more to high technology production processes and service-oriented activities.

These structural shifts, together with reduced material intensity and improved cleaner-production practices, have contributed to an overall slowdown in industry-related pollution generated by the developed economies and to greater resource efficiency. Technological advance and environmental regulation have also contributed to stable or declining levels of some polluting emissions such as sulphur dioxide and some heavy metals, notably in North America and Western Europe.

Developing countries are still on a rising curve of production and pollution. Rapid industrialization, and the construction of large, material-intensive metropolitan centres, and related transport and distribution networks, mean that these countries are replicating the resource use patterns typical of the earlier phases of development in the industrialized

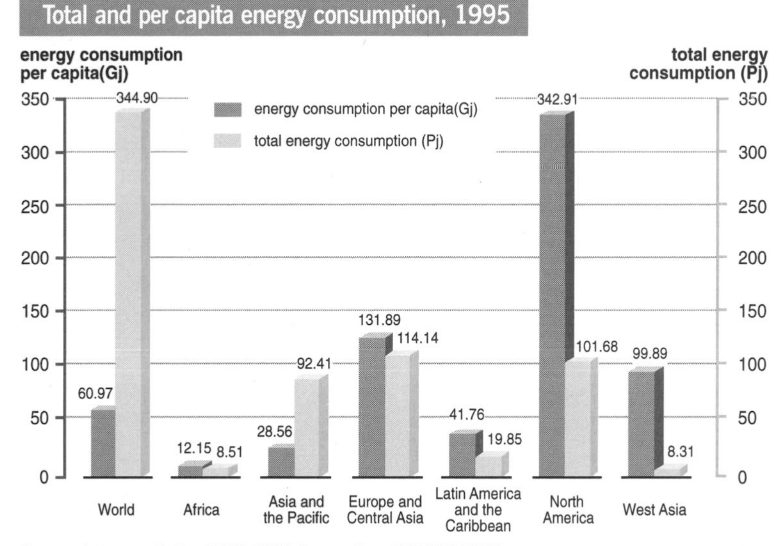

Total and per capita energy consumption, 1995

Source: data compiled by UNEP GRID Geneva from UNSTAT 1997

In 1995, the high-income countries, home to 20 per cent of the world's population, accounted for about 60 per cent of world commercial energy use

Farmers have traditionally satisfied increasing demand by ploughing new land but opportunities for expansion are now limited

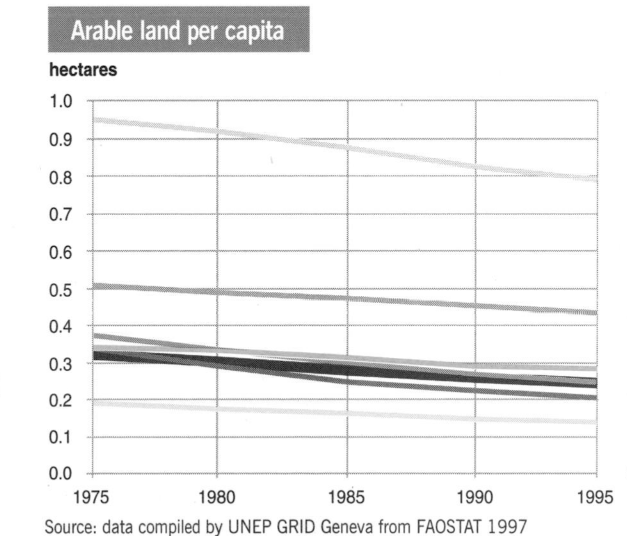

Arable land per capita

Source: data compiled by UNEP GRID Geneva from FAOSTAT 1997

world. The environmental efficiency now being sought in industrialized countries is often seen as a luxury in developing countries.

At the same time, agriculture remains important in the economy of most developing countries, contributing a higher proportion of national GDP, and providing more employment, than in the developed world. However, the area of arable land available per head of population is decreasing in all regions as populations grow (see graph on page 5): the global availability of cropland has now fallen by some 25 per cent over two decades, from 0.32 hectares per capita in 1975 to 0.24 hectares in 1995 (FAOSTAT 1997). Farmers have traditionally satisfied increasing demand by ploughing new land but opportunities for expansion are now limited. Raising productivity has therefore been central to increased grain production. Breeders have boosted the yield potential of cereals substantially, and the currently controversial use of genetically-modified strains may do so further. Fertilizer use continues to rise in many developing countries, though there is concern about diminishing returns from increased applications and the threat of nitrate pollution of freshwater supplies. Irrigation has also been a key to increased grain yields, expanding at 2.3 per cent a year from 1950 to 1995 (FAOSTAT 1997).

World population will reach 6 000 million during 1999 – but the rate of growth has begun to slow

Population

The world population has more than doubled since 1950 and will reach 6 000 million during the year of this report's publication (see graph below). It reached 1 000 million in 1804. It took 123 years to add another 1 000 million; 33 years to reach 3 000 million in 1960, 14 years to reach 4 000 million, 13 years to reach 5 000 million in 1987 and 12 years to reach 6 000 million in 1999. The rate of population growth, though now beginning to fall, still adds nearly 80 million people a year (United Nations Population Division 1998a).

World population

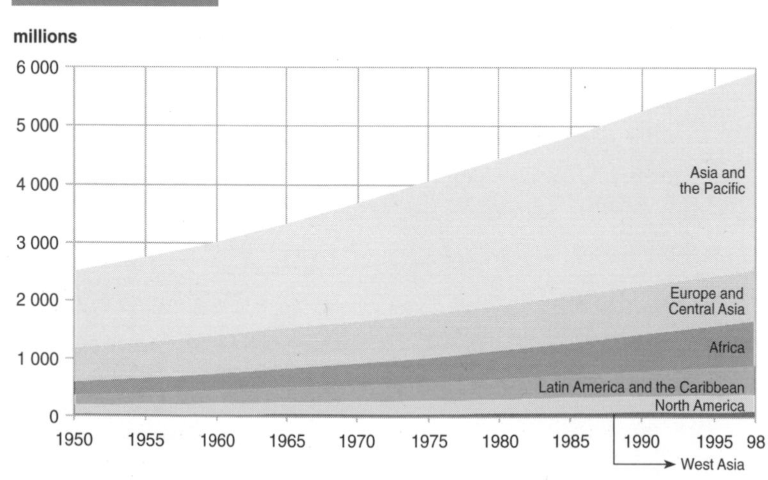

Source: compiled by UNEP GRID Geneva from United Nations Population Division 1998a

The demands placed on the environment to provide resources for human activities and to absorb wastes have grown steadily with rising population and increasing per capita consumption. The rate of growth of population has declined significantly in recent years, thanks to falling fertility in most regions, and the most recent population forecasts from the United Nations indicate that, under a medium-fertility scenario, the global population is likely to peak at about 8 900 million in 2050 (United Nations Population Division 1998a). This projection assumes that all developing countries will achieve replacement fertility levels (2.1 children per woman) over the next half century. Currently, the highest fertility rates tend to be found in countries suffering from poverty, food insecurity and natural resource degradation. Since falling fertility is correlated with rising income and improvements in areas such as health care, employment and women's education, a transition to stable population numbers in these regions cannot be taken for granted. If fertility rates were to exceed the medium scenario by just half a child per couple, the world population would rise to some 27 000 million people (United Nations Population Division 1998b).

Given that many natural resources (such as water, soil, forests and fish stocks) are already being exploited to or beyond their limits, in at least some regions, the efforts required to meet the needs of an additional 3 000 million people in the next 50 years will be immense, even at present consumption levels. If poverty is to be reduced and economic benefits distributed more equitably, then a further major increase in production will be required, not to mention significant modifications to economic, social and political systems. Whether the planetary environment can meet these demands, and under what conditions, is an open question.

Political organization

Political regimes often affect the environment. During the colonial era, for example, political systems changed land-use patterns in many regions. The colonization process exploited natural resources for export, established large monocultures and opened up a largely unexploited domain. The transition from colonies to new states shifted control of land tenure to national authorities. Newly established governments frequently paid more attention to rapid economic development than to fair and equitable access to natural resources by the poor. The situation was reminiscent of early European development where landlords and the bureaucracy denied the basic right to land to the poor, leading to catastrophic consequences in the 18th and 19th centuries in Western Europe and in the 20th century in Eastern Europe.

From the 1950s onwards, centrally-planned regimes sought fast economic growth through state-managed industrialization plans. Systems of quotas and production targets were driven by political decisions rather than market efficiency and this led to excessive resource use and waste. The legacy of these forms of industrial production in the former Soviet Union and Eastern Europe has not only been economic dislocation but daunting environmental problems such as the death of the Aral Sea, nuclear contamination, and high levels of air and water pollution (see Chapter 2, Europe).

Since 1989, most such regimes have begun to move towards market-based systems of economic organization and economic liberalization, often accompanied by democratization. While market systems have been inherently efficient at economic organization, environmental costs have traditionally been excluded from the decision-making process. This has allowed unsustainable exploitation of natural resources as well as unsustainable demands on natural pollution sinks. However, some valuation of environmental resources and services is gradually being established in the market through the regulation and assignment of property rights. The good example is the successful system of sulphur dioxide emission trading in the United States which has helped achieve substantial reductions in emissions.

Another potentially worrying trend has been the shift of economic power and decision-making through globalization. At the national level, governments have been the primary mechanisms for the defence of the common good in the environmental and social areas, and for raising resources through taxation to be redistributed to these ends. With globalization and the shift of many activities to the international level, national governments are losing influence. Multinational corporations and institutional investors have become increasingly powerful internationally. Although their first priority is profit, many leading international corporations and banks are adding environmental and social value to economic value as corporate priorities, and leading significant initiatives towards more sustainable development. NGOs have also become increasingly influential.

However, these trends address only part of the problem. There is increasing evidence that there is a real need to introduce stronger global coordination and governance structures to protect the global commons, and to better means of financing global environmental action (these issues are the subject of specific recommendations in Chapter 5). In all developed and some developing countries, 20 to 45 per cent of GDP is transferred to the central government as taxes and other revenue, representing a significant effort to meet the collective needs of society for security and welfare (World Bank 1998). In comparison, global contributions to the United Nations and other international organizations are minimal, even though the need for global political, social and environmental security is growing. As a greater proportion of wealth creation by the private sector is globalized and escapes national taxation, the base of economic activity supporting national environmental and social action, as a proportion of total activity, will shrink. The lack of international sources of funds for environmental protection is one reason why global environmental stewardship is falling so far behind development.

Conflict, peace and security

Serious armed conflicts continued, with heavy loss of life, during the 1990s. Serious conflicts have plagued countries in Africa, Central Asia, West Europe and West Asia over the past few years. Loss of life in war is accompanied by increased pressure on ecosystems. Resource productivity collapses in war-affected areas, and there is a danger that environmental damage will affect much wider areas than those directly involved in the conflict. This was the case in both the Second Gulf War and the recent conflict in Yugoslavia. In the latter, the destruction of chemical and petrochemical complexes in Serbia led to the pollution of the Danube river, causing problems in the downstream countries of Bulgaria and Romania. The flow of refugees to neighbouring Balkan countries also led to environmental problems and the spread of disease.

War-related refugees are often compelled to

extract fuelwood and freshwater resources at an unsustainable rate in order to survive. The number of people receiving refugee assistance worldwide reached an all-time high of 27.4 million in 1995, before dropping to 22.7 million in 1997 (UNHCR 1998). In 1999, these numbers were swelled by refugees fleeing Kosovo during the conflict in Yugoslavia.

In addition to the environmental stress caused by warfare, there is now increasing concern that environmental degradation and resource shortages may actually cause armed conflict. Examples of environmental degradation capable of escalating into violence include severe water shortages, widespread desertification, health-threatening toxic contamination, and refugee flight from environmental wastelands. Even within nations, increasing demands for limited natural resources create domestic tensions, as well as intensifying the pressure between private and public interests. National security is now increasingly dependent on environmental security.

On the positive side, military expenditures have fallen in most areas of the world (see graph below). In 1997, world military expenditure was about US$740 000 million – the equivalent of US$125 per capita. It fell by an average of 4.5 per cent a year during the decade 1988–97 (SIPRI 1998). The ratio of global military expenditure to gross national product fell to 3 per cent in 1994, a new low for the entire period since 1960, compared to 5.5 per cent in 1984. For developing countries, the ratio fell consistently over the decade, from 6.1 per cent in 1984 to 2.6 per cent in 1994, except for an increase in 1990 (USACDA 1997). This has positive implications for the redirection of military finance towards social expenditure although there is little evidence that savings in military expenditure have been used to finance environmental action. However, the environment has certainly benefited from a reduced military consumption of minerals and petroleum.

Over the past 10 years, military expenditures have been greatly reduced in nearly all regions. This has also reduced the military consumption of minerals and petroleum

Military expenditures

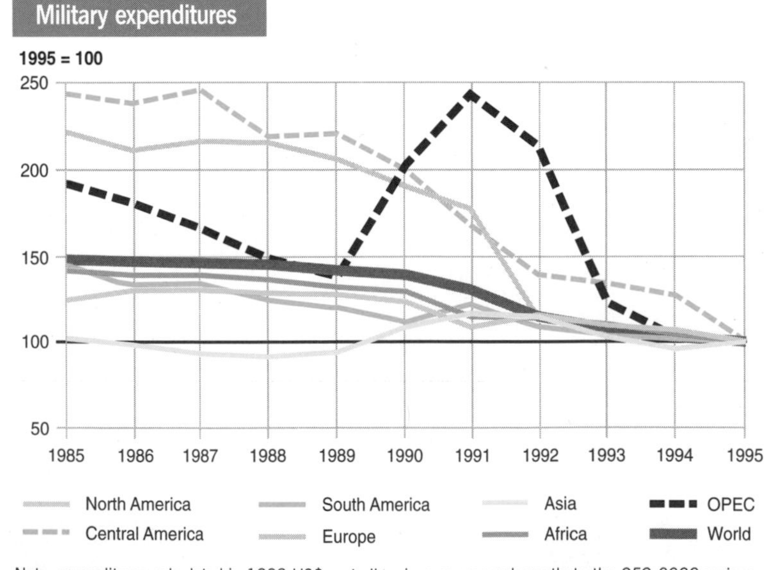

Note: expenditures calculated in 1993 US$; not all regions correspond exactly to the *GEO-2000* regions

Source: BICC 1998

Regionalization

Since many environmental problems extend beyond national boundaries, the development of regional levels of governance, through regional conventions, economic cooperation organizations, and even unions of governments, is creating institutional structures able to respond to transboundary environmental issues. An example is the unification of Europe and the expansion of the European Union currently under consideration. The Maastricht and Amsterdam treaties have placed sustainable development squarely on the European Union policy agenda, as has *Agenda 2000*. The current trend in the European Union to go for 'framework' legislation is having an impact on the environment at national and regional levels. The environmental steps required of countries before they can join will strengthen the region's response to its environmental problems. The process of European unification also provides an example of the efforts that will be needed to reduce global inequalities between nations. The Pacific Islands are another region where a complementary set of regional organizations and regional conventions are creating a strong framework of regional environmental legislation and collaboration.

The implications of these major economic, demographic and political trends, and their interactions in the global environment, are being addressed in the current debates over global levels and patterns of consumption, and over poverty reduction. *Agenda 21* explicitly recognized the complex nature of these issues. On the one hand, a wealthy minority of the world's population is consuming at an unsustainably high level, causing disproportionate damage to global ecosystems, while protecting only their local environment. On the other hand, a poor, larger and rapidly-growing proportion of the world's population is being forced by poverty to degrade the natural resource base on which it is directly dependent. In addition, a vast global 'middle class' is expected to be created by continued economic growth and globalization. What will be the environmental impacts of 3 000–4 000 million consumers, with rising incomes, who all want to live an affluent lifestyle? What will happen as their success contrasts increasingly with the lot of the very poor? Since some planetary resources may be too limited to support this increased consumption, how will the resulting tensions be resolved?

Areas of danger and opportunity

Globalization and the private sector

There are a number of important environmental dimensions to globalization. Stratospheric ozone depletion, climate change driven by global warming, and the worldwide spread of persistent organic pollutants are obvious examples. There is also an accelerated process of biological globalization. The increase in trade, transport and travel has created many new opportunities, both deliberate and accidental, for organisms to move around the world and invade new environments. Some of these introductions bring net benefits but many are of aggressive invasive species that upset the local ecological balance and crowd out other more desirable, useful or perhaps unique species. These changes can degrade ecosystems and lead to significant losses in biological diversity, ecosystem resilience and productivity.

A related problem is the loss in genetic diversity of many crop plants and domesticated animals under market and commercial pressures to maximize productivity and profit. Many varieties and breeds, often evolved over centuries of local selection, have desirable features that adapt them to particular local environments, resist specific diseases or environmental extremes, or give them unique features, but that do not lend themselves to mass marketing. Today, strong pressures for the globalization of agriculture are eliminating much of this traditional diversity. Yet the future of sustainable agriculture may well lie with a much greater level of adaptation to local conditions in order to maximize all forms of productivity as well as a wider range of environmental services. Excessive globalization today could destroy much of the potential for better agriculture tomorrow.

Over the past 25 years, financial markets have grown and become internationally integrated. International capital flows have expanded rapidly, particularly foreign direct investment in developing and transition countries, which nearly tripled in the first six years of this decade (World Bank 1997a). The importance of the private sector in this globalization is illustrated by the fact that, in 1996, foreign exchange trading by the big investors amounted to some US$350 million million (Martens and Paul 1998), more than ten times the world's GDP of about US$30 million million (World Bank 1998). Total revenue of the top 500 companies was about US$11 million million, 50 per cent each for industry and services (Fortune 1998). Private foreign investment, concentrated in a limited number of developing countries, was about US$250 000 million, compared to overseas development assistance (ODA) of less than US$50 000 million (see graph below). These figures demonstrate the overriding importance of the private

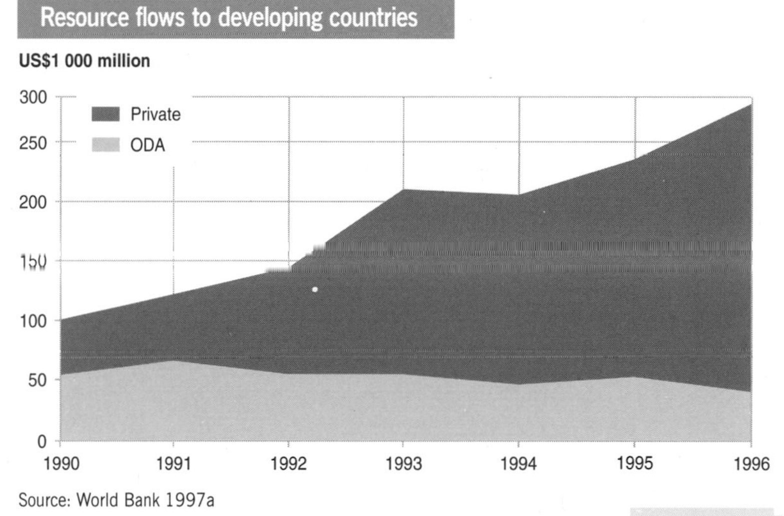

Source: World Bank 1997a

sector in the world's economy and, consequently, in environmental issues.

In 1996, private foreign investment was about five times larger than overseas development assistance

This massive increase in private financial flows has been paralleled by a decrease in ODA by governments leading, in effect, to a decreased capacity of public sector and multilateral agencies to deliver public goods such as environmental health. Transnational corporations, private banks and pension funds cannot be expected to cover major public health and environmental infrastructure costs. In addition, public agencies in most countries are hampered by their tremendous debt burden, and environmental agencies and activities are often among the first to be cut back in order to manage deficits and interest payments (Martens and Paul 1998). The international financial situation thus has a direct impact on the ability of countries to address important environmental and social issues.

Equally, important factors in globalization are the rise of transnational corporations and the consequent diffusion of new technologies and common work practices. The operations of these giant corporations, together with the revolution in communications technology and transportation, have enabled the growth of truly global systems of production and distribution. Globalization has been accelerated by important policy and institutional changes, including the dismantling of trade barriers and capital controls, and the creation of a multinational trade regime regulated by the World Trade Organization (WTO). Partly as a result of public pressure, the private sector is now taking an increasingly responsible attitude towards the sustainable management of the global environment. However, this will not be sufficient in itself to address all global environmental problems.

The International Centre for Trade and Sustainable Development

The International Centre for Trade and Sustainable Development (ICTSD) was established in Geneva in September 1996 to contribute to a better understanding of development and environment concerns in the context of international trade. It exists to help NGOs secure information on, and communicate their concerns to, key policy-making fora on trade and development – WTO, UNCTAD, other intergovernmental organizations and national governments. At the same time, the Centre aims to increase awareness among policy-makers and trade officials of the NGO community's work on trade, development and environment. It helps those involved in environment and development work understand how the trade policy-making process operates, and what negotiations and issues it deals with, and encourages decision-makers to understand the importance of integrating sustainable development imperatives into their priorities. ICTSD's role is to facilitate cooperation and integration through objective information dissemination, policy dialogues and research support.

The Centre is governed by a small coalition of well-known organizations: Consumer Unity and Trust Society (India), Fundación Futuro Latinoamericano (Ecuador), the International Institute for Sustainable Development (Canada), the Swiss Coalition of Development Organizations and the World Conservation Union (IUCN).

Source: ICTSD 1998

Trade

International trade has grown much faster than global GDP over the past 25 years and the international community is committed to further trade liberalization. The Uruguay Round GATT negotiations in 1994 explicitly incorporated the issue of environment into the future work programme of WTO. The aim of the WTO Committee on Trade and Environment has since been to promote mutually-supportive trade and environment policies. Trade-driven economic growth has brought increasing wealth to many countries, and helped to finance environmental protection measures. However, trade can and does harm the environment. Where environmental issues are not incorporated in economic prices and decision-making, trade can magnify unsustainable patterns of economic activity and resource exploitation. Conflicts between trade liberalization and environmental protection have already arisen, and the disturbing and short-sighted emerging pattern seems to be that national environmental protection measures are being challenged on the grounds that they erect barriers to trade. As an example, WTO recently ruled that the United States could not discriminate against imports of shrimp caught without the use of turtle excluder devices which allow sea turtles to escape from shrimp nets (WTO 1998). Similar attempts to protect dolphins and sea birds from the effects of industrial-scale fishing practices have also been struck down (GATT 1991).

Other international organizations, including OECD, UNEP and UNCTAD, are addressing trade and environment, and a number of new organizations have been created to further understanding of trade and environment issues, including the the International Centre for Trade and Sustainable Development (see box left).

Given the expected growth in world trade in the coming decades, and the pressure for action to counter increasing environmental depredation, future conflicts seem more, not less, likely to arise. In a 1998 speech to WTO, the Executive Director of UNEP denounced the dichotomy between trade liberalization and protectionism as obsolete. The real challenge will be to ensure that future trade liberalization is pursued with a view to maximizing overall human welfare, he said. This must include effective and cost-efficient management of the environmental resources and environmental quality on which human livelihoods and human health depend (Töpfer 1998).

International debt

One of the signs of imbalance in the international economic system is the excessive level of international debts accumulated by many nations. Deteriorating terms of trade for developing countries exporting agricultural and other commodities have made it increasingly difficult for these countries to reimburse their debts. From the environmental perspective, the need to pay off these debts has driven many developing countries to sell off their natural resources, particularly timber and minerals, for whatever price they could obtain, often in environmentally-destructive ways. Export cash crops have been favoured over food production for local consumption. Environmental standards have been kept low or non-existent to help attract foreign investment. Structural adjustment programmes have required reductions in government expenditures, with the environment being one of the easiest areas to cut. The indebted countries have thus been pushed towards further environmental deterioration.

Demographic changes

In parallel with these changes there have been profound demographic shifts, as people have migrated, and continue to migrate, from rural to urban areas in search of work and new opportunities. Since 1950, the number of people living in urban areas has jumped from 750 million to more than 2 500 million people. Currently, some 61 million people are added to cities each year through rural to urban migration, natural increase within cities, and the transformation of villages into urban areas (United Nations Population Division 1997). Urbanization creates new needs and aspirations, as people work, live, move and socialize in different ways, and require different products and services. Urban environmental impacts and demands are also different.

By 2025, the total urban population is projected to double to more than 5 000 million people, and 90 per cent of this increase is expected to occur in developing countries (United Nations Population Division 1997). Most of the world's children born in the 21st century will grow up in cities, with their perceptions and consumption behaviour shaped by an urban environment. The innate environmental sensitivity of people raised on the land or close to nature is being lost.

The demographic shift that has not been considered or allowed, because of its political sensitivity, is the globalization of population movements. The free movement of capital is now seen as normal, and the uninhibited free trade in goods and services is the goal of governments through WTO. However, genuine globalization should also imply a

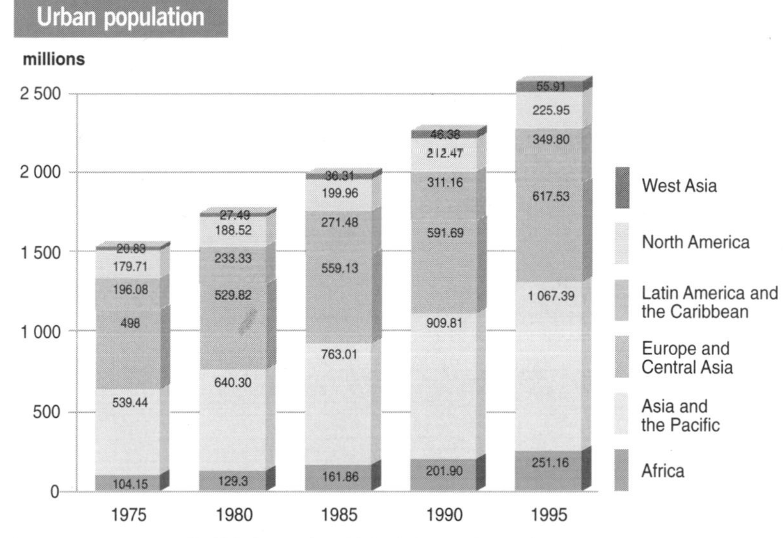

Source: compiled by UNEP GRID Geneva from United Nations Population Division 1997 and WRI, UNEP, UNDP and WB 1998

Some 61 million people are added to cities each year through rural to urban migration, natural increase within cities, and the transformation of villages into urban areas

third condition: that all people should be able to move freely to live and work wherever they like. This is the one change that would allow optimization of the population to environmental carrying capacity, and a rapid reduction in the economic and social disparities between countries that are so destabilizing at present. Efforts in the European Union to eliminate barriers to internal population movements are a precursor of what will be required.

Consumer culture

The global market and the purchasing power of an increasingly wealthy and urban population are driving the homogenization of lifestyles and popular culture. The late 20th century 'consumer society' can be characterized by a growing emphasis on the individual, a search for wider opportunities and experiences, a desire for comfort and autonomy, and personal material accumulation. The advent of international advertising, electronic communications and wide access to the mass media have fed a worldwide public appetite for new and more products, and for travel. Rising affluence has fuelled the 'Western' model of consumption, and its emulation all over the world. And, though developing countries still account for less than 20 per cent of global GDP, many of their people are joining the consumer society. Per capita incomes are rising, and habits of diet, mobility and resource consumption are changing to reflect industrial country patterns.

This consumer lifestyle as presently practised is

environmentally wasteful and inefficient, requiring large quantities of resources per capita and generating wastes that create further environmental problems when they are disposed of and released into the environment. Yet this does not have to be the case. Technological change can reduce resource use many times without lowering the standard of living. Efforts to increase environmental efficiency, reduce waste and introduce recycling are growing and spreading. It is widely recognized, at least by many NGOs and the wealthiest governments in OECD, that a tenfold reduction in resource consumption in industrialized countries is a necessary long-term target if adequate resources are to be released for the needs of developing countries (von Weizäcker and others 1995, OECD 1998). An extract from the 1997 Carnoules Declaration calling for nations to adopt this approach is reproduced in the box on the left.

Alongside the consumer culture, the world has other value systems and lifestyles which may be less visible and invasive but which represent the rich diversity of human experience and fulfilment. Many of these are more respectful of the environment, and provide options worth considering in the move towards more sustainable forms of society. The poor are also cut off from the consumer society, which is still largely irrelevant to their struggle for existence. A lifestyle that excludes one-third of the world's population, however dominant it may appear at the moment, should not be regarded as the supreme achievement of 20th-century civilization.

The 1997 Carnoules Statement

The following is extracted from the Factor 10 Club's 1997 Carnoules Statement to government and business leaders:

'Within one generation, nations can achieve a ten-fold increase in the efficiency with which they use energy, natural resources and other materials.

The Factor 10 Club, an international body of senior government, non-government, industry, and academic leaders working out of Germany's Wuppertal Institute, believes that such a goal is within the reach of technology and, with appropriate policy and institutional changes, could be brought within the reach of economics and politics. In the process, we should see a steady improvement in the quality of life of communities, new opportunities and improved competitiveness for business, expanded possibilities for employment, and an increased potential for wealth creation and its more equitable distribution ...

Most significantly, we enter the new Millennium with the transition already under way. During the past few decades, economic and technological changes have resulted in a reduction in the demand for energy and some materials per unit of production. The link between growth and its impact on the environment has also been severed. Indeed, a new economy has begun to emerge, one that is more efficient and potentially more sustainable. It is marked by people producing more goods, more jobs, and more income while using less energy and resources for every unit of production ...

Rising levels of consumption by the rich and a doubling of the world's population over the next 40-50 years would require a factor 4 increase in food production, a factor 6 increase in energy use and at least a factor 8 growth in income. If this is to be achieved without pushing the planet beyond certain critical thresholds that we are only now beginning to understand, governments must support policies that encourage industry and society to achieve ever-greater levels of energy and resource productivity and dematerialization.

We call upon governments, industry and international and non-governmental organizations, to adopt a Factor 10 increase in energy and resource productivity as a strategic goal for the new Millennium.

Some governments and international and business bodies have already begun to move in this direction. Austria and the Netherlands, for example, adopted this strategic goal in 1995. In Germany, the national parliament (Bundestag) holds a regular enquete on material flows in the German economy, in order to establish a basis for further policy decisions. The World Business Council for Sustainable Development and the United Nations Environment Programme have jointly called for a factor 20 increase in eco-efficiency. In cooperation with the newly founded Factor 10 Institute, the Ministry of Economics in Vienna is now preparing a countrywide information campaign to help small and medium-sized businesses to design eco-intelligent products. The Canadian government has established a Commissioner for the Environment and Sustainable Development who will review government policies and programmes against sustainability criteria and report annually to Parliament. The OECD is exploring Factor 10 as a potential thrust. In the United States, the President's Council on Sustainable Development has taken an active interest in Factor 10 and eco-efficiency.'

Source: Carnoules 1998

Technology

Demand for technical innovation – driven by economic growth, industrialization and social development – has been met through a sharp rise in the numbers of practising scientists and engineers – a 15-fold increase in the past 50 years (Hammond 1998) and greatly increased communication within the research and development community. More than two million research papers are now published each year and the number is rising. In the industrialized economies particularly, technological innovations have led to greater efficiency in energy and materials use, with many products being reduced in size and weight through the use of lighter materials such as aluminium in place of steel, and plastics in place of metals. Improved technologies mean that recycling rates for many key raw materials have also increased. In addition, demand has shifted away from heavy goods towards less-material intensive products, consumer goods and service industries.

These trends constitute the phenomenon of dematerialization, driven in part by relative price changes and substitution (Bagnoli and others 1996), which has significantly slowed the rate of increase in the use of many (though not all) raw materials in the industrialized countries. This means that economic growth can, to some extent, be delinked from growth in resource use. Per capita use rates for steel, timber and copper, for example, have generally stabilized or even declined in the OECD countries. Resource intensity (the quantity of energy and materials required for constant economic output) has fallen by about 2 per cent per year since 1970 (Glyn 1995). In absolute terms, however, consumption of energy and most raw materials continues to rise in countries which still have population growth. Resource intensity in developing countries is still high, though there is evidence that efficiency is improving. Some Asian economies are becoming fuel-efficient at lower levels of per capita income than was the case in the developed world.

Cleaner technologies have played a critical role in many of the successes in pollution control in industrialized countries recorded since the 1970s. Flue scrubbers at power stations, waste recovery and recycling systems, and catalytic converters fitted to vehicles are now mainstream technologies in the developed countries. As regulation and compliance regimes have gradually been tightened, a major global environmental market has emerged for the environmental technologies and services required to meet these new standards.

Transport

During the 20th century, there has been a shift away from rail and water transport in favour of road and air. The single most dramatic change has been the rise in personal mobility in developed countries, encouraged by cheap oil, affordable motor cars and lifestyles built around commuting, out-of-town shopping, dispersed families and leisure activities. Since World War II, the number of vehicles on the road has risen from about 40 million to some 680 million (International Road Federation 1997). The fastest growth is now found in the developing world, though car ownership is still low (see bar chart).

If current rates of expansion continue, there will be more than 1 000 million vehicles on the road by 2025. Transport now accounts for one-quarter of world energy use, and about one-half of the world's oil production; motor vehicles account for nearly 80 per cent of all transport-related energy. Transport is thus a major contributor to greenhouse gas emissions and urban air pollution. Transport infrastructure – roads, car parks, airports, rail lines – is also responsible for substantial land use, habitat degradation and fragmentation. The transport sector has, so far, proved highly resistant to attempted policy reforms.

Numbers of motor vehicles

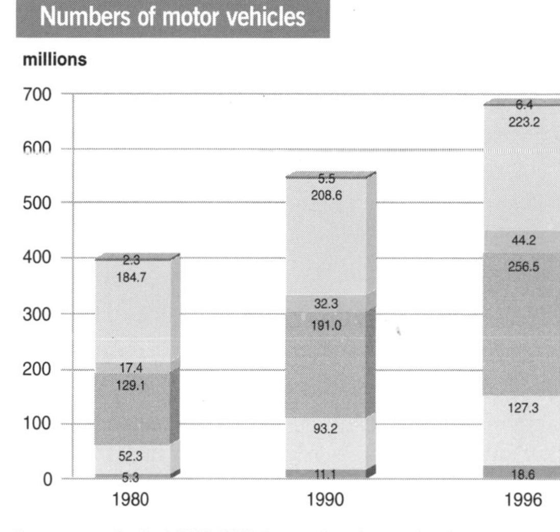

Source: compiled by UNEP GRID Geneva from International Road Federation 1997

Note: 1980 figure for North America is estimated

If current rates of expansion continue, there will be more than 1000 million vehicles on the road by 2025

Improvements in fuel efficiency and vehicle emission reductions have consistently been offset or outpaced by volume growth. However, the economic costs of congestion and pollution, in terms of lost production and health care, are increasingly recognized. A recent study for the United States estimated the total annual costs of congestion (including lost productivity, wasted fuel and increased accident insurance) at US$340 per capita (FHA 1990).

Air transport is also growing very fast. Dense air traffic is now leading to long delays on some flights, particularly in Europe, and this in turn is encouraging the use of high-speed trains (see Chapter 2, Europe).

Responses

Environmental policy issues

Regulation is still the core instrument of environmental policy. The industrialized countries enacted a 'first generation' of legislation in the 1960s and 1970s, aimed principally at protecting human health from the impacts of air, water and soil pollution. In the 1990s, many countries reformed their sectoral environmental approaches into better integrated strategic policies. Comprehensive environmental protection laws are now in place in the newly industrializing countries and other developing countries. Legislation concentrates on standards, bans, permits and quotas and, in some cases, specification of technologies or technical standards to be used in industry. These instruments have long been favoured because they promise certainty of outcome – though without costly monitoring and enforcement, this promise may not be realized.

Recognizing that regulation will be ineffective if poorly drafted, or not supported by adequate inspection and enforcement agencies, some developing country governments, notably in the Asia and Pacific region, have also increased their spending on environmental personnel. For example, the annual increase in the number of officials in central environment agencies in the Asian newly industrializing economies during 1989–94 was 7.6 per cent, compared with 4.7 per cent in all government organizations (ESCAP/ADB 1995). Command-and-control legislation has its limitations, however – notably the time needed to draft, enact and implement adequate laws, the inflexibility of regulation and possible cost-inefficiencies in implementation on the ground. In addition, problems with inspection and enforcement, especially in rapidly developing nations, appear to be worsening as limited capacity and resources face an explosion of industrial activity and urbanization.

The use of economic instruments for environmental management is gaining acceptance in the OECD countries and, increasingly, elsewhere. The most recent survey by the OECD Environment Directorate indicates that the number of economic instruments used by member states for environmental protection has increased by nearly 50 per cent over 1987 levels (OECD 1997).

There is also a clear trend towards the integration of environmental policy-making into the broader sphere of sustainable development. Following the Rio Earth Summit, many countries established their own national councils for sustainable development to address sustainability issues and coordinate national responses to the United Nations Commission on Sustainable Development (UNCSD). These bodies have been set up in more than 130 countries since 1992, and more than 50 countries have initiated official government mechanisms to formalize participation with the public and other stakeholders.

This increasing involvement of civil society alongside the public and private sectors is a significant new development in environmental governance. In North America, for instance, there were civil society consultations in late 1996 leading to the Hemispheric Summit of the Americas on Sustainable Development. Civil society groups in many parts of the world are involved in community indicator networks, watershed-based initiatives, efforts by the International Council for Local Environmental Initiatives, Habitat's network of community initiatives, and the environmental activities of indigenous people's and women's movements. This localization of national and global initiatives is an appropriate way to tackle many types of environmental problems and should become increasingly important in the future.

UN Development Assistance Framework

The United Nations Development Assistance Framework (UNDAF) is a key component of the Secretary-General's reform proposal. Action 10 (a) of his *Renewing the United Nations: a programme for reform* states:

'In order to achieve goal-oriented collaboration, programmatic coherence and mutual reinforcement, the United Nations programmes of assistance will be formulated and presented as part of a single United Nations Development Assistance Framework with common objectives and time-frame. Programme funds managed by each of the programmes and funds will be included in the document, but remain clearly identifiable. Preparations would entail collaborating programming and close consultation with Governments, including compatibility with Country Strategy Notes wherever they exist.'

In August 1997, the United Nations Development Group initiated a pilot phase to test the operationalization of Action 10 (a) in 18 countries. The experience gained will be used to guide the implementation of the UNDAF process in other countries.

Source: UNDG (undated)

The UNDAF process at work in Mozambique

The UNDAF pilot exercise in Mozambique provided an effective framework for increased interagency collaboration in the country. The UNDAF preparation process, led by the UN Resident Coordinator and the country team, involved a series of consultations where representatives from all UN agencies provided input on how to enhance collaboration and programme coordination. The UNDAF, based on the Government's Country Strategy Note and UN Common Country Assessment, helped focus United Nations development assistance on three strategic objectives:

- increasing access to and quality of basic social services, infrastructure development and employment generation;
- promoting good governance and strengthening the capacity of civil society organizations;
- promoting the sustainable management of natural resources.

The United Nations country team is working closely with the World Bank, bilateral donors and NGOs to improve overall coordination of development activities in Mozambique.

On the other hand, a worrying trend has emerged with the recent decline in environmental expenditures in many countries in the face of budgetary constraints. More positively, intergovernmental processes have not lost momentum. The series of United Nations conferences and summits on key issues of development, notably those concerned with environment and development (UNCED 1992), with small island developing states (SIDS 1994), population (ICPD 1994), human settlements (Habitat II 1996) and food security (WFS 1996), all explicitly addressed the role of natural resource conservation and environmental quality in achieving broadly-based development goals. A high degree of international consensus now exists on the principles and frameworks for action, although the practicalities of implementation remain an immense challenge. In a new pilot initiative, the United Nations Development Assistance Framework (UNDAF) has been launched in an attempt to provide a common framework for all development funds, programmes and agencies of the UN system (see box left). The exercise also aims to improve coordination of follow-up action to global conferences and relevant decisions of the General Assembly.

One challenge is to develop integrated approaches to planning and analysis. A key constraint to the emergence of strong sustainability institutions is the fragmentation of research into disciplines, government units into sectors, and so on. Designing frameworks for linking across subjects and sectors, over various spatial scales, regions and themes, to give a more integrated perspective, is becoming essential to a full understanding of the planetary and human environments, as the range of subjects treated in this report illustrates. Just as important is the development of a cadre of trained professionals in integrated environmental assessment, equipped with appropriate analytical tools and models, and supported by global observing systems and other data collection processes. These will be important ingredients in the institutional preparations for sustainable development.

The concept of development

The concept of development was defined in the 1950s and 1960s as a largely economic process, in which wealth would trickle down and improve human welfare. Today this has now given way to much broader definitions of development. UNDP has focused on human development and issued a series of *Human Development Reports* exploring critical issues such as gender inequality, growth, poverty and consumption patterns (UNDP 1998 and previous years). It also calculates a human development index based on life expectancy, adult literacy, school enrolment and GDP per capita (see bar chart). The Rio Earth Summit in 1992 defined sustainable development, through its action plan *Agenda 21*, as a multifaceted process involving the full range of environment and development issues and requiring the participation of governments, international organizations and major groups. The World Bank has expanded its definition of wealth to include produced assets, natural capital, and human and social capital, with the latter generally being the major component of national wealth (World Bank 1997b).

Efforts to develop indicators of sustainable development have raised the challenge of defining

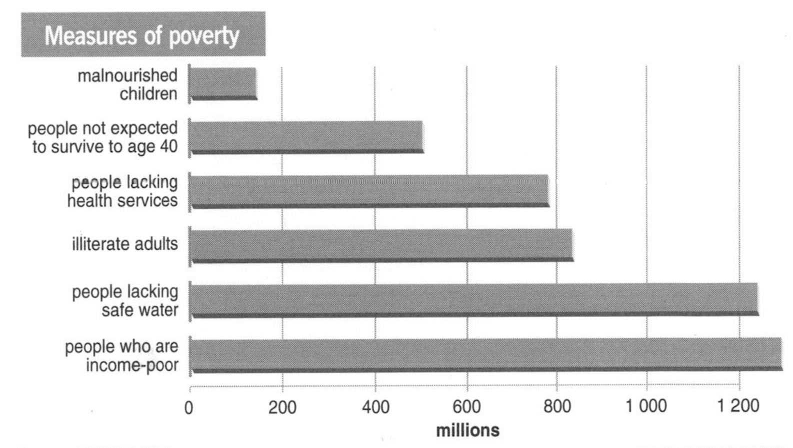

Source: UNDP 1997

Many millions of people still suffer from different forms of poverty; more than 1 300 million are 'income-poor' and have to live on less than US$1/day

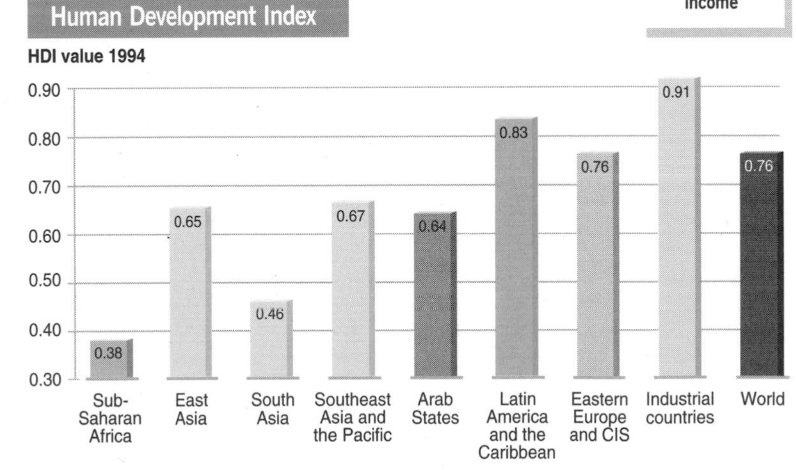

Source: UNDP 1997

Note: not all regions correspond with those of *GEO-2000*

UNDP's Human Development Index is a composite index based on life expectancy, educational attainment and income

such a broad subject through quantitative measures. The countries pilot testing national indicators of sustainable development in support of the UNCSD programme on indicators find they need at least 50 indicators to cover the major dimensions of sustainability (Government of the Czech Republic 1998). The UNCSD study calls for economic, social, environmental and institutional indicators covering driving forces, states and responses across all the programme areas of *Agenda 21*, with 134 indicators identified in the first phase. These activities have made clear how many dimensions there are to development. They also highlight the need to develop clearer targets and goals, and new indicators for the less tangible aspects of development including individual welfare, community cohesion, institutional development, knowledge and culture.

Science and research

The role played by scientists in advising intergovernmental policy processes on sustainable development has expanded rapidly. A number of international scientific research programmes for the global environment, such as the International Geosphere-Biosphere Programme, the World Climate Research Programme and the International Human Dimensions of Global Change Programme, are addressing the challenging questions raised by global change and human pressures. At the level of operational information gathering, an increasing number of national institutions and experts are contributing to the global monitoring efforts of the different global environmental observing systems: the Global Climate Observing System, the Global Ocean Observing System and the Global Terrestrial Observing System.

Scientists are also playing an increasing role in policy advice via participation in bodies such as the Intergovernmental Forum on Chemical Safety, the Intergovernmental Forum on Forests and the subsidiary scientific and technical bodies under the climate change, biodiversity and desertification conventions. These bodies provide scientific and technical input to intergovernmental negotiations and the implementation of multilateral environmental agreements.

A third role which has grown rapidly in importance is that of the independent scientific assessment processes; notable examples include the Intergovernmental Panel on Climate Change and the Joint Group of Experts on Scientific Aspects of Marine Environmental Protection (Fritz 1998). These activities are being stimulated by the concept of sustainable development, with its emphasis on the integration of environmental, economic and social concerns. A concept of 'new national systems of innovation' is emerging, which favours more interaction among universities, scientific research organizations, government agencies and the private sector. The net result should be an improvement in the scientific basis of policy-making (UNCSD 1998a).

While such research and improved environmental monitoring must continue, the need for further study should not be taken as an excuse for postponing action on critical environmental problems. In almost all areas, there is enough knowledge to initiate actions such as reducing harmful subsidies or organizing public/private partnerships for resource management. New information from research can then help to refine policy action. There is a particular need for more observations and information to improve the monitoring of policy effectiveness and to strengthen accountability.

Business and industry

Recent years have brought greater recognition of the complexity of environmental issues, and some withdrawal of government from the detailed oversight of industrial operations. Instead of legislative micro-management, objectives are set and the details of implementation left to industry. The response has been a trend towards greater corporate responsibility, realized through self-regulation, corporate environmental policies, voluntary codes of practice (such as the chemical industry's Responsible Care Programme), and the use of environmental audits and open reporting. Such initiatives are becoming more mainstream, particularly now that the total quality management concept has been extended to the environmental sphere through systems such as the European Union's Environmental Management and Audit Scheme (EMAS), the British Standards Organization BS7750 and the ISO 14 000 series of management standards.

Many companies are pioneering cleaner production systems under the rubric of 'industrial ecology', which aim to close substance loops and thereby reduce or eliminate toxic pollution and waste generation. Cleaner production has proved popular in industry, at least partly because the costs of this approach tend to diminish over time, while the costs of controlling pollution and cleaning up after the event become increasingly high as new regulations are introduced (see graph). A range of new tools, such as the eco-compass (diagram below), has been produced to aid in the ecodesign of new products and improve the environmental performance of existing products.

Many of these initiatives are being undertaken in partnership with national governments or international organizations. During the 1998 UNCSD session, the Secretary-General of the International Chamber of Commerce reaffirmed the organization's commitment to bringing the 'financial, managerial and technical expertise' of industry and business to sustainable development. However, industry representatives emphasized that the sector is looking to governments to establish appropriate frameworks, such as legal and fiscal incentives, which would encourage speedier introduction of clean technologies and other measures (UNCSD 1998b).

Industry's environmental performance is being increasingly held to account by the general public. NGOs are emerging as unofficial industry 'watchdogs', in countries such as the Philippines as well as in more traditionally activist regions such as North America. Greater public scrutiny is being facilitated by right-to-know legislation enacted in many places, including Canada, the United States and the European Union.

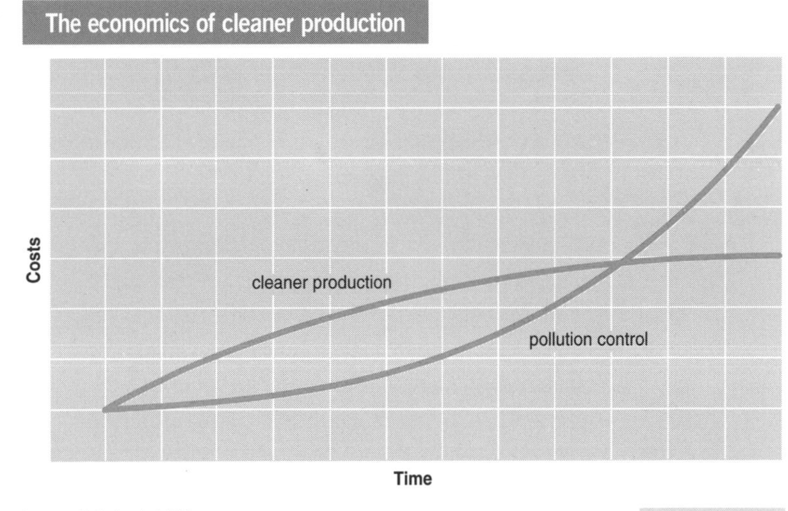

Source: Rabobank 1998

The Århus Convention on Access to Information, Public Participation in Decision Making and Access to Justice in Environmental Matters adopted in June 1998 is a good example.

Despite progress, however, there is often a gap between the environmental concern and performance of leading multinationals and large companies, and that of small and medium sized enterprises (SMEs). The largest companies have both the resources to invest in environmental action and the visibility to motivate such action. Small companies, which represent a major

While the costs of pollution control continue to escalate with time, those of cleaner production diminish

The eco-compass

The eco-compass developed by Dow Europe is a useful tool for assessing the environmental impact of a product. The assessment is made by constructing a series of concentric hexagons, with each corner representing a different environmental dimension. These are (moving clockwise from the top of the diagram):

- service extension (for example making products last longer);
- revalorization (re-manufacturing, reuse and recycling possibilities);
- resource conservation (renewability of materials used);
- energy (consumed per unit of production);
- material intensity (weight of resources used per unit of production);
- health and environment (risks to people and ecosystems).

The concentric hexagons represent scores of 0-5, starting with 0 at the centre and 5 at the perimeter. All uses of the eco-compass must start with a baseline product, which is given a score of 2 on all six dimensions. The product to be compared is then evaluated on a factor basis for each dimension – for example, if the manufacture of the baseline product uses 100 kWh of energy per unit of production, and the new product uses only 25 kWh, the new product scores a factor of four. When scores are plotted for all six dimensions, the eco-compass takes on a new shape, making it easy to compare its environmental performance with that of the baseline.

service extension
revalorization
resource conservation
energy
material intensity
health and environment
0 1 2 3 4 5
base product
new product

Source: Fussler 1996, Rabobank 1998

part of industrial activity around the world, have neither. How to get the positive experience of businesses at the cutting edge of environmental involvement to filter down to the mass of industrial activity in SMEs below them is one of the unresolved challenges of the moment, though attempts are being made (EEA 1998).

Service industries are also becoming more engaged in environmental issues. A recent survey of financial institutions worldwide indicates that the sector now has a high awareness of environmental issues. Respondents to the survey believed that environmental issues will become increasingly integrated with core business activities over the next decade, and that financial institutions will be more likely to look for transactional opportunities with environmentally-related businesses (UNEP 1995).

The UNEP Financial Services Initiative on the Environment

The UNEP Financial Services Initiative on the Environment is intended to help integrate environmental considerations into all aspects of the financial service sector's operations. A core part of the initiative is to foster endorsement of both the UNEP Statement by Financial Institutions on the Environment and Sustainable Development (written in 1992, revised 1997) and the Statement of Environmental Commitment by the Insurance Industry (1995), which commit signatories to incorporating environmentally-sound practices into their operations. More than 100 financial institutions and 80 insurance companies, from more than 25 countries, have signed their respective statements.

Banks and other lending institutions are now considering 'environmental risks' alongside more traditional banking risks, and many lenders now operate screening practices as part of their asset management (OECD 1998 and 1997).

The insurance industry is another sector taking an active interest in environmental and sustainability issues. Liability for clean-up costs is one risk, and climate change from global warming is seen as a potentially serious threat to the industry's financial stability. Economic damage from weather-related disasters exceeded US$200 000 million during 1990–96, four times the total losses for the previous decade (Worldwatch Institute 1997). In 1995 the insurance industry, aided by UNEP, produced a Statement of Environmental Commitment which promised – or warned of – greater attention to environmental risks in core activities such as loss prevention, product design, claims handling and asset management (UNEP 1998).

Employment policies and changes

High technology is promoting the emergence of 'post-industrial' economies in the wealthier OECD countries, which are characterized by shifts away from heavy industry and manufacturing, towards services, high-technology industries such as software, and the cultivation of high value-added niche markets. The resulting decentralization, labour mobility, personal flexibility and higher skill levels requires a transformation of work force skills and working habits. Unfortunately, the industrial transition is occurring more rapidly than the adjustments that the labour market can make, leading to considerable structural unemployment, labour unrest and social tensions. Nevertheless, opportunities abound to reduce the environmental pressures created by more traditional work patterns dependent on commuting by private car and working in energy-intensive commercial buildings. For example, traffic congestion and air pollution might be reduced through encouragement of telecommuting, or on-line working. Another trend is the expansion of employment opportunities in the relatively new sector of environmental technologies and services, which have risen dramatically since the 1980s. In 1990, the environmental equipment and services industries and environment-related activities in the OECD countries employed nearly 8 million people (OECD 1996).

Consumer awareness and information

Thanks to widespread economic growth, 3 000 to 4 000 million people have experienced substantial improvements in their incomes and standards of living since the 1960s (UNDP 1997). Overall consumption, unsurprisingly, has risen dramatically to what is probably an unsustainable level. Nevertheless, there are encouraging signs of real interest among consumers in more environmentally-sustainable products and services, and a growing number of initiatives by business and NGOs to supply this new market. For example, concerns over declining fish stocks have prompted Unilever and the World Wide Fund for Nature to form the Marine Stewardship Council, which will establish industry-wide principles for sustainable fishing (see box right). Fish harvested according to the Council's standards will be eligible for certification, or eco-labelling, so that consumers can choose to buy the more sustainable product. A similar certification scheme has been launched by the Forest Stewardship Council to guide consumers towards wood products from sustainably managed forests. A number of cooperative organizations have sprung up to promote the 'Fair Trade' movement, which aims to achieve fair prices for small farmers who produce coffee, fruits or vegetables using environmentally-friendly methods. Such products are beginning to move from niche markets to the mainstream: Fair Trade coffee, for example, now commands five per cent of the UK market (IIED 1997). A small but influential number of companies has taken the decision to 'green' their product lines, for example, through the use of organic cotton in clothing.

There are also the other third of the world's people, the poor who have not benefited by improved standards of living. They need to be empowered with the knowledge and minimal resources necessary to ameliorate their own situation in ways that are also environmentally sound and sustainable. The benefits of environmental science and management should not be reserved only for the well-to-do and educated, but should also be translated into forms accessible to all the world's population.

Agenda 21 and subsequent declarations stress the critical role of education in instilling greater understanding of the concept of sustainability in the next generation. 'Greener' and more integrated educational systems can foster appreciation of the ways in which economic, social and ecological systems are interdependent. In an increasingly urbanized society, the formal educational system is called upon to replace environmental learning that once took place through direct contact with nature. Progress so far appears disappointing, however. It is difficult to introduce new subjects into school curricula, and limited change is visible in most university curricula (UNU 1998), although there has been rapid growth in the specialized environmental courses now offered by universities throughout the world. The mass media have played an important role in raising public awareness on the environment, especially in relation to disasters, but environmental coverage is still superficial and scattered.

Towards sustainable fish and forests

The Marine Stewardship Council

The World Wide Fund for Nature (WWF) formed a conservation partnership with Unilever in 1996 to create market incentives for sustainable fishing by establishing an independent Marine Stewardship Council (MSC). The MSC's mission is to work for sustainable marine fisheries by promoting responsible, environmentally-appropriate, socially beneficial and economically-viable fisheries practices while maintaining the biodiversity, productivity and ecological processes of the marine environment.

The Council is establishing a set of globally-agreed principles and criteria for sustainable fishing, developing a process for international implementation and conducting test cases for certification of fisheries.

Only fisheries meeting these standards will be eligible for certification by independent, certifying firms accredited by the MSC. Products from fisheries certified to MSC standards will be marked with an on-pack logo. This will allow consumers to select fish products that they know come from sustainable, well-managed sources, thus creating a market incentive for industry to shift to sustainable fishing practices.

The Forest Stewardship Council

The Forest Stewardship Council (FSC) was founded in [illegible] to support environmentally-appropriate, socially beneficial and economically-viable management of the world's forests. It is an association of Members consisting of representatives from environmental and social groups, the timber trade and the forestry profession, indigenous people's organizations, community forestry groups and forest product certification organizations from around the world.

The FSC is introducing an international labelling scheme for forest products, which provides a credible guarantee that the product comes from a well-managed forest. All forest products carrying the FSC logo have been independently certified as coming from forests that meet the FSC Principles and Criteria of Forest Stewardship. Forest inspections are carried out by FSC-accredited certification bodies. Certified forests are visited on a regular basis to ensure they continue to comply with the Principles and Criteria.

The FSC also supports the development of national and local standards that implement the international Principles and Criteria of Forest Stewardship at the local level. These standards are developed by national and regional working groups which work to achieve consensus amongst those involved in forest management and conservation in each part of the world. FSC has prepared guidelines to help working groups develop regional certification standards.

Sources: MSC 1999 and FSC 1998

Conclusions

We live in a world of accelerating global change, where internationally-coordinated environmental stewardship is lagging behind economic and social development. It is a world where environmental gains through new technology and policies are overtaken by the pace and scale of population growth and economic development.

The globalization of the economy and society is accompanied by the globalization of resource management and environmental problems but the institutional response to this is lagging behind.

It is clear that environmental management cannot be separated from the improved management of human society. The global human ecosystem is threatened by grave imbalances in productivity and in the distribution of goods and services. A significant proportion of humanity still lives in dire poverty, and projected trends are for increasing discrepancies between those that benefit from economic and technological development, and those that do not. This unsustainable progression of extremes of wealth and poverty threatens the stability of the whole human system, and with it the global environment.

Environmental governance at all levels requires a new partnership between governments and civic society that can foster the eradication of poverty and an equitable distribution of environmental costs and benefits. Signs of such new partnerships and the development of regionally-conducive frameworks – such as in the European Union – are emerging but too often remain restricted to wealthier regions and to multinationals which are under public scrutiny from pressure groups in the developed world.

The environment cannot be separated from the human condition but is one essential complement of sustainable human development. The processes of globalization that are so strongly influencing the evolution of society need to be directed towards resolving rather than aggravating the serious imbalances that divide the world today. All the partners involved – governments, intergovernmental organizations, the private sector, the scientific community, NGOs and other major groups – need to work together to resolve this complex and interacting set of economic, social and environmental challenges in the interests of a more sustainable future for the planet and human society.

References

Adriaanse, A., Bingezu, S., Hammond, A., Moriguchi, Y., Rodenburg, E., Rogich, D., and Schultz, H. (1997). *Resource Flows: The Material Basis of Industrial Economies.* A joint publication of the World Resources Institute (WRI), the Wuppertal Institute, the Netherlands Ministry of Housing, Spatial Planning and Environment, and the National Institute for Environmental Studies (RIVM). WRI, Washington DC, United States

Bagnoli, P., McKibben, W. and Wilcoxen, P. (1996). *Global Economic Prospects: Medium Term Projections and Structural Change*, Brookings Discussion Paper in International Economics No. 121 (also United Nations University Center for Advanced Studies Working Paper No. 1). The Brookings Institution, Washington DC, United States

BICC (1998). *BICC Yearbook: Conversion Survey 1997.* Bonn International Centre for Conversion, Bonn, Germany http://bicc.uni-bonn.de/milex/milexdata.html

Carnoules (1998). *Carnoules Statement 1997* http://www.baltic-region.net/science/factor10.htm

CDIAC (1998). *Revised Regional CO_2 Emissions from Fossil-Fuel Burning, Cement Manufacture, and Gas Flaring: 1751–1995.* Carbon Dioxide Information Analysis Center, Environmental Sciences Division, Oak Ridge, Tennessee, United States. http://cdiac.esd.ornl.gov/cdiac/home.html

EEA (1998). *Environmental Management Tools for SMEs: a handbook.* European Environment Agency, Copenhagen, Denmark

ESCAP/ADB (1995). *State of the Environment in Asia and the Pacific 1995.* United Nations Economic and Social Commission for Asia and the Pacific, and Asian Development Bank. United Nations, New York, United States

FAOSTAT (1997). *FAOSTAT Statistics Database.* Food and Agriculture Organization of the United Nations, Rome, Italy http://www.fao.org

FHA (1990). *Estimates of Urban Roadway Congestion: 1990.* DOT-T-94-01. Federal Highway Administration, US Department of Transportation, Washington DC, United States

Fortune (1998). Fortune's 500/Global 500. *Fortune,* Special Issue 1998 http://www.pathfinder.com/fortune/global500/

Fritz, J. S. (1998). *Report on International Scientific Advisory Processes on the Environment and Sustainable Development.* UNEP/DEIA/TR.98-1. UNEP, Nairobi, Kenya http://www.unep.ch/earthw/sciadv.htm

FSC (1998). Forest Stewardship Council, http://www.fscoax.org/

Fussler, C., with James, P. (1996). *Driving Eco-innovation.* Pitman, London, United Kingdom

GATT (1991). *Tuna/Dolphin Report.* GATT Dispute Resolution Panel, Geneva, Switzerland, September 1991

Gies, F. and G. (1994). *Cathedral, Forge and Waterwheel: Technology and Invention in the Middle Ages.* HarperCollins, New York, United States

Glyn, A. (1995). Northern Growth and Environmental Constraints. In Bhaskar, V. and Glyn, A. (eds.), *The North, The South: Ecological Constraints and the Global Economy.* Earthscan, London, United Kingdom

Government of the Czech Republic (1998). *Fourth International Workshop on Indicators of Sustainable Development: Report.* 19–21 January 1998, Prague, Czech Republic. Charles University Environmental Center, Prague, Czech Republic http://www.czp.cuni.cz/csd/

Hammond, A. L. (1998). *Which World? Scenarios for the 21st Century.* Island Press, Washington DC, United States

ICTSD (1998). *International Centre for Trade and Sustainable Development.* ICTSD, Geneva, Switzerland http://www.ictsd.org

IIED (1997). *Unlocking Trade Opportunities: Changing Consumption and Production Patterns.* International Institute for Environment and Development, London, United Kingdom

International Road Federation (1997). *World Road Statistics,* 1997 Edition. IRF, Geneva, Switzerland, and Washington DC, United States

Lacey, R., and Danziger, D. (1999). *The Year 1000: what life was like at the turn of the First Millennium.* Little, Brown, London, United Kingdom

Martens, J. and Paul, J. A. (1998). The coffers are not empty: financing for sustainable development and the role of the United Nations. *Global Policy Forum,* July 1998 http://www.globalpolicy.org/socecon/global/paul.htm

MSC (1999). Marine Stewardship Council, http://www.msc.org/

OECD (1996). *Environmental Performance in OECD countries: progress in the 1990s.* OECD, Paris, France

OECD (1997). *Evaluating Economic Instruments.* OECD, Paris, France

OECD (1998). OECD Environment Ministers share goals for action. News Release, 3 April 1998. OECD, Paris, France

Rabobank (1998). *Sustainability: choices and challenges for future development.* Rabobank International, Leiden, the Netherlands

SIPRI (1998). *SIPRI Yearbook 1997.* Swedish International Peace Research Institute, Stockholm, Sweden

Töpfer, K. (1998). Statement by Klaus Töpfer, Executive Director, UNEP, to the WTO Symposium on Trade, Environment and Sustainable Development, Geneva, 17 March 1998

UNCSD (1998a). *Science and Sustainable Development.* E/CN.17/1998/6/Add.3. United Nations Commission on Sustainable Development, United Nations, New York

UNCSD (1998b). United Nations Commission on Sustainable Development (UN CSD), 20 April-1 May 1998, New York, United States http://www.un.org/esa/sustdev/csd.htm

UNDG (undated). *A Framework for Change: a report from the United Nations Development Group.* UNDG, New York, United States

UNDP (1997). *Human Development Report 1997.* Oxford University Press, New York, United States, and Oxford, United Kingdom

UNDP (1998). *Human Development Report 1998.* Oxford University Press, New York, United States, and Oxford, United Kingdom

UNEP (1995). *Global Survey: Environmental Policies and Practices of the Financial Services Sector.* UNEP, Geneva, Switzerland

UNEP (1998). *Financial Services and the Environment: Questions and Answers.* UNEP/ROE/98/3. UNEP Regional Office for Europe, Geneva, Switzerland

UNHCR (1998). *State of the World's Refugees, 1997-1998: A Humanitarian Agenda.* United Nations High Commissioner for Refugees, Geneva, Switzerland http://www.unhcr.ch/sowr97/statsum.htm

United Nations Population Division (1997). *World Population Prospects 1950-2050 (The 1996 Revision).* United Nations, New York, United States

United Nations Population Division (1998a). *World Population Prospects 1950-2050 (The 1998 Revision).* United Nations, New York, United States

United Nations Population Division (1998b). *World Population Projections to 2150.* United Nations, New York, United States

UNSTAT (1997). *1995 Energy Statistics Yearbook.* United Nations Statistical Division, New York, United States

UNU (1998). *Preparing for a Sustainable Future. Higher Education and Sustainable Human Development: Strategy for Future Action.* United Nations University, Paper prepared for the World Conference on Higher Education and Sustainable Human Development, Paris, 5-9 October 1998 (draft)

USACDA (1998). *Word Military Expenditures and Arms Transfers 1996.* US Arms Control and Disarmament Agency, Washington DC, United States
http://www.acda.gov

WBCSD (1997). *Signals of Change.* World Business Council for Sustainable Development, Geneva, Switzerland

von Weizsäcker, E., Lovins, A., and Lovins, H. (1995). *Faktor Vier.* Droemer Knaur, München, Germany

World Bank (1997a). *Global Development Finance 1997.* The World Bank, Washington DC, United States

World Bank (1997b). *Expanding the Measure of Wealth: Indicators of Environmentally Sustainable Development by the World Bank.* Environmentally Sustainable Development Studies and Monograph Series No. 17. The World Bank, Washington DC, United States

World Bank (1998). *World Development Indicators, 1998.* The World Bank, Washington DC, United States

Worldwatch Institute (1997). *Vital Signs 1997-98.* Worldwatch Institute, Washington DC, United States

WTO (1998). *United States Import Prohibition of Certain Shrimp and Shrimp Products.* 15 May 1998, Report of the Panel, World Trade Organization, Geneva, Switzerland

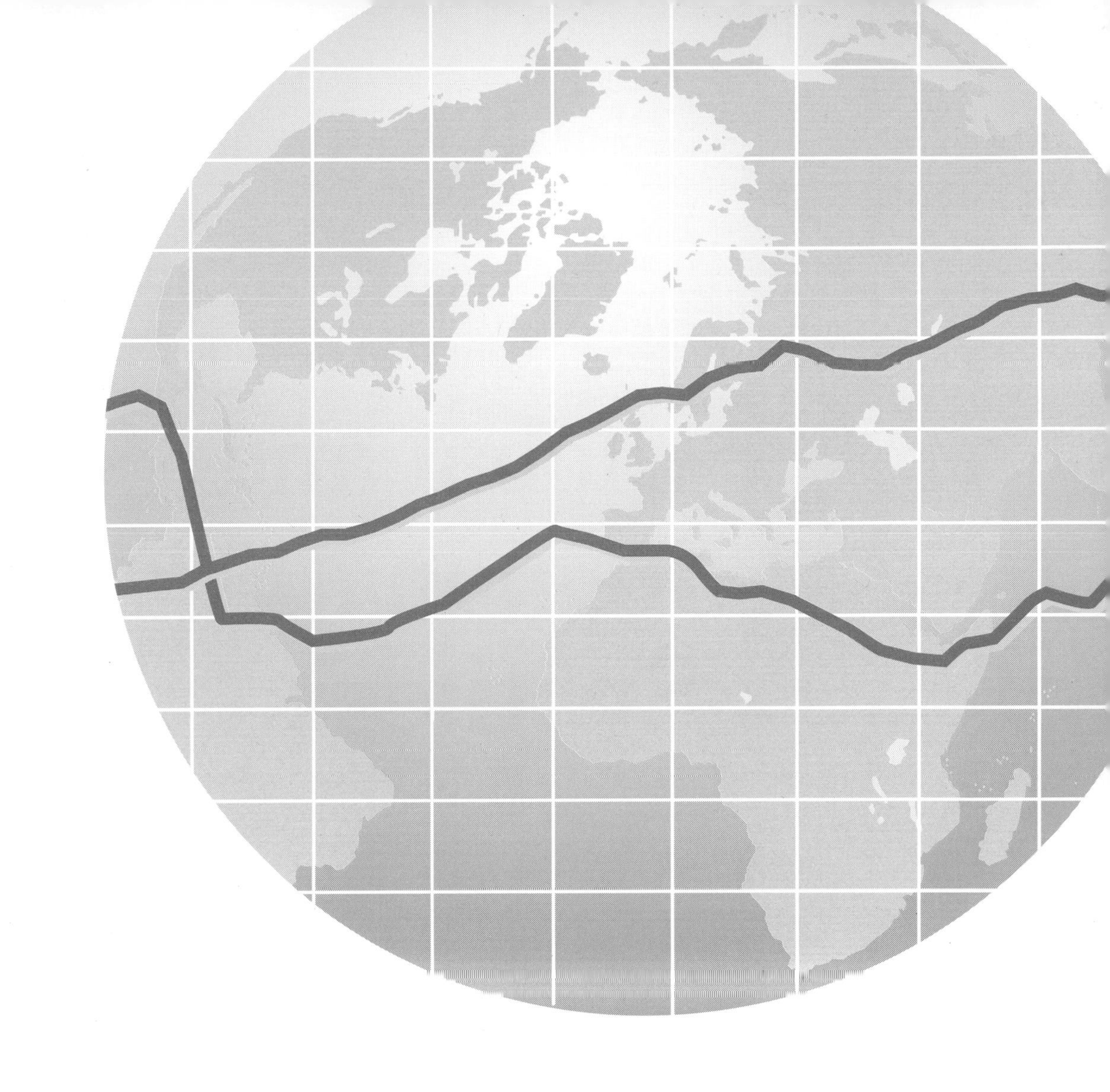

Chapter Two

The State of the Environment

Global and Regional Synthesis

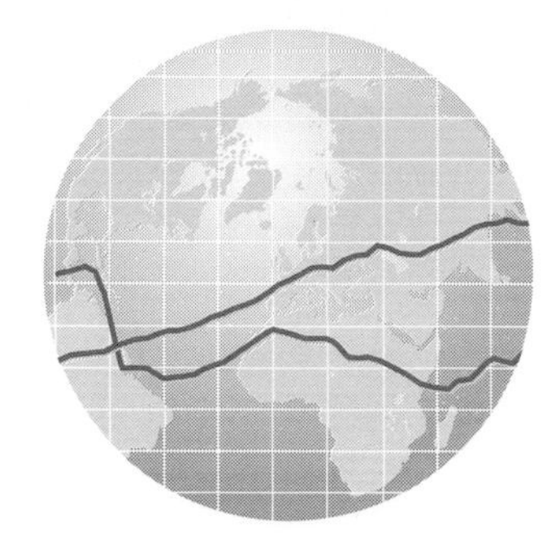

KEY FACTS

- Global emissions of CO_2 reached a new high of nearly 23 900 million tonnes in 1996 – 400 million tonnes more than in 1995 and nearly four times the 1950 total.
- Human activities now contribute more to the global supply of fixed nitrogen than do natural processes: we are fertilizing the Earth on a global scale and in a largely uncontrolled experiment.
- Without the Montreal Protocol, levels of ozone-depleting substances would have been five times higher by 2050 than they are today, and surface UV-B radiation levels would have doubled at mid-latitudes in the northern hemisphere.
- Losses from natural disasters over the decade 1986–95 were eight times higher than in the 1960s.
- The estimated health cost of the 1997–98 forest fires to the people of Southeast Asia was US$1 400 million
- The countries projected to suffer from serious shortfalls in food supply are also those faced with rapidly growing populations and urbanization, low productivity agriculture, high debt and insufficient wealth to import food.
- In 1996, 25 per cent of the world's approximately 4 630 mammal species and 11 per cent of the 9 675 bird species were at significant risk of total extinction.
- If present consumption patterns continue, two out of every three persons on Earth will live in water-stressed conditions by the year 2025.
- More than half the world's reefs are potentially threatened by human activities, with up to 80 per cent at risk in the most populated areas
- Urban air pollution problems are reaching crisis dimensions in many cities in the developing world.

This chapter provides an overview of the state of the environment at the end of the second millennium. Most of the analysis is regional but it begins with an overview of issues that are of global significance for the environment: climate change, stratospheric ozone depletion, nitrogen loading, toxic chemicals and hazardous waste, natural disasters, *El Niño*, forest fires and biomass burning, and human health and the environment. The section continues with a synthesis of the sectoral issues that are examined in detail by region later in the chapter: land and food, forests, biodiversity, freshwater, coastal and marine areas, the atmosphere and urban areas. The chapter then describes these sectoral issues by region. The policies and other measures being used to address these issues are discussed in Chapter 3.

Global issues

Climate change

Annual global emissions of carbon dioxide from the burning of fossil fuels, cement manufacture and gas flaring reached a new high of nearly 23 900 million tonnes in 1996 (CDIAC 1999). This was some 400 million tonnes more than in 1995 and nearly four times the 1950 total. Only in some countries in Europe and Central Asia has there been a significant drop in emissions during the past decade, mainly as a result of the economic crises in Eastern and Central Europe. Atmospheric concentrations of CO_2 in 1997 reached more than 360 parts per million (ppm), the highest level in 160 000 years (Keeling and Whorf 1998).

In assessing the possible impact of rising atmospheric concentrations of CO_2 and other greenhouse gases (GHGs), the WMO/UNEP Intergovernmental Panel on Climate Change (IPCC) concluded in its 1995 report that 'the balance of evidence suggests that there is a discernible human influence on global climate' (IPCC 1996a). Recent research suggests that climate change would have complex impacts on the global environment. The IPCC mid-range scenario projects an increase in global mean temperature of 2.0 °C, within a range of 1.0 to 3.5 °C, by the year 2100, the largest warming in the past 10 000 years. Average sea level is projected to rise by about 50 cm, within a range of 15 to 95 cm, by the year 2100. A 50-cm rise in sea level would lead to the displacement of millions of people in low-lying delta areas and a number of small island states could be wiped out (IPCC 1996b).

In a warmer world there would be higher agricultural production in the high latitudes of the northern and southern hemispheres but reduced production in the tropics and sub-tropics where there is already food deficiency. The species composition of forests and other terrestrial ecosystems is likely to change – entire forest types may disappear. Although forest productivity could increase, the standing biomass of forests may not increase because of more frequent outbreaks and extended ranges of pests and pathogens, and increasing frequency and intensity of fires. Climate change could influence lakes, streams and wetlands through altered water temperatures, flow regimes and water levels. Increases in the variability of water flow, particularly the frequency and duration of large floods and droughts, would tend to reduce water quality and biological productivity and habitat in freshwater ecosystems (IPCC 1998).

In addition to these environmental effects, climate change may have direct and indirect health impacts. Greater frequency and severity of heat waves, and changes in agriculture and food production, could affect nutritional status and vector distributions (Lindsey and Birley 1996). The expansion of warmer areas may increase and extend the ranges of mosquito and other vector populations, affecting the incidence of vector-borne diseases and re-introducing malaria to Europe (Bradley 1996).

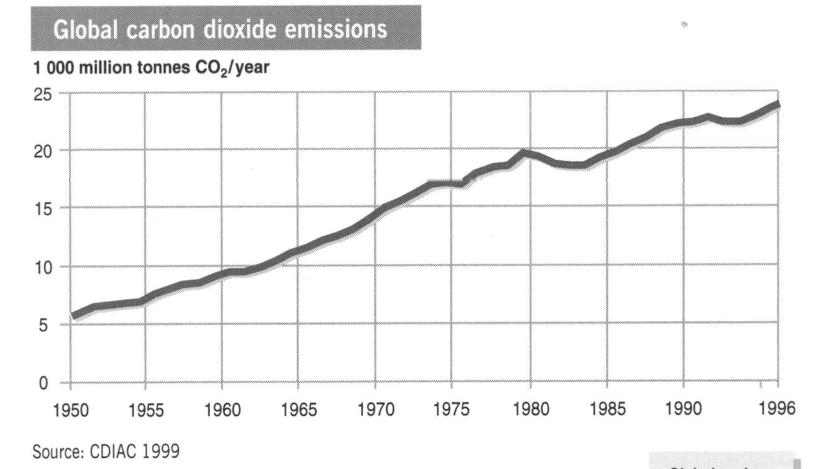

Source: CDIAC 1999

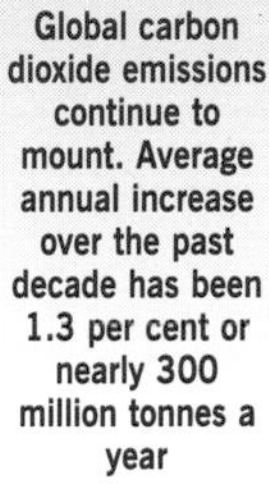
Global carbon dioxide emissions continue to mount. Average annual increase over the past decade has been 1.3 per cent or nearly 300 million tonnes a year

Despite the improved ability of climate models to simulate observed trends, there are still considerable uncertainties in key factors, including the magnitude and patterns of natural variability, the effects of human influence, and the rates of carbon sequestration. There are also new questions to be resolved. For example, is the observed increasing magnitude of *El Niño* events during recent decades related to human-induced climate change? To what extent do reductions in sulphur emissions, required to reduce the acid rain problem, offset warming by greenhouse gases by reducing sulphate aerosols in the atmosphere?

A key factor in assessing the consequences of climate change is the inertia of the climate system: climate change occurs slowly and once a significant change has occurred it will not disappear quickly. Hence, even if a stabilization of greenhouse gas concentrations is achieved (see box on page 26), warming could continue for several decades, and sea

Only in Europe have per capita emissions of carbon dioxide declined over the past 20 years. Emissions are much higher in North America than in other regions

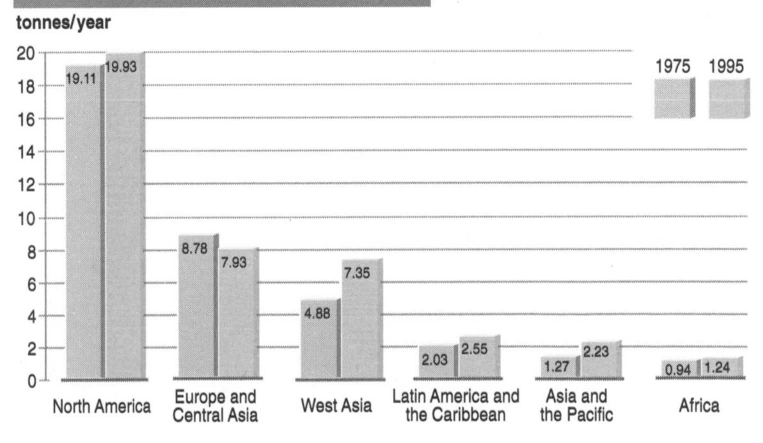

Source: compiled by UNEP GRID Geneva from CDIAC 1998 and WRI, UNEP, UNDP and WB 1998

At what level should greenhouse gas concentrations be stabilized?

According to the IPCC (1996a), stabilization of CO_2 at 450 ppm and other GHGs at levels somewhat above the present concentrations will lead to an increase of the global mean temperature by 1.5–4.0 °C, and stabilization at 550 ppm CO_2 will lead to an increase of 2.0–5.5 °C. Carbon cycle models show that immediate stabilization of the atmospheric CO_2 concentration at its present level of about 360 ppm could be achieved only if emissions are immediately reduced by 50–70 per cent and further reduced thereafter. If stabilization at below 550 ppm were to be aimed for, the annual mean per capita CO_2 emission for the whole world would need to be approximately 5 tonnes during the next century and below 3 tonnes by 2100. Current levels are about 4 tonnes/capita as a world average, with a maximum emission of nearly 20 tonnes/capita in North America and a minimum of less than 1 tonne/capita in many parts of Africa.

levels could continue to rise for centuries.

Future GHG emissions will be a function of global energy demand, and the rate of development and introduction of carbon-free and low carbon energy technologies. Several variables make predictions of future emissions uncertain: economic growth rates, energy prices, the adoption of effective energy policies and the development of efficient industrial technologies. Meeting the targets for emission reductions agreed at Kyoto, itself a formidable challenge for some countries, is only a first step in bringing under control what is generally agreed to be the most critical environmental problem that the world faces. But even meeting all the targets agreed at Kyoto will have an insignificant effect on the stabilization levels of carbon dioxide in the atmosphere.

Stratospheric ozone depletion

Global consumption of chlorofluorocarbons (CFCs), the most prevalent ozone-depleting substances (ODS), fell from 1.1 million tonnes in 1986 to 160 000 tonnes in 1996 (see graph below), thanks to an almost complete phase out by industrialized countries (UNEP 1998a). Several factors contributed to the success of policies directed at reducing the consumption of ODS: damage to the ozone layer could be ascribed to a single group of substances, alternative substances and processes were developed at acceptable costs, a scientific assessment was introduced to make adjustments to the Montreal Protocol as required, the Protocol contained flexible implementation schemes and evaluation procedures, and the principle of 'common but differentiated' responsibilities was recognized for the developed and developing countries.

CFC production has fallen from a peak of more than 1 million tonnes a year to 160 000 tonnes in 1996 as a result of the Montreal Protocol

Global CFC production

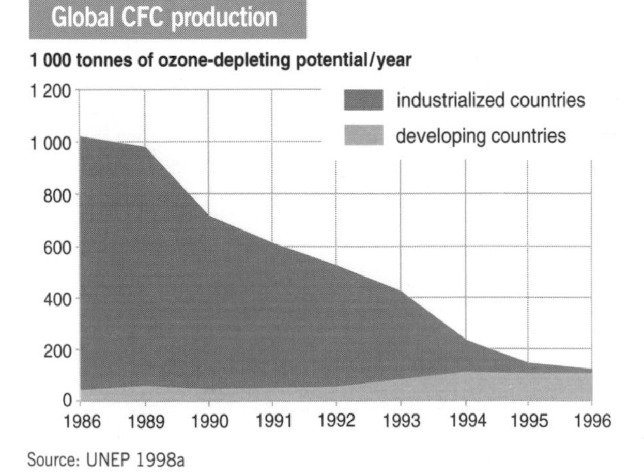

Source: UNEP 1998a

One measure of the Protocol's success is that the ozone layer is now expected to recover to pre-1980 levels by the year 2050. Without the Protocol, levels of ODS would have been five times higher then than they are today, and surface UV-B radiation levels would have doubled at mid-latitudes in the northern hemisphere (UNEP 1999).

The total combined abundance of ODS in the lower atmosphere peaked in about 1994 and is now slowly declining (WMO, UNEP, NOAA, NASA and EC 1998). While total chlorine is declining, total bromine is still increasing, as is the abundance of CFC substitutes. If reductions in the use of ODS continue as envisaged in the Montreal Protocol, then concentrations of these substances in the stratosphere should have peaked between 1997 and 1999, and should begin to decline during the next century. The rate of decline in stratospheric ozone levels at mid-latitudes has already started to slow. The unusually low ozone values above the Arctic in late winter/spring observed in six out of the past nine years could have been accentuated by the unusually cold and prolonged stratospheric winters experienced during those six years (WMO, UNEP, NOAA, NASA and EC 1998).

Despite significant progress in bringing the problem of ozone-layer depletion under control, a number of outstanding challenges remain (see box right). Past (and continuing) emissions of ODS will result in increases in UV-B radiation that are likely to lead to increases in the incidence and severity of a variety of short- and long-term human health effects, particularly on the eyes, the immune system and the skin. Recent evaluations of UV-related excess skin cancer risks in Europe caused by ozone depletion suggest that, even though stratospheric ozone concentrations should reach a minimum around the year 2000 (which assumes that the measures in force are fully implemented), excess skin cancer incidence is not expected to begin to fall until about 2060, because of the time lags involved.

The response of terrestrial ecosystems to increased UV-B is evident primarily in interactions among species rather than in the performance of individual organisms. Recent studies indicate that increased UV-B affects the balance of competition among higher plants, the degree to which higher plants are consumed by insects and the susceptibility of plants to pathogens (UNEP 1998b). Increased UV-B can be damaging for crop varieties but this may be offset by protective and repair processes.

In terms of overall impact, ozone depletion interacts with the climate change process. Stratospheric loss of ozone has caused a cooling of the global lower stratosphere: changes in stratospheric ozone since the late 1970s may have offset about 30 per cent of the warming effect of other greenhouse gases over the same period (WMO, UNEP, NOAA, NASA and EC 1998). There are also complex interactions between ozone depletion, climate change and the abundance of methane, nitrous oxide, water vapour and sulphate aerosols in the atmosphere. For example, carbon is an important element in the absorption of UV radiation. Climate change and acid rain have led to decreases in the dissolved organic carbon concentration in many North American lakes (Schindler and others 1996). As organic carbon levels have decreased, UV radiation has been able to penetrate much more deeply into surface waters, resulting in greater UV-B exposure of fish and aquatic plants.

While the potential impact of stratospheric ozone depletion means there is no room for complacency, the cooperative measures that followed the identification of the problem remain an outstanding and encouraging example of the ability of the international community to act in unison in protecting the global environment.

Current ozone losses and UV-B increases

	ozone loss (%)	UV-B increase (%)
Northern hemisphere, mid-latitudes, winter/spring	6	7
Northern hemisphere, mid-latitudes, summer/autumn	3	4
Southern hemisphere, mid-latitudes, year-round	5	6
Antarctic spring	50	130
Arctic spring	15	22

Note: figures are approximate and assume other factors, such as cloud cover, are constant

Source: WMO, UNEP, NOAA, NASA AND EC (1998)

Challenges in the protection of the ozone layer

CFC production in developing countries, notably Brazil, China, India, Republic of Korea, Mexico and Venezuela, more than doubled between 1986 and 1996, while consumption rose by some 10 per cent (UNEP 1998a). Because production levels in the years 1995–97 will be used as base levels to determine the timing of phase out in developing countries, scheduled to begin in mid-1999 with elimination due by 2010, the current high production will inflate the allowed levels of production for years to come. The Russian Federation will not eliminate its production of CFCs before the year 2000, and some of the European transition economies are experiencing economic and technical difficulties with CFC substitution (UNEP 1998c).

Halon production, mostly for use in fire-fighting equipment, is rising again, primarily in developing countries. For example, production of halons in China grew nearly fourfold between 1991 and 1996 (UNEP 1998a). This trend is of particular concern since a given amount of halons can destroy up to ten times more ozone than the same amount of CFCs.

CFC elimination is being undermined by a rise in illegal trading. Substantial demand still exists in the developed world, mostly to service existing refrigeration and cooling equipment. Illegally-imported virgin CFCs are cheaper than legally recycled CFCs or new CFCs obtained from limited existing stocks. The incentives for smuggling are therefore high. Estimates of the size of the global CFC black market range from 20 000 to 30 000 tonnes annually.

Nitrogen loading

Evidence is mounting that human activities are seriously unbalancing the global nitrogen cycle. Nitrogen is abundant in the atmosphere but must be fixed by micro-organisms in the soil, water and in the roots of nitrogen-fixing plants before it is available for use by plants and the animal life dependent on them. The advent of intensive agriculture, fossil fuel combustion and widespread cultivation of leguminous crops has led to huge additional quantities of nitrogen being deposited into terrestrial and aquatic ecosystems. Human activities have at least doubled the amount of nitrogen available for uptake by plants (Vitousek and others 1997) and now contribute more to the global supply of fixed nitrogen than do natural processes: we are fertilizing the Earth on a global scale and in a largely uncontrolled experiment.

The principal form of anthropogenic nitrogen, accounting for some 60 per cent of the total, is inorganic nitrogen fertilizer. Global fertilizer use soared from less than 14 million tonnes in 1950 to 145 million tonnes in 1988; by 1996 it had fallen back to about 135 million tonnes (FAOSTAT 1997). Consumption is now stable or declining in the industrialized countries but demand is still rising in the developing world. The major driving force is increasing food production, driven in turn by increasing human population and the growing demand for livestock products, particularly in developing countries.

Global fertilizer consumption

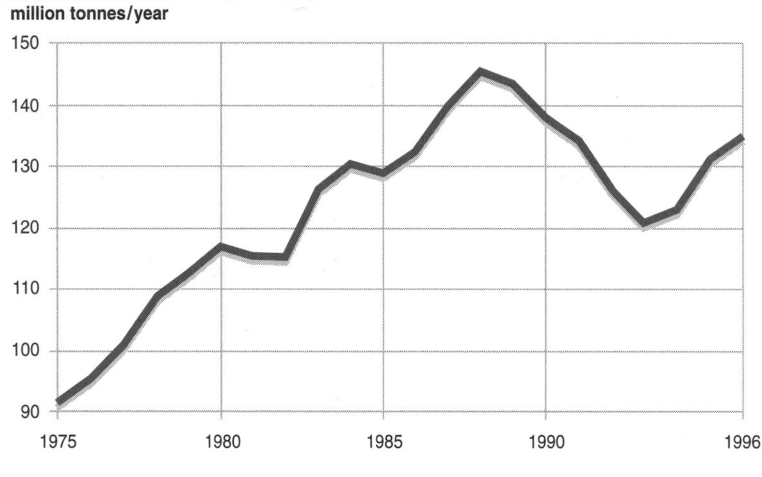

Source: FAOSTAT 1997

Global fertilizer use is less than it was in the late 1980s but consumption is still increasing in the developing countries

Typically, less than half of the nitrogen applied is taken up by plants – the rest is lost to the air, dissolved in surface waters or absorbed into groundwater. The cultivation of leguminous crops such as soybeans, peas and alfalfa accounts for about 25 per cent of anthropogenic nitrogen, and fossil fuel burning for about 12 per cent (Vitousek and others 1997). Other sources include biomass burning, draining wetlands (resulting in the release of organic nitrogen in the soil) and conversion of woodland to cropland.

The huge increase in nitrogen loading of the environment has had a number of consequences. There has been a large rise in the nitrogen levels of drinking water supplies, resulting mainly from agricultural run-off and wastewater. In some major rivers of the northeastern United States, for example, nitrate concentrations have risen up to tenfold since the beginning of the century, necessitating costly purification systems to protect human health (Carpenter and others 1998). Globally, human activities have increased the amount of riverine transport of dissolved inorganic nitrogen by a factor of 2–4 (Seitzinger and Kroeze 1998). Nitrogen-based trace gases emitted during fossil fuel combustion (notably from automobiles) are major contributors to atmospheric pollution. Nitric oxide is an important precursor of ground-level ozone, the component of photochemical smog that is most dangerous to human health and crop productivity. It can also be transformed into nitric acid and, together with sulphuric acid resulting from sulphur emissions, washed out of the atmosphere as acid rain. Acidification of forests, soils and surface waters is increasingly the result of nitrogen emissions in industrialized countries, as sulphur emissions are brought under control.

Rising nitrogen loads combined with phosphorous have led to exuberant and unwanted plant and algal growth in many freshwater habitats and coastal areas throughout the world. In the United States, eutrophication – rapid plant growth in water resulting in oxygen deprivation for other species – accounts for about half of the impaired lake area and 60 per cent of the impaired river reaches (Carpenter and others 1998). Large areas of northern Europe, where intensive agriculture and high fossil fuel combustion coincide, are now in a state of nitrogen saturation: no more nitrogen can be taken up by plants, and additional deposits are simply dispersed into surface water, groundwater and the atmosphere without playing any role in the biological systems for which they were intended.

Excess levels of nitrogen can reduce plant diversity by enhancing the growth of plants best able to utilize it at the expense of others. In large areas of northern Europe, for example, high levels of nitrogen deposition have resulted in the conversion of heathlands rich in biodiversity into grasslands containing relatively few species (Wedin and Tilman 1996).

Nitrogen deposition is also causing more fundamental damage to ecosystems. Elevated nitrogen levels in soils increase the leaching of minerals such as potassium and calcium, which promote plant growth and are essential as a buffer against acidity. As soil acidity is increased, aluminium ions are mobilized and can reach concentrations sufficient to damage tree roots or kill fish if the aluminium washes into waterways (Kaiser 1996).

There is compelling evidence that nutrient enrichment is at least partly to blame for damage to estuaries and coastal seas, and some of the decline in coastal fisheries production. In brackish water, nitrogen is usually the limiting nutrient for algal activity and plant growth. River discharges laden with nitrogen-rich sewage and fertilizer run-off therefore tend to stimulate algal blooms, which can lead to oxygen starvation in coastal waters at lower depths. This has caused significant fish kills in the Baltic Sea, Black Sea and Chesapeake Bay (Vitousek and others 1997). Biodiversity can also be reduced as 'nuisance' algae come to dominate marine ecosystems. The world's oceans are being plagued by a rising incidence of algal blooms – known as brown or red tides (see box on page 151).

There is a growing consensus among researchers that the scale of disruption to the nitrogen cycle may have global implications comparable to those caused by disruption to the carbon cycle. On the positive side, it appears possible that the nitrogen and carbon cycles are interacting. Since nitrogen is normally a limiting factor in plant growth, increased available nitrogen may be enhancing overall plant growth which, in turn, would enhance the Earth's carbon storage potential. This extra vegetation may explain the puzzle of the world's 'missing' carbon – the difference between the amount of carbon emitted and the amount known to be accumulating in the atmosphere each year (Vitousek and others 1997).

On the negative side, nitrogen emissions to the atmosphere are contributing to global warming. Nitrous oxide is a potent greenhouse gas, accounting for about 6 per cent of the enhanced greenhouse effect. It is long-lived in the lower atmosphere, and concentrations are currently increasing at the rate of 0.2 to 0.3 per cent per year. In the upper atmosphere, the gas also contributes to ozone depletion. Most of the atmospheric nitrous oxide is of biological origin, being produced by bacteria in soils and surface waters. Recent increases in emissions are attributed to human activities, in particular related to agriculture and land use (Environmental Pollution 1998).

Current trends suggest that nitrogen-related problems are likely to worsen. The world's rising demand for food makes it likely that fertilizer use will increase (despite research on genetically-modified nitrogen-fixing crops) and the transportation boom shows no sign of slackening. Far greater efforts will have to be devoted to developing more efficient methods of plant nutrient management (FAO 1998). If enhanced energy efficiency measures or shifts to cleaner fuels are undertaken to curb carbon emissions, the benefits in terms of reduced nitrogen emissions may prove equally great.

Toxic chemicals and hazardous waste

Exposure to chemical agents in the environment – in air, water, food and soil – has been implicated in numerous adverse effects on humans from cancer to birth defects. The 'old' poisons, such as lead and mercury, some industrial solvents and some pesticides, are still of concern in many parts of the world but there is a reasonable level of understanding of their effects and the measures needed to protect human health and the environment from them (although such measures are not always adequately implemented). There is far less knowledge about the toxicological effects of a number of new chemicals coming onto the market. These may be present in household products, cosmetics and even pharmaceuticals.

In addition, exposure to hazardous chemicals can result from industrial and transportation accidents and from inadequate management and disposal of wastes, particularly hazardous wastes (see box below).

Hazardous waste production

The output of hazardous wastes worldwide was about 400 million tonnes a year in the early 1990s, of which some 300 million tonnes were produced by OECD countries (UNEP 1994a), mainly from chemical production, energy production, pulp and paper factories, mining industries, and leather and tanning processes.

Progressively tighter regulatory controls have increased the costs of waste disposal in many countries. Export to developing countries with less stringent controls and a lower public awareness of the issue has been one way in which some companies have side-stepped these regulations. Officially, fewer than 1 000 tonnes a year are traded to developing countries but illegal traffic in hazardous waste poses a potentially serious threat to the environment and human health (de Nava 1996).

One way of combating such trade is through the system called Prior Informed Consent (PIC) for Certain Hazardous Chemicals in International Trade. Operated by FAO and UNEP, PIC is a procedure that helps participating countries learn more about the characteristics of potentially hazardous chemicals that may be shipped to them, initiates a decision-making process on the future import of these chemicals and helps disseminate this decision to other countries (IRPTC 1999). The aim is to promote a shared responsibility between exporting and importing countries in protecting human health and the environment (see 'The Rotterdam Convention', page 202).

Policy makers are also focusing on a more integrated approach to waste management, one that uses cleaner production concepts to minimize the volume of wastes generated by manufacturing processes (UNEP 1998d).

UNEP survey of international trade in selected POPs, 1990–94

product	*no. of countries where use is banned*	*no. of countries where import is banned*	*no. of countries reporting production*	*production reported (tonnes)*	*no. of countries reporting exports*	*exports reported (tonnes)*	*no. of countries reporting imports*	*imports reported (tonnes)*
Aldrin	26	52	1	2.1	0	-	1	50.1
Chlordane	22	33	0	-	>2	?	4	227.8
DDT	30	46	3	2 070	2	356.4	3	62
Dieldrin	33	54	1	3.1	1	8 kg	2	36.5
Endrin	28	7	0	-	0	-	1	1 000 litres
Heptachlor	23	34	0	-	0	-	3	435.1
Hexachlorobenzene	13	4	0	-	1	35.8	4	1.1
PCB	2	5	0	-	?	?	1	?
Toxaphene	18	1	1	241.4	0	-	2	277.4

Notes: Survey based on responses from 60 governments representing 75 per cent of worldwide chemical trade. The '?' refers to a statement 'yes' made by the submitting country. However, in the case of PCBs an export figure of 739.6 tonnes has been entered but without a year attribution; and an amount of 12 451 tonnes has been imported into a country in 1994 but for a grouped entry of PCBs, PBBs and PCTs.

Source: UNEP 1996a

Two groups of hazardous chemicals – heavy metals and persistent organic pollutants (POPs) – are currently receiving particular attention. Although the emissions of some of these substances are falling, concentrations in the environment are of concern, both near highly contaminated areas and as a result of widespread distribution through the food chain (UNEP 1996a).

Exposure to heavy metals has been linked with developmental retardation, various cancers, and kidney damage. Exposure to high levels of mercury, gold and lead has also been associated with the development of auto-immunity, in which the immune system starts to attack its own cells, mistaking them for foreign invaders (Grover-Kerkvliet 1995). Several studies have shown that lead exposures can significantly reduce the IQ of children (Goyer 1996). In some countries, heavy metal emissions are falling as a result of the removal of lead from petrol, improvements in wastewater treatment and incinerators, and improved industrial technologies. Significant further improvements could be achieved if the available technologies were more widely applied (EEA 1998).

POPs are fat-soluble toxic chemicals that do not easily degrade, persist for many years in the environment, concentrate up the food chain, and accumulate in animal and human tissues. They often end up thousands of kilometres from where they are used or released. The growing evidence that some POPs can have serious human health effects has pushed governments to collective action (see Chapter 3). Although POPs include a wide range of chemicals, much recent research and regulatory action focuses on the industrial PCBs, polychlorinated dioxins and furans (unwanted by-products of various industrial processes) and pesticides such as DDT, chlordane and heptachlor. Despite restrictions on the use of these chemicals in many developed countries, they are still manufactured there for export and remain widely used in developing countries.

Concern over the impact of POPs on the environment and human health has increased further with the emergence of scientific findings that suggest that certain POPs (and also some organo-metallic compounds) – called endocrine disrupters because they interact with the endocrine, or hormone system – may be playing a role in a range of problems from reproductive and developmental abnormalities to neurological and immunological defects in humans and other animals (Colborn and others 1996).

It is estimated that hundreds of thousands of

people die every year from acute exposure to toxic chemicals but precise figures are not available. In some developing countries, poisoning is among the most frequent cause of mortality in hospital patients.

There is particular and growing concern about the threats that chemicals pose to children's health. The main problems include both acute exposure leading to poisoning and chronic, low-level exposures causing functional and organic damage during periods of special vulnerability, when neurological, enzymatic, metabolic and other systems are still developing. Exposure of unborn children to toxic chemicals may produce irreversible effects. For example, low levels of mercury have severe effects on the foetuses of pregnant mothers who ingest contaminated food. Recent research suggests that these chemicals may affect the ability of children to learn, integrate socially, fend off disease and reproduce (Colborn 1997).

Natural disasters

Natural disasters include earthquakes, volcanic eruptions, fires, floods, hurricanes, tropical storms, cyclones, landslides and other events that cause loss of life and livelihoods. It is estimated that almost 3 million people have perished as a result of natural disasters in the past three decades while tens of millions have suffered hardship (UN 1997).

Things appear to be getting worse, in two ways: natural disasters appear to be becoming more frequent and their effects more severe. The Munich Reinsurance Company estimates that total global economic losses from natural disasters for the two years 1997 and 1998 reached US$120 000 million. Allowing for inflation, losses over the decade 1986–95 were eight times higher than in the 1960s (Munich Re 1997 and 1998).

Overall, the poor are the most likely to suffer from major disasters and the least likely to be insured against loss. In 1997, Asia suffered 33 per cent of the world's catastrophic events, 67 per cent of the casualties and 28 per cent of the economic losses. However, only 0.2 per cent of those losses were covered by insurance policies. The global insurance industry paid out US$4 500 million for disaster-related damage in 1997, and 66 per cent of the claims were made in the United States (Munich Re 1998). Thus insured losses and repayments are concentrated mainly in the richer industrialized countries.

While the consequences of most natural disasters are generally confined to one or a few countries or to even smaller areas, some may affect large parts or even the whole of the planet. The debris from very large volcanic eruptions, for example, can spread around the entire globe, and the *El Niño* phenomenon (see page 32) can have effects many thousands of kilometres away from the region in the Pacific Ocean where it originates.

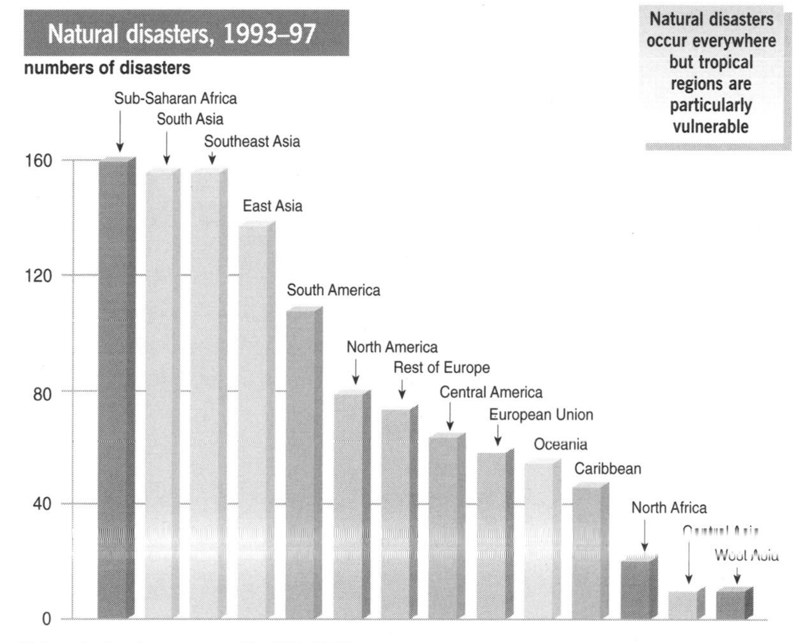

Note: not all regions correspond to *GEO-2000* regions

Source: CRED 1999

Global warming models indicate that rising global temperatures are likely to affect many atmospheric parameters including precipitation and wind velocity, and raise the incidence of extreme weather events, including storms and heavy rainfall, cyclones and drought. It may or may not be just coincidence that the Munich Reinsurance Company recorded more than 700 'large loss events' in 1998, compared with between 530 and 600 during previous recent years. The most frequent natural catastrophes were windstorms (240) and floods (170), which accounted for 85 per cent of the total economic losses (Munich Re 1998).

Volcanic eruptions and earthquakes are confined to seismically-active areas and remain stable in incidence. The incidence of other natural disasters such as storms and floods, however, is increasing in frequency and magnitude (Munich Re 1997), and some of these natural phenomena – particularly floods – are being exacerbated or triggered by human degradation of the environment

Major disasters: the past three years

Major disasters over the past three years include:

- a cyclone in the Indian province of Gujarat in June 1998 killed more than 10 000 people
- Hurricane George caused damage estimated at US$10 million in the Caribbean in September 1998
- Hurricane Mitch led to more than 9 000 deaths in Nicaragua and Honduras in October 1998, and caused major setbacks to development plans
- flooding of the Yangtze river in China between late June and mid-August 1996 affected 20 million people and caused economic losses of more than US$20 000 million
- flooding in Central Europe in 1997 caused economic damage estimated at US$2 900 million in Poland and US$1 800 million in the Czech Republic – in Eisenhüttenstadt the previous flood record of 1854 was exceeded by 62 cm
- severe floods were also recorded in 1997 in Kenya, Myanmar, Somalia, the United States and along the Pacific coast of Latin America
- earth tremors caused major destruction in many towns and villages in central Italy in 1997, as did mudslides in 1998
- in 1997, earthquakes claimed the lives of more than 2 300 people in Iran

Sources: Munich Re 1997 and 1998

and disturbance of formerly stable ecosystems.

The vulnerability of rural and urban populations to natural disasters is also growing, due to population increase and inadequately planned urbanization. The number and density of people living in cities within earthquake and tropical cyclone zones have risen dramatically in the past two decades. In many developing regions, population pressures and poverty are forcing farmers to cultivate marginal and vulnerable areas in flood plains or on hill slopes. Poor planning decisions have also led to siting potentially-hazardous facilities, such as nuclear power plants, chemical factories and major dams, in earthquake zones and densely-populated areas.

Deforestation can, in the short-term, lead to increased run-off and soil erosion, mudslides and flash flooding. Poor forest management has exacerbated flash floods across the world, such as those witnessed in the Philippines which killed more than 5 000 people in 1991, and the mudslides in southern Italy in 1998.

Urban development, settlement, drainage of wetlands for agriculture, and canalization of rivers for irrigation or navigational purposes have removed the normal flood plains of numerous rivers throughout Asia, Europe and, in particular, the Americas. In the absence of natural absorption basins, rivers rise higher, flow faster and flood more violently. For example, the flooding of the Odra river in Central Europe in the summer of 1997 is estimated to have cost nearly US$6 000 million in economic losses in affected countries, with the most severe damage in Poland which lost 2 000 km of railway line, 3 000 km of roads, 900 bridges and 100 000 houses (Munich Re 1997). The 1996 flooding of the Yangtze River in China killed more than 2 700 people, left two million people homeless, drowned tens of thousands of animals, destroyed crops over some 20 million hectares of farmland, and resulted in a 4–6 per cent loss in GDP. China experienced severe flooding again in 1998.

El Niño

El Niño is the term used to describe a phenomenon which starts with the surface warming of an area of the eastern Pacific close to the equator (see map right) and whose effects spread over most of the world. *El Niños* are not natural disasters – indeed, some of their effects may be beneficial – but natural variations in climate. They normally occur every three to five years, lasting 6–18 months, and peak around Christmas time, which is why Peruvian fishermen called the phenomenon '*El Niño*' (boy child). Between *El Niño*s there are often periods marked by a cooling of the surface waters of the same area of the Pacific, a phenomenon called *La Niña*. *El Niño*s are also marked by fluctuations of atmospheric pressure which parallel those of the surface sea temperature in the eastern equatorial Pacific. The whole cycle is called the *El Niño* Southern Oscillation (ENSO).

*El Niño*s have far-reaching effects. The build up of warm water along the west coast of South America prevents the normal upwelling of cold water from the ocean depths. In the western Pacific, the normally rain-bearing cloud systems shift eastward toward the central and eastern Pacific, bringing heavy rainfall to these areas while countries in the western Pacific, such as Australia, Indonesia and Papua New Guinea, experience drought. The effects of the changes in wind speed and direction, sea surface temperatures and the depths of the warm water often extend into temperate latitudes. For example, most *El Niño* winters are mild over western Canada and parts of the northern United States, and wet over the southern United States from California to Florida. Southern China is subject to storms and southern Africa has a tendency to drought (WCN 1998a).

The 1997/98 *El Niño* was one of the strongest on record, developing more quickly and with higher temperature rises than ever recorded. The episode developed rapidly throughout the central and eastern tropical Pacific Ocean in April–May 1997. During the second half of the year, it became more intense than the major *El Niño* of 1982/83, with sea-surface temperature anomalies across the central and eastern Pacific of 2–5 °C above normal.

The warming effect of *El Niño* was a major factor contributing to the record high global temperature in 1997. The estimated global mean surface temperature for land and marine areas averaged 0.44 °C above the 1961–90 base period mean. The previous warmest year was 1995, with an anomaly of +0.38°C (WCN 1998b).

By mid-January 1998, the volume of *El Niño*'s warm water pool had decreased by about 40 per cent since its maximum in early November 1997 but its surface area in the Pacific was still about 1.5 times the size of the continental United States. This warm pool had so much energy that its impacts dominated world climate patterns up to mid-1998.

The *El Niño* of 1982/83 was estimated to have been responsible for 2 000 deaths and about US$13 000 million worth of damage worldwide (WCN 1998c). The *El Niño* of 1997/98 has been blamed for extreme rainfall and flooding in equatorial central and eastern Africa, and severe storms along the California coast and in the southeastern United States. Severe droughts occurred in northeastern Brazil, parts of southern Africa and Indonesia, and were responsible for drought-related famine in Papua New Guinea (see also box right).

The 1997/98 *El Niño* was the first to have been widely predicted, thanks to the comprehensive *El Niño* observing network which now spans the Pacific Ocean, and a network of observational satellites. The former includes ships, drifting buoys and sea-level gauges on many Pacific islands, all relaying their observations to meteorological centres in real time. In addition, several satellites measure the temperature and elevation of the sea surface.

In 1997, information from these systems was analysed by several teams of forecasters, many of whom predicted that a major *El Niño* was on the way. As a result, scientists have compiled a complete picture of an *El Niño* which can be used as a benchmark against which to measure future ones.

An outstanding question centres on whether there is a causal link between *El Niño* and global warming – it is unclear whether global warming is increasing the incidence or severity of *El Niño*.

El Niño: sea temperature anomalies in January 1998

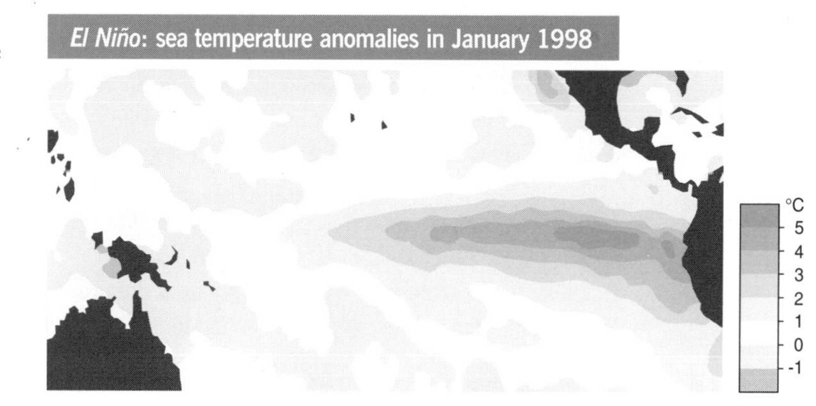

Source: NOAA 1998

Sea surface temperature anomalies in January 1998, at the height of the 1997/98 _El Niño_

Some impacts of the 1997/98 _El Niño_

South America

- Guyana, severely affected by drought, began water conservation measures
- the coasts of Ecuador and northern Peru received 350–775 mm of rain during December 1997 and January 1998, compared to the normal 20–60 mm
- torrential rains soaked southern Brazil, south-eastern Paraguay, most of Uruguay and adjacent parts of northeastern Argentina
- rain on Colombia's Pacific coast increased the threat of landslides while inland forest fires destroyed about 150 000 hectares
- the sea level in the Colombian Pacific rose 20 cm

Africa

- unusually warm weather was reported in most of South Africa, southern Mozambique, and the central and southern portions of Madagascar
- heavy rain fell across central and southern Mozambique, the northern half of Zimbabwe and parts of Zambia, causing flash floods in places
- Kenya was particularly hard hit by flooding, where many villages were cut off, and the main Nairobi-Mombasa road was made impassable

Asia and the Pacific

- in Indonesia and the Philippines, long-term dryness persisted over the region
- tropical storms Les and Katrina caused heavy rain in northern Australia
- torrential rains in southern China

North America

- unusual jet stream patterns over North America led to severe storms over the eastern North Pacific and the west coast of the United States.

Source: WCN 1998d

Forest fires and biomass burning

During 1996–98, fire swept through forests in Australia, Brazil, Canada, China's northeastern Inner Mongolian region, France, Greece, Indonesia, Italy, Mexico and several other countries in Latin America, the Russian Federation, Turkey and the United States. Satellite photos showed that about 3.3 million hectares of Brazilian forest were devastated as a result of the fires. More than 3 million hectares of forest in Mongolia were burnt in 1996. The fires in Southeast Asia in 1997 were the worst in 15 years, with at least 4.5 million hectares burnt, and smoke and haze affecting some 70 million people (Liew and others 1998). The fires in Indonesia threatened at least 19 protected areas, many of which are rich in biodiversity (WWF 1998).

The forests of Southeast Asia and of the Brazilian Amazon were especially vulnerable to fire in 1997 and 1998 because of a severe drought probably related to the strong *El Niño* of the same period (see page 32) and/or changing global weather patterns. After the severe *El Niño* of 1982, the largest fires then on record raged across Kalimantan. The 1997 and 1998 fires were far more extensive and coincided with an even more severe *El Niño*.

In many countries, vegetation, forests, savannahs and agricultural lands are burnt to clear land and change its use. Forest clearing accelerates as populations expand and pressures to exploit natural resources increase. Much of the expansion into forested areas uses the cheapest form of cover removal: fire. Thus increased pressure for development has led to much of the recent fire damage in tropical rain forests as loggers, cattle farmers and peasants take advantage of the dry season to clear land for farming.

In Indonesia and South America, much of the blame for starting fires fell on small farmers. But only 12 per cent of the forest cleared in the Amazon is actually used for arable farming. The remaining 88 per cent is used for pasture. New areas are usually made accessible for ranching and agriculture as a result of the construction of logging roads to extract mahogany (WWF 1997).

The health impacts of forest fires can be serious and widespread. Estimates for the fall-out from fires in Southeast Asia suggest that 20 million people were in danger of respiratory problems. The estimated health cost to the people of Southeast Asia was US$1 400 million, mostly related to short-term health problems (EEPSEA/WWF 1998). In 1997, smoke and air pollution from fires in Guatemala, Honduras and Mexico drifted across much of the southeastern United States, prompting Texas officials to issue a health warning to residents.

Another major consequence of forest fires is their potential impact on global atmospheric problems, including climate change. Only in the past decade have researchers realized the important contributions of biomass burning to the global budgets of carbon dioxide, methane, nitric oxide, tropospheric ozone, methyl chloride and elemental carbon particulates.

The extent of biomass burning has increased significantly over the past 100 years. It is now recognized as a significant global source of atmospheric emissions, contributing more than half of all the carbon released into the atmosphere (see table left). The burning of tropical savannahs is estimated to destroy three times as much dry matter per year as the burning of tropical forests (Andreae 1991).

Forest fires are dealt with in more detail in the regional sections on Asia and the Pacific and on Latin America that follow.

Biomass burning

source of burning	*biomass burned (million tonnes dry matter/year)*	*carbon released (million tonnes carbon/year)*
Savannahs	3 690	1 660
Agricultural waste	2 020	910
Tropical forests	1 260	570
Fuelwood	1 430	640
Temperate and boreal forests	280	130
Charcoal	20	30
WORLD TOTAL	8 700	3 940
For comparison:		
Global carbon emissions (1996) from fossil fuel burning, cement manufacture and gas flaring		6 518

Sources: Andreae 1991, CDIAC 1999

Human health and the environment

The World Health Organization estimates that poor environmental quality contributes to 25 per cent of all preventable ill-health in the world today (WHO 1998).

Traditional problems, such as contaminated water, poor sanitation, smoky indoor air and exposure to mosquitoes and other animal disease vectors, are still the primary environmental factors in ill health (see table on page 36). Across the world, insufficient water supplies, inadequate sanitation and poor hygiene are primarily responsible for global outbreaks of cholera and other diarrhoeal diseases, which claim three million lives each year (WHO 1997a).

Vector-borne diseases, affecting more than 700 million in total a year, are considered the most sensitive to climatic and environmental conditions. Malaria, the best-known vector-borne disease, has been declared 'public enemy number one' by WHO and affects more than 500 million people in 90 countries, causing 1.5–2.7 million deaths per year (WHO 1997a).

In many developing regions, these traditional environmental health problems are now exacerbated by emerging problems of pollution from industry and agriculture (Smith 1997). Chemical agents, particularly air-borne ones, are considered to be major factors in causing and worsening tuberculosis, bronchitis, heart disease, cancers and asthma. Tuberculosis, the single largest cause of death in adults from infectious diseases, was responsible for three million deaths in 1996, 95 per cent of which occurred in the developing world (WHO 1997a). Exposure to pesticides, fertilizers and heavy metals poses health risks through soil, water, air and food contamination. Global pesticide use has resulted in 3.5–5 million acute pesticide poisonings per year (WHO 1990). Recent epidemiological studies have suggested a link between organochlorine pesticides and cancer, including lymphoma and breast cancer (Zahm and Devesa 1995)

The emergence of some 30 new diseases, including AIDS, Ebola and haemorrhagic illnesses, in the past 20 years has become a growing public health issue. Demographic changes, in particular rapid, unplanned urbanization, have resulted in conditions that encourage the spread of diseases such as dengue fever. The two primary mosquito vectors of the dengue, *Aedes aegypti* and *Aedes albopictus*, have adapted from their natural forest environments, where they breed in tree holes containing rainwater, to the urban environment where they breed in drains, water cans, discarded tyres, pots and bottles (Gubler and Clark 1994).

Urbanization (see page 47) can be an important source of health problems: nearly half of the world's

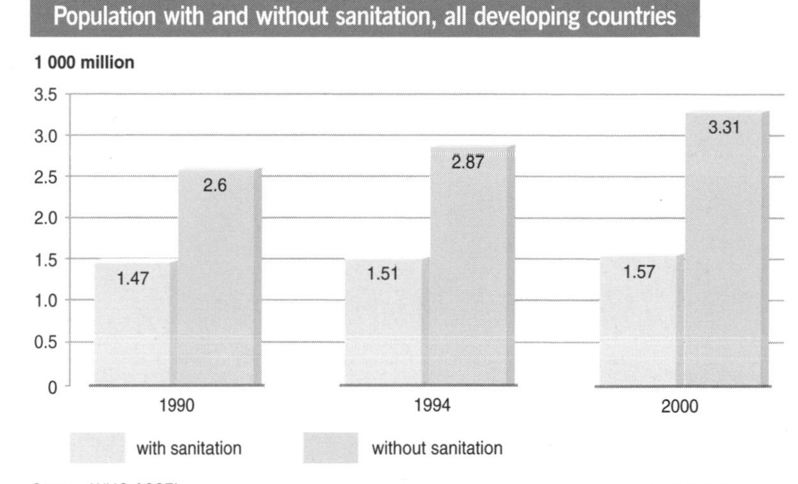

Source: WHO 1997b

Inadequate sanitation is one of the prime causes of disease. In the developing countries, the provision of sanitation is not keeping up with population growth

population will live in urban centres by the end of the 20th century but currently 30–60 per cent of the urban population are in low-income countries, and lack adequate housing with sanitary facilities, drainage systems, and piping for clean water (UNCHS 1996). This number is expected to increase since local and city authorities often lack the resources, knowledge, trained personnel and financial capacity needed to meet their responsibilities in providing services and amenities essential for healthy living. Increased exposure to biological and chemical health risks in urban areas is particularly harmful to children. Children suffer the greatest number of deaths due to diarrhoeal diseases (2.5 million deaths per year) and this number is likely to increase in populations of the urban poor. Prevalence of asthma, often exacerbated by air pollutants, has also increased among children (Woolcock and Peat 1997). Studies suggest a quantitative relationship between atmospheric carcinogen levels and lung cancer, and WHO has estimated that 50 per cent of the global burden of chronic respiratory illnesses is associated with air pollutants (WHO 1997b).

As the global population continues to grow, there is increasing pressure to develop agriculture, roads and transportation systems in previously unsettled areas. This kind of land conversion can encourage the spread of diseases harmful to human health. For example, leishmaniasis, an infectious disease transmitted through a sandfly bite, has increased to 12 million cases (WHO 1998) each year alongside increasing land development in Africa, Latin America and West Asia (WHO 1997b). Forest clearance in

Environmental factors affecting health

	Polluted air	*Poor sanitation and waste disposal*	*Polluted water or poor water management*	*Polluted food*	*Unhealthy housing*	*Global environmental change*
Acute respiratory infections	●				●	
Diarrhoeal diseases		●	●	●		●
Other infections		●	●	●	●	
Malaria and other vector-borne diseases		●	●		●	●
Injuries and poisonings	●		●	●	●	●
Mental health conditions					●	
Cardiovascular diseases	●					●
Cancer	●			●		●
Chronic respiratory diseases	●					

Source: WHO 1997a

particular is associated with higher incidence of diseases such as malaria.

People in developing countries face larger amounts of indoor pollutants, such as sulphur and nitric oxides and arsenic compounds, due to greater exposure to open fires which burn biomass, coal or wood as fuel. Indoor pollution, a more severe threat to women and children who spend more time indoors, causes respiratory disorders and is also linked to heart and lung disease mortality. A study in India and Nepal demonstrates that cardiovascular disease is more common among women who have been exposed to indoor pollutants (WHO 1992).

Regional synthesis

The sections that follow provide an overview of the sectoral issues discussed in detail by region in the rest of this chapter. Seven sectoral issues are covered:

- land and food;
- forests;
- biodiversity;
- freshwater;
- marine and coastal areas;
- atmosphere; and
- urban areas.

Where facts are not referenced in this synthesis account, a fuller description (with references) will be found in the appropriate regional section.

Land and food

The Earth could, in theory, support far more than its present population but the distribution of good soils and favourable growing conditions does not match that of the population. The problem is being exacerbated by increasing land degradation caused by deforestation, poor management of arable and pasture land, including over-use of fertilizers and pesticides, the clearance of marginal land for cultivation, poor management of watersheds and water resources, uncontrolled dumping of wastes, deposition of pollutants from the air and poor land-use planning. Although land degradation is occurring all over the world, the problem is particularly serious where local food production cannot provide an adequate diet or even enough for bare survival. Their low agricultural yields and the pressures of high population growth have forced millions of small farmers to clear forests and cultivate fragile marginal lands, causing soil erosion and deepening rural poverty.

There is a lack of reliable data on land degradation but it is likely that soil degradation has affected some 1 900 million hectares of land worldwide (UNEP/ISRIC 1991). The largest area affected, about 550 million hectares, is in Asia and the Pacific. In China alone, between 1957 and 1990, the area of arable land was reduced by an area equal to all the cropland in Denmark, France, Germany and the Netherlands combined, mainly because of land degradation (ESCAP 1993).

In Africa, an estimated 500 million hectares of land have been affected by soil degradation since about 1950 (UNEP/ISRIC 1991) – including 65 per cent of the region's

Desertification

Desertification is a significant threat to the arid, semiarid and dry sub-humid areas of the globe – the 'susceptible drylands' which cover 40 per cent of the Earth's land surface. Soil degradation in the drylands affects or puts at risk the livelihoods of more than 1 000 million people who are directly dependent on the land for their habitat and source of livelihood.

Dryland soils are particularly vulnerable because they recover only slowly from disturbance. With a limited supply of water, new soil forms very slowly, salts once accumulated tend to remain where they are, and soils that are dry, poorly held together and sparsely covered by vegetation are susceptible to erosion. Infrequent rains are particularly erosive, especially where vegetation cover is sparse. Susceptible areas include the savannahs of Africa, the Great Plains and the Pampas of the Americas, the Steppes of southeast Europe and Asia, the outback of Australia and the margins of the Mediterranean.

Some 1 035 million hectares, or 20 per cent of the world's susceptible drylands, are affected by human-induced soil degradation (UNEP/ISRIC 1991). Of this total, 45 per cent is affected by water erosion, 42 per cent by wind erosion, 10 per cent by chemical deterioration and 3 per cent by physical deterioration of the soil structure. Water erosion is the dominant form of degradation in semi-arid areas (51 per cent of total degradation) and dry sub-humid regions (also 51 per cent), and wind erosion is dominant in the arid zone (60 per cent).

One major consequence of desertification is the development crisis affecting many dryland countries. Drylands still provide much of the world's grain and livestock, and form the habitat that supports the last remaining big game animals. The human population of the drylands lives in increasing insecurity as productive land per capita diminishes.

agricultural land (Oldeman 1994). Crop yields in Africa could be halved within 40 years if degradation of cultivated land continues at present rates (Scotney and Dijkhuis 1989). Land degradation affects about 300 million hectares of land in Latin America, as a result of soil erosion, loss of nutrients, deforestation, overgrazing and poor management of agricultural land (UNEP/ISRIC 1991). In Europe, some 12 per cent of the land area (115 million hectares) is affected by water erosion and some 4 per cent (42 million hectares) by wind erosion; in North America about 95 million hectares are affected by degradation, mainly erosion (UNEP/ISRIC 1991).

FAO projections for food supplies by region (FAO 1996) suggest that future problems will be concentrated in sub-Saharan Africa and South Asia, and that chronic under-nutrition is expected to affect 11 per cent of the population, or 637 million people in these countries in the year 2010. The countries projected to suffer from serious shortfalls in food supply are also those faced with rapidly growing populations and urbanization, low productivity agriculture, high debt and insufficient wealth to import food. Food availability in all other regions is projected to be adequate by the year 2010, as agricultural production growth is expected to keep pace with growing food requirements.

A particular problem, not only in developed countries, is the rising demand for meat, fish, poultry and dairy products, which encourages farmers to raise livestock. Growing fodder and feed crops for animals can displace subsistence food crops and is a less efficient use of land.

Since the 1970s, the FAO has been assessing the actual and potential areas of cultivated land in 117 developing countries to see which are or could become self-sufficient in food. There are some 2 500 million hectares of land in these countries that could be cultivated, of which about 760 million hectares are already under cultivation. By the year 2000, the FAO estimated that 64 countries would be facing a critical situation and, using traditional subsistence agriculture, 38 of these would be unable to support even half their projected populations (FAO 1995).

Global production of crops and livestock grew by 2.6 per cent in 1996. In developing countries,

Calorie intake has generally increased over the past two decades but there were downturns in both Europe and Central Asia (as a result of political upheaval) and and in West Asia (as a result of war) during 1990–95

Calorie intake per capita

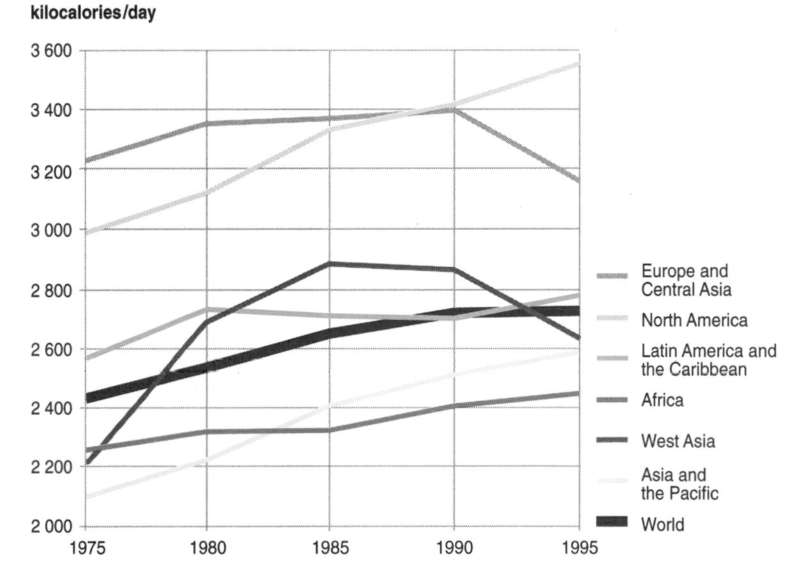

Source: compiled by UNEP GRID Geneva from FAOSTAT 1997 and WRI, UNEP, UNDP and WB 1998

production in that year grew by 2.9 per cent, compared to 5.2 per cent in 1995, 5 per cent in 1994 and 4 per cent in 1993.

Forests

Eighty per cent of the forests that originally covered the Earth have been cleared, fragmented or otherwise degraded (WRI 1997). Most of the remaining forest is located in just a few places, mostly in the Amazon Basin, Canada, Central Africa, Southeast Asia and the Russian Federation. These large blocks of ecologically-intact natural forest are valuable because they house indigenous cultures, shelter global biodiversity, provide ecosystem services, store carbon, contribute to local and national economic growth, and meet recreational needs. Yet logging, mining and other large-scale development projects threaten 39 per cent of the remaining natural forests, with those in South and Central America, western North America and the boreal regions of the Russian Federation most at risk (WRI 1997).

Between 1990 and 1995, 56 million hectares of forests were lost, a net loss of 65 million hectares in developing countries being partially offset by an increase of nearly 9 million hectares in the developed world

Worldwide, there are now some 3 500 million hectares of forest, about half in the tropics and the rest in the temperate and boreal zones (FAO 1997a). The great majority are natural and semi-natural; forest plantations make up only about 5 per cent of the total. Slightly more than half of the total area is in developing countries.

Despite increased public awareness and a large number of initiatives, deforestation is still continuing in most of Africa, Latin America, and Asia and the Pacific. During 1980–90 alone, the Latin American region lost 62 million hectares (6.0 per cent) of its natural forest, the largest loss in the world during those years, with a further 5.8 million hectares a year lost during 1990–95 (FAOSTAT 1997).

Globally, between 1990 and 1995, 56 million hectares of forests were lost, a total loss of 65 million hectares in developing countries being partially offset by an increase of nearly 9 million hectares in the developed world.

Deforestation has been arrested – and even reversed – in North America and Europe. For example, Europe's forest area has increased by more than 10 per cent since the early 1960s as a result of tree planting and, in part, natural regeneration of marginal lands (EEA 1995). In West Asia, where forest is scarce, vigorous replantation and reforestation programmes have turned the tide in some, though not all, countries.

The underlying driving forces behind deforestation are poverty, population and economic growth, urbanization and expansion of agriculture lands. Clearance for agriculture is the largest cause of tropical deforestation; logging, however, is responsible for an estimated one-third of the total, the proportion rising to about one-half in Asia, and possibly higher still in parts of South America (FAO 1997a).

The demand for wood continues to rise; world production of wood products, including fuelwood and charcoal as well as commercial timber products, is 36 per cent higher than in 1970. Wood remains the main and often the only source of energy for many people in large areas of the developing world. In Africa, where 90 per cent of the population depends on firewood and other biomass for energy, the production and consumption of firewood and charcoal doubled between 1970 and 1994 and is expected to rise by another 5 per cent by 2010 (FAO 1997a). Commercial wood production is still dominated by the developed world, though developing countries increased their share of industrial roundwood output from 17 per cent in 1970 to 33 per cent in 1994 (FAO 1997a). The industrial countries are largely self-sufficient in timber and pulp products, with the important exception of Japan. Wood in Europe is produced mainly from managed forests and plantations but logging from natural or virgin forests remains common in North America. The biggest projected demand for

Change in forest extent, 1990–95

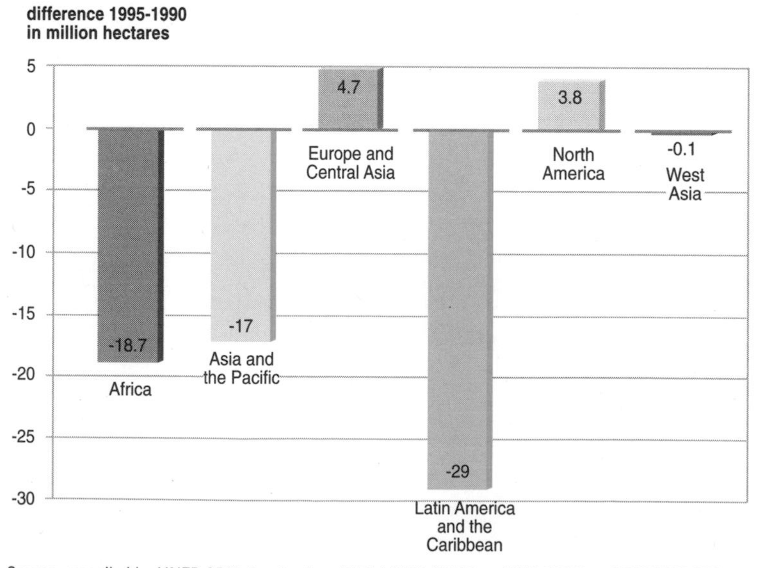

Source: compiled by UNEP GRID Geneva from WRI, UNEP, UNDP and WB 1998 and FAOSTAT 1997

commercial wood will come from Asia, where demand is rising most rapidly but reserves are already inadequate.

There are no quantitative assessments of the condition and health of forests on a global level. However, significant degradation still continues worldwide. About 60 per cent of all forests in Western and Central Europe, and large areas around industrial installations in East Europe and Central Asia, are either seriously or moderately degraded, mainly as a result of pollution. In some areas of Europe, however, there have been some improvements in forest condition, which are being interpreted as a response to improvements in air quality (UNECE/CEC 1997).

Africa's forests are being degraded by droughts, fuelwood extraction, civil wars and the refugees that result, untimely bush fires and the advance of agriculture. Over-exploitation has resulted in forests of critically low quality in many parts of the world. For example, only about 10 per cent of the remaining forests in parts of the Mekong basin are now commercially valuable (MRC/UNEP 1997) and changes in the structure and composition of large forest areas in Latin America have led to irreversible losses in biodiversity (WRI 1997).

A growing realization of the scale of loss and degradation of forests has mobilized media attention and public concern, changed policies, legislation and institutional arrangements, and focused local, national and international efforts on promoting sustainable forest management. Timber harvesting from natural forests in some countries has been reduced for environmental reasons, and more emphasis is being placed on increasing the efficiency of harvesting of forest products and manufacturing operations, expanding the area of tree plantations, rehabilitating degraded lands and reducing demand through wood substitutes and other alternatives. Non-wood forest products, important for household use and income and, in some cases, as export products, are increasingly being recognized as a substantial component of the forest economy in some regions. Growing awareness of the social and environmental functions of trees and forests has led to the planting or expansion of urban and community forests.

Another major change is an acknowledgement that achievement of sustainable forest management on a global level necessitates the involvement of a range of interest groups and accommodation of multiple interests in forestry planning (see Chapter 4).

Biodiversity

All species, as well as all individuals within a species, have a finite life span and thus changes in biodiversity are inevitable. Accelerated and enhanced reduction in diversity at gene, species and ecosystem level, however, is not only intrinsically undesirable but a significant threat to human material welfare because it implies reduced ability of ecosystems to provide key products and services.

The total number of species on the Earth is very large: around 1.7 million have been described but many more are believed to exist – estimates range from 5 to nearly 100 million; 12.5 million has been proposed as a reasonable working estimate (see table). The most species-rich environments on Earth are moist tropical forests which extend over some 8 per cent of the world's land surface and probably hold more than 90 per cent of the world's species. Overall, the regions richest in biodiversity are Africa, Asia and the Pacific, and Latin America.

The conservation status of most species is not known in detail but two large animal groups – mammals and birds – have been comprehensively assessed and may be representative of the status of biodiversity in general. In 1996, 25 per cent of the world's approximately 4 630 mammal species, and 11 per cent of the 9 675 bird species were assessed as globally threatened – that is, at significant risk of total extinction (IUCN 1996). Countless other species,

Known and estimated total numbers of species

	known number of species	estimated total number of species
Insects	950 000	8 000 000
Fungi	70 000	1 000 000
Arachnids	75 000	750 000
Nematodes	15 000	500 000
Viruses	5 000	500 000
Bacteria	4 000	400 000
Plants	250 000	300 000
Protozoans	40 000	200 000
Algae	40 000	200 000
Molluscs	70 000	200 000
Crustaceans	40 000	150 000
Vertebrates	45 000	50 000
World total (all groups)	1 700 000	12 500 000

Source: WCMC 1992

although not yet globally threatened, now exist in reduced numbers and as fragmented populations, and many of these are threatened with extinction at national level.

Most of the threatened species are land-based, with more than half occurring in forests (Collar and others 1994) but evidence is growing of the vulnerable nature of freshwater habitats and marine habitats such as coral reefs. For example, in the United States, freshwater species – nearly 70 per cent of the mussels, 50 per cent of the crayfish and 37 per cent of the fishes are threatened – are at greater risk than terrestrial species (Master and others 1998).

Food plants exemplify the most fundamental values of biodiversity. Originally, plants were consumed directly from the wild, and gathering of wild produce continues throughout the world today. Only a few of the many species of flowering plants have been treated as direct food sources though others provide food for animals which in turn are hunted or farmed by people.

Around 200 species have been domesticated as food plants, and of these about 20 are crops of major international economic importance. Relatively few botanical families account for the world's main food plants: *Gramineae* (grasses, including cereals) and *Leguminosae* (legumes, including peas, beans and lentils) are foremost among these.

The evolution of food crops over many centuries of domestication has increased the range of genetic diversity but development and promotion of high-yielding cultivars for modern intensive agriculture is now rapidly reversing this trend, leading to a dangerous reliance on genetically uniform crops, often ones that need high inputs of fertilizer and pesticides to perform effectively. As intensive agriculture has spread widely, many local varieties have been displaced and some have disappeared entirely. Wild relatives of cultivated species are also often threatened with extinction as a result of habitat change.

An increasingly restricted genetic base appears to underlie periodic production failure in economically important crops, leading to increased yield variability and increased synchronicity of variation across large areas; for example, a 15 per cent reduction in maize harvest in 1970 in the United States was attributed to widespread cultivation of a blight-susceptible variety (WCMC 1992).

Wood, still mainly harvested from wild sources, is one of the most important commodities in international trade. Potentially valuable timber resources in many parts of the world are being degraded through excess harvesting, inadequate management and habitat loss. For example, of more than 600 large tree species in Ghana, around 60 are used in the timber trade and some 25 species have been identified as of conservation concern because of over-exploitation or rarity (WCMC 1992). Recent analysis (Oldfield and others 1998) of around 10 000 tree species (out of a possible world total of 100 000) found that nearly 6 000 met the criteria for threatened status defined by IUCN, with 976 categorized as Critically Endangered, 1 319 as Endangered and 3 609 as Vulnerable. Habitat loss or modification is the underlying source of risk, particularly for restricted range species but felling was the individual threat most often cited (for 1 290 species).

The conservation status of trees

			number of tree species
estimated world total			100 000
species assessed			10 091
globally threatened			5 904
of which,	Critically Endangered	976	
	Endangered	1 319	
	Vulnerable	3 609	
extinct			95

Note: the 95 extinct tree species includes 18 that still exist but not in the wild

Source: WCMC Species Database, data available at http://wcmc/org/uk, and Oldfield and others 1998

At the broadest level, biodiversity loss is driven by economic systems and policies that fail to value properly the environment and its resources, legal and institutional systems that promote unsustainable exploitation, and inequity in ownership and access to natural resources, including the benefits from their use. While some species are under direct threat, for example from hunting, poaching and illegal trade, the major threats come from changes in land use leading to the destruction, alteration or fragmentation of habitats. For example, Niger has lost 80 per cent of its freshwater wetlands during the past two decades (UNDP 1997), two-thirds of Asian wildlife habitats have been destroyed with the most acute losses in the Indian sub-continent, China, Vietnam and Thailand (Braatz 1992) and, in the Latin American region, the average annual deforestation rate during 1990–95 was 2.1 per cent in Central America and more than 1 per cent in Paraguay, Ecuador, Bolivia and Venezuela (FAO 1997a).

A further major cause of biodiversity loss is the widespread introduction of animal and plant species outside their natural range, resulting in change at the community and ecosystem level, and sometimes total destruction of some of the species originally present (UNEP 1995). For example, some 18 per cent of 119 threatened mammals in Australia and the Americas, and 20 per cent of world threatened birds, were affected by introduced predators or competitors in 1992 (WCMC 1994). The effects of alien species are especially pronounced in closed systems such as lakes and islands. For example, at least 60 per cent of the cichlid fishes in Lake Victoria are estimated to be extinct as a result of the introduction of Nile perch (Keenleyside 1991).

Environmental pollution is an increasingly major threat to biodiversity in many countries. Pesticide residues have reduced the population of several bird species and other organisms. Air and water pollution stress ecosystems and reduce populations of sensitive species, especially in coastal zones and wetlands. Rapid environmental change, such as *El Niño* events, can also have significant impacts on natural habitats, as can the longer-term effects of climate change, for example reductions in the volumes of water bodies after persistent dry weather. The effect of events such as forest fires can be multiplied many times wherever habitats are already fragmented and species depleted.

The conservation of biodiversity is often regarded as less important than the short-term economic or social interests of the sectors that influence it most heavily. A major requirement is to incorporate biodiversity concerns into other policy areas.

Freshwater

Global freshwater consumption rose sixfold between 1900 and 1995 – at more than twice the rate of population growth. About one-third of the world's population already lives in countries with moderate to high water stress – that is, where water consumption is more than 10 per cent of the renewable freshwater supply (see maps on page 42). The problems are most acute in Africa and West Asia but lack of water is already a major constraint to industrial and socio-economic growth in many other areas, including China, India and Indonesia (Roger 1998). In Africa, 14 countries are already subject to water stress or water scarcity, and a further 11 countries will join them in the next 25 years (Johns Hopkins 1998). If present consumption patterns continue, two out of every three persons on Earth will live in water-stressed conditions by the year 2025 (WMO and others 1997). The declining state of the world's freshwater

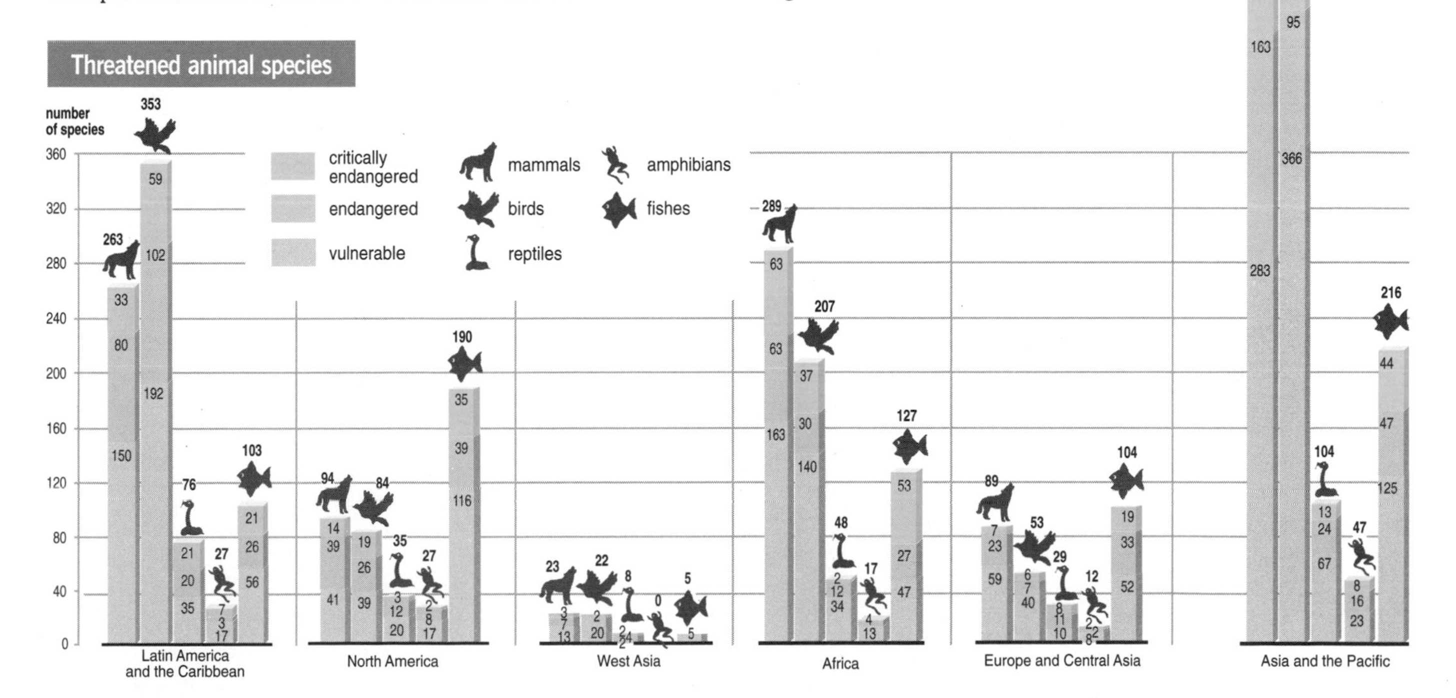

Bar chart shows total number of threatened animal species by region. Most are land-based but freshwater habitats and marine habitats such as coral reefs are increasingly vulnerable

Source: WCMC/IUCN 1998

Global water stress, 1995 and 2025

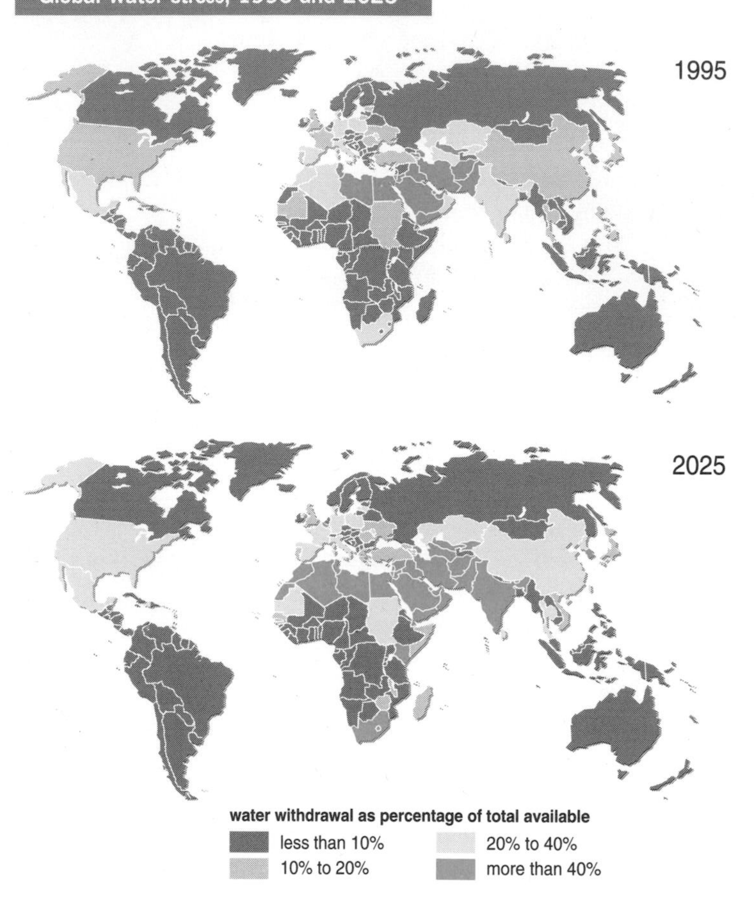

Note: water stress is defined as follows:
low, less than 10% of total available is withdrawn
moderate, 10–20% of total available is withdrawn
medium-high, 20–40% of total available is withdrawn
high, more than 40% of total available is withdrawn

Source: WMO and others 1996

By the year 2025, as much as two-thirds of the world population may be subject to moderate to high water stress

resources, in terms of quantity and quality, may prove to be the dominant issue on the environment and development agenda of the coming century.

About 20 per cent of the world's population lacks access to safe drinking water and about 50 per cent lacks adequate sanitation. In many developing countries, rivers downstream of large cities are little cleaner than open sewers. Levels of suspended solids in Asia's rivers, for example, almost quadrupled since the late 1970s and rivers typically contain four times the world average and 20 times OECD levels. The faecal coliform count in Asia's rivers is 50 times higher than the WHO guidelines. People using such water for washing, bathing or drinking are at high risk. In Latin America as a whole, only about 2 per cent of sewage receives any treatment. Worldwide, polluted water is estimated to affect the health of about 1200 million people and to contribute to the death of about 15 million children under five every year (ICWE 1992).

Sewage pollution is also common in groundwater in many developing countries. For example, groundwater in Merida, Mexico, has been severely affected by the influx of stormwater and sewage, and there is a risk that the contamination will spread to the wells which supply the city. Similar problems have occurred in Sri Lanka and many Indian cities, and are expected in Jakarta and Manila, which have 900 000 and 600 000 septic tanks respectively (UNEP 1996b).

While sewage pollution is the largest and most common problem, it is not the only one. The intensive use of pesticides and fertilizers has led to chemicals being leached into freshwater supplies in many places. Nitrate pollution from excess fertilizer use (see page 27) is now one of the most serious water quality problems. Maximum allowable levels of nitrates in drinking water are exceeded in some places in every country in Europe (OECD 1994) and in many countries in other regions. Even in the United States, more than 40 million people obtained their drinking water in 1994 from a system in which there were violations of health-based standards, mainly those relating to nitrates. In some parts of Africa, nitrate loads in some suburban groundwater wells are 6–8 times WHO acceptable levels. Not only are nitrates dangerous to human health, leading to brain damage and even death in some infants (OECD 1994), but they stimulate rapid algal growth in waterways, leading to eutrophication in both inland waters and the sea. The red tides in the Gulf of Mexico and elsewhere are the direct result of the over-use of fertilizers in agriculture.

Industrial wastes are significant sources of water pollution, often giving rise to contamination with heavy metals (lead, mercury, arsenic and cadmium) and persistent organic compounds. A study of 15 Japanese cities, for example, showed that 30 per cent of all groundwater supplies are contaminated by chlorinated solvents from industry; in some cases, the solvents from spills travelled as far as 10 km from the source of pollution (UNEP 1996b).

Over-abstraction has also affected the quality of groundwater. This has led to seawater intrusion along shorelines, causing salinization of coastal agricultural lands. As a result, some arable land, such as that on the Batinah coastal plain of Oman, has been completely lost (UNEP/ESCWA 1992). It is estimated that the saline interface between the sea and groundwater advances at an annual rate of 75–130 metres in Bahrain (UNEP/ESCWA 1991). In Madras, India, salt water intrusion has moved 10 km inland, rendering many irrigation wells useless (UNEP 1996b). Salt water intrusion is of particular concern in small island states, where the limited groundwater supply is surrounded by salt water.

Inland water bodies have suffered in many areas from industrial pollution and poor land management. In Scandinavia, for example, hundreds of lakes, particularly small ones, still suffer from acidification and it will take a long time for water quality to return to normal (EEA 1997). All the major rivers in the European part of the former Soviet Union and in Siberia have been diverted into chains of artificial lakes. In most, lake bed sediments are highly polluted, and high inputs of phosphorus and other nutrients have often led to eutrophication. The Aral Sea – which lost one-third of its area, two-thirds of its water and almost all its native organisms as a result of the diversion of its input waters for irrigation (UNEP 1994b) – will probably never recover. Fisheries in the Black Sea have collapsed, and rapidly rising water levels in the Caspian Sea have inundated many surrounding villages and towns. The causes of the latter are unknown but climate change may be implicated (WCN 1997).

Worldwide, agriculture accounts for more than 70 per cent of freshwater consumption, mainly for irrigation of agricultural crops. In Africa and Asia, agriculture accounts for nearly 80 per cent. Agricultural demand for water is projected to increase sharply, since much of the additional food that will be needed to feed the world population in the future is expected to come from an increase in irrigated land. In regions where water is in short supply, however, there may be a good case for buying in staple foods and using the irrigation water saved for domestic and industrial purposes.

Household demand, particularly in urban areas, is rising rapidly, particularly among wealthy consumers, in developed and developing countries, with an abundance of household appliances and garden irrigation. Europe and North America are the only regions currently using more water in industry than in agriculture. On current trends, industrial water use will more than double by the year 2025 with a four-fold increase in pollutant emissions to watercourses (WMO and others 1997). In some countries, industrial water demand will rise even more sharply. Industrial water use in China, for example, is projected to increase more than fivefold by the year 2030 (Brown and Halweil 1998).

Groundwater supplies about one-third of the world's population, and is the only source of water for rural dwellers in many parts of the world. Excessive withdrawal of groundwater, in quantities greater than the ability of nature to renew the aquifers, is now widespread in parts of the Arabian Peninsula, China, India, Mexico, the former Soviet Union and the United States. The water table has dropped by tens of metres in many places where there is intensive groundwater use. An estimated 65 per cent of public water supplies in Europe come from groundwater sources, and groundwater withdrawal in the European Union rose by 35 per cent between 1970 and 1985 (EEA 1995). Falling water tables have also exacerbated land subsidence in many regions as well as saltwater intrusion into groundwater. Parts of California's San Joaquin Valley, for example, have sunk by 8 metres since the 1920s, causing land fissures and disruption to roads, railways and housing.

Groundwater resources in West Asia in general and on the Arabian Peninsula in particular are in a critical condition because the volumes withdrawn far exceed natural recharge rates, threatening water distribution systems that have been used for thousands of years.

Limited availability, contamination and increased water demand have made groundwater withdrawals more costly, and this has contributed to greater social inequity. In Gujarat, India, for example, excessive groundwater withdrawal has caused the water level in aquifers to fall by 40 metres in some cases (UNEP 1996b). This has deprived many poor farmers of freshwater, since they cannot afford to sink boreholes to the required depth. Wealthier farmers are able to move further inland and buy new land.

There are many natural constraints to access to freshwater, such as the uneven distribution of water in different regions, and the variable effects of weather. Water managers are also becoming increasingly concerned about the unpredictable effects of climatic variability on water resources, including those associated with *El Niño* and anthropogenic climate change.

It is also becoming clear that good water management can solve many of the problems of pollution and scarcity. Most of the citizens of Jordan and Israel, for example, two of the most 'water-scarce' countries in the world, have access to adequate supplies of safe water, largely as a result of an effective irrigation strategy.

Marine and coastal areas

The oceans are the largest ecosystems on Earth. They are as rich and diverse as any terrestrial ecosystem yet are still largely unexplored. While the deep ocean is mainly unpolluted, evidence is emerging of environmental degradation in some areas, and a decline in many marine species. The coastal marine environment, by contrast, is clearly being affected by the modification and destruction of habitats, over-fishing and pollution. Many of these impacts can be traced back to land-based human activities located far from the sea. Enclosed seas are the most endangered. The Aral Sea is effectively dead and semi-enclosed seas such as the Mediterranean, the Black Sea and the Baltic are highly polluted. Coastal lagoons are globally polluted.

More than one-third of the world's population lives within 100 km of a seashore (Cohen and others 1997) – 50 per cent of the population in North America and 60 per cent in Latin America, where 60 of the largest 77 cities lie on the coast. By 2000, nearly 500 million people will be concentrated in urban conglomerations along the shores of Asia (WRI/UNEP/UNDP 1994).

The natural environment of coastal areas, which includes wetlands, estuaries, mangroves and coral reefs, is being degraded by agricultural and urban development, industrial facilities, port and road construction, dredging and filling, tourism and aquaculture. Dam construction, even located far inland, can alter water flow patterns that support important fisheries, as well as cutting off the supply of sediment necessary to maintain deltas and coastlines.

The many people living in coastal zones, and even those located far inland, generate large quantities of wastes and other polluting substances that enter the seas directly or through coastal watersheds, rivers and precipitation from polluted air. While coastal pollution is gradually being controlled in many industrialized countries, it is still rising rapidly as a result of population growth, urbanization and industrial development in developing regions. For example, 38 per cent of Africa's coastline and 68 per cent of its marine protected areas are under a high degree of threat from development.

Many coastal waters carry excessive sediment and are contaminated by microbes and organic nutrients. Nitrogen, resulting from sewage discharges, agricultural and urban run-off, and atmospheric precipitation, is a particular problem (see page 27). The destruction of wetlands and mangroves, which act as natural filters for sediment, excessive nitrogen and wastes, has also accelerated nutrient build-up. Additional pollution sources are oil leaks and accidental spills from shipping, discharge of bilge water, oil drilling and mineral extraction. Some persistent pollutants are even reaching deep ocean waters.

Worrying evidence is emerging of the accelerating destruction of the world's coral reefs by pollution. More than half the world's reefs are potentially threatened by human activities, with up to 80 per cent at risk in the most populated areas (WRI, ICLARM, WCMC and UNEP 1998).

There have been some, albeit isolated, improvements in the state of the coastal and marine environment. Examples include improved bathing beaches in many regions, the clean-up of some rivers in western Europe, and a decline in DDT levels in the Baltic Sea and off the Pacific Coast of North America, resulting in the recovery of some animal and bird

Shrimp farming

Between 1980 and 1990, the production of farmed prawn and shrimp grew 600 per cent, with about 75 per cent coming from Asia. Annual production worldwide is now more than 1 million tonnes. While shrimp farming was initially seen as a way of reducing harvesting pressure in heavily used natural fisheries and the collateral damage done to other species, environmental problems associated with the industry, including habitat conversion, damage to wild populations and effluents, have led to a reappraisal of the industry.

While most shrimp farms are on salt flats and similarly suitable land, an increasing proportion are being put on wetlands and areas of former mangrove forests. Globally, shrimp farming accounts for considerably less than 10 per cent of the total loss of mangroves but this proportion is increasing. Damage to wild populations of shrimp is restricted primarily to the South American fisheries where farmers prefer to raise larvae caught in the wild rather than those raised in a hatchery. The inflated price for wild caught larvae has caused much damage to wild populations. Finally, the widespread over-fertilization and seeding of ponds and the increasing use of antibiotics and other chemicals have led to severe problems with effluents.

The environmental record of the large producers is improving fast and most problems are now caused by smaller producers. Small-scale shrimp farming is important, however: it provides millions of jobs and is an important stimulus to local economies. Efforts to improve the technology, especially the success of hatching larvae and other methods of feeding, will lead to large environmental gains, as will the trends for smaller producers to form cooperatives. Conversion of wetland and mangrove habitat for shrimp production should, however, be strictly controlled.

Source: Boyd and Clay 1998

populations. However, much more needs to be done to swing the global balance from destruction towards recovery, including more effort to address the problem of marine debris which threatens marine wildlife.

There is a growing understanding of the possible impact of climate change on the marine environment, for example through more evaporation from warmer seas increasing atmospheric humidity and thus reinforcing the greenhouse effect (Epstein 1997). Until recently, attention has focused on the impact on small island states and low-lying countries of a rise in sea level and an increase in the frequency or intensity of storms resulting from climate change. There could, however, be more complex effects. For example, if warming continues, freshwater from melted Arctic ice may form a cap on the Norwegian and Greenland Seas, resulting in changes to deep ocean circulation patterns that might divert to the south the waters of the Gulf Stream that presently keep western Europe warm in the winter (Broecker 1997).

Surface warming and increased thermal stratification may also reduce phytoplankton productivity, which forms the basis of the entire marine food chain. A build up of carbon dioxide in the atmosphere can lead to increased acidity of the surface ocean (Epstein 1997) which, together with UV-B penetration, can also reduce phytoplankton productivity; it can also change the carbonate content in surface waters, which could interfere with coral growth. Extensive coral bleaching (see page 337) has also recently been associated with the warming of surface waters (Pomerance 1999).

Over the past half-century, the world's fishing fleets have been industrialized, in response to growing demand and high subsidies, through the introduction of high-technology fishing gear, sonar fish tracking systems, and on-board processing and refrigeration which enable boats to stay at sea for many weeks. The global marine fish catch rose from some 50 million tonnes in 1975 to more than 97 million tonnes in 1995 (see bar chart). This increase masks a complicated picture in which new species of fish, and new fishing grounds, have been successively exploited and depleted. Aquaculture output, meanwhile, has grown dramatically, now accounting for almost 20 per cent of all fish and shellfish production (FAO 1997b). Repeated failures to implement measures to control over-fishing mean that approximately 60 per cent of the world's ocean fisheries are now at or near the point at which yields decline (Grainger and Garcia 1996) and many

Global marine fish catch

million tonnes/year

Region	1975	1980	1985	1990	1995
West Asia	0.36	0.29	0.32	0.36	0.43
North America	3.76	4.90	6.16	7.27	6.22
Latin America and the Caribbean	6.29	9.18	13.27	15.71	21.11
Europe and Central Asia	12.11	12.41	12.89	20.91	18.50
Asia and the Pacific	22.94	25.18	29.74	34.72	40.94
Africa	6.75	7.44	9.17	10.96	10.28

Source: compiled by UNEP GRID Geneva from FAO 1997c

Global marine fish catch has grown considerably over the past two decades but the rate of growth has begun to slow. The catch in Africa, Europe and North America had begun to decline by 1990

local fishing communities have suffered catastrophic reductions in their annual harvest.

The fishing industry is also degrading marine habitats and species, often in the most biologically-productive and commercially-valuable marine habitats, such as mangroves and coral reefs. Intensive forms of aquaculture are generating additional environmental problems in the form of severe local water pollution and destruction of coastal ecosystems.

Nearly 1 000 million people depend on fish for their primary source of protein, and demand for food fish is projected to increase from about 75 million tonnes in 1994/95 to 110–120 million tonnes in 2010. With careful management, the marine catch could be sustainably increased by about 10 million tonnes a year. However, if no effective action is taken soon, production could decline. According to FAO, most of the projected increase in demand for food fish can be met only through continued increases in aquaculture (FAO 1997b).

Atmosphere

Strenuous efforts have begun to abate atmospheric pollution in many industrialized countries but urban air pollution problems are reaching crisis dimensions in most cities of the developing world. Acid rain remains a problem, with critical loads (the threshold at which acid deposition causes damage) frequently exceeded over large parts of North America, Europe and Southeast Asia (Kuylenstierna, Cinderby and Cambridge 1998). The precipitation of atmospheric pollutants at sea is the major source of open ocean

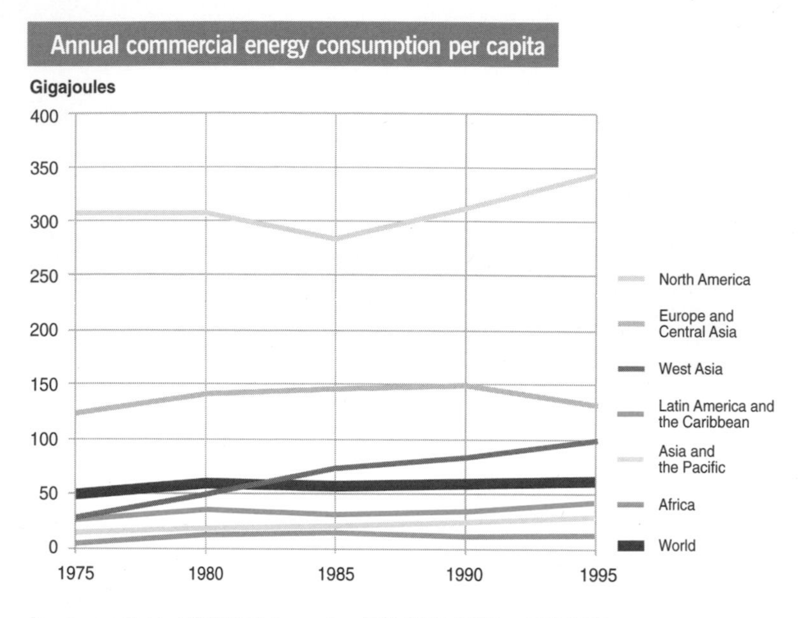

Source: compiled by UNEP GRID Geneva from WRI, UNEP, UNDP and WB 1998

Future levels of atmospheric pollution will be governed largely by the use of fossil fuel energy – but in Europe and the United States, SO_2 emissions have declined since 1980 despite increasing energy consumption during 1980–90

pollution, and the identification of processes that transport toxic chemicals from warm regions into the Arctic (see page 176) shows how the atmosphere links the global environment into a single integrated system.

The Convention on Long Range Transboundary Air Pollution has resulted in significant reductions in emissions of acidifying gases in Europe and North America – for example, between 1985 and 1994, SO_2 emissions in Western, Central and Eastern Europe fell by 50 per cent in line with the Convention on Long Range Transboundary Air Pollution protocols (Olendrzynski 1997). However, emissions in other regions, especially in parts of Asia, are a major and growing problem. For example, if current trends continue, emissions of sulphur dioxide from coal burning in Asia will surpass emissions from North America and Europe combined by the year 2000 and continue to grow thereafter, unlike emissions in North America and Europe which are expected to fall (see table right). Impacts have already been observed: for example, the World Bank has estimated China's overall annual forest and crop losses due to acid rain at US$5 000 million (World Bank 1997); in Japan, many monitoring sites recorded annual sulphur dioxide deposition at levels equal to or greater than those in Europe or North America; and in the Republic of Korea winter rain acidity has been nearly as high as pH4 (Shrestha and Iyngararasan 1998).

There is growing understanding of the links between atmospheric problems such as local air pollution, acid rain, global climate change and stratospheric ozone depletion. Isolated responses to one environmental problem may in fact worsen another. For example, catalytic converters on cars decrease nitric oxide emissions and help to reduce acid rain and urban smog but they release higher levels of nitrous oxide, which is a potent greenhouse gas and a contributor to stratospheric ozone depletion. Sulphate aerosols in the upper atmosphere contribute to acid rain but may offset greenhouse warming – and thus reducing sulphur emissions from power plants by switching to low-sulphur coal or using scrubbers may exacerbate the problem of climate change (IPCC 1996a).

Atmospheric pollution is a relatively minor problem in the African and West Asian regions. In Africa, there are problems in the urbanized and industrialized areas in the north and south, for example from vehicles, which are often old and use leaded fuel, and from some manufacturing, mining and industrial activities and power stations. Biomass burning is an additional problem in Africa. If the projected demand for transport and electricity in Africa is met with current technologies, emissions from vehicles will rise fivefold and from power stations elevenfold by 2003 (World Bank 1992). In West Asia, air pollution is a problem mainly in the larger cities, exacerbated by the high temperatures and levels of sunlight.

Despite improvements in the levels of some atmospheric pollutants in North America and Western Europe, resulting from the effective implementation of control measures, and pollution reductions in Eastern Europe and Central Asia, mainly from economic restructuring, significant problems remain. For example, critical loads for acid deposition are still being exceeded for more than 25 per cent of ecosystems in Western and Central Europe, and emissions of nitrogen oxides in North America increased by about 10 per cent from the 1980s to the

SO_2 emissions from fossil fuel burning

	1980	*1990*	*1995*	*2000*	*2010*
	(millions of tonnes of sulphur dioxide)				
Europe	59	42	31	26	18
United States	24	20	16	15	14
Asia	15	34	40	53	79

Note: per capita emissions in Asia are still many times lower than those in Europe or the United States

Source: Worldwatch Institute 1998

1990s (International Joint Commission 1997). These problems are likely to worsen as the economies in Eastern Europe and Central Asia grow stronger, and with the continuing increase in car use in these regions, in the rest of Europe and in North America.

In Latin America, the main anthropogenic source of atmospheric emissions is deforestation. Biomass burning and the establishment of new types of vegetation cover in the Amazon basin have significant ecological implications for the region, the continent and the globe (LBA 1996). Some parts of the region also suffer from air pollution from industry and from large cities. The situation may worsen as a result of the deregulation and privatization of the energy sector in, for example, Argentina, Brazil and Colombia – where there may be a trend away from biomass and hydropower to more use of fossil fuels (Rosa and others 1996).

The Asian and Pacific region has experienced significant growth in atmospheric pollution, resulting from the heavy use of coal and high sulphur fuels, traffic growth and forest fires. The most serious problems are in urban areas and the developing countries in the region. Japan, however, has reduced sulphur emissions through gains in efficiency, heavier reliance on oil and nuclear power, and stringent pollution control laws.

In all regions, future levels of atmospheric pollution will be governed largely by the use of energy from fossil fuels. The Intergovernmental Panel on Climate Change has predicted that global economic output may double between now and 2050, with energy demand reaching nearly three times that in 1990 (IPCC 1995). If the developing countries follow the conventional development path, there would be a massive increase in the emission of atmospheric pollutants. However, this need not be the case – as has been proved by some developed countries. For example, in Europe, sulphur emissions peaked in the 1970s (see page 262) and subsequently declined steadily, despite increasing energy consumption. Similarly, the mechanisms developed for the Kyoto Protocol could help developing countries restrict their emissions of greenhouse gases.

Urban areas

Almost 3 000 million people, about half of the world's population, live in urban areas and about 160 000 more join them every day. Cities affect far more than the areas they occupy: their 'ecological footprints' can be enormous because of their huge demands for energy, food and other resources, and the regional and global impacts of their wastes and emissions to soil, air and water. London's ecological footprint, for example, considering only its consumption of food and forest products, and the area needed to assimilate its emissions of carbon dioxide, is calculated to be 125 times greater than the area of the city itself (IIED 1995).

Growth of urban populations

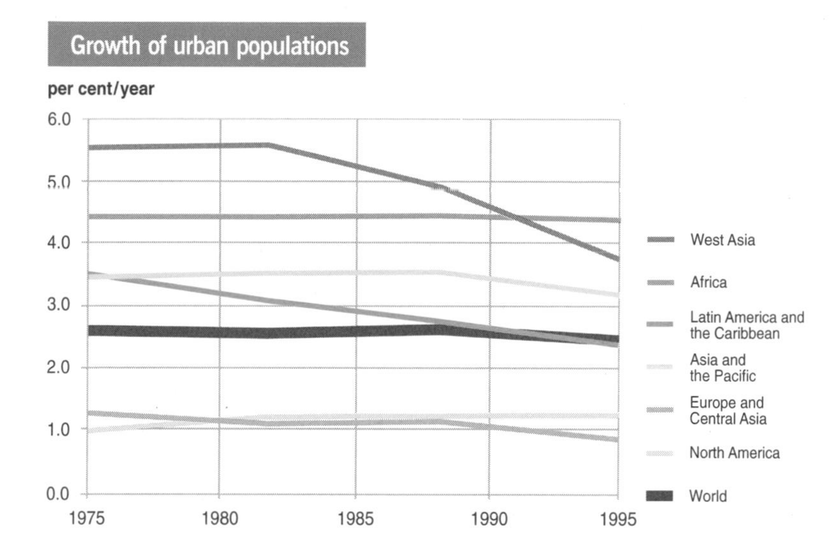

Source: compiled by UNEP GRID Geneva from United Nations Population Division 1997

While rates of urban growth have slowed in all regions except North America, they are still adding some 160 000 people to the world's urban population every day

Urbanization has been one of the most striking developments of the 20th century. In Africa, for example, only 5 per cent of the population lived in urban areas at the beginning of the century, about 20 per cent in the 1960s and about 35 per cent in 1995. Africa's annual urban growth rate is now the highest in the world, at more than 4 per cent (see graph). The urban population of the Asia-Pacific region, now about 35 per cent of the total population, grew by 3.2 per cent a year between 1990 and 1995, compared with 0.8 per cent a year for the rural population (United Nations Population Division 1997). About 70 per cent of the population of North America, Europe and Latin America now lives in cities, and worldwide 326 cities have populations of more than 1 million, compared with 270 in 1990 (WRI, UNEP, UNDP and WB 1996). In Western Europe (EEA 1998) and North America (WRI, UNEP, UNDP and WB 1996), in contrast with most other regions, there is a move out of large cities into suburbs and smaller urban centres.

Most of the growth in the world's population is taking place in developing countries and most of the projected increase of 1 000 million people between 1999 and around 2010 is likely to be absorbed by cities

Industrial estates in urban areas

Industrial estates are now common features of the urban landscape. The International Development Research Council records more than 12 000 estates around the world, ranging in size from 1 to more than 10 000 hectares.

Industrial estates are primarily designed to improve production efficiency through the co-location of manufacturing plant and services but many of them also pose a substantial threat to the environment. Their size and number are expanding, particularly in countries undergoing fast industrialization.

There is a clear need for sound environmental management on industrial estates. Fortunately, the proximity of industries that cause environmental damage can be turned to advantage. If environmental concerns are integrated into estate development at all stages, cumulative damaging effects can be avoided. For example, permits can be granted to ensure that only compatible industries are allowed to locate in a given area. Estates with sound environmental management include energy efficiency measures, resource conservation, waste minimization, cleaner production and information centres, and preparedness and planning for accidents.

The most advanced estates form a miniature 'industrial ecosystem' in which individual manufacturing processes optimize the consumption of energy and raw materials and the effluents of one process serve as the raw material for another process. The industrial district of Kalundborg in Denmark is a prime example of this industrial symbiosis. For 15 years, industries on the site have exchanged by-products such as surplus energy, waste heat and other materials. In one example, waste heat (in the form of cooling water) from the Asnaes Power Plant provides district heating to houses and buildings in the municipality of Kalundborg.

Source: UNEP 1997

in these countries – cities already faced with enormous backlogs in housing and infrastructure development, and struggling with increasingly overcrowded transportation systems, insufficient water supplies, deteriorating sanitation and environmental pollution. In spite of this, people continue to migrate to cities in the hope of a better life, often as a result of the devastation of rural economies by land degradation.

At least 600 million urban dwellers in Africa, Asia and Latin America live in squatter settlements and shanty towns, in housing of such poor quality and with such inadequate provision for water, sanitation, drainage and garbage removal that their lives and health are under continuous threat (UNCHS 1996). The number of people living in such conditions is likely to expand very rapidly; while large cities in some developing countries have been growing at rates of up to 10 per cent per annum, slums and squatter settlements in some of them are growing twice as fast. An increasing number of the urban poor, probably more than 100 million, are homeless, a serious problem in developed as well as in developing countries (UNCHS 1996).

Among the most serious environmental problems in cities are air and water pollution, solid waste accumulation and disposal (including toxic and hazardous wastes), and noise. Many cities are also at risk from natural hazards or hazards whose origin may be natural but where human actions have significantly increased the level of risk (see page 31).

Most urban air pollution comes from the combustion of fossil fuels, in motor vehicles and for industrial processes, heating and electricity generation, but some also comes from incinerators, petrochemical plants and refineries, metal smelters and the chemicals industry. Some primary pollutants can combine to form even more damaging secondary pollutants. For example, ozone and other photochemical oxidants are formed when hydrocarbons react with nitrogen oxides and oxygen in the presence of sunlight. Tropospheric ozone is one of the major components of urban smog, a growing problem in cities around the world.

While urban air pollution is coming under control in some countries, the situation is deteriorating rapidly in many heavily industrialized cities in developing countries. In China, for example, smoke and small particles from burning coal cause more than 50 000 premature deaths and 400 000 new cases of chronic bronchitis a year in 11 of its largest cities (World Bank 1997), and private car circulation has been restricted in some cities in South America and Europe in attempts to reduce harmful levels of air pollution. Worldwide, more than 1 000 million urban residents are exposed to health-threatening levels of air pollution (Schwele 1995).

Many cities are facing serious shortages of safe water as a result of over-exploitation of resources and pollution. Bangkok's water table, for example, has fallen 25 metres since the late 1950s and saltwater has penetrated its wells (WWF 1990). Daily demand for water in Beijing increased almost 100 times between 1950 and 1980 (WRI, UNEP and UNDP 1992). Urban demand for water in Latin America is likely to rise fivefold during the next four decades (WRI, UNEP and UNDP 1994).

Recent trends in urbanization reflect economic and political changes. Within the context of the structural changes in the world economy, some regions and cities have proved more flexible than nations in adapting to changing economic conditions. Well-managed urbanization can lead to improvements in the living standards of the world's population. However, the transition to an urbanized world has profound implications for the state of the world environment.

References

Andreae, M. O. (1991). In Levine, J.S. (ed.). *Global Biomass Burning*. MIT Press, Cambridge, United States

Boyd, C.E., and Clay, J.W. (1998). Shrimp Aquaculture and the Environment. *Scientific American*, June 1998

Braatz, S. (1992). *Conserving Biological Diversity. A Strategy for Protected Areas in the Asia-Pacific Region*. World Bank Technical Paper No. 193. World Bank, Washington DC, United States

Bradley, D. (1996). Human Health and Tropical Development in Health and the Environment: The Linacre Lectures. In McMichael, A. (ed.), *Climate Change and Human Health: an assessment prepared by a Task Group on behalf of the World Health Organization and the United Nations Environment Programme*. Oxford University Press, Oxford, United Kingdom, and New York, United States

Broecker, W. S. (1997). Thermohaline circulation, the Achilles Heel of our climate system: Will man-made CO_2 upset the current balance? *Science* 278, 1582–1588

Brown, L., and Halweil, B. (1998). *China's Water Shortage*. Worldwatch Press Release, 22 April 1998. Worldwatch Institute, Washington DC, United States

Carpenter, S., N.F. Caraco, D.L. Correll, R.W. Howarth, A.N. Sharpley and V.H. Smith (1998). Nonpoint pollution of surface waters with phosphorus and nitrogen. *Ecological Applications*, No. 3, Summer 1998

CDIAC (1998). *Revised Regional CO_2 Emissions from Fossil-Fuel Burning, Cement Manufacture, and Gas Flaring: 1751–1995*. Carbon Dioxide Information Analysis Center, Environmental Sciences Division, Oak Ridge, Tennesse, United States.
http://cdiac.esd.ornl.gov/cdiac/home.html

CDIAC (1999). *Revised Regional CO_2 Emissions from Fossil-Fuel Burning, Cement Manufacture, and Gas Flaring: 1751–1996*. Carbon Dioxide Information Analysis Center, Environmental Sciences Division, Oak Ridge, Tennesse, United States.
http://cdiac.esd.ornl.gov/cdiac/home.html

Cohen, J.E., Small, C., Mellinger, A., Gallup, J., and Sachs, J. (1997). Estimates of coastal populations. *Science*, 278, 5341, 1211–1212

Colborn, T. (1997). *Our Planet*, Vol. 8, No. 6

Colborn, T., Dumanoski, D., and Myers, J.P. (1996). *Our Stolen Future: are we threatening our fertility, intelligence and survival? – a scientific detective story*. Little, Brown, London, United Kingdom

Collar, N. J., Crosby, M. J. and Stattersfield, A. J. (1994). *Birds to Watch 2: The World List of Threatened Birds*. BirdLife International, Cambridge, United Kingdom

CRED (1999). EMDAT Database of the Centre for Research on the Epidemiology of Disasters, Catholic University of Louvain, Belgium
http://www.md.ucl.ac.be/entites/esp/epid/misson/intro_uk.htm

de Nava, C.C. (1996). World wide Overview of Hazardous Wastes. *Toxicology and Industrial Health*, Vol.12, No. 2a

EEA (1995). *Environment in the European Union 1995: Report for the Review of the 5th Environmental Action Programme*. Office for Official Publications of the European Communities, Luxembourg

EEA (1997). *Air Pollution in Europe in 1997*. Office for Official Publications of the European Communities, Luxembourg

EEA (1998). *Europe's Environment: The Second Assessment*. Office for Official Publications of the European Communities, Luxembourg, and Elsevier Science, Oxford, United Kingdom

EEPSEA/WWF (1998). *Haze damage from 1997 Indonesian fires exceeds US$1.3 billion*. Press Release by the World Wide Fund for Nature (WWF) Indonesia Programme and the Economy and Environment Program for Southeast Asia (EEPSEA), 24 February 1998
http://www.geocities.com/RainForest/2701/eepsea1.htm

Environmental Pollution (1998). First International Nitrogen Conference. Papers published in *Environmental Pollution*, 102, 1
http://www.minvrom.nl/environment/nitrogen/409.htm
http://www.hbz-nrw.de/elsevier/02697491/sz984251/

Epstein, P. R. (1997). Climate, Ecology and Human Health. *Consequences*, Vol. 3, No. 2

ESCAP (1993). *State of Urbanization in Asia and the Pacific*, United Nations, New York, United States

FAO (1995). *Dimensions of Need*. FAO, Rome, Italy

FAO (1996). *Food, Security and Nutrition*. World Food Summit, FAO, Rome, Italy

FAO (1997a). *State of the World's Forests*. FAO, Rome, Italy

FAO (1997b). *Yearbook of Fishery Statistics*. FAO, Rome, Italy

FAO (1997c). *Fishstat-PC*. FAO, Rome, Italy. http://www.fao.org

FAO (1998). *Guide to Efficient Plant Nutrition Management*. FAO, Rome, Italy

FAOSTAT (1997). *FAOSTAT Statistics Database*. FAO, Rome, Italy.
http://www.fao.org

Gilbert, C. (1997). *Indonesia's Peat Smoulders Underground*. Environment News Service, IGC Networks Headlines Digest
http://www.concentric.net/~blazingt/info/sarawak.htm

Glover-Kerkvliet, J. (1995). Environmental Assault on Immunity. *Environmental Health Perspectives*, Vol. 103, No. 3

Goyer, R.A. (1996). Results of lead research: prenatal exposure and neurological consequences. *Environmental Health Perspectives*, Vol. 104, No. 10

Grainger, R., and Garcia, S. (1996). *Chronicles of Marine Fishery Landings (1950–1994): Trend Analysis and Fisheries Potential*. FAO Fisheries Technical Paper 359. FAO, Rome, Italy

Gubler, D., and Clark, G. (1994). Community based integrated control of the Aedes aegypti: a brief overview of current programs. *American Journal of Tropical Medicine and Hygiene*, Vol. 50, No. 6

ICWE (1992). International Conference on Water and the Environment: development issues for the 21st century, 26–31 January 1992, Dublin, Ireland. ICWE Secretariat, WMO, Geneva, Switzerland

IIED (1995). *Citizens Action to Lighten Britain's Ecological Footprint*. International Institute for Environment and Development, London, United Kingdom

International Joint Commission (1997). *The IJC and the 21st Century. Response of the IJC to a Request by the Governments of Canada and the United States for Proposals on How to Best Assist Them to Meet the Environmental Challenges of the 21st Century*. International Joint Commission, Washington DC, United States, and Ottawa, Canada

IPCC (1995). *Climate Change 1994: Radiative Forcing of Climate Change and an Evaluation of the IPCC IS92 Emission Scenarios*. Houghton, J., Meiro Filho, L.G., Bruce, J., Lee., H., Callander, B.A., Haites, E., Harris, N., and Maskell, K. (eds.), UNEP/WMO. Cambridge University Press, Cambridge, United Kingdom

IPCC (1996a). *Climate Change 1995: The Science of Climate Change*. Houghton, J., Meiro Filho, L.G., Callander, B.A., Harris, N., Kattenberg, A., and Maskell, K. (eds.), UNEP/WMO. Cambridge University Press, Cambridge, United Kingdom

IPPC (1996b). *Climate Change 1995: Impacts, Adaptations and Mitigation of Climate Change: Scientific-Technical analyses. Contribution of Working Group II to the Second Assessment Report of the Intergovernmental Panel on Climate Change*. Watson, R.T., Zinyowera, M.C., and Moss, R.H. (eds.), WMO/UNEP. Cambridge University Press, Cambridge, United Kingdom

IPCC (1998). *The Regional Impacts of Climate Change: An Assessment of Vulnerability. A special report of IPCC Working Group II*. Cambridge University Press, Cambridge, United Kingdom

IRPTC (1999). *Prior Informed Consent for Certain Hazardous Chemicals in International Trade*
http://irptc.unep.ch/pic/

IUCN (1996). *1996 IUCN Red List of Threatened Animals*. IUCN, Gland, Switzerland

Johns Hopkins (1998). Solutions for a Water-Short World. *Population Report,* Vol. XXVI, No. 1, September 1998. Johns Hopkins Population Information Program, Baltimore, Maryland, United States http://www.jhuccp.org/popreport/m14sum.stm

Kaiser, J. (1996). Acid Rain's Dirty Business: Stealing Minerals from Soil. *Science*, Vol. 272, 198, 12 April 1996

Keeling, C.D. and Whorf, T.P. (1998). *Atmospheric CO_2 concentrations – Mauna Loa Observatory, Hawaii, 1958-1997* (revised August 1998). NDP-001. Carbon Dioxide Information Analysis Center, Oak Ridge National Laboratory, Oak Ridge, Tennessee, United States http://cdiac.esd.ornl.gov/cdiac/home.html

Keenleyside, M.H.A. (1991). *Chichlid Fishes: Behaviour, Ecology and Evolution.* Chapman and Hall, London, United Kingdom

Kuylenstierna, J.C.I., S. Cinderby and H. Cambridge (1998). Risks from Future Air Pollution. In Kuylensierna, J. and Hicks, K. (eds.). *Regional Air Pollution in Developing Countries.* Stockholm Environment Institute, York, United Kingdom

LBA (1996). *The large scale biosphere-atmosphere experiment in Amazonia*. INPE, São Paulo, Brazil

Liew, S.C., Lim, O.K., Kwoh, L.K., and Lim, H. (1998). Study of the 1997 forest fires in South East Asia using SPOT quicklook mosaics. *Proceedings, 1998 International Geosience and Remote Sensing Symposium,* Vol. 2, p. 879-881, Seattle, Washington, United States

Lindsey, S. and Birley, M. (1996). Climate Change and Malaria Transmission. *Annals of Tropical Medicine and Parasitology*, Vol. 90, No. 6

Master, L.L., Flack, S.R. and Stein, B.A. (eds., 1998). *Rivers of Life: Critical Watersheds for Protecting Freshwater Biodiversity*. The Nature Conservancy, Arlington, Virginia, United States

MRC/UNEP (1997). *Mekong River Basin Diagnostic Study: Final report.* Mekong River Commission (MRC) and UNEP, Bangkok, Thailand

Munich Re (1997 and 1998). *Annual Review of Natural Catastrophes, 1997 and 1998.* Münchener Rückversicherungs-Gesellschaft (Munich Reinsurance Company), Munich, Germany

NOAA (1998). http://nic.fb4.noaa.gov:80/ products/analysis_monitoring/enso_advisory/ advfig1.gif

OECD (1994). *Towards Sustainable Agricultural Production – cleaner technologies*. OECD, Paris, France

Oldeman, L.R. (1994). Global Extent of Soil Degradation. In *Soil Resilience and Sustainable Land Use* (eds. D.J. Greenland and I. Szabolcs), p. 99-118. CAB International, Wallingford, United Kingdom

Oldfield, S., Lusty, C. and MacKinven, A. (1998). *The World List of Threatened Trees*. WCMC and IUCN. World Conservation Press, Cambridge, United Kingdom

Olendrzynski, K. (1997). Emissions. In *Transboundary Air Pollution in Europe,* edited by E. Berge. MSC-W Status Report 1997. Norwegian Meteorological Institute, Oslo, Norway

Pomerance, R. (1999). *Coral Bleaching, Coral Mortality and Global Climate Change.* Report to the US Coral Reef Task Force Meeting in Hawaii, 5-6 March 1999

Reuters (1998). 16 April 1998

Roger, P. (1998). *Role of Governments in Regulating Industrial Water Activities*. Background Paper No. 16, Commission on Sustainable Development, Sixth Session, 20 April–1 May 1998

Rosa, L. P., M. T. Tolmasquim, E. La Rovere, L. F. Legey, J. Miguez, R. Schaeffer (1996). *Carbon dioxide and methane emissions: a developing country perspective.* COPPE/UFRJ, Rio de Janeiro, Brazil

Schindler, D.W., Curtis, J.P., Parker, B.R. and Stainton, M.P. (1996). Consequences of climate warming and lake acidification for UV-B penetration in North American boreal lakes. *Nature,* 379, 705–708

Schwele, D. (1995). Public Health Implications of Urban Air Pollution in Developing Countries. *Proceedings of the 10th World Clean Air Congress*, Espoo, Finland, 28 May - 2 June 1995

Scotney, D.M. and Djikhuis, F.H. (1989). *Recent Changes in the Fertility Status of South African Soils.* Soil and Irrigation Research Institute, Pretoria, South Africa

Seitzinger, S.P., and Kroeze, C. (1998). Global distribution of nitrous oxide production and N inputs in freshwater and coastal marine ecosystems. *Global Biogeochemical Cycles,* 12, 93–113

Shrestha, S., and Iyngararasan., M. (1998). An Overview of of Acid Rain Impacts in the Asia and Pacific. In Kuylensierna, J., and Hicks, K. (eds.). *Regional Air Pollution in Developing Countries.* Stockholm Environment Institute, York, United Kingdom

Smith, K.R. (1997). Development, Health and Environmental Risk Transition. In Shahi, G.S. (ed.), *International Perspectives on Environment, Development and Health*. Springer, New York, United States

UN (1997). *Environment and Sustainable Development: International Decade for Natural Disaster Reduction*. Report of the Secretary-General, 3 November 1997. United Nations, New York, United States

UNCHS (1996). *An Urbanizing World: Global Report on Human Settlements 1996*. Oxford University Press, Oxford, United Kingdom, and New York, United States

UNDP (1997). *Human Development Report 1997*. Oxford University Press, New York, United States, and Oxford, United Kingdom

UNECE/CEC (1997). *Forest Condition in Europe, 1977.* Federal Research Centre for Forestry and Forest Products, Germany

UNEP (1994a). *UNEP Data Report.* UNEP, Nairobi, Kenya

UNEP (1994b). *The Pollution of Lakes and Reservoirs*. UNEP Environment Library No. 12, UNEP, Nairobi, Kenya

UNEP (1995). *Global Biodiversity Assessment.* United Nations Environment Programme. Cambridge University Press, Cambridge, United Kingdom

UNEP (1996a). *UNEP Survey on Sources of POPs.* Report prepared for an IFCS Expert Meeting on Persistent Organic Pollutants, Manila, the Philippines, 17-19 June 1996. UNEP, Geneva, Switzerland

UNEP (1996b). *Groundwater: a threatened resource*. UNEP Environment Library No. 15, UNEP, Nairobi, Kenya

UNEP (1997). *The Environmental Management of Industrial Estates.* UNEP IE Technical Report No. 39. UNEP, Paris, France

UNEP (1998a). *Production and Consumption of Ozone Depleting Substances 1986–1996*. Ozone Secretariat, United Nations Environment Programme, Nairobi, Kenya http://www.unep.org/unep/secretar/ozone/pdf/Prod-Cons-Rep.pdf

UNEP (1998b). *Environmental Effects of Ozone Depletion: 1998 Assessment*. Ozone Secretariat, United Nations Environment Programme, Nairobi, Kenya

UNEP (1998c). *Report of the Technology and Economic Assessment Panel, 1998.* Ozone Secretariat, UNEP, Nairobi, Kenya

UNEP (1998d). *Cleaner Production: a guide to sources of information*. UNEP, Paris, France

UNEP (1999). *Synthesis of the Reports of the Scientific, Environmental Effects, and Technology and Economic Assessment Panels of the Montreal Protocol. A Decade of Assessments for Decision Makers Regarding the Protection of the Ozone Layer: 1988-99.* Ozone Secretariat, UNEP, Nairobi, Kenya

UNEP/ESCWA (1991). *The National Plan of Action to Combat Desertification in Bahrain*. UNEP, Bahrain

UNEP/ESCWA (1992). *The National Plan of Action to Combat Desertification in Oman*. UNEP, Oman

UNEP/ISRIC (1991). *World Map of the Status of Human-Induced Soil Degradation (GLASOD). An Explanatory Note,* second revised edition (edited by Oldeman, L.R., Hakkeling, R.T., and Sombroek, W.G.). UNEP, Nairobi, Kenya, and ISRIC, Wageningen, Netherlands

United Nations Population Divison (1997). *Urban and Rural Areas, 1950-2030 (the 1996 Revision),* on diskette. United Nations, New York, United States

Vitousek, P. M., J. Aber, R. W. Howarth, G. E. Likens, P. A. Matson, D. W. Schindler, W. H. Schlesinger, and G. D. Tilman (1997). Human alteration of the global nitrogen cycle: causes and consequences. *Ecological Applications* 7, 737–750

WCMC (1992). *Global Biodiversity: Status of the Earth's Living Resources*. Groombridge, B. (ed.). Chapman and Hall, London, United Kingdom

WCMC (1994). *Biodiversity Data Source Book.* Groombridge, B. (ed.). World Conservation Press, Cambridge, United Kingdom

WCMC/IUCN (1998). WCMC Species Database, data available at http://wcmc/org/uk, assessments from the 1996 IUCN Red List of Threatened Animals

WCN (1997). Caspian Sea Levels: explaining the changes. *World Climate News,* No. 10, January 1997

WCN (1998a). The Impacts of *El Niño* Events. *World Climate News,* No. 13, June 1998

WCN (1998b). The 1997–1998 *El Niño. World Climate News,* No. 13, June 1998

WCN (1998c). Major *El Niño* Event. *World Climate News,* No. 12, January 1998

WCN (1998d). Regional Impacts of the 1997–1998 *El Niño. World Climate News,* No. 13, June 1998

Wedin, D.A., and Tilman, D. (1996). Influence of nitrogen loading and species composition on the carbon balance of grasslands. *Science,* Vol. 274, p. 1720

WHO (1990). *Public health impacts of pesticides used in agriculture.* WHO, Geneva, Switzerland

WHO (1992). *Indoor air pollution from biomass fuel.* WHO/PEP/92.3A. WHO, Geneva, Switzerland

WHO (1997a). *World Health Report, 1997: Conquering Suffering, Enriching Humanity.* WHO, Geneva, Switzerland

WHO (1997b). *Health and environment in sustainable development, five years after the Earth Summit.* WHO, Geneva, Switzerland

WHO (1998). *The World Health Report 1998: Life in the 21st Century, A Vision for All*. WHO, Geneva, Switzerland

WMO and others (1997). *Comprehensive Assessment of the Freshwater Resources of the World*. WMO, Geneva, Switzerland

WMO, UNEP, NOAA, NASA AND EC (1998). *Scientific Assessment of Ozone Depletion: 1998. Volumes I and II.* Global Ozone Research and Monitoring Project - Report No. 44. WMO, Geneva, Switzerland

Woolcock, A., and Peat, J. (1997). Evidence for the increase in asthma worldwide. In Chadwick, D. and Cardew, G. (eds.), *The Rising Trends in Asthma.* Ciba Foundation Symposium 206. Wiley, Chichester, United States

World Bank (1992). *Development and Environment, World Development Report*. Oxford University Press, Oxford, United Kingdom, and New York, United States

World Bank (1997). *Clear Water, Blue Skies: China's Environment in the New Century*. China 2020 Series. World Bank, Washington DC, United States

Worldwatch Institute (1998). *Vital Signs 1998*. Worldwatch Institute, Washington DC, United States

WRI (1997). *The Last Frontier Forests: Ecosystems and Economies on the Edge*. D. Bryant, D. Nielsen and L. Tangley (eds.). WRI, New York, United States

WRI, ICLARM, WCMC and UNEP (1998). *Reefs at Risk: a map-based indicator of threats to the world's coral reefs. WRI,* Washington DC, United States

WRI, UNEP and UNDP (1992). *World Resources 1992-93: A Guide to the World Environment*. Oxford University Press, New York, United States, and Oxford, United Kingdom

WRI, UNEP and UNDP (1994). *World Resources 1994-95: A Guide to the World Environment*. Oxford University Press, New York, United States, and Oxford, United Kingdom

WRI, UNEP, UNDP and WB (1996). *World Resources 1996-97: A Guide to the Global Environment* (and the *World Resources Database* diskette). Oxford University Press, New York, United States, and Oxford, United Kingdom

WRI, UNEP, UNDP and WB (1998). *World Resources 1998-99: A Guide to the Global Environment* (and the *World Resources Database* diskette). Oxford University Press, New York, United States, and Oxford, United Kingdom

WWF (1990). *Atlas of the Environment.* Lean, G., Hinrichsen, D. and Markham, A. (eds.). Arrow, London, 1990

WWF (1997). *Rain Forests on Fire* http://www.worldwildlife.org/new/fires/report2.htm

WWF (1998). *The Year The World Caught Fire* http://www.panda.org/news/features/01-98/story3.htm

Zahm, S. and Devesa, S. (1995). *Childhood cancer: overview of incidence, trends and environmental carcinogens*, Environmental Health Perspectives, Vol 103, Supplement 6

Africa

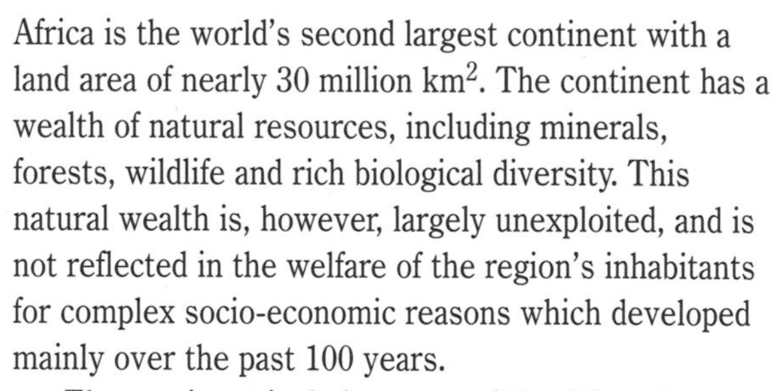

KEY FACTS

Reducing the poverty of the poor majority of Africans is the overriding priority. This poverty is a major cause and consequence of the environmental degradation and resource depletion which threaten economic growth. New approaches that put the poor at the top of the environment and development agenda could tap and release the latent energy and talents of Africans to bring about development that is economically, socially and environmentally sustainable.

- Africa remains under-populated: its population density of 249 people per 1 000 hectares is low compared to the world average of 442.
- Africa is the only continent on which poverty is expected to rise during the next century.
- An estimated 500 million hectares of land have been affected by soil degradation since about 1950, including as much as 65 per cent of agricultural land.
- As a result of declining food security, the number of undernourished people in Africa nearly doubled from 100 million in the late 1960s to nearly 200 million in 1995.
- Africa lost 39 million hectares of tropical forest during the 1980s, and another 10 million hectares by 1995.
- Fourteen countries are subject to water stress or water scarcity, and a further 11 will join them by 2025.
- Africa emits only 3.5 per cent of the world's total carbon dioxide now and this is expected to increase to only 3.8 per cent by the year 2010.
- While the large external debts of many African countries are a major concern, many of the same countries also have growing 'environmental debts' where the cost of remedial action will be far greater than preventive action.

Africa is the world's second largest continent with a land area of nearly 30 million km^2. The continent has a wealth of natural resources, including minerals, forests, wildlife and rich biological diversity. This natural wealth is, however, largely unexploited, and is not reflected in the welfare of the region's inhabitants for complex socio-economic reasons which developed mainly over the past 100 years.

The continent includes some of the driest deserts, largest tropical rain forests and highest equatorial mountains in the world. But key natural resources are unevenly distributed. For example, more than 20 per cent of the remaining tropical forest is in a single country, the Democratic Republic of the Congo, while a major share of the continent's water resources are in a few large basins such as the Congo, Niger, Nile and Zambezi river systems.

Many of the events that have shaped Africa's geo-political, socio-economic and environmental development over the past century are related to the colonization of the region and its subsequent partition in 1885 among several European countries. During the first half of the 20th century, the colonial authorities imported economic development policies and patterns which largely neglected the adverse impacts on the poor majority of people and on the environment. On achieving independence during and after the 1960s, African governments inherited and maintained centralized economic and sectoral institutions and narrowly focused economic growth policies, usually with the encouragement and support of international aid agencies. These national and international 'development' policies, in combination with rapid population growth and increased poverty, had progressively adverse impacts on the state of the environment throughout the continent.

Since the 1970s, the environment and key natural resources in most African countries have been increasingly threatened by escalating and unsustainable pressures from fast-growing populations and cities as well as expanding agricultural and industrial activities. Significant economic and environmental damage has also resulted from civil conflicts and war caused in part by the arbitrary division of territory and peoples, as well as inequitable development patterns set during colonial times. In the push for accelerated economic growth after independence, many national development projects as well as international aid and lending policies failed to take into account the adverse impacts of their activities on the environment and natural resource base.

Throughout Africa, reducing the poverty of the poor majority of people is the overriding priority for governments. This poverty is a major cause and consequence of the environmental degradation and resource depletion which threaten present and future economic growth. Improving the health, income and living conditions of the poor majority remains the top political and policy imperative if Africa is to move toward development that is economically, socially and environmentally sustainable.

Social and economic background

Africa has undergone major social, economic and political transformations. At the turn of the 20th century, the total population was only 118 million, 7.4 per cent of the world population. In the following 50 years, the population grew slowly, as high fertility rates were offset by high death rates due to poor health conditions, infectious diseases, civil wars and the struggle against colonialism. When mortality rates began to decline sharply from the 1950s onwards, due to improved health conditions associated with economic development, there was a dramatic population increase. By 1997, the population was estimated at 778.5 million, more than 13 per cent of the world population (United Nations Population Division 1996). It is projected that by the year 2025, the population in Africa will almost double to 1 453 million, representing about 18 per cent of the projected world population (United Nations Population Division 1996).

Population

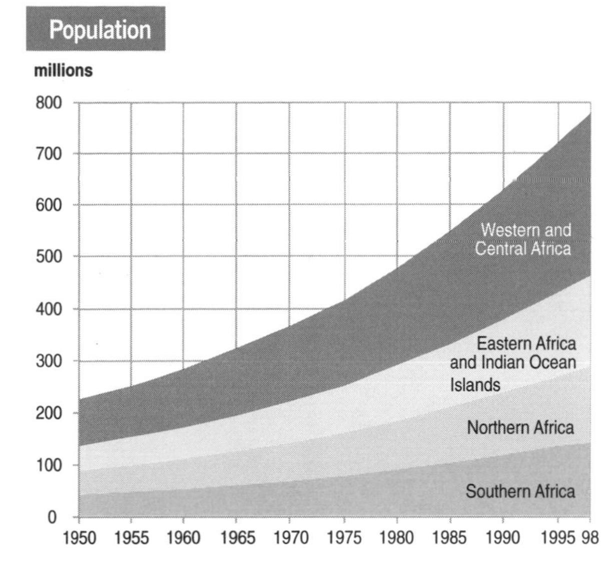

Source: compiled by UNEP GRID Geneva from United Nations Population Division 1996

Although Africa has more than 13 per cent of the world population, population densities in most of the region are low compared to those in other parts of the world

Despite such rapid population growth, Africa remains under-populated: its population density of 249 per 1 000 hectares is low compared to the world average of 442 or the 1 130 found in Asia (WRI, UNEP, UNDP and WB 1998). However, wide variations of population density occur within and between countries. Mauritius has the highest population density in Africa, at 5 562 per 1 000 hectares, while Namibia's 19 people per 1 000 hectares is the lowest (WRI, UNEP, UNDP and WB 1998).

Fertility rates in Africa are projected to decline from 6.5 during 1975–80 to 5.3 during 1995–2000 (WRI, UNEP, UNDP and WB 1998). Western and Central Africa has the highest fertility rate of 6.6, while Southern and Northern Africa have the lowest rates of 4.1 and 4.2, respectively. Epidemic diseases have had a serious impact on the African population. In recent years, HIV/AIDS has become one of the major causes of death. In 1996, about 14 million people in sub-Saharan Africa had HIV/AIDS, about 64 per cent of the worldwide total (AIDS Analysis Africa 1996).

Poverty and environmental degradation 'are linked in a vicious circle in which people cannot afford to take proper care of the environment' (SARDC, IUCN and SADC 1994). Poverty has been and remains a major cause and consequence of environmental degradation

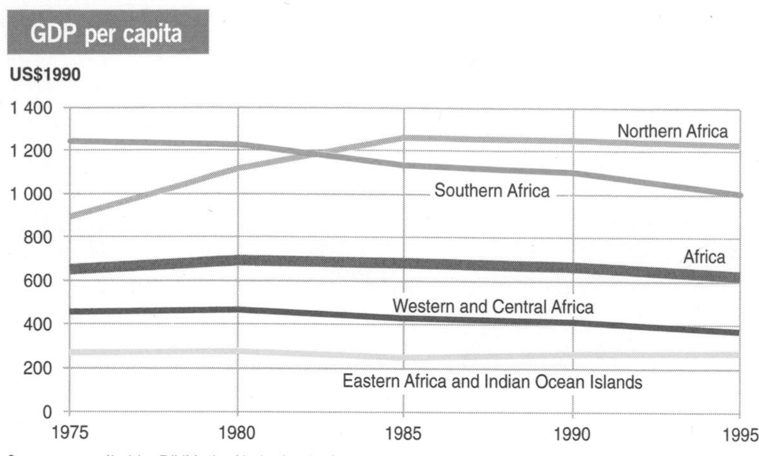

Source: compiled by RIVM, the Netherlands, from World Bank and UN data

Although per capita GDP has been declining in most African countries since 1980, there have been some signs of economic recovery since 1995

and resource depletion. Currently almost 40 per cent of people in sub-Saharan Africa live below the poverty line, and both income poverty and human poverty are increasing (UNDP 1997). According to current projections, Africa is the only continent on which poverty is expected to rise during the next century (UNDP 1998).

The human condition in Africa remains as daunting as ever. Of 45 countries on the UN list of Low Human Development Indicators, 35 are in Africa (UNDP 1997). Reducing poverty and improving human development are major challenges for the continent.

Although the 1980s were considered a 'lost' decade for both economic and environmental improvement in Africa (UNEP 1991), with either negative or sluggish economic growth, there have been signs of economic recovery since the mid-1990s. In 1996, GDP grew by 4–5 per cent for the second year in a row, higher than population growth, and nearly three-quarters of sub-Saharan countries had more than 3 per cent growth (GCA 1997). Rates, however, varied from –15.4 per cent in Burundi to 37.3 per cent in Equatorial Guinea and 16.1 per cent in Malawi (UNECA 1997). This overall good performance, in every sub-region except Western and Central Africa, was the result of better weather, a more favourable international environment and improved macroeconomic policies. Agriculture was a notable contributing factor to higher growth (GCA 1997).

During the 1980s and 1990s, many countries embarked on economic reform through structural adjustment programmes. While economic liberalization may have helped fuel economic recovery, there are indications that economic growth will worsen rather than improve environmental conditions. For this reason there is no substitute for explicit environmental policy actions (WRI, UNEP, UNDP and WB 1996).

The debt burden has been a major constraint for many nations, which have had to spend more on servicing their debt than providing basic social services. In 1997, Africa's total debt stock stood at US$349 000 million, or 67.5 per cent of GDP, with a debt service ratio of 21.3 per cent of exports plus remittances (UNECA 1998). External debt varies widely. For example, Nigeria, Côte d'Ivoire, the Sudan, the Democratic Republic of the Congo and Angola account for nearly half the debt for sub-Saharan Africa (United Nations 1996). Although the debt issue is being addressed by the international community, the relief will be selective and will take a number of years to have effect (UNDP 1997).

Africa's share in world trade is small and is shrinking due to fierce competition from other regions, which enjoy faster and more sustained economic growth. In 1995, the continent's terms of trade had fallen to 89 per cent of the 1987 baseline index (GCA 1997). Nevertheless, exports and imports significantly influence the regional economy, with exports alone accounting for 25 per cent of the regional GDP and imports providing 20 per cent of the domestic supply. Imports have increased from US$91 600 million in 1990 to US$125 200 million in 1996, making Africa one of the most open regions in the world (UNECA 1997).

Although considerable improvements have been made in reducing political instability and civil unrest, which are themselves a manifestation of intense competition for declining opportunities and resources, much more needs to be done to attain and sustain socio-economic growth, durable peace and equitable income distribution. There are now strong signs of a return to peace and security, and progress towards democratic governance and popular participation. Civil wars in Angola, the Sudan and the Democratic Republic of the Congo have still to be resolved. Resettlement and reconstruction are still slow in countries such as Burundi, Liberia, Rwanda and Somalia. Another predicament relates to refugees and other displaced persons (see box right). Most African countries have generously shared their limited resources with refugees, at times to the detriment of the environment.

Growing fiscal constraints and competition for

ever-dwindling public resources have seen the environment being sacrificed in terms of budgetary allocations for the more pressing demands of health and education. As a result, donor funding is sustaining most environmental management programmes.

Land and food

Land is the critical resource and the basis for survival for most people in Africa. Agriculture contributes about 40 per cent of regional GDP and employs more than 60 per cent of the labour force (World Bank 1998). The contribution of agriculture to national GDP is generally highest in Eastern, and Western and Central Africa. In Ethiopia and Somalia, for example, the agricultural sector provides more than 60 per cent of national GDP.

Land degradation is a serious problem throughout Africa, threatening economic and physical survival. Key issues include escalating soil erosion, declining fertility, salinization, soil compaction, agrochemical pollution and desertification. An estimated 500 million hectares of land have been affected by soil degradation since about 1950 (UNEP/ISRIC 1991), including as much as 65 per cent of agricultural land (Oldeman 1994). Soil losses in South Africa alone are estimated to be as high as 400 million tonnes annually (SARDC, IUCN and SADC 1994). Soil erosion affects other economic sectors such as energy and water supply. In a continent where too many people are already malnourished, crop yields could be cut by half within 40 years if the degradation of cultivated lands were to continue at present rates (Scotney and Dijkhuis 1989).

Recurrent droughts are also a major factor in the degradation of cultivated land and rangelands in many parts of Africa. The two problems are often interlinked. While drought increases soil degradation problems, soil degradation also magnifies the effect of drought (Ben Mohamed 1998).

In many countries, a combination of inequitable land distribution, poor farming methods and unfavourable land tenure and ownership systems have led to declining productivity on grazing lands, falling crop yields and diminishing returns from the water supplied. In Uganda, much land is owned and used according to customary practices which provide little or no incentive to protect and conserve it, leading to mismanagement and degradation (NEMA 1996).

In Southern Africa, escalating land degradation over the past decade has been caused by increased livestock. Overgrazing causes more than half the soil degradation in the sub-region. In Namibia, livestock production subsidies actually encourage farmers to raise more livestock than if they had to meet the full costs themselves (Byers 1997). With new economic policy changes under way in the region of the Southern African Development Community (SADC), including the removal of such subsidies, stocking rates are expected to decline over the next decade. Declining agricultural yields in the SADC region are also attributed to water erosion which is responsible for about 15 per cent of land degradation. About 2 per

Refugees and the environment in Tanzania

The Rwandan refugee crisis in mid-1994 led to the influx of more than 600 000 people into the Ngara District of northwest Tanzania. Considerable environmental damage was caused by refugees harvesting firewood and building poles, poaching in the Burigi and Biharamulo Game Reserves, and the use of cheap refugee labour in charcoal and timber operations. Refugees also put some 15 000 hectares of land under cultivation in Ngara alone.

UNHCR and its local and international partners established a range of projects to improve the situation. In the emergency phase, the German aid agency GTZ set up simple cookstove projects and began marking important trees for protection around the camps. Local NGOs were co-opted to produce tree seedlings. Then CARE International (with IFAD funds) set up a fully-fledged environmental programme with refugees and local people that included large-scale firewood supply, tree seedling production, environmental education, agroforestry and soil stabilization measures, in addition to the existing programmes on stove dissemination and tree protection. More than 1.5 million new trees were planted, improved cooking stoves were adopted by 85 per cent of the refugees, wood harvesting was reduced by more than 60 per cent, and poaching was eventually stopped.

UNHCR worked closely with district authorities and government natural resource personnel to establish environmental task forces that were able to promote technical debate, help in conflict resolution and avoided duplication of activities.

The region benefited from the interest of many donors and development organizations, including UNDP, USAID, CARE, GTZ, IFAD, ACCORD and Help Age, an expansion of existing Dutch-supported District Rural Development programmes, and new support to local NGOs working in the environment sector.

Many lessons were learnt: some of the key issues are pre-emptive site planning, establishing inter-agency coordination from the start, and promoting better cooking techniques to reduce demand for fuelwood. It also became clear that nature is a great healer. The power of the Tanzanian landscape to recover from the blows dealt it by the refugees has been impressive. Only a year after the refugees went home, the Ngara area was reverting to natural woodland.

Source: based on material supplied by UNHCR

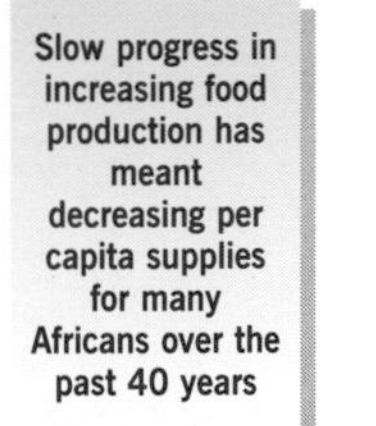
Slow progress in increasing food production has meant decreasing per capita supplies for many Africans over the past 40 years

Per capita food production

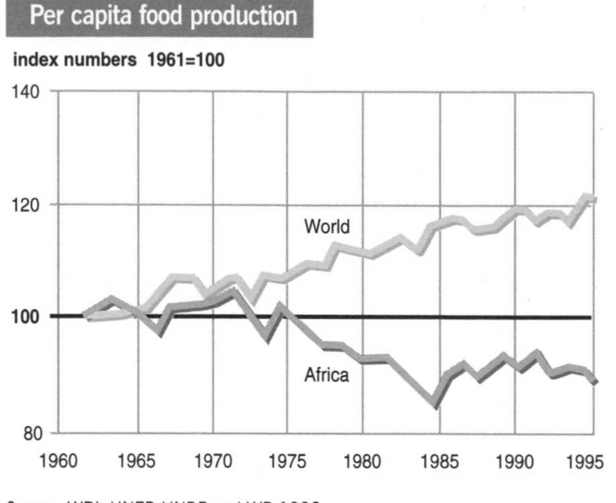

Source: WRI, UNEP, UNDP and WB 1998

cent of the soils in Southern Africa are also damaged by physical degradation such as the sealing and crusting of topsoil, leading to a reduction of available soil water, the compaction of topsoil and waterlogging (Byers 1997).

In Western and Central Africa, a combination of rising population growth, inappropriate agricultural practices such as shifting cultivation and suppression of fallow, variable climatic conditions, persistent drought and overgrazing are major causes of land degradation. In Northern Africa, land degradation is particularly acute in the desert fringes of Algeria, the Eastern Rift and High Atlas regions in Morocco, and the mountainous regions of Tunisia.

Only North Africa has been able to make major increases in the per capita calorie supply. Land degradation and drought were important causes of the decline in Southern Africa

Nearly two-thirds of African land is arid or semi-arid. The continent is the most seriously affected by desertification which threatens more than one-third of Africa's land area, particularly in Mediterranean Africa, the Sudano-Sahelian region and Southern Africa (Darkoh 1993). In Northern Africa alone, more than 432 million hectares (57 per cent of total land) are threatened by desertification (CAMRE/UNEP/ACSAD 1996). Although overgrazing has long been considered the primary cause of desertification in Africa, it is now thought that rainfall variability and long-term droughts are more important determinants (UNEP 1997).

Calorie intake per capita

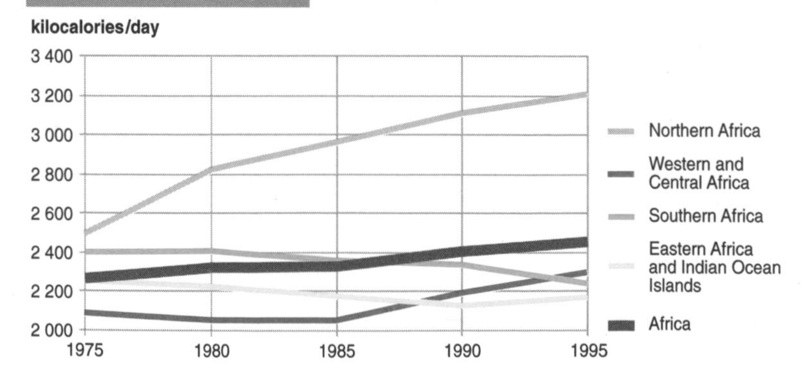

Source: compiled by UNEP GRID Geneva from FAOSTAT 1997 and WRI, UNEP, UNDP and WB 1998

Recurrent drought also has a severe impact on food security. In the 1994–95 cropping season in Southern Africa, cereal harvests declined by 35 per cent compared to the previous year. The maize harvest alone fell by 42 per cent due to drought (SADC 1995). Drought was equally devastating during the 1991–92 cropping season in the SADC region, where cereal production was nearly halved, with more than 20 million out of 85 million people affected by food shortages (Lone, Laishley and Bentsi-Enchill 1993). In spite of new measures to minimize the impact of drought, such as developing crop cultivars and animal breeds that are drought-tolerant, the recurrent droughts in Southern Africa are expected to continue to lower yields for another decade or more. However, the SADC countries aim to improve food security by promoting regional comparative advantages through trade in food commodities.

Although a net food exporter before 1960, Africa has become more dependent on food imports and food aid over the past three decades. During 1974–90, food imports in sub-Saharan Africa rose by 185 per cent and food aid by 295 per cent (UNDP 1997). In 1995, food imports accounted for 17 per cent of total food needs in the region. That rate is projected to at least double by 2010 (Nana-Sinkam 1995). In Western and Central Africa, food already constitutes more than 30 per cent of the value of imports (UNCTAD 1996).

Land degradation is a major factor in constraining food production in Africa to only a 2 per cent average annual increase. As this is much lower than the average population growth rate, per capita food production has been falling (see graph top left), and household and national food security is at risk in many countries. Other factors that lower food self-sufficiency and security in Africa include pests and diseases, inappropriate food production and storage practices, inadequate food processing technologies, civil wars and the low economic status of the women who produce the bulk of the food. Unless urgent and effective land conservation and watershed management measures are taken, food insecurity will continue to be a critical local, national and regional problem.

As a result of declining food security, the number of undernourished people in Africa nearly doubled from 100 million in the late 1960s to nearly 200 million in 1995. Projections indicate that the region will be able to feed only 40 per cent of its population by 2025 (Nana-Sinkam 1995). Yet the agricultural potential of the continent remains largely untapped. Although there are an estimated 632 million hectares of arable land in Africa, only 179 million hectares are actually cultivated (FAOSTAT 1997). As with other natural resources, the arable land is unevenly distributed. More than 246 million hectares of the as yet uncultivated arable land, representing nearly 40 per cent of the remaining total in the region, is found in only three countries (the Democratic Republic of the Congo, Nigeria and the Sudan).

The poverty of Africa's poor is both a cause and a consequence of accelerating soil degradation and declining agricultural productivity. Poverty reduction is thus the major challenge for those responsible for policy and decision making on the protection and sustainable use of land resources in Africa.

Forests

The forests of Africa cover 520 million hectares and constitute more than 17 per cent of the world's forests. They are largely concentrated in the tropical zones of Western and Central, Eastern and Southern Africa. With more than 109 million hectares of forests, the Democratic Republic of the Congo alone has more than 20 per cent of the region's forest cover, while Northern Africa has little more than 9 per cent (FAO 1997a), principally along the coast of the western Mediterranean countries. Forests include dry tropical forests in the Sahel, Eastern and Southern Africa, humid tropical forests in Western and Central Africa, diverse sub-tropical forest and woodland formations in Northern Africa and the southern tip of the continent, as well as mangroves in the coastal zones.

Forests play an important economic role in many countries. Forest products provide 6 per cent of GDP in the region, the highest in the world. But the share of forest products in trade is only 2 per cent, lower than the world average of 3 per cent (FAO 1998).

Africa's forests are threatened by a combination of factors including agricultural expansion, commercial harvesting, increased firewood collection, inappropriate land and tree tenure regimes, heavy livestock grazing, and accelerated urbanization and industrialization. Drought, civil wars and bush fires also contribute significantly to forest degradation (FAO 1997a and 1998). Inappropriate agricultural systems such as the *chitemene*, a system of shifting cultivation practised in parts of Southern and Central Africa, and *tavy* slash-and-burn agriculture in Madagascar, are responsible for considerable forest losses. Until recently, Southern Africa was losing more than 200 000 hectares of forests a year to shifting cultivation (Chidumayo 1986), although this is now starting to decline as farmers change to more settled agricultural practices.

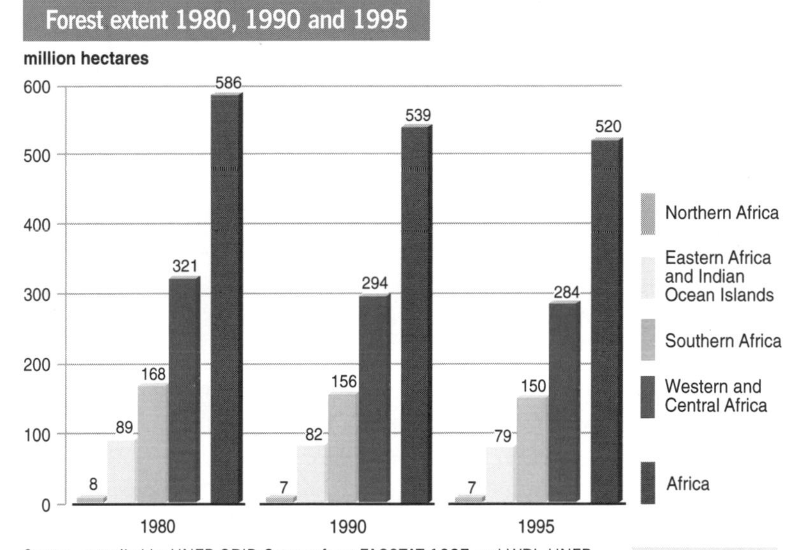

Source: compiled by UNEP GRID Geneva from FAOSTAT 1997 and WRI, UNEP, UNDP and WB 1998

During 1990–95 the annual rate of deforestation in Africa was about 0.7 per cent, a slight decline from 0.8 per cent during 1980–90

Throughout Africa, there has been an increasing demand for wood products, especially firewood, charcoal and roundwood. As a result the consumption of forest products nearly doubled during 1970–94. The production and consumption of firewood and charcoal rose from 250 to 502 million m^3 during the same period. Recent projections estimate that consumption will rise by another 5 per cent by 2010 (FAO 1997a). More recently, new economic reform measures have removed subsidies on energy alternatives which further increased the demand for firewood. At least 90 per cent of Africans depend on firewood and other biomass for their energy needs (FAO 1997a).

In Western and Central Africa, much of the tropical humid forests have already undergone substantial commercial harvesting. The total volume of wood exploited annually in the sub-region is more than 200 million m^3. Nearly 90 per cent is consumed as firewood and charcoal, and only 2 per cent as industrial roundwood (FAO 1997a). However, as it produces only a small proportion of the world's industrial roundwood, Africa is a net importer of industrial wood. Five countries in Northern Africa – Algeria, Egypt, Libya, Morocco and Tunisia – together account for 60 per cent of the imports. With the exception of a few countries such as Kenya, Nigeria, South Africa, Swaziland, Tanzania and Zimbabwe, all sub-Saharan African countries import all their paper (FAO 1997a).

Large-scale oil exploration and mining in Western and Central Africa have also led to the loss of forest resources, especially in Cameroon, the Congo, Gabon and Nigeria.

Wildlife also contributes to forest degradation and loss in Africa, particularly elephants in areas such as the Sengwa, Hwange, Mana Pools, Luangwa Valley and Chobe national parks in Southern Africa, where they destroy forests by knocking down trees and 'simplifying' the habitat and ecological processes.

During 1990–95 the annual rate of deforestation in Africa was about 0.7 per cent, a slight decline from 0.8 per cent during 1980–90 (FAOSTAT 1997). The highest rates were recorded in the moist western parts of the continent. During the 1980s, Africa lost an estimated 47 million hectares of forest. By 1995 another 19 million hectares had been lost (FAO 1997a), an area the size of Senegal. Losses have been particularly high in countries such as Uganda, where forest and woodland cover shrunk from an estimated 45 per cent of total land area in 1900 to only 7.7 per cent by 1995 (Ministry of Natural Resources, Uganda, 1995).

Elephants contribute to forest degradation by knocking down trees and habitat 'simplification'

Tree plantations and agroforestry are increasingly important aspects of forest rehabilitation, especially in non-tropical Northern and Southern Africa. Although providing significant amounts of timber, firewood and other useful products, afforestation rates throughout Africa are far less than the rate of deforestation (FAO 1997a).

The pressures on African forests will inevitably continue rising to meet the needs of fast-growing populations in rapidly urbanizing and industrializing countries, especially if most of their people remain poor.

Biodiversity

Africa has a large and diverse heritage of flora and fauna, including major domesticated agricultural crops such as sorghum and millet. The continent is home to more than 50 000 known plant species, 1 000 mammal species, and 1 500 bird species. Traditionally, African societies depended on many of these indigenous species for survival and developed strategies to protect and conserve them for the benefit of their own and future generations. In some cultures, areas that were particularly rich in biodiversity were often designated as sacred groves and protected areas.

The first national parks in Africa were created in the first half of the 20th century, including the Kruger National Park in South Africa in 1928 and the Toubkal Nature Reserve in Morocco in 1944. In 1938 the Arab countries convened a symposium on nature conservation which resulted in the designation of many of the existing protected areas in their countries (UNESCO 1954).

Eastern Africa has the highest numbers of endemic species of mammals (55 per cent), birds (63 per cent), reptiles (49 per cent) and amphibians (40 per cent), whereas species endemism is relatively low in Northern Africa. Madagascar is the most endemic-rich country in Africa, and the sixth in the world for higher vertebrates (mammals, birds and amphibians), with more than 300 endemic species, and the third-most plant-rich country in Africa after the Democratic Republic of the Congo and Tanzania (WCMC 1992). One of the six most significant concentrations of plants in the world is the Cape Floral Kingdom (WWF 1996).

Savannahs, the richest grasslands in the world, are the most extensive ecosystem in Africa. They support many indigenous plants and animals as well as the world's largest concentration of large mammals such as elephants, buffalo, rhinoceros, giraffes, lions, leopards, cheetah, zebras, hippopotami, kudus, waterbucks and oryx.

This large and diverse biological heritage is at risk in all regions of Africa (see illustration). Some species have already been reported as extinct, including four antelope species in Lesotho and Swaziland, the blue wildebeest in Malawi, the tssessebe in Mozambique, the endemic bluebuck from the south-western Cape in South Africa and the kob in Tanzania (Stuart, Adams and Jenkins 1990). Many other species are now under threat of extinction. In Mauritania, an estimated 23 per cent of the mammals are now at risk (WCMC 1992). In Western and Central Africa, the endangered species include timber plants such as *Guarea excelsa, Milicia excelsa, Nauclea diderric,* such medicinal species as *Voacanga africana, Zanthoxyhmm zanthoxyloides* and *Brucea guineensis,* and mammal species such as the chimpanzee, the Senegal hartebeest (*Alcelaphus bucelaphus*), elephants (*Loxodanta africana*) and one of the three manatee species (*Trichechus senegalensis*). In Eritrea, 22 plant species are reportedly threatened with extinction (Eritrea Agency for the Environment 1995).

The number of threatened species may be higher than the illustration on this page shows because species diversity in Africa is not yet fully documented.

African wetlands also have a rich biological diversity, with many endemic and rare plant species as well as wildlife such as migratory birds. Wetlands are found in most African countries, the largest including the Okavango Delta, the Sudd in the Upper Nile, the Lake Victoria and Chad basins, and the floodplains and deltas of the Congo, Niger and Zambezi rivers. Despite being among the most biologically-productive ecosystems in Africa, wetlands are often regarded locally either as wasteland, habitats for pests and threats to public health or as potential areas for agriculture. As a result many wetlands are being lost. During the past two decades, for example, Niger lost more than 80 per cent of its freshwater wetlands (Niger Ministry of Environment and Hydraulics 1997). Coastal wetlands in Egypt and Tunisia and freshwater wetlands in the Sudan are also under increasing threat. Freshwater ecosystems found in lakes, rivers and wetlands may be the most endangered ecosystems of all. They have already lost a greater proportion of their species and habitats than terrestrial or marine ecosystems, and are in danger of further losses from dams, pollution, overfishing and other threats (WRI, UNEP, UNDP and WB 1998).

One of the three manatee species (*Trichechus senegalensis*) is an endangered species off the west coast of Africa

Environmental pollution is an increasingly major threat to biodiversity in many countries. Pesticide residues have reduced the populations of several bird species and other organisms. Both air and water pollution stress ecosystems and reduce populations of sensitive species, especially in the coastal zones where there is a high population density and industrial activity. As the region continues to industrialize, the adverse impact of pollution on biodiversity will become even more severe unless cleaner production

Africa's large and diverse biological heritage is at risk in all the sub-regions

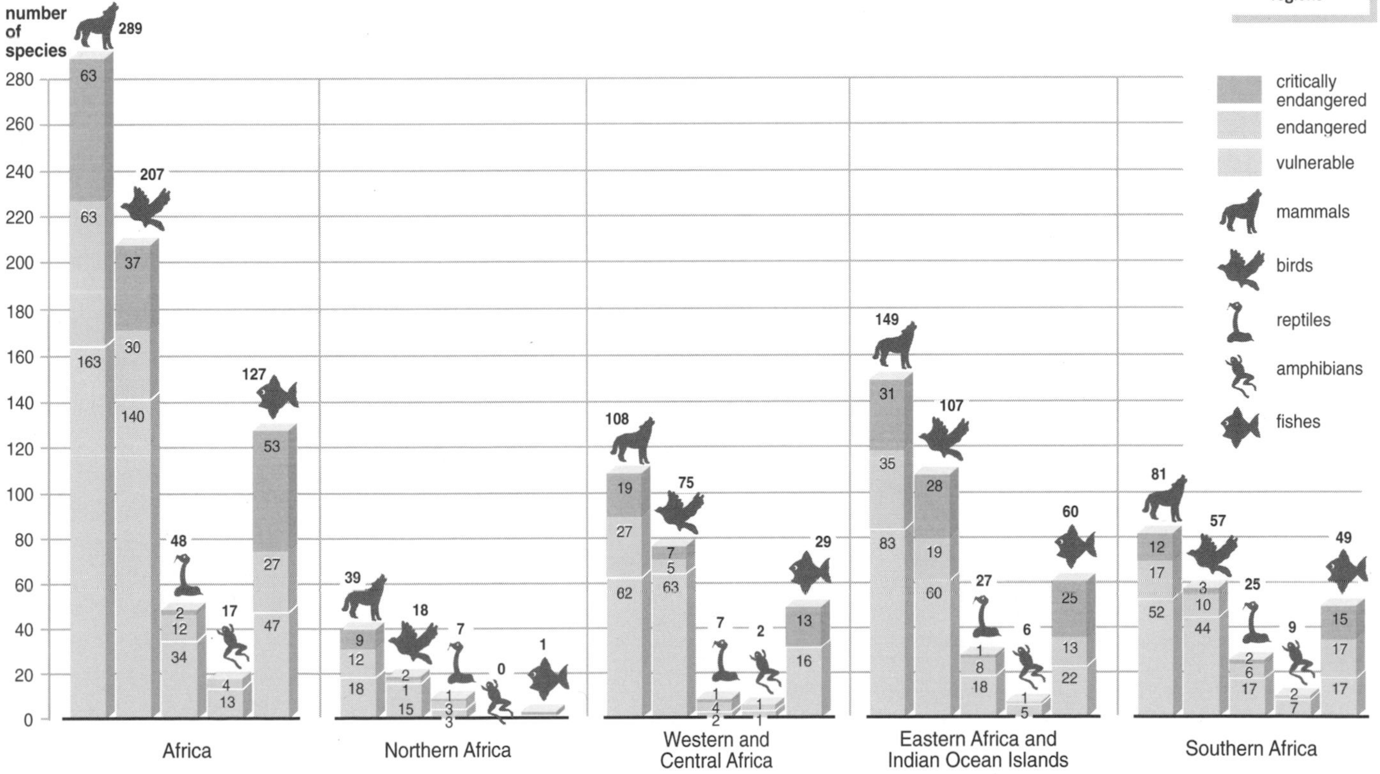

Source: WCMC/IUCN 1998

technologies are adopted.

The introduction of exotic species over the past century also contributed to biodiversity loss as some alien species 'out-competed' native vegetation. For example, parts of the *fynbos* in South Africa and eastern highland grasslands in Zimbabwe were invaded by exotic Australian *Acacia* and *Pinus* species, and threatened the survival of the indigenous *Restio, Erica* and *Protea* species (Geldenhuys 1996). Island species, such as those found in the Indian Ocean, are particularly vulnerable to extinction caused by competition or predation (WCMC 1992). However, the introduction of undesirable exotic species is declining, a positive trend that will probably continue as regulations on the importation of biological resources become increasingly stringent.

There are some 3 000 protected areas in Africa but neither their size nor their number are likely to increase because of intense competition for land

Civil conflict and war have also led to significant ecological damage and biodiversity losses in and outside protected areas, as well as to the marginalization of environmental management institutions and conservation programmes. By 1991, the wildlife populations of national parks and reserves in Angola had been reduced by civil war to only 10 per cent of their 1975 levels (Huntley and Matos 1992). Similar losses are likely to have occurred in the Great Lakes region during the past five years but have yet to be quantified.

Climate change is the latest emerging threat to biodiversity in Africa. It has already been identified as a contributing cause in the decline of amphibian populations, due to drastic reductions in the volume of water bodies after persistent dry weather in combination with intensified human activities along the shorelines.

The 3 000 or so protected areas in Africa total nearly 240 million hectares (see illustration). Neither the size nor number of protected areas is likely to increase in the future because of increasingly intense competition for land to meet the needs of expanding populations, cities, agriculture and industry.

Size and number of protected areas

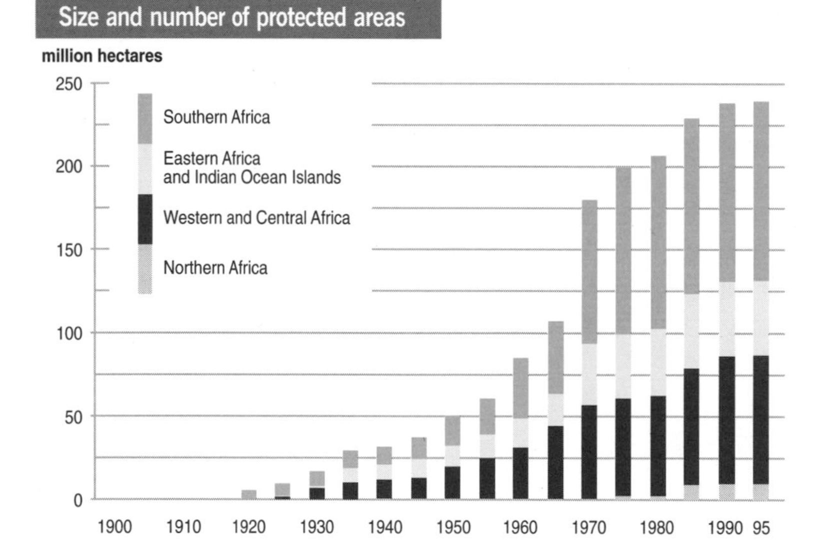

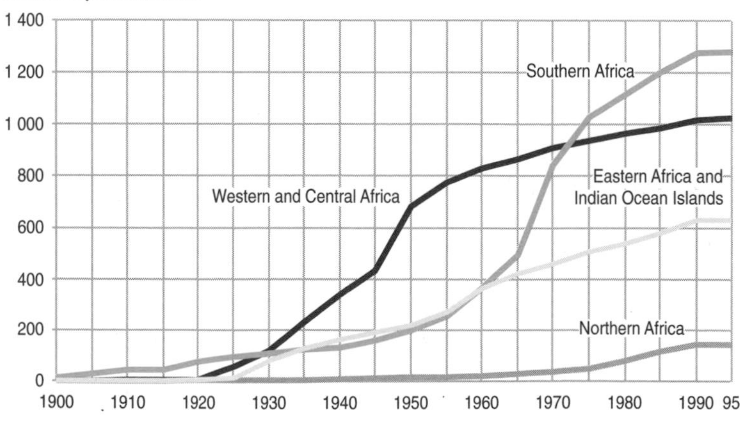

Note: includes all nationally-designated protected areas as well as IUCN categories I–VI.
Source: WCMC 1998

Freshwater

While Africa uses only about 4 per cent of its renewable freshwater resources (WRI, UNEP, UNDP AND WB, 1998), water is becoming one of the most critical natural resource issues. The continent is one of the two regions in the world facing serious water shortages (Johns Hopkins 1998). Africa has abundant freshwater resources in large rivers and lakes such as the Congo, Nile and Zambezi river basins, and in Lake Victoria, the second largest freshwater lake in the world. However, there are great disparities in water availability and use within and between African countries because the water resources are so unevenly distributed. For example, the Congo River watershed contains 10 per cent of Africa's population but accounts for about 30 per cent of the continent's annual run-off (Johns Hopkins 1998). Other contributing factors are the inadequate assessment and underdevelopment of water resources, the lack of technical and institutional infrastructure as well as the lack of investment in water resource development.

Most freshwater comes from seasonal rains, which vary with climatic zone. The greatest rainfall occurs along the equator, especially the area from the Niger delta to the Congo River basin. The Sahara desert has virtually no rain. Northern and Southern Africa receive 9 and 12 per cent respectively of the region's rainfall (FAO 1995). In Western and Central Africa, rainfall is exceptionally variable and unpredictable. While the Sahelian countries have limited supplies of freshwater, most countries in the humid tropical zone have abundant water. The availability of water varies considerably even within countries and the situation is further complicated by frequent droughts as well as inappropriate water management programmes.

Groundwater resources are crucial for many countries and people in Africa, particularly during the dry season and in large arid zones. Groundwater is a main source of water in many rural areas, including for nearly 80 per cent of the human and animal populations in Botswana (Government of Botswana 1993) and at least 40 per cent in Namibia (Heyns 1993). In Libya, groundwater accounts for 95 per cent of the country's freshwater withdrawals (FAO 1997b), while in areas such as the Pangani Basin of Tanzania groundwater is a significant source for irrigated agriculture (World Bank and DANIDA 1995). In many parts of the continent, groundwater resources have not yet been fully explored and tapped.

The demand for water is increasing rapidly in most countries due to population growth and economic development. Although some African countries have high annual averages of available water per capita, many others already or soon will face water stress (1 700 m^3 or less per person annually) or scarcity conditions (1 000 m^3 or less per person annually). Currently, 14 countries in Africa are subject to water stress or water scarcity, with those in Northern Africa facing the worst prospects (Johns Hopkins 1998). A further 11 countries (see map) will join them in the next 25 years (Johns Hopkins 1998).

In the SADC region, water demand is projected to rise by at least 3 per cent annually until the year 2020, a rate about equal to the region's population growth rate (SARDC, IUCN and SADC 1994). It has been estimated that by 2025 up to 16 per cent of Africa's population (230 million people) will be living in countries facing water scarcity, and 32 per cent (another 460 million) in water-stressed countries (Johns Hopkins 1998). Africa's share of water on a per capita basis is estimated to have declined by as much as 50 per cent since 1950 (Bryant 1994).

Rising demand for increasingly scarce water resources is leading to growing concerns about future access to water, particularly where water resources are

Water stress and water scarcity in the year 2025

water scarcity in 2025
less than 1 000 m^3/person/year

water stress in 2025
1 000 to 1 700 m^3/person/year

Source: Johns Hopkins 1998

By the year 2025, 25 African countries will be subject to water scarcity or water stress

shared by two or more countries. About 50 rivers in Africa are shared by two or more countries. Access to water from any of these shared rivers could provoke conflict, particularly in the Nile, Niger, Volta and Zambezi basins (Johns Hopkins 1998).

As in other dry regions, agriculture is the largest user of water in Africa, accounting for 88 per cent of total water use (WRI, UNEP, UNDP and WB 1998). However, with only 6 per cent of cropland under irrigation, there is considerable potential to increase food production through irrigation, and demand for water for irrigation will continue to grow. Some 40–60 per cent of the region's irrigation water is currently lost through seepage and evaporation. This contributes to serious environmental problems such as soil salinization and waterlogging, although water 'lost' in this way may end up in aquifers whence it can be pumped to irrigate nearby fields.

Freshwater fisheries are the main source of income and protein for millions of Africans. The annual freshwater fish catch is estimated at about 1.4 million tonnes, with Egypt alone contributing about 14 per cent (FAO 1997c). However, the damming of the Nile, and the disposal of untreated sewage and industrial effluents, has endangered species and reduced the fish catch in many regions, including the Nile Delta and Lake Chad (Johns Hopkins 1998).

The main threats to water quality in Africa include eutrophication, pollution and the proliferation of invasive aquatic plants such as the water hyacinth

(*Eichhornia crassipes*) and *Salvinia molesta* weeds. The water hyacinth has seriously affected most water bodies in the region, including Lake Victoria, the Nile River and Lake Chivero. As no effective means of controlling this weed has yet been found, the water hyacinth will continue to disrupt water transport, water supplies to urban areas, the fishing industry, power generation and the livelihoods of many local communities.

Industrial wastes are still discharged without treatment into rivers and lakes in most African countries, causing a major and persistent health problem. Saltwater intrusion into surface and groundwater sources is also a major problem, especially along the Mediterranean coast and on oceanic islands such as the Comoros which are highly dependent on groundwater resources and at risk from sea-level rise. A related problem is the high level of dental and skeletal fluorosis that occurs in several areas, particularly on Africa's east coast.

With recurring droughts and chronic water shortages in many areas, most countries and people already pay an increasingly high price for water and for the lack of water. The poor, especially women and children, usually pay the highest price in cash terms to buy small amounts of water. They also expend more in calories carrying water from distant sources, suffer more in impaired health from contaminated or insufficient water, and also lose more in diminished livelihoods and even lost lives.

More than 300 million people in Africa still lack reasonable access to safe water. Even more lack adequate sanitation (UNDP 1996). In sub-Saharan Africa, only about 51 per cent of the population have access to safe water, and 45 per cent to sanitation (UNDP 1997). There are, however, great variations throughout the continent. In Libya and Mauritius almost all the population has access to safe water and sanitation, compared to only about one-quarter in Chad, Ethiopia and Madagascar (UNDP 1997). Urban residents generally have better access to safe water and sanitation than those living in rural areas. For example, in 1994, only 30 per cent of the rural population in Uganda had access to safe water compared to 60 per cent in urban centres (Ministry of Natural Resources, Uganda, 1995).

Marine and coastal areas

Africa's coastal ecosystems and marine biodiversity contribute significantly to the economies of many coastal countries, mainly through fishing and tourism. For example, in Namibia the fisheries sector contributes more than 35 per cent of GDP and employs more than 12 000 people (Namibia Foundation 1994). The marine fishing industry makes an important contribution to the balance of trade in countries such as Morocco, which had the highest average annual marine catches on the continent at 844 000 tonnes in 1995 (FAO 1997c). In Southern Africa, the annual marine fisheries catch was estimated as 1.25 million tonnes in 1995, with a potential sustainable catch of 2.7–3.0 million tonnes (FAO 1997c and SADC 1996). However, the sub-region has experienced major changes in the composition and total landings of fish. Once regarded as one of the richest fishing grounds in the world, the catches on the west coast have declined sharply from the 3 million tonnes harvested during the 1950s and 1960s (FAO 1993).

Coastal zones are also important for the tourists they attract and the revenue they generate, particularly for countries such as Egypt, Gambia, Kenya, Mauritius, Morocco, Seychelles, Tanzania and Tunisia. Tourism is heavily dependent on the quality of the coastal environment, and coastal zone degradation therefore has serious implications for the industry. This is particularly true in small island countries, such as Mauritius and Seychelles, that are economically dependent on tourism. At the same time, the unmanaged growth of the tourism industry can have a detrimental effect on the coastal environment and resources (World Bank 1995a).

Coastal and marine resources have not been adequately assessed, and are under increasing threat from development-related activities. Habitat conversion and degradation, overexploitation, pollution and sedimentation, coastal erosion, eutrophication, species introductions and climate change are considered the major causes of marine biodiversity loss (World Bank 1995b).

Up to 38 per cent of the African coastline of 40 000 km is considered to be under a high degree of threat from developments which include cities, ports, road networks and pipelines, including 68 per cent of marine protected areas (WRI, UNEP, UNDP and WB 1996, and World Bank 1995a).

Urbanization of the coastal zone, particularly where it is poorly controlled, is creating concern. It is projected that Western and Central African coastal populations will double to 50 million in the next 25 years, leading to a continuous chain of cities along the

1 000-km Gulf of Guinea (World Bank 1995a) which will exceed the carrying capacity of the coastal corridor. The expanding land and sea-based activities along the Mediterranean coast of North Africa and in the Red Sea over the past 20 years also pose increasing threats to coastal ecosystems. Some 40–50 per cent of the Mediterranean population is already concentrated along the coast, and this population is expected to double by the year 2025 (UNEP 1996).

Marine pollution from major coastal cities is common and has even reached toxic levels in some cases. In 1990 coastal cities and towns in Southern Africa discharged more than 850 million litres of industrial and human wastes into the sea daily through more than 80 pipelines, largely without any treatment (Cock and Koch 1991). In 1992, the lack of adequate infrastructure in Maputo caused significant coastal sewage and pollution problems, while in Angola untreated industrial waste pumped into the Bay of Luanda resulted in bacterial contamination (IUCN 1992). There are no immediate prospects of reducing the coastal pollution problems faced by many African countries.

Africa's coastal ecosystems are also threatened by industrial pollution, mining and oil exploration activities. Although the level of industrial development in Africa is still relatively low, the rate is accelerating along the coastal zone (World Bank 1995a). Most industries still discharge their untreated wastes directly into rivers and, ultimately, the oceans. The Mediterranean basin is now one of the most polluted, semi-enclosed basins in the world. But pollution also affects unenclosed seas. In 1993 industrial waste was found in the coastal waters near major centres along the entire coastline, stretching from Dar Es Salaam and Maputo on the east coast, to Durban and Cape Town in South Africa, and to Walvis Bay in Namibia and Boa do Cacuaco, 15 km north of Luanda in Angola (SARDC, IUCN AND SADC 1994). In the Indian Ocean there are increasing risks of pollution from oil spills because this is the main transportation artery for oil from the Middle East to Europe and America, with an estimated 470 million tonnes transported annually (Salm 1998). Similar risks apply in Northern Africa as more than 100 million tonnes of oil are transported through the Red Sea annually with insufficient maritime traffic regulations (World Bank 1996a). Petrochemical complexes add to the problem. For example, three major complexes at Annaba, Arzew and Skikda in Algeria discharge large quantities of chromium, mercury, oils, phenols, acids, chlorine and urea into the sea (World Bank 1995c). Similar situations exist in Egypt, Libya, Morocco and Tunisia.

In many areas, coastal erosion is a growing problem, driven by natural processes which are exacerbated by the upstream construction of dams and the development of other forms of coastal infrastructure such as artificial lagoons and the clearing of mangrove systems. In the longer term, climate change is also a

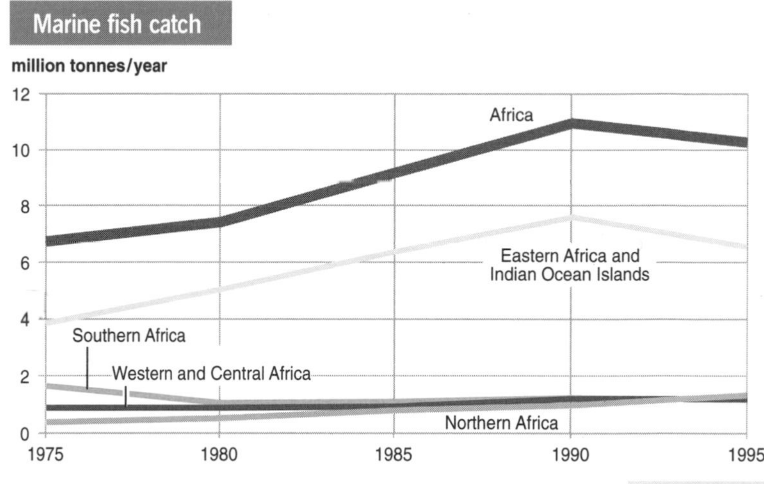

Source: compiled by UNEP GRID Geneva from FAO 1997c

Total marine fish catch grew by more than 50 per cent during 1975–90 but there was a downturn in Eastern Africa after 1990. The catch in Southern Africa is far below the 3 million tonnes of the 1950s

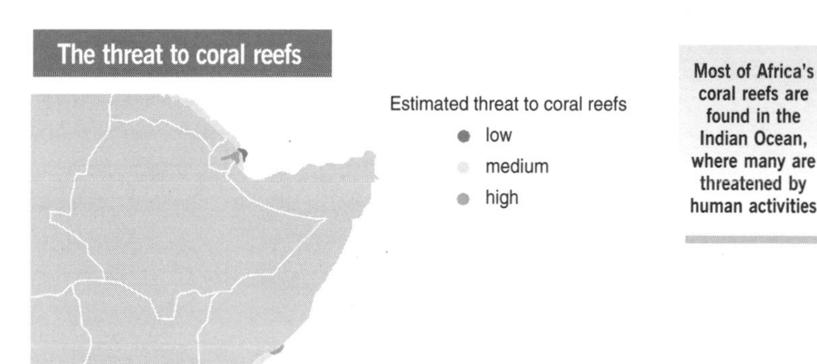

Source: WRI, ICLARM, WCMC and UNEP 1998

Most of Africa's coral reefs are found in the Indian Ocean, where many are threatened by human activities

major threat to critical coastal ecosystems such as the Nile, the Niger and other low-lying deltas and oceanic islands, particularly in the Indian Ocean, which may be inundated by rising sea levels.

Coral reefs are increasingly under threat from human activities (see map on page 63), particularly from coastal development and overexploitation as well as blast fishing and land-based pollution. The Indian Ocean contains about 15 per cent of the world's mapped coral reefs, of which more than one-half is estimated to be at risk from human activities. Coral reefs in the northern Red Sea (in the Gulf of Aqaba and near the Gulf of Suez) and along the coast of Djibouti are also considered to be under a high degree of threat (WRI, ICLARM, WCMC and UNEP 1998). Unprecedented coral bleaching following *El Niño* was reported in the Indian Ocean during the first half of 1998, due to extremely high ocean temperatures (NOAA 1998). Such stress weakens corals and can ultimately lead to their death (WRI, ICLARM, WCMC and UNEP 1998).

In Western and Central Africa, and in Eastern Africa, traditional biomass fuels provide most of the energy used – elsewhere commercial fuels have taken over

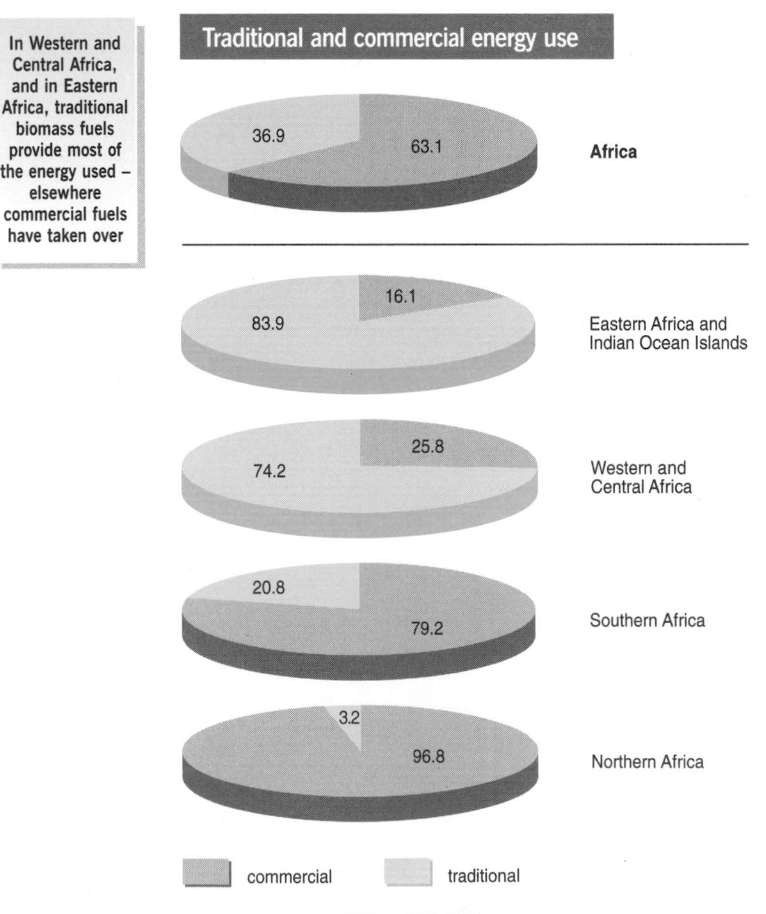

Source: WRI, UNEP, UNDP and WB 1998

Atmosphere

Atmospheric pollution has emerged as a problem in most African countries only in the past few decades. Its severity and impacts are still largely unknown, although it is believed that vegetation, soils and water in some areas have been adversely affected by gaseous pollutants and acid rain.

The main sources of atmospheric pollution are bush fires, vehicle emissions, manufacturing, mining and industry. Major industrial sources include thermal power stations, copper smelters, ferro-alloy works, steel works, foundries, fertilizer plants, and pulp and paper mills. If the projected growth of demand for vehicular transport and electricity is to be met with current technologies, emissions from thermal power stations are projected to increase elevenfold and from vehicles fivefold by 2003 (World Bank 1992). The use of leaded fuel in vehicles is also a major concern. Lead pollution is worsened by the region's ageing vehicles, most of which are more than 15 years old. They are also said to emit five times more hydrocarbons and carbon monoxide, and four times more nitrogen oxides, than new ones (World Bank 1995c).

Air pollution is most significant in the more urbanized and industrialized countries of Northern and Southern Africa. In Southern Africa, air pollution is largely from thermal power stations. About 89 per cent of electricity generation in the SADC region is from coal, mostly produced in South Africa where it accounted for 97 per cent of total electricity generation in 1994 (Sivertsen and others 1995). As South African coal contains about 1 per cent sulphur, the country emits more sulphur dioxide than any other in the SADC region and is ranked as the 15th largest emitter of greenhouse gases in the world (USAID 1997). During 1990–91, South Africa contributed 66 per cent of all sulphur emissions in the SADC region, whereas Lesotho, Swaziland and Mozambique jointly contributed only 0.9 per cent (Sivertsen and others 1995). As this dependence on coal-based thermal power will persist for years, sulphur dioxide pollution will remain a problem unless measures are taken to reduce the levels of sulphur in coal or provide incentives for developing alternative energy sources such as hydropower, wind, geothermal and solar.

Mining is a major source of income and also of air pollution in Southern Africa. The mining industry employs more than 800 000 people, generates 60 per cent of foreign exchange earnings and contributes about 11 per cent of GDP in the SADC region (SADC 1992). Sulphur emissions from mining are estimated at one million tonnes per year and are a growing concern, particularly among people with respiratory problems (SADC 1992).

Indoor air pollution caused by the widespread use of biomass as a cooking fuel is also a major contributor to the high incidence of respiratory diseases because of the exposure to smoke and other pollutants in a confined space. In sub-Saharan Africa, biomass use is expected to provide nearly 80 per cent of the total energy used even in 2010. In Northern Africa, the corresponding figure is much lower (see pie charts left); even in 1995, traditional fuel use was only some 3 per cent of the total (WRI, UNEP, UNDP and WB 1998).

In West Africa, the Harmattan winds often result in high atmospheric dust loading and poor visibility, and contribute to respiratory and other diseases. The continual build-up of mineral dust concentrations since the 1960s is likely to have a climatic impact through a land-atmosphere feedback mechanism (Ben Mohamed and Frangi 1986, Ben Mohamed 1985 and 1998).

Despite these problems, most African states have few or no specific air quality standards. City dwellers, in particular, are exposed to respiratory diseases such as asthma, bronchitis and emphysema as a result of industrial emissions and vehicle exhaust fumes (UNECA 1996). Heat islands in urban areas have also been shown to affect weather and local climate (Hewehy 1993).

Africa's emissions of the greenhouse gases that cause climate change are still low, estimated to be only 7 per cent of global emissions (World Bank 1998). Africa presently emits only 3.5 per cent of the world's total carbon dioxide. South Africa alone contributes 44 per cent of the region's emissions. Total carbon dioxide emissions in the region are expected to increase to 3.8 per cent of the world total by the year 2010 due to increased industrialization and urbanization (Energy Information Administration 1997). As they serve as a sink for carbon dioxide and mitigate greenhouse gas emissions, Africa's vast forest reserves play a key role in alleviating and balancing the emissions of the industrialized world. However, this crucial function is threatened by accelerating deforestation.

Climate change, resulting in sea-level rise and flooding or erosion of low-lying coastal areas and lagoons, will have serious adverse impacts on ecosystems, water resources, coastal zones and

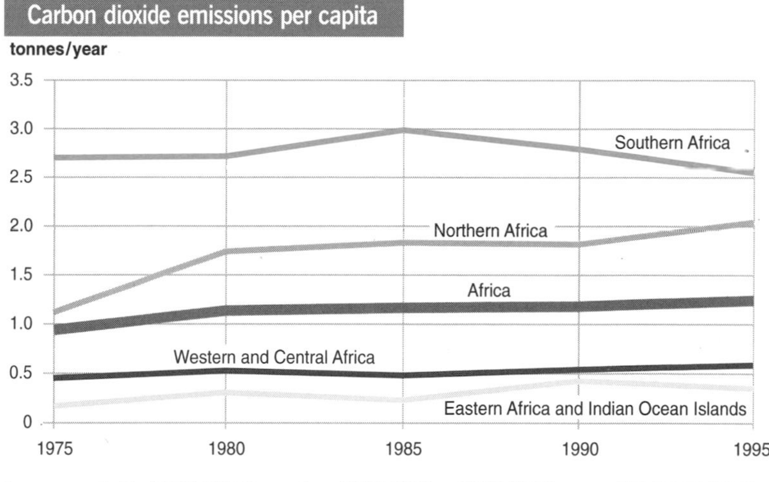

Source: compiled by UNEP GRID Geneva from CDIAC 1998 and WRI, UNEP, UNDP and WB 1998

Africa emits only 3.5 per cent of the world's total carbon dioxide –total emissions are expected to increase to 3.8 per cent of the world total by the year 2010

human settlements, particularly in the countries of Western and Central Africa, the Nile Delta and the Indian Ocean island states. Poverty makes many African peoples and countries particularly vulnerable to the impacts of climate change, especially in areas dependent on rain-fed agriculture. This vulnerability is increased by recurrent natural disasters such as drought, floods and cyclones. Increases in water stress and drought may also increase the incidence of vector-borne diseases and hunger. In 1998, the *El Niño* is thought to have been the cause of serious floods in Southern and Eastern Africa and exacerbated outbreaks of cholera, malaria and Rift Valley fever in Kenya and Somalia (CARE 1998).

Urban areas

At the start of the 20th century, 95 per cent of Africans lived in rural areas. Even in the 1960s, Africa remained the least urbanized continent, with an urban population of 18.8 per cent. By 1996 this had doubled, and at least 43 per cent of the population is expected to live in urban areas by 2010 (United Nations Population Division 1997). Average annual urban growth rates in Africa during 1970–2000 were the highest in the world at more than 4 per cent, and these are projected to decrease slightly to about 3 per cent during 2020–25 (United Nations Population Division 1997).

In Northern Africa, more than half the population now lives in cities while in Southern and in Western and Central Africa urbanization levels are still about 33–37 per cent. Eastern Africa is the least urbanized sub-region, with 23 per cent (United Nations Population Division 1997). The differences in urbanization rates are even greater among countries. In a few, such as Algeria, the Congo, Djibouti, Libya, Mauritania, South Africa and Tunisia, more than 50 per cent of the population now lives in urban areas, while in Rwanda and Burundi the urbanization levels are only 6 to 8 per cent (United Nations Population Division 1997).

Major cities in Africa are experiencing rapid growth. Nairobi, Dar es Salaam, Lagos and Kinshasa grew sevenfold during 1950–80, mainly because of rural-urban migration (Johns Hopkins 1998). During 1950–95 the population of Cairo quadrupled from 2.4 million to 9.7 million. Lagos in Nigeria is now even bigger with 10.3 million inhabitants (United Nations Population Division 1997). In 1997, the largest cities in 24 African countries had populations of more than one million each (UNDP 1997), nearly half of them in Western and Central Africa. Rapid urbanization is expected to continue for decades.

Urban residents make heavy demands on the environment as they generally consume more resources than rural dwellers, and generate large quantities of solid waste and sewage. In Northern Africa, at least 20 per cent and as much as 80 per cent of urban solid wastes are disposed of by dumping in open spaces. Rapid urbanization in Lagos increased solid waste generation sixfold to about 3.7 million tonnes a year in 1990, plus another half a million tonnes of largely untreated industrial waste because

Urbanization continues growing steadily in Africa. It is much higher in Northern Africa than elsewhere, and is still quite low in Eastern Africa and the Indian Ocean Islands

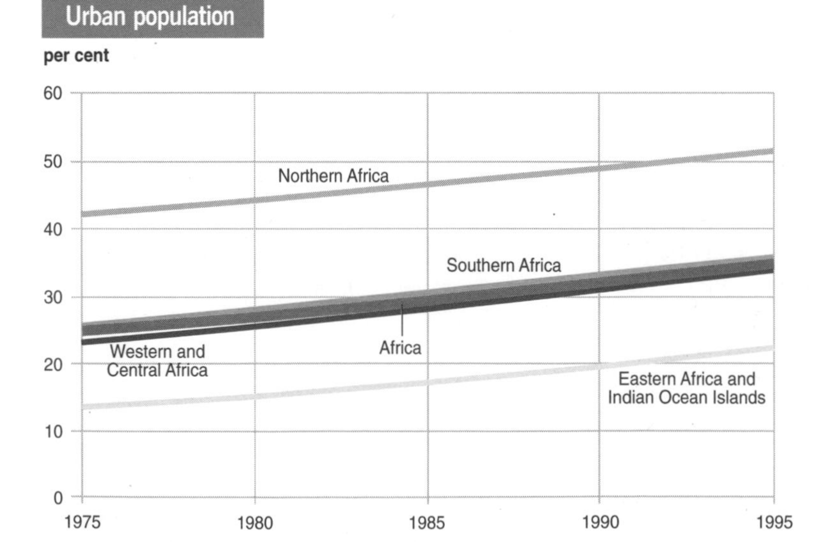

Source: compiled by UNEP GRID Geneva from United Nations Population Divison 1997 and WRI, UNEP, UNDP and WB 1998

Solid waste production, wastewater treated and garbage collection

	solid waste per capita (kg/year)	*wastewater treated (%)*	*households with garbage collection (%)*
Abidjan	365	58	70
Ibadan	401	-	40
Kinshasa	438	3	0
Bujumbura	511	4	41
Lomé	693	-	37
By comparison:			
Toronto	511	100	100

Source: Habitat 1997

90 per cent of the industries in Nigeria lack pollution control facilities (IMO 1995). The 1.3 million inhabitants of Lusaka produce 1 400 tonnes of solid waste daily, of which 90 per cent is not collected because the local authority has too few staff, funds and equipment. As only 36 per cent of Lusaka's residents have sewerage services, most of the rest use pit latrines, a common situation throughout Africa (Agyemang and others 1997).

The lack of adequate solid waste disposal and sewerage services causes serious public health problems in many cities, causing many diseases including often-fatal water-borne diseases such as cholera and dysentery. In 1994, 61 960 cases of cholera resulting in 4 389 deaths were reported in Angola, the Democratic Republic of the Congo, Malawi, Mozambique and Tanzania (WHO 1995). Another 171 000 cases of dysentery with at least 600 deaths were reported in Malawi, Mozambique and Zimbabwe (Holloway 1995). Poor drainage in some urban areas contributes to the spread of malaria which kills more that 1.5 million people in Africa annually (Tavengwa 1995).

The concentration of industries in or near cities is also a major source of environmental pollution and resource depletion. In 1994, the spill of toxic chemicals from a pulp and paper company into the Usuthu river in Swaziland killed many fish (Mavimbela 1995). In Mozambique, more than 126 factories in and around Maputo discharge their wastes directly into the environment (Couto 1995). In Tanzania, textile mills are reported to release dyes, bleaching agents, alkalis and starch directly into Msimbazi Creek in Dar es Salaam (Bwathondi, Nkotagu and Mkuula 1991).

An estimated 81 per cent of urban residents have access to safe water and 66 per cent to sanitation facilities. The situation is worse in the rural areas where only 47 per cent of the people have access to safe water (World Bank 1996b). However, the urban statistics combine the richest and poorest residents in a single average which disguises the daily reality of the poor majority in large slums who lack reasonable access to safe water. For their small share of water, the urban poor pay an unfair price, usually at least four and sometimes as much as ten times more per litre than the metered rates of those living in the elite residential areas (Serageldin 1995).

Conclusions

Over the past 100 years the state of the terrestrial, freshwater and marine environments has declined in virtually all respects. Environmental degradation and resource depletion have escalated particularly over the past three decades due to the cumulative impacts of rapid growth in population, intensive agriculture, urbanization and industrialization. The priority list of environmental challenges includes land degradation, deforestation, declining biodiversity and marine resources, water scarcity, and deteriorating water and air quality.

A major reason for these adverse environmental trends in Africa is that most people and countries are poor. Their poverty is both a cause and a consequence of environmental degradation. However, the main cause of many environmental problems is the persistence of economic, agricultural, energy, industrial and other sectoral policies which largely neglect – and fail to avoid – harmful impacts on the environment and natural resource base.

Another reason for poverty results from the political instability of many countries over the past few decades. Where such instability has resulted in civil war, the human and environmental effects have been even more devastating.

Poverty also exists in spite of the wealth of Africa's natural resources. Many Africans are unable to benefit from this wealth, partly because it is very unevenly distributed across the continent and partly for complex reasons connected with Africa's socio-economic history over the past 100 years.

A number of key issues dominate Africa's environmental problems. These include:

- Increased food insecurity resulting from rapid population growth, degradation of agriculture and arable lands, and mismanagement of available water resources combined with poor economic policies to support food production. Land degradation is also a serious environmental problem. However, Africa owns vast areas of unexploited arable land which could be exploited in the future through the integrated management of land, water and human resources.
- African forests are shrinking as a result of deforestation. Unless energy alternatives to firewood, and other sources of income for people whose lives depend on forests, can be found, deforestation will continue.
- The richness of African biodiversity requires greater protection and a sustainable use that will ensure the income of those who depend on it. There is a need to maximize biodiversity landscape protection, to give priority to biodiversity areas close to areas of high population density, and to give balanced attention to such regions as the arid and semi-arid areas.
- Africa's freshwater problems are acute and worsening. Most arise from the poor management of water resources, lack of financial resources required for sustainable development and efficient utilization of resources, absence of effective regional and basin development plans and shared management, and under-estimation of the groundwater potential to supplement irrigation and drinking water supplies.
- Coastal and marine resources are subjected to increasing pressure and are being degraded as a result of increased urbanization and overexploitation coupled with mismanagement. There is an urgent need for integrated coastal zone management.
- Air pollution has now emerged as an environmental issue of concern in most major cities of Africa. Regulatory measures and environmental standards need to be introduced to combat the problem. African emissions of greenhouse gases remain modest but should not be left unregulated.
- The African urbanization rate is the highest in the world. This is resulting in a deteriorating urban environment. Most of these problems, however, are common, predictable and are inevitable consequences of rapid urbanization. They can be resolved through efficient and effective urban

management systems. The challenge is to adopt city planning, development and management approaches that conform to the principles of sustainable urban development.

In general, regular monitoring, assessments and public reports of the state of the environment are particularly crucial prerequisites for better policy and decision-making. They are also cost-effective. While the large external debts of many African countries are a major concern, many of the same countries also have growing 'environmental debts' where the cost of remedial action will be far greater than preventive action. A 1990 World Bank study estimated the longer-term losses to Nigeria of not acting to prevent environmental degradation at more than US$5 100 million per year, which then represented more than 15 per cent of GDP (World Bank 1990). Some environmental losses are irreversible. Groundwater polluted by industrial and agricultural chemicals cannot readily be cleaned. Top soil washed or blown away in a few years takes centuries to replace. Extinct plant and animal species are lost forever, as are their potential health, economic and other benefits (UNEP 1993).

On the eve of the 21st century, African countries are reconsidering some conservation practices and community-based approaches which prevailed in their societies in the 19th century. For example, inter-cropping and agroforestry were traditional farming practices well into the 20th century. Using community-based approaches, several African governments have achieved notable successes in wildlife management.

Data on environmental conditions and trends need to be improved but also combined with existing economic and social data to provide a better basis for sustainable development planning and decision-making. Many African governments still prepare separate reports on the national economy, on health and social conditions, and on the state-of-the-environment while neglecting the linkages between these issues.

The key challenge is to reduce poverty. New approaches that put the poor at the top of the environment and development agenda could tap and release the latent energy and talents of Africans to bring about development that is economically, socially, environmentally and politically sustainable.

References

Agyemang, O., and others (1997). *An Environmental Profile of the Greater Lusaka Area: Managing the Sustainable Growth and Development of Lusaka.* Lusaka City Council, Lusaka, Zambia

AIDS Analysis Africa (1996). Southern Africa Edition, Vol. 7, No. 2 (Aug/Sept 1996), Whiteside and van Niftrik Publications, Howard, South Africa

Ben Mohamed, A. (1985). *Turbidity and humidity parameters in Sahel: possible climatic implications.* Sp. Env. Report No. 16, WMO-No. 547, WMO, Geneva, Switzerland

Ben Mohamed, A. (1998). Wind erosion in Niger: extent, current research, and ongoing soil conservation activities. In M.K. Sivakumar, M. Zobisch, S. Koala and T.P. Maukonen (eds.), *Wind Erosion in Africa and West Asia: Problems and Control Strategies.* Proceedings of the ICARDA, ICRISAT, UNEP, WMO Expert Group Meeting, 22–25 April 1997, Cairo, Egypt. ICARDA, Aleppo, Syria

Ben Mohamed, A., and J.P. Frangi (1986). Results from ground-based monitoring of spectral aerosol optical thickness and horizontal extinction: some characteristics of dusty Sahelian atmospheres. *Journal Clim. Appl. Metor.,* 25, 1807-1815

Bryant, E. (1994). Water: tapping Africa's most basic resource. *Africa Farmer,* 12 (July 1994), 25-28

Bwathondi, P.O.S., Nkotagu, S.S. and Mkuula, S. (1991). *Pollution of the Msimbazi Valley,* National Environmental Management Council, Dar es Salaam, Tanzania

Byers, B.A. (1997). *Environmental Threats and Opportunities in Namibia: A Comprehensive Assessment.* Department of Environmental Affairs, Windhoek, Namibia

CAMRE/UNEP/ACSAD (1996). *State of Desertification in the Arab Region and the Ways and Means to Deal with it* (in Arabic with English summary). Damascus, Syria.

CARE (1998). *El Niño: El Niño Flooding Part of Equation as Infectious Diseases Spread in Kenya and Somalia.* http://www.care.org/newscenter/elnino/nino128.html

Chidumayo, E.N. (1986). *Species Diversity in some Zambian Forests.* University of Zambia, Lusaka

CDIAC (1998). *Revised Regional CO_2 Emissions from Fossil-Fuel Burning, Cement Manufacture, and Gas Flaring: 1751–1995.* Carbon Dioxide Information Analysis Center, Environmental Sciences Division, Oak Ridge, Tennessee, United States. http://cdiac.esd.ornl.gov/cdiac/home.html

Cock, J., and Koch, E., (eds., 1991). *Going Green: People, Politics and the Environment in South Africa.* Oxford University Press, Cape Town, South Africa

Couto, M. (1995). *Pollution and its Management.* SARDC, Harare, Zimbabwe

Darkoh, M. B. K. (1993). Desertification: the Scourge of Africa. *Tiempo,* Issue No. 8, April 1993 http://www.cru.uea.ac.uk/tiempo/floor0/archive/t8art1.htm

Energy Information Administration (1997). *International Energy Outlook.* http://www.eia.doe.gov/oiaf/ieo97/home.html

Eritrea Agency for the Environment (1995). *National Environmental Management Plan for Eritrea.* Eritrea Agency for the Environment, Asmara, Eritrea

FAO (1993). *Yearbook – Fishery Statistics: Catches and Landings,* Vol. 76. FAO, Rome, Italy

FAO (1995). *Irrigation in Africa in Figures,* Water Report 7. FAO, Rome, Italy

FAO (1997a). *State of the World's Forests 1997.* FAO, Rome, Italy

FAO (1997b). *Irrigation in the Near East Region in Figures,* Water Report 9. FAO, Rome, Italy

FAO (1997c). *Fishstat-PC.* FAO, Rome, Italy. http://www.fao.org.

FAO (1998). *State of the World's Forests 1998.* FAO, Rome, Italy

FAOSTAT (1997). *FAOSTAT Statistics Database.* FAO, Rome, Italy. http://www.fao.org

GCA (1997). *African Social and Economic Trends.* Annual Report 1996 of the Global Coalition for Africa, Washington DC, United States

Geldenhuys, C. (1996). *Past, Present and Future Forest Management in the Southern African Region with Special Emphasis on the Northern Regions of Namibia.* Directorate of Forestry, Windhoek, Namibia

Government of Botswana (1993). Botswana Country Paper. In *Proceedings of the Workshop on Water Resources Management in Southern Africa,* Victoria Falls, Zimbabwe, 5–9 July 1993

Habitat (1997). *Global Urban Indicators Database.* Habitat, Nairobi, Kenya

Hewehy, M. A. (1993). *Impacts of Air Pollution on Cultural Resources in Cairo.* 13th Annual Meeting, International Conference of IAIA, Shanghai, China

Heyns, P. (1993). Water Management in Namibia. In *Proceedings of the Workshop on Water Resources Management in Southern Africa,* Victoria Falls, Zimbabwe

Holloway, A. (1995). Challenges for Long-term Disaster Reduction: Elements of Telemedicine, Harare, Zimbabwe, unpublished

Huntley, B.J. and Matos, E. (1992). *Biodiversity: Angolan Environment Status Quo Assessment Report.* IUCN Regional Office for Southern Africa, Harare, Zimbabwe

IMO (1995). *Global Waste Survey: Final Report.* IMO, London, United Kingdom

IUCN (1992). *Angola: Environmental Status Quo Assessment Report.* IUCN Regional Office for Southern Africa, Harare, Zimbabwe

Johns Hopkins (1998). Solutions for a Water-Short World. *Population Report,* Vol. XXVI, No. 1, September 1998. Johns Hopkins Population Information Program, Baltimore, Maryland, United States http://www.jhuccp.org/popreport/m14sum.stm

Lone, S., Laishely, R. and Bentsi-Enchill, N.K. (1993). *Africa Recovery,* Briefing Paper No. 9. United Nations Department of Public Information, New York, United States

Mavimbela, S. (1995). *Pollution and its Management.* SARDC, Harare, Zimbabwe

Ministry of Natural Resources, Uganda (1995). *The National Environmental Action Plan for Uganda.* NEAP Secretariat, Ministry of Natural Resources, Kampala, Uganda

Namibia Foundation (1994). *Marine Fisheries in Regional Context.* Namibia Brief No. 18, Windhoek, Namibia

Nana-Sinkam, S. C. (1995). *Land and Environmental Degradation and Desertification in Africa.* Joint UNECA/FAO publication, Addis Ababa, Ethiopia, February 1995 http://www.fao.org/desertification/DOCS/361117/36111700.htm#TOP

NEMA (1996). *State of the Environment Report for Uganda*. National Environmental Management Authority, Kampala, Uganda

Niger Ministry of Environment and Hydraulics (1997). *Plan National pour l'Environnement et le Developpement Durable du Niger.* Niamey, Niger

NOAA (1998). *1998 Coral Bleaching in Indian Ocean Unprecedented.* National Oceanographic and Atmospheric Administration Press Release, 1 July 1998, Washington DC, United States

Oldeman, L.R. (1994). Global Extent of Soil Degradation. In *Soil Resilience and Sustainable Land Use* (eds. D. J. Greenland and I. Szabolcs), p. 99-118. CAB International, Wallingford, United Kingdom

SADC (1992). *Environmental Effects of Mining in the SADC Region*. Mining Sector Coordinating Unit, Lusaka, Zambia

SADC (1995). *Climate Change.* Proceedings of the First SADC Conference held in Windhoek, Namibia, 2-6 March 1992. SADC ELMS, Maseru, Lesotho

SADC (1996). *Food, Agriculture and Natural Resources*. SADC Annual Consultative Conference, Johannesburg, South Africa

Salm, R.V. (1998). The Status of Coral Reefs in the Western Indian Ocean with notes on the related ecosystems. In *The International Coral Reef Initiative (ICRI),* Western Indian Ocean and Eastern Africa Regional Workshop Report, 29 March-2 April 1996, Mahé, Seychelles. UNEP, Nairobi, Kenya

SARDC, IUCN and SADC (1994). *State of the Environment in Southern Africa*. SARDC, IUCN and SADC, Harare, Zimbabwe, and Maseru, Lesotho

Scotney, D.M. and Djikhuis, F.H. (1989). *Recent Changes in the Fertility Status of South African Soils.* Soil and Irrigation Research Institute, Pretoria, South Africa

Serageldin, Ismail (1995). *Toward Sustainable Management of Water Resources.* World Bank, Washington DC, United States

Sivertsen, B., Matale, C. and Pereira, L.M. (1995). *Sulphur Emissions and Transfrontier Air Pollution in Southern Africa*. SADC, Maseru, Lesotho

Stuart, S.N., Adams, R.J. and Jenkins, M.O. (1990). *Biodiversity in Sub-Saharan Africa and its Islands: Conservation, Management and Sustainable Use.* IUCN, Gland, Switzerland

Tavengwa, T. (1995). SADC Unites in War Against Malaria: WHO Lays Platform of Action. In *The Herald,* 7 December 1995, Harare, Zimbabwe

UNCTAD (1996). *Annual Report.* UNCTAD, Geneva, Switzerland

UNDP (1996). *Human Development Report 1996*. Oxford University Press, New York, United States, and Oxford, United Kingdom

UNDP (1997). *Human Development Report 1997*. Oxford University Press, New York, United States, and Oxford, United Kingdom

UNDP (1998). *Combatting poverty.* http://www.undp.org/undp/rba/undp_af/poverty.htm

UNECA (1996). *Urban Environment and Health in ECA Member States.* UN Economic Commission for Africa, Addis Ababa, Ethiopia

UNECA (1997). *Report on the Economic and Social Situation in Africa.* UN Economic Commission for Africa, United Nations, Addis Ababa, Ethiopia

UNECA (1998). *Africa Economic Report 1998.* UN Economic Commission for Africa, United Nations, Addis Ababa, Ethiopia http://www.un.org/depts/eca/divis/espd/aer98.htm#ia5

UNEP (1991). *Regaining the Lost Decade: A Guide to Sutainable Development in Africa.* UNEP, Nairobi, Kenya

UNEP (1993). *Accelerating the Transition to Sustainable Development: Implications of Agenda 21 for West Africa*. UNEP, Nairobi, Kenya

UNEP (1996). *The State of the Marine and Coastal Environment in the Mediterranean Region.* MAP Technical Reports Series No. 100. UNEP, Athens, Greece

UNEP (1997). *World Atlas of Desertification,* second edition. Edward Arnold, London, United Kingdom

UNEP/ISRIC (1991). *World Map of the Status of Human-Induced Soil Degradation (GLASOD). An Explanatory Note,* second revised edition (edited by Oldeman, L.R., Hakkeling, R.T., and Sombroek, W.G.). UNEP, Nairobi, Kenya, and ISRIC, Wageningen, Netherlands

UNESCO (1954). *Symposium on the Protection and Conservation of Nature in the Near East.* UNESCO, Cairo, Egypt

United Nations (1996). *UN System-Wide Special Initiative on Africa.* UN, New York, United States

United Nations Population Division (1996). *Annual Populations 1950-2050 (the 1996 Revision),* on diskette. United Nations, New York, United States

United Nations Population Divison (1997). *Urban and Rural Areas, 1950-2030 (the 1996 Revision),* on diskette. United Nations, New York, United States

USAID (1997). *Climate Change Action Plan*. USAID, Washington DC. United States

WCMC (1992). *Global Biodiversity: Status of the Earth's Living Resources*. Chapman and Hall, London, United Kingdom

WCMC (1998). WCMC Protected Areas Database http://www.wcmc.org.uk/protected_areas/data

WCMC/IUCN (1998). WCMC Species Database, data available at http://wcmc/org/uk, assessments from the 1996 IUCN Red List of Threatened Animals

WHO (1995). *Cholera – 1994: Situation in the African Region*. WHO, Harare, Zimbabwe

World Bank (1990). *Towards the Development of an Environmental Action Plan for Nigeria*. Report No. 9002-UNI. World Bank, Washington DC, United States

World Bank (1992). *Development And Environment, World Development Report*. Oxford University Press, Oxford, United Kingdom, and New York, United States

World Bank (1995a). *Africa: A Framework for Integrated Coastal Zone Management.* World Bank, Washington DC, United States

World Bank (1995b). *Towards Environmentally Sustainable Development in Sub-Saharan Africa: A World Bank Agenda.* World Bank, Washington DC, United States

World Bank (1995c). *Middle East and North Africa: Environmental Strategy Towards Sustainable Development.* Report No. 13601-MNA. World Bank, Washington DC, United States

World Bank (1996a). *The Experience of the World Bank in the Legal, Institutional and Financial Aspects of Regional Environment Programs: Potential Applications of Lessons Learned for the ROPME and PERSGA Programs*. World Bank, Washington DC, United States

World Bank (1996b). *African Water Resources: Challenges and Opportunities for Sustainable Development*. World Bank, Washington DC, United States

World Bank (1998). *The World Bank and Climate Change: Africa* http://www.worldbank.org/html/extdr/climchng/afrclim.htm

World Bank and DANIDA (1995). *Rapid Water Resources Assessment Report*, Vol. 2, Basin Report. World Bank and DANIDA, Dar es Salaam, Tanzania

WRI, ICLARM, WCMC and UNEP (1998). *Reefs at Risk: a map-based indicator of threats to the world's coral reefs.* Washington DC, United States http://www.wri.org/indictrs/rrstatus.htm#world

WRI, UNEP, UNDP and WB (1996). *World Resources 1996-97: A Guide to the Global Environment* (and the *World Resources Database* diskette). Oxford University Press, New York, United States, and Oxford, United Kingdom

WRI, UNEP, UNDP and WB (1998). *World Resources 1998-99: A Guide to the Global Environment* (and the *World Resources Database* diskette). Oxford University Press, New York, United States, and Oxford, United Kingdom

WWF (1996). http://www.livingplanet.org/resources/factsheets/general/27biodisa.htm

Asia and the Pacific

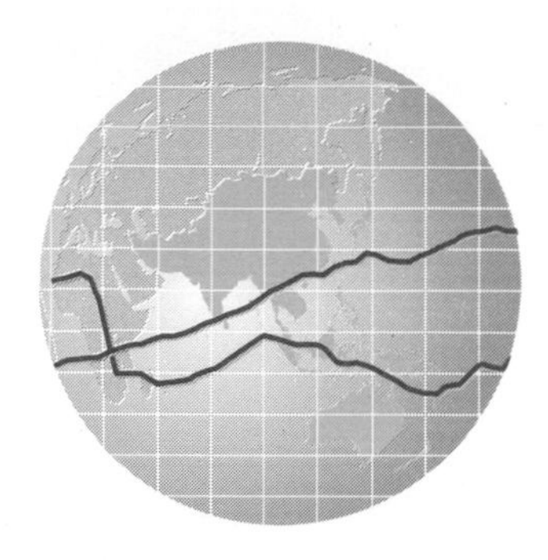

KEY FACTS

A 'business-as-usual' scenario suggests that continued rapid economic growth and industrialization may result in further environmental damage and that the region may become more degraded, less forested, more polluted and less ecologically diverse in the future. Asia's particular style of urbanization – toward megacities – is likely to further exacerbate environmental and social stresses

- Some 75 per cent of the world's poor live in Asia.
- There is great pressure on land resources in the region in which some 60 per cent of the world population depends on 30 per cent of its land area.
- The limiting factor to producing more food in the future will be freshwater supplies, especially in populous and arid areas.
- About one million hectares of Indonesia's national forests have been destroyed by fires that burned for several months from September 1997. Fires also burnt more than 3 million ha of forests in Mongolia in 1996.
- Increasing habitat fragmentation in Southeast Asia has depleted the wide variety of forest products that used to be the main source of food, medicine and income for indigenous people
- At least one in three Asians has no access to safe drinking water and at least one in two has no access to sanitation.
- Demand for primary energy in Asia is expected to double every 12 years while the world average is every 28 years.
- While the proportion of people living in urban centres is still lower than in developed countries, it is rising rapidly, and is focused on a few urban centres.

Social and economic background

In the past 100 years, most countries in the region have undergone unparalleled social, political and economic transformations. Colonialism, which dominated much of the region, was replaced by other political systems. Economies, which were largely agrarian, became industrialized, export-oriented and better integrated with global markets. Agriculture was intensified to increase production for home consumption and export.

Rapid industrialization and economic growth have changed virtually every dimension of life, especially in East and Southeast Asia. Yet, by many measures – of health, education, nutrition, as well as income – the quality of life within the region remains poor for most people. At least one in three Asians has no access to safe drinking water and at least one in two has no access to sanitation (ADB 1997). Average cereal consumption is one-third that in the developed countries, and average calorie intake, though rising, is low in most sub-regions (see graph top right). Literacy rates tend to be low, particularly for women (ADB 1997). Poverty is a major problem: some 75 per cent of the world's poor live in Asia (UNESCAP/ADB 1995).

In most countries, economic development and industrialization have taken a heavy toll on the environment. At the turn of the century, environmental degradation in the region was largely due to poor farming methods, colonial expansionist land practices in South and Southeast Asia, and foreign invasions and mineral exploitation in China. Japan and China were among the first industrialized economies in the region. Following World War II, there was rapid growth of the commercial and services sectors in Japan and improvements in health, education, housing and nutrition. In Southeast Asia, rapid economic growth began in the early 1980s. Change came later to South Asia where Structural Adjustment Programmes and economic liberalization have only recently begun to be implemented.

The economies of Australia and New Zealand are based more on natural resources than other industrialized countries in the region. Agriculture and mining account for most export earnings derived from trade in commodities in Australia (Commonwealth of Australia 1998). Pacific island states rely heavily on their natural resources directly through agricultural production, forestry and fisheries and indirectly through tourism. The Exclusive Economic Zone (EEZ) of these small countries comprises a large proportion of their total area and, for some, offshore marine resources are almost the only basis for economic development.

In the past decade, economic growth rates have varied dramatically. The fastest growing economies were in China and Thailand, both of which registered about 8 per cent average annual growth during 1985–95. In 1995 GDP per capita ranged from US$14 791 for Australasia and the Pacific, to US$1 183 for Southeast Asia to only US$484 for South Asia (see graph below right).

Economic growth has been largely fuelled by industrialization and international trade. During 1980–95, the share of the industrial and services sectors in the region's total GDP increased significantly while the agricultural sector declined, except in most Mekong basin countries. In 1960, 75 per cent of working Asians were employed in agriculture; by 1990 this had fallen to 62 per cent. Over the same period, the share of people working in industry grew from around 15 to 21 per cent (ADB 1997). The dramatic economic growth rates of the early part of the 1990s have been followed by equally dramatic economic slumps, particularly in Southeast Asia. Average economic growth rates in the Pacific island countries also appear to have turned sharply negative since the East Asian financial crisis.

Economic growth has been accompanied by some improvement in health and education, although the

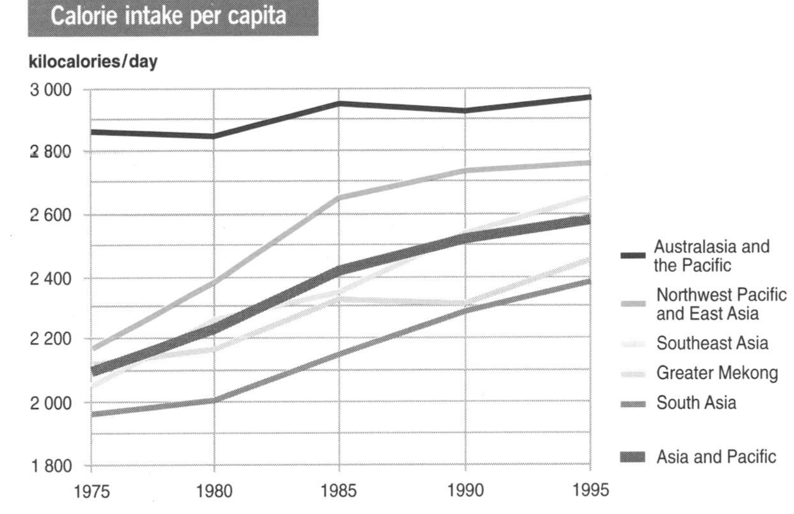

Source: compiled by UNEP GRID Geneva from FAOSTAT 1997 and WRI, UNEP, UNDP and WB 1998

Though calorie intake has risen rapidly in the region, it is still about one-third lower than in developed countries

region lags far behind the developed world. Adult literacy varies from only 15 per cent female literacy in Afghanistan and 41 per cent male literacy in Nepal to more than 95 per cent for both sexes in countries such as the Republic of Korea and Japan (UNESCO 1995). The literacy rate is also high in Australia, although some groups still have poor literacy skills (Commonwealth of Australia 1998).

Life expectancies in some countries are now comparable with those of middle- and high-income countries. There has also been a marked decline in infant mortality, from 68 per 1000 live births in 1990 to 59 per 1000 in 1995, although great disparities still exist in the region (UNESCAP/ADB 1995).

The region has huge variations in wealth and economic growth. GDP/capita for Australasia and the Pacific was more than seven times the regional average in 1995

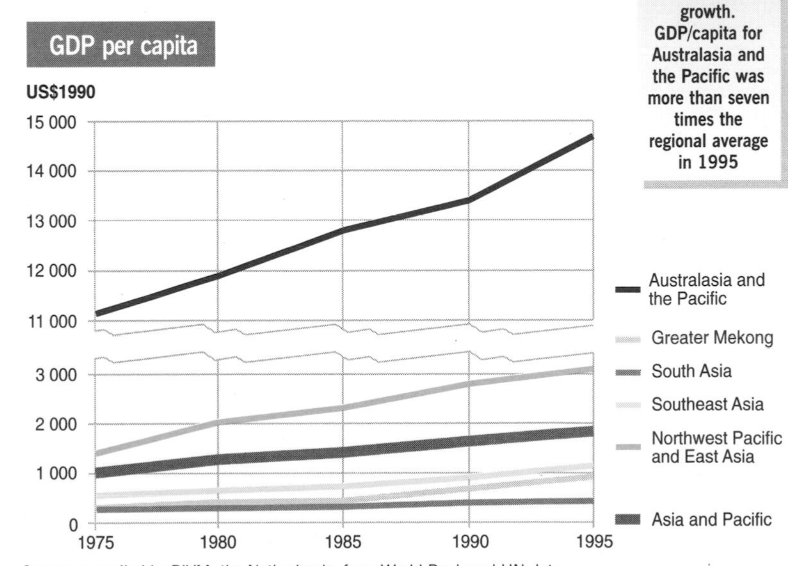

Source: compiled by RIVM, the Netherlands, from World Bank and UN data

Annual growth of per capita GDP, 1975–95

Australasia and the Pacific	1.42%
Greater Mekong	5.62%
Northwest Pacific and East Asia	3.80%
South Asia	1.54%
Southeast Asia	4.08%
ASIA AND PACIFIC	**3.09%**
(World	1.17%)

Source: compiled by RIVM, the Netherlands, from World Bank and UN data

The advantages of rapid economic growth have not filtered down to all levels of society. Poverty remains a significant problem, particularly in South Asia where there are more than 515 million of the region's 950 million poor people (UNDP 1997) and around 39 per cent of the population is below the poverty line, with numbers still increasing.

Rural poverty, together with rapid industrialization, has led to rates of rural-urban migration that are significantly higher than the global average, although at around 33 per cent the proportion of urban population in the region was lower than the global average of 45 per cent in 1995 (United Nations Population Division 1997).

Population densities in South Asia are among the highest in the world, and there is great pressure on land resources throughout the region – in which some 60 per cent of the world population depends on 30 per cent of its land area (UNESCAP/ADB 1995). Bangladesh had 922 persons per km^2 in 1995 (WRI, UNEP, UNDP and WB 1998). The combination of rural poverty and population pressure has forced people to move to ecologically-fragile areas. In addition, the number of landless people is increasing.

The large and growing populations found in most sub-regions are one of the causes of environmental degradation and pollution

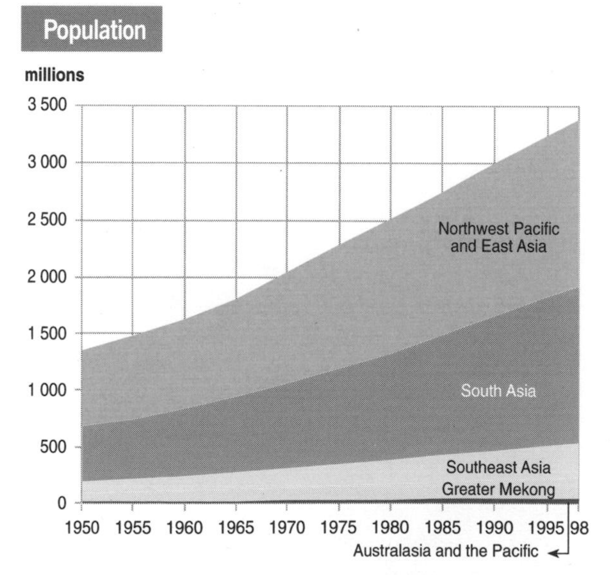

Source: compiled by UNEP GRID Geneva from United Nations Population Division 1996

Population growth rates have declined in recent years and the region's overall growth rate is now the same as the world average of 1.4 per cent. The highest sub-regional figures are 1.9 and 1.7 per cent a year for South and Southeast Asia respectively (United Nations Population Division 1996).

The combination of high population density and growth, rapid industrialization and urbanization, and poverty has taken its toll on the region's natural resource base, accelerated environmental degradation and led to a substantial increase in air and water pollution. Other significant environmental problems include land degradation caused by deforestation and inappropriate agricultural practices, water loss, and mangrove clearance for aquaculture. Estimates of the economic costs of environmental degradation in Asia range from 1 to 9 per cent of national GNPs (ADB 1997).

In addition, the natural disasters which regularly hit the region, especially the South Pacific island states, can have extremely damaging impacts on both the environment and fragile economies (see box).

Environmental experts are not agreed about the relative significance of the various causes of environmental degradation. Rapid population growth is often blamed but the damage continues even in countries with low and declining population growth rates (ADB 1997). Economic growth is also blamed but rising incomes eventually lead to improvements as growing popular demand for a better environment forces a favourable policy shift. This point is not far off for Asia's more prosperous newly industrialized economies (ADB 1997).

Asia's environmental crisis may be mainly a result of market and policy failures, neglect and institutional weaknesses. Only a few countries in East Asia, including China, have succeeded in implementing policies to reverse the trend of continuing degradation.

The most important external force shaping the region's future will be the increased integration of the world economy. Global trade provides access to the information, ideas, technologies, and the other critical

Natural disasters

Cyclones, floods, storm surges, earthquakes, droughts, landslides and volcanic eruptions affect many countries in the region causing great loss of life and extensive damage to property and infrastructure. These disasters seriously affect the pace of development. One cyclone in Bangladesh in November 1970 caused almost half a million casualties, with colossal damage to property and infrastructure. In 1976, a single earthquake in China took nearly 300 000 lives.

Trends in disaster events are disturbing. Statistics show that during the period 1900–91, there have been more than 3 500 disasters – roughly 40 a year – and that they have killed more than 27 million people.

There is also evidence that the frequency of disasters is increasing. For the top 10 most disaster-prone countries in the region – Australia, Bangladesh, China, India, Indonesia, the Islamic Republic of Iran, Japan, New Zealand, Philippines and Viet Nam – there were a total of 1 312 disasters during the 25 years 1966–90, which killed 1.7 million people and affected more than 2 000 million. The frequency of disasters in this period was 52.5 a year, compared to only 24.8 a year over the period 1900–91.

It is possible that global climate change may result in similar or even worse trends in the future. An increase in the occurrence of natural disasters during the next decades could have serious economic effects. For example, the cyclones of 1990 and 1991 in Samoa caused a total estimated loss of US$416 million, about four times the GDP of that country. Samoa has yet to recover from these cyclones. Precautionary measures and mitigation plans need be strengthened in all countries of the region.

Sources: CRED 1991 and 1993

resources that are the backbone of economic progress. Greater mobility of international capital will also mean less scope for autonomy in macro-economic policy. In the future, as recent events in Southeast and East Asia have shown, global capital markets will react to changes in fiscal and monetary policies more quickly and more severely than in the past.

A 'business as usual' scenario suggests that continued rapid economic growth and industrialization may result in further environmental damage and that the region may become more degraded, less forested, more polluted and less ecologically diverse in the future. Asia's particular style of urbanization – toward megacities – is likely to further exacerbate environmental and social stresses (ADB 1997).

Land and food

At the turn of the century, most countries (except Australia and New Zealand) were agriculturally-based, practising traditional subsistence farming, including shifting cultivation and nomadic livestock grazing. As the region's population increased, the need for greater food production put increasing pressure on land resources. The area of cropland, which stood at about 210 million ha in 1900, expanded during 1980–95 from 426 to 453 million ha, largely at the expense of forest cover which decreased by 42.6 million ha (8.3 per cent) over the same period.

Agricultural activities have often suffered from and caused environmental degradation. Bringing marginal land into production is a case in point. In Japan and the Republic of Korea, for example, urban and industrial developments on flat coastal areas have encroached on arable land and led to increasing cultivation of forested hill slopes. In Southeast Asia the introduction of the cash economy has induced some hill tribes to convert unsuitable upland areas to intensive commercial cropping. Elsewhere, particularly in Malaysia and Indonesia, rural people who relied on traditional shifting cultivation have been forced to move to marginal lands and, as a result of declining productivity, are having to adopt shorter fallow periods. By comparison, in the 1970s and early 1980s, New Zealand sheep farmers facing declining export returns responded to government subsidies by clearing patches of steep marginal forest land to boost production (Roper 1993, New Zealand Ministry for the Environment 1997). The results of these actions included degradation of watersheds through accelerated erosion, and increased sedimentation and flooding downstream.

In Southeast Asia, land conversion was intensified by increased commercial logging and the introduction of commercial crops (Kummer 1993, Uhlig 1984). For instance, arable land increased five-fold in Malaysia during 1900–50 to accommodate rubber and oil palm plantations (ADB 1994). Agricultural land use in Southeast Asia has expanded only slightly from 16.8 per cent of total land area in 1975 to 19.6 per cent in 1992 (ADB 1995).

About 50 per cent of land cover in Australia has been changed by complete clearing, thinning of vegetation, overgrazing, changed fire regimes and other habitat modifications. In New Zealand, large areas of forest had been cleared and virtually all grazeable land converted to pasture by 1920 to provide wool, meat and dairy exports. Today, around 50 per cent of land in both countries is used for grazing (Commonwealth of Australia 1996, New Zealand Ministry for the Environment 1997).

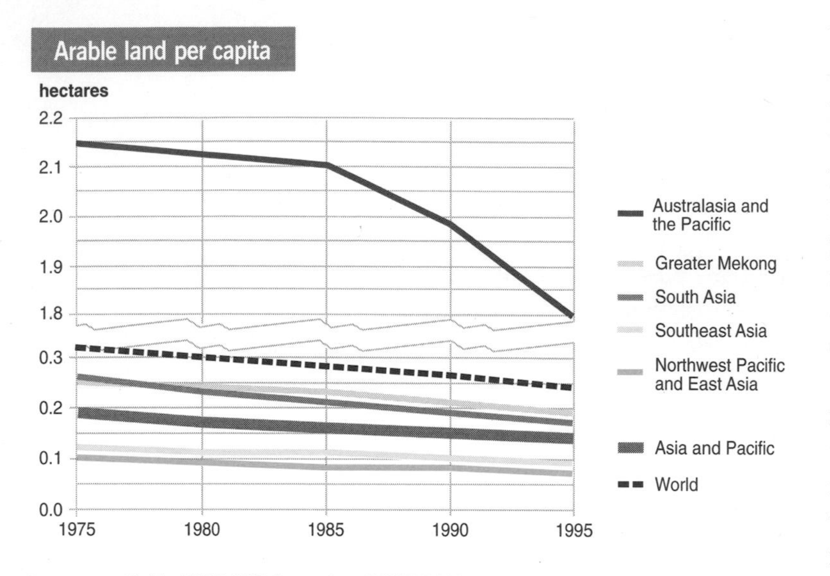

Source: compiled by UNEP GRID Geneva from FAOSTAT 1997

The Asia and Pacific region has much less arable land per capita than the world average, even though Australia has more than any other country in the world

In East Asia, recent trends have been different. Rapid economic growth after World War II caused a dramatic increase in the conversion of natural and agricultural areas to urban and industrial infrastructures, particularly in Japan and the Republic of Korea. In China, the extent of arable land also decreased – between 1957 and 1990 by an area equal to all the cropland in France, Germany, Denmark and the Netherlands combined (UNESCAP/ADB 1995) – but this was largely due to land degradation. Mongolia is unique in the sub-region with 75 per cent of its total land area occupied by nomadic livestock herds and an overgrazing problem.

Overall there is less land per person in the Asia-Pacific region today than in other parts of the world, and population density is highest in South Asia. In 1995, whilst Australia had most cropland per capita (2.66 ha) in the world, much of the region, particularly in South and East Asia, had well below the world average of 0.24 ha. At least 10 major countries in the region have less than 0.10 ha and, as the graph above shows, so do two sub-regions: Southeast Asia (0.09 ha) and Northwest Pacific and East Asia (0.07 ha).

Maintaining, let alone improving, the situation is becoming increasingly difficult as populations continue to expand and agricultural land is lost to urban, industrial and transport infrastructure, particularly in coastal areas. Attempts to compensate by expanding agricultural production into other areas, as in previous decades, are constrained by the decreasing availability of suitable land and widespread land degradation.

The increasing scarcity of agricultural land has been moderated by dramatic improvements in agricultural yields with the development of high-yielding crop varieties, irrigation systems and increased agrochemical inputs. For example, grain yield in China nearly quadrupled between 1952 (1 300 kg/ha) and 1996 (4 600 kg/ha). Similarly grain production per capita increased from 300 kg per year in 1952 to around 400 kg per year in 1996 (State Statistical Bureau, various years) and the country is currently self-sufficient in this commodity. Between 1980 and 1990, food production in Southeast Asia grew faster than anywhere in the world (UNCTAD 1994). Over the past decade, food production continued to increase throughout the region but has not matched the growth rate of the previous decades. The limiting factor to producing more food in the future will be freshwater supplies, especially in populous and arid areas. In Southeast Asia, there is already little potential for additional large-scale water development schemes.

Southeast Asia, Australia and New Zealand are all net exporters of agricultural commodities. Australia feeds the equivalent of about an extra 50 million people with cereals (WRI, UNEP and UNDP 1994). However, the region as a whole is a net food importer, and food security is a high priority in most countries, particularly in East and South Asia (UNESCAP/ADB 1995).

The region will rely increasingly on imported food. By 2010, Asia's share of world cereal imports is estimated to rise to about 42 per cent from its current level of 33 per cent (ADB 1997), and this may put pressure on world food balances and affect world food prices. In South Asia, structural adjustment programmes, including trade liberalization and the removal of subsidies from forest products and some food and cash crops, will encourage intensive production and increased exports (Dutta and Rao 1996). This pattern of growth in agribusiness is likely to place an increasing strain on rural production resources.

The combination of rapid urban and industrial growth, extensive deforestation and unsustainable agriculture, including inadequate soil conservation, cultivation of steep slopes and overgrazing, has had a devastating impact on land resources. According to GLASOD, of the world's 1 900 million ha of land affected by soil degradation during the past 45 years, the largest area (around 550 million ha) is in the Asia-Pacific region (UNEP/ISRIC 1991). For Asia this constitutes about 20 per cent of total vegetated land. Dry parts of the region are particularly vulnerable, and it is estimated that 1 320 million people (39 per cent of the region's population) live in areas prone to drought and desertification (UNEP 1997). The more recent Assessment of Soil Degradation in South and South-East Asia (ASSOD 1997, see map) found that agricultural production is substantially reduced by degradation in dry areas. Nearly 180 million ha in China, including 90 per cent of China's extensive grasslands (SEPA 1998), 110 million ha in India and 62 million ha in Pakistan are degraded, representing 56, 57 and 86 per cent respectively of susceptible drylands (UNEP 1997).

Soil erosion has reduced agricultural potential in many countries. In India, for instance, as much as 27 per cent of the soil has been affected by severe erosion (ADB 1997), water being one of the principal causes of the removal of nutrient-rich topsoil, particularly in the Himalayas. In the Islamic Republic of Iran, 45 per cent of agricultural land is affected by light to moderate water erosion (FAO, UNDP and UNEP 1994). Wind erosion is also extensive and severe, affecting about 25 million ha in India and Pakistan, particularly the drybelt stretching from Central Iran to the Thar desert, and another 75 million ha in China (UNEP 1997). Woods (1983), in assessing the extent and severity of Australia's land degradation in 1977, estimated that about 38 per cent of agricultural lands required treatment for wind and/or water erosion. More recent national-scale information on erosion is currently being prepared.

Irrigated agriculture has degraded existing arable lands and resulted in vast expanses of salinized and

Severity of soil degradation in South and Southeast Asia

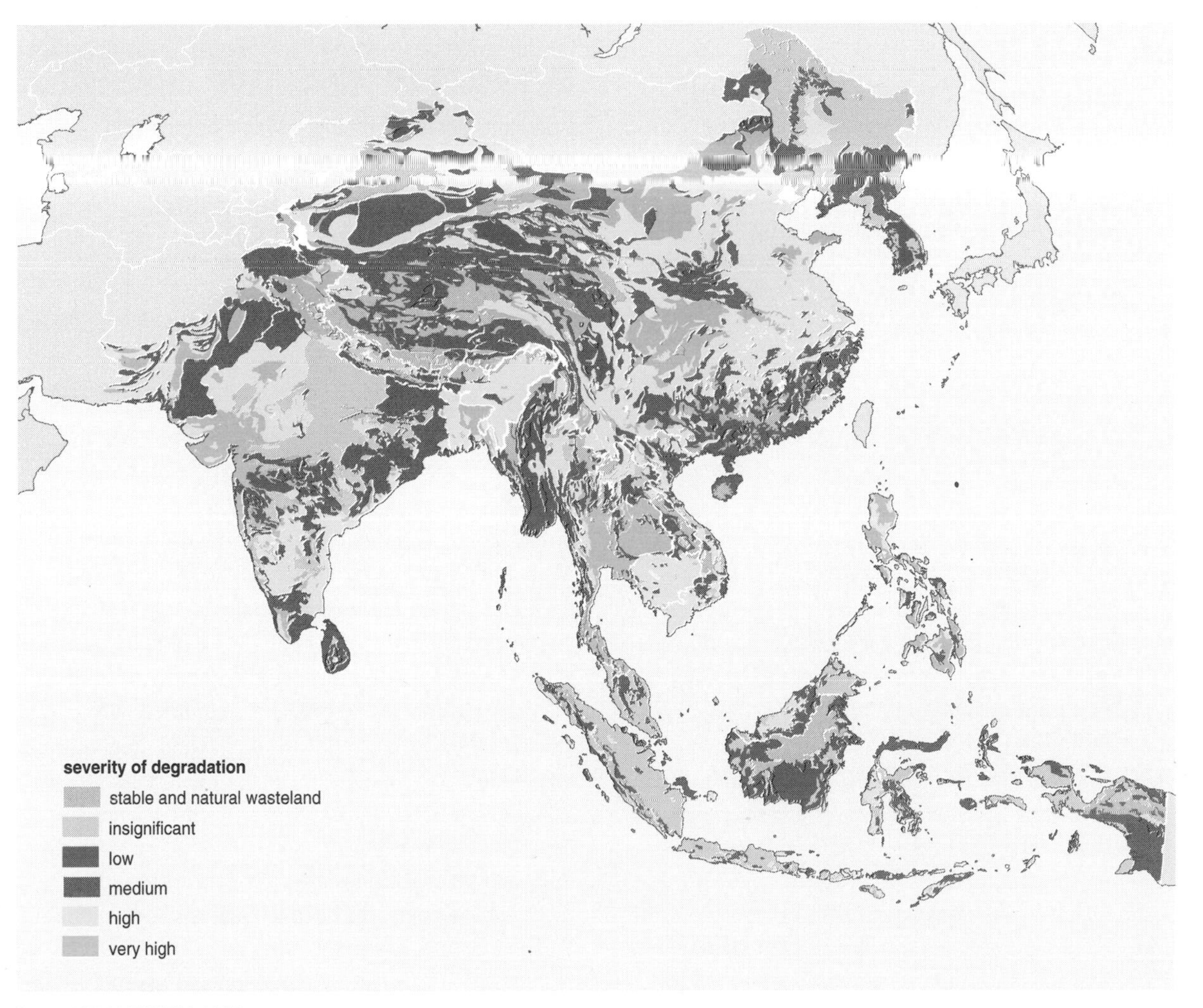

Source: ISRIC/UNEP/FAO 1997

More than 350 million ha – some 53 per cent of all land in the ASSOD area are desertified

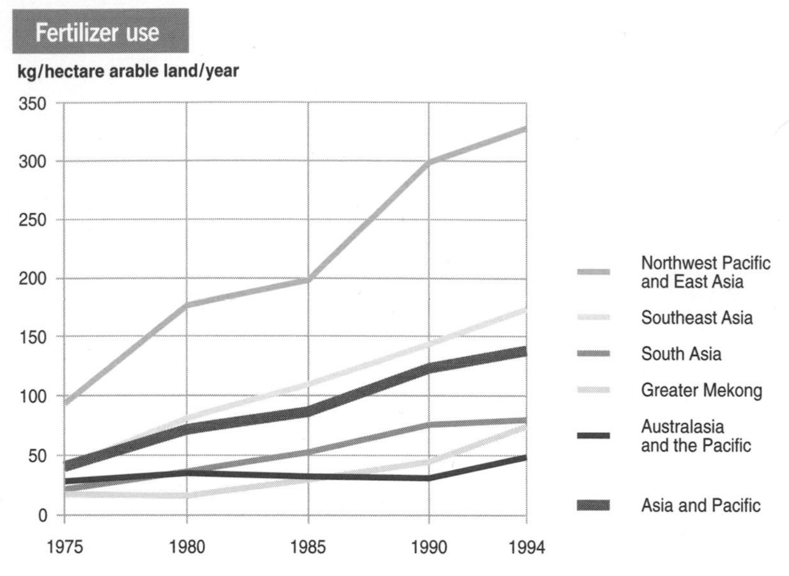

Source: Compiled by UNEP GRID Geneva from FAOSTAT 1997

The Northwest Pacific and East Asian sub-region has the highest rate of fertilizer use in the world, though some countries have now begun to reduce their usage

waterlogged soils. The Asia-Pacific region is responsible for around 75 per cent of all human-induced salinization in arid, semi-arid and dry sub-humid areas – the susceptible drylands – of the world (UNEP 1997). In the mid-1980s, Pakistan, India and China could alone account for about 50 per cent (30 million ha) of the world's irrigated land damaged by salinization (Postel 1989). In Pakistan, salt build-up in the soil is known to reduce crop yields by 30 per cent (Worldwatch Institute 1997). Estimates of secondary salinity (dryland and irrigated) in Australia vary from three to nine million hectares (SCARM 1998). This has reduced productivity and sometimes increased erosion in these areas (Commonwealth of Australia 1996). Not all water-related problems are associated with irrigation, though. In the Mekong Basin, natural leaching of rock salt into soils and saltwater intrusion in the delta area have degraded some of the most productive agricultural areas (MRC/UNEP 1997a).

Excessive agrochemical inputs in parts of the region are causing further degradation and soil pollution. In Australia, for example, some 30 million ha of soils within the higher rainfall, improved pasture and cropping areas have been acidified, and have a pH_{water} of less than 5.5 (SCARM 1998). Acidification can lead to toxic soils, poorer water and nutrient uptake by plants, and thus reduced yields (SCARM 1998). Japan and the Republic of Korea are now cutting back on the use of agrochemicals. At the same time, maintenance of soil fertility is a crucial issue. In the Mekong Basin, soil productivity is expected to continue its decline with the use of increasingly intensive agricultural practices (MRC/UNEP 1997).

Urbanization and industrial development, including the construction of dams and mining, have continued to contribute to land degradation in the region. For example, mineral exploitation has already degraded some 2 million ha of land in China and continues to affect 40 000 ha more each year. The long-term impacts of nuclear weapons' testing and the hazardous and toxic materials left behind after military activities have been particular concerns for the South Pacific nations. In addition, some of these countries are regularly exposed to tropical cyclones which inflict damage on infrastructure and crops as well as hindering crop growth due to residual salt and the loss of topsoil (SPREP 1993).

With roughly 60 per cent of the world's population depending upon only one-third of the world's land area, the region is hard put to provide the basic needs of its expanding population. The major challenge is to optimize land use for competing needs.

Forests

Primary forests in the region have been substantially depleted. Excessive cutting for timber and clearing for agriculture, including commercial plantation crops, have been the two major direct causes of deforestation (FAO 1997a, UNESCAP/ADB 1995). Commercial logging by the top five producers in the region – China, India, Indonesia, Japan and Malaysia – produces more than 200 million m^3 of roundwood annually (ASEAN 1997, and WRI, UNEP, UNDP and WB 1998). Forty per cent of Australia's forests have been cleared, and only about 25 per cent of the original forest estate in Australia remains relatively unaffected by clearing or harvesting (Commonwealth of Australia 1996). Many of the remaining forests in the Mekong basin countries have been logged so extensively that they are now of critically low quality. For example, only about 10 per cent of the remaining forests in the Lao People's Democratic Republic are commercially valuable (MRC/UNEP 1997a). Illegal and unmonitored logging is also a significant cause of deforestation. Commercial logging in the larger Pacific Islands has been largely driven by offshore demand, particularly in Asia, and deforestation rates have recently approached 2 per cent in countries such as Samoa (Government of Western Samoa 1994).

Fuelwood harvesting, irrigation and hydroelectric power development, mining, the expansion of urban and industrial infrastructure and railways, diseases, invasive species and cyclones have also contributed significantly to deforestation. Fires have also had a substantial impact (Gadgil and Guha 1992). About one million ha of Indonesia's national forests have been destroyed by forest fires (see box on page 90) that have burned almost continuously for several months beginning in September 1997 (EEPSEA/WWF 1998). Less well known are the fires that burnt more than 3 million ha of forests in Mongolia in 1996 (FAO 1997b).

Wars have taken a further toll. Much of Japan's forests were destroyed in World War II. In the former Korea (before separation), forest resources were excessively removed during the last years of Japanese colonial rule and were severely damaged by the Korean War in 1950–53 (OECD 1997). War in Indochina during the 1960s and early 1970s was extremely detrimental: about 2 million ha of forest in Viet Nam were destroyed through bombing and spraying of defoliant (WCMC 1994) and the toxic after-effects of residual dioxin held back forest regeneration for several years, especially in mangrove areas. Similarly, there was substantial loss of forest cover in northern Laos due to bombing during the war (DAI 1995).

In keeping with the rest of the region, New Zealand's forests have also had a chequered history (see box).

Between 1850 and 1980, nearly 24 per cent (224 million ha) of forests were removed throughout the region. With the exception of China, where only 7–8 per cent of land remained under forest cover by the middle of the century, rates of deforestation have increased significantly since 1930 and are now estimated to be 0.6 per cent a year, rising to 1.6 per cent a year in the Mekong basin (see chart above). In the process of deforestation, two-thirds of wildlife habitats have been destroyed (IUCN 1986), and vast expanses of naturally-fragile land, particularly upper water catchment areas, have been exposed to soil erosion. Six countries (China, Indonesia, Malaysia, Myanmar, Philippines and Thailand) account for three-quarters of recent deforestation in the region.

Average per capita forest cover for the region was 0.17 ha in 1995, considerably lower than the world average of 0.61 ha. Though there are large variations within the region, the 555 million ha of forests that remained in 1995 seem incapable of satisfying the needs of the population, and domestic wood shortages are beginning to appear, notably in Philippines,

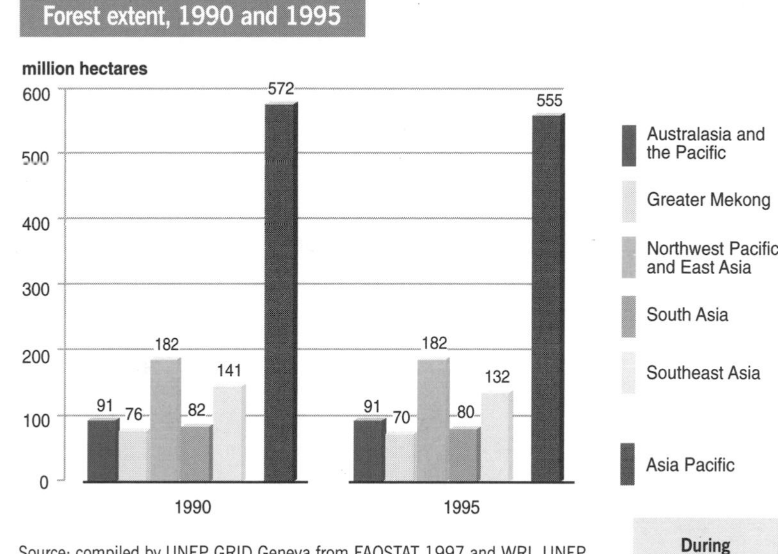

Source: compiled by UNEP GRID Geneva from FAOSTAT 1997 and WRI, UNEP, UNDP and WB 1998

During 1990–95, the region's forests were reduced by 17 million hectares. Deforestation was fastest in the Mekong (1.6 per cent a year) and in Southeast Asia (1.3 per cent a year)

Thailand and South Asia. In Nepal, for example, nearly 90 per cent of all the energy consumed is still in the form of traditional fuel (WRI, UNEP, UNDP and WB 1995). In East Asia, most timber is now imported from abroad, though there has been a recent policy shift to boost self-sufficiency in some countries such as Japan and the Republic of Korea.

Forest clearance in New Zealand

The last major phase of forest clearing in New Zealand took place during 1900–20. In earlier centuries, Maori fires had reduced natural forest cover from 85 to about 53 per cent of the land area. European settlers then reduced it further to 23 per cent. In 1919 the Forest Service was set up to create a sustainable timber supply and to protect the remaining alpine forests to avoid erosion and for flood control. Because the indigenous trees were slow growing, the Forest Service logged these, where practicable, and replaced them with fast-growing exotic trees (mostly *Pinus radiata* from California). Private companies and landowners also began planting exotic forests and, by 1996, these covered 1.6 million hectares. Public opposition to the logging of state-owned indigenous forests mushroomed in the 1970s. In 1986, as part of broader reforms, the Forest Service was disbanded. The state's exotic forests were sold off and most of its 4.9 million hectares of indigenous forests were designated for protection under the newly established Department of Conservation.

By the late 1980s, New Zealand's environmentalists had turned their attention to logging in privately-owned indigenous forests. This led, in 1993, to an amendment to the Forests Act of 1949 which banned most indigenous timber production unless it was under a certified sustainable management plan. Today, less than 3 per cent of New Zealand's indigenous forests are logged, accounting for less than 1 per cent of total timber output. Another major development was the New Zealand Forest Accord signed in 1992 by environmentalists and major forestry companies. The Accord prevents exotic forests from being planted at the expense of regenerating indigenous vegetation or significant wildlife habitat. Although it does not bind all forest owners, the Accord has significantly modified the behaviour of the larger companies.

Source: New Zealand Ministry for the Environment 1997

Large numbers of animal species are threatened in the region, which includes 4 of the the world's 12 mega-diverse countries

Several countries have now introduced sustainable forest and agricultural management policies and increased the extent of protected areas. Commercial logging bans in Cambodia, Lao People's Democratic Republic, Thailand, Viet Nam have slowed but not halted deforestation. For example, before the logging ban in 1988, forest depletion in Thailand averaged more than 480 000 ha per year. During 1993–95 it was down to some 100 000 ha per year (MOSTE Thailand 1997).

Asia's dominance of world trade in tropical hardwoods is likely to decline. At current rates of harvesting, remaining timber reserves in Asia will last for fewer than 40 years (ADB 1994). Continued development of urban and industrial infrastructure in forested areas may increase opportunities for forest exploitation by providing easy access for logging and encroachment (EA 1997). Forest fires are also contributing significantly to forest destruction and will continue doing so unless major efforts are made to stop them.

However, the region leads in the establishment of forest plantations. Nine of the top 15 developing countries for forest plantation establishment are in the region – Bangladesh, China, India, Indonesia, the Republic of Korea, Myanmar, Thailand, Philippines and Viet Nam (FAO 1997a). In China, for example, government afforestation campaigns increased forest cover from 12 per cent in the 1980s to almost 14 per cent (34.25 million ha) by 1996 (SEPA 1996a). In addition, a forest network has been established covering 16 million ha of farmland and, under China's *Agenda 21*, a further 29 million ha will be reforested by 2010, increasing forest cover to 17 per cent of total land area. In Australia, the Plantation 2020 Vision aims to treble the area of Australia's plantation estate from 1 to 3 million ha by 2020 (Plantation 2020 Vision Implementation Committee 1997).

Biodiversity

The region includes parts of three of the world's eight biogeographic divisions, namely the Palaearctic, Indo-Malayan and Oceanian realms. The region also includes the world's highest mountain system (Himalayas), the second largest rain forest complex and more than half the world's coral reefs. The Southeast Asian sub-region is noted as the centre of diversity of wild and domestic cereals and fruit species (ASEAN 1997) .

Of the 12 'mega-diverse' countries identified by McNeeley and others (1990), four are in this region, namely Australia, China, Indonesia and Malaysia. China is ranked third in the world for biodiversity with

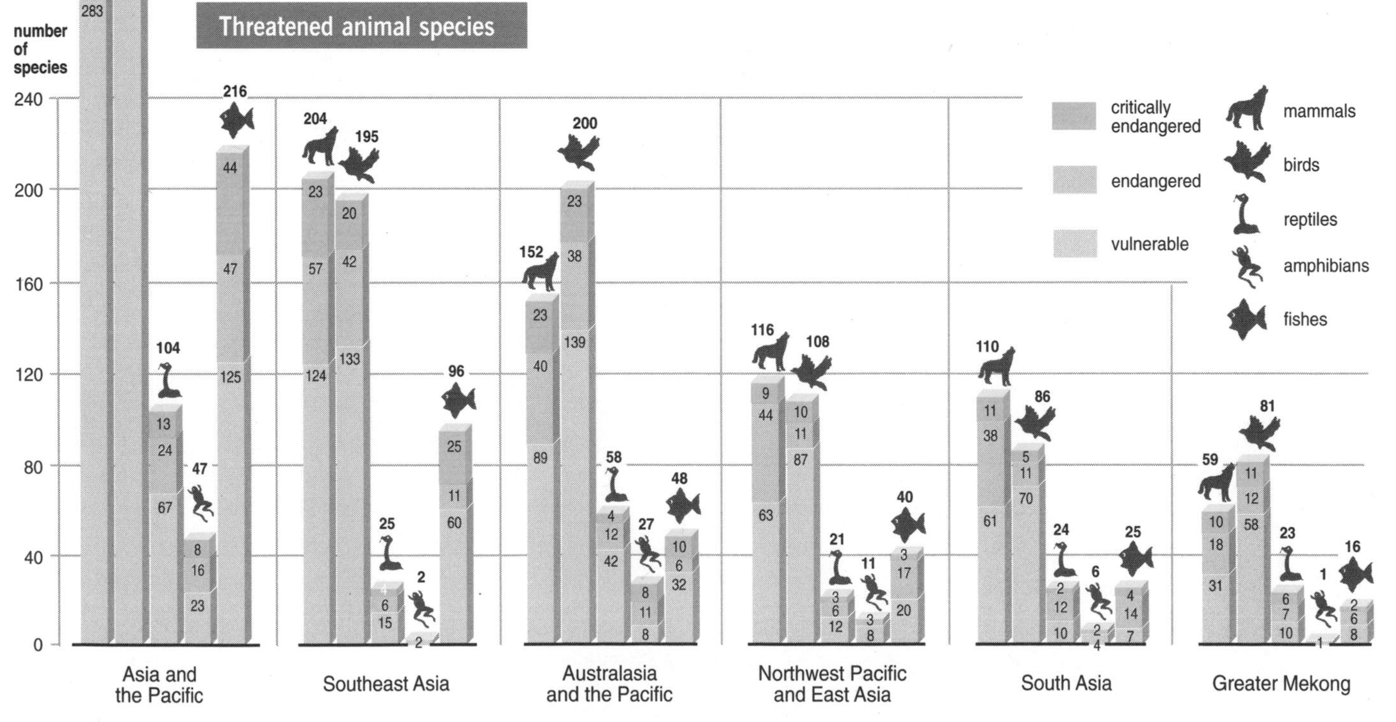

Threatened animal species

Source: WCMC/IUCN 1998

more than 30 000 species of advanced plants and 6 347 kinds of vertebrates, representing 10 and 14 per cent respectively of the world total (SEPA 1996b). Australia has an estimated one million species of which about 85 per cent of flowering plants, 84 per cent of mammals, more than 45 per cent of birds, 89 per cent of reptiles, 93 per cent of frogs and 85 per cent of inshore, temperate-zone fish are endemic (Commonwealth of Australia 1996).

In the past half century, the rich biological resources of the region have been increasingly exploited both for international trade and to sustain the growing population. The direct harvesting and export of natural products, particularly timber and fish, the expansion of agriculture into primary forests, wetlands and grasslands, and the replacement of traditional native crops with high-yielding exotic species have had severe impacts on the region's biodiversity. In addition, urbanization, industrialization, pollution, mining, tourism, introduced species, hunting, illegal trade in endangered species and the lack of proper management practices have taken their toll. In the past decade, demand on biological resources increased sharply due to rapid economic and population growth.

Increasing habitat fragmentation in Southeast Asia has depleted the wide variety of forest products that used to be the main source of food, medicine and income for indigenous people (MRC/UNEP 1997a). Destruction has been less severe in the Mekong basin, particularly along inaccessible national borders. A major concern in South Asia, particularly in the Indian subcontinent, has been the loss of biodiversity brought about by compounding the long-term pressures on grasslands with rapid growth in human and livestock populations (WCMC 1992).

Modern agriculture has also reduced genetic diversity. In Indonesia, for example, some 1 500 varieties of rice disappeared during 1975–90 (WRI, UNEP and IUCN 1992) and similar trends have been observed throughout the region. By 2005, India is expected to produce 75 per cent of its rice from just 10 varieties compared with the 30 000 varieties traditionally cultivated (Ryan 1992).

Hunting, poaching and illegal trade in endangered species have a widespread impact on biodiversity in many countries. In the Lao People's Democratic Republic and Viet Nam, for example, wildlife meat is considered a delicacy and a wide range of species are hunted for this purpose (MRC/UNEP 1997a). Poaching and the illegal harvesting and trade of medicinal plants and animals has increased both in Mongolia and the Republic of Korea from where they are exported to the lucrative black markets of Pacific neighbours (JEC 1997). In Pakistan, falcons are smuggled to the Middle East, lizards and snakes are killed for their skins, and crocodile hunting is still a popular sport and recreational activity (Government of Pakistan 1994).

Introduced species (see box) have been another significant cause of biodiversity loss, especially in Oceania. The brown tree snake, which attacked many native bird populations (see box on page 144), is a prime example.

Coastal biological resources have been depleted by commercial fishing, including poison and blast fishing. In New Zealand waters there are recent indications of destruction of seamount ecosystems by fishing trawlers seeking deep water fish, such as orange

Australia: changes to major ecosystems, 1788–1995

- About 40 per cent of all forests have been cleared
- More than 60 per cent of coastal wetlands in southern and eastern Australia have been lost
- Nearly 90 per cent of temperate woodlands and mallee have been cleared
- More than 99 per cent of temperate lowland grasslands in south-eastern Australia have been lost
- About 75 per cent of rainforests have been cleared
- Up to 85 per cent of some seagrass beds have died in recent decades

The world's worst record of mammal extinctions ...

- 10 out of 144 marsupial species and 8 out of 53 native rodent species became extinct over the past 200 years

And, what is believed to be, the current status of land animals and plants ...

- 5 per cent of higher plants
- 23 per cent of mammals
- 9 per cent of birds
- 7 per cent of reptiles
- 16 per cent of amphibians
- 9 per cent of freshwater fish

extinct, endangered or vulnerable

In addition, many species have been imported and are creating great damage. These include rabbits (approximately 200 million), foxes (5 million), cats (12 million), goats, Buffel grass, Rubber vine, Para grass, the giant sensitive plant, Siam weed, and the fungus *Phytophthora cinnamomi* — which is a pathogen threatening entire native plant communities in some areas of southern Australia. In addition at least 55 species of marine fish and invertebrates, plus several seaweeds, have been introduced, either intentionally for aquaculture or accidentally in ships' ballast water or encrusted on their hulls. These are damaging marine and coastal environments.

Source: Commonwealth of Australia (1996)

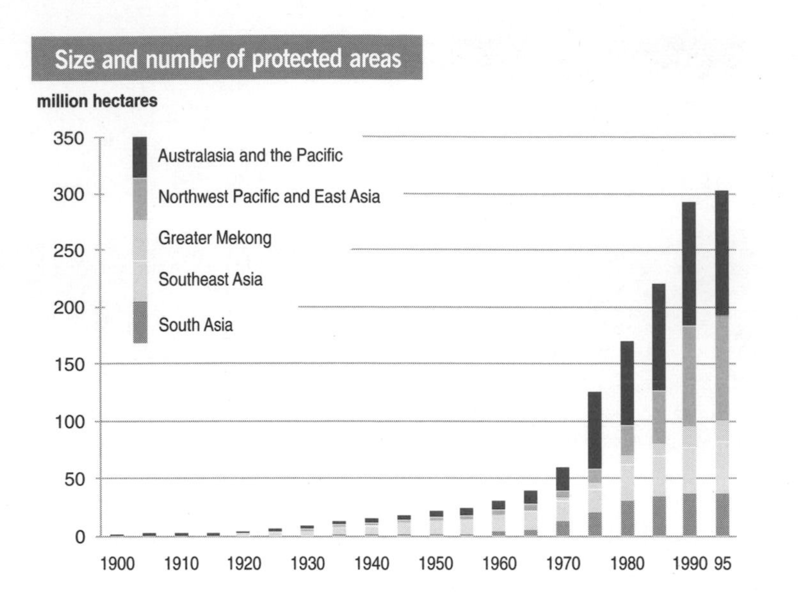

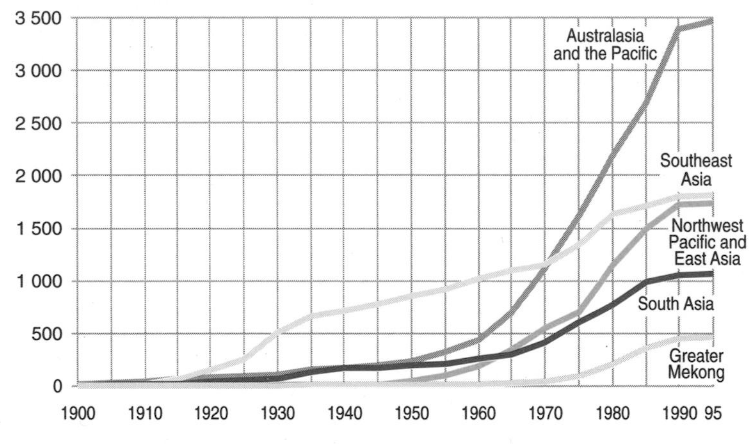

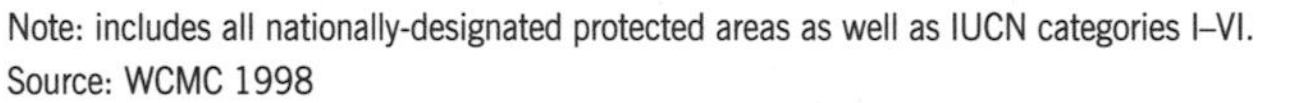
Note: includes all nationally-designated protected areas as well as IUCN categories I–VI.
Source: WCMC 1998

Australasia and the Pacific has the greatest number of protected areas but Northwest Pacific and East Asia has the largest protected area

roughy (Jones 1992, Probert 1996). In addition, pollution from shipping, in particular oil, and in some areas the discharge of toxic wastes, have adversely affected the marine environment (UNESCAP 1990 and 1995). Coastal ecosystems, particularly coral reefs and associated fish life, have been degraded by the combined effects of agricultural run-off and siltation, urban sewage, industrial pollution and, in countries such as Maldives (Government of Maldives 1994), by tourism. Mangrove destruction damages the spawning grounds for numerous aquatic species and has often triggered other forms of coastal ecosystem degradation. The Chakaria Sundarbans in eastern Bangladesh, which have been almost completely cleared for aquaculture (UNESCAP/ADB 1995), and Thailand's mangrove forests, where more than half of the total area (some 208 220 ha) disappeared between 1961 and 1993 (GESAMP 1993), are two notable examples. Land reclamation and other coastal developments have also been responsible for the destruction of wildlife habitat and some fine natural coastlines in the Republic of Korea (Government of Republic of Korea 1994).

The true extent of habitat change and species loss in the region has not yet been fully quantified because current data are inadequate or out of date (Dearden 1996, UNESCAP/ADB 1995), However, it is estimated that about two-thirds of Asian wildlife habitats have already been destroyed and 70 per cent of the major vegetation types in the Indo-Malayan realm (covering South Asia, the Mekong basin and Southeast Asia) have been lost, with a possible associated loss of up to 15 per cent of terrestrial species (Braatz 1992 and MacKinnon 1994). Dry and moist forests have suffered 73 and 69 per cent losses respectively, and wetlands, marsh and mangroves have been reduced in extent by 55 per cent (Braatz 1992). Overall habitat losses have been most acute in the Indian sub-continent, China, Viet Nam and Thailand (Braatz 1992).

The 'hot spots' (where the disappearance of already-threatened moist tropical forest would cause the greatest losses of biodiversity) include the remaining forests in Philippines, peninsular Malaysia, northwestern Borneo, the eastern Himalayas, the Western Ghats in India, southeastern Sri Lanka and New Caledonia (UNESCAP/ADB 1995).

Many species are threatened. Of the 640 species listed for protection under the Convention on International Trade in Endangered Species of Wild Fauna and Flora (CITES), 156 are found in China and around 15 to 20 per cent of the country's fauna and flora species are endangered. In Australia, approximately 5 per cent of angiosperms and 9 per cent of terrestrial vertebrates are rated either as endangered or vulnerable (Commonwealth of Australia 1996). In New Zealand, the threatened species list has continued to grow, partly because of improved knowledge and partly because some species and habitats are continuing to decline. Some 800 species and 200 sub-species of animals, plants and fungi are now threatened. One of the worst affected groups is New Zealand's endemic land and freshwater birds, three-quarters of which (37 out of 50 species) are now threatened (New Zealand Ministry for the Environment 1997).

To date, a few countries have designated more than 15 per cent of their territory as protected areas but lack of resources, weak policy enforcement, weak institutional capacity and poor interagency cooperation within the region are limiting their effectiveness. For example, the Asian Wetland Bureau has estimated that 15 per cent of all wetland habitat in South Asia is afforded some legal protection but only 10 per cent is totally protected. Furthermore, the degree of protection in South Asia is greater than in either Southeast Asia or East Asia (Samar 1994).

Few countries in the region have a complete listing of species and there is little information on most of the ecosystems. With such a generally poor understanding of the existing biodiversity, it is neither possible to assess accurately how much of it is threatened, nor what effects various human activities, in concert, may be having on particular species or ecological communities. This constitutes a major impediment to the conservation and management of biodiversity in the region.

Freshwater

Freshwater withdrawal from rivers, lakes, reservoirs, underground aquifers and other sources increased more in Asia during the past century than in other parts of the world, from 600 cubic kilometres in 1900 to approximately 5 000 cubic kilometres by the mid-1980s (da Cunha 1989). In Beijing, for example, the daily demand for water increased almost 100 times between 1950 and 1980 (WRI 1990). One consequence of this rapid rise in demand has been a proliferation of dams and reservoirs – between 1950 and 1986, the number of large dams increased from 1 562 to 22 389 (ICOLD 1984 and 1989).

Agriculture, mainly irrigation, accounts for the major part of water withdrawals. In the more industrialized countries, agriculture accounts for up to 50 per cent of withdrawals but this rises to more than 90 per cent in all South Asian countries except Bhutan, and reaches 99 per cent in Afghanistan (WRI, UNEP, UNDP and WB 1998).

In common with other parts of the world, the exploitation of water resources in the region has caused major disruption to hydrological cycles. Water development programmes for hydropower and to meet domestic and industrial requirements, coupled with deforestation in important watersheds, have reduced river levels and depleted wetlands. In addition, the mismanagement of water resources and increased irrigation has used groundwater reserves faster than they can be replenished, thus depleting aquifers and lowering water tables. Other activities, including the removal of vegetation from stream banks and flood control channelling, have changed the natural character of watercourses and estuaries.

Contamination by pollutants has also seriously degraded water quality, thereby reducing the amount of clean water available. The overall result has been to decrease the annual per capita availability of freshwater in the developing countries of the region from 10 000 m^3 in 1950 to approximately 4 200 m^3 by the early 1990s (see bar chart below).

Water availability varies greatly within the region. Within Southeast Asia alone, annual per capita internal renewable water resources range from about 172 m^3 a year in Singapore to more than 21 000 m^3 in Malaysia (WRI, UNEP, UNDP and WB 1988). Singapore is currently meeting its freshwater demands by importing some of its supply from Malaysia. In China, water resources are estimated at 2 348 m^3/capita (SEPA 1997). Supplies in India, the Islamic Republic of Iran, the Republic of Korea, Pakistan and Thailand are considerably below this, at between 1400 and 1900 a year. At the other end of the spectrum, Bhutan and Lao People's Democratic Republic have around 50 000 m^3/capita and Papua New Guinea an enormous 174 000 m^3/capita a year (WRI, UNEP, UNDP and WB 1998).

The region's water resources are under increasing pressure. Some arid countries, such as Afghanistan and the Islamic Republic of Iran, already have chronic water shortages. Most developing countries in the

Renewable freshwater resources

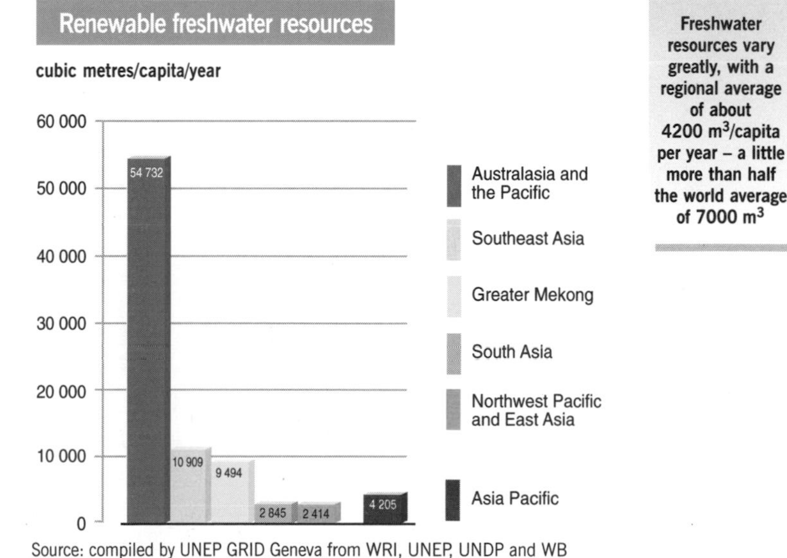

Source: compiled by UNEP GRID Geneva from WRI, UNEP, UNDP and WB 1998

Freshwater resources vary greatly, with a regional average of about 4200 m^3/capita per year – a little more than half the world average of 7000 m^3

region have experienced growing water scarcity, deteriorating water quality, and sectoral conflicts over water allocation. In many parts of the region, misuse and overexploitation of water resources has resulted in the depletion of aquifers, falling water tables, shrinking inland lakes and diminished stream flows, even to ecologically-unsafe levels. Many of the Pacific islands lack adequate water storage capacity, resulting in water scarcity despite the high rainfall. This problem was highlighted by the 1997–98 'drought' in the sub-region. Drinking water supplies in coastal areas are supplemented by the coconut.

Water quality has been steadily degraded by a combination of factors including sewage and industrial effluent, urban and agricultural run-off and saline intrusion. Whilst many water quality problems are ubiquitous, others are either localized or more common in specific parts of the region.

Levels of suspended solids in Asia's rivers almost quadrupled since the late 1970s (ADB 1997, GEMS 1996) and rivers typically contain 4 times the world average and 20 times OECD levels (GEMS 1996). Sedimentation is closely tied to erosion levels and poses critical problems for most of the Mekong River, although total suspended solids (TSS) loads are not as high as in some other Asian rivers. For example, GEMS data indicate that TSS concentration in the Mekong is approximately 294 mg per litre, compared to the Ganges River at 1 130 mg per litre (MRC/UNEP 1997a). Suspended solid levels are highest in China (ADB 1997). Sediment from erosion also continues to foul rivers in Australia (Commonwealth of Australia 1996) and New Zealand, though the removal of sheep from steep pastures is now reducing sedimentation rates in some catchments in the latter (Smith and others 1993).

Water pollution caused by organic matter, pathogenic agents, and hazardous and toxic wastes is another serious problem. Biological oxygen demand (BOD) in Asian rivers is 1.4 times the world average. BOD levels declined in the early 1980s but increased in the 1990s because of increased organic waste loading. Asia's rivers contain three times as many bacteria from human waste (faecal coliform) as the world average and more than ten times OECD guidelines (ADB 1997). The reported median faecal coliform count in Asia's rivers is 50 times higher than the WHO guidelines (ADB 1997). Faecal coliform counts in Southeast Asia are the worst in the region (ADB 1997).

Asia's record with regard to safe water supply is poor. One in three Asians has no access to a safe drinking water source that operates at least part of the day within 200 metres of the home (ADB 1997). Access to safe drinking water is worst in South and Southeast Asia. Almost one in two Asians has no access to sanitation services and only 10 per cent of sewage is treated at primary level (ADB 1997). Effluent flows straight into surface or groundwater.

Dirty water and poor sanitation cause more than 500 000 infant deaths a year in the region, as well as a huge burden of illness and disability (WHO 1992). According to WHO, diarrhoea associated with contaminated water poses the most serious threat to health in the region and the region accounted for about 40 per cent of the total global diarrhoea episodes in the under-fives during 1990.

A number of toxic pollutants also affect human health. For example, Asia's surface water contains 20 times more lead than surface waters in OECD countries, mainly from industrial effluents (ADB 1997). The worst lead contamination in the region is in Southeast Asia (ADB 1997). Bangladesh and some adjacent parts of India suffer from arsenic contamination of groundwater (see box). Dioxin contamination is becoming an emerging issue in Japan (NLA 1997).

Agrochemical inputs including fertilizers and pesticides, and animal wastes from livestock, are a growing source of freshwater pollution. Excessive levels of nitrates from agricultural run-off are a major cause of eutrophication throughout the region (UNESCAP/ADB 1995). Levels of nutrients, particularly phosphorus, remain unacceptably high in

Arsenic pollution in Bangladesh

In Bangladesh, almost 97 per cent of the population have access to drinking water through more than 4 million tubewells and regard this source as a safeguard against diarrhoeal diseases. However, high arsenic concentrations have now been identified in large numbers of the wells in rural areas. Of 20 000 tubewells tested to date, 19 per cent were found to be contaminated by unacceptable levels of arsenic (more than 0.05 mg per litre) and 2 200 cases of arsenicosis have been identified. The cause of arsenic contamination in groundwater has not yet been established with certainty but natural geological changes are presumed to be the primary cause.

Source: MoHFW 1998

Australian rivers, lakes and reservoirs (Commonwealth of Australia 1996). In New Zealand, the increase in dairying and fertilizer use is intensifying pollution in shallow lakes, streams and groundwater (Smith and others 1993). During the 1990s, freshwater resources in the Mekong basin suffered moderate to severe eutrophication. Eutrophication of surface water is also becoming a serious problem in Southeast Asia. The region as a whole had more lakes and reservoirs with eutrophication problems (54 per cent) than Europe (53 per cent), Africa (28 per cent), North America (48 per cent) and South America (41 per cent) (UNEP 1994).

Water pollution issues vary greatly. In Southeast Asia, the industrial sector is the main source of water pollution but untreated domestic wastewater as well as chemical residues and animal wastes increasingly threaten water quality in most major rivers. In the Mekong basin, organic matter, microbes and toxic metals have polluted freshwater bodies, though most water quality problems result from natural processes (MRC/UNEP 1997a). In Japan, heavy metal and toxic chemical pollution has been reduced but surface waters are affected by organic pollution (OECD 1994). In New Zealand, the number of sewage treatment plants grew from 5 to 258 during 1950–96, reducing sewage pollution (New Zealand Ministry for the Environment 1997).

Demand for water will increase throughout the region into the next century. By 2025, India is expected to be water stressed with per capita water availability decreased to some 800 m^3; China will reach the water-stress threshold before 2025 (WMO and others 1997). Southeast Asia still has adequate supplies to cope with demand over the next decade (ASEAN 1997).

While agriculture will continue to use most water, freshwater demand is growing fastest in the urban and industrial sectors. As a consequence, a major freshwater issue in many countries will be how to allocate scarce water resources among competing sectors.

Without better management practices, groundwater depletion is likely to be aggravated. At present rates of exploitation, for instance, the aquifer in Male, Maldives, is expected to be exhausted in the next few years (Government of Maldives 1994). In Mongolia, groundwater availability will be a main concern owing to the scarcity of surface water supplies.

The future quality of freshwater is one of the most pressing environmental problems in many parts of the region. Growing populations and water contamination from a wide range of sources imply reduced per capita availability in the future. The challenge will be to use dwindling supplies to satisfy a wide range of demands, and to increase national, sub-regional and regional cooperation to avoid conflict over the shared use of water resources.

Marine and coastal areas

With more coastline than any other region in the world (some Pacific islands consist of nothing more than a coastal zone), the region's rich marine resources have for long been central to its development. At the turn of the century, many countries relied almost exclusively on inland and marine fisheries as their sole source of protein and, in some cases, foreign exchange. In the latter half of the century, the rapid expansion of fisheries, coupled with population and industrial growth, resulted in increased migration to coastal cities and the expansion of coastal settlements. Today, about one-quarter of the world's 75 largest cities are situated along the region's coastline. This has resulted in increased domestic and industrial effluent, more areas of landfill, increased dredging, and the erosion of coastlines and coastal habitats. In addition, pollution from upstream, and the expansion of aquaculture production at the expense of mangrove forests, have further degraded marine and coastal resources.

In many parts of the region, economic development has been most active in coastal zones, putting enormous pressures on coastal ecosystems. For example, during the 1960s, heavy industries concentrated along Japanese coastal areas caused extreme water pollution, especially in semi-closed areas, which damaged fishery resources and resulted in red tides (JEC 1997). Strict laws and standards since the 1970s have successfully improved the quality of coastal waters, although eutrophication in areas such as Tokyo Bay is still serious despite the development of sewage systems. In addition, impoverished coastal inhabitants have exploited coastal resources with little regard for sustainability and this has resulted in the loss of critical ecosystems. Coastal erosion, resulting from increased land subsidence from groundwater extraction, and off-shore mining of sand and dredging are two other notable problems in some places. Long stretches of Australia's 70 000-km coastline are far from major population centres and among the least-polluted places on Earth. However, human activities have caused extensive losses of saltmarshes, mangroves and seagrass beds, particularly near urban areas (Commonwealth of Australia 1996).

Marine-based tourism is also increasing. In Maldives, for instance, marine-based tourism now contributes more than 19 per cent of the country's GDP and 30 per cent of government revenue (Government of Maldives 1998). It has also caused environmental degradation, particularly through the construction of hotels, beach clubs and marinas involving infilling, dredging and the resuspension of contaminated silts.

The introduction of modern fishing has transformed fisheries in the region. Marine fisheries production increased by an average of 2.9 per cent a year during 1975–95 (see graph right). By 1990, the region accounted for 38 per cent of the world marine fish catch and 8 Asia-Pacific countries were among the world's top 15 fishing nations. However, by the early 1990s, traditional marine fish stocks had reached full exploitation in many areas (FAO 1991, ASEAN 1997) and overfishing was threatening the diversity and quantity of fish (UNESCAP/ADB 1995). Stocks that have been severely affected include all the assessed fish stocks in the Northwestern Pacific off the Asian coast and Skipjack tuna and inshore stocks of demersal fish off Southeast Asia (FAO/RAPA 1994). In Australia, most major seafood species are now fully fished and some species, including the southern bluefin tuna, have been overfished (Commonwealth of Australia 1996, BRS 1997). The world's largest tuna fishery, in the Western and Central Pacific, is in relatively good health, with the exception of one species (the bigeye).

Destructive fishing techniques, pollution, sedimentation and the crown-of-thorns starfish have destroyed many reefs in East Asia

The threat to coral reefs

Source: WRI, ICLARM, WCMC and UNEP 1998

The island states of the South Pacific have so far not played a major role in offshore fishery industries, even in their own EEZs, though a number of prospects for them to participate more actively have been identified (UNEP 1985). Some nations have already managed to generate revenue from the licence fees charged to foreign commercial fishing companies for fishing rights in their EEZs. However, implementation and management issues such as surveillance and enforcement capacities still need to be addressed. More importantly, the levels of harvest and ecologically-destructive fishing technologies need to be controlled to prevent depletion of fish stocks.

Aquaculture expansion over the past decade has affected many of the region's coastal areas through habitat conversion, introduction of exotic species, increased use of chemicals (pesticides, antibiotics and hormones), and in other ways. In 1992, the region provided 87 per cent of global aquaculture produce (FAO/RAPA 1994). As already mentioned, aquaculture has been a major cause of the destruction of more than 3 million ha of Southeast Asia's mangrove forests. In the Mekong basin, mangrove forests have also been degraded drastically, both in area and quality, particularly in the southern Mekong Delta. In Viet Nam, for example, mangrove forests shrank from 400 000 ha to 252 000 ha between 1950 and 1983 (MoSTE Viet Nam 1997). The once extensive mangrove forests along the coast of Thailand were reduced from nearly 368 000 ha in 1961 to 160 000 ha in 1996 (OEPP 1998). Substantial areas of mangroves have also been lost in South Asia and southern Japan (EA 1997). The widespread felling of mangroves has reduced coastal protection from cyclones and storm surges, increased seawater intrusion and acidified surface waters (Lean and others 1990). Indirectly, it has also affected commercial demersal fisheries that rely on mangroves as nursery areas.

Aquaculture production will become increasingly important since many marine fish stocks have reached their maximum exploitative level. However, in Southeast Asia, with the exception of Cambodia, where the cutting of mangroves for charcoal production and the conversion of forest to shrimp farms are recent and growing concerns (MRC/UNEP 1997b), the rate of mangrove forest depletion and coastal environmental degradation may slow down as a result of protective measures (ASEAN 1997).

Coral reefs have come under similar pressures (see map left). In addition to mangroves, excessive land reclamation and coastal development have harmed extensive areas of coral reef in the south of Japan since 1972 when Okinawa returned to Japanese control. At the same time, the crown-of-thorns starfish has destroyed some 90 per cent of the reef habitat and prevented reef recovery (EA 1997, Mezaki 1988). In the Philippines, the combined effects of sediments, industrial and domestic pollutants, and destructive fishing techniques have damaged some 70 per cent of coral reefs. This has affected local self-sufficiency in many communities since Filipinos derive nearly half of their protein from fish (FAO 1993a). Reefs in the Mekong basin have been similarly affected. Australia has the largest area of coral reefs in the world and, by international standards, they are still in good condition. However, they are now exposed to significant pressures, with those close to population centres and tourism activities showing the most signs of damage. The Great Barrier Reef is the best-known example (see box on page 88).

The crown-of-thorns starfish has destroyed many reef habitats in the region and prevented reef recovery

Coastal and marine water pollution has increased throughout the region, mainly due to direct discharges from rivers, increased surface run-off and drainage from expanding port areas, oil spills and other

Marine fish catch

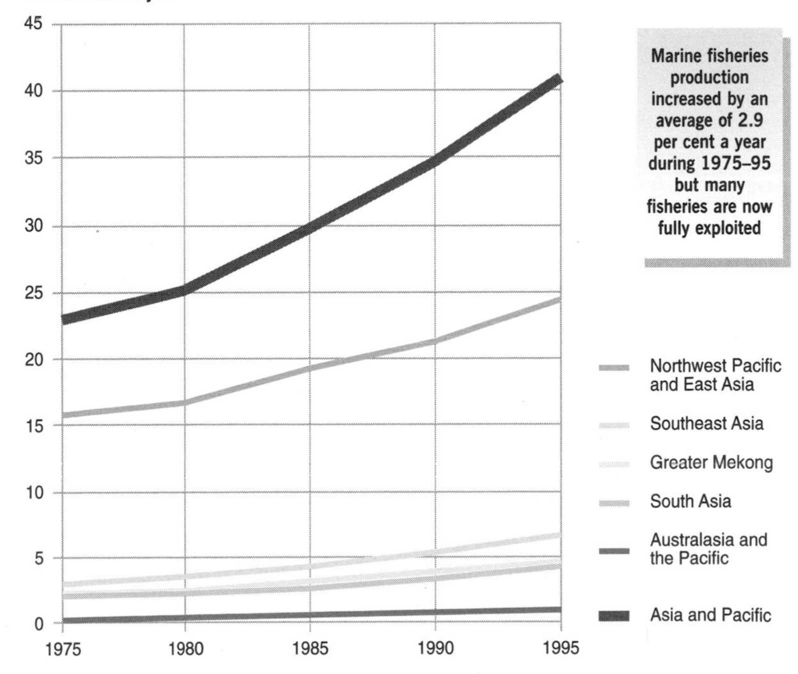

Marine fisheries production increased by an average of 2.9 per cent a year during 1975–95 but many fisheries are now fully exploited

Source: compiled by UNEP GRID Geneva from FAOSTAT 1997

contaminants from shipping, and domestic and industrial effluent. Offshore mineral exploration and production activities are further sources of pollutants.

Two-thirds of the world's total sediment transport to oceans occurs in Southeast Asia (GEMS 1996) – the combined result of active tectonics, heavy rainfall, steep slopes and erodible soils disturbed by unsound agricultural and logging practices. In Malaysia, the damage to fisheries from siltation alone exceeds the damage caused by bacterial contamination from sewage (FAO 1993b). In addition, Asian rivers are generally heavily contaminated with untreated sewage and industrial effluent.

Oil pollution is a significant problem along major shipping routes and an increasing number of accidents have occurred in recent years. The open ocean of the Sea of Japan/East Sea contains nearly twice as much oil as the surface of the Northwestern Pacific Ocean. Coastal installations are also a major source. In the port of Chittagong in Bangladesh, about 6 000 tonnes of crude oil are spilled a year and crude oil residue and wastewater effluent from land-based refineries amount to about 50 000 tonnes a year (Khan 1993).

One of the most serious issues is the decline in water quality caused by rising levels of nutrients from land-based sources. Expansion of intensive agricultural practices has resulted in increased agrochemical pollution, particularly in developing countries of the region. Fertilizer consumption rose by 340 per cent over the period 1975–95 (see figure on page 78) and pesticide use has increased fourfold since 1977 (Holmgren 1994). Inorganic nitrogen and phosphorus are the major pollutants of China's near-shores seas. In South Asia, shellfish and finfish have been contaminated by increasing pesticide pollution. Australia's marine waters are generally low in nutrients and therefore productivity. However, rising levels of nutrients is considered to be one of the most serious marine issues. Each year, Australia's sewerage systems discharge about 10 000 tonnes of phosphorus and 100 000 tonnes of nitrogen, much of which enters the sea (Commonwealth of Australia 1996).

Red tides are caused by phytoplankton blooms which deplete oxygen in coastal waters, causing the mass death of aquatic organisms. In addition, the algae may produce toxins which cause shellfish poisoning and present a serious health hazard to consumers. Red tides have become a major concern in several countries, including Philippines (UNESCAP/ADB 1995), Australia (Hallengraeff 1995), New Zealand, Japan (OECD 1994), the Republic of Korea (Government of Republic of Korea 1998) and China (Zhang and Zou 1997) and seem to be increasing in frequency. The repeated blooms of toxic algae that have killed marine life and made hundreds of people ill are considered the most dramatic development in New Zealand's coastal waters since 1990 (Chang 1993, Robertson and Murdoch 1998).

It has been suggested that changes in sea surface conditions may have contributed to the spread of toxic algae and invasive seaweeds in New Zealand waters (Chang 1993, Hawes 1994). Average temperatures in the oceans around Australasia are generally rising and there is evidence that sea level in the sub-region has risen about 2 mm a year over the past 50 years (IPCC 1998). The potential impacts of climate change and sea-level rise are considered among the greatest environmental threats for the island states in the region and particularly for the atolls of the South Pacific. The predicted changes such as increased frequency and severity of tropical cyclones, coastal

The Great Barrier Reef

The Great Barrier Reef – the largest system of coral reefs in the world – is about 2 500-km long and comprises 2 900 separate reefs and 940 islands. Its high species diversity includes more than 400 species of coral, 4 000 species of molluscs, 1 500 species of fish, 6 species of turtles, 35 species of sea birds and 23 species of sea mammals. The region is Australia's premier marine tourism destination. About two million people visit the reef and its adjacent coast annually and the number of visitors is increasing by 10 per cent each year. The combined value of tourism and fishing on the reef is estimated at US$1 000 million a year.

The Great Barrier Reef is one of the least-disturbed coral reef systems and much of it is still in a relatively good condition. The main pressures on the reef include:

- declining water quality in inshore areas, due mainly to elevated sediments and nutrients from changes in land use in coastal catchments;
- fishing (particularly trawling of the sea floor and overfishing of reef species);
- coral mortality caused by outbreaks of the crown-of-thorns starfish (the causes of which still remain unknown but which have damaged nearly 20 per cent of the reefs in the past 30 years);
- storms;
- the potential threat of oil and chemical spills, and ballast water introductions from shipping; and
- the effects of tourism.

Source: Commonwealth of Australia (1996)

inundation and flooding as well as saltwater contamination of drinking water supplies could have profound consequences for agriculture, forestry, coastal development and human health (IPCC 1998).

By the year 2000, nearly half the world's coastal population (477.3 million people) is expected to be housed in urban conglomerations along Asian shores (WRI, UNEP and UNDP 1994). It is clear that the uncontrolled exploitation of the region's coastal resources requires a more effective management system, and that there is an urgent need to move from information gathering to concrete actions aimed at management and problem-solving.

Atmosphere

In the past quarter of a century, atmospheric pollution increased significantly in much of the region, largely as a result of escalating energy consumption due to economic growth and greater use of motor vehicles. The use of poor quality fuels with a high sulphur content such as coal, inefficient methods of energy production and use, traffic congestion, poor automobile and road conditions, leaded fuel and inappropriate mining methods have exacerbated the situation. Forest fires are also contributing significantly to air pollution. Significant health threats also exist from the use of low-quality traditional solid fuels, such as wood, crop residues and dung, for cooking and heating in lower-income urban households and rural areas.

Per capita commercial energy use more than doubled in most parts of the region between 1975 and 1995 (see graph above right). In 1995, the region accounted for 26.8 per cent of the world's commercial energy consumption. While global energy consumption fell by 1 per cent per year between 1990 and 1993, Asia's energy consumption grew by 6.2 per cent a year (ADB 1997). Fossil fuels now account for about 80 per cent of energy generation in the region, with coal accounting for about 40 per cent. The region also accounted for about 41 per cent of global coal consumption in 1993 (EIA 1995).

With the increase in the use of relatively high carbon content fuels such as coal and oil, emissions of CO_2 also increased fast – at twice the average world rate of 2.6 per cent a year during 1975-95 (CDIAC 1998). Since the 1970s, industrial emissions of CO_2 have grown 60 per cent faster in Asia than anywhere else (ADB 1997). China and Japan are the first and second largest CO_2 emitters respectively in the region

Annual commercial energy consumption per capita

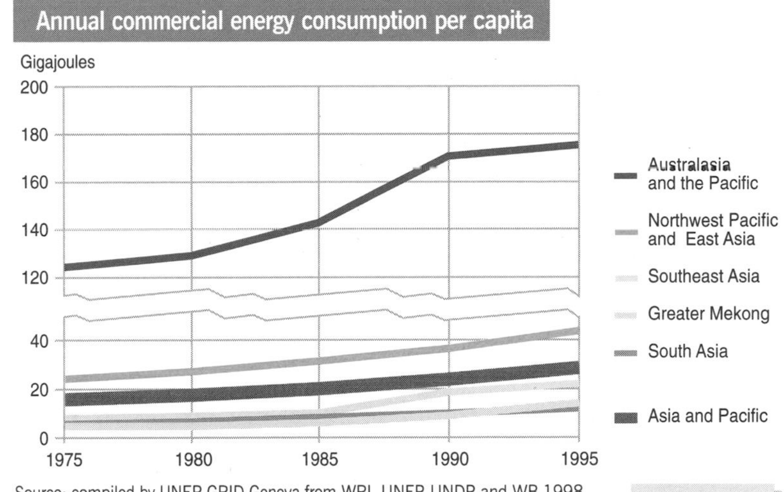

Source: compiled by UNEP GRID Geneva from WRI, UNEP, UNDP and WB 1998

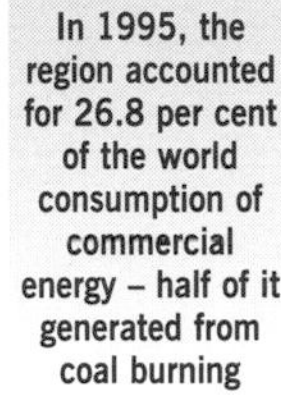
In 1995, the region accounted for 26.8 per cent of the world consumption of commercial energy – half of it generated from coal burning

(WRI, UNEP, UNDP and WB 1998). However, CO_2 emissions per capita are low, little more than than half the world average (see graph below) and only 11.2 per cent of the level in North America in 1995. Past land clearing has also contributed a significant proportion of CO_2 emissions in some countries.

Sulphur dioxide emissions in Asia increased from 11.25 million tonnes of sulphur equivalent in 1970 to 20 million tonnes in 1986 – at least four times the rate of any other region (Hameed and Dignon 1992). Nitrogen oxide (NO_x) emissions from fossil fuel combustion increased by about 70 per cent (Hameed and Dignon 1992). However, total emissions were significantly less than those of North America and Europe during the same period.

While per capita emissions of carbon dioxide are little more than half the world average, they grew twice as fast as the world average during 1975–95

Carbon dioxide emissions per capita

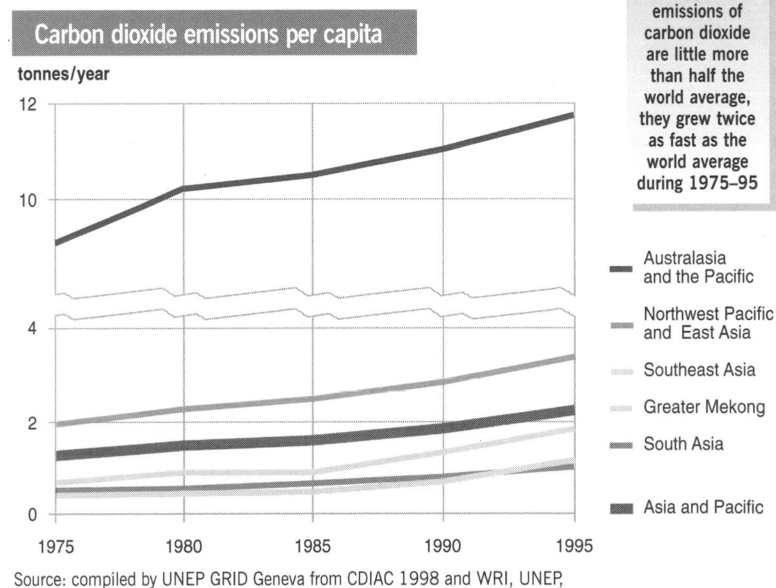

Source: compiled by UNEP GRID Geneva from CDIAC 1998 and WRI, UNEP, UNDP and WB 1998

The severity of air pollution varies considerably across Asia. Even major urban areas of Australia, where the concentration of air pollutants is generally low, occasionally experience levels of pollution that exceed air quality goals (Commonwealth of Australia 1996, NSW EPA 1997).

Two of Asia's giant economies, China and India, rely heavily on coal. Ninety per cent of China's 18 million tonnes of SO_2 emitted into the atmosphere annually come from coal burning (State Planning Commission 1997). Overall, Asian emissions of SO_2 are at least 50 per cent higher than those of North America, Africa and Latin America (ADB 1997). Three of Asia's 11 megacities (see table opposite) exceed WHO guidelines for acceptable SO_2 levels (WHO and UNEP 1992).

With increasing SO_2 emissions, acidification is an emerging issue. The most sensitive areas are in south China, the southeast of Thailand, Cambodia and south Viet Nam (Hettelingh and others 1995). On the other hand, there is no evidence of significant acid deposition in Australia, which is not subjected to emissions from neighbouring countries and where fossil fuels have a low sulphur content (Commonwealth of Australia 1996).

Transportation contributes the largest share of air pollutants to the urban environment. The total number of registered vehicles in the region in 1996 was about 127 million, 4.24 per cent more than in the previous year (International Road Federation 1997). In Seoul, car ownership doubled in a single year between 1991 and 1992 (Ministry of Environment, Republic of Korea, 1990 and 1995). Lead pollution is a particular problem in megacities of Southeast Asia. The introduction of unleaded fuels is reducing average lead levels, although the rate of decline is slower in Asia than elsewhere.

Ten of Asia's 11 megacities exceed WHO guidelines for particulate matter by a factor of at least three (WHO and UNEP 1992). Levels of smoke and dust, a major cause of respiratory diseases, are generally twice the world average and more than five times as high as in industrial countries and Latin America (ADB 1997). Recent forest fires in Indonesia are a further, notorious source of air particulates (see box below).

Recent studies show that smoke and dust particles can significantly damage human health. According to WHO estimates, Bangladesh, India, Nepal and Indonesia together account for about 40 per cent of the global mortality in young children caused by pneumonia (WHO 1993). In China, smoke and small particles from burning coal cause more than 50 000 premature deaths and 400 000 new cases of chronic bronchitis a year in 11 of its large cities (World Bank 1997a). The negative impacts of domestic burning of solid fuels are not confined to developing countries. Winter air pollution, mostly from coal and wood-burning fires in private homes, is a persistent problem in New Zealand (New Zealand Ministry for the Environment 1997).

Some countries have managed to gain partial control over air quality deterioration. During the past

Smoke haze over Indonesia on 19 October 1997

Source NASA 1997

Indonesian forest fires and air pollution

The Indonesian forest fires that began to burn in September 1997 in Kalimantan and Sumatra have greatly increased pollution levels in Southeast Asia, releasing an estimated 110-180 million tonnes of CO_2 to the atmosphere (*Bangkok Post*, 27 September 1997). The area affected by both CO_2 and other air pollutants from the fire spread east-west for more than 3 200 km, covering six Southeast Asian countries and affecting perhaps 70 million people. Smoke reached as far south as Darwin, Australia. Peak levels of particulates in Kuala Lumpur, Singapore, and many Indonesian cities exceeded 6 000 $\mu g/m^3$ (World Bank 1997b). The air pollution index (API), which is a measure of SO_2, NO_2, CO, ozone and dust particles, reached a critical level of 288 $\mu g/m^3$ on 26 September 1997 in Betong district in Southern Thailand. In the Malaysian state of Sarawak, the API hit a record 839 $\mu g/m^3$ on 23 September 1997. Levels of 100–200 $\mu g/m^3$ are considered 'unhealthy'; levels of more than 300 are equivalent to smoking 80 cigarettes a day and are 'hazardous'.

two decades, Japan successfully reduced emissions of SO_2, NO_x and CO through technological innovation, institutional development, and cooperation by all levels of government and industry. SO_2 emissions, for instance, decreased nearly 40 per cent between 1974 and 1987 (WRI, UNEP and UNDP 1992). Similar air pollution problems in the Republic of Korea have been reduced since the 1980s by increasing the use of low-sulphur oil and liquified natural gas (Government of Republic of Korea 1998).

Demand for primary energy in Asia is expected to double every 12 years while the world average is every 28 years. High carbon-content fuels are likely to continue to dominate the region's energy market. Coal will remain the fuel of choice throughout much of the region, because of its abundance and easy availability, especially in China, India and Mongolia, and demand is projected to increase by 6.5 per cent a year (World Bank 1997c).

By the year 2000, SO_2 emissions from coal burning in Asia are expected to surpass the emissions of North America and Europe combined (World Bank 1997a) and, if current trends in economic development continue without effective SO_2 control measures, will more than triple within the next 12 years. This is likely to result in a significant increase in acid deposition problems, especially within East Asia. The Korean peninsula will be seriously affected by cross-border acid rain. Mongolia may receive acid rain from its north-western border with Russia. In addition, urban air pollution will be aggravated by increasing emissions from transport. A study of Nepal, for instance, estimates that total emissions will increase fivefold by 2013, about two-thirds of which are likely to come from the transport sector (Shrestha and others 1996).

While the region's contribution to the greenhouse effect and total world emissions of atmospheric pollutants are currently limited, both are increasing fast. Air quality is proving detrimental to human health in many parts of the region. These trends are likely to continue.

Urban areas

Levels of urbanization in the region are relatively low. Some 23.6 per cent of Asians lived in urban areas in 1975, increasing to 34 per cent in 1995 – less than half the values found in North America, Europe and Latin America. However, with rapid economic development, particularly over the past thirty years, urban populations have increased fast with most of the urban population concentrated in a few cities. The impacts of rapid urbanization include encroachment on agricultural and forest lands, urban air and water pollution (and associated diseases), unavailability of safe drinking water and the overexploitation of groundwater causing urban land settlement and subsidence, seawater intrusion, increasing traffic congestion, noise pollution and significant increases in solid municipal and industrial wastes.

Asia's urban population was slightly more than 1067 million in 1995 (see bar chart on page 92), having grown at an average annual rate of 3.2 per cent during 1990–95, compared with just 0.8 per cent growth in rural populations. Of the 369 cities in the world with more than 750 000 residents, 160 are in Asia and the Pacific, compared to 79 in Europe, 64 in North America, 35 in Africa and 31 in South America. In 1994, 9 of the world's 14 largest urban centres (megacities with more than 10 million residents) were in the Asia-Pacific region, including the largest, Tokyo.

Over the past decade, urban growth rates ranged from 0.4 per cent in Japan to 7.5 per cent in Afghanistan (United Nations Population Division 1997). India and Pakistan are home to the largest and fastest growing cities in the sub-region; Karachi and Mumbai are growing at 4.2 per cent per annum, followed by Delhi at 3.8 per cent per annum.

The urban population in Southeast Asia tends to

Air quality in 11 megacities

City	SO_2	SPM	Lead	CO
Bangkok	●	●●●	●●	●
Beijing	●●●	●●●	●	●
Calcutta	●	●●●	●	●
Delhi	●	●●●	●	●
Jakarta	●	●●●	●●	●●
Karachi	●	●●●	●●●	●
Manila	●	●●●	●●	●
Mumbai	●	●●●	●	●
Seoul	●●●	●●●	●	●
Shanghai	●●	●●●	●	●
Tokyo	●	●	●	●

●●● Serious problem. WHO guidelines exceeded by more than 100 per cent
●● Moderate to heavy pollution. WHO guidelines exceeded by up to 100 per cent
● Low pollution. WHO guidelines are normally met or may be exceeded from time to time by a small amount

Source: WHO and UNEP 1992

be concentrated in the highly industrialized capitals (ASEAN 1997). Some capitals grew at a phenomenal speed: Jakarta, for example, grew to 8 million residents in 15 years, one-tenth the time it took New York City to reach the same population (UNESCAP/ ADB 1995). Urban growth rates in the sub-region may now be down to 3.5 per cent (ASEAN 1997).

In East Asia, China was one of the first countries to establish large cities. In Japan, 50.3 per cent of the total population lived in urban areas by 1950, increasing to 78 per cent in 1996 (World Bank 1998). In Mongolia, the urban population grew from 21 per cent in 1956 to 54 per cent by 1994, with 27.5 per cent of the Mongolian population concentrated in the capital city, Ulaanbaatar (State Statistical Office of Mongolia 1996). In the Republic of Korea, Seoul's population increased tenfold during 1950–90 and now contains more than one-quarter of the country's total population. China's urban population rose from 192 million to 377 million between 1980 and 1996, making it the country with the highest total urban population in the world (World Bank 1998). The most phenomenal increase in city dwellers occurred in the Democratic People's Republic of Korea after the end of the Korean War (1950–53); the urban population increased from 17.7 per cent in 1953 to 61.2 per cent in 1995 (United Nations Population Division 1997)).

Urban population

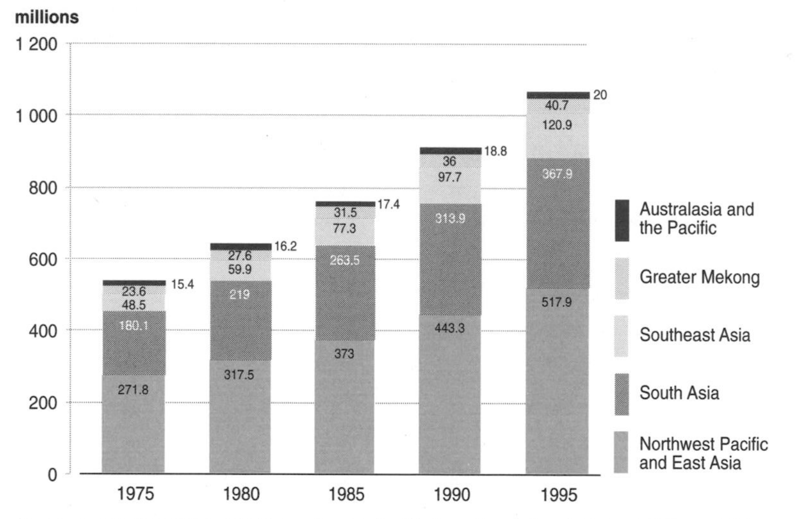

Source: compiled by UNEP GRID Geneva from United Nations Population Division 1997 and WRI, UNEP, UNDP and WB 1998

Asia's urban population was slightly more than 1067 million in 1995, having grown at an average annual rate of 3.2 per cent during 1990–95

Slums are growing in many cities. In Colombo, for instance, some 50 per cent of the population resides in slums and squatter areas (Government of Sri Lanka 1994) whilst a similar percentage applies to the entire urban populations in Indonesia and Bangladesh. The Republic of Korea is an exception with a remarkably low proportion of about 1 per cent (UNESCAP 1993).

Except in a few countries, including Mongolia (MNE 1996) and the Pacific Island states, traffic congestion is another serious problem, causing air pollution and extending travel times, thereby affecting human health and incurring economic losses through delays.

Growing populations have frequently outpaced the development of urban infrastructure. Access to safe drinking water in urban areas ranges from 35 per cent in Indonesia to 66 per cent in Nepal to 100 per cent in Maldives and Singapore (ASEAN 1997, WRI, UNEP, UNDP and WB 1996, and Government of Maldives 1998). Access to sanitation ranges from 62 per cent in Pakistan to 100 per cent in Maldives (WRI, UNEP, UNDP and WB 1996, Government of Maldives 1998, ASEAN 1997). In China, more than 300 cities have experienced water shortages (State Planning Commission 1995) and only 20 per cent of urban sewage receives concentrated treatment (State Statistical Bureau 1997). In the face of intensifying urbanization, Cambodia and the Lao People's Democratic Republic have particularly severe problems with limited infrastructure and resources (MRC/UNEP 1997a).

Some parts of East Asia provide a good level of services to urban residents. For instance all urban populations in Japan have access to health care, safe water and sanitation (World Bank 1997a).

All urban areas in Australia and New Zealand have adequate transport infrastructure, sewerage systems, stormwater drainage systems, piped water, electricity and waste disposal services, but in recent years these have come under strain – especially in the largest cities (Commonwealth of Australia 1996). For example, Auckland has experienced water shortages, floods and power cuts (New Zealand Ministry for the Environment 1997).

There are approximately 270 000 indigenous Australians. They comprise about 18 per cent of Australians living in remote settlements but less than 2 per cent of those living in cities. The quality of drinking water in remote communities is generally poorer than in metropolitan areas, reflecting poor water treatment and poor source-water quality and availability. On almost every health measure, indigenous Australians suffer poorer health than other Australians; death rates, for example, are two to four times those of the total Australian population (Commonwealth of Australia 1996).

The total waste generated in the region amounts to 2 600 million tonnes a year, of which solid waste accounts for 700 million tonnes and industrial activities generate 1 900 million tonnes (UNESCAP/ADB 1995). The East Asia sub-region generated 46 per cent (327 million tonnes) of the region's total municipal solid waste in 1992; this proportion is projected to increase to 60 per cent by 2010 (UNESCAP/ADB 1995). The Republic of Korea produced a 50 per cent increase in industrial waste in the period 1991–95 alone (Government of Republic of Korea 1998). In New Zealand, many of the country's estimated 7 800 contaminated sites are in urban industrial areas (New Zealand Ministry for the Environment 1997).

A large percentage of industrial wastes in Southeast Asia, including hazardous chemicals, are discharged without treatment. These wastes affect not only the health of workers who handle them but also residents living near factories. However, many countries now have effective legislation for the safe handling, treatment and disposal of these substances (ASEAN 1997).

Many urban waste disposal systems are inadequate. Disposal of untreated wastewater is spreading water-borne diseases and damaging marine and aquatic life. In response, investment in domestic wastewater treatment systems has been accelerated in many Southeast Asian countries, including Malaysia. High rates of urbanization in the island states of the South Pacific has also resulted in serious waste management and pollution problems, particularly with respect to their impacts on groundwater resources. Environmentally-safe disposal of solid waste and sewage is a major concern for the island states of the region where land and therefore available disposal sites are limited and sewerage systems are lacking.

In most countries, the urban population is likely to grow threefold in the next 40 years (UNESCAP/ADB 1995). China alone is expected to have 832 million urban residents by 2025.

As urban areas, especially megacities, expand further, increases in traffic congestion, water and air pollution, and slums and squatters settlements can be expected. Most large Asian cities already face an acute shortage of safe drinking water and a fivefold increase in demand is anticipated within the next 40 years (UNESCAP/ADB 1995). Public expenditure on water and sanitation is around one per cent of GDP for most countries of the region, and is likely to rise.

In East Asia, many governments are attempting to reduce the growth of their primary cities by curbing rural-urban migration. A new trend for Chinese cities is represented by Dalian, Zhuhai and Xiamen, Zhangjiagang, Shenzhen and Weihai, the Environmental Star Cities, where great efforts are being made to emphasize urban environmental planning and pollution prevention amid economic development (SEPA 1998).

Urbanization is one of the most significant issues facing Asia and the Pacific. How to deal with increasing amounts of urban and industrial waste is a major concern for most of the region. While the proportion of people living in urban centres is still lower than that in developed countries, it is rising rapidly, and is focused on a few urban centres.

References

ADB (1994). *The Environment Program: Past, Present and Future.* Asian Development Bank, Manila, Philippines

ADB (1995). *Key Indicators for Developing Asian and Pacific Countries.* Asian Development Bank, Manila, Philippines

ADB (1997). *Emerging Asia: Changes and Challenges.* Asian Development Bank, Manila, Philippines

ASEAN (1997). *First ASEAN State of the Environment Report.* ASEAN Secretariat, Jakarta, Indonesia

ASSOD (1997). *ASSOD: The New Assessment of Soil Degradation in South and South-East Asia.* ISRIC, Wageningen, the Netherlands

Braatz, S. (1992). *Conserving Biological Diversity. A Strategy for Protected Areas in the Asia-Pacific Region*. World Bank Technical Paper No. 193. World Bank, Washington DC, United States

BRS (1997). *Status of Fisheries Reports 1997*. Resource Assessments of Australian Commonwealth Fisheries. Bureau of Resource Sciences, DPIE, Canberra, Australia

CDIAC (1998). *Revised Regional CO_2 Emissions from Fossil-Fuel Burning, Cement Manufacture, and Gas Flaring: 1751–1995.* Carbon Dioxide Information Analysis Center, Environmental Sciences Division, Oak Ridge, Tennessee, United States. http://cdiac.esd.ornl.gov/cdiac/home.html

Chang, F. Hoe. (1993). The early 1993 shellfish poisonings and toxic algal blooms in New Zealand – an update. *Water and Atmosphere,* 1, 4, 8-9

Commonwealth of Australia (1996). *Australia: State of the Environment 1996*. State of the Environment Advisory Council and Department of the Environment, Sport and Territories. CSIRO Publishing, Collingwood, Australia

Commonwealth of Australia (1998). *1998 Yearbook Australia*. Australian Bureau of Statistics, Canberra, Australia

CRED (1991). Centre for Research on the Epidemiology of Disasters, Disaster Events Database, *CRED Disasters in the World*, November 1991, Brussels, Belgium

CRED (1993). Centre for Research on the Epidemiology of Disasters, Disaster Events Database, Disaster Ranking over 25 years. *CRED Bulletin*, January 1993, Brussels, Belgium

da Cunha, L.V. (1989). *Sustainable Development of Water Resources*, International Symposium on Integrated Approach to Water Pollution, Lisbon, Portugal

Dearden, P. (1996). Biodiversity in the Highlands of Northern Thailand: Some Research Approaches. In *Biodiversity in Asia: Challenges and Opportunities for the Scientific Community*, Proceedings of a Conference on Prospects of Cooperation on Biodiversity Activities, Chiang Rai, Thailand, 15–19 January 1996

DAI (1995). *East Asia Country Environmental Profiles*. Development Alternatives, Inc. for USAID Regional Support Mission for East Asia, Bangkok, Thailand

Dutta, K., and Rao J. M. (1996). Growth, distribution and environment: sustainable development in India. *World Development*, 24, 2, 287-305

EA (1997). *Quality of the Environment in Japan 1997.* Japanese Environment Agency, Tokyo, Japan

EEPSEA/WWF (1998). *Haze damage from 1997 Indonesian fires exceeds US$1.3 billion.* Press Release by the World Wide Fund for Nature (WWF) Indonesia Programme and the Economy and Environment Program for Southeast Asia (EEPSEA), 24 February 1998 http://www.geocities.com/RainForest/2701/eepsea1.htm

EIA (1995). *International Energy Annual: 1993*. Energy Information Agency, US Department of Energy, Washington DC, United States

FAO (1991). *Recent Developments in World Fisheries*. FAO, Rome, Italy

FAO (1993a). *Marine Fisheries and the Law of the Sea: A Decade of Change*. FAO Fisheries Circular No. 853. FAO, Rome, Italy

FAO (1993b). Indo-Pacific Fisheries Commission: Papers Presented at the Seventh Session of the Standing Committee on Resources Research and Development. Bangkok, Thailand

FAO (1997a). *State of the World's Forest 1997*. FAO, Rome, Italy

FAO (1997b). *Provisional Outlook to 2010*. FAO, Rome, Italy

FAO/RAPA (1994). *Selected Indicators of Food and Agricultural Development in Asia and the Pacific Region, 1983–93*. Publication 1994/24. FAO/RAPA, Bangkok, Thailand

FAOSTAT (1997). *FAOSTAT Statistics Database.* FAO, Rome, Italy. http://www.fao.org

FAO, UNDP and UNEP (1994). *Land Degradation in South Asia: its severity, causes and effects upon the people*. World Soil Resources Report No. 78. FAO, Rome, Italy

Gadgil, M. and Guha, R. (1992). *Ecological History of India*. Oxford Universty Press, Delhi, India

GEMS (1996). *Annotated Digital Atlas of Global Water Quality.* GEMS Water Collaborating Center, Ontario, Canada. Available on diskette and at http://www.cciw.ca/gems/intro.html

GESAMP (1993). Impact of oil and related chemicals and wastes on the marine environment. GESAMP Reports and Studies No. 50. IMO, London, United Kingdom

Government of China (1994). *China's Agenda 21*. Beijing, China

Government of Maldives (1994). *State of the Environment Maldives 1994.* Ministry of Planning, Human Resources and the Environment, Male, Republic of Maldives

Government of Maldives (1998). *Statistical Year Book of Maldives 1997*. Male, Republic of Maldives

Government of Pakistan (1994). *Country Report on State of Environment in Pakistan,* presented at the Regional Meeting on the State of the Environment in Asia and Pacific, Myanmar

Government of Republic of Korea (1994). *Environmental Protection in Korea*. Ministry of Environment, Kwacheon, Republic of Korea

Government of Republic of Korea (1998). *Environmental Protection in Korea 1997*. Ministry of Environment, Kwacheon, Republic of Korea

Government of Sri Lanka (1994). *State of the Environment of Sri Lanka*. Ministry of Environment and Parliamentary Affairs, Colombo, Sri Lanka

Government of Western Samoa (1994). *Western Samoa: National Environment and Development Management Strategies.* SPREP, Apia, Western Samoa

Hallengraeff, G.M. (1983). Marine phytoplankton communities in the Australian region: current status and future threats. In L. Zann (ed.), *State of the Marine Environment for Australia: Technical Annexe 1– The Marine Environment.* Great Barrier Reef Marine Park Authority, for the Department of the Environment, Sport and Territories, Ocean Rescue 2000 Program, Townsville, Australia

Hameed, S. and Dignon, J. (1992). Global Emissions of Nitrogen and Sulfur Oxides in Fossil Fuel Combustion 1970–86. *J. Air Waste Management Assoc.,* 42, 159-63

Hawes, I. (1994). Sea lettuce; a 'blooming' nuisance. *Water and Atmosphere,* 2, 4, 20–22

Hettelingh, J. P., Chadwick M., Sverdrup H. and Zhao D., (1995). Chapter 6 in *RAINS-ASIA: An Assessment Model for Acid Deposition in Asia*. The World Bank, Washington DC, United States

Holmgren S. (1994). *An environmental assessment of the Bay of Bengal region*. Bay of Bengal Programme, BOPG/REP/67. BOBP, Madras, India

ICOLD (1984). *World Register of Dams – 1984.* Central Office, International Commission on Large Dams, Paris, France

ICOLD (1989). *World Register of Dams – 1988 update.* Central Office, International Commission on Large Dams, Paris, France

International Road Federation (1997). *World Road Statistics* 1997 Edition. IRF, Geneva, Switzerland, and Washington DC, United States

IPCC (1998). *The Regional Impacts of Climate Change: An Assessment of Vulnerability. A special report of IPCC Working Group II.* Intergovernmental Panel on Climate Change. Cambridge University Press, Cambridge, United Kingdom

IUCN (1986). *Review of Protected Area Systems in the Indo-Malayan Realm.* IUCN, Gland, Switzerland

JEC (1997). *Asian Environmental Report 1997/ 98.* Japan Environmental Council, Toyoshinsya, Japan

Jones, J. B. (1992). Environmental impact of trawling on the seabed: a review. *New Zealand Journal of Marine and Freshwater Research*, 26, 59-67

Khan, M. A. (1993). *Problems and Prospect of Sustainable Management of Urban Water Bodies in the Asia and Pacific Region*. Bangkok, Thailand

Kummer, D. M. (1993). *Trends in Land Use and its Impact: an Attempt at Sub-global Explanation, in Rural Land Use in Asia and Pacific*, a report of an APO symposium, 29 September–6 October 1992. Asian Productivity Organization, Tokyo, Japan

Lean, G., Hinrichsen D. and Markham A. (1990). *Atlas of the Environment*. WWF and Arrow Books, London, United Kingdom

MacKinnon, J. (1994). Analytical Status Report of Biological Conservation in Asia-Pacific Region. In *Biodiversity Conservation in Asia-Pacific: Constraints and Opportunities*, Proceedings of a Regional Conference

McNeeley, J. A., Miller, K. R., Reid, W. V., Mittermeier, R. A., and Werner, T. B. (1990). *Conserving the World's Biological Diversity.* WRI, World Conservation Union, World Bank, WWF-US and Conservation International, Washington DC, United States, and Gland, Switzerland

Mezaki, S. (1988). *Sango no Umi.* Koubunken (in Japanese)

Ministry of Environment, Republic of Korea (1990 and 1995). *Environment Statistical Yearbooks 1990 and 1995.* Ministry of Environment, Kwacheon, Republic of Korea

MNE (1996). *Nature and Environment in Mongolia*. Ministry of Nature and Environment, Ulaanbaatar, Mongolia (in Russian)

MoHFW (1998). *Arsenic Contamination Mitigation Project.* Ministry of Health and Family Welfare, Bangladesh

MoSTE Thailand (1997). *Thailand's Action for Sustainable Development.*Thailand's Country Report to the UN Commission on Sustainable Development. Ministry of Science, Technology and Environment, Bangkok, Thailand

MoSTE Vietnam (1997). *State of the Environment Report of Viet Nam 1994.* Ministry of Science, Technology and Environment, Hanoi, Viet Nam

MRC/UNEP (1997a). *Mekong River Basin Diagnostic Study: Final Report.* Mekong River Commission, Bangkok, Thailand

MRC/UNEP (1997b). *Greater Mekong Sub-region: State of the Environment Report*. Mekong River Commission, Bangkok, Thailand

NLA (1997). *Water Resources in Japan 1997.* National Land Agency, Japan (in Japanese)

NASA (1997). http://jwocky.gsfc.nasa.gov/uvaer/indonesia/indo292.gif

New Zealand Ministry for the Environment (1997). *The State of New Zealand's Environment 1997*. GP Publications, Wellington, New Zealand

NSW EPA (1997). *New South Wales State of the Environment 1997.* NSW Environment Protection Authority, Sydney, Australia

OECD (1994). *OECD Environmental Performance Reviews: Japan*. OECD, Paris, France

OECD (1997). *OECD Environmental Performance Reviews: Korea Republic*. OECD, Paris, France

OEPP (1998). *Thailand's State of the Environment Report 1995–96.* Office of Environmental Policy and Planning, Bangkok, Thailand

Plantation 2020 Vision Implementation Committee (1997). *Plantation for Australia: The 2020 Vision*. Ministerial Council on Forestry, Fisheries and Aquaculture; Standing Committee on Forestry; Plantations Australia; Australian Forest Growers; and National Association of Forest Industries. MCFFA/SCF Secretariat, Department of Primary Industries and Energy, Canberra, Australia

Postel, S. (1989). *Water for Agriculture: facing the limits.* Worldwatch Paper 93. Worldwatch Institute, Washington DC, United States

Probert, P. K. (1996). Trawling the depths: Deep-sea fishing off New Zealand may have long-lasting effects. *New Zealand Science Monthly*, 7, 9610

Robertson, D. and Murdoch, R. (1998). *Fish kills off Kaikoura linked to toxic algae.* Media Release, 2 March 1998. National Institute of Water and Atmospheric Research (NIWA), Wellington, New Zealand

Roper, B. (1993). The end of the golden weather: New Zealand's economic crisis. In Roper, B. and Rudd, C. (eds.). *State and Economy in New Zealand*. Oxford University Press, Auckland, New Zealand

Ryan, J. C. (1992). Conserving Biological Diversity. In: L. Brown and others. *State of the World 1992*. W. W. Norton, New York, United States

Samar, S. (1994). The Biological Value of the Asia-Pacific Region. In *Biodiversity Conservation in Asia-Pacific: Constraints and Opportunities*, Proceedings of a Regional Conference

SCARM (1998). *Sustainable Agriculture. Assessing Australia's Recent Performance*. A report to the Standing Committee on Agriculture and Resource Management of the National Collaborative Project on Indicators for Sustainable Agriculture. SCARM Technical Report 70. CSIRO Publishing, Collingwood, Australia

SEPA (1996a). *Report on the State of the Environment in China.* State Environmental Protection Administration of China, Beijing, China

SEPA (1996b). *Country Study Report on Biodiversity in China.* State Environmental Protection Administration of China, Beijing, China

SEPA (1997). *National Report on Sustainable Development, 1997.* State Environmental Protection Administration of China, Beijing, China

SEPA (1998). *Report on the State of the Environment in China 1997.* State Environmental Protection Administration of China, China Environmental Science Press, Beijing, China

Shrestha, R. M. and Malla, S. (1996). Air pollution from energy use in a developing country city: the case of Kathmandu Valley, Nepal. *Energy - the international journal*, 21, 9, 785-794

Smith, C. M., Wilcock, R. J., Vant, W. N., Smith, D. G. and Cooper, A. B. (1993). *Towards Sustainable Agriculture in New Zealand: Freshwater Quality in New Zealand and the Influence of Agriculture.* MAF Policy Technical paper 93/10. Ministry of Agriculture and Fisheries, Wellington, New Zealand

SPREP (1993). *Taule'alo, Tu'u'u Itei. Western Samoa SOE Report,* SPREP, Apia, Western Samoa

State Planning Commission (1995). *Report on China's Population, Resources and Environment.* China Environmental Sciences Press, Beijing, China

State Planning Commission (1997). *China's Energy Development Report.* Economic Management Press, Beijing, China

State Statistical Bureau (various years). *China Statistical Yearbooks.* China Statistical Publishing House, Beijing, China

State Statistical Office of Mongolia (1996). Mongolian Economy and Society in 1995. Ullanbaatar. Mongolia

Uhlig, H. (ed., 1984). *Spontaneous and Planned Settlement in Southeast Asia.* Institute of Asian Affairs, Hamburg, Germany

UNCTAD (1994). *UNCTAD Commodity Year Book.* UNCTAD, New York, United States

UNDP (1997). *Human Development Report 1997.* Oxford University Press, New York, United States, and Oxford, United Kingdom

UNEP (1985). *Environment and Resources in the Pacific.* Dahl, A. L. and Carew-Reid, J. (eds.). *UNEP Regional Seas Reports and Studies* No. 69. UNEP, Nairobi, Kenya

UNEP (1992). *Marine Pollution from Land-based Sources: Facts and Figures.* UNEP Industry and Environment, Paris, France

UNEP (1994). *The Pollution of Lakes and Reservoirs.* UNEP Environment Library No. 12. UNEP, Kenya, Nairobi

UNEP (1997). *World Atlas of Desertification.* Second Edition. Arnold, London, United Kingdom

UNEP/ISRIC (1991). *World Map of the Status of Human-Induced Soil Degradation (GLASOD). An Explanatory Note,* second revised edition (edited by Oldeman, L.R., Hakkeling, R.T., and Sombroek, W.G.). UNEP, Nairobi, Kenya, and ISRIC, Wageningen, Netherlands

UNEP/ISRIC/FAO (1997). *Soil Degradation in South and Southeast Asia: the Assessment of the Status of Human-induced Soil Degradation in South and Southeast Asia.* Prepared for UNEP by G.W.J. Van Lynden and L.R. Oldeman, ISRIC, Wageningen, Netherlands

UNESCAP (1990). *State of the Environment in Asia and the Pacific 1990.* UNESCAP, Bangkok, Thailand

UNESCAP (1993). *The State of Urbanization in Asia and the Pacific 1993.* UN, New York, United States

UNESCAP/ADB (1995). *State of the Environment in Asia and the Pacific 1995.* United Nations Economic and Social Commission for Asia and the Pacific, and Asian Development Bank. United Nations, New York, United States

UNESCO (1995). UNESCO'S Statistics on Education 1995. Unesco, Paris, France http://www.unesco.org/unesco/educprog/stat.95/english/Asia.WSTIND2.e.html

United Nations Population Division (1996). *Annual Populations 1950-2050 (the 1996 Revision),* on diskette. United Nations, New York, United States

United Nations Population Division (1997). *Urban and Rural Areas, 1950-2030 (the 1996 Revision),* on diskette. United Nations, New York, United States

WCMC (1992). *Global Biodiversity: Status of Earth's Living Resources.* World Conservation Monitoring Centre, Cambridge, United Kingdom

WCMC (1994). *The Socialist Republic of Vietnam: an Environmental Profile.* World Conservation Monitoring Centre, Cambridge, United Kingdom

WCMC (1998). WCMC Protected Areas Database http://www.wcmc.org.uk/protected_areas/data

WCMC/IUCN (1998). WCMC Species Database, data available at http://wcmc/org/uk, assessments from the 1996 IUCN Red List of Threatened Animals

WHO (1992). *Our Planet, Our Health.* Report of the WHO Commission on Health and the Environment. WHO, Geneva, Switzerland

WHO (1993). *The Work of WHO in the South-East Asia Region, 1 July 1991–30 June 1993. WHO,* New Delhi, India

WHO and UNEP (1992). *Urban Air Pollution in Megacities of the World.* Blackwell, Oxford, United Kingdom

WMO and others (1997). *Comprehensive Assessment of the Freshwater Resources of the World.* WMO, Geneva, Switzerland

Woods, L. E. (1983). *Land Degradation in Australia.* AGPS, Canberra, Australia

World Bank (1997a). *Environment matters: towards environmentally and socially sustainable development.* The World Bank, Washington DC, United States

World Bank (1997b). *1997 World Development Indicators.* The World Bank, Washington DC, United States

World Bank (1997c). *Can the Environment Wait? Priorities for East Asia.* The World Bank, Washington DC, United States

World Bank (1998). *1998 World Development Indicators.* The World Bank, Washington DC, United States

WRI, ICLARM, WCMC and UNEP (1998). *Reefs at Risk: a map-based indicator of threats to the world's coral reefs. WRI,* Washington DC, United States

WRI, UNEP and IUCN (1992). *Global Biodiversity Strategy: Guidelines for Action to Save, Study and Use Earth's Biotic Wealth Sustainably and Equitably.* WRI. Washington

WRI, UNEP and UNDP (1990). *World Resources 1990-91: A Guide to the World Environment.* Oxford University Press, New York, United States, and Oxford, United Kingdom

WRI, UNEP and UNDP (1992). *World Resources 1992-93: A Guide to the World Environment.* Oxford University Press, New-York, United States, and Oxford, United Kingdom

WRI, UNEP and UNDP (1994). *World Resources 1994-95: A Guide to the World Environment*. Oxford University Press, New York, United States, and Oxford, United Kingdom

WRI, UNEP, UNDP and WB (1996). *World Resources 1996-97: A Guide to the Global Environment* (and the *World Resources Database* diskette). Oxford University Press, New York, United States, and Oxford, United Kingdom

WRI, UNEP, UNDP and WB (1998). *World Resources 1998-99: A Guide to the Global Environment* (and the *World Resources Database* diskette). Oxford University Press, New York, United States, and Oxford, United Kingdom

Zhang, C. and Zou, J. (1997). Recent Progress of Study on Harmful Algal Blooms in China: an Overview. In Zhou, J. (ed.). *Sources, Transport and Environmental Impact of Contaminants in the Coastal and Estuarine Areas of China*. SCOPE China Publication Series 4. China Ocean Press, Beijing, China

Europe and Central Asia

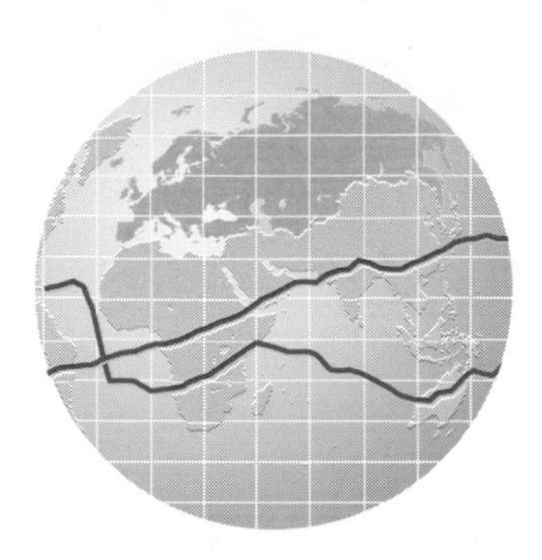

KEY FACTS

There have been important improvements in some, though not all, environmental parameters in Western Europe. In the other sub-regions, political change has resulted in sharp though probably temporary reductions in industrial activity, reducing many environmental pressures.

- GDP per capita in Western Europe countries is typically ten times higher than in the rest of the region.
- In Western Europe, sulphur dioxide emissions fell by more than one-half during 1980–95 but the sub-region still produces nearly 15 per cent of global CO_2 emissions.
- Forest area in Western and Central Europe has grown by more than 10 per cent since the 1960s – but nearly 60 per cent of forests are seriously or moderately damaged by acidification, pollution, drought or forest fires.
- In many countries in the region, half the known vertebrate species are under threat.
- One of the most serious forms of river pollution is high concentrations of nutrients, causing eutrophication in the lakes and seas into which they discharge.
- Most stocks of commercially exploited fish in the North Sea are in a serious condition – the North Sea fishing fleet needs to be reduced by 40 per cent to match fish resources.
- Road transport is now the main source of urban air pollution.
- Of the ten countries in the world with the highest SO_2 emissions per capita, seven are in Central Europe, one in Eastern Europe and two in North America.
- About 60 per cent of large cities in the region are overexploiting their groundwater resources.
- As of 1 January 1999, 360 cities had joined the European Sustainable Cities and Towns campaign.

The late 20th century has been yet another dramatic period in Europe's turbulent history. Until the late 1980s, the region was marked by sharp political and socio-economic divisions between market economies in the west and centrally-planned economies in the rest of the region, with very limited cooperation and often deep conflict between east and west.

In Western Europe, the material standard of living has improved steadily since 1945, along with growing agricultural and industrial production. Signs of severe environmental degradation became increasingly obvious during the 1960s and 1970s, and most countries responded by developing environmental policies – initially directed at local and regional air and water pollution problems. These policies, in combination with factors such as the relatively high price of energy during the oil crisis years, have improved the situation – for example sulphur dioxide emissions fell by more than one-half between 1980 and 1995 (EMEP/MSC 1998). But there has been less progress in other areas: for example, Western Europe is responsible for nearly 14 per cent of the world's carbon dioxide emissions (CDIAC 1998).

Development under the centrally-planned economies in Central and Eastern Europe and Central Asia was understood mainly in terms of growth of physical production (especially in the industry and energy sectors) and this resulted in the severe exploitation of renewable and non-renewable resources. Heavy industry, resource extraction, energy production and the military sector were all associated with high levels of environmental pollution. Extreme specialization was an important element of central planning, resulting in a relatively large demand for transport which increased environmental pressures in some areas. But there were also some positive elements for the environment: the wide use of public transport rather than private cars, strong state systems of nature protection, re-usable packaging for foodstuffs, some sustainable farming and forestry practices, and the separate collection of garbage for recycling in some countries. High educational levels were also a positive force.

One of the most influential changes during the past decade has been the increase in European integration. At the same time, the European Union is expanding, and trade between countries within the region is also growing. Some changes, such as the harmonization of Central and Eastern European legislation to European Union law and a possible shift from medium-distance air travel to high-speed trains, may be beneficial; others, such as increasing car use, are more likely to be harmful to the environment.

Although European integration is generally regarded as a positive development, it may threaten the environment in several ways. The desire of people in the transition countries, especially the young, to attain the living standards and consumption levels of Western countries, with pressures to develop the economy first and solve environmental problems later, may have serious repercussions. And 'blindly' adapting to Western resource management techniques may well result in the loss of traditional, more sustainable approaches that still exist in some parts of Central and Eastern Europe and Central Asia.

Social and economic background

The population of Europe and Central Asia is now about 872 million (United Nations Population Division 1997) – double that at the beginning of the century. Population growth has declined in all sub-regions (see graph above) but relatively high consumption patterns mean that even a slow population growth increases pressure on the environment. In addition, the average size of households is becoming smaller, resulting in greater per capita resource use (EEA 1998a).

Central Asia has a higher rate of population growth than the other sub-regions due to high fertility rates. Providing adequate working and living conditions for

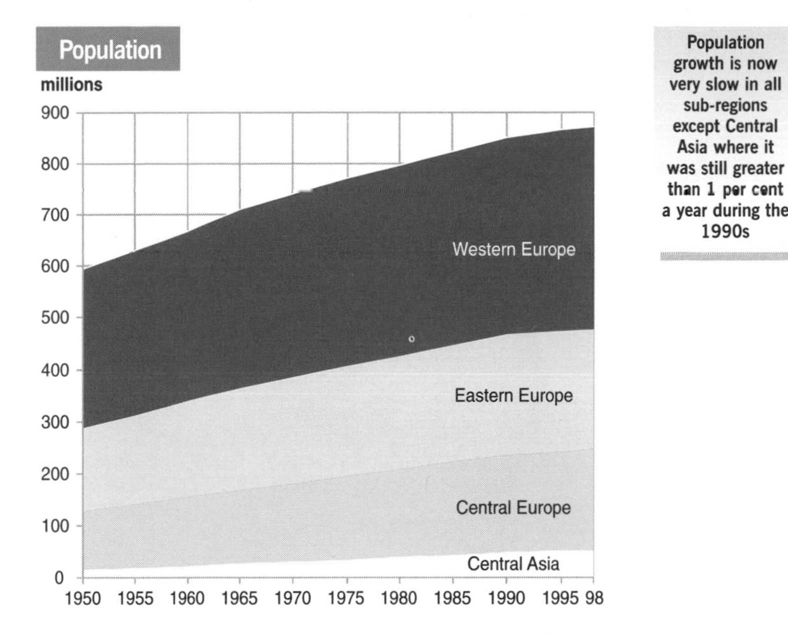

Source: compiled by UNEP GRID Geneva from United Nations Population Division 1996

Population growth is now very slow in all sub-regions except Central Asia where it was still greater than 1 per cent a year during the 1990s

this growing population in a sustainable way may be a major development challenge for this region in the future. The migration of hundreds of thousands of people within Eastern Europe and Central Asia has created a number of environment-related problems in recipient countries (IOM 1998).

Per capita GDP in Western Europe is up to ten times larger than in the other sub-regions

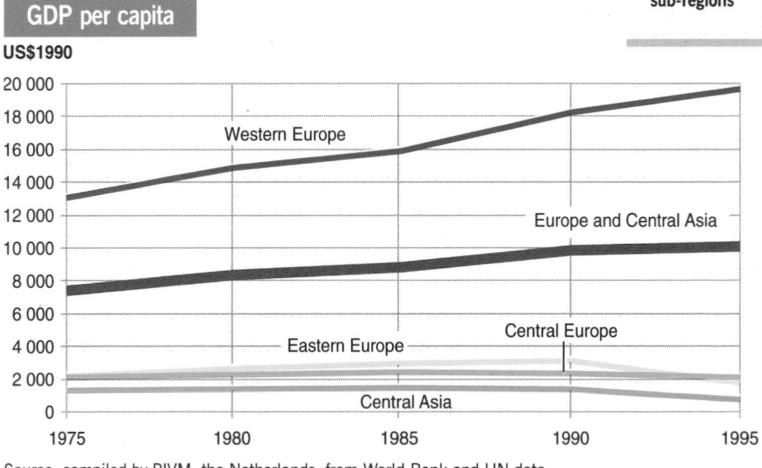

Source: compiled by RIVM, the Netherlands, from World Bank and UN data

The spread of incomes across the region reported in *GEO-1* is still present. GDP per capita in Western European countries is up to ten times higher than in the rest of the region (see graph).

The economies of Western Europe have recovered from the recession of the early 1990s and are currently growing at around 2.5 per cent per year. An

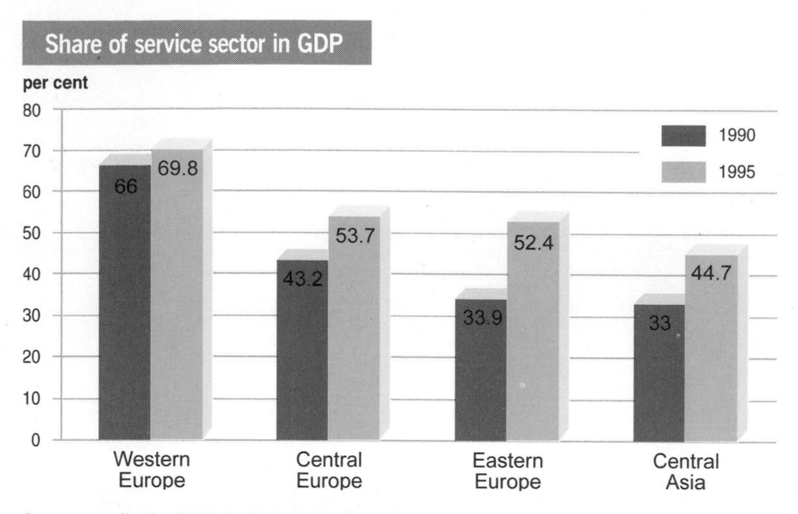

Source: compiled by RIVM, the Netherlands, from World Bank 1997

The service sector is growing rapidly in importance in all the European sub-regions

important factor has been the realization of the Single Market. Unemployment, however, is still relatively high and between 1990 and 1995 grew from 7.8 to 10.2 per cent (UNECE 1996). All Western European countries are currently experiencing relatively rapid growth in the service sector. Although this could result in less environmental pressure than similar growth dominated by industrial activities, the effect may be smaller than expected due to the significant pressures related to transport and tourism. Moreover, a service economy based on increased imports of agricultural and industrial products from other parts of the world simply shifts environmental pressures to other regions.

Life expectancy in Western Europe is still considerably higher than in the other three sub-regions

In Eastern Europe and Central Asia, GDP fell by some 40 per cent as a result of the economic collapse of around 1990. In combination with high inflation rates, this led to a dramatic increase in poverty levels, especially among the older generation and people living in the older industrial regions. Many countries in Central Europe (where GDP/capita fell by only some 7 per cent during 1990-95) seem to be beginning to resolve some of their political and institutional problems, with economic growth resulting from price liberalization, privatization, the reform of tax, legal and financial systems, and international trade. Most countries in Eastern Europe and Central Asia are still experiencing significant economic problems, although there are indications of economic recovery in some countries (World Bank 1996, EBRD 1996 and 1997). The share of the services sector in GDP is increasing throughout the region (see bar chart left), mostly due to the decline in the industrial sector, which traditionally played a large role in total economic activity.

A further striking difference between Western Europe and the rest of the region is in life expectancy. Contributory factors to the generally poorer overall health of the population in Central and Eastern Europe are thought to include the socio-economic situation, lifestyle (including smoking and diet), medical care and environmental factors such as urban pollution and drinking water quality. During the past five years, the health situation in Eastern Europe has worsened (see graph), most markedly with a significant drop in life expectancy for men. Life expectancy in Central Asia appears to be improving – probably due to more effort being devoted to medical care after independence (World Bank 1997 and United Nations Population Division 1996).

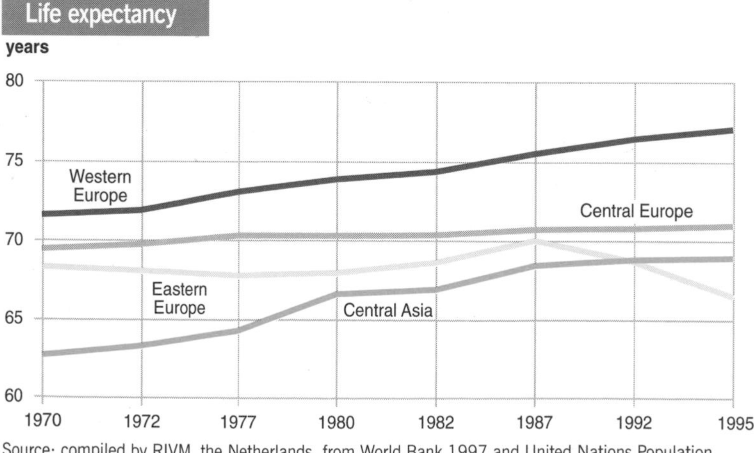

Source: compiled by RIVM, the Netherlands, from World Bank 1997 and United Nations Population Division 1996

Driving forces

The European Environment Agency's recent report on the pan-European region: *Europe's Environment: the Second Assessment* (EEA 1998a) concludes that industry, transport, energy and agriculture are the key sectoral driving forces that impact on Europe's environment. In addition, tourism is playing an increasing role – unless properly managed, it can impose a substantial burden on fragile ecosystems, wildlife habitats and coastal regions.

The relative contribution of industry to many environmental problems, while still very important, has decreased over the past decade. In Western Europe, emissions of pollutants to air and water are falling as environmental objectives are increasingly integrated into decision making. In the other sub-regions, a considerable decrease in environmental pressures has resulted from the drop in industrial

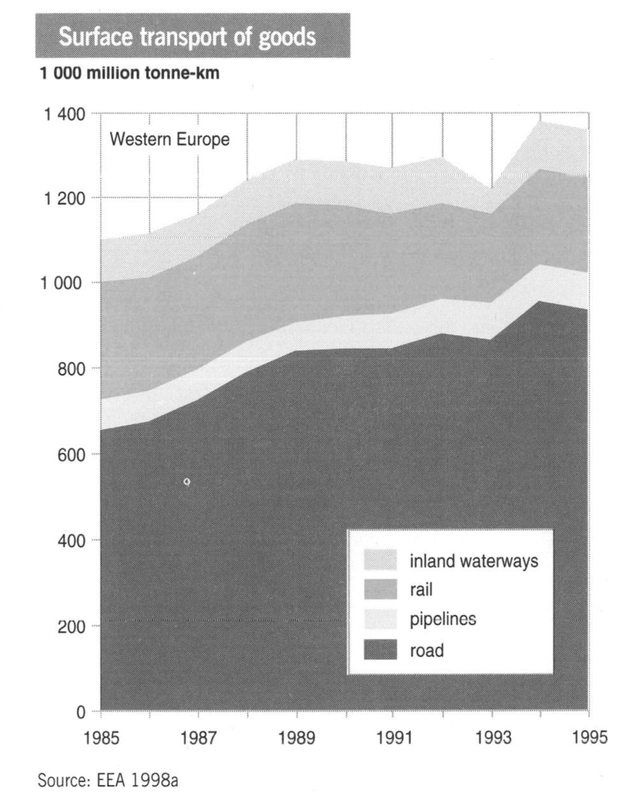

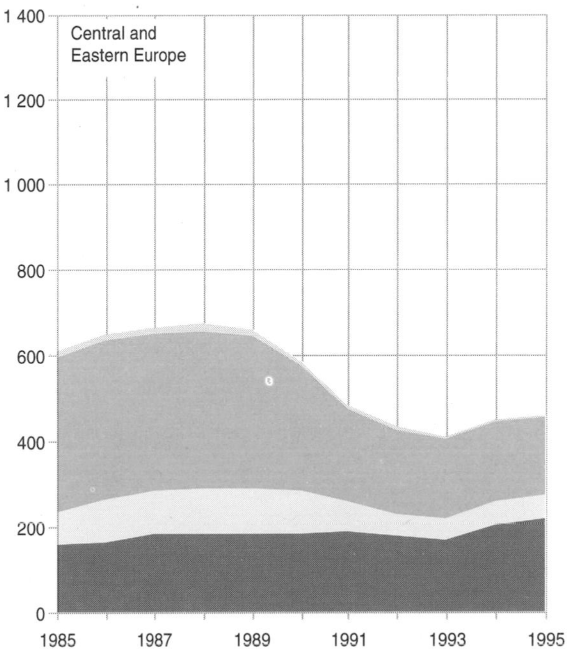

Source: EEA 1998a

In Western Europe, most goods are moved by road. In Central and Eastern Europe, rail transport has been more important but the transport of goods by road is growing

activity, especially in old, energy-intensive, heavy industries. Many highly polluting industrial plants, however, are still operating.

Transport plays an important role in climate change, acidification, summer smog and urban environmental problems. Throughout the region, the environmental impact of transport is increasing as technology and environmental policy are failing to keep up with the pace of growth. In Western Europe, for example, total mobility increased by about 3.6 per cent per year between 1985 and 1995 but fuel efficiency improved at only about 1 per cent a year (Schipper and others 1993, CE 1997). Some underlying factors are that private car use is growing at the expense of public transport, cars are getting larger, and there are fewer people per car. Air travel in Western Europe is growing more rapidly than any other transport mode (82 per cent in 10 years). In Western Europe, most goods are moved by road. In Central and Eastern Europe, rail transport has been more important but the transport of goods by road is growing (see graphs above). Similar trends are found for passenger travel. In Central and Eastern Europe and Central Asia, there has been a rapid expansion of the car fleet over the past few years; in the Baltic States, for instance, car ownership increased from 118 cars to around 150 cars per 1000 of population between 1989 and 1993 (IEA 1996).

The rapid growth in passenger and freight traffic is partly a consequence of rapid integration processes but the related growth of environmental pollution, noise and health problems makes a timely transition to more sustainable transportation and settlement patterns imperative.

Energy use is a basic driving force behind several environmental problems such as climate change, acidification and pollution by heavy metals and particulates. Transport of oil and gas can result in spills and leakages. Other energy sources such as hydropower and nuclear power can also cause significant environmental impacts.

In Western Europe, energy use grew relatively slowly in the early 1990s as a result of economic recession but is now increasing more rapidly as GDP grows by 2 to 3 per cent per year while energy efficiency is only improving by about 1 per cent per year (EEA 1998a). Relatively low energy prices have provided only limited stimulus for energy efficiency improvement. In Central and Eastern Europe and Central Asia, energy use has fallen significantly as a result of restructuring but is expected to rise again as economic recovery takes off (see graph on page 102).

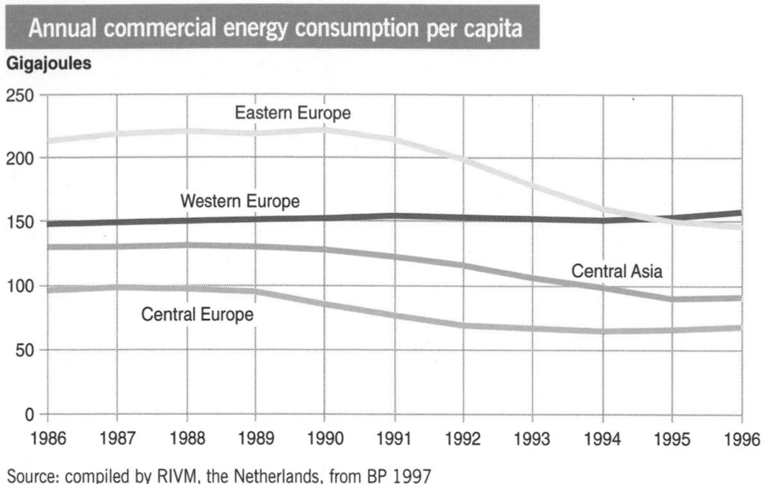

Source: compiled by RIVM, the Netherlands, from BP 1997

Per capita energy consumption has fallen in all sub-regions except Western Europe since 1990 but is expected to rise again with economic recovery

Considerable scope exists for improvement in energy efficiency throughout the region, especially in Eastern Europe and Central Asia.

Important changes are taking place with regard to agriculture. In Western Europe, agricultural subsidies have been significantly reduced and policies are beginning to direct more attention to environmental requirements and the need for more sustainable agriculture. Environmental considerations, however, have yet to be fully integrated with the Common Agricultural Policy (CAP) of the European Union.

In most of the rest of the region, agriculture has been based on large-scale farming associated with negative environmental effects such as wind erosion in Central Asia and nitrate pollution in Central and Eastern Europe. Economic recession has resulted in major reductions in agricultural inputs such as fertilizers, with corresponding environmental benefits. In some areas, farmers are now beginning to concentrate more on improving quality than on maximizing production (Bouma and others 1998). New techniques aimed at increasing productivity may, however, result in similar environmental pressures to those experienced in Western Europe.

Land and food

Western and Central Europe today contain virtually no untouched natural areas. In these sub-regions, about one-third of the land is forested, coverage ranging from about 6 per cent in Ireland to 66 per cent in Finland (EEA 1995). During this century, agricultural yields in Western Europe have increased substantially as a result of intensification, decreasing the total area needed for agricultural purposes. As a result, the total forest area is now slightly larger than it was a hundred years ago. In many countries subsidies for afforestation, which are aimed primarily at increasing timber production, also have environmental and social aims. Most forest expansion, however, has been on poorer land.

In Central and Eastern Europe and in Central Asia, agriculture was dominated by large-scale collective farming for much of the century. There has also been a decline in the area of agricultural land in Central Europe since around 1990. Significant changes in land use may be taking place on marginal land where many areas, small and large, are being abandoned (Bouma and others 1998).

In Eastern Europe, almost half the land area is still covered by forests – and, with about a further 20 per cent covered by natural grasslands, about 70 per cent of the total can be regarded as non-domesticated. However, over the century almost 10 per cent of this area has been converted into cropland or pasture land (Klein Goldewijk and Battjes 1997).

Throughout most of the region, the dominant pressures on land use are from agriculture and forestry, although urbanization, transport and tourism, with their associated infrastructures, are becoming increasingly important.

During the first half of the century, Central Europe and the former Soviet Union were major food exporters while most Western European countries were importers. The practice of subsidizing agricultural prices in Western Europe began after World War II. The CAP of the EEC was aimed mainly at increasing agricultural productivity in order to achieve self-sufficiency in food, providing a fair and stable income for farmers, and reasonable prices for consumers. While achieving these objectives, however, the CAP increased pressure on land resources and produced large food surpluses. Intensive agriculture, with its high input of fertilizers and pesticides, jeopardized soil and water resources as well as natural and semi-natural habitats (Mannion 1995).

In Eastern Europe and Central Asia, collective agriculture was also heavily subsidized but was unable to provide adequate supplies of food. Climatic variations led to large fluctuations in yields. Poor facilities for food storage and distribution led to large losses. Many attempts to extend the area under cultivation caused extensive ecological destruction. For example, in the mid-1950s, vast areas of natural grassland (dry steppes) in northern Kazakhstan were ploughed up and later suffered from severe wind erosion (State Committee of Kazakhstan 1993).

In Central Europe, per capita food production was higher than in the former Soviet Union but the impact of heavy machinery and the widespread use of chemicals and fertilizers contributed to the degradation of soil structure, soil erosion and acidification. By the mid-1980s in Hungary, for example, about 50 per cent of farmland was affected by acidification and 17 per cent by severe soil erosion (Government of the Hungarian Republic 1991).

There have been major developments in agriculture throughout the region during the past few years. Major reform of the CAP started in 1992, with measures designed to compensate farmers for using less intensive farming methods (converting arable land to meadows and pastures, preserving habitats and biodiversity, afforestation and long-term set-aside) and to promote organic agriculture (EEA 1995). Increasing production is no longer the prime requirement, and subsidies are gradually being replaced by direct payments for the implementation of the new policies. However, environmental considerations are still only a small part of the CAP (EEA 1998a).

The pressure on land resources in Central and Eastern Europe and Central Asia has also started to decrease but for different reasons. The collapse of centrally-planned economies and the ending of state subsidies to large collective farms were among the major reasons for a sharp decrease in the use of agricultural chemicals, the abandonment of large irrigation projects, falling livestock numbers and a general reduction in agricultural land, all of which resulted in a sharp decline in agricultural production.

While some of these developments are encouraging from the point of view of a shift to more sustainable patterns of land use, there are a number of other areas of concern.

Soil erosion has always been a serious problem, particularly in the Mediterranean region, including Turkey. Of Europe's total land area (from the Atlantic to the Urals), 12 per cent is affected by water erosion and 4 per cent by wind erosion, generally as a result of unsustainable agricultural practices (UNEP/ISRIC 1991).

Salinization and waterlogging are serious problems, especially in areas where large but poorly-managed irrigation systems were constructed. Large-scale irrigation projects in Eastern Europe and Central Asia have resulted in soil pollution from excessive

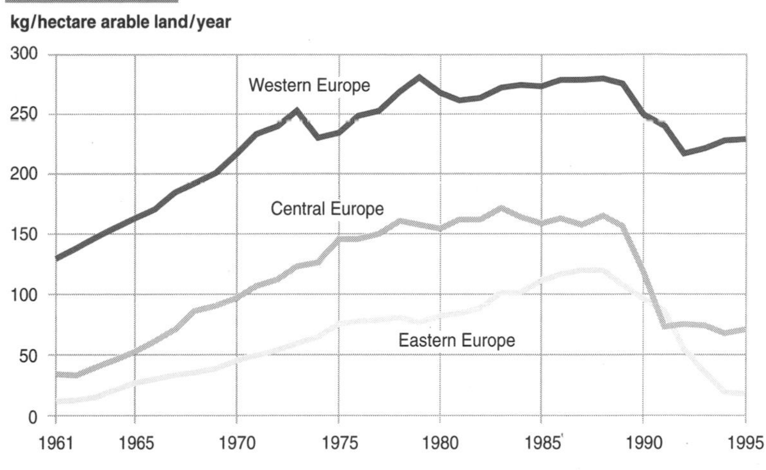

Source: compiled by Jonathan Clarke from FAOSTAT 1997

Note: figures for Central Asia are not shown

Fertilizer use has fallen in all sub-regions but most markedly in Central and Eastern Europe

fertilizer and pesticide use, salinization, and especially waterlogging due to high percolation of water from unprotected channel beds (State Committee of Turkmenistan on Statistics, 1994). Land in the Aral Sea basin has suffered particularly severe degradation as a result of such unsustainable projects.

Pollution of land by excessive use of fertilizers and pesticides and by contaminants such as heavy metals, persistent organic pollutants and artificial radionuclides is widespread. Many of these pollutants end up in surface and groundwater but they can also be taken up directly by crops and affect soil structure. Heavy metal pollution of soils around urban centres also poses health problems for city dwellers who may have their recreational activities and gardens near those sites, often even growing their own food in suburban vegetable gardens.

The declining use of fertilizers

Kazakhstan is a good example of how changes in agricultural organization have led to declining use of agricultural inputs in Central Asia. In 1995 only 386 state farms remained out of an original 764, while use of mineral fertilizers was only 16 per cent of the 1993 level, and the area to which mineral fertilizers were applied diminished to just 9 per cent of 1993 levels (Ministry of Ecology and Bioresources of Kazakhstan 1996).

Similar changes are taking place in Central Europe. Fertilizer use in Albania fell from 176 to just 26 kg/ha between 1990 and 1994. In Poland the figures were 235 kg/ha in 1985 and 100 kg/ha in 1994. In the Czech Republic (perhaps one of the most successful transition economies) fertilizer use fell from 346 kg/ha in 1985 to 107 in 1994 (FAOSTAT 1997).

Soil acidification has been a problem mainly in Western and Central Europe. Emissions of acidifying compounds have been reduced significantly in recent years but critical loads for acidification are still being exceeded in more than 10 per cent of the land area of Western and Central Europe (RIVM/CCE 1998). In Eastern Europe, acidification is concentrated in specific areas. For example, emissions from power plants and metallurgical plants in Norilsk (Central Siberia) affect taiga and tundra ecosystems for many kilometres around the plants. The sensitivity of terrestrial ecosystems to acidifying deposition is relatively high in most of Siberia, especially in the western part. Specific areas with high deposition of acidifying compounds can also be found in Central Asia but in general the soil there is much less vulnerable to acidification.

More than 300 000 contaminated sites have been identified in Western Europe (EEA 1998a). It is generally assumed that there are numerous contaminated sites in Central and Eastern Europe and Central Asia, including many hundreds of abandoned military bases with high levels of soil contamination, mainly with oil products, heavy metals and sometimes radioactive compounds, but a comprehensive survey is still lacking. Large areas of land in the Russian federation, Ukraine and especially Belarus were affected by radioactive fallout from the Chernobyl nuclear disaster in 1986, with serious consequences for food production as well as health, and there are continuing restrictions on access in some areas around the plant.

Oil pollution has affected large areas in West Siberia, mainly because of oil and gas extraction, and along the Caspian shores in Azerbaijan and Kazakhstan. Oil leaks are a major problem with more than 23 000 leakages from pipelines alone in 1996, mainly as a result of corrosion (State Committee of the Russian Federation on Environmental Protection 1997). Ageing of the oil pipeline infrastructure linking Siberia with Europe poses a threat of further soil pollution and the frequency of accidents can be expected to increase.

Over the next ten years, implementation of the new CAP in Western Europe and falling subsidies in Central and Eastern Europe and Central Asia are likely to reduce pressure on agricultural land. In the West, the arable area will probably continue to decrease, with remaining fields becoming increasingly productive, a gradual restoration and interlinking of the small remaining areas of natural habitat, and further reforestation.

Agriculture in the Central European countries seeking to join the European Union will face intense competition which may result in a decrease in the number of farms and thus reduce pressure on land resources. At the same time, as the economic situation improves, with more use of private cars and demand for one-family houses on individual plots, growing suburbanization may result in an increase in demand for land and a decrease in prime agricultural land.

In Eastern Europe and Central Asia, long-lasting environmental disasters will continue to affect land use, and the Aral Sea will probably continue to shrink, with serious consequences for agricultural production in the basin.

Forests

Most of the indigenous forests in Europe and Central Asia, in particular in Western and Central Europe, disappeared long ago as a result of human activities. In Western Europe, for instance, only about 1 per cent of the forest area is original forest. However, large areas of natural temperate forest can still be found in some of the Nordic and Baltic countries and in parts of the Russian Federation.

Over the past century, the total forested area within the region declined from 45 to 42 per cent, mainly as a result of deforestation in Eastern Europe (Klein Goldewijk and Battjes 1997). In contrast, the area covered by forest in Western and Central Europe grew as a result of major reforestation programmes. In some countries, these programmes achieved impressive results – Hungary, for example, has increased its forest area by 0.6 million hectares over the past 50 years (Ministry for Environment and Regional Policy of Hungary 1994) and Ukraine increased its forested area by 1.5 million hectares or 21 per cent over the past 30 years (Ministry of Nature Protection of Ukraine 1994).

The total forest area in Western and Central Europe has grown by more than 10 per cent since the 1960s as a result of replanting and the natural regeneration of marginal areas (EEA 1995). More recently, between 1990 and 1995, the total forest area in Europe grew very marginally (see bar chart opposite); the increase during this period was more than 10 per cent in Armenia, Greece, Ireland, Kazakhstan and Uzbekistan, where large-scale plantation projects were carried out. Only 3 out of the 54 countries in the region as a whole lost forest area in this period, and even these losses were insignificant (FAOSTAT 1997).

Intensive forestry, as generally practised in Western Europe, cannot provide the same biodiversity as natural forests. The use of fast-growing species, especially in the Nordic countries, has somewhat relieved the pressure on existing forests. But this has led to the loss of a vast number of species which used to inhabit indigenous forests but cannot survive in monoculture plantations. Forestry practice in the Baltic States and many other parts of Central and Eastern Europe and Central Asia has remained relatively small-scale and less technologically advanced than in Western Europe. This has been highly beneficial for the preservation of species diversity. On the other hand, there has been large-scale clear cutting of the remaining indigenous forests, especially in Siberia. All in all, there is little diversity in European forests today, with just a few species dominating, including *Pinus sylvestris* (24 per cent) and *Picea abies* (23 per cent) (EEA 1995).

The increasing globalization of world markets, with more forest products at competitive prices appearing from other parts of the globe, may help to conserve Western European forests (FAO 1997a). While the objectives of forest management in Europe are changing towards sustainable management as a central goal, rather than the more traditional objective of sustainable yield, most forest areas are still under the type of management that takes little account of general biodiversity concerns (EEA 1998a).

In many of the transition countries, forestry agencies and policies are rapidly being re-structured as land privatization creates large numbers of private forest owners. However, there is little certainty about the long-term environmental implications of the privatization of forest land and its effects on biodiversity and land use (FAO 1997a). Some countries, such as Belarus, the Russian Federation and Ukraine are reluctant to privatize forest lands.

In the Russian Federation, the decline in timber production is linked with the general decline of industry and a sharp decrease in timber supplies to the former Soviet republics. There has been severe over-harvesting in some areas, for example in Siberia, which is near the huge timber markets that exist in some far eastern countries. In some Eastern European and Central Asian countries, military conflicts,

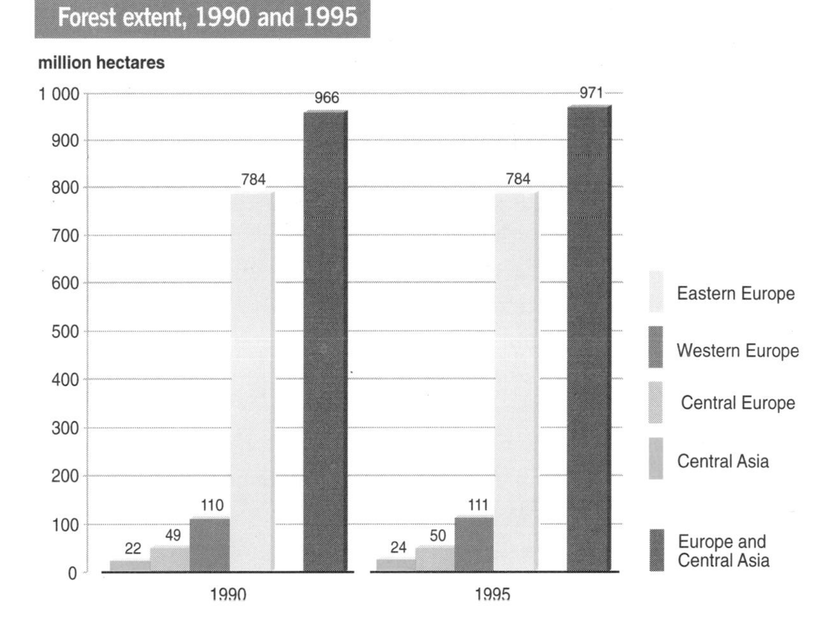

Source: compiled by UNEP GRID Geneva from FAOSTAT 1997, FAO 1997a and WRI, UNEP, UNDP and WB 1998

Data on forest extent show a small growth in all sub-regions during 1990-95

economic blockades (Ministry of Nature Protection of the Republic of Armenia 1993) and rural poverty (which results in large demands for wood as a primary source of energy, Guseinova 1997) have caused local deforestation. In some areas, the removal of tree cover on steep mountain slopes has triggered erosion and contributed to avalanche damage in winter (UNDP 1995).

Forests throughout much of the region are affected by air pollution, pest outbreaks, drought and fires. In most of Europe but particularly in Central Europe where air pollution is highest, crown condition during the past decade has been declining, with a quarter of the trees examined being defoliated by more than 25 per cent. While forest damage has developed less drastically than feared in the early 1980s, there is severe and in some cases catastrophic damage in some places. About 60 per cent of total forests in Western and Central Europe are either seriously or moderately damaged. The reduced vitality of forests probably has several different causes, either separately or in combination, including acidification, other pollutants, drought and forest fires. In some areas and in different species and age groups, however, there has been some improvement in forest condition, which is interpreted as a response to improvements in air quality (EC/UNECE 1997).

Few specific data are available on forest health in large parts of Eastern Europe and Central Asia but vast areas of forest around industrial centres are known to be affected by acidification (RIVM/UNEP 1999) and the contamination of soils with heavy metals (see The Polar Regions, page 176). For example, 1.3 million hectares of forest in the Russian Federation are affected by industrial emissions (Federal Service of the Russian Federation on Hydrometeorology and Environmental Monitoring 1997), although even this huge area represents only 0.1 per cent of the Russian Federation's forest resources.

Following the Chernobyl disaster in 1986, it is estimated that more than 7 million hectares of forest and other wooded land were contaminated by radioactivity, preventing forest work and increasing the danger of secondary radioactive pollution in case of fires (FAO 1997a).

In many European countries, half of the known vertebrate species are threatened and more than one-third of Europe's bird species are in decline

Forested areas in Western Europe and in the Baltic States will probably continue to expand, especially as the new CAP releases more land from intensive agricultural production. With diminishing emissions of acidifying gases, pressure on forest stands will diminish although acidic forest soils will need much more time to recover. In Central Europe, stricter forest and nature protection regulations will probably be introduced, and this should result in less deforestation. As in Western Europe, forested areas may gradually expand and there may also be an increase in the extent of protected forests.

Provided adequate policies are established and enforced, Russian forests could provide a sustainable supply of large volumes of wood for domestic and world markets as well as providing a significant global sink for future increases in the world's carbon dioxide emissions at very moderate cost (Isaev 1995)

Pressure on forest resources in Armenia, Georgia and Central Asia, where forests play a crucial role in preventing serious land degradation in mountainous or arid regions, is likely to continue.

Biodiversity

A century ago, Europe and Central Asia still contained

Threatened animal species

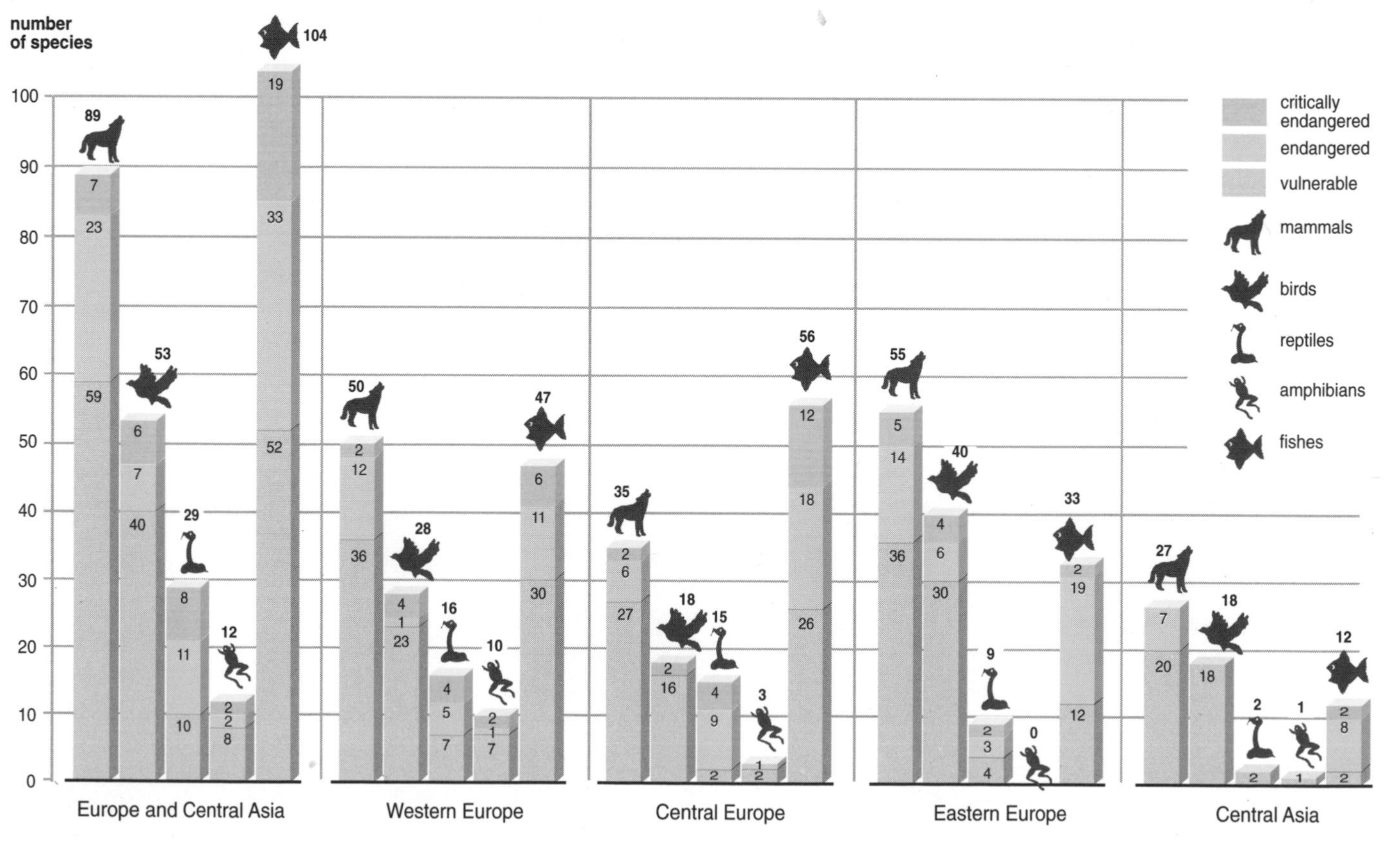

Source: WCMC/IUCN 1998

vast and relatively untouched natural areas but railways were beginning to open up the interior to development. After World War I, further pressures on natural ecosystems resulted from the rapid increase in the road network and the number of private cars.

The pressure on natural ecosystems accelerated after World War II. At the same time, international programmes aimed at coordinating efforts for nature preservation were initiated as realization of the importance of nature preservation grew among the international community. These efforts resulted in considerable growth in the extent of protected areas (see illustration right) but implementation of protection measures within these areas was uneven and unable to reverse the general decline.

An important development, initiated in the 1970s, was the creation of the IUCN's (World Conservation Union) Red Data book and national Red Data books for rare and endangered animal and plant species. These have played an important role in raising public awareness of the importance of the preservation of endangered species, long considered one of the key objectives of biodiversity preservation.

The main current pressures on biodiversity in Europe and Central Asia are from land-use changes, pollution, changes in agricultural, forest and water management, the introduction of alien species and breeds, over-exploitation of resources and tourism (EEA 1998a). Climate change could also become an important (if not dominant) source of pressure (Alcamo and others 1998, RIVM/UNEP 1997).

In many European countries, half of the known vertebrate species are threatened (see bar chart left). Of special importance for Western Europe is the problem of protection of sites used by migratory birds. At present more than one-third of Europe's bird species are in decline, mainly due to damage to their habitats by land-use changes and increasing pressure from agriculture and forestry (Tucker and Heath 1994, Tucker and Evans 1997).

Over the past decade it has become increasingly clear that natural and semi-natural habitats are becoming too small to sustain certain species. This problem of habitats is especially acute in Western Europe, where the continuing development of infrastructure in terms of roads and high-speed railway lines is a major threat.

In Central and Eastern Europe, in contrast to Western Europe, many natural, indigenous habitats still survive. During the socialist era, many factories were concentrated in industrial areas, which caused serious local pollution problems but diminished pressure elsewhere. The lack of modern roads and railways and the poor development of infrastructure were beneficial for wildlife, since there was little fragmentation of natural landscapes. On the other hand, a number of large-scale development projects were highly destructive to wildlife, such as drainage activities in Eastern Europe in the 1960s and 1970s which transformed the last remaining large natural wetlands into agricultural land (Ministry of Natural Resources and Environmental Protection of Belarus 1998).

In Central Asia, over-irrigation in the Aral Sea basin led to the almost complete destruction of unique valley and delta forests, once a habitat for a vast variety of bird, fish and mammal species. The diversity of species in these areas has been greatly

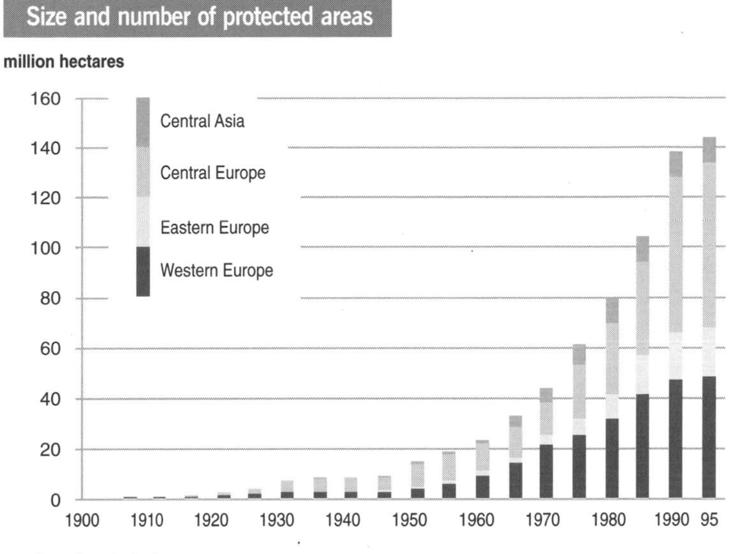

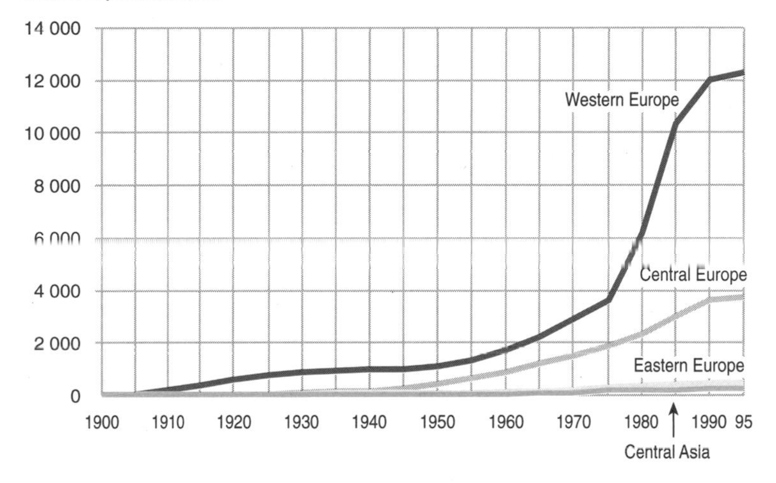

Source: WCMC 1998

The size and number of protected areas has grown sharply since World War II but protection measures within these areas have been unevenly implemented

reduced (State Committee of the Republic of Uzbekistan on Nature Protection 1995).

The major debate on biodiversity conservation has now shifted from the protection of endangered species and genetic diversity to the protection of habitats. Of greatest importance to biodiversity are the shrinking natural and semi-natural habitats. An important pan-European issue is the linking of different protected areas into networks by creating bio-corridors or passageways for wildlife. There is an urgent need to provide better protection of entire landscapes and their associated characteristic land management systems. For example, the preservation and proper maintenance of small woods and hedges around agricultural fields provides the vital shelter needed for the survival of many species of birds and small field animals. The modernization of farming systems, usually involving the enlargement of agricultural fields, often results in the destruction of such wildlife havens.

Wetlands were among the first habitats to be protected, because of their importance to wildlife preservation. In Europe and Central Asia about 300 wetland sites are protected under the Ramsar Convention, in addition to some 70 world natural heritage sites and biosphere reserves, also important for wildlife preservation (EEA 1995). Despite such protection, wetland loss is continuing, especially in Southern Europe but also in many agricultural and urbanized areas in northwestern and Central Europe. Other important habitats and landscapes under pressure include sand dunes, old and semi-natural woodlands, and semi-natural agricultural areas.

There is little sign of any reversal in the decline of species and loss of habitat in Western Europe. The situation in the rest of the region offers more opportunities and challenges. The fall of the 'iron curtain' revealed a relatively untouched and scarcely populated border zone between the NATO and the former Warsaw Pact countries. In addition, large relatively natural areas still exist in Central and Eastern Europe and Central Asia. However, the system of nature protection is now under enormous pressure in these sub-regions as many different demands compete for very limited funds. In the region as a whole, the conservation of biodiversity is still regarded as less important than the short-term economic and social interests of the sectors that influence it most heavily (Conservation Foundation, NGO Eco-Accord and CEU 1998). A major obstacle to securing conservation goals remains the need to incorporate biodiversity considerations into other policy areas (EEA 1998a).

Freshwater

Water resources in Europe and Central Asia have been profoundly influenced over the past century by human activities, including the construction of dams and canals, large irrigation and drainage systems, changes of land cover in most watersheds, high inputs of chemicals from industry and agriculture into surface and groundwater, and depletion of aquifers. As a result, problems of overuse, depletion and pollution have become evident – and more and more conflicts are developing between various uses and users.

Although most people in Europe and some in Central Asia enjoy adequate supplies of freshwater, the distribution of resources is uneven (see bar chart right). Within the overall sub-regional pattern, there are many local variations: the most abundant supplies are in northern Europe and the remote Asian parts of the Russian Federation, and there is relative scarcity, on a per capita basis, in Mediterranean regions, where agriculture competes for freshwater with growing tourism and industrial use, and in the drylands of Central Asia (WRI, UNEP, UNDP and WB 1998) where poorly-designed irrigation systems have caused problems in areas which would naturally have adequate supplies.

Europe (but not Central Asia) and North America are the only global regions where more water is used for industry (55 per cent on average in Europe) than for agriculture (31 per cent) or domestic purposes (14 per cent). Agriculture, however, is the dominant water use in most Mediterranean countries (almost 60 per cent on average) and in Central Asia (more than 90 per cent) (WRI, UNEP, UNDP and WB 1998). In Western Europe, total demand for water increased from about 100 km^3/year in 1950 to about 560 km^3/year in 1990 (EEA 1995) but has since declined slightly as a result of improved water management, more recycling and a shift from water-consuming industries.

In Central and Eastern Europe, water consumption has declined during the past decade, mainly as a result of economic restructuring. The overall drop in water consumption is mainly a result of reduction of water abstraction for industrial purposes; urban demand is steadily growing, driven by the rising urban population and increased per capita consumption as standards of living improve (EEA 1998a). There are few data on corresponding trends in Central Asia. The dominant source of freshwater within the region is surface water.

The most important freshwater pollutants are nitrate, pesticides, heavy metals and hydrocarbons, and the most important consequences of this pollution are eutrophication of surface waters and possible effects on human health. Over-use, resulting in lowering of the water table, is causing salt water to intrude into groundwater in coastal regions (UNEP/ISRIC 1991, Szabolcs 1991).

In many Western and Central European countries, groundwater samples have been found to contain nitrate at above the maximum admissible concentration in water intended for human consumption set in the European Union Drinking Water Directive (EEA 1998b).

Proper access to safe drinking water in Eastern Europe and Central Asia is often limited by the poor quality of surface and groundwater, shortages of chemicals for treatment, and the poor state of distribution mains and networks. The situation is worst in the Central Asian regions near the Aral Sea (UNDP 1996). Infectious intestinal diseases, often caused by poor drinking water, are among the main causes of infant mortality in the southern regions of the Russian Federation and the Central Asian states (UNICEF 1998).

Despite the introduction of water quality targets in the European Union and the attention to water quality in the Environmental Action Programme for Central and Eastern Europe, the state of many rivers remains poor. One of the most serious forms of river pollution in Europe is high concentrations of nutrients, causing eutrophication in the lower reaches of rivers and the lakes and seas into which they discharge. There have, however, been some improvements in the most seriously polluted rivers since the 1970s, reflecting decreases in discharges resulting from improvements in wastewater treatment and emission controls, helped by the reduced use of phosphorus in detergents (EEA 1998a). In general, an improvement in European rivers can be observed for phosphorus and organic matter, with no clear trend for nitrate (see graphs on page 110). In the Rhine river basin, for example, co-ordinated actions have resulted in a significant decrease in organic matter and phosphorus pollution since the mid-1980s.

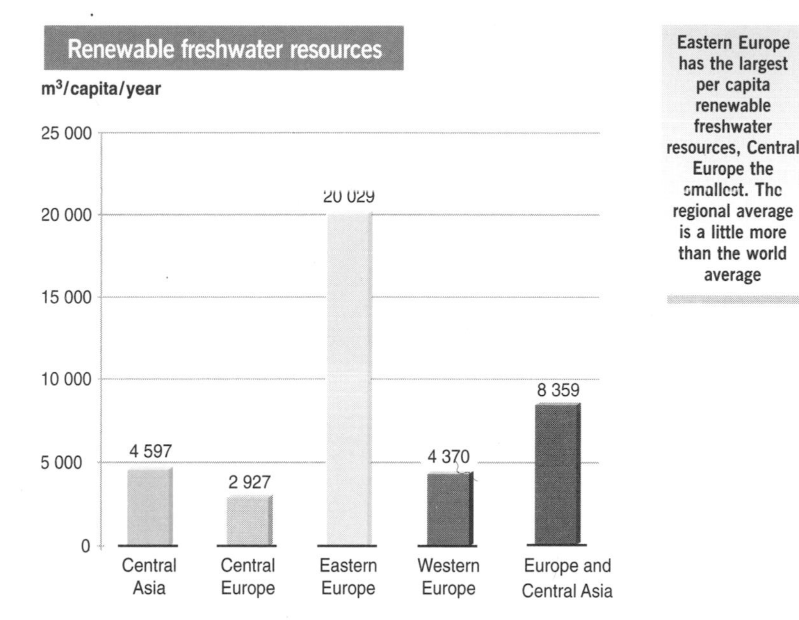

Source: compiled by UNEP GRID Geneva from WRI, UNEP, UNDP and WB 1998

Eastern Europe has the largest per capita renewable freshwater resources, Central Europe the smallest. The regional average is a little more than the world average

The pollution of the Danube with phosphorus compounds has decreased since 1990 (Ministry of Environment Protection of Romania 1996) as a result of the decline in industrial production and fertilizer use in some of the countries through which it flows.

Water quality in Europe's natural and man-made lakes appears to be improving but water quality in many lakes is still poor and well below that in natural lakes in a good ecological state. In Scandinavia, for example, hundreds of lakes, particularly small ones, still suffer from acidification and, although sulphur deposition is falling, it will take a long time for water quality to return to normal (EEA 1997). In the former Soviet Union, many artificial reservoirs were constructed during the 1930s and in particular after

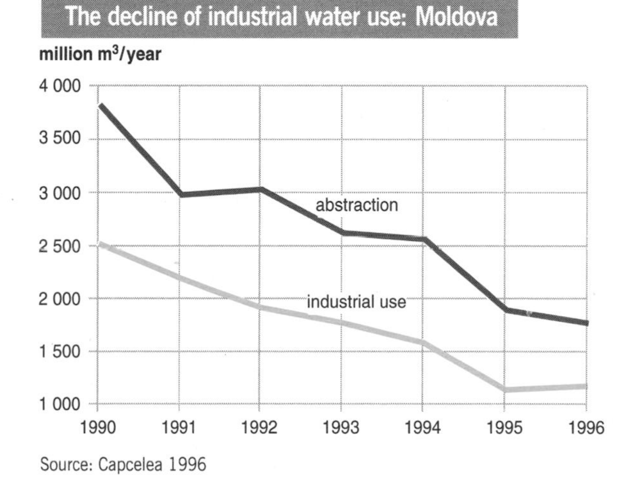

Source: Capcelea 1996

Decline in industrial water use in Central and Eastern Europe is well illustrated by these figures on total abstraction and industrial water use from the Republic of Moldova

World War II. All the major rivers in the European part of the former Soviet Union and in Siberia were diverted into chains of artificial lakes. In most cases, lake bed sediments are highly polluted, and high inputs of phosphorus and other nutrients have often led to eutrophication (Federal Service of the Russian Federation for Hydrometeorology and Environmental Monitoring 1997).

Europe's groundwater is endangered in many ways. Significant pollution by nitrate, pesticides, heavy metals and hydrocarbons has been reported from many countries. Trends in nitrate pollution are unclear. Although smaller quantities of pesticides are being used, environmental impacts are not necessarily diminishing because the range of pesticides in use is also changing (EEA 1998b).

In the Russian Federation, 1 400 areas with polluted groundwater have been identified, most of which (82 per cent) are west of the Urals. In 36 per cent of the cases, pollution is due to industry, in 20 per cent to agriculture (fertilizers and wastes from farm animals), in 10 per cent to municipal landfills and in 12 per cent to mixed sources (Ministry of Nature Protection of the Russian Federation 1996). In Ukraine, especially in the eastern industrial areas, the pollution of aquifers by heavy metals, mainly from the mining and chemical industries, is so serious that many wells can no longer be used as a source of drinking water (Ministry of Nature Protection of Ukraine 1994). Few wells around the Aral Sea still provide safe drinking water (Ministry of Ecology and Bioresources of Kazakhstan 1996).

The demand for clean water is expected to grow throughout Europe and Central Asia. This may aggravate the already tense water supply situation in areas which already face problems, such as the Mediterranean countries, particularly during dry summers. A major challenge will be to reduce the vast losses from distribution networks, especially in Central and Eastern Europe and Central Asia, where losses can exceed 50 per cent (Statistical Committee of the CIS 1996).

Water quality in the European Union will probably gradually improve as increasingly stringent legislation and regulations are implemented. In particular, the European Commission's Urban Waste Water Treatment Directive (91/271/EEC), Nitrates Directive (91/676/EEC) and the proposed Water Framework Directive should result in substantial improvements.

In Central Europe, the resumption of economic growth may result in increasing water pollution, reversing the improvements that have resulted from the sharp decline in industrial activity and reductions in the use of fertilizers and pesticides.

The situation in Eastern Europe will depend on economic growth and the development of industry, the main consumer (and polluter) of water resources. Water pollution problems may persist and worsen as economies recover, with industrial enterprises placing low emphasis on prevention measures and governments taking insufficient measures to enforce pollution reduction strategies.

A main issue for the future in Central Asia is the resolution of the allocation of water rights and water prices between upstream and downstream users (Dukhovny and Sokolov 1996).

The Aral Sea is probably now irrecoverable but it

Graphs show declining levels of phosphorus (PTOT) and organic matter (BOD) in European rivers, with no clear trend for nitrate (NO_3N). Values are relative to 1975 (1975=0)

Phosphorus, nitrate and organic matter in three European rivers

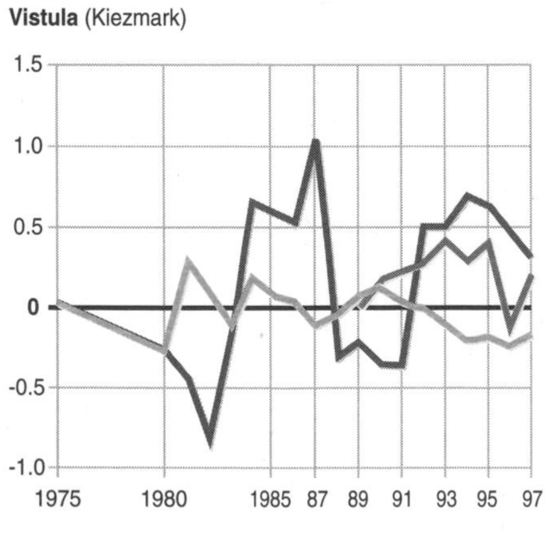

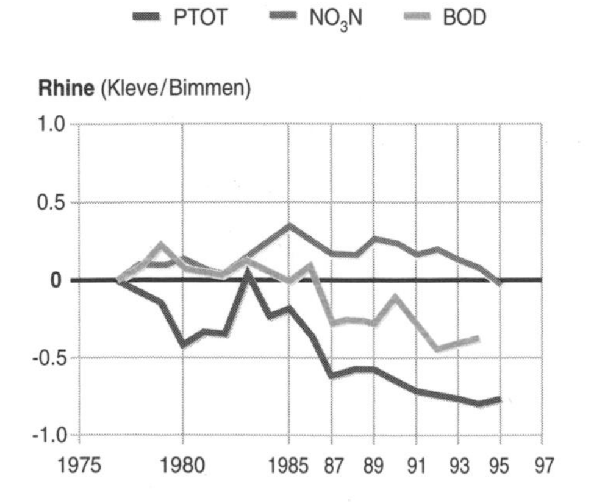

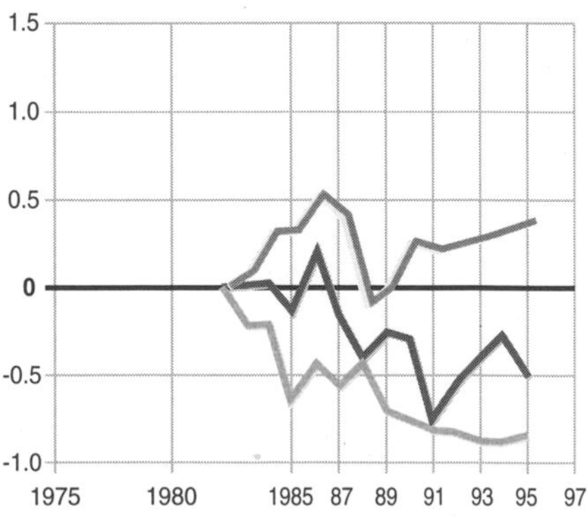

Source: EEA-ETC/IW 1996

may be possible to increase run-off in the rivers that feed it, and thus at least stabilize its water level, by improving water-use techniques in the Aral Sea basin.

Marine and coastal areas

A characteristic feature of the region is that much of it is surrounded by semi-enclosed and enclosed seas, such as the Mediterranean, Black, Azov, Aral, Caspian, Baltic and White Seas. These seas have limited or non-existent water exchange with the open seas, and are thus sensitive to the build-up of pollutants.

Contemporary pollution patterns were already established by the turn of the century; for example, industry was growing, mainly in Western European countries bordering the North Sea basin, oil fields were starting to develop in the Baku region on the Caspian Sea, and metallurgical and chemical factories were being built near the Sea of Azov.

After World War II, the rapid development of heavy industries, increasing use of chemicals in agriculture and rapid population growth led to increasing pressure on all the seas in and around Europe and Central Asia. These seas are also affected by pollutants and nutrients from agricultural and industrial sources generated far upstream, particularly along the Danube, Dnieper, Oder, Rhine, Vistula and Volga. Atmospheric deposition is also an important source of sea pollution.

With about one-third of Europe's population living within 50 km of coastal waters, urban and industrial development and tourism are resulting in growing pressures on already hard-pressed areas. The major issues of concern are eutrophication, contamination, mainly with heavy metals, persistent organic pollutants (POPs) and oil, overfishing and degradation of coastal zones (EEA 1998a).

According to the European Environment Agency (EEA 1998a), the most affected seas are the North Sea (overfishing, high nutrient and pollutant inputs), the Baltic Sea (high nutrient and pollutant inputs), the seas around the Iberian peninsula (overfishing and pollution from heavy metals), parts of the Mediterranean Sea (high nutrient inputs, coastal degradation, overfishing and the disposal of plastics), and the Black Sea (rapid increase of nutrient inputs and overfishing). In addition, the Arctic seas suffer from pollution from oil products, POPs and other materials. The Arctic Seas are dealt with in detail in the section on the polar regions (see page 176).

Enclosed seas (such as the Caspian and Aral Seas) are facing rapid changes in sea level. Overfishing is also a serious problem in many seas.

North Sea

Two major problems affect the North Sea: eutrophication and overfishing. Phosphorus and nitrate discharges into the North Sea are increasing, mainly due to the run-off of surplus nutrients from agriculture, resulting in eutrophication of coastal waters. There was some reduction in the fishing fleet in 1995 and 1996 but most of the stocks of commercially exploited fish in the North Sea are in a serious condition. It has been estimated that the North Sea fishing fleet should be reduced by 40 per cent to match available fish resources (ICES 1996).

While other pollutants such as oil, heavy metals and persistent organic substances have been detected, particularly near point sources of emission, concentrations in biota and sediments are generally low. Total discharges of contaminated water from oil production facilities are increasing, however, as fields are getting older and more come into production. The concentration of oil in the water is still low, and dispersion and dilution is rapid (SFT 1996 and 1997).

Baltic Sea

Pollution of this shallow sea has been a serious problem throughout the second half of the 20th century. The high population in its basin – 77 million people (EEA 1995) – has resulted in high inputs of

Some North Sea fish stocks are at historically low levels and most are overexploited

North Sea fish stocks

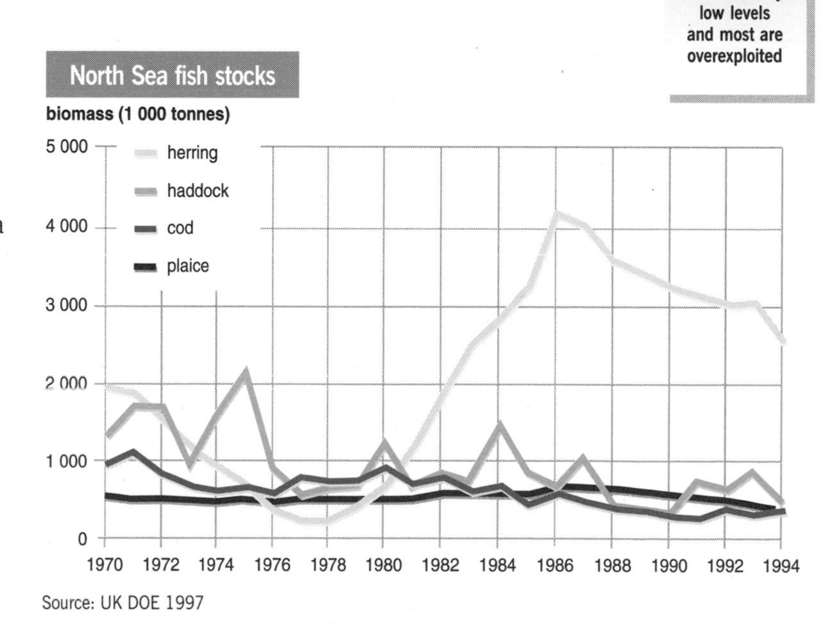

Source: UK DOE 1997

pollutants from surrounding countries, especially from Poland and the former Soviet Union. Until the late 1980s, major cities on its shores such as St Petersburg (4 million inhabitants) and Riga (800 000 inhabitants) had inadequate wastewater treatment facilities, and emissions from many industrial enterprises, including chemical factories, added to the problem (Mnatsakanian 1992).

Eutrophication is widespread, with algal blooms becoming more frequent and concentrations of toxic organic compounds growing (EEA 1995). Eutrophication has affected fish stocks, which have also suffered from overfishing and possibly, in the case of Baltic salmon, from organochloride pollutants (ICES 1994). Action to prevent the discharge of pollutants, by the Helsinki Commission, has, however, stabilized and even reduced the levels of some pollutants since the early 1990s (Ministry of Transport of the Republic of Latvia and Latvian Hydrometeorological Agency 1994).

Inputs of pollutants to the Baltic Sea will probably continue to fall as a result of major programmes to reduce emissions and bring production facilities in line with modern requirements under way in Estonia, Latvia, Lithuania, Poland and, to a certain extent, the Russian Federation.

Mediterranean Sea

The Mediterranean Sea is subject to pollution (including chemical and bacterial contamination and the spread of pathogenic micro-organisms) and eutrophication, mainly from inputs from rivers, especially along the African shores, the southern coasts of France and the North Adriatic (EEA 1995). The problems are mainly in semi-enclosed bays, some of which still receive large amounts of untreated sewage. Discharge of nitrogen and phosphorus is probably the cause of the phytoplankton blooms, the ‘red tides’ that are now frequent in certain parts of Mediterranean (UNEP/MAP 1996). The rapid growth of tourism is a major threat to the environment and biodiversity in much of the area. The sea has the highest species diversity among the European seas: some fish species are being overexploited, while others are thought to be within safe biological limits (FAO 1997b).

Black Sea and Azov Sea

The ecosystems of the Black Sea and the adjacent Sea of Azov have experienced drastic changes during the past ten years. About 170 million people live in the catchment area of the Black Sea (EEA 1995). Most pollutants enter it from international rivers (mainly the Danube but also the Dnieper, Dniester and Don), which bring down nutrients, oil, heavy metals, pesticides, surfactants and phenols. Eutrophication and overfishing, together with increased numbers of a species of jellyfish accidentally introduced in the 1980s, have led to a drastic decline in fish catches. Falling industrial activity since the early 1990s has resulted in less pollution of coastal waters (Ministry of Nature Protection of the Russian Federation 1996) but the construction of oil terminals in the Russian Federation and Georgia and the expected increase of oil tanker traffic are likely to increase pollution again.

The adjacent shallow Sea of Azov which, at the beginning of the century, had one of the richest fisheries in the world, has been under constant pressure from major industrial activities and high use of agrochemicals. During the past 30 years it has suffered from eutrophication and become practically devoid of fish (EEA 1995).

Aral Sea and Caspian Sea

The shrinking of the Aral Sea, caused by the construction of dams and irrigation networks which drastically diminished run-off in its basin (see *GEO-1*), has continued despite the international attention which the problem has received. Measures to return irrigation waters to the sea have not been sufficient to counter-balance evaporation, and the release of salt dusts along its shores continues.

The Caspian Sea possesses 85 per cent of the world’s stock of sturgeon and is the source of 90 per cent of all black caviar. The sea is contaminated by a number of chemicals, including phenols, oil products

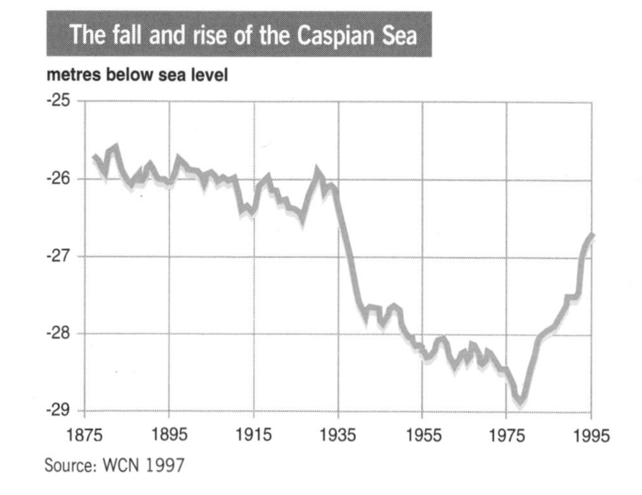

Source: WCN 1997

and surfactants (Azerbaijan State Committee for the Environment and UNDP 1997). The water level of the Caspian fell during the 1960s and 1970s as its natural fluctuations were compounded by the construction of dams which slowed down water turnover. This was followed by rapid construction on exposed shores in the Russian Federation and the bordering Central Asian republics. But in 1978 the water level started to rise again and has now risen by some 2.5 metres (see graph below left). The sea level has now stabilized and may even be falling. While the changed levels of the Aral Sea are certainly due to mismanagement of local water resources, the causes of the changes to the Caspian Sea level are more complicated; global warming may be a contributory factor (WCN 1997).

These changes in level have serious effects on settlements, infrastructure and land resources, and especially on oil and gas extraction facilities (Dukhovny and Sokolov 1996). The problems are complex and potentially serious. On the political front, there is the question of establishing ownership of and responsibility for infrastructure if repairs are needed. Livelihoods and the local economy, particularly fisheries and the oil industry, are threatened, and there are environmental risks to fish stocks and water supplies, for example from leakages from broken pipes.

Pacific Seas

Pollution along the Pacific coast of the Russian Federation, mainly by oil and oil products, heavy metals and pesticides, is concentrated mainly in ports and bays. In addition, pulp and paper factories are significant sources of pollution. Serious potential threats are presented by off-shore oil drilling at the northern tip of Sakhalin Island (Ministry of Nature Protection of the Russian Federation 1996), the dumping of liquid radioactive wastes in the open parts of the Sea of Japan and rusting nuclear submarines in naval bases along the coast north of Vladivostok.

Nuclear pollution in the Arctic

From the 1960s onwards, Soviet naval authorities dumped liquid wastes and buried solid nuclear wastes and obsolete reactors from submarines and nuclear icebreakers in the shallow waters along the eastern coast of the Novaya Zemlya archipelago and in the Barents and Kara Seas (Governmental Commission on Radioactive Waste Pollution of the Seas 1993). Although the wastes were often buried in special protective covers designed to prevent contact with sea water for several hundred years, at least in theory, these sites, together with the Russian Federation's ageing nuclear submarine fleet, represent an important potential threat of nuclear contamination (AMAP 1997).

Atmosphere

Until the 1960s, coal was the primary source of energy in most parts of the region for electricity generation, industry and domestic heating. Air purification devices were practically non-existent. This led to high levels of air pollution, particularly in cities, with soot, dust, sulphur dioxide (SO_2) and nitrogen oxides (NO_x). Winter smogs, particularly the notorious episodes in London during the 1950s, had serious effects on health and also on building materials and historic monuments.

In Western Europe, after World War II, industry was restructured and oil, gas and nuclear power were increasingly used for energy production. This, together with the introduction of low sulphur fuels, natural gas, electricity and district heating schemes for domestic heating, contributed to the virtual disappearance of winter smog. Road transport, on the other hand, has grown inexorably and is now the main source of urban air pollution.

In Central Europe, in the post-war period, a policy of self-reliance on domestic energy resources and the forced development of heavy industry resulted in the exploitation of local energy resources, often of poor quality and high sulphur content (for example, brown coal and oil shales). Power plants were built in clusters near coal mines to reduce transportation costs. This high concentration of plants caused major pollution problems, mainly with SO_2. The Black Triangle area – at the borders between the former German Democratic Republic, Czech Republic and Poland – as well as the Upper Silesia region in Poland and the Ostrava basin in the Czech Republic suffered most.

During the post-war period, Eastern European and Central Asian countries relied increasingly on oil, gas, hydro and nuclear power for electricity generation, resulting in less air pollution than from the generally dirtier fuels used in Central Europe. But there were other sources of air pollution, such as the production of ferrous and non-ferrous metals, pulp and paper, and chemicals, which were often located close to cities.

Throughout the century, there has been a gradual shift in industrial emission 'hot spots' from northwestern Europe towards the east and south (EEA 1995). During the past ten years, the levels and patterns of air pollution in Europe have changed as a result of the adoption of important agreements aimed at reducing emissions and the dramatic changes occurring in Central and Eastern Europe and Central Asia (EEA 1997).

The most drastic improvement in urban air quality throughout Western and Central Europe during the past 10 years has been the decline in pollution from SO_2. However, research suggests that about 25 million urban dwellers in Europe are still exposed at least once a year to levels above the WHO Air Quality Guidelines for health protection, mainly due to winter smog episodes in Central and northwestern Europe (EEA 1998a). Summer smogs, too, are of continuing concern in many cities: the number of people exposed to summer smog conditions above WHO guidelines is 37 million (EEA 1998a). Of the ten countries in the world with the highest SO_2 emissions per capita, seven are in Central Europe, one in Eastern Europe and two in North America.

Europe is responsible for approximately one-third of global greenhouse gas emissions. In Western Europe, per capita emissions of carbon dioxide (the main greenhouse gas) fell slightly between 1990 and 1995, mainly due to economic recession, the restructuring of industry in Germany, and the switch from coal to natural gas for electricity generation. Emissions fell much more in Central and Eastern Europe during the same period, mainly as a result of economic re-structuring and the related drop in economic activity. CO_2 emissions are expected to start to rise again in all the sub-regions in the near future (RIVM/UNEP 1999). Emissions of most other greenhouse gases (methane, nitrous oxide and CFCs) have also fallen (EEA 1998a). Emissions of CFC replacement gases, in particular HCFCs and HFCs (both greenhouse gases) are, however, increasing.

Emissions of acidifying substances in the region as a whole have decreased substantially. Between 1985 and 1994, SO_2 emissions in Western Europe, Central Europe and Eastern Europe fell by 50 per cent as a result of the Convention on Long-range Transboundary Air Pollution protocols (Olendrzynski 1997). The main reasons for these reductions were the installation of low-sulphur coal and flue-gas desulphurization equipment at large point sources in Western Europe and the renewal of power plants and economic restructuring in Eastern Europe.

Significant reductions in ammonia emissions have also been achieved, resulting from changes in agricultural policy in Western Europe and reduced agricultural activity in Central and Eastern Europe. NO_x emissions have also been lower. Total nitrogen emissions (NO_x plus ammonia) fell by 19 per cent between 1990 and 1995, the largest falls occurring in Central and Eastern Europe. The transport sector has become the largest source of NO_x in Europe, contributing 60 per cent of the total in 1995. The use of vehicle exhaust catalysts is helping to reduce emissions in Western Europe but relatively slowly because of the low turnover rate of the vehicle fleet (EEA 1998a). In Central and Eastern Europe, emissions of NO_x from stationary sources fell due to the economic recession but this has been partially nullified by the sharp growth in the use of private cars, especially in large cities. During the recession years of 1990–94, the number of private cars in the Russian Federation increased by 143 per cent, in Ukraine by 130 per cent, in Kazakhstan by 123 per cent, and in Armenia by 110 per cent (Statistical Committee of the CIS 1996).

As a result of these reductions in emissions, the area of Europe where the deposition of acidifying compounds exceeds critical loads for ecosystems has been significantly reduced. Nevertheless, in Western and Central Europe, critical loads are still being exceeded for more than 10 per cent of ecosystems (EMEP/MSC 1998).

Ozone concentrations in the troposphere over Europe (the layer of the atmosphere from the ground to 10–15 km) are typically three to four times their pre-industrial levels. Tropospheric ozone is the main contributor to the summer smogs that occur over large parts of Europe every year and which have been causing respiratory problems for several decades. The problem is most severe in parts of Western and Central Europe, and results mainly from emissions of the main precursor gases (NO_x and non-methane volatile organic compounds) from industry and vehicles. Although emissions of these precursors fell in 1994 by 14 per cent in comparison with 1990, ozone concentrations remain high and often well above threshold limits set by the WHO. In the European Union, for example, about 330 million people are exposed at least once a year to levels that exceed these threshold limits (Malik and others 1996).

According to the Kyoto agreements, greenhouse gas emissions in Western Europe should be reduced to

8 per cent below 1990 levels by 2010. Under 'business-as-usual' conditions, however, it is highly unlikely that this target will be met. Nevertheless, the technical potential for emission reductions is large enough in principle to allow the Kyoto target to be reached. Achieving this will be major challenge for Western Europe in the coming decade. Most of the Central European countries committed themselves to reductions of between 5 per cent and 8 per cent which will probably require additional measures to be taken. The Russian Federation and Ukraine have to stabilize their emissions in 2010 compared to 1990. According to current expectations, this goal will be met without additional environmental policies (RIVM/UNEP 1999).

With progress in reducing emissions of SO_2, emissions of nitrogen are gradually becoming a more important acidification factor (EEA 1997). It is unlikely that the European Union's Fifth Environmental Action Plan target of a 30 per cent reduction of emissions of NO_x by the year 2000 will be met, mainly due to the expected growth in road traffic, and further reductions will be required beyond 2000 to reduce acidification and tropospheric ozone (EEA 1998a).

Most of the recent air quality improvements in Central and Eastern Europe and Central Asia have been due to economic decline. Many air pollution problems are likely to persist, and worsen as economies recover, with industrial enterprises disregarding air pollution prevention measures, using the harsh economic situation or the fact that total pollution has already decreased as a justification for their lack of action. The generally weak environmental protection bodies in many countries are unlikely to be able to enforce effective air pollution reduction strategies in the near future, and measures aimed at recovery from near or complete economic collapse are likely to take precedence over those aimed at protecting or improving the environment.

Urban areas

The built-up areas in Europe and Central Asia have grown dramatically over the past hundred years – currently almost three-quarters of the population in the region live in cities. Rapid city growth has had many ecological and environmental health consequences. For example, major industrial areas, originally developed in open country on the outskirts of cities, have been surrounded by residential areas, whose inhabitants often suffer from health and pollution problems. Although the patterns of city growth in Western Europe have differed from those in the East, the general direction of development and the environmental consequences have often been similar.

In Western Europe, the 1960s and 1970s were periods of rapid suburbanization and decline of city centres, while in Central and Eastern Europe there was massive urbanization. Suburban growth is now starting to gain momentum in parts of Central and Eastern Europe where economic transformation is enabling wealthy people and the growing middle classes to buy suburban family houses and commute to work by car. An important recent development in the eastern part of the region has been the large migration of Russian-speaking people, mainly from Armenia, Georgia and Central Asia, to cities in European Russia, which has caused additional pressure on already vulnerable social systems from the point of view of new housing, job creation and medical care (IOM 1998).

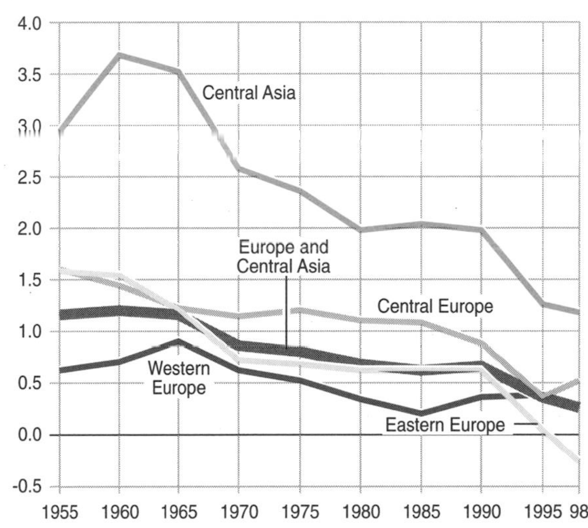

Source: compiled by UNEP GRID Geneva from United Nations Population Division 1997 and WRI, UNEP, UNDP and WB 1998

Some three-quarters of the region's population live in cities, and rates of urban growth have now slowed to almost zero except in Central Asia; in Eastern Europe, they are now negative

Overall, air quality in most cities has improved over the past few decades. Ozone, however, remains a major problem in some Western European cities. Transport has become the major contributor to several air pollution problems in Western Europe. Despite rigorous and effective measures to reduce car emissions, most air pollution in major cities still comes from automotive sources, and the number of cars continues to increase. At the same time, there have been some improvements in transport-related air quality; for example, atmospheric concentrations of lead are falling due to the reduction of the lead content in petrol (EEA 1997).

In Eastern and Central Europe and Central Asia, the worst pollution in cities occurred during the 1970s and 1980s, when industrial production increased with little regard for environmental consequences. Although there were fewer cars than in the West, they were mainly domestically-produced models with high emission rates. Although emissions from stationary sources have fallen considerably since 1990, there has been some growth in urban mobility and car ownership; this is expected to accelerate in the coming decade. Emissions are likely to grow as a result, despite the introduction of cleaner cars (EEA 1998a).

In general, in practically all major cities, automotive sources are replacing stationary sources as the dominant contributor to air pollution, resulting in a reduction of winter smog but an increase of summer smog.

Problems with municipal waste have increased, with waste quantities per capita in Western Europe rising by 35 per cent since 1980. Most waste is dealt with by the cheapest available method: in OECD Europe during 1991–95, 66 per cent of municipal waste went to landfills, 18 per cent was incinerated, 9 per cent was recycled, 6 per cent was composted and 1 per cent was treated in other ways (OECD 1997). Recycling of waste in most Western European countries is increasing.

The sharp growth in the number of vehicles in the region is now the major cause of urban air pollution

Urban wastewater treatment standards vary markedly across the region. Most of the population in northern Europe now lives in houses or flats connected to a sewer. In many cities in southern and eastern Europe, however, water receives no or only limited treatment. In most Central and Eastern European cities, wastewater is still collected together with rainwater and discharged to water bodies without treatment, causing eutrophication, especially in some urban estuaries (EEA 1998a).

About 60 per cent of large cities in Europe are overexploiting their groundwater resources, and water availability may increasingly constrain urban development in some areas. Leakages from water mains of up to 50 per cent are common (EEA 1998a). Many cities in Eastern Europe and Central Asia have for many decades suffered from poor quality drinking water due to pollution of surface and groundwater sources, obsolete water purification techniques and the poor state of water mains. These problems were aggravated after the beginning of economic transition when many local municipalities lacked funds to improve the drinking water supply. For example, in the Russian Federation, in 1995, about 22 per cent of drinking water samples did not meet chemical standards and almost 9 per cent exceeded acceptable bacteriological levels (Ministry of Nature Protection of Russian Federation 1996).

Urban noise is an important problem throughout the region. The maximum acceptable noise level is regularly exceeded in most cities. In Europe overall, about 10 million people are exposed to environmental noise levels that may cause hearing loss (OECD/ ECMT 1995).

In spite of progress in some environmental areas, the large cities of the region will continue to present major environmental challenges. Their 'ecological footprints', the ecological productive areas needed to support their populations with renewable and non-renewable resources and to absorb their emissions and wastes, are large and growing. Many city authorities are exploring ways of achieving sustainable growth in the context of Local *Agenda 21* policies, which require the implementation of measures aimed at reducing use of water, energy and materials, and better planning of land use and transportation. As of 1 January 1999, 360 cities, some 334 of which are in Western Europe, have already joined the European Sustainable Cities and Towns Campaign (ESCT 1998).

The growth in vehicles

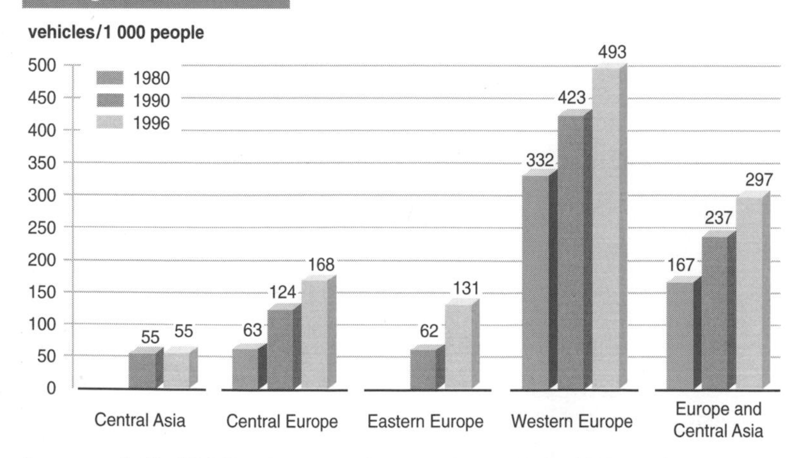

Source: compiled by UNEP GRID Geneva from International Road Federation 1997; data for Eastern Europe and Central Asia for 1980 are not available

References

Alcamo, J., Leemans, R., and Kreileman, E. (eds., 1998). *Global Change Scenarios of the 21st Century.* Elsevier Science, Oxford, United Kingdom

AMAP (1997). *AMAP Assessment Report: Arctic Pollution Issues.* Arctic Monitoring and Assessment Programme, Oslo, Norway

Azerbaijan State Committee for the Environment and UNDP (1997). *State of the Environment Report 1997.* Baku, Azerbaijan

Bouma J., Varallyay, G. and Batjes, N. H. (1998). Principal land use changes anticipated in Europe. *Agriculture, Ecosystems & Environment,* 67, 103-119

BP (1997). *BP Statistical Review of World Energy.* British Petroleum Co., London, United Kingdom

Capcelea, A. M. (1996). *The Republic of Moldova on the Route to Sustainable Development.* Stiinta, Chisinau, The Republic of Moldova

CDIAC (1998). *Revised Regional CO_2 Emissions from Fossil-Fuel Burning, Cement Manufacture, and Gas Flaring: 1751–1995.* Carbon Dioxide Information Analysis Center, Environmental Sciences Division, Oak Ridge, Tennessee, United States. http://cdiac.esd.ornl.gov/cdiac/home.html

CE (1997). *Efficiency and Sufficiency; towards sustainable energy and transport.* Centre for Energy Conservation and Environmental Technology, Delft, the Netherlands

Conservation Foundation, NGO Eco-Accord and CEU (1998). International environmental collaboration. In Stewart J. M. (ed.). *Russia: a case study.* Conservation Foundation London Initiative on the Russian Environment. Prepared for the Environment for Europe Conference, Aarhus, Denmark, 23–28 June 1998

Dukhovny, V. and Sokolov, V. (1996). Development of the Caspian-Aral Seas Program: an ICWC Perspective. In Glantz M. and Zonn I. (eds.). *Scientific, Environmental and Political Issues in the Circum-Caspian Region.* Proceedings of the NATO advanced research workshop, May 1996

EBRD (1996). *Transition Report.* European Bank for Reconstruction and Development, London, United Kingdom

EBRD (1997). *Transition report update.* European Bank for Reconstruction and Development, London, United Kingdom

EC/UNECE (1997). *Forest Condition in Europe,* 1997 Technical Report. European Commission and the United Nations Economic Commission for Europe, Brussels, Belgium, and Geneva, Switzerland

EEA (1995). *Europe's Environment: the Dobříš Assessment.* European Environment Agency, Copenhagen, Denmark

EEA (1997). *Air Pollution in Europe in 1997.* European Environment Agency, Copenhagen, Denmark

EEA (1998a). *Europe's Environment: The Second Assessment.* European Environment Agency, Copenhagen, Denmark

EEA (1998b). *Groundwater Quality and Quantity.* To be published in the EEA Environmental monograph series. European Environment Agency, Copenhagen, Denmark

EEA-ETC/IW (1996). *Surface Water Quantity Monitoring in Europe.* EEA Topic Report No. 3/1996. European Environment Agency, Copenhagen, Denmark

EMEP/MSC (1998). EMEP Meteorological Synthesizing Centre (MSC). *Data on transboundary air pollution.* MSC, Oslo, Norway

ESCT Campaign (1998). European Sustainable Cities and Towns Campaign, Rue de Treves 49-51, Box 3, B-1040 Brussels, Belgium http://www.who.dk/healthy-cities/esctc.htm

FAO (1997a). *State of the World's Forests.* FAO, Rome, Italy

FAO (1997b). *Review of the State of the World's Fisheries: Marine Fisheries.* FAO circular, No 920. FAO, Rome, Italy

FAOSTAT (1997). *FAOSTAT Statistics Database.* FAO, Rome, Italy. http://www.fao.org

Federal Service of the Russian Federation on Hydrometeorology and Environmental Monitoring (1997). *Review of Environmental Pollution in the Russian Federation in 1996.* Moscow, Russia (In Russian)

Government of the Hungarian Republic (1991). *Hungarian National Report to UNCED.* Budapest, Hungary

Governmental Commission on Radioactive Waste Pollution of the Seas (1993). *Facts and Problems Linked with Dumping of Radioactive Wastes in the Seas around the Russian Federation* (edited by A. V. Yablokov). Governmental Commission on Radioactive Waste Pollution of the Seas, Moscow, Russia (in Russian)

Guseinova, F. (1997). *State of the environment in Azerbaijan.* Report prepared by the Green Movement of Azerbaijan in 1997

ICES (1994). *Report on the study group on occurrence of M-74 in fish stocks.* International Council for Exploration of the Seas, Report C. M. 1994/ENV, No. 9. Copenhagen, Denmark

ICES (1996). *The 1996 report of the ICES Advisory Committee on Fishery Management.* International Council for Exploration of the Seas, Coop. Res. Rep. No. 221. Copenhagen, Denmark

IEA (1996). *World Energy Outlook,* 1996 edition. International Energy Agency, Paris, France

International Road Federation (1997). *World Road Statistics,* 1997 edition. IRF, Geneva, Switzerland, and Washington DC, United States

IOM (1998). *International Organization for Migration.* Report on Migration of Population in CIS Countries. IOM, Geneva, Switzerland

Isaev, A. O. (1995). *Ecological problems of CO_2 absorption by re-forestation in Russia.* Centre for Russian EcoPolicy, Moscow, Russia (in Russian)

Klein Goldewijk, C. G. M. and Battjes, J. J. (1997). *A Hundred Year (1890-1990) Database for Integrated Environmental Assessment (HYDE version 1.1).* National Institute of Public Health and the Environment, Bilthoven, the Netherlands

Malik, S., Simpson, D., Hjellbrekke, A.-G and ApSimon, H. (1996). *Photochemicals model calculations over Europe for Summer 1990: model results and comparison with observations.* EMEP/MSC-W Report 2/96. DNMI, Oslo, Norway

Mannion, A. M. (1995). *Agriculture and Environmental Change.* John Wiley, Chichester, United Kingdom

Ministry for Environment and Regional Policy of Hungary (1994). *Environmental Indicators of Hungary*. Budapest, Hungary

Ministry of Ecology and Bioresources of Kazakhstan (1996). *National Report on the State of the Environment in Kazakhstan in 1995*. Almaty, Kazakhstan (in Russian)

Ministry of Ecology and Bioresources of Kazakhstan (1997). *Ecological Bulletin of the Republic of Kazakhstan for 1996*. Almaty, Kazakhstan (in Russian)

Ministry of Environment Protection of Romania (1996). *Environment Protection Strategy of Romania*. Bucharest, Romania

Ministry of Natural Resources and Environmental Protection of Belarus (1998). *State of the Environment: Country Overview – Belarus TACIS* (available from the European Environment Agency, Copenhagen, Denmark)

Ministry of Nature Protection of the Republic of Armenia (1993). *National Ecological Report of Armenia*. Yerevan, Armenia

Ministry of Nature Protection of the Russian Federation (1996). *National Report on the State of the Environment in the Russian Federation in 1995*. Moscow, Russia

Ministry of Nature Protection of Ukraine (1994). *State of the Environment and Activities in Ukraine*. Kiev, Ukraine

Ministry of Transport of the Republic of Latvia and Latvian Hydrometeorological Agency (1994). *Environmental Pollution in Latvia: Annual Report for 1993.* Riga, Latvia

Mnatsakanian, R. (1992). *Environmental Legacy of the Former Soviet Republics*. Centre for Human Ecology, University of Edinburgh, Edinburgh, United Kingdom

OECD (1997). *Environmental Data Compendium 1997*. Organization of Economic Cooperation and Development, Paris, France

OECD/ECMT (1995). *Urban Travel and Sustainable Development*. Organization of Economic Cooperation and Development, Paris, France

Olendrzynski, K. (1997). Emissions. In *Transboundary Air Pollution in Europe,* edited by E. Berge. MSC-W Status Report 1997. Norwegian Meteorological Institute, Oslo, Norway

RIVM/CCE (1998). *Data on area with exceedances of critical loads.*RIVM (National Institute of Public Health and the Environment) and Coordination Centre for Effects, Bilthoven, the Netherlands

RIVM/UNEP (1997). *The Future of the Global Environment: a Model-based Analysis Supporting UNEP's First Global Environment Outlook.* Bakkes, J. A. and van Woerden, J. W. (eds.). RIVM 402001007 and UNEP/DEIA/TR.97–1. RIVM, Bilthoven, the Netherlands, and UNEP, Nairobi, Kenya

RIVM/UNEP (1999). *Energy-related policy options in Europe and Central Asia 1990–2010 – environmental impacts of scenarios with and without additional policies*. Van Vuuren, D.P., and Bakkes, J.A (eds.). *GEO-2* Technical Background Report Series (UNEP.DEIA&EW.TR/99-4). UNEP, Nairobi, Kenya

Schipper, L., Fugueroa, M. J., Price, L. and Espey, M. (1993). Mind the gap – the vicious circle of automobile fuel use. *Energy Policy,* 21, 1173-1190

SFT (1996 and 1997). *Environmental surveys in the vicinity of petroleum installations on the Norwegian shelf*. Reports for 1994 and 1995, State Pollution Control Authority, Norway, report No. 96.15 and No. 97.13, Oslo, Norway

State Committee of Kazakhstan (1993). National Report of Kazakhstan on the State of the Environment. *Eurasia Journal*, No. 6, 1993 (in Russian)

State Committee of the Republic of Uzbekistan on Nature Protection (1995). *National Report on the State of the Environment and Use of Natural Resources in the Republic of Uzbekistan in 1994.* Tashkent, Uzbekistan (in Russian)

State Committee of the Russian Federation on Environmental Protection (1997). *National Report on the State of the Environment in the Russian Federation in 1996*. Moscow, Russia (in Russian)

State Committee of Turkmenistan on Statistics (1994). *Nature Protection and Management of Natural Resources in Turkmenistan in 1993*. Ashgabat, Turkmenistan (in Russian)

Statistical Committee of the CIS (1996). *Environment in CIS Countries*. Moscow, Russia (in Russian)

Szabolcs, I. (1991). Salinisation potential of European Soils. In Brower, F. M., Thomas, A., and Chadwick, M. J. (eds.). *Land use changes in Europe: a process of change, environmental transformation and future patterns*. Kluwer Academic Publishers, Dordrecht, the Netherlands

Tucker, G. M. and Evans, M. (1997). *Habitats for Birds in Europe: a conservation strategy for the wider environment*. BirdLife Conservation Series 6. BirdLife International, Cambridge, United Kingdom

Tucker, G. M. and Heath, M. F. (1994). *Birds in Europe, their Conservation Status*. BirdLife International, Cambridge, United Kingdom

UK DOE (1997). Environment Protection Statistics and Information Management Division, UK Department of the Environment: http://www.environment.detr.gov.uk/epsim/indics/

UNDP (1995). *UNDP Human Development Report – Georgia*. Government of the Republic of Georgia and UNDP, Tbilisi, Republic of Georgia

UNDP (1996). *UNDP Human Development Report – Kazakhstan*.Government of the Republic of Kazakhstan and UNDP, Almaty, Kazakhstan

UNECE (1996). Secretariat for the Economic Commisssion for Europe, Geneva, Switzerland. *Economic Bulletin for Europe,* Vol. 48

UNEP/ISRIC (1991). *World Map of the Status of Human-Induced Soil Degradation (GLASOD). An Explanatory Note,* second revised edition (edited by Oldeman, L.R., Hakkeling, R.T., and Sombroek, W.G.). UNEP, Nairobi, Kenya, and ISRIC, Wageningen, Netherlands

UNEP/MAP (1996). *The State of the Marine and Coastal Environment in the Mediterranean Region*, MAP Technical report series 100, UNEP-Athens, Greece

UNICEF (1998). *State of the World's Children 1998*. UNICEF, New York, United States. http://www.unicef.org/sowc98

United Nations Population Division (1996). *Annual Populations 1950-2050 (the 1996 Revision),* on diskette. United Nations, New York, United States

United Nations Population Divison (1997). *Urban and Rural Areas, 1950-2030 (the 1996 Revision),* on diskette. United Nations, New York, United States

WCMC (1998). WCMC Protected Areas Database http://www.wcmc.org.uk/protected_areas/data

WCMC/IUCN (1998). WCMC Species Database, data available at http://wcmc/org/uk, assessments from the 1996 IUCN Red List of Threatened Animals

WCN (1997). Caspian Sea Levels: explaining the changes. *World Climate News*, No. 10, January 1997

World Bank (1996). *Annual Report 1996.*World Bank, Washington DC, United States

World Bank (1997).*World Development Indicators 1996/1997.*World Bank, Washington DC, United States

WRI, UNEP, UNDP and WB (1998). *World Resources 1998-99: A Guide to the Global Environment* (and the *World Resources Database* diskette). Oxford University Press, New York, United States, and Oxford, United Kingdom

Latin America and the Caribbean

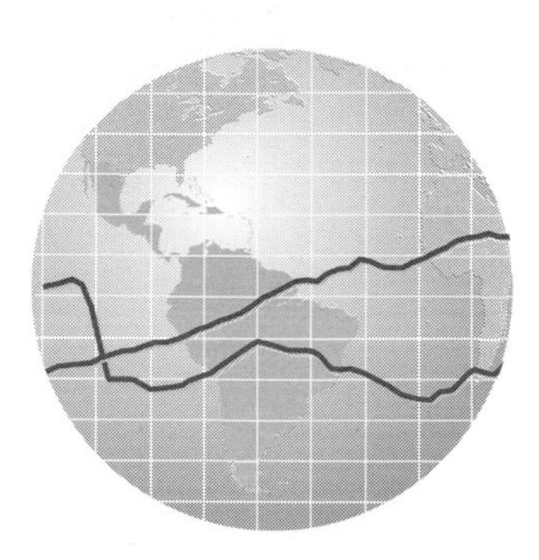

KEY FACTS

Two major environmental issues stand out in the region. The first priority is to find solutions to the problems of the urban environment, which now houses nearly three-quarters of the region's population. The second priority is to find ways of promoting the sustainable use of tropical forests and biodiversity.

- The income of the richest 20 per cent of the population is 19 times more than that of the poorest 20 per cent, compared to a figure of just 7 for the industrial countries.
- The environmental costs of improved farm technologies have been very high. During the 1980s Central America increased production by 32 per cent but doubled its consumption of pesticides.
- The natural forest cover continues to decrease in all countries. A total of 5.8 million hectares a year was lost during 1990–95, resulting in a 3 per cent total loss for the period.
- Most forests in eastern and southern Amazonia are subjected to severe dry seasons each year, particularly during *El Niño* events. These forests are on the edge of the rainfall regime that is necessary for them to resist fire.
- It is estimated that 1 244 vertebrate species are now threatened with extinction.
- A large decrease in the marine fisheries catch is expected as a result of the 1997–98 *El Niño*.
- Many countries have substantial potential for curbing carbon emissions, given the region's renewable energy sources and the potential of forest conservation and reforestation programmes to provide valuable carbon sinks.
- In São Paulo and Rio de Janeiro, air pollution is estimated to cause 4 000 premature deaths a year.

Social and economic background

The Latin American region contains 15 per cent of the world's land area (20 million km^2) and 7.7 per cent of its population (484 million); it generates 5.7 per cent of world GDP (World Bank 1997). Brazil is the largest country with 8.5 million km^2 and 159 million people, followed by Argentina (2.8 million km^2 and 34 million) and Mexico (1.9 million km^2 and 91 million). Mexico is included here in the sub-region Meso-America (see page xxxiii), and the term Central America is used to mean Meso-America less Mexico.

In the period 1940–80, the region's population grew from 160 to 430 million people and its total energy consumption increased fourfold (CEPAL 1996). Profound social and economic changes over the past 20 years have led to significant impacts on the region's natural resources. In most countries, dictatorships have given way to civilian democracies, inflation has been reduced, foreign investment has increased and free-market reforms are under way. The democratization process has also opened up new opportunities for public participation. Protectionist barriers have been removed unilaterally, or as part of regional accords such as Mercosur, leading exports to grow at 6 per cent a year during the early 1990s, compared with 1.8 per cent a year during the mid-1980s. The restructuring of the state in search of simpler and more agile forms of government, economic growth, the liberalization of the economy and the privatization of state enterprises are now the major political themes of the region. These reforms appear to be laying a foundation for a rate of progress that seemed impossible during the 'lost decade' of the 1980s. However, there are many conflicting trends.

Latin America's GDP is now more than US$1 600 000 million. All countries showed an increase in UNDP's index of human development for the period 1960–94, as well as widespread improvement in quality of life. But despite these positive trends the region is still characterized by an unequal distribution of wealth. The expected triumph of free-market reforms over poverty has yet to be delivered. On the contrary, the number of people below the poverty line had reached 160 million by 1995 (World Bank 1996). The gap between incomes is widening further, real wages have fallen and unemployment is now higher than in 1990. The income of the richest 20 per cent of the population is 19 times more than that of the poorest 20 per cent, compared to a figure of just 7 for the industrial countries (UNDP 1997).

The growing poverty gap is also having a profound impact on health in the region. Indigenous and other marginalized urban groups often suffer from a lack of basic services (potable water and sanitation) and social discrimination which further exacerbates the situation. Problems such as malnutrition and iodine deficiency are most serious among these populations (as high as 47 and 20 per cent, respectively, in Bolivia), as are diseases such as cholera that also stem from lack of potable water and sewage treatment systems (PAHO 1994). Despite the endemic presence of Chagas' disease (more than 20 per cent of the population is infected by *Trypanosoma cruzi* in several countries), malaria and dengue fever, life expectancy increased by 38 per cent during 1960–94 (UNDP 1997) and infant mortality decreased by 45 per cent during 1980–90 (PAHO 1994). Vaccine-preventable diseases also declined. Although infectious disease is still a

GDP per capita

US$1990

Latin America and the Caribbean
South America
Meso-America
Caribbean

Source: compiled by RIVM, the Netherlands, from World Bank and UN data

Although GDP/capita has increased for the region as a whole, the number of people below the poverty line is still rising

Economic conditions in the Caribbean

Over the past decade, the Caribbean countries have undertaken a number of economic reforms, with mixed results. For most countries, growth rates were positive during the 1990s with most economies rebounding in 1996 and 1997 due to the improved performance of exports in general, and tourism and free trade zones in particular.

Those countries where tourism and financial services are well developed have the highest per capita income (Caribbean Development Bank 1997). Thirteen are classified as middle-income countries and nine have per capita incomes above the average for middle-income countries. The Cayman Islands and the British Virgin Islands recorded per capita GDPs of US$35 930 and US$26 957 respectively in 1996. However, dependence on preferential trading arrangements, tourism and official development assistance has made most States vulnerable to external developments.

The region has benefited from preferential trade schemes adopted by the United States, Canada and the European Union. In the case of the European Union, the Lomé Convention has provided free access to the European market for some products and has also provided financial and technical assistance. Some Caribbean countries have had easier access to European Union markets than lower-cost competitors elsewhere in the region.

Tourism accounts for one-quarter of foreign exchange earnings and provides one-fifth of all jobs (McElroy and Albuquerque 1998). Agriculture is still a significant export earner and means of livelihood in several countries. Sugar and bananas are the most important agricultural products.

However, economic growth has failed to keep pace with population growth in many countries and widespread poverty still exists: some 38 per cent of the total population – more than seven million people – are classified as poor. With the urban population forecast to rise from 62 per cent in 1995 to 69 per cent by 2010 (United Nations Population Division 1997), urban poverty will become an increasing concern.

Natural disasters have a negative impact on the sub-region. Most islands lie within the hurricane belt and are vulnerable to frequent damage. Recent major natural disasters include hurricanes Gilbert (1988) and Hugo (1989), the eruptions of the Soufriere Hills Volcano in Montserrat (1997) and the Piparo Mud Volcano in Trinidad (1997), as well as drought conditions in Cuba and Jamaica during 1997–98, attributed to the *El Niño*. More recently hurricane Georges devastated large areas, as did hurricane Mitch.

significant cause of mortality in Latin America, the most common causes of death are cardiovascular disease and malignant neoplasms. In fact, the region suffers from the ailments of both the developing and the industrialized world, although the more developed nations often have higher rates of cardiovascular disease, cancer and obesity, whereas the less developed countries have higher rates of malaria, Chagas' disease and dengue fever. In addition, mortality due to violence, accidents and AIDS are increasing in many countries (PAHO 1994).

Environmental emergencies have had a significant impact on the well-being of Latin Americans. Earthquakes, forest fires, volcanic eruptions, hurricanes and other events often devastate local infrastructure and destroy crops, causing further setbacks in the development process. The *El Niño* event of 1997–98 caused drought in Amazonia and many parts of Meso-America, and led to the death of thousands of cattle, crop losses and widespread forest fires. At one point, for example, the State of Sonora in Mexico had only 2.5 per cent of its normal water withdrawal capacity and enough water to serve its population for a mere month (La Nación 1998a). Forest fires have caused serious health problems, airport closures and destroyed hundreds of thousands of hectares of natural forest (La Nación 1998b). While most such disasters cannot be prevented, up-to-date environmental information, widespread preparation and education can reduce their impacts.

Extensive areas of South America and Meso-America have been affected by land degradation

The region's central challenge is now to build a political consensus that will maintain stability and economic growth while addressing the growing social and environmental problems. All those most concerned – governments, politicians, industrial management and labour leaders – seem aware of the seriousness of the environmental issues that are discussed below. There is also growing public awareness of the impacts of economic activities on the environment.

Land and food

Latin America has the world's largest reserves of cultivable land. The agricultural potential of the region is estimated at 576 million hectares (Gomez and Gallopin 1995). During 1980–94, the area under cultivation and permanent pasture increased and the forested area decreased (FAO 1997a and 1997b).

Almost 250 million hectares of land in South America are affected by land degradation while 63 million hectares are affected in Meso-America (see diagram below left). Soil erosion constitutes the major threat (68 per cent and 82 per cent of the affected land in South America and Meso-America respectively), while chemical degradation (mainly loss of nutrients) covers an area of 70 million hectares in South America and 7 million hectares in Meso-America (UNEP/ISPRIC 1991). In South America, some 100 million hectares of land have been degraded as a result of deforestation and some 70 million hectares of land have been overgrazed. The major cause of land degradation in Meso-America is poor management of agricultural land. Oldeman (1994) estimates that in South America 45 per cent of cropland, 14 per cent of permanent pastures and 13 per cent of forest and woodlands are affected by land degradation. In Meso-America, 74 per cent of cropland, 11 per cent of permanent pastures and 38 per cent of forested areas are estimated to be affected by land degradation.

In the Caribbean, inappropriate use of land for rapid and unplanned urbanization has led to the irretrievable loss of valuable land which should have been kept for agriculture, watershed protection and biodiversity conservation.

Expansion of permanent pastures into previously forested areas is still the main source of deforestation

Land areas and degradation

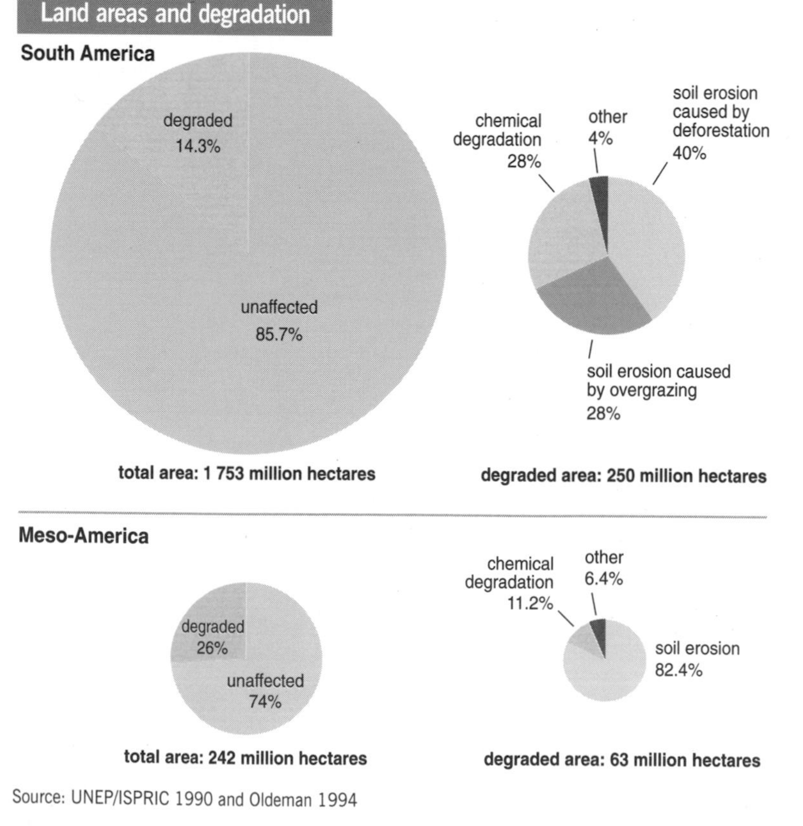

Source: UNEP/ISPRIC 1990 and Oldeman 1994

Losses from desertification

Total losses from desertification in the region may be as much as US$975 million a year. If losses due to drought as well as desertification are included, annual losses may exceed US$4 800 million. According to UNEP figures, US13 000 million would be necessary to restore degraded land and thus prevent these losses. However, due to a lack of comparability in current data and doubts about the socio-economic benefits of anti-desertification initiatives, many policy makers are reluctant to allocate funds for recuperation.

The social costs of drought and desertification may be even higher: millions of people move from the countryside to cities (in many cases in other countries), settling in the periphery of urban areas, perpetuating and aggravating urban poverty. When only men migrate to cities, leaving behind their wives and children, the latter become even more vulnerable. Women are often not recognized as legitimate partners by community and government authorities. Thus drought and desertification not only exacerbate poverty but also aggravate social breakdown and political instability.

Source: FGEB 1994

in the Brazilian Amazon (Nepstad and others 1997) although much of this area is initially used as cropland. Soybean production, mainly for export, has been the main driving force of the agricultural frontier expansion in northern Argentina, eastern Paraguay and central Brazil (Klink, Macedo and Mueller 1995). Farming technology has improved agricultural yields throughout the region. The environmental costs of these improved technologies, however, have been very high. During the 1980s Central America increased production by 32 per cent and its cultivated area by 13 per cent but doubled its consumption of pesticides (FAO 1997a).

In addition, sheep and cattle ranching have led to overgrazing and subsequent desertification (see box above), particularly in the Argentinean steppes where 35 per cent of the pasture land has been lost (Winograd 1995). In Central America, steep slopes, intense rainfall and poor agricultural practices have made erosion the principal cause of the loss of agricultural potential. Severe inequality in land distribution associated with insecure land tenure is also leading to the over-exploitation of resources for short-term benefits (Fearnside 1993 and Jones 1990).

If appropriate soil conservation measures are not adopted (including the implementation of new criteria for crop selection), the degradation of arable land will continue, endangering food production and affecting food security. It is also expected that trade integration initiatives, such as Mercosur, will have a major impact on production systems, favouring crops with high international prices and low labour requirements such as strawberries and cut flowers (Gligo 1995).

Forests

Natural forest covers 47 per cent of the total land area of the region. Almost all (95 per cent) are tropical (852 million hectares), located in Central America, the Caribbean and tropical South America. The remaining resources, covering some 43 million hectares, are found in temperate South America, mainly in Argentina, Chile and Uruguay (FAO 1997b). The northern Amazon Basin and the Guyana shield are home to the largest tract of intact, roadless forest in the world (WRI 1997). The Amazon basin is also important in global metabolism, accounting for approximately 10 per cent of net terrestrial primary production (LBA 1996).

Natural forest cover continues to decrease due to clearance for cropland and stock farming, construction of roads, dams and other infrastructure, and mining (FAO 1997b). The Latin American region lost 61 million hectares (6 per cent) of its forest cover during 1980–90, the largest forest loss in the world during these years.

The extent of natural forest cover continues to decrease in all sub-regions. More than 90 million ha were lost during 1980–95, resulting in an 8.7 per cent total loss for the period

The natural forest cover continues to decrease in all countries. A total of 5.8 million hectares a year was lost during 1990–95, resulting in a 3 per cent total loss for the period (FAO 1997b). The highest average rate

Forest extent 1980, 1990 and 1995

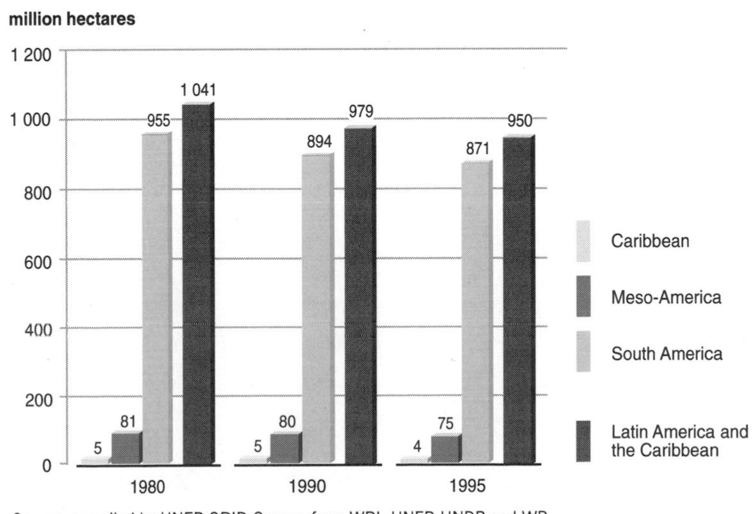

Source: compiled by UNEP GRID Geneva from WRI, UNEP, UNDP and WB 1998 and FAO 1997a and 1997b

of annual deforestation was in Central America (2.1 per cent). Bolivia, Ecuador, Paraguay and Venezuela had deforestation rates greater than 1 per cent a year for the same period (FAO 1997b). In Paraguay, for example, forest cover decreased in the eastern region from 8.8 million hectares (55 per cent coverage) in 1945 to 2.9 million hectares (18 per cent coverage) in 1991. In the western region, the decrease was from 16.8 million hectares (70 per cent coverage) to 10.8 million hectares (45 per cent coverage). The estimated deforestation rate for 1992 was 200 000 hectares a year (Stöhr 1994).

Brazil lost approximately 15 million hectares of forest area in the period 1988–97 (see graph below). Though deforestation in the Brazilian Amazon nearly

Latest figures on deforestation in the Brazilian Amazon show a substantial decline from the all-time high of 1994–95

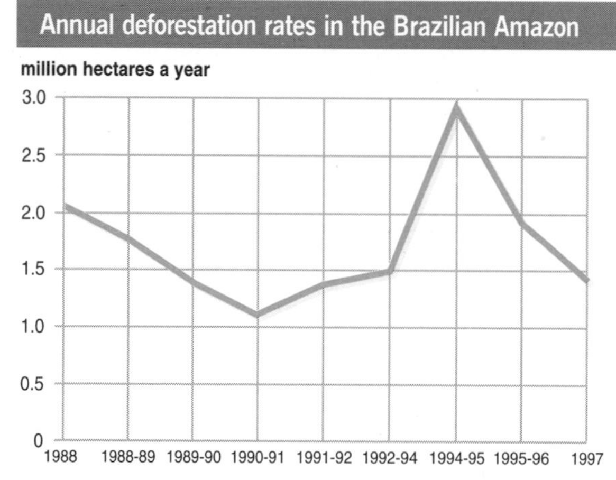

Notes: Data for 1993 and 1994 are estimates based on the mean rate of deforestation during 1992–94. The 1997 rate is estimated from an analysis of 47 Landsat images.

Source: INPE/IBAMA 1998

doubled between 1994 and 1995, with 2.9 million hectares of forest cleared in 1995, the greatest extent in recorded history, there have since been substantial declines – to about 1.8 million hectares in 1996 and an estimated 1.3 million hectares in 1997 (INPE/ IBAMA 1998). Of the eight countries in the world that still have more than 70 per cent of original forest cover, six are located in South America (Brazil, Colombia, French Guyana, Guyana, Suriname and Venezuela).

The expansion of the agricultural frontier has been one of the main causes of deforestation. Traditional slash-and-burn practices have been the primary means of advancing the agricultural frontier in many countries. However, modern agriculture, mining and the need for new roads and settlements are responsible for the largest forest clearings. Two other factors are also becoming important threats: logging (harvesting for the forest products industry) and fire caused by drought and human carelessness. In Bolivia, Guyana and Suriname, a drive to exploit natural resources, mainly brought about by an economic crisis, has accelerated the fragmentation of pristine forests over the past decade. Selective logging has changed the structure and composition of much of the remaining forested areas, particularly in southeastern Amazonia and along river courses, leading to irreversible losses in biodiversity (WRI 1997). An increasing number of countries are considering granting extensive forest concessions to forestry companies. In Guyana, one company has been granted nearly 6 million hectares, and countries such as Suriname, Bolivia and Venezuela are following suit by opening up large areas of primary forests to forest harvest (Bryant, Nielsen and Tangley 1997). The increasing pressures for forest concession in South America may worsen deforestation and forest degradation throughout the region.

In the Caribbean, large tracts of forest have been lost because of direct forest exploitation, as well as the conversion of forested areas into cropland and permanent pasture. Historically, forest clearance for sugar and banana plantations has been common in nearly all Caribbean countries. Fragmentation has also affected many of the natural forests in the Caribbean.

Damage by forest fires has increased, causing great losses to the economies of Central American countries (CCAD and IUCN 1996). A combination of both logging and drought is increasing the flammability of Amazonian forests. Logging increases flammability by opening up the leaf canopy, allowing sunlight to penetrate to the fuel layer on the ground, and by increasing the fuel load through the production of woody debris. Even virgin forests become flammable when drought is severe. Most forests in eastern and southern Amazonia (half of the 400 million hectares of closed canopy forest in Brazilian Amazonia) are subjected to severe dry seasons each year (see box), and more particularly during *El Niño* events. These forests are on the edge of the rainfall regime that is necessary for them to resist fire (Nepstad and others 1997).

Production and trade of forest products varies widely. Fuelwood accounts for 78 per cent of the region's production and industrial roundwood for 16 per cent. However, the trade of products from natural

Forest fires in the Amazon

Severe seasonal droughts associated with *El Niño* events and selective harvesting of timber are increasing the flammability of large areas of forest in the Amazon region. Forest ground fires can kill up to 50 per cent of a forest's above ground biomass, with large but poorly understood effects on forest fauna. Surface fires increase forest flammability and thus lead to dangerous positive feedback in which Amazon landscapes become successively more flammable with each burning season. These fires are usually not included in deforestation monitoring programmes, and may increase the area of forest affected by human activity by 60 per cent. Surface fires may also release significant amounts of carbon into the atmosphere.

In many regions of Amazonia, the rains that fell in 1998 were sufficient to extinguish the fires of the 1997 burning season but did not recharge the soil moisture that had been lost. In early 1998, a combination of prolonged drought and expanding slash-and-burn agriculture led to large forest fires which drew worldwide attention to the region. According to recent reports (United Nations Disaster Assessment Coordination 1998 and Barbosa 1998), 14 per cent of the Brazilian State of Roraima was burned, an area of approximately 3.3 million hectares of which 1 million hectares was forest. The UN disaster assessment task force estimated losses of 14 000 head of cattle, 700 silos and 100 rural houses, directly affecting 12 000 people (of which about 7 000 were indigenous peoples).

These fires may be a harbinger of a much larger forest fire problem in Amazonia, in which severe seasonal drought exceeds the capacity of deep Amazonian soils to buffer forests against the leaf-shedding that increases their vulnerability to fire. The forest area that could become vulnerable to fire in the 1998 dry season was estimated to be more than twice the size of Roraima, and ten times the size of Costa Rica.

Source: Moreira 1998

forests may be affected as major import countries insist on timber certification. The focus on endangered species can also affect trade; Brazil, for example, has placed a ban on mahogany harvesting (IBAMA 1998). Non-timber forest products, and non-timber gathering, still constitute the main source of cash income for many poor farmers throughout tropical South America.

The need for forest conservation has been placed high on the political agenda in many countries. Another positive development is the use of incentives for promoting the establishment of forest plantations. Recent policy reforms in Guatemala, Paraguay and Uruguay are expected to stimulate the reforestation of thousands of hectares.

Despite all these efforts, the region's forest resources are still under extreme and competing pressures. On the one hand, large population groups are heavily dependent on forests for food, especially in tropical South America (FAO 1997b) and there has been heavy encroachment of forests by the rural poor in their search for land for agricultural use. On the other hand, strong external and internal pressures are being put on countries with extensive tropical forests to try to conserve and protect these unique ecosystems.

Biodiversity

The tropical, sub-tropical and temperate habitats of the Latin America region are exceptionally rich in biodiversity. The neotropical ecological zone contains 68 per cent of the world's tropical rain forests (FAO 1997b). The region contains 40 per cent of the plant and animal species of the planet, and is considered to have the highest floristic diversity in the world (Heywood 1995). The warm Amazonian valleys, the high and cold Andean mountains, the Brazilian Atlantic forest, and the dry forests of Meso-America are home to some of the world's richest ecosystems. Arid and semi-arid vegetation occurs in the mountainous areas running from southern Ecuador to Chile, in northern Colombia, Venezuela, Argentina and northeastern Brazil. Brazil, Paraguay and Bolivia share some of the world's most important continental wetlands, including 400 000 km^2 of marshlands (the *pantanal* and *chaco*), renowned for their diversity.

The main problem is how to avoid habitat destruction and the consequent extinction of species, many of which are not yet described by science. The expansion of agriculture into semi-arid regions, forest cutting and depletion of wetlands have reduced the populations of many species. Loss of habitat has been the greatest threat. Habitat conversion has been severe in the Central American forests, the *chaco* forest, the savannah ecosystems of the Brazilian *cerrado* – which houses the largest diversity of all savannah floras in the world – and the Mediterranean-type shrublands of the Pacific coast (Dinerstein and others 1995). Mexico hosts 51 per cent of all migratory bird species from its northern neighbours, and the loss of critical overwintering sites due to deforestation and other land use changes may threaten the survival of these populations (Robinson 1997 and Greenberg 1990).

No systematic evaluation of habitat turnover and

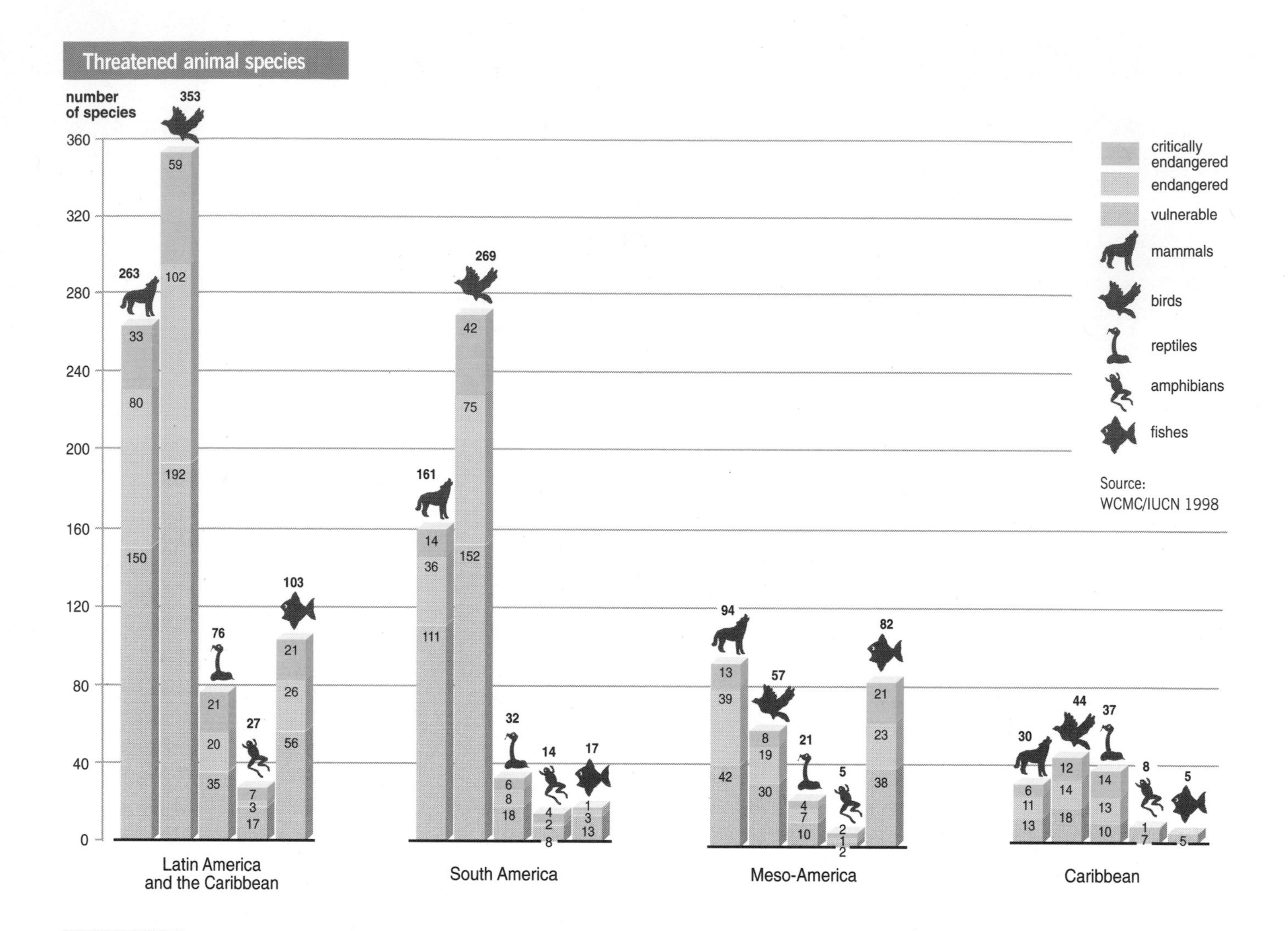

Many of the region's animal species are now vulnerable, endangered or critically endangered

species depletion has been attempted but what figures are available suggest a significant impact; several hundred vertebrate species are now threatened with extinction (Baillie and others 1996). The intensification of agricultural practices, forest replacement with plantations, new technologies for cultivating dry lands (a major reservoir of biodiversity) and the modification of the coastline suggest that these trends may worsen in the near future.

The biota of all countries are threatened. Brazil has the second largest number of threatened bird species (103 species) in the world, and Peru and Colombia occupy the fifth place with 64 species each (Baillie and others 1996). A third of Chilean vertebrates (marine fishes excluded) are threatened (Simonetti and others 1995). Brazil also has 71 threatened mammal species (the fourth highest in the world). More than 50 per cent of Argentinean mammals and birds are also threatened. Areas with large numbers of threatened birds tend also to have large numbers of threatened mammals. This indicates that the two groups may be susceptible to similar threats.

Ecosystems and their vegetation are similarly threatened. In central Chile, for example, it is estimated that 30 per cent of the *maulino* forest in the Cordillera de la Costa was replaced by pine plantations during 1978–87 (CODEFF 1987).

The amount of land under some form of conservation and protection is continuing to rise, with some 6.6 per cent of the region's land under categories of strict protection. However, many types of ecosystems are still under-represented or not represented in protected areas (Dinerstein and others 1995). Furthermore, many protected areas, despite their declared legal status, are really only protected on paper, and lack any real means of preventing degradation. Central America is recognizing the social value of biodiversity for local communities as it re-evaluates its biodiversity and natural resources as the basis for the generation of new products and hence socio-economic development (CCAD and IUCN 1996).

Despite much publicized support for biodiversity conservation, the lack of governmental and institutional support for research and development in biodiversity suggests that trends of declining biological diversity will continue unabated over the next decades.

Freshwater

The Latin American region is extremely rich in water resources: the Amazon, Orinoco, São Francisco, Paraná, Paraguay and Magdalena rivers carry more than 30 per cent of the world's continental surface water. Nevertheless, two-thirds of the region's territory is classified as arid or semi-arid. These areas include large parts of central and northern Mexico, northeastern Brazil, Argentina, Chile, Bolivia and Peru.

Demand for water is growing rapidly as populations and industrial activity expand and irrigated agriculture (the largest use) continues to increase (WRI, UNEP, UNDP and WB 1996). Many current patterns of water withdrawals are clearly unsustainable, such as pumping from aquifers at rates far greater than they are recharged.

Despite the advances of the past ten years, access to safe water remains an important issue. Many people still lack an adequate water supply and a sewage system. In 1995, around 70 per cent of the Central American population had access to a piped public water supply but in Latin America as a whole as little as 2 per cent of sewage receives any treatment (World Bank 1997). If action is not taken in the near future, these problems could present severe health and environmental risks.

Where industry, mining and use of agricultural chemicals are expanding, rivers become contaminated with toxic chemicals and heavy metals. Virtually all countries in Latin America have artisanal mining activities, of which gold is the most mined mineral. It is estimated that as many as one million artisanal miners are producing some 200 tonnes of minerals annually (Veiga 1997). Nevertheless, mercury emissions have been reduced from the high levels observed in the late 1980s as a result of a reduction of informal mining activities due to scarcity of easily exploitable ores, better organization of mining activities (largely by NGOs), and the high cost of mercury which has led many miners to recycle. However, probably about as much mercury is still emitted as gold is produced. Since the beginning of the new gold boom in Latin America at the end of the 1970s, around 5 000 tonnes of mercury may have been discharged into the forests and urban environment (Veiga 1997).

One of the origins of groundwater pollution is seepage from improper use and disposal of heavy metals, synthetic chemicals and hazardous wastes. The quantity of such compounds reaching groundwater from waste dumps appears to be doubling every 15 years in Latin America (UNDP 1995). Aquifer depletion and salt water intrusion are also important sources of groundwater contamination.

Size and number of protected areas

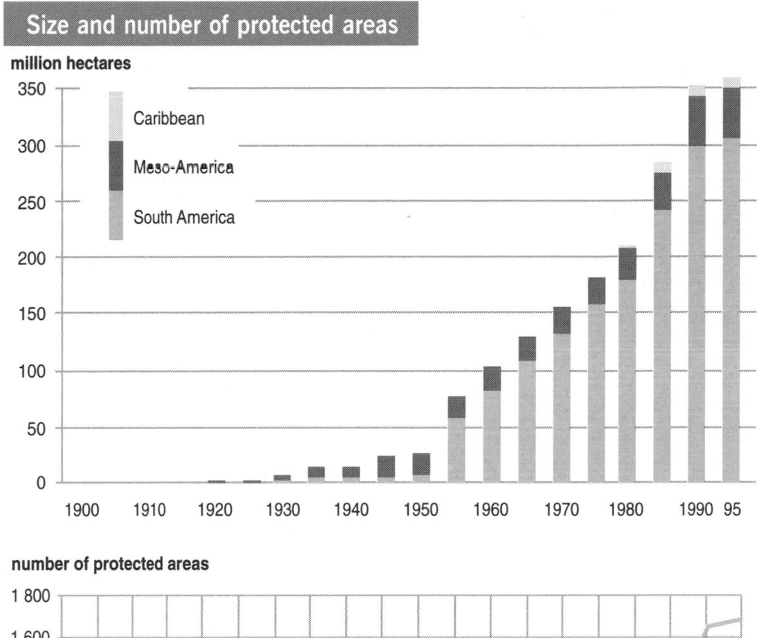

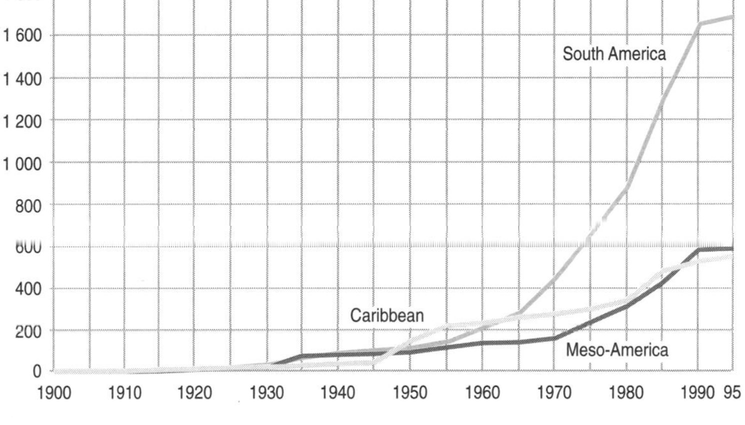

Source: WCMC 1998

While the number and area of protected sites in Latin America continue to grow, many are protected only on paper and have no real resistance to degradation.

The *Hidrovia* project

The Plata Basin countries – Argentina, Bolivia, Brazil, Paraguay and Uruguay – have agreed to ensure permanent navigation of the Paraguay-Paraná waterway in view of its importance to the region's economic development. The proposed *Hidrovia* Project involves a 3 282-km long waterway starting in Puerto Cáceres, Brazil, and ending in Nueva Palmira, Uruguay. This project will harness two rivers to the needs of modern trade; the Paraná and Paraguay, tributaries of the River Plate, which drains South America's largest river basin after the Amazon.

Feasibility studies are being made of two different schemes, Modules A and B. Module A is a short-term, more restricted project, consisting mainly of dredging from Santa Fé (Argentina) to Corumbá (Brazil) and Puerto Quijarro (Bolivia), including the Tamengo Canal, and sign posting from Corumbá to Nueva Palmira (Uruguay). This module only marginally includes the Pantanal region. Module B is a more ambitious, long-term project, which includes environmental impact assessments (already completed and under analysis by participating governments), dredging, course changes, correction and stabilization of navigation channels, regulation of water resources, as well as construction of water engineering structures from Cáceres, Brazil, to Nueva Palmira, Uruguay. In this case the scheme directly affects the *Pantanal*, the world's greatest wetlands. Scientists are worried that altering the river's course might alter the *Pantanal's* water cycle, drying up some regions and flooding others. Environmentalists fear for the area's animals and plants. The Brazilian government agreed in 1997 to prevent any further construction activities affecting the *Pantanal* region, and restrain itself to maintenance activities on the Brazilian side of the waterway.

Sources: Bucher and others 1993, CIHPP 1995

The sediments produced by erosion, and the discharge of domestic, industrial and agrochemical wastes, are the principal causes of water quality deterioration. The Alcehuate in El Salvador and the Virilla in Costa Rica are just two examples of rivers heavily polluted by agro-industrial activities and metropolitan development. As industry, irrigation and population expand, so do the environmental and economic costs of providing additional water supplies. Some countries, such as Mexico and Peru, now use more than 15 per cent of their total freshwater reserves per year.

The main cause of water pollution is the direct discharge of untreated domestic and industrial wastes to surface water bodies, which contaminates not only the water bodies but adjacent groundwater aquifers. The geographical distribution of water pollution in the region is dominated by flows from large metropolitan areas. The major contributing factors are: the concentration of population and industrial production in large metropolitan centres; growth in conventional sewerage systems which has not been accompanied by corresponding treatment facilities; intensification of agricultural land use close to metropolitan areas; changes in economic structure, with increased emphasis on manufacture; concentrated run-off from paved areas in the growing cities; and the need for artificial regulation of stream flows. As a result, the quality of water bodies near large metropolitan areas has been seriously compromised. A secondary source of point pollution comes from mining.

The costs of supplying water to the cities are continually rising, with dramatic examples from large and growing urban areas. In Mexico City, water is pumped over elevations exceeding 1 000 metres into the Valley of Mexico, and in Lima upstream pollution has increased treatment costs by about 30 per cent (World Bank 1997). Investment in sanitation and water offer high economic, social and environmental returns, but the next four decades will see urban population rising threefold and domestic demand for water increase fivefold in Latin America (WRI, UNEP and UNDP 1994).

Water availability has been a fundamental factor in

the development of irrigation throughout the region. An area of 697 000 km^2 is currently irrigated, corresponding to 3.4 per cent of the region's territory (World Bank 1996) but salinization and waterlogging are eating away the productivity of 40 years of irrigation investments in countries such as Mexico, Chile and Argentina (Winograd 1995).

After the hydroelectric projects that dominated the region in the 1970s, such as Itaipu, Salto Grande and Yaciretá in the Plata Basin, and Tucuruí and Balbina in the Amazon Basin, the current trend in South America is the construction of *hidrovias* or waterways. Two ambitious projects are under way in the region, the Paraná-Paraguay and the Araguaia-Tocantins waterways, that are planned to harness five river systems over a total length of 8 000 km to improve continental navigation networks (see box opposite).

During the past decade, environmental problems related to water have affected both urban and rural areas. In the arid and semi-arid areas, there has been increased competition for scarce water resources. Using polluted water for drinking and bathing spreads infectious diseases such as cholera, typhoid and gastroenteritis. Several countries have had recent outbreaks of these diseases, which affect the urban poor in particular.

In the Caribbean, housing developments continue to be sited in sensitive areas such as on steep hillslopes in the upper parts of water catchment areas, and too close to sensitive groundwater aquifers. Freshwater resources are thus being damaged at the same time as demand for water is increasing.

Marine and coastal areas

The marine and coastal systems of the region support a complex interaction of distinct ecosystems, with an enormous biodiversity, and are among the most productive in the world. Several of the world's largest and most productive estuaries are found in the region, such as the Amazon and Plata Rivers on the Atlantic

Many of the region's coral reefs are under threat; the Caribbean sub-region is the most affected, with 29 per cent of its reefs at high risk

The threat to coral reefs

Source: WRI, ICLARM, WCMC and UNEP 1998

Marine fish catch

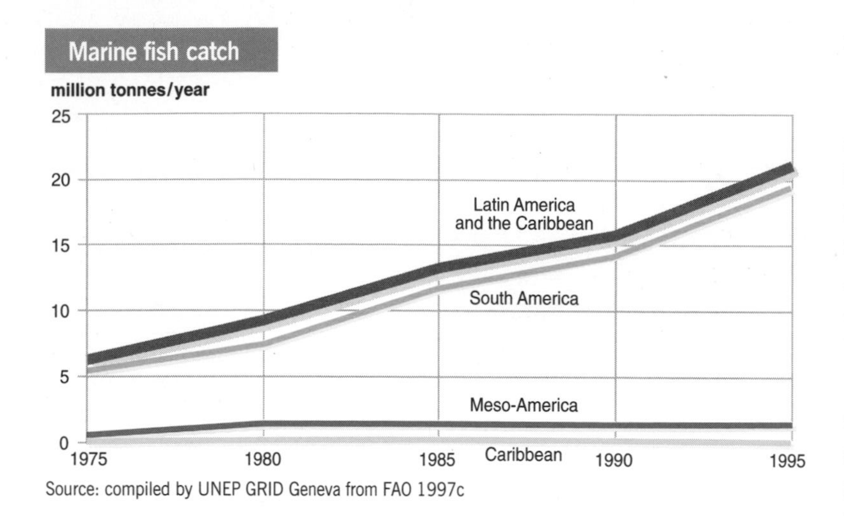

Source: compiled by UNEP GRID Geneva from FAO 1997c

The marine fishery catch has been growing fast in South America but the 1997–98 El Niño is expected to produce a large decrease

coast, and the Guayaquil and Fonseca on the Pacific. The coast of Belize has the second largest barrier reef in the world. The waters off Chile and Peru support one of the top five commercial fisheries and the world's fastest growing fishery is off the coast of Argentina and Uruguay (IDB 1995). The region's coastal zone is 64 000 km long and encompasses 16 million km^2 of maritime territory. For countries such as the island nations of the Caribbean, Panama and Costa Rica, this territory represents more than 50 per cent of the total area.

The total marine catch in 1995 was about 21 million tonnes (see graph above), some 20 per cent of the world catch. Over the decade 1985–95, many South American countries doubled or tripled their marine catch, and Colombia increased its catch fivefold. During 1970–83 Peru's catch fell from 12 to 2 million tonnes due to *El Niño* events but had increased to nearly 9 million tonnes by 1995 (IDB 1995). A large decrease is expected as a result of the 1997–98 *El Niño*, the most severe event recorded to date (see box right). The effects of fisheries on marine biodiversity and resource sustainability are also of major concern, since more than 80 per cent of the commercially exploitable stocks in the southwestern Atlantic and 40 per cent in the southeastern Pacific are either fully fished, over-fished or depleted (FAO 1997c).

A wide range of fisheries activities (industrial, artisanal and recreational) coexist in the Caribbean sub-region. Overall landings from the main fisheries rose from around 177 000 tonnes in 1975 to a peak of 256 000 tonnes in 1985 before declining to around 136 000 tonnes in 1995. According to an FAO assessment, some 35 per cent of the region's stocks are overexploited (FAO 1997c). The sub-region also has the highest percentage discard, mostly as by-catch of shrimp trawling.

Mariculture is less important than in some other tropical regions but is growing in countries such as Ecuador where a significant shrimp mariculture industry has developed, mostly in converted mangrove areas. Latin America produced 21.6 per cent of the world's farmed shrimp in 1995. Aquaculture in Chile is growing at more than 30 per cent a year, compared with 9.5 per cent worldwide. Activities are concentrated in salmon farming, induced by favourable export markets, and are generating some US$450 million a year in export earnings. In 1997 salmon exports were more 145 000 tonnes (Ministerio de Economía 1997), and this trend is expected to continue (Instituto de Fomento Pesquero 1998).

The need for integrated coastal management in the Caribbean

Industries dealing with horticulture and aquaculture, oil, lumber, chemicals, textiles, vehicle repairs and ship building have all added large quantities of hazardous materials to rivers, estuaries, wetlands and coastal areas, and have had major impacts on the aquatic and marine environments (Davidson 1990). Agrochemical residues are also found in estuarine and coastal sediments as well as in coastal waters. Land reclamation for residential, industrial, agricultural and tourism purposes has caused the degradation of coastal and marine ecosystems of the sub-region.

Limited infrastructure serving the tourism industry and coastal settlements has further contributed to pollution problems in coastal waters. In addition to locally-generated waste, the increasing popularity of the Caribbean as a destination for cruise ships and yachts has led to an increase in the volume of waste being discharged directly into the natural environment, because port reception facilities for ship-generated solid waste are generally inadequate. In densely developed coastal areas, the risk of sewage pollution of coastal waters is high because of the height of the groundwater table and the absorptive capacity of the soils. In countries such as Barbados, Jamaica and Haiti, the degradation of protective reef systems by sewage-induced eutrophication has contributed to coastal erosion and the destruction of beaches.

Clearly, careful planning and management of all sectoral activities simultaneously will result in greater overall benefits than pursuing sectoral development plans independently of one another. Integrated coastal management approaches are required, combining all aspects of the human, physical and biological aspects of the coastal zone within a single management framework.

El Niño

The term *El Niño* was originally applied to a tepid current found along the coast of Peru and Ecuador every Christmas. The term is now used to describe the exceptionally warm and long-lived currents that occur every two to seven years, beginning in the summer and lasting for as long as 22 months. *El Niño* effects are propagated across the world's weather systems. In South America, they have caused catastrophic floods in southern Brazil and along the Pacific coasts of Ecuador, Peru and Chile; severe droughts in Brazil's northeastern region and the Altiplano areas of Peru and Bolivia; and dramatic falls in the Pacific coast fish catch. The 1997–98 El Niño resulted in economic losses that exceeded the combined international non-reimbursable, non-military, development assistance for the same period.

Tourism accounts for about 12 per cent of GDP in the region, mainly in coastal areas. Some 100 million tourists visit the Caribbean annually, contributing about 43 per cent of the Caribbean's combined GNP and one-third of export earnings (WTTC 1993). By the year 2005, scuba diving tourism alone could generate revenues of approximately US$1 200 million in the Caribbean (WTO 1994). In addition to generating employment (10 million were employed in tourism in 1993), tourism investments lead to important land use changes in coastal areas. Many rural coastal areas are experiencing a gradual shift from dependence on local fisheries and agriculture towards the provision of tourism services and related activities (WTTC 1993).

The region's ports are the second leading destination for containerized US exports, and the Panama canal is a major focus of maritime trade. The total tonnage passing through the region's ports increased from 3.2 to 3.9 per cent of the world total during 1980–90, and a marked increase is expected as a result of trade liberalization and privatization of the region's ports (UNCTAD 1995). Expanding ports and maritime trade are often accompanied by intensified transportation corridors in coastal ocean areas, as is happening off Argentina, Brazil, Ecuador and Uruguay.

All these activities are bringing rapid and often drastic transformation to coastal and marine areas. Land conversion is causing degradation of coastal habitats, including mangroves, estuaries and coral reefs. Mangroves, for example, have been disappearing fast over the past 20 years, and as much as 65 per cent of Mexico's mangroves have already been lost (Suman 1994). Coastal water quality has been declining throughout the region, due to increasing discharges of untreated municipal waste.

The reefs of the Caribbean and adjacent waters constitute about 12 per cent of the world total, and are good indicators of the severe damage that has been inflicted on the environment. Today 29 per cent of the sub-region's reef areas (see map on page 129) are considered at high risk due to increased run-off and sedimentation caused by deforestation, nutrient contributions from sewage from hotels and shipping, coastal construction and mining (Bryant and others 1998). Declining coastal water quality, reef degradation and beach erosion are linked in a cycle which threatens public health, shore-front properties and tourism.

Poor and landless people have settled in flood-prone coastal areas in countries such as Brazil, Ecuador, Guyana and Honduras, increasing coastal pollution, over-harvesting, and conflicts over access to traditional fishing areas (IDB 1995).

Atmosphere

Most Latin American countries find it difficult to obtain reliable information from which to prepare emission inventories of greenhouse gases (GHG). Emission factors specific to a particular region or system are scarce, forestry and land use change are difficult to characterize, and many of the data either do not exist or must be derived from related statistics or even from anecdotal evidence. The trends emerging from completed (Uruguay and Argentina) and preliminary (Costa Rica, Mexico and Venezuela) inventories suggest that more than 50 per cent of emissions come from industrial production and energy generation. In Brazil and Chile, gross emissions of GHG due to energy consumption are considerably

Per capita emissions of carbon dioxide are well below the 12 tonnes estimated for the high-income economies, and also below the world average of 4.0 tonnes

Carbon dioxide emissions per capita

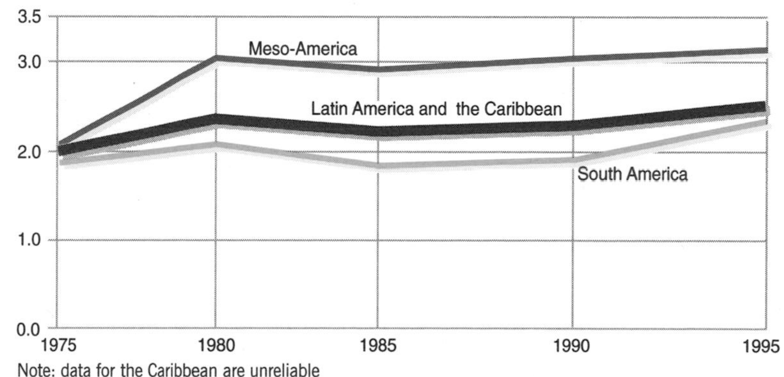

Note: data for the Caribbean are unreliable
Source: compiled by UNEP GRID Geneva from WRI, UNEP, UNDP and WB 1998 and CDIAC 1998

lower than emissions from deforestation, land use change and agriculture (Bonduky and others 1995).

The region is responsible for 4.3 per cent of the world's total carbon dioxide emissions from industrial processes, and for 48.3 per cent of emissions from land-use changes. Methane emissions from anthropogenic sources correspond to 9.3 per cent of the world's total. The average per capita emission of carbon dioxide for 1995 was 2.55 tonnes, well below the 11.9 tonnes estimated for the high-income economies, and also below the world average of 4.0 tonnes (CDIAC 1998).

The main anthropogenic source of emissions is deforestation, and Amazonia is an important natural source of methane and nitrogen oxides. Conversion of primary tropical forests to agriculture and to secondary vegetation is a significant change on a global scale. In the Amazon basin, which covers nearly 7 million km^2, biomass burning and the establishment of new types of vegetation cover will have significant ecological implications for the region, the continent and the planet (LBA 1996).

In 1993, about 70 per cent of Brazilian electricity came from hydropower (Rosa and others 1996). In Central America more than 50 per cent of the energy produced is generated by hydropower. However, there is increasing conflict over access to and use of water because of the vulnerability of hydropower to climatic variability. Uruguay, for example, generates most of its energy from hydropower but severe droughts over the past few years have given rise to water allocation problems which have affected agricultural production. In addition, there is a trend in, for example, Argentina, Brazil and Colombia to move from renewable forms of energy to fossil fuels, in both the electric power and the transportation sectors, as a result of the deregulation of the energy sector (Rosa and others 1996). Deregulation and privatization of energy could increase emissions since market forces will probably not favour biomass and hydropower. Private investment tends to prefer fossil fuel power plants to hydroelectric plants because capital costs are lower and the return on the investment is faster – even though energy costs are higher (Tolmasquim 1996).

Many countries have substantial potential for curbing carbon emissions, given the region's renewable energy sources of biomass and hydropower, and the potential of forest conservation and reforestation programmes to provide valuable carbon sinks. The use of ethanol as a substitute for gasoline can also reduce carbon dioxide emissions.

Urban areas

The region has a highly urban population. In 1950, 43 per cent of the total population lived in urban areas but this had risen to 73.4 per cent by 1995 (see graph opposite). Most of the urban population lives in large cities such as Mexico City (16.5 million people), São Paulo (16 million) and Rio de Janeiro (10 million). Buenos Aires, with nearly 12 million people, and Santiago, with 5 million, house 34 per cent of the population of Argentina and Chile (WRI 1996). While urbanization itself does not necessarily have negative socio-economic or environmental impacts, unplanned urban growth has led to the development of outer and inner city slums, many of which lack basic services.

Increasing population density and economic activity have led to increased pollution in many cities. Santiago, for example, is now one of the most polluted urban areas in the world; the main sources of air pollution are urban transport, and small and medium-sized industries (IMO 1995). Air pollution is causing severe respiratory problems among city dwellers, with higher rates of pneumonia than in many other cities, and many premature deaths from respiratory diseases. Treatment costs are high and there are productivity losses due to absenteeism (O'Ryan 1994). Mexico City, São Paulo and Bogota are also suffering from severe air pollution. The Brazilian programme of adding alcohol to gasoline has, despite reducing carbon dioxide emissions by some 30 per cent and decreasing air pollution, not been sufficient, and São Paulo now restricts private car circulation, as do Mexico City and Santiago. In São Paulo and Rio de Janeiro, 27 million people are exposed to high levels of particulate air pollution estimated to cause 4 000 annual cases of premature mortality (CETESB 1992).

Lead emissions are also a major problem. The main sources of exposure are emissions from vehicles that use leaded gasoline, industrial production, particularly of paints and batteries, and food. The effects of leaded gasoline are felt most in urban environments. Residents of areas with high levels of traffic generally have a much higher level of lead in the blood than those exposed to less traffic. However, over the past decade or so, the lead content of gasoline has been decreasing in most countries, and lead-free gasoline has been introduced. Countries with the largest share of lead-free gasoline are Brazil (100 per cent), Costa Rica (100 per cent), Guatemala (80 per cent) and Mexico (46 per cent) (Christopher and others 1996).

Until the mid-1970s, poverty was generally more common in rural areas than in urban ones. In the 1990s, however, regional statistics show that 65 per cent of poor households are in urban areas (World Bank 1996). During 1990–94, the level of urban poverty fell from 33 to 24 per cent in Chile, and from 12 to 6 per cent in Uruguay (CEPAL 1996). In Buenos Aires, 17 per cent of the population live in households with unsatisfied basic needs (overcrowded households, inadequate housing, poor access to clean drinking water and sanitary infrastructure) compared to 22 per cent in the 1980s (La Serna and others 1997).

Unplanned growth in urban areas has most effect on the poor, who often lack proper water supplies and sanitation services, even though these are generally well developed in urban areas. Unplanned growth has other environmental effects associated with inadequate means of disposing of wastewater, a demand for water that exceeds supply, and the pollution of groundwater.

The topographic position of most Latin American cities makes it difficult to use conventional methods of water and sewage treatment, and collection of solid wastes. In Central America, a study of 158 urban centres with more than 10 000 residents showed that industrial and domestic wastes were not treated before disposal (Incer 1994). Caracas, La Paz, São Paulo, Rio de Janeiro and Lima, among others, have serious sanitation problems. In Brazil, sewerage networks reach only 49 per cent of the urban population. Lack of water and sanitation in Brazil's urban areas is estimated to cause some 8 500 cases of premature mortality per year (Barros and others 1995).

The disposal of solid waste is also problematic. Only 30 years ago, the generation of solid waste per capita was 0.2–0.5 kg a day while it is now 0.5–1.00 kg a day. The problem lies not only in the quantity but also in the quality and composition of the wastes, which have changed from being dense and almost entirely organic to voluminous and increasingly non-biodegradable, with a higher percentage of toxic substances. In 1995, Latin America was generating approximately 275 000 tonnes a day of solid urban waste. A fleet of 30 000 trucks and 350 000 m^3 of land a day would be needed to collect and bury this garbage in a sanitary fashion. In reality, the region's waste disposal facilities are composed of 35 per cent sanitary landfills and 25 per cent semi-controlled landfills (PAHO 1995). The other disposal facilities do not comply with minimum norms and are best regarded as no more than rubbish dumps.

Progress has been made but only in a few large cities that, because they are so big, distort the statistics and lead to unwarranted optimism. In reality, the situation in other cities is grim. In Brazil, a national survey found that 57 per cent of cities have open garbage dumps, 14 per cent have controlled landfills, and 28 per cent have sanitary landfills or other adequate methods of waste disposal. In Chile,

Urban population

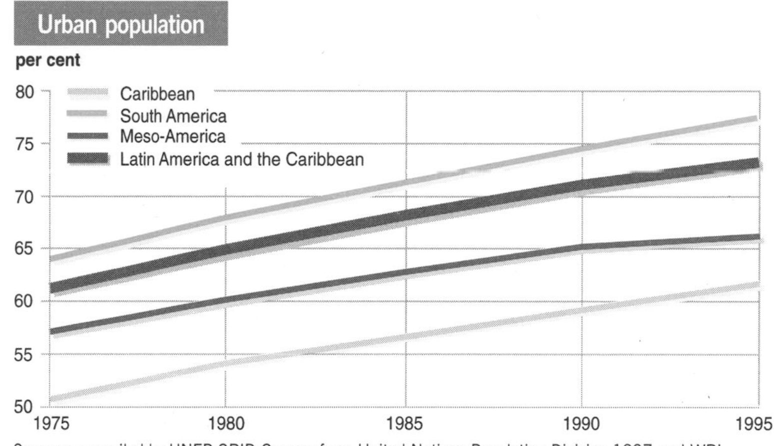

Sources: compiled by UNEP GRID Geneva from United Nations Population Division 1997 and WRI, UNEP, UNDP and WB 1998

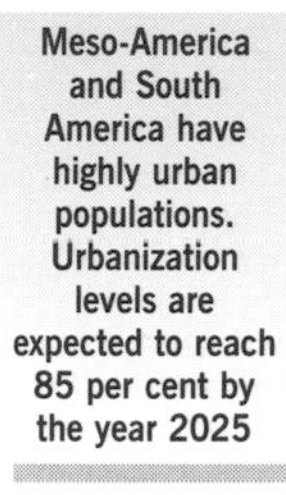
Meso-America and South America have highly urban populations. Urbanization levels are expected to reach 85 per cent by the year 2025

Urban population growth rates are falling sharply, mainly because of the already very high levels of urbanization in the region

Growth of urban populations

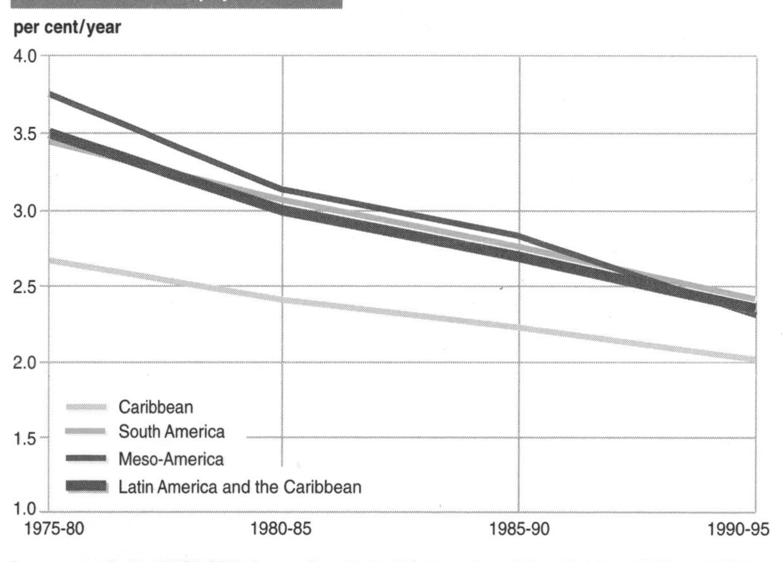

Source: compiled by UNEP GRID Geneva from United Nations Population Division 1997 and WRI, UNEP, UNDP and WB 1998

78 per cent of waste disposal facilities are sanitary, as are 30 per cent in Mexico. In other countries, such as Bolivia, Ecuador, Peru and most Central American nations, there are no sanitary landfills outside the capital cities, although both Bolivia and Colombia have interesting programmes for medium-sized cities (PAHO 1995).

In most Caribbean countries, increased population and per capita incomes, altered patterns of consumption, and the rapid development of the tourism and industrial sectors have led to fast rates of growth in waste generation. In 1994, the volume of waste disposed of at the major landfills in Trinidad and Tobago was 44 per cent more than in 1979 whereas population increase over the same period was only 30 per cent (Goddard 1997). The quality and composition of the waste has also altered significantly, becoming much less biodegradable. Increasing amounts of plastics, aluminium, paper and cardboard packing cases are being used and disposed of by households and businesses.

Solid waste production, wastewater treated and garbage collection

	solid waste per capita (kg/year)	wastewater treated (%)	households with garbage collection (%)
Brasilia	182	54	95
Havana	584	100	100
La Paz	182	0	92
San Salvador	328	2	46
Santiago	182	5	57
By comparison:			
Toronto	511	100	100

Source: Habitat 1997

Urban populations are growing very rapidly. For example, in the Brazilian states of Rondonia, Para and Mato Grosso, and in Santa Cruz de la Sierra in Bolivia, the urban population has been doubling every decade (WRI 1997). In the Brazilian cities of Manaus and Belém, the population grew by more than 65 per cent in the past decade (IBGE 1991). Although the rates of growth of large cities declined in the 1990s, the problem is now moving to the medium-sized cities. Increases in income in countries such as Chile, for example, have changed consumption patterns in favour of more intensive use of the environment. Higher income is accompanied by increased consumption and more waste. In Santiago, for example, areas with higher income levels produce 1 kg of waste per inhabitant a day, while the poor areas generate 0.5 kg (Escudero 1996).

It seems likely that over the next decade water problems will be moderated but not solved by investment in sanitation programmes. Urban air pollution, in spite of special governmental programmes, may remain high because of the increase in the number and use of private cars, the main source of air pollutants. No significant reductions are expected in industrial wastes and emissions because of the magnitude of the problem, and the time and resources needed to implement effective policies.

Conclusions

Recent trends in Latin America and the Caribbean point to high growth rates, new opportunities for public participation, and an improvement in certain aspects of the quality of life. Nevertheless, the region is still characterized by high (and growing) inequalities, both in the distribution of wealth and in access to opportunities. While there are significant improvements in some health indicators, the growing poverty gap is leading to a resurgence of infectious disease.

The environmental costs of regional economic expansion are already extremely high, and seem to be growing. The major issues are:

- an accelerating over-exploitation of land and marine resources;
- nutrient depletion and soil erosion;
- overgrazing and subsequent desertification;
- continuing deforestation;
- loss of biodiversity and habitat degradation;
- groundwater contamination and depletion;
- increasing conflict over access to and use of water;
- air pollution;
- heavy metal contamination; and
- urban waste disposal problems.

If today's central challenge in Latin America and the Caribbean is to build a political consensus that will maintain stability and economic growth, the accelerating social and environmental problems listed above must also be strongly addressed. Current improvements in accessing environmental information are an important asset in this respect.

The first priority is to find solutions to the problems of the urban environment, which now houses nearly three-quarters of the region's population. Improved housing, sanitation, transportation and employment in large urban areas are badly needed.

The second priority is to find ways of promoting the sustainable use of tropical forests and biodiversity. There are many examples of what should not be done but forest conservation and reforestation has at last become a political priority in many countries.

References

Baillie, J., Groombridge, B., Gärdenfors, U., and Stattersfield, A.J. (eds., 1996). *1996 IUCN Red List of Threatened Animals.* IUCN, Gland, Switzerland

Barbosa, R. I. (1998). *Avaliaçao preliminar da área dos sistemas naturais atingida por incêndios no estado de Roraima.* Instituto Nacional de Pesuisa da Amazonia, Manaus, Brazil

Barros, M. E., Sergio F. Piola e Solon M. Vianna (1995). *Política de Saúde no Brazil: Diagnóstico e Perspectivas.* IPEA, Brasília, Brazil

Bonduki, Y., K. Bowers, B. Braatz, M. Perdomo, N. Pereira, A. M. Segnini (1995). Latin American Greenhouse Gas Emissions and Mitigation Options. In *Revista de la Facultad de Ingenieria – Universidad Central de Venezuela*, Vol. 10, No. 1-2

Bryant, D., D. Nielsen and L. Tangley (1997). *The last frontier forests: ecosystems and economics on the edge.* World Resources Institute, Washington DC, United States

Bucher, E. H., A. Boneto, T. P. Boyle, P. Canevari, G. Castro, P. Huszar, and T. Stone (1993). *Hidrovia – An Initial Environmental Examination of the Paraguay-Paraná Waterway*. Wetlands for the Americas Manomet, Massachusetts, United States, and Buenos Aires, Argentina

Caribbean Development Bank (1997). *Annual Report 1997.* CDB, St Michael, Barbados

CCAD and IUCN (I 996). *Reducción del efecto invemadero mediante la limitación y absorbción del CO_2 en América Central.* Propuesta Plan de Prevención y Combate de Inciendos Forestales en América Central. Comisión Centroamericana de Ambiente y Desarollo, Consejo Centroamericana de Bosques y Areas Protegidas, Unión Mundial para la Naturaleza, Oficina para Mesoamérica, San José, Costa Rica

CDIAC (1998). *Revised Regional CO_2 Emissions from Fossil-Fuel Burning, Cement Manufacture, and Gas Flaring: 1751–1995.* Carbon Dioxide Information Analysis Center, Environmental Sciences Division, Oak Ridge, Tennessee, United States. http://cdiac.esd.ornl.gov/cdiac/home.html

CEPAL (1996). *1980-1995 – 15 ans de desempeno economico.* Comisión Economica para America Latina y el Caribe, Naciones Unidas, Santiago, Chile

CETESB (1992). *Relatorio de Qualidadde do Ar em São Paulo.* Compahia de Tecnologia de Saneamento Ambiental, São Paulo, Brazil

Christopher, P. H., M. Hernández-Avila, D.P. Rall (1996). *El Plomo en América – Estrategias para la prevención*

CIHPP (1995). *Hidrovia Paraguay-Paraná: Secretaria Ejecutiva.*Comité Intergubernamental de la Hidrovía Paraguay-Paraná, Maldonado, Uruguay

CODEFF (1987). *Evaluación de la destrucción y disponibilidad de los recursos forestales nativos en la VII y VIII regiones*. Comite Nacional Pro Defensa de la Fauna y Flora, Santiago, Chile

Comisión Nacional de Población de México (1998). *Estadísticas ambientales 1996.* Comisión Nacional de Población de México, Mexico City, Mexico

Davidson, L. (1990). *Environmental Assessment of the Wider Caribbean Region*. UNEP Regional Seas Reports and Studies No. 121. UNEP, Nairobi, Kenya

Dinerstein, E., D. M. Olson, D. J. Graham, A. L. Webster, S. A. Primm, M. P. Bookbinder, G. Ledec (1995). *A Conservation Assessment of the Terrestrial Ecoregions of Latin America and the Caribbean.* World Bank, Washington. United States

Escudero, Juan y Sandra Lerda (1996). Implicaciones ambientales de los cambios en los patrones de consumo en Chile. In Sunkel, O. (ed.). *Sustentabilidade Ambiental del Crecimiento Economico Chileno*. Universidad de Chile, Santiago, Chile

FAO (1997a). *FAOSTAT Statistics Database.* FAO, Rome, Italy. http://www.fao.org

FAO (1997b). *State of the World's Forests 1997.* FAO, Rome, Italy

FAO (1997c). *FAO Fishstat-PC.* FAO, Rome, Italy

Fearnside, P.M. (1993). Deforestation in Brazilian Amazonia: the effect of population and land tenure. In *Ambio,* 22, 537–45

FGEB (1994). *Anales del Taller Latino Americano de la Desertificación.* Fundación Grupo Esquel Brazil, Fortaleza, Ceará, Brazil, 1994

Gligo, N. (1995). The Present State and Future Prospects of the Environment in Latin America and the Caribbean. In *CEPAL Review* 55, April 1995

Goddard, G. (1997). Background Paper on Solid Waste Management in Trinidad and Tobago, Port of Spain. Internal report of the Environmental Management Authority of Trinidad and Tobago. Port of Spain, Trinidad and Tobago

Gómez, I. A and G. C. Gallopin (1995). Potencial Agrícola de la América Latina. In *El Futuro Ecológico de un Continente: Una Vision Prospectiva de la América Latin.* Editorial de la Universidad de las Naciones and Fondo de Cultura Económica, Mexico

Greenberg, R. (1990). *Southern Mexico: Crossroads for Migratory Birds.* Smithsonian Migratory Birds Center, National Zoological Park, Washington DC, United States

Heywood, V. H. (1995). *Global Biodiversity Assessment*. Cambridge University Press, Cambridge, United Kingdom

IBAMA (1998). *Mogno no Brasil*. IBAMA, Brasília, Brazil

IBGE (1991). *1960, 1970, 1980 and 1991 Censo Demográfico.* Fundação Instituto Brasileiro de Geografia e Estatística, Rio de Janeiro, Brazil

IDB (1995). *Coastal and Marine Resources Management: Strategy Profile.* Inter-American Development Bank, Washington DC, United States

IMO (1995). *Global Waste Survey – Final Report*. International Maritime Organization, Manila, Philippines

Incer, J. (1994). Deterioro ambiental en Centroamérica y sus efectos sobre la salud. In *Conferencia Centroamericana sobre Ecologia y Salud,* Programa Medio Ambiente y Salud en el Istmo Centroamericano (MASICA), Comisión Centroamericana de Ambiente y Salud, Oficina Panamericana de la Salud, San José, Costa Rica

INPE/IBAMA (1998). *Deforestation in Amazonia 1995-1997.* INPE/IBAMA, Brasilia, Brazil

Instituto de Fomento Pesquero (1998). Mercado del sector pesquero. In *Boletín Trimestral* 42, March 1998

Jones, J. (1990). *Colonization and Environment: land settlement in Central America*. United Nations University Press. Tokyo, Japan

Klink, C. A, R. H. Macedo, C. C. Mueller (1995). *Bit by bit the Cerrado loses space*. WWF and Pró-Cer (Sociedade de Pesquisas Ecológicas do Cerrado), Brasilia, Brazil

La Nación (1998a). Humo asfixia al humo. 20 May 1998, San José, Costa Rica

La Nación (1998b). Lucha Contra el Fuego. 21 May 1998, San José, Costa Rica

La Serna, Carlos y Claudio Tecco (1997). El Caso de Argentina. In Raul Urzua and Diego Palma (eds.), *Pobreza Urbana y Descentralización: Estudio de casos*. Centro de Analisis de Politicas Publicas, Universidad de Chile, Santiago, Chile

LBA (1996). *The large scale biosphere-atmosphere experiment in Amazonia*. INPE, São Paulo, Brazil

Ministerio de Economía, Subsecretaría de Pesca, *Informe sectorial pesquero*, Enero-Diciembre 1997, Chile

Moreira, A.G. (1998). Woods Hole Research Center, Brazil http://www.whrc.org

Nepstad, D.N., C.A. Klink, C. Uhl, I.C. Vieira, P. Lefebvre, M. Pedlowski, E. Matricardi, G. Negreiros, I.F. Brown, E. Amaral, A. Homma and R. Walker (1997). Land-use in Amazonia and the Cerrado of Brazil. In *Ciencia and Cultura – Journal of the Brazilian Association for the Advancement of Science,* 49, 1/2, 73-86

Oldeman, L.R. (1994). Global Extent of Soil Degradation. In *Soil Resilience and Sustainable Land Use* (eds. D.J. Greenland and I. Szabolcs), p. 99-118. CAB International, Wallingford, United Kingdom

O'Ryan, R. (1994). *Sustainable Development and the Environment in Chile: a Review of the Issues.* Universidad de Chile, January 1994

PAHO (1995). *El Manejo de Residuos Sólidos Municipales en América Latina y El Caribe*. Serie Ambiental No 15, Pan American Health Organization, Washington DC, United States

PAHO (1994). *Health Conditions in the Americas, 1994.* www.paho.org/english/country.htm

Robinson, S.K. (1997). The Case of the Missing Songbirds. In *Consequences,* 3, 1, 2–15

Rosa, L. P., M. T. Tolmasquim, E. La Rovere, L. F. Legey, J. Miguez, R. Schaeffer (1996). *Carbon dioxide and methane emissions: a developing country perspective.* COPPE/UFRJ, Rio de Janeiro, Brazil

Simonetti, J., A.M. Arroyo, A.S. Spotorno and E. Lozada (1995). *Diversidad Biologica en Chile*. CONICYT, Santiago, Chile

Stöhr, Gerhard (ed.) (1994). *Paraguay: Perfil del Pais con informaciones y comentarios relacionados al desarrollo economico y social.* GTZ GmbH, Asuncion, Paraguay

Suman, Daniel (1994). *El ecosistema de manglar en America Latina y la cuenca del Caribe: su manejo y conservacion.* Rosenstiel School of Marine and Atmospheric Science, University of Miami, Miami, Florida, United States

Tolmasquim, M. T. (1996). CO_2 emissions from energy systems: comparing trends in Brazil with trends in some OECD countries. In Rosa, L. P., M. T. Tolmasquim, E. La Rovere, L. F. Legey, J. Miguez, R. Schaeffer (1996). *Carbon dioxide and methane emissions: a developing country perspective.* COPPE/UFRJ, Rio de Janeiro, Brazil

UNCTAD (1995). *Review of Maritime Transport 1994.* UNCTAD, New York, United States, and Geneva, Switzerland

UNDP (1997). *Human Development Report 1997*. Oxford University Press, New York, United States, and Oxford, United Kingdom

UNEP/ISRIC (1991). *World Map of the Status of Human-Induced Soil Degradation (GLASOD). An Explanatory Note,* second revised edition (edited by Oldeman, L.R., Hakkeling, R.T., and Sombroek, W.G.). UNEP, Nairobi, Kenya, and ISRIC, Wageningen, Netherlands

United Nations Disaster Assessment Coordination (1998). *Incêndios no estado de Roraima: Agosto 1997 – Abril 1998.* Brasilia, Brazil

United Nations Population Division (1996). *Annual Populations 1950-2050 (the 1996 Revision),* on diskette. United Nations, New York, United States

United Nations Population Division (1997). *Urban and Rural Areas, 1950-2030 (the 1996 Revision),* on diskette. United Nations, New York, United States

Veiga, M. M. (1997). *Introducing New Technologies for Abatement of Global Mercury Pollution in Latin America.* UNIDO/UBC/CETEM/CNPq, Rio de Janeiro, Brazil

WCMC (1998). WCMC Protected Areas Database http://www.wcmc.org.uk/protected_areas/data

WCMC/IUCN (1998). WCMC Species Database, data available at http://wcmc/org/uk, assessments from the 1996 IUCN Red List of Threatened Animals

Winograd, M. (1995). *Indicadores Ambientales para Latinoamérica y el Caribe: Hacia la sustentabilidad en el uso de tierras.* GASE, Proyecto IICA/GTZ, OEA and WRI. San José, Costa Rica

World Bank (1996). *Social Indicators of Development 1996.* World Bank, Washington DC, United States

World Bank (1997). *World Development Report: the state in a changing world*. Oxford University Press, Oxford, United Kingdom, and New York, United States

WRI (1997). *The Last Frontier Forests: Ecosystems and Economies on the Edge*. D. Bryant, D. Nielsen and L. Tangley (eds.). WRI, New York, United States

WRI, ICLARM, WCMC and UNEP (1998). *Reefs at Risk: a map-based indicator of threats to the world's coral reefs.* Washington DC, United States

WRI, UNEP and UNDP (1994). *World Resources 1994–95.* Oxford University Press, New York, United States, and Oxford, United Kingdom

WRI, UNEP, UNDP and WB (1996). *World Resources 1996-97: A Guide to the Global Environment* (and the *World Resources Database* diskette). Oxford University Press, New York, United States, and Oxford, United Kingdom

WRI, UNEP, UNDP and WB (1998). *World Resources 1998-99: A Guide to the Global Environment* (and the *World Resources Database* diskette). Oxford University Press, New York, United States, and Oxford, United Kingdom

WTO (1994). *Tendencias del mercado turistico (Americas) : 1980-1993.* Comisión de la OMT para las Americas, Madrid, Spain

WTTC (1993). *Travel and Tourism: A New Economic Perspective*. The 1993 WTTC Report, Research Edition, World Travel and Tourism Council, London, United Kingdom

North America

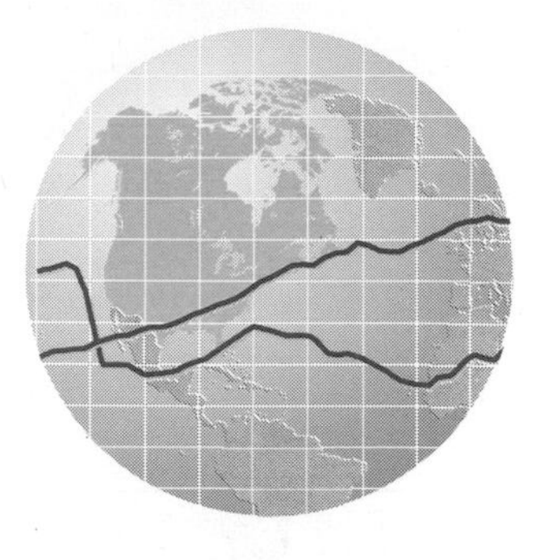

KEY FACTS

North Americans use more per capita energy and resources than any other region. This causes acute problems for the environment and human health. The region has succeeded, however, in reducing some environmental impacts.

- The North American region is at a critical environmental cross-roads: important decisions have now to be made that will determine whether the region's economic activity and patterns of production and consumption will become more sustainable.
- There is continuing concern about the effects of exposure to pesticides and other toxic compounds on human health and the environment in general.
- Emissions of CO, VOCs, particulates, SO_2 and lead have been markedly reduced over the past 20 years.
- Fuel use is high – in 1995 the average North American used more than 1 600 litres of fuel (compared to about 330 litres in Europe).
- The oxygen-depleted 'dead zone' that now appears off the US Gulf Coast each summer – at the peak of fertilizer run-off from the Corn Belt – is the size of New Jersey.
- Global warming could move the ideal range for many North American forest species some 300 km to the north, undermining the utility of forest reserves established to protect particular plant and animal species.
- The impact of development on critical biological resources is an important issue across the region. Changes to ecosystems caused by the introduction of non-indigenous species are of particular concern.
- Fish stocks off the east coast have nearly collapsed. The Atlantic finfish catch declined from 2.5 million tonnes in 1971 to less than 500 000 tonnes in 1994.

Trends in environmental quality in North America are mixed. On the positive side are improvements in some aspects of air and water quality, and reduced levels of soil erosion in much of the region. On the negative side are sharp declines in fish stocks in major marine fisheries, continued logging pressures on old-growth forests, growing invasions of exotic species and other threats to biodiversity, and increasing outbreaks of toxic organisms in estuaries and coastal zones associated with excess run-off of nutrients. Success in reducing emissions of some toxic industrial materials must be compared with continued high levels of industrial use of such materials and slow progress in cleaning up toxic waste sites. Success in phasing out production of CFCs and other ozone-depleting gases must be balanced against failure to reduce emissions of carbon dioxide, the primary greenhouse gas, and hence the region's growing contribution to the risk of climate change.

Beyond these specific issues, and underlying all of them, is the scale of economic activity in North America. The large and robust North American economy brings many benefits, stimulating job creation, and increasing welfare and opportunities for the region's inhabitants. But it also puts increasing stresses on regional environmental quality and has major impacts across the global environment. The United States and Canada have among the highest per capita consumption of energy and other natural resources in the world, and they contribute a disproportionate share of global emissions of greenhouse gases. North American consumption also provides strong incentives for increasing international trade, leading to increased industrialization and resource use throughout the world. Thus the footprint of North America's impact on the environment stretches well beyond the region itself.

Discussion of Mexico is included in the North American chapters on topics heavily affected by cross-border issues such as conservation of biodiversity and migratory species, transportation management, watershed management and air pollution. For other issues Mexico is included in the Latin America section of the report. Thus, unless explicitly indicated, North America here refers to Canada and the United States.

Social and economic background

The North American region is characterized by continuing economic growth with strong, market-oriented economies. The region's economic vitality is creating new regional and global opportunities but also exacerbating some existing environmental stresses and creating new ones.

North America has a population of approximately 304 million which is growing at an annual rate of 0.8 per cent (United Nations Population Division 1996). The region is a magnet for immigration, which contributes substantially to its growth. Approximately three-quarters of the population is urban, living in cities, suburbs or large metropolitan areas (United Nations Population Division 1997).

As a region rich in some fossil fuels (though not oil) and in hydropower resources, North America maintains some of the lowest energy prices in the world. Low energy costs have favoured the development of energy-intensive economies and promoted widespread reliance on automobiles. Although technological changes have increased the energy efficiency of many industrial processes, along with the fuel efficiency of automobiles, these trends have been more than offset by rising use and changing consumption patterns (see graph right). North Americans drive further than they did a decade ago, for example, and the sale of small vans and light trucks as family vehicles has increased enormously, resulting in rising energy use for transportation. The result is both additional pressure on urban air quality and, notably, on the global climate.

Energy use is not the only aspect of the existing pattern of production and consumption with serious environmental consequences. The consumer culture and suburban lifestyles of many Canadian and US inhabitants and low – in fact declining – prices for most natural resources have led to very high per capita uses of such resources and correspondingly large amounts of industrial and post-consumer wastes. High levels of resource use also have impacts on coastal, freshwater, forest and other ecosystems. In addition, political trends have favoured diminished state intervention in markets, accompanied by deregulation, privatization and reductions in government expenditures in both North American

Population

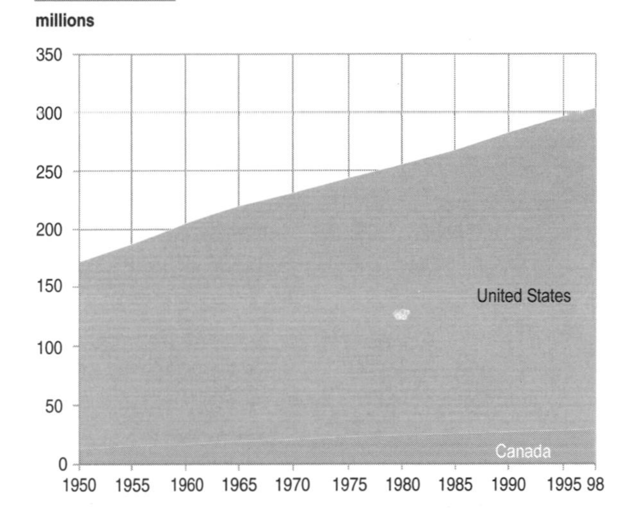

Source: United Nations Population Division 1996

North America's population of 304 million is growing at 0.8 per cent a year. Immigration contributes substantially to growth

Annual commercial energy consumption per capita

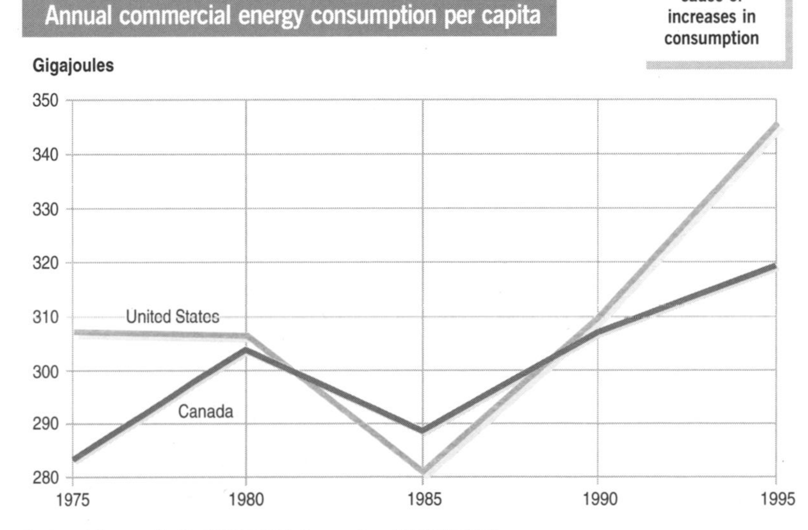

Source: data compiled by UNEP GRID Geneva from UNSTAT 1997

Energy production in North America continues to rise – low energy prices are one cause of increases in consumption

countries. As a result, many types of policy interventions, such as higher energy taxes, are considered socially and politically unacceptable. This is in sharp contrast to Europe, where taxes are higher and the price paid by the consumer for petrol is two to four times more than in North America. The higher taxes may be reflected in per capita consumption of fuel: in 1995, approximately 329 litres were used per person in Europe in contrast with 1 642 litres per person in the United States and Canada (WRI, UNEP, UNDP and WB 1998).

Environmental health problems, especially those associated with agricultural and industrial pollutants, continue to be an issue of concern. Agricultural production contributes significantly to the use and release of toxic material: approximately 26 000 tonnes of pesticides are used annually in agriculture in the Great Lakes basin (International Joint Commission 1997, WWF 1997). The Natural Resource Defense Council reports that US pesticide use reached an all-time high in 1995. More than 540 000 tonnes of pesticides were used, reversing a downward trend in the previous few years (Natural Resource Defense Council 1996). The estimate excludes 'inert' ingredients such as petroleum, benzene and other toxic compounds, which can comprise more than 50 per cent of the volume of formulated pesticides. The estimate also excludes use of non-conventional pesticides such as wood preservatives and disinfectants, which the US Environmental Protection Agency (EPA) previously estimated at more than 450 000 tonnes a year. There is continuing concern about the effects of exposure to these and other compounds on human health and the environment in general (International Joint Commission 1997, Council on Environmental Quality 1997, Colburn and others 1996). The Great Lakes Science Advisory Board of the International Joint Commission concluded that certain chemicals in the environment may affect the endocrine systems of wildlife and humans. Endocrine disrupters, a group of chemicals that includes some persistent organic pollutants (POPs), may block or mimic the natural action of hormones such as oestrogen, and disrupt the regulation of sexual and reproductive development (US EPA 1997a).

The impact of development on critical biological resources is an important issue across the region. Changes to ecosystems through the introduction of invasive non-indigenous species are of particular concern. Increased air traffic and changes in global trade have added to this problem. There are many ways in which alien species can disrupt entire ecosystems: by replacing native species, changing existing water and nitrogen-cycling regimes, depriving indigenous animals of their normal diets, introducing new pathogens against which native species have no defences, and changing the genetic make-up of native species by mating with them (Powledge 1998). In the United States, approximately 15 per cent of 4 500 established exotic species cause serious economic or ecological harm (US Congress 1993). In Canada, more than 500 species of introduced plants have become agricultural weeds (OECD 1995). Estimates of economic losses for the United States alone range up to several thousand million dollars a year (Jenkins 1996).

Issues of environmental equity or justice have emerged in recent years as a result of evidence showing that the impact of pollution and resource degradation often falls disproportionately on poor or racially-distinct neighbourhoods or indigenous communities. And in a largely urban society it is easy to forget that environmental resources are still an essential part of subsistence for some groups and communities within North America; for them, as for many others, environmental degradation is far more than an aesthetic issue or a loss of recreational opportunities. Public opinion polls indicate that environmental quality and environmental protection are considered important issues across the region. This concern for environmental protection is manifested in the implementation of waste management practices. For example, by 1995 40 US States had comprehensive recycling or waste reduction laws, and 44 had legislated or announced goals of 20 to 70 per cent for recycling or waste reduction (Council on Environmental Quality 1997). Resolving the contradictions between environmental values and economic and social pressures for increased production and consumption of natural resources is one of the challenges facing North America today. How the region meets this challenge matters greatly, and not just to its inhabitants. Because of the region's economic and political influence as well as its sheer size, North America has a strong influence on both economic trends and environmental policies around the world.

Land and food

North America's land base is more than 19.3 million km^2, or approximately 14 per cent of the world's land area (FAO 1997a). Currently, about 27 per cent of the land is devoted to agriculture, with 2.3 million km^2 in crops and 2.7 million km^2 in permanent pasture (OECD 1997). Most of North America's agricultural land is in the United States: approximately 82 per cent of all cropland and 90 per cent of all permanent pastures is found south of the US/Canadian border.

At the turn of the century, land could be easily acquired in North America, and waves of new settlers were drawn to the region's productive agricultural lands. As settlements spread across the continent, forests and grasslands were converted to agricultural uses. The vast natural grasslands of North America's Great Plains were gradually transformed into agricultural lands to cultivate grain or support livestock. With the advent of the tractor and the market forces created by World War I, the largest expansion of tilled areas for wheat occurred in the Great Plains. The area planted to wheat increased by 60 000 km^2. Between 1921 and 1929 another 61 000 km^2 of grassland in the southern plains were converted to wheat. However, overgrazing, poor farming techniques and drought conditions led to massive soil erosion on the Plains during the Dust Bowl years of the 1930s (Southwick 1996). Huge dust storms extending from Canada to Mexico affected nearly 4 million km^2, until they were eventually brought under control by soil conservation and farm rehabilitation programmes (MacNeill 1989, Mannion 1991).

After World War II, intensified and modernized agricultural practices led to a large rural-to-urban migration and a decline in traditional, small-scale family farming practices. Fewer and larger farming enterprises came to manage the production of large fields of monocultures or intensively-rotated crops. Several factors led to the intensification and modernization of agriculture including the movement of people away from agricultural pursuits; replacement of human labour with mechanization; temporarily improved pest control technologies; and specialization within the production sector. Canada and the United States eventually became the world's leading sources of surplus foodgrain, exporting 132 million tonnes yearly during the 1980s, compared to about 5.5 million tonnes before World War II (World Commission on Environment and Development 1987).

The intensification of agricultural production in the United States and Canada has increased productivity by a factor of three or four since the 1950s (Lipske 1993) but these gains have also increased environmental stresses. The heavy use of pesticides and fertilizers (see bar chart below) resulted in run-off that became and remains a major source of water pollution. Approximately 950 000 km^2 of land in the United States and Canada are affected by soil degradation, primarily water and wind erosion

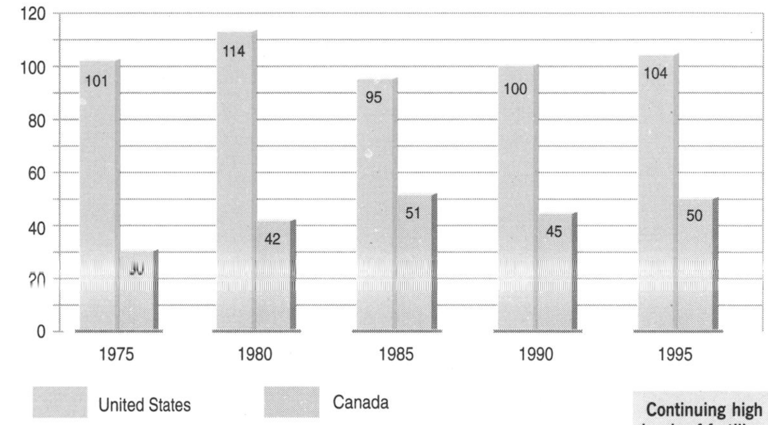

Source: data compiled by UNEP GRID Geneva from FAO 1997a

Continuing high levels of fertilizer use, particularly in the United States, are causing serious problems related to nitrogen run-off

(UNEP/ISRIC 1991). Farmers have responded by applying increasing amounts of fertilizers but studies have shown continued declines in the average rates of change in productivity (Batie 1993). In 1997 the United States used 20 million tonnes of fertilizer, a figure slightly less than the all-time high of 21.5 million tonnes in 1981 (FAO 1997c). Despite efforts to reduce negative environmental impacts, agricultural production accounts for a significant use and release of toxic material (International Joint Commission 1997), and OECD studies list nitrate pollution as one of the most serious water quality problems in North America (OECD 1994). As the amount of fertilizer used approaches the physiological capacity of crops to absorb nutrients, the excess nitrogen poses a threat to ecosystem health. Along with phosphorus, another key fertilizer ingredient, nitrogen promotes overgrowth of algae in rivers, lakes and bays. As the algae die and decay, they use up large amounts of the water's oxygen, depriving other species of the oxygen they need to survive. In the case of the US Gulf Coast, the oxygen-depleted 'dead zone' (see box on page 151) that now appears each summer – at the peak of fertilizer run-off from the Corn Belt – is nearly the size of El Salvador (Tolman 1995).

The United States and Canada are involved in an issue of emerging international concern – the impact of certain chemical pesticides that are classed as persistent organic pollutants (POPs). POPs are toxic substances composed of organic chemical compounds and are of particular concern because they are toxic to humans and animals; do not degrade readily in the environment; tend to bioaccumulate; and often change from solid to gaseous phase, travelling long distances in the air before being redeposited in the environment (US EPA 1997a). Research on POPs has focused on 12 chemicals – the 'dirty dozen' – nine of which are pesticides and include DDT, chlordane and heptachlor (WWF 1998). Scientific evidence continues to show that some POPs cause genetic, reproductive and behavioural abnormalities in wildlife and humans, and may be associated with increased incidence in humans of cancer and neurological deficits (US EPA 1997a). The pesticides are used on crops such as cotton, vegetables, fruits and nuts. Although specific data on the location and amount used worldwide are difficult to obtain, most of the nine pesticides are still in use or exist in many countries (WWF 1998). All of the 12 POPs are either banned from use or regulated in North America (FASE 1996). Recent efforts to find and promote alternatives to POPs include newer and pest-specific pesticides and biological control methods (US Congress 1995a, National Research Council 1996). An international agreement on POPs is expected to be adopted in the year 2000.

A further source of concern in North America is the long-range transport of airborne emissions from other countries or regions, such as Eastern Europe, the Russian Federation and Asia, that may lead to deposition in North America. These pollutants bioaccumulate in wildlife and humans that consume the wildlife, particularly affecting the health of aboriginal and northern Canadian communities.

The expansion of farmland, especially into wetlands, destroyed habitats and contributed to the loss of biodiversity. Since the 1950s, however, the conversion of wetlands to agricultural land has declined. In the mid-1950s, agriculture, with government encouragement, was responsible for approximately 87 per cent of wetland conversions. In contrast, during 1982–92 about 57 per cent of total wetland losses were attributed to urban development and only 20 per cent to agriculture. Since 1992, approximately 4 000 km^2 of wetlands have been placed into the US Department of Agriculture's Wetland Reserve programme. These lands are mostly under permanent easement and are not available for conversion to farmland. A shift toward organic farming and other low-impact agricultural techniques, including conservation tillage and integrated pest management, also helps to moderate some environmental stresses.

With mounting concern over the health and environmental effects of agricultural chemical use, organic agriculture is expanding. According to the Organic Farming Research Foundation (Organic Farming Research Foundation 1996), 49 per cent of US organic farmers intended to increase their organic areas over the next two to three years, and only 4 per cent were planning to decrease them. In recognition of this trend, in 1998 the US Department of Agriculture's Agricultural Marketing Service launched a proposal to establish a National Organic Program which would establish national standards for the organic production and handling of agricultural products, and would include a National List of synthetic substances approved for use in the production and handling of organically-produced products. It would also establish an accreditation programme for State officials and private entities that wish to be accredited to certify farm, wild-crop harvesting and handling operations that comply with the programme's requirements. The programme would also include labelling requirements for organic products and products containing organic ingredients, and enforcement provisions (US Department of Agriculture 1998).

Hazardous waste management has received attention in both Canada and the United States, at least in part from public pressure about sites posing threats to public health, as well as from concerns about potential liability for damage caused by hazardous waste (OECD 1995 and 1996). By September 1995, the US EPA had identified 1 374 National Priority List sites – sites where contamination presents the most serious threat to human health and the environment (Council on Environmental Quality 1997). Work was being conducted at 93 per cent of these sites, with permanent clean-up in progress at 60 per cent of the sites. An additional 15 622 sites remained which were classified as potentially hazardous or low priority. In Canada, in 1993, provinces reported on close to 4 800 contaminated sites that posed a threat to the environment (OECD 1995).

Over the next 10 years, changing trade patterns may intensify environmental impacts on agricultural land. Recent reductions in agricultural subsidies and tariff protection may increase US and Canadian production. As global trade barriers are reduced, there may be increasing pressure on export agriculture to satisfy a growing demand for food. For instance, a greater demand for meat by Asian countries could lead to an expansion in livestock production (Government of Canada 1996). If agricultural production increases, water and land use pressures may increase.

Forest extent, 1990 and 1995

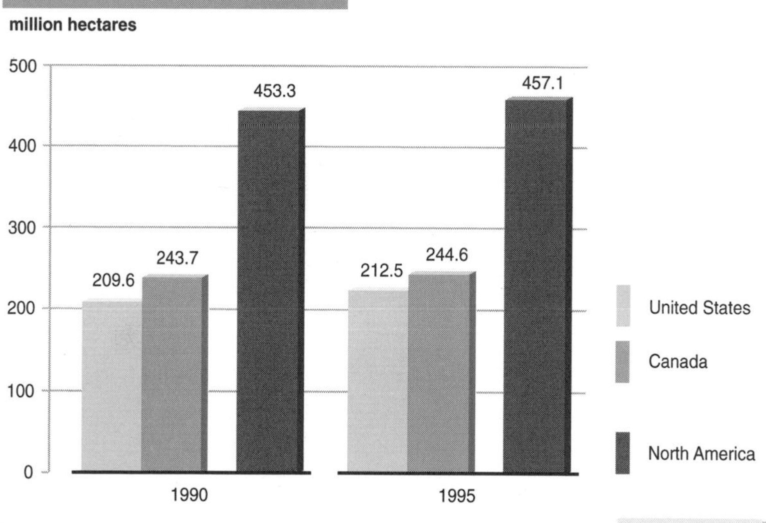

Source: FAO 1997a

Forests

Forests are one of the most prominent features of North America, covering about 25 per cent of the land area (FAO 1997b). North American forests constitute a rich resource, providing economic and recreational benefits, as well as watershed protection, wildlife habitat, and many other ecological services. The 460 million hectares of forests in Canada and the United States which are managed for commercial purposes comprise about 13 per cent of the world's total (FAO 1997b).

The past 100 years have brought both gains and losses to North America's regional forest cover, with substantial regrowth in the eastern United States and Canada. Overall, the extent of forest cover has stabilized in many parts of the region.

The quality of North America's forests, however, has deteriorated. While the world's second largest expanse of intact, natural forest is in the boreal regions of North America, forests in the United States are becoming increasingly fragmented and biologically impoverished, invaded by exotic species, or lacking in the characteristics that support viable populations of indigenous species (Bryant and others 1997). *GEO-1* mentioned the harvesting of high quality, old growth timber stands in the northern Pacific region, including the states of Washington and Oregon, which fuelled a spirited public debate over forest policy in both Canada and the United States. In the intermountain West, pine and mixed conifer stands are in an unhealthy condition as a result of decades of overgrazing, heavy harvesting and fire suppression (Council on Environmental Quality 1997).

There is increasing debate about the use of forests for commercial logging, recreation and conservation. An estimated 57 per cent of North American

North America is home to 13 per cent of the world's forests. While their area is increasing, their quality is still deteriorating

The Model Forest Program

The movement to establish sustainable forestry practices has gained momentum throughout North America. One outcome has been the International Model Forest Program, which originated in Canada in 1992. The objectives of this programme are to accelerate the implementation of sustainable development in forestry practices; to apply new and innovative approaches, procedures, techniques and concepts in the management of forests; and to test and demonstrate best sustainable forestry practices using the most advanced technology and forestry practices available.

In 1998, there were 14 Model Forests in North America, with 11 in Canada and 3 in the United States (Natural Resources Canada 1998). One of the successes of the Canadian programme has been building partnerships between aboriginal groups, industrial partners and educational institutions (International Model Forest Network 1997). Building these partnerships, however, proved more difficult than expected, and in some cases took more than two years (Natural Resources Canada 1997). Though there has been progress in developing sustainable forest management tools, there is little evidence of practical application. The challenge is thus to translate management decisions to on-the-ground actions (Natural Resources Canada 1997). The US Model Forest Program has focused on international outreach, developing internet materials and educational activities (International Model Forest Network 1997).

(including Mexico) forests are considered commercially productive (Commission for Environmental Cooperation 1999), and the forest industry directly and indirectly employs an estimated 1 in 17 Canadians (Natural Resources Canada 1998). North America is a leading producer and consumer of timber, pulp and newsprint, and is highly competitive in world markets for timber products. The forest products industry of Canada and the United States is technologically advanced, and capable of removing and processing timber with great efficiency.

Over the past ten years, the industry has achieved notable success in reducing water pollution from pulp and paper mills, and interest in sustainable forestry practices is growing. Changes have also been made to encourage greater public participation in forest management and to involve partners and interested groups in open and transparent decision making (Natural Resources Canada 1997). New forest community partnerships have been established among governments, industry, labour, environment groups, private woodlot owners, aboriginal people, academia and others in trying to achieve a more holistic approach to forest management that balances the environmental, economic, social and cultural demands placed on forests.

The next ten years may see an intensification of the debate over forest practices and the use of forests. Worldwide demand for forest products is expected to grow, increasing pressure for commercial production from North American forests. Recreational demands are also expanding, and in many areas may constitute a higher-value economic use of forests, one that is in principle compatible with conservation goals.

Global warming could move the ideal range for many North American forest species some 300 km to the north (Council on Environmental Quality 1997). The rate of climate change will influence the stress experienced by forest species and an increase in fires, droughts and pest populations would produce adverse effects on forest systems. Forest reserves, established to protect particular plant and animal species, may no longer be located in areas where the climate is hospitable to those species (Council on Environmental Quality 1997).

Biodiversity

Because ecosystems extend across borders, Mexico is included as part of North America in this section. North America's biodiversity increases along a latitudinal gradient from north to south. Some 7 807 plants, 233 mammals and 160 birds are endemic or unique to North America (Commission for Environmental Cooperation 1999). Endemism is particularly high in Mexico, where as many as 40–50 per cent of flowering plants, more than half of the reptiles and amphibians, nearly 50 per cent of the fish, and 33 per cent of the mammals are considered endemic (World Bank 1995). In contrast, Canada has relatively low endemism, with unique species mainly limited to islands and areas that escaped glaciation (Government of Canada 1996). Endemism in the United States is highest on islands, especially the Hawaiian Islands, where 44 per cent of the higher plants, 95 per cent of the molluscs, 43 per cent of the birds and 30 per cent of inshore fishes are unique to those islands (Allison and others 1995, Hourigan and Reese 1987). Much less is known about species richness and endemism in marine habitats.

Over the past century, habitat destruction, over-zealous hunting or harvesting, and competition from introduced species has led to the decline and extinction of many North American species (Langner and Flather 1994). The past century also witnessed a decrease in the genetic diversity of agricultural crops and livestock (Government of Canada 1996). In more recent decades, threats attributed to hunting and over-harvesting have diminished in comparison to those

Stopping the brown tree snake

The brown tree snake, a nocturnal reptile native to Papua New Guinea, was accidentally introduced to Guam in the 1940s. It is now an uncontrollable pest and has wiped out Guam's native bird population. It is also a serious health risk for humans. Hawaii, one of the hubs of Pacific travel, is under constant threat of invasion from foreign species, of which the brown tree snake is the most urgent.

Hawaii is using a number of prevention tools to reduce the risk of invasion:

- integrated planning throughout the area to prevent the entrance of all brown tree snakes, and information exchange by print and the Internet about habits and control methods;
- coordination of public policy, designation of the snake as a pest prevention priority, and provision of funding for inspection of carriers leaving Guam and arriving in high-risk ports;
- training quarantine officials and others to recognize the snake, and training Snake Watch Attack teams on each island; and
- promoting media coverage of the threat, publicizing the issue in schools and educating travellers.

from habitat destruction, degradation and fragmentation. Indeed, habitat loss and alteration have become a major threat to the continued diversity of wildlife in North America.

Wetland habitats, which are essential to many forms of wildlife, are particularly threatened. Since the 17th century, many of Canada's wetlands have been lost or severely degraded. Drainage for agriculture accounted for about 85 per cent of these losses, and for 80 per cent of the estuary marches in the Fraser River Delta, 70 per cent of prairie potholes, 68 per cent of southern Ontario wetlands, and 65 per cent of Atlantic coastal salt marshes (Rubec 1994). Similarly, in the United States, more than half of the wetlands have been drained, dredged or modified in some way. Most of this wetland loss (48 million hectares) has occurred in the 48 conterminous states, with Alaska losing only a fraction of its original 68 million hectares (OECD 1996). Overall, though, wetlands still comprise some 1.27 million km^2, 13 per cent of Canada's area; this amounts to one-quarter of the world's wetlands (Government of Canada 1996).

There are 50 wetlands of international importance in North America (the Ramsar sites that parties to the Convention on Wetlands of International Importance especially as Waterfowl Habitat agree to establish as wetland reserves). They cover 14.2 million hectares, of which 13 million hectares are in Canada (Ramsar 1997). Wetland loss has contributed, among other things, to the long-term declines of some duck populations (Caithamer and Smith 1995). Over the past decade, awareness of wetland loss has been growing and efforts to protect these habitats have increased. But losses continue to outpace the gains made through wetland restoration projects. Coastal marshes and other aquatic ecosystems are particularly prone to degradation because the concentration of settlements in coastal areas and around rivers and lakes continues to grow (Langner and Flather 1994).

The Migratory Bird Treaty Act, enacted by the United States in 1918, helped in the recovery of many North American birds suffering from population declines due to excessive hunting (Harrington 1995). More recent cooperative efforts, including the North American Breeding Bird Survey (begun in 1966 by both the United States and Canada), have led to better information on status and trends in North American migratory species (LaRoe and others 1995). Problems with migratory species remain, however, with habitat loss one of the biggest threats to migratory populations in North America. Many birds, bats, butterflies and sea mammals summer in the northern reaches of North America and winter in its southern climes, whereas others migrate exclusively within the tropics. Mexico hosts 51 per cent of all migratory bird species from its northern neighbours, and the loss of critical over-wintering sites, due to deforestation and other land-use changes, may threaten the survival of these populations (Robinson 1997, Greenberg 1990). The transformation or degradation of breeding habitat in the north, along with deprivation of vital stopover areas along developed coastal areas, is also contributing to a decline in migratory songbirds (Temple 1998, Robinson 1997, Terborgh 1989). Other migrants are under multiple pressures. The extraordinary migration of the Monarch butterfly, for example, may be threatened by coastal development in California, the deforestation of its habitat in Mexico's cloud forests, and threats to milkweed habitat in Canada (Malcolm 1993, Schappert 1996).

In recent decades, the deliberate or inadvertent introduction of exotic species has emerged as a growing threat to native biodiversity that has both economic and biological costs. In the United States, for example, at least 4 500 species of foreign origin have established populations, with approximately 15 per cent of these species causing harm (US Congress 1993). High-impact species, such as the gypsy moth, imported fire ants (*Solenopsis invicta*), purple

The zebra mussel invasion

The zebra mussel is a small freshwater mollusc native to Russia that was introduced to North America from Europe in the 1980s through ship ballast water. It invaded southern Canada and the Great Lakes and expanded into North America's inland waters at an alarming rate. It is now found in two-thirds of all US waterways. This invasive species is causing the decline of many aquatic species and communities, and large-scale changes in local food webs. It attaches itself to other mussel species, for example, reducing their populations, and it filters out and removes phytoplankton and zooplankton that are the base of the food chain. It also causes substantial economic damage by clogging the water-intake structures of power plants and municipal water treatment plants, and encrusts the bottoms of commercial and recreational boats (Institute of Water Research 1997, Sea Grant Minnesota 1997).

Threatened animal species

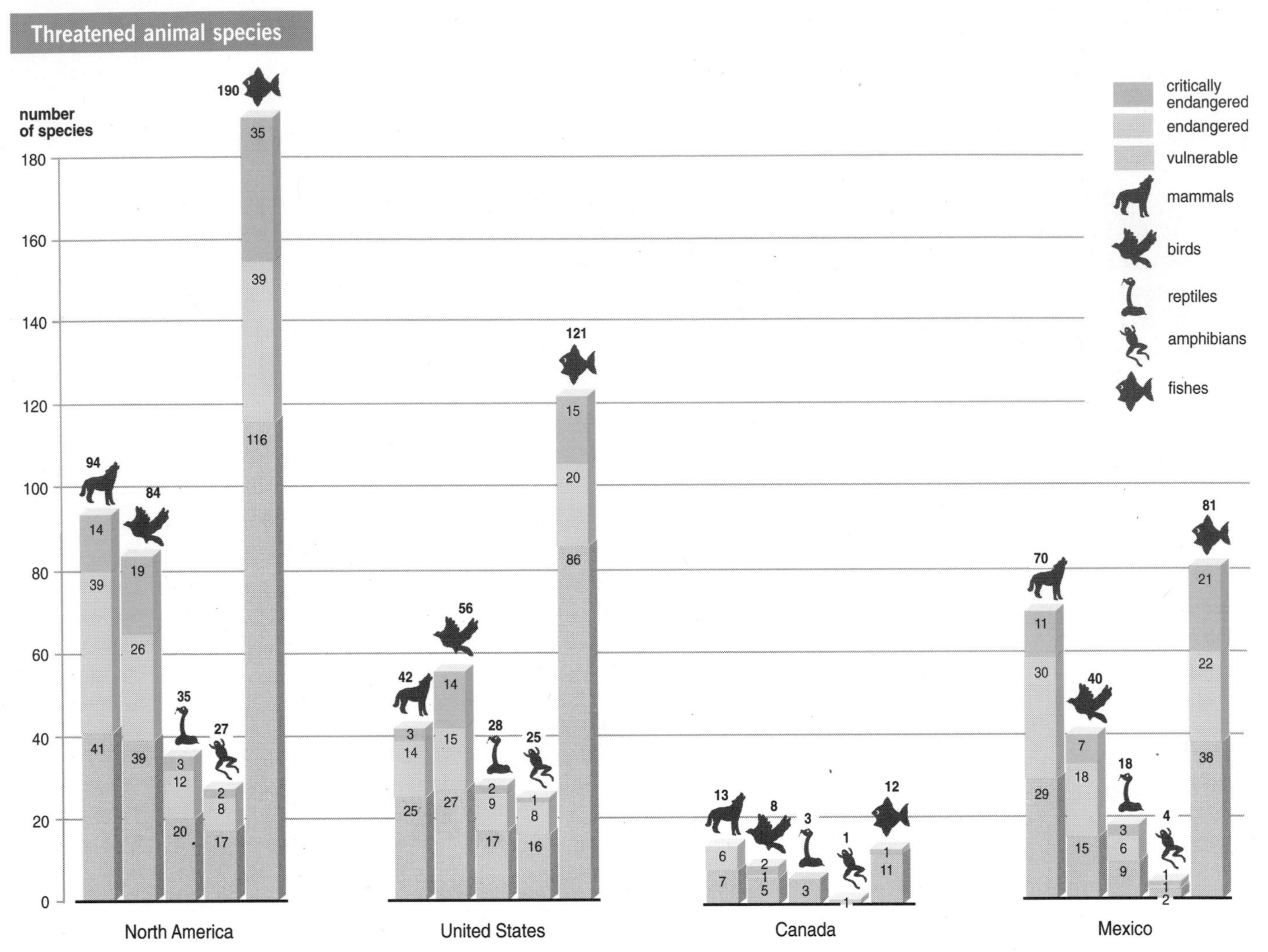

Source: WCMC/IUCN 1998

There are an estimated 430 threatened species of mammals, birds, reptiles, amphibians and fish in North America, and one-third of the region's freshwater fish stocks are threatened.

loosestrife (*Lythrum salicaria*) and the zebra mussel (see box on page 145), can affect many national interests including agriculture, industry, human health and the protection of natural areas.

In November 1990, the Non-indigenous Aquatic Species Act was passed by the US Congress, and in May 1993, the first ballast water law in the world was adopted. The ballast water law requires ships entering the Great Lakes with ballast water to exchange that water on the high seas. While ballast water exchange reduces the risk of introducing invasive species, it does not eliminate it. Recommendations for future strategies to eliminate undesirable introductions include alternatives in ship design and treatment of ballast water with, for example, heat, ultraviolet light or ozone to remove foreign organisms (Mills and others 1998).

Overall, there are an estimated 430 threatened species of mammals, birds, reptiles, amphibians and fishes in North America (see bar chart above). One-third of the region's freshwater fish stocks are threatened or rare. The United States had the greatest diversity of freshwater mussels in the world but 55 per cent of the species are now either extinct or threatened with extinction (Williams and Neves 1995). Concern with vanishing plant species has also become a regional and global concern, particularly in relation to genetic stock. A 1997 report assessing the condition of approximately 20 500 species of native US plants and animals gives about two-thirds of the species satisfactory marks while about one-third are of conservation concern (Stein and Flack 1997). Organisms with especially poor marks include animals depending on freshwater habitats such as mussels, crayfish, fishes and amphibians. Flowering plants also receive low scores, with one-third of their species – a total of some 5 144 species – in trouble.

Limited data on the status of marine species are available from the US National Marine Fisheries Service within the National Oceanic and Atmospheric Administration (National Marine Fisheries Service 1997). At least 85 species of marine mammals are found in US marine waters, including the Atlantic Coast, Gulf of Mexico and Pacific Coast. Eighteen of these species are listed as threatened or endangered under the US Endangered Species Act. Increases in populations have resulted from the prohibition of commercial whaling but decreases have resulted from the by-catch associated with commercial fishing, illegal killings, strandings, entanglement, disease and exposure to contaminants (Kinsinger 1995).

Throughout North America, some 2.5 million km^2 of land, freshwater and marine areas have been set aside as national parks and other types of protected areas (see illustration right) that help to sustain and preserve rare, threatened or vulnerable ecosystems, and the species and genetic resources that they harbour (Commission for Environmental Cooperation 1999). This amounts to approximately 9 per cent of North America's total land area, and the number and 
amount of protected lands are growing. In Mexico, more than 10 new biosphere reserves were decreed in the past decade (Secretaría de Medio Ambiente Recursos Naturales y Pesca 1996), while in Canada protected areas have increased by 15 per cent since 1990 (Government of Canada 1996) and will continue to increase with the implementation of Marine Protected Areas and National Marine Conservation Areas. The US system of protected areas doubled in size in 1980, with the creation of Tongass National Park in Alaska.

An encouraging trend is the growing recognition of the need to protect representative areas of all the region's diverse ecosystems. Eco-regional assessment, however, has not turned up good news. In an examination of eco-regions of North America, the World Wildlife Fund found that eco-regions in the United States and southern Canada are under severe threat (Ricketts and others 1997). The most critical eco-regions include temperate broadleaf and mixed forests, temperate grasslands, savannahs and shrublands; 60 per cent of the eco-regions classified as critical or endangered are part of these habitat types.

Over the next 10 years, with the implementation of new legislation such as the Marine Protected Areas and the National Marine Conservation Areas, the number of protected areas may increase. Despite efforts to protect biodiversity, however, human activities are likely to continue to encroach on ecosystems and to jeopardize the habitats of threatened species in North America. The expansion of regional and international trade may increase the introduction of exotic species into North America. On a longer time scale, climate change may also force rapid adaptation, alter communities of plants and animals, and their migration patterns, and cause extinctions.

The need to conserve biodiversity requires the development of an analytical framework for monitoring its status and for establishing priorities. One North American example is the Canadian Biodiversity Strategy developed by the Canadian Government (Environment Canada 1994). This strategy sets out national goals and strategic directions for the conservation and sustainable use of biodiversity, and

Size and number of protected areas

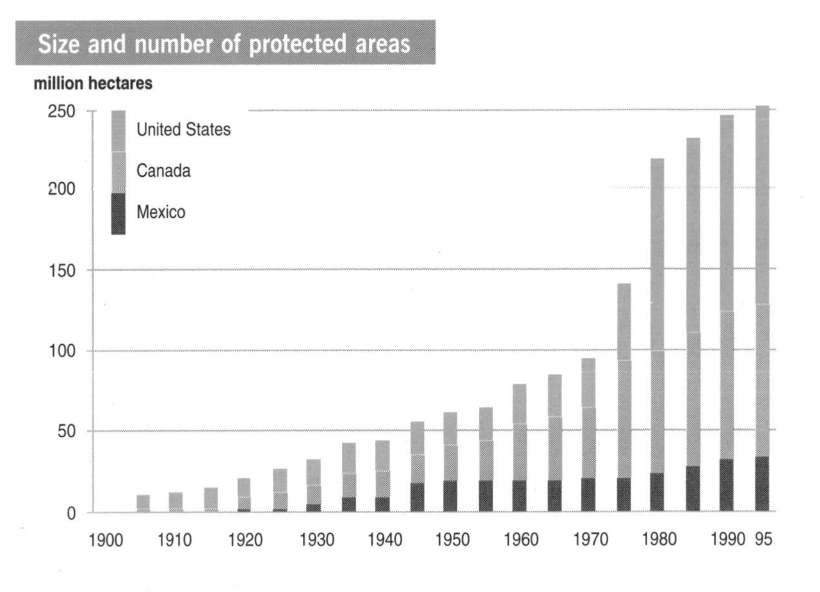

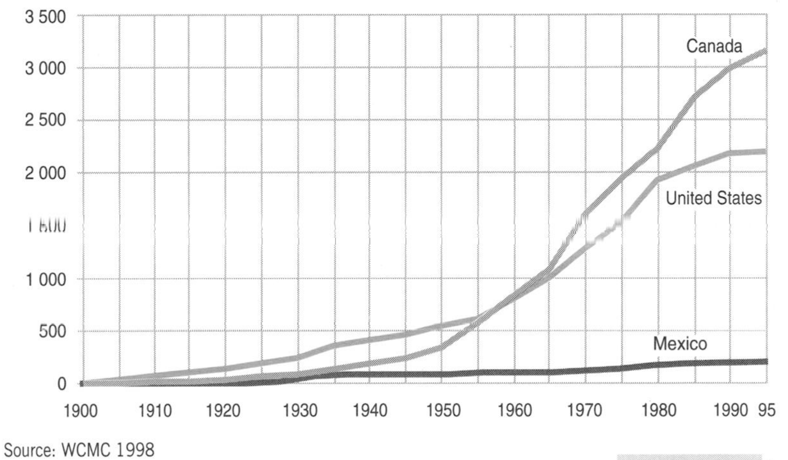

Source: WCMC 1998

Both the number and area of protected sites in North America continue to grow

acts as a planning framework for federal, provincial and territorial governments. There are also many initiatives in the area of environmental indicator development that lend support to the implementation of the strategy.

Freshwater

As *GEO-1* demonstrated, North America has an abundant supply of freshwater resources but it is unequally distributed across the region. Surface and groundwater sources together provide an annual 5 308 km^3 of renewable and fossil water to the two countries, which is about 13 per cent of the global total (WRI, UNEP, UNDP and WB 1998). On a per capita basis, Canada has 10 times more water resources than the United States. However, water scarcities occur in many parts of North America, including some parts of Canada's prairie provinces and the US southwest (OECD 1995, 1996).

Over the past 100 years, the demand for water has increased steadily in North America. This is partly a result of population growth and increasing municipal demands for water. It is also related to North America's energy-intensive industrial development and the dramatic expansion of irrigated agriculture. The latter has occurred mainly in the United States where the area of irrigated land has risen from 1.5 million hectares in 1890 to approximately 21 million hectares in 1995 (Council on Environmental Quality 1997). Dams and diversion projects have flourished over the past century, as communities and economic sectors have sought access to secure water supplies. Meanwhile, water has been pumped from some underground aquifers faster than natural recharge rates, depleting an important resource and causing water tables in the United States to fall by up to 120 cm a year in some irrigated regions (Pimentel and others 1997). Cotton farming has had major effects on water supplies in northern Texas and parts of New Mexico, for instance. These areas were traditionally used for cattle ranching but large-scale agriculture was made possible with the advent of groundwater irrigation. Cotton farming then increased demands for water from the Ogallala aquifer and led to severe groundwater depletion (Kasperson and others 1996).

The United States uses much more of its water for agriculture (irrigation) than Canada. Water use for power generation is high in both countries

Freshwater withdrawals by sector

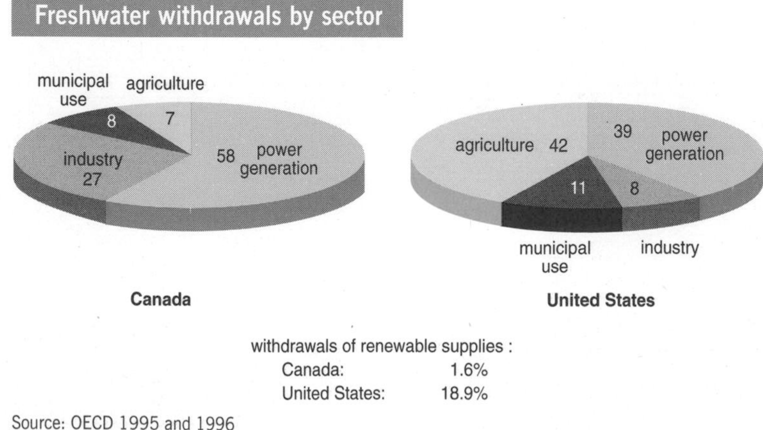

Source: OECD 1995 and 1996

During the 1990s, several measures have been taken to increase water use efficiency. As a result, per capita water use has not risen much in North America. In fact, Canada reported a 3.3 per cent decrease in daily municipal per capita water use between 1991 and 1994 – equivalent to a reduction of 22 litres per person per day (Government of Canada 1996). Nevertheless, Canada and the United States still rank among the world's largest consumers of water: average annual per capita withdrawal of water resources was 1 798 m^3 for North America (1991 data), in contrast with 645 m^3 for the world (1987), 625 m^3 for Europe (1995), and 202 m^3 for Africa (1995) (WRI, UNEP, UNDP and WB 1998).

Although the municipal supply, demand and quality of water receive much attention, the sectors that use most water in North America are agriculture and power generation (see pie charts left). In the United States, each accounts for about 40 per cent of total water withdrawal. In Canada, the figures are 58 per cent for power generation and 7 per cent for agriculture (OECD 1996 and 1995). Recently, however, these withdrawals have been declining while domestic use has been rising – it has almost doubled since 1960, reflecting population growth and urban expansion (OECD 1996).

The growth of municipal and industrial demands for water has led to conflicts over the distribution of water rights. Water resources are now a major constraint to growth and increased economic activities envisioned by planners, especially in the west and southwestern arid lands of the United States (Council on Environmental Quality 1997). Agricultural consumption accounts for a large share of water use in these areas. At the same time, demands for recreation, aesthetic enjoyment and wildlife habitat have become increasingly important in the management of North America's water resources. Both commercial and recreational fishing are also important water uses.

Conflicting demands for water have prompted many to favour the establishment of water management boards, in combination with water conservation measures.

The domestic water and sewer systems introduced for urban areas early in this century greatly improved the quality of drinking water and reduced the incidence of water-borne disease. Over the past decade, the quality of drinking water in North America has been improved still further. In 1994, more than 80 per cent of community water systems, serving 240 million people, reported no violations of health-based standards. Even so, in the same year more than 40 million people in the United States obtained their drinking water from a system in which there were violations of health-based standards (Council on Environmental Quality 1997). Canada also enjoys relatively high-quality water (Government of Canada 1996). However, some groundwater, which more than six million Canadians rely on for their water supplies, may be contaminated as a result of poor earlier management of wastes or industrial chemicals. Because groundwater moves slowly, detection of such contamination is often long delayed – sometimes until well after the source of the contamination has disappeared. Well water is often affected by faecal coliform bacteria and nitrates, which may be present in 20–40 per cent of all rural wells in Canada (Government of Canada 1996).

The International Joint Commission states that boundary areas are vulnerable to impairment from toxic chemical use: 'The Great Lakes region, acting as a sink for many persistent, bioaccumulative compounds, is the most prominent example. While there has been progress in curbing use of the most harmful compounds and in restoring contaminated areas since the 1970s, releases persist. A 1995 analysis by Environment Canada showed that Great Lakes basin industries released 173 092 tons of materials listed on the Canadian National Pollutant Release Inventory or the US Toxic Release Inventory in one year. When air releases originating on both sides of the border within the "one-day airshed" of the basin were taken into account, the total nearly doubled to 319 098 tons' (International Joint Commission 1997, Environment Canada 1995).

In some parts of North America, especially in older cities where sanitary and storm water networks are combined and become overcharged in wet weather, wastewater is still discharged into water bodies without treatment (OECD 1996). Many rural and indigenous populations, unconnected to municipal supplies and dependent on well water, experience water quality problems. As recently as the mid-1980s, half the homes on indigenous people's reserves in Canada were without running water, sewers or septic tanks. Significant progress has been made, with about 96 per cent of homes in 1996–97 having some form of potable water supply, and almost 92 per cent of homes with sewage disposal facilities (Minister of Public Works and Government Services, Canada, 1998). More specific examples were cited in *GEO-1*.

Significant achievements have been made in reducing industrial pollutants in the United States – for example, through the Federal Water Pollution Control Act of 1956, as amended by bills such as the Clean Water Act of 1977. However, as the scope of economic activity widens, new pollutants are introduced to water supplies. Agrochemical run-off is the main source of water pollution in agricultural regions of North America, contributing 60 per cent of the total impaired stream length and 57 per cent of impaired lake surface in the United States (OECD 1996). Pesticide and herbicide run-off has contaminated groundwater in many areas and has been registered in most water bodies, including the Great Lakes and the St Lawrence, the Susquehanna and Colorado rivers. Nitrogen and phosphorus levels that exceed national standards have been found in the surface and groundwater of areas devoted to intensive agriculture, resulting in over-fertilization and the eutrophication of water bodies.

Fish consumption advisories provide another measure of water quality. In 1995, consumer advisories to limit consumption of certain fish species increased by 14 per cent over the previous year; advisories were issued in 1995 for 1740 water bodies in 47 states, an increase of 209 warnings from 1994 (Council on Environmental Quality 1997). Mercury accounted for more than two-thirds of the warnings. Warnings for PCBs rose 37 per cent, for chlordane 16 per cent and for DDT (which has been banned in the United States since 1972) by 3 per cent. These increases may result from the increased number of surveys being carried out by States and therefore do not necessarily indicate worsening conditions. However, they do show where local water quality problems exist (Council on Environmental Quality 1997).

Over the next 10 years, water use will continue to rise. Expanding populations will require more water to serve domestic, commercial, recreational and manufacturing needs (International Joint Commission 1997). Climate change is expected to further increase the demand for irrigation water in some areas of North America, especially the Great Plains.

Marine and coastal areas

North America's coastline is at least 400 000 km long and is marked by a diversity of ecosystems, including estuaries, bays, inlets, barrier islands, fjords, tidal flats, lagoons, salt marshes, mangrove swamps, coral reefs, deltas and dunes. These areas support a profusion of marine resources, many of which are harvested commercially, as well as recreational and tourist activities. More than 50 per cent of the US population lives in coastal areas and by the year 2025 this will reach 75 per cent (National Oceanic and Atmospheric Administration 1998a). Coastal areas are thus economically significant.

Yet the ocean's living resources and the benefits derived from them are threatened by fisheries operations, chemical pollution and eutrophication, alteration of physical habitat, and invasions of exotic species. New threats may be caused by ozone depletion and human-induced climate change (National Oceanic and Atmospheric Administration 1998a, National Research Council 1995).

Finfish catches in Canada have declined dramatically over the past 40 years, causing severe economic and social problems

At the beginning of this century, settlements along North America's coasts were characterized by small fishing communities, and a few cities situated on important transportation routes. Marine resources were abundant. Indeed, the vast stocks of cod on the Atlantic's Grand Banks, the most important cod-fishing ground in the world, attracted European fishing vessels long before immigrants settled in the region. The low level of technology precluded over-exploitation of fish stocks, whereas marine mammals were more susceptible and in great demand. In the Arctic, commercial whaling had greatly reduced some species even by the early 1900s.

During the 1950s and 1960s, new technologies increased the harvesting capability of the fishing industry, and fishing intensified. The industry based itself in fewer and larger ports and in the hands of a few large companies. Total fish catches in North America rose from 3.9 million tonnes in 1961 to a peak of 7.56 million tonnes in 1987 (FAO 1997a).

In the North Atlantic, 21 of the 43 groundfish stocks in Canadian waters are now in decline and 16 more have shown no recent signs of growth (OECD 1995). Ground fish, such as cod, haddock, redfish and several species of flatfish, are most affected by pressure from overfishing, and ground fish stocks off the east coast, especially cod, have nearly collapsed. The Atlantic finfish catch, of which groundfish form the bulk, declined from 2.5 million tonnes in 1971 to less than 500 000 tonnes in 1994. Other factors reinforcing these declines include changes in ocean characteristics, particularly temperature and salinity, increased predation by growing seal populations, failures within Canada's domestic fishery management system, and foreign overfishing outside Canada's 200 mile limit (OECD 1995). The alarming economic, social and environmental implications of declining cod stocks had been known for many years but action was taken only in the 1990s. Canada placed a two-year moratorium on Northern Cod, which has since been extended indefinitely. In 1993, the United States imposed stricter limits and shorter seasons. The collapse of the industry caused severe economic dislocation for those whose income and way of life depended on the sea. It is still unclear to what degree cod populations are recovering and what level of harvest is sustainable for the near future. Concerns also exist for the fate of the west coast salmon fishery, with little known for certain about the impact of habitat loss on the Pacific Coast salmon resource (Commissioner of the Environment and Sustainable Development 1998). The situation is further complicated because the salmon migrate between the north Pacific and their spawning rivers, in both Canada and the United States. Fishing boats from both countries chase the migrating fish in ocean waters

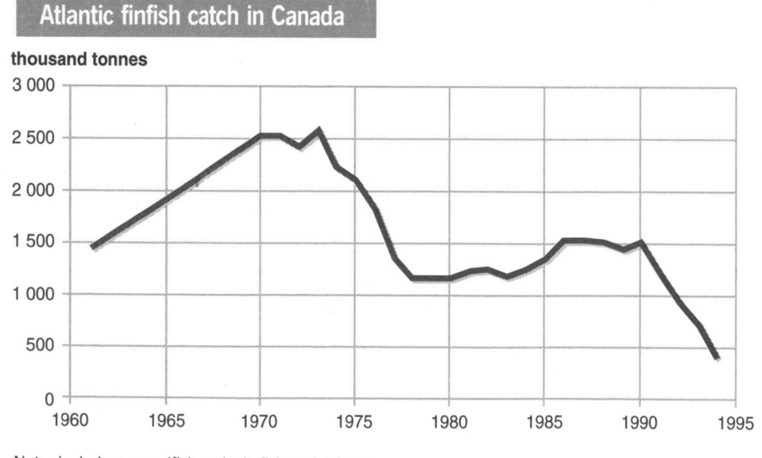

Note: includes groundfish, pelagic fish and salmon

Source: *Statistics Canada.* National Accounts and Environment Division, System of National Accounts, 1996

and high catches have contributed to declining stocks. Data suggest that, for a variety of reasons of which overfishing is only one, stocks of salmon and steelhead in the Columbia River basin have declined by 80 per cent from historic levels, and salmon stocks in California are down by 65 per cent (OECD 1996). To prevent overfishing, Canada and the United States have tried to implement a treaty entitling each country to a catch proportionate to the number of fish spawned in its own rivers. But with neither country able to agree on an equitable split, no agreement has been effective and overfishing continues (Canadian Department of International Affairs and Foreign Trade 1996).

Partly in response to the decline in fish stocks, Canada has implemented a new approach to fisheries management, which includes diversification, promotion of aquaculture and improved aquaculture technology. The production value for five main aquaculture species in Canada grew from an estimated C$7 million in 1984 to an estimated C$460 million in 1997 (farm gate value, Price Waterhouse Cooper 1998). Aquaculture has also grown fast in the United States, with a fourfold increase in the output of fish, shellfish and aquatic plants during the 1980s (US Congress 1995b). While the industry has been aided by improvements in fish husbandry and genetics, harvesting methods, and better processing and transportation systems, environmental issues are becoming important (National Research Council 1992, Stickney 1994). These include pollution from fish waste and feeds; disease transfer from cultured to wild stock; interference with recreational and commercial fishing and boating; reduction of genetic diversity when cultured stock breed with wild stock; competition between native wild species and escaped cultured species for habitat and food resources; and aesthetic impacts from noise, smells and unsightly constructions (Stickney 1994).

Despite management successes, many fish stocks in the United States are threatened. Of the 727 marine species covered under federal management in the nation's 200-mile offshore exclusive economic zone, sufficient information to determine fishery status was available for only 279 species, less than two-fifths of the total (National Oceanic and Atmospheric Administration 1998b). Of these, 86 (31 per cent) were listed as overfished, 183 (66 per cent) were listed as not overfished, and 10 (3 per cent) were considered to be approaching an overfished condition. The overfished species included some of the most valuable commercial fish and shellfish.

Fertilizer run-off creates 'dead zone' in the Gulf of Mexico

The Gulf of Mexico is of great importance for tourism, fisheries, shipping, and oil and gas exploitation, and for wetlands that provide habitat for 75 per cent of North America's migratory waterfowl (OECD 1996). The Gulf of Mexico also is a 'hot spot' for threats to its marine ecosystems. It receives excess nutrients from fertilizer run-off carried by the Mississippi River, which drains 40 per cent of the continental United States. The Gulf of Mexico also receives nutrients from domestic sewage systems, materials flushed out by floods and carbonaceous materials from marsh erosion along the coast. These factors have contributed to the creation of the hypoxic zone, a concentrated area of algae blooms that consume oxygen when they decompose. This has led to the death or displacement of fish in 1 688 km^2 of bottom area off the coast of Louisiana and Texas (Ocean Planet 1995). In addition, some 57 per cent of the shellfish growing area within the Gulf of Mexico has been closed because of health risks (OECD 1996).

Over the past ten years, outbreaks of harmful micro-organisms in coastal waters have become more frequent. Excess nutrients – phosphorus and nitrogen – from agricultural and other human activities are thought to contribute to such outbreaks. *Pfiesteria piscicda*, a toxic dinoflagellate, has been implicated as a cause of fish kills involving millions of fish at many sites along the North Carolina coast as well as smaller fish kills involving thousands of fish in several tributaries of the Chesapeake Bay (US EPA 1997b). The shellfish industry has been severely affected by agricultural run-off, and pesticides have been found in high concentrations in the shellfish inhabiting marine lagoons and estuaries in the Gulf of Mexico (see box above). The Chesapeake Bay suffers from chronic overloading of nutrients, most of which originate from intensively-farmed cropland and livestock production.

While the primary problem in the Gulf of Mexico is hypoxia, there are additional concerns about pesticide and heavy metal pollution accumulation in fish and shellfish. The US EPA Monitoring and Assessment Program for estuaries (EMAP-E) found that about 10 per cent of marine catfish, 2 per cent of Atlantic croaker and 2 per cent of commercial shrimp examined from Gulf of Mexico estuaries had elevated levels of arsenic in edible tissue. However, most of this arsenic is probably in a form that is not toxic to humans. Approximately 1–2 per cent of marine catfish had elevated levels of cadmium, selenium or zinc. About 2 per cent of shrimp had elevated levels of chromium and selenium. Two per cent of Atlantic croaker had concentrations of chromium of more than 2 parts per million. While these numbers show that background levels of contaminants in fish and shellfish are low, higher concentrations of contaminants may be expected near the sources of contamination (US EPA 1997c).

Marine resources are affected in less visible ways by the loss of habitat. The Gulf of Mexico, one of North America's most productive marine areas, is heavily affected by coastal development and human activity. Poor water quality arising from human activities is damaging wetland and seagrass habitat, and coral reefs. The Florida Keys Reefs, extending from Miami to the Dry Tortugas, may support more marine fish species than any other coastal region of the mainland United States and are a major tourism attraction, with more than a million divers visiting the area each year. Yet polluted waters from Florida Bay, and anthropogenic nutrients from storm run-off, sewage and agricultural sources threaten the health of these reefs (WRI, UNEP, UNDP and WB 1998). Agricultural and urban run-off has also led to beach closings: during 1996 beaches were closed or health warnings against swimming were issued 2 596 times. Since 1988, more than 18 590 beach closures and swimming warnings have been put in place across the nation (US Natural Resource Defense Council 1997).

Over the next 10 years, North Americans will continue to be drawn to coastal areas to live and to participate in recreational and tourist activities. Domestic and international demand for fish and fish products is likely to continue to grow. But increased and intensifying human activity will aggravate the environmental problems already suffered by marine and coastal ecosystems. Growing oil imports may increase the incidence of accidental oil spills. Aquaculture itself poses environmental risks, especially as the industry expands, through pollution of the surrounding area from fish faeces, uneaten food and other organic debris, and through the accidental escape of non-indigenous species (Iwama 1991). Threats to human health from more frequent outbreaks of toxic micro-organisms in coastal waters may also increase.

Atmosphere

The dynamic socio-economic transformations occurring in North America over the past century have led to dramatic changes in the atmosphere, including local air pollution and urban smog, transboundary pollution problems such as acid precipitation, and global impacts such as stratospheric ozone depletion and global climate change. These changes have had profound impacts on human and environmental health in North America, as well as on human populations and the environment worldwide.

The release of contaminants into the atmosphere followed the introduction of motor vehicles and industrial expansion across the region within the past century. Although pollution was traditionally concentrated in larger cities and industrial areas, the explosive growth of automobile use facilitated the dispersion of economic activities and human settlements. By the 1960s, the effects of pollution on both local and regional air quality were acute in some parts of North America, with effects on human health, particularly the respiratory system, and the quality of ecosystems (Dockery and others 1996, US EPA 1996).

Low fuel costs and the development of an energy-intensive economy have resulted in the burning of large amounts of fossil fuels in North America, particularly in the United States. After a decline in CO_2 emissions in the early 1980s due to oil price increases, emissions continued to climb, from 1 368 million tonnes in 1984 to 1 607 million tonnes 10 years later. The United States is the world's largest emitter of greenhouse gases – and also emits more per capita than any other country in the world.

Acid precipitation is a serious transboundary air pollution concern in North America. It results from emissions of SO_2 and NO_x, largely from industries and power plants in the US midwest, carried northward by prevailing winds. Thousands of lakes in southeast Canada and northeast United States have become so acidic that they no longer support healthy fish populations. The problem was not addressed until the mid-1970s, by which time precipitation acidity over eastern North America was ten times the pre-industrial value. Changes in industrial processes, fuels and legislation, as well as bilateral cooperation between Canada and the United States, have resulted in declining emissions; SO_2 emissions were reduced by 54 per cent in eastern Canada between 1980 and 1995, and US utility emissions of SO_2 declined by a similar amount. NO_x emissions, however, increased approximately 10 per cent from the 1980s to the 1990s, and only 10 per cent of the lakes in Quebec and the Atlantic Provinces showed reduced acidity by 1994 (International Joint Commission 1997).

Smog is also a serious transboundary air pollution issue with major environmental and human health effects. Canada and the United States have agreed to develop a Joint Plan of Action on Transboundary Air Pollution that will address the major components of smog – ground-level ozone and particulates – and which will include the negotiation of a new ozone annex to the bilateral Air Quality Agreement in 1999. Ground-level ozone is a secondary pollutant formed by reactions between NO_x and volatile organic compounds (VOCs), particularly during the summer months. Pollutants from Mexican cities, some of which are subject to severe smog, are often blamed for non-attainment of air quality standards in nearby US cities.

Over the past decade, there has been a notable decline in the North American production of chlorofluorocarbons (CFCs), the most important ozone-depleting gases, in response to the Montreal Protocol on Substances that Deplete the Ozone Layer, which entered into force in 1989. As a result of cooperation among governments, CFC producers and industry, atmospheric concentrations of CFCs have levelled off (Elkins and others 1993). Nevertheless, CFC production is still legal in developing countries, including Mexico, and a thriving black market has developed for CFCs in North America. This has become a potentially important emerging environmental issue.

There have also been some improvements in local and regional air quality over the past decade, although significant problems remain. In the United States, except for increased NO_x emissions of about 14 per cent, emissions of CO, VOCs, particulates and SO_2 decreased between 1970 and 1994 (Council on Environmental Quality 1997, US EPA 1995). Lead emissions had the most spectacular decline (98 per cent over the same period), due to the adoption of unleaded fuels. But despite declining emissions, air quality is still a public health concern. Particulate pollution is causing increased hospital admissions for the treatment of respiratory and heart diseases, and respiratory infections are causing absence from both schools and work (US EPA 1996, Shprentz 1996). Similarly, high levels of ozone are blamed for irritating the respiratory tract and impairing lung function, causing coughing, shortness of breath and chest pain. In an analysis of ozone health impacts in 13 cities, the American Lung Association estimated that high ozone levels were responsible for 10 000 to 15 000 extra hospital admissions, and 30 000 to 50 000 additional emergency visits to hospitals during the 1993–94 ozone season (Ozkaynak and others 1996).

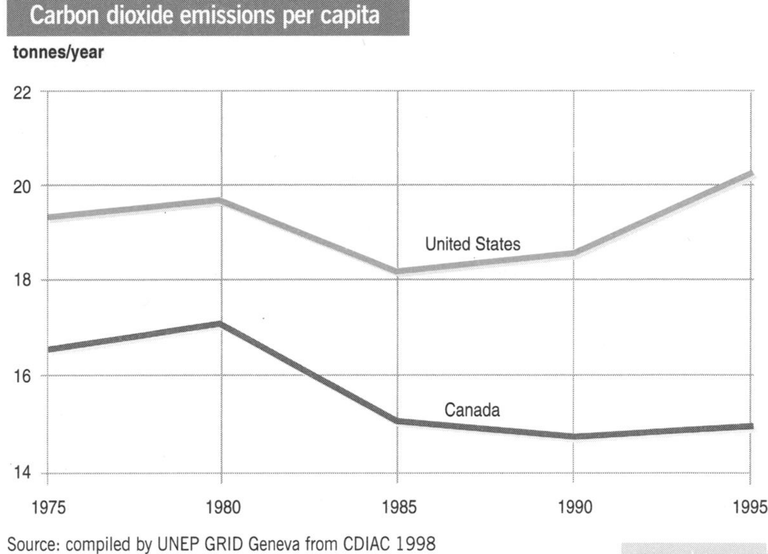

Source: compiled by UNEP GRID Geneva from CDIAC 1998

After a decline in CO_2 emissions in the early 1980s due to oil price increases, US (but not Canadian) per capita emissions continued to climb. The United States emits more per capita than any other country in the world

Over the next 10 years, air quality may improve in some cities but is likely to decline further in others, particularly those with growing populations and increased automobile use. Greenhouse gas emissions in both Canada and the United States in the year 2000 are expected to exceed 1990 levels and to continue to rise as energy consumption increases and automobile transportation expands. By supporting the adoption of the Kyoto Protocol to the Climate Change Convention, both Canada and the United States have shown that they intend to address their high levels of greenhouse gas emissions. The protocol specifies that Canada should reduce its emissions by 6 per cent and the United States by 7 per cent below 1990 levels, during the period 2008–12. However, by exceeding 1990 emissions in the year 2000, Canada and the United States will not meet the 'aim' of the Convention of returning emissions in 2000 to 1990 levels. Higher than expected economic growth, lower energy prices, slower gains in energy efficiency and slower adoption of renewable energy sources have raised US emissions more quickly than anticipated even a few years ago (US Department of Energy 1997).

Urban areas

North America urbanized rapidly early in the 20th century under the combined pressures of population increases, immigration from other regions, and rural/urban migration. Later, the automobile and the rapid development of rail and road transportation networks led to a process of suburbanization, as wealthier inhabitants escaped the congestion and pollution of inner cities. By 1970 this settlement pattern, dependent on the automobile, accounted for between one-half and two-thirds of the US population (Greenwood and Edwards 1979). Large cities emerged along the eastern and western seaboards of the United States and along the Canadian shores of Lake Ontario.

Production of municipal waste

	year	*total (1000 tonnes)*	*per capita (kg)*
Canada	1992	18 110	630
United States	1994	189 696	730
North America	-	238 316	620

Source: OECD 1997

In 1980, approximately 76 per cent of the population in Canada and the 74 per cent of the population in the United States lived in urban areas (United Nations Population Division 1997). Over the past 30 years, the rise in the percentage of population inhabiting urban centres has slowed significantly: the United Nations estimate for urban population for North America in the year 2000 is 77 per cent. Nevertheless, by the year 2020, the urban population in Canada is expected to be 81 per cent and in the United States 85 per cent (United Nations Population Division 1997).

Canada and the United States make up one of the wealthiest urban-industrial regions in the world and have been able to mitigate the most severe environmental impacts of dense population settlements. Substantial political and economic resources have been used to provide the infrastructure and technology for the delivery of potable water, adequate sanitation, wastewater removal, and solid and hazardous waste disposal. Many urban areas have managed to regulate and stabilize local air pollution problems despite the rise in vehicle emissions associated with urban sprawl and increased commuter distances.

Yet the scale of economic growth associated with North American cities, and their reliance on high levels of energy and other resources, contribute significantly to many of the region's pollution and waste problems. North Americans, in fact, are the largest producers of municipal solid waste in the world. Between 1980 and 1995, the average North American produced 620 kg of such waste per year, compared with 430 kg per year for the average European (OECD 1997). Waste reduction, reuse and recycling efforts in the United States are gradually reducing the quantities of municipal waste being incinerated and landfilled: the proportion of waste recovered tripled between 1970 and 1993 to 22 per cent. Recycling rates of products such as glass and paper, however, are still low in comparison with most OECD countries (OECD 1996).

Conclusion

Over the next 10 years, economic growth is expected to be steady, with a continued growth in resource consumption at least in proportion to population. If the commitments under the Kyoto Protocol are met, however, there would have to be greater efficiency in the use of energy and other resources, as well as significant fuel switching to cleaner and less carbon-intensive fuels such as natural gas (which contains less carbon per energy unit than other fuels).

The North American region is at a critical environmental cross-roads: important decisions have now to be made that will determine whether the region's economic activity and patterns of production and consumption will become more sustainable. These decisions will affect both regional and global environmental quality.

References

Allison, A., Miller, S.E., and Nishida, G.M. (1995). Hawaii Biological Survey. In LaRoe, E.T., Farris, G.S., Puckett, C.E., Doran, P.D., and Mac, M.J. *Our Living Resources: a Report to the Nation on the Distribution, Abundance, and Health of US Plants, Animals, and Ecosystems.* US Department of the Interior, National Biological Service, Washington DC, United States, p. 362

Baillie, J., Groombridge, B., Gärdenfors, U., and Stattersfield, A.J. (eds., 1996). *1996 IUCN Red List of Threatened Animals.* IUCN, Gland, Switzerland

Batie, S.S. (1993). *Soil Erosion: Crisis in America's Croplands?* The Conservation Foundation, Washington DC, United States

Bryant, D., Nielsen, D., and Tangley, L. (1997). *The Last Frontier Forests: Ecosystems and Economies on the Edge.* WRI, Washington DC, United States

Caithamer, D. F., and Smith, G. W. (1995). North American Ducks. In LaRoe, E.T., Farris, G.S., Puckett, C.E., Doran, P.D., and Mac, M.J. *Our Living Resources: a Report to the Nation on the Distribution, Abundance, and Health of US Plants, Animals, and Ecosystems.* US Department of the Interior, National Biological Service, Washington DC, United States, pp. 34-7

Canadian Department of Foreign Affairs and International Trade (1996). *Pacific Salmon Treaty.* Ottawa, Canada

CDIAC (1998). *Revised Regional CO_2 Emissions from Fossil-Fuel Burning, Cement Manufacture, and Gas Flaring: 1751–1995.* Carbon Dioxide Information Analysis Center, Environmental Sciences Division, Oak Ridge, Tennessee, United States. http://cdiac.esd.ornl.gov/cdiac/home.html

Colburn, T., Dumanoski, D., and Myers, J.P. (1996). *Our Stolen Future.* Dutton, New York, United States

Commission for Environmental Cooperation (1999). *On Track? Sustainability and the State of the North American Environment.* Commission for Environmental Cooperation, Montreal, Canada

Commissioner of the Environment and Sustainable Development (1998). *Report of the Commissioner of the Environment and Sustainable Development to the House of Commons.* Ottawa, Ontario, Canada

Council on Environmental Quality (1997). *Environmental Quality, The 25th Anniversary Report of the Council on Environmental Quality.* Executive Office of the President, US Government Printing Office, Washington DC, United States

Dockery, D., and others (1996). Health Effects of Acid Aerosols on North American Children: Respiratory Symptoms. In *Environmental Health Perspectives,* 104(5), 503

Elkins, J.W. and others (1993). Decrease in the Growth Rates of Atmospheric Chlorofluorocarbons 11 and 12. In *Nature,* 364, 780

Environment Canada (1994). *Canadian Biodiversity Strategy.* Environment Canada, Ottawa, Canada. http://www.ec.gc.ca/cepa/ip02/e02_01.html

Environment Canada (1995). *Industrial Releases Within the Great Lakes Basin: An Evaluation of NPRI and TRI Data.* Environment Canada, Ottawa, Canada

FASE (1996). *Exporting Risk, Pesticide Exports from US Ports, 1992–1994.* Foundation for Advancements in Science and Education, Los Angeles, United States

FAO (1997a). *FAOSTAT Statistics Database.* FAO, Rome, Italy. http://www.fao.org.

FAO (1997b). *State of the World's Forests 1997.* FAO, Rome, Italy

FAO (1997c). *FAO Fertilizer Yearbook 1997.* FAO, Rome, Italy

Gordon, D. (1995). Regional and Global Protected Area Statistics and Information on the 1996 United Nations List of National Parks and Protected Areas. Paper presented at the IUCN Commission on National Parks and Protected Areas, North American Regional Meeting, Banff National Park, Alberta, Canada, 14–19 October 1995

Government of Canada (1996). *The State of Canada's Environment – 1996.* Print and CD-ROM, Environment Canada, Ottawa, Canada. http://www1. ec.gc.ca/~soer/

Greenberg, R. (1990). *Southern Mexico: Crossroads for Migratory Birds.* Smithsonian Migratory Birds Center, National Zoological Park, Washington DC, United States

Greenwood, N.J. and Edwards, J.M.B. (1979). *Human Environments and Natural Systems.* Wadsworth Publishing Company, Inc., Belmont, California, United States

Harrington, B.A. (1995). Shorebirds: East of the 105th Meridian. In LaRoe, E.T., Farris, G.S., Puckett, C.E., Doran, P.D., and Mac, M.J. *Our Living Resources: a Report to the Nation on the Distribution, Abundance, and Health of US Plants, Animals, and Ecosystems.* US Department of the Interior, National Biological Service, Washington DC, United States, pp 57–60

Hourigan, T.F,. and Reese, E.S. (1987). Mid-ocean isolation and the evolution of Hawaiian reef fishes. In *Trends Ecol. E,* 2, 187–191.

Institute of Water Research (1997). *Zebra Mussels and Aquatic Nuisance Species.* Institute of Water Research, Michigan State University, Ann Arbor Press, Inc., Ann Arbor, United States

International Joint Commission (1997). *The IJC and the 21st Century. Response of the IJC to a Request by the Governments of Canada and the United States for Proposals on How to Best Assist Them to Meet the Environmental Challenges of the 21st Century.* International Joint Commission, Washington DC, United States, and Ottawa, Canada

International Model Forest Network (1997). *Annual Report 1996-97.* International Model Forest Network, Ottawa, Canada

Iwama, G.K. (1991). Interactions between Aquaculture and the Environment. In *Critical Reviews in Environmental Control,* 21, 2, 177–216

Jenkins, P. (1996). Free Trade and Exotic Species Introductions. In Sandlund, O.T., and others (eds.), *Proceedings of the Norway/UN Conference on Alien Species.* Directorate for Nature Management/Norwegian Institute for Nature Research, Trondheim, Norway

Kasperson, J.X., Kasperson, R.C., and Turner II, B.L. 1996. Regions at Risk. In *Environment* 38, 10, 4–15 and 26–29

Kinsinger, A. (1995). Marine Mammals. In LaRoe, E.T., Farris, G.S., Puckett, C.E., Doran, P.D., and Mac, M.J. *Our Living Resources: a Report to the Nation on the Distribution, Abundance, and Health of US Plants, Animals, and Ecosystems.* US Department of the Interior, National Biological Service, Washington DC, United States, pp. 94–96

Langner, L.L., and Flather, C.H. (1994). *Biological Diversity: Status and Trends in the United States.* US Department of Agriculture, Forest Service, Rocky Mountain Forest and Range Experiment Station, Fort Collins, United States

LaRoe, E.T., Farris, G.S., Puckett, C.E., Doran, P.D., and Mac, M.J. *Our Living Resources: a Report to the Nation on the Distribution, Abundance, and Health of US Plants, Animals, and Ecosystems.* US Department of the Interior, National Biological Service, Washington DC, United States

Lipske, M. (1993). Natural Farming Harvests: New Support. In J. L. Butler and C. Schmidt (eds.), *Midwest Regional Environmental Issues Manual: Bringing Environmental Issues Closer to Home,* pp. 73–5). Saunders College Publications, Forth Worth, United States

MacNeill, J. (1989). Strategies for Sustainable Economic Development. *Scientific American,* 261, 3, 154–65.

Malcolm, S. B. (1993). Conservation of Monarch Butterfly Migration in North America: An Endangered Phenomenon. In S. B. Malcolm and M. P. Zalucki (eds.), *Biology and Conservation of the Monarch Butterfly*. California Natural History Museum of Los Angeles County, Los Angeles, United States

Mannion, A. M. (1991). *Global Environmental Change: A Natural and Cultural Environmental History.* Longman Scientific and Technical with John Wiley and Sons, Inc., New York, United States

Mills, E.L., Hall, S.R., and Pauliukonis, N.K. (1998). Exotic Species in the Laurentian Great Lakes; From Science to Policy. In *Great Lakes Research Review,* 3, 2, 1–7

Minister of Public Works and Government Services, Canada. (1998). *Canada and Freshwater* (Monograph No. 6 in Sustainable Development in Canada Monograph Series). Ottawa, Canada.

National Marine Fisheries Service (1997). *Report to Congress on the Status of Fisheries in the United States.* Washington DC, United States

National Oceanic and Atmospheric Administration (1998a). Population: Distribution, Density and Growth. In *State of the Coast Report.* NOAA, Silver Spring, United States

National Oceanic and Atmospheric Administration (1998b). *Ensuring the Sustainability of Ocean Living Resources* (Year of the Ocean Discussion papers). NOAA, Silver Spring, United States

National Research Council (1992). *Marine Aquaculture: Opportunities for Growth.* National Academy Press, Washington DC, United States

National Research Council (1995). *Understanding Marine Biodiversity: A Research Agenda for the Nation.* National Academy Press, Washington DC, United States

National Research Council (1996). *Ecologically Based Pest Management: New Solutions for a New Century.* National Academy Press, Washington DC, United States

Natural Resource Defense Council (1996). *US Pesticide Use at All-time High.* http://www.nrdc.org/search/fzintr.html

Natural Resource Defense Council (1997). *Testing the Waters VII (Ocean Update July 1997).* NRDC, New York, United States

Natural Resources Canada (1996). *Model Forest Network, Year in Review: 1994–95.* Natural Resources Canada, Canadian Forest Service, Science Branch, Ottawa, Canada

Natural Resources Canada (1997). *The Sustainable Management of Forests* (Sustainable Development in Canada Monograph Series, Monograph No. 1). Ottawa, Canada

Natural Resources Canada (1998). *The Canadian Forest Service.* Ottawa, Canada

Ocean Planet (1995). Smithsonian Travelling Exhibition. *Threats to the Health of the Oceans.* Smithsonian Institute, Washington DC, United States
http://seawifs.gsfc.nasa.gov/ocean_planet.html

OECD (1994). *Towards Sustainable Agricultural Production: Cleaner Technologies.* OECD, Paris, France

OECD (1995). *Environmental Performance Reviews: Canada.* OECD, Paris, France

OECD (1996). *Environmental Performance Reviews:United States.* OECD, Paris, France

OECD (1997). *OECD Environmental Data: Compendium 1997.* OECD, Paris, France

Organic Farming Research Foundation (1996). http://www.panna.org/panna/

Ozkaynak, H., and others (1996). *Ambient Ozone Exposure and Emergency Hospital Admissions and Emergency Room Visits for Respiratory Problems in 13 US Cities.* American Lung Association, Washington DC, United States

Pimentel, D. Houser, J., Periss, E., White, Fang H., Mesnick, L., Barsky, T., Tariche, S., Schreck, J., and Alpert, S. (1997). Water Resources: Agriculture, the Environment, and Society. In *BioScience,* 47, 2, 97-106

Price Waterhouse Coopers (1998). Northern Aquaculture Statistics – the Year in Review. *Northern Aquaculture,* July 1998

Powledge, F. (1998). Biodiversity at the Crossroads. In *BioScience,* 48, 5, 347–52

Ramsar (1997). *List of Wetlands of International Importance.* Ramsar Convention Bureau, Gland, Switzerland.

Ricketts, T., Dinerstein, E., Olson, D., Loucks, C., Eichbaum, W., Kavanagh, D., Hedao, Pl, Hurley, P., Carney, K., Abell, R., and Walters, S. (1997). *A Conservation Assessment of the Terrestrial Eco-regions of North America. Volume I. The United States and Canada.* WWF, Washington DC, United States

Robinson, S.K. (1997). The Case of the Missing Songbirds. In *Consequences,* 3, 1, 2–15.

Rubec, C.D.A. (1994). Canada's Federal Policy on Wetland Conservation: a global model. In Mitsch, W.J. (ed.), *Global Wetlands: Old World and New World.* Elsevier Science, Amsterdam, the Netherlands

Schappert, P. (1996, unpublished draft). Distribution, Status and Conservation of the Monarch Butterfly, *Danaus plexippus (L.),* in Canada. Commission for Environmental Cooperation, Montreal, Canada

Sea Grant Minnesota (1997). *Exotic Species.* http://www.d.umn.edu/seagr/exotic/z_overview.htm.

Secretaría de Medio Ambiente Recursos Naturales y Pesca. (1996). *Programa de Áreas Naturales Protegidas de México 1995-2000.* Instituto Nacional de Ecología, SEMARNAP, Mexico City, Mexico

Shprentz, D. (1996). *Breathtaking: Premature Mortality Due to Particulate Air Pollution in 239 American Cities.* Natural Resources Defense Council, New York, United States

Southwick, C.H. (1996). *Global Ecology in Human Perspective.* Oxford University Press, New York, United States

Stein, B.A., and Flack, S.R. (1997). *1997 Species Report Card: The State of US Plants and Animals.* The Nature Conservancy, Arlington, United States

Stickney, R.R. (1994). *Principles of Aquaculture.* John Wiley and Sons, Inc., New York, United States

Temple, S. A. (1998). Easing the Travails of Migratory Birds. In *Environment,* 40, 1, 7–9 and 28–32

Terborgh, J. (1989). *Where Have All The Songbirds Gone?* Princeton University Press, Princeton, United States

Tolman, J. (1995). Poisonous Runoff from Farm Subsidies. In *Wall Street Journal,* 8 September 1995

UNEP/ISRIC (1991). *World Map of the Status of Human-Induced Soil Degradation (GLASOD). An Explanatory Note,* second revised edition (edited by Oldeman, L.R., Hakkeling, R.T., and Sombroek, W.G.). UNEP, Nairobi, Kenya, and ISRIC, Wageningen, Netherlands

UNSTAT (1997). *1995 Energy Statistics Yearbook.* United Nations Statistical Division, New York, United States

United Nations Population Division (1996). *Annual Populations 1950-2050 (the 1996 Revision),* on diskette. United Nations, New York, United States

United Nations Population Divison (1997). *Urban and Rural Areas, 1950-2030 (the 1996 Revision),* on diskette. United Nations, New York, United States

US Congress (1993). *Harmful Non-Indigenous Species of the United States* (OTA-F-565). Office of Technology Assessment, US Government Printing Office, Washington DC, United States

US Congress (1995a). *Biologically Based Technologies for Pest Control* (OTA-ENV-636). Office of Technology Assessment, US Government Printing Office, Washington DC, United States

US Congress (1995b). *Current Status of Federal Involvement in US Aquaculture* (OTA Background Paper). Office of Technology Assessment, US Government Printing Office, Washington DC, United States

US Department of Agriculture (1998). Agricultural Marketing Service. http://www.ams.usda.gov/nop/index.htm

US Department of Energy (1997). Energy Information Administration, press release, 12 November 1997

US EPA (1995). *National Air Quality and Emissions Trends Report.* US EPA, OAQPS, Research Triangle Park, North Carolina, United States

US EPA (1996). *Review of National Ambient Air Quality Standards for Particulate Matter: Policy Assessment of Scientific and Technical Information* (Report No. EPA-452/R-96-013). US EPA, Washington DC, United States. http://www.epa.gov/gumpo/emap/module2.html

US EPA (1997a). *Emerging Global Environmental Issues.* Office of International Activities, Washington DC, United States

US EPA (1997b). *Pfiesteria piscicida.* Office of Wetlands Oceans and Watersheds, Washington DC, United States. http://www.epa.gov/OWOW/estuaries/pfiesteria/fact.html#5

US EPA (1997c). *Monitoring and Assessment Program for Estuaries (EMAP-E).* http://www.epa.gov/gumpo/emap/module2.html

WCMC (1998). WCMC Protected Areas Database http://www.wcmc.org.uk/protected_areas/data

WCMC/IUCN (1998). WCMC Species Database, data available at http://wcmc/org/uk, assessments from the 1996 IUCN Red List of Threatened Animals

Williams, J. D., and Neves, R. J. (1995). Freshwater Mussels: A Neglected and Declining Aquatic Resource. In LaRoe, E.T., Farris, G.S., Puckett, C.E., Doran, P.D., and Mac, M.J. *Our Living Resources: a Report to the Nation on the Distribution, Abundance, and Health of US Plants, Animals, and Ecosystems.* US Department of the Interior, National Biological Service, Washington DC, United States, p. 362

World Bank (1995). *Mexico Resource Conservation and Forest Sector Review* (Report No. 13114-ME): Natural Resources and Rural Poverty Operation Division, Country Department II, Latin America and the Caribbean Regional Office, World Bank

World Bank (1997). *1997 World Development Indicators.* World Bank, Washington DC, United States

World Commission on Environment and Development (1987). *Our Common Future.* Oxford University Press, Oxford, United Kingdom

WRI, ICLARM, WCMC and UNEP (1998). *Reefs at Risk: a map-based indicator of threats to the world's coral reefs.* WRI, Washington DC, United States

WRI, UNEP, UNDP and WB (1998). *World Resources 1998-99: A Guide to the Global Environment* (and the *World Resources Database* diskette). Oxford University Press, New York, United States, and Oxford, United Kingdom

WWF (1997). *Reducing Reliance on Pesticides in the Great Lakes Basin.* WWF, Washington DC, United States

WWF (1998). *Background Paper on Persistent Organic Pollutants (POPs).* WWF, Washington DC, United States

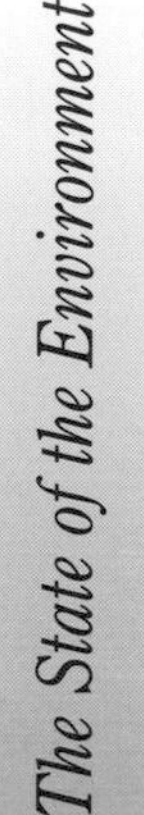

West Asia

KEY FACTS

The region is facing a number of major environmental issues. The scarcity and degradation of water and land resources is the most pressing. Degradation of the marine and coastal environment, loss of biodiversity, industrial pollution and management of hazardous wastes also threaten socio-economic development in the region.

- Land degradation has been a serious problem over the past decade, and the region's rangelands – important for food security – are deteriorating, mainly as a result of overstocking what are essentially fragile ecosystems.
- Thanks to reforestation programmes, forest areas have changed little over the past decade but the cost of imported forest products is high.
- Over the next decade, urbanization, industrialization, a growing population, abuse of agrochemicals, uncontrolled fishing and hunting, war chemicals and military manoeuvres in the desert are expected to increase pressures on the region's fragile ecosystems and their endemic species.
- Groundwater resources in West Asia in general and on the Arabian Peninsula in particular are in a critical condition because the volumes withdrawn far exceed natural recharge rates. Unless improved water management plans are put in place, a series of water-related issues will interact to cause major environmental problems in the future.
- Some 1.2 million barrels of oil are spilled into the Persian Gulf annually. The level of petroleum hydrocarbons in the area exceeds that in the North Sea by almost three times and is twice that of the Caribbean Sea.
- The oil-producing countries generate from 2–8 times more hazardous waste per capita than does the United States.

The West Asia region occupies an area of about 3.95 million km^2 (CAMRE/UNEP/ACSAD 1996) and comprises two sub-regions: the Arabian Peninsula (Bahrain, Kuwait, Oman, Qatar, Saudi Arabia, the United Arab Emirates and Yemen) and the Mashriq (Iraq, Jordan, Lebanon and Syria, and West Bank and Gaza). It is surrounded by four marine water bodies: the Mediterranean Sea, the Red Sea, the Arabian Sea and the Persian Gulf.

The arid and semi-arid climate is characterized by low, erratic, unpredictable rainfall and high evaporation rates. Seventy-two per cent of the area receives less than 100 mm of annual rainfall, 18 per cent receives 100–300 mm and less than 10 per cent receives more than 300 mm. Most of the rainfall occurs during winter (CAMRE/UNEP/ACSAD 1996).

Social and economic background

The discovery of oil in the early 1930s heralded a new economic and environmental chapter in the region's history. The eastern areas of the Arabian Peninsula and northern Iraq emerged as the main sources of fossil fuel (oil and gas) in the world. With this came a period of rapid socio-economic transformation with unprecedented rates of urbanization, hastily-planned industrialization, mass immigration towards the oil-rich states from other parts of the region, as well as an influx of expatriates from outside the region. The combined effects of these influences together with rapidly transformed life styles and consumption patterns have been overwhelming.

Fossil fuels are still the main source of wealth and the region's GDP can fluctuate widely – as it did in the 1980s – due to changes in oil prices. In 1995 total GDP reached US$257 900 million, of which US$218 500 million (85 per cent) are the share of the Gulf Cooperation Council countries (the GCC includes Bahrain, Kuwait, Oman, Qatar, Saudi Arabia and the United Arab Emirates) which contain only 30 per cent of the region's population (UNESCWA 1997). Per capita GDP in GCC countries averaged US$8 579, with the lowest in Oman (US$6 223) and the highest in the United Arab Emirates (US$18 122). Per capita GDP in Mashriq countries averaged US$674 in 1995, with the lowest in Iraq (US$70) and the highest in Lebanon (US$2 950). However, economic progress during the past 30 years, coupled with increasing population pressures, has led to extensive degradation of the region's natural resources.

During the past half century, population increased nearly fivefold (see figure opposite), from just under 20 million in 1950 to 92 million in 1998 (United Nations Population Division 1996). Population growth during the period 1990–95 was slightly less than 3.0 per cent annually and has started to decline in many countries. The average population growth rate is expected to drop to 2.66 per cent by 2010, ranging from 1.3 per cent in Lebanon to 4.0 per cent in West Bank and Gaza (United Nations Population Division 1997).

The population is young and the working population (age group 15–65 years) constitutes only 54.6 per cent of the total (World Bank 1997). The urban population increased from 55.3 per cent in 1980 to 66.5 per cent in 1995, thus putting urban areas under severe pressure. Many countries are also facing serious problems of unemployment, illiteracy, poverty and inadequate basic services even though the human development index has risen, in some cases quite sharply, over the past three decades.

The impact of the events of the past 30 years or so on the environment has been considerable. The most pressing environmental concerns are:

- water resources (quantity and quality);
- the degradation of marine and coastal environments; and
- land degradation and desertification.

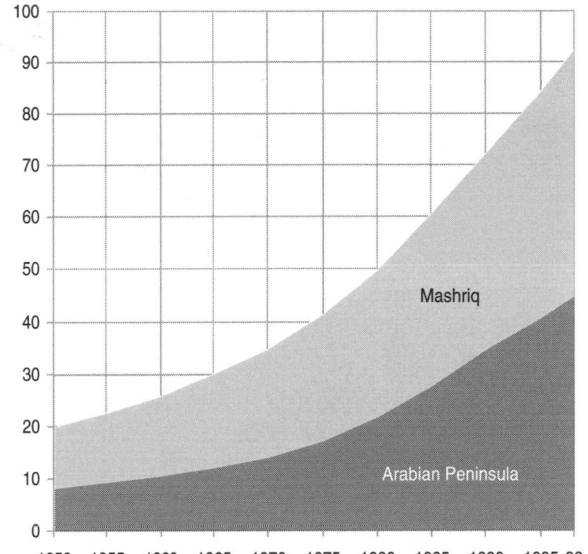

Source: data compiled by UNEP GRID Geneva from United Nations Population Division 1996

During the past half century, population increased nearly fivefold. Population growth during the period 1989–95 averaged 3.8 per cent annually but has now started to decline in many countries

Other major environmental issues include deteriorating conditions in human settlements and urban sprawl; the loss of biodiversity; industrial pollution; inappropriate management of toxic chemicals and hazardous waste; and degradation of the cultural heritage.

These rapid and profound changes have produced serious environmental management problems. State environmental authorities are generally young, inadequately staffed and short of the expertise needed not only to address current issues but also a substantial backlog of environmental problems, both in depleting natural resources and in polluting the environment. This situation is beginning to change. Environmental issues are gradually coming to the forefront of national concerns. There are also encouraging signs of an emerging awareness among the public of the need to protect the environment. The past two decades have seen the emergence of environmental NGOs that are beginning to promote popular support for national efforts to protect the environment. The business community has also begun to take its environmental responsibilities more seriously.

Land and food

Traditionally, grazing and subsistence farming were the main forms of agriculture but, by the middle of

Population growth, the degradation of farmland and urban expansion are reducing the area of arable land available per capita

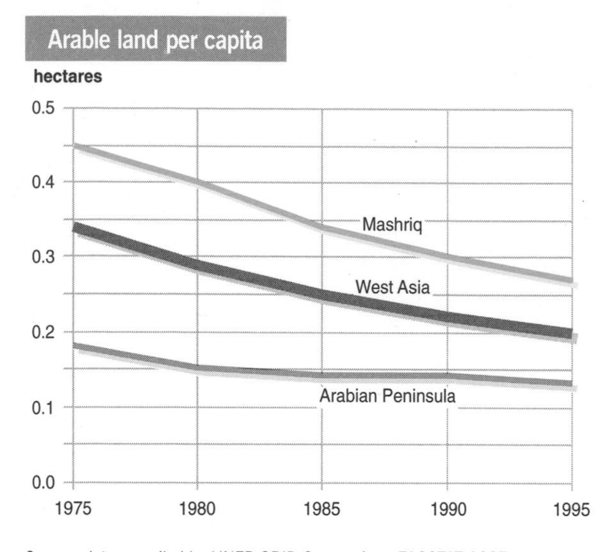

Source: data compiled by UNEP GRID Geneva from FAOSTAT 1997

this century, modern agricultural systems had been introduced to increase food production. Marginal lands and some rangelands were put under cultivation to cope with increasing food demand. The aridity of the environment, deforestation, overgrazing and the extension of cereal crops onto rangelands have led to the deterioration of natural vegetation cover there and accelerated desertification (Nahal 1995). During the 1980s, population growth and other demographic changes led to urbanization, increasing food demand, the intensification of land exploitation and a diminishing per capita availability of cultivated land (see figure above) in all countries except Saudi Arabia, Qatar and Lebanon (FAOSTAT 1997).

Much of the West Asian region is desert, has been desertified or is vulnerable to desertification. Overall, only 4.4 per cent is not at risk

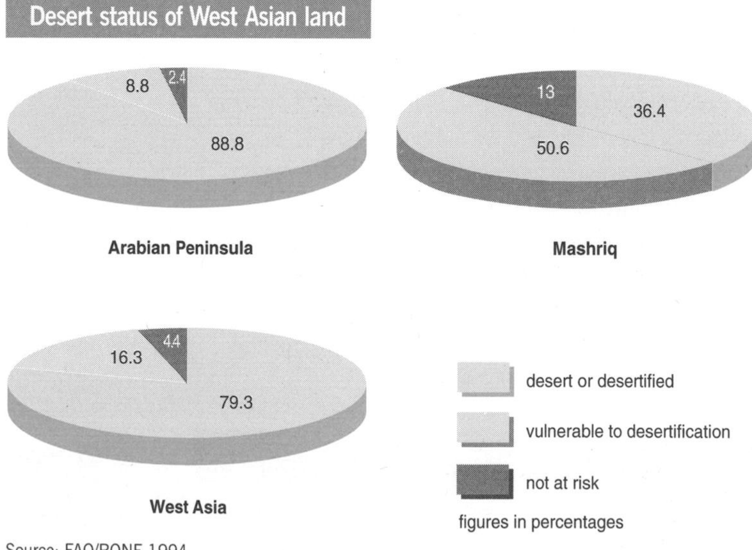

Source: FAO/RONE 1994

Rangelands occupy much of West Asia and are the main form of vegetation cover. The rangeland area varies with annual rainfall fluctuations but in 1994 was estimated at about 150 million hectares or 38 per cent of the total area of the region (FAOSTAT 1997).

In the past, nomadic tribes developed a number of forms of rangeland protection (known as the Al-hema, Hamiyah and Sann systems) which are among the oldest in the world. These systems set aside large tracts of rangeland to be used as reserves during periods of stress. During the 1950s, two major events caused many nomads to abandon their traditional grazing systems: new land-use laws were introduced in several countries which designated rangelands as public property; and agricultural machinery was widely introduced, with some rangelands being ploughed up for cereal farming, particularly barley. As a result of these and other trends, rangelands are deteriorating throughout the region (see box opposite).

Land degradation has been a dominant problem throughout the past decade. Most land is either desertified or vulnerable to desertification (see figure below). The percentage of desertified land ranges from 10 in Syria to nearly 100 in Bahrain, Kuwait, Qatar and the United Arab Emirates. In Jordan, Iraq, Syria and the countries of the Arabian Peninsula, desertification has affected wide areas of rangelands. In Lebanon degradation is serious on steep mountainous land. Salinity is also a serious problem in Bahrain, Iraq, Jordan, Oman, Syria and the United Arab Emirates (CAMRE/UNEP/ACSAD 1996).

The following paragraphs summarize the key issues affecting land and food in West Asia:

- Overgrazing and fuelwood gathering have led to deterioration and desertification of more than 36 million hectares of rangelands in Jordan, Iraq and Syria (AOAD 1995).
- Wind erosion affects 28.1 per cent (1.1 million km^2) of the total area, mainly in GCC countries, Iraq and Syria. Water erosion affects large areas in all Mashriq countries and Saudi Arabia, including 1 260 hectares in Lebanon, more than 1 million hectares in Syria and up to 21 per cent of Iraq. Annual soil loss due to water erosion amounts to 200 tonnes/hectare in the mountainous area of Jordan (CAMRE/UNEP/ACSAD 1996), and reaches similar values on deforested hill slopes in Syria.
- Poor irrigation techniques have resulted in salinization, alkalinization and nutrient depletion in

The deteriorating rangelands

Drought, overgrazing, collecting and uprooting woody species for use as fuel, ploughing and the mismanagement of water resources are the principal causes of rangeland deterioration. Desirable range species have been slowly disappearing and several important medicinal and forage plant species have been destroyed. More than 30 per cent of grazing land in Saudi Arabia is deteriorating as a result of overgrazing, the gathering of woody plants and dryland farming (El-Khatib 1974).

Sheep density on some rangelands is more than one mature head per hectare, some four times the natural carrying capacity (Le Houerou 1995). This is made possible by heavily subsidized complemenatary feeding but leads to impoverished rangelands which become invaded by unpalatable species. Dust storms, sand dunes, desert pavements and various forms of wind and water erosion follow. Deterioration of rangelands and the increasing dependency of the sheep industry on the world grain market could have serious long-term effects, jeopardizing food security in the region.

Governments have taken several steps to restore ecological balance in the rangelands. For example, laws and decrees have been enacted to prevent cultivation of rangelands and a number of protected rangeland areas have been declared – there are now more than 60 in Syria and Jordan alone (ACSAD 1997a). However, most of these plans have failed to produce significant results; rangelands are continuing to deteriorate mainly because these ecosystems are so fragile and because the grazing stock far exceeds the carrying capacity of the land.

large areas. The percentage of irrigated land that is salinized by irrigation is estimated to be 33.6 in Bahrain, 3.5 in Jordan, 85.5 in Kuwait and 5.9 in Syria (FAO 1997a).

- Fertile agricultural land around major cities has been lost to urbanization, industrial establishments and transportation infrastructure. One result is that the food gap in the region increased from US$10 700 million in 1993 to US$11 800 million in 1994 (FAO/UNESCWA 1994; UNESCWA 1997).
- Deterioration of rangeland and farm productivity is forcing farmers to abandon agricultural land and migrate to cities, increasing pressure on services and infrastructure. It is estimated that the cost of soil degradation in Syria is equivalent to about 12 per cent of the value of the country's agricultural output or about 2.5 per cent of total GNP (Ministry of State for Environmental Affairs, Syria, 1997).

Land degradation is expected to continue unless countries undertake more mitigation measures. Fortunately, most countries have now launched national action plans to combat desertification.

Forests

Natural forests used to cover much of the north of the region but there has been a long history of overexploitation and degradation. Extensive land clearing for human settlements and agriculture, grazing by goats, sheep and other animals for thousands of years, illicit felling, burning for charcoal production, fires and inappropriate agricultural practices have now virtually exterminated the natural forests, including much of the former forests of pistachio, oak, juniper and cedar trees that were found in the north of the region. The greatest loss was during World War I, when the best trees from Lebanon

Land use, 1982–84 and 1992–94

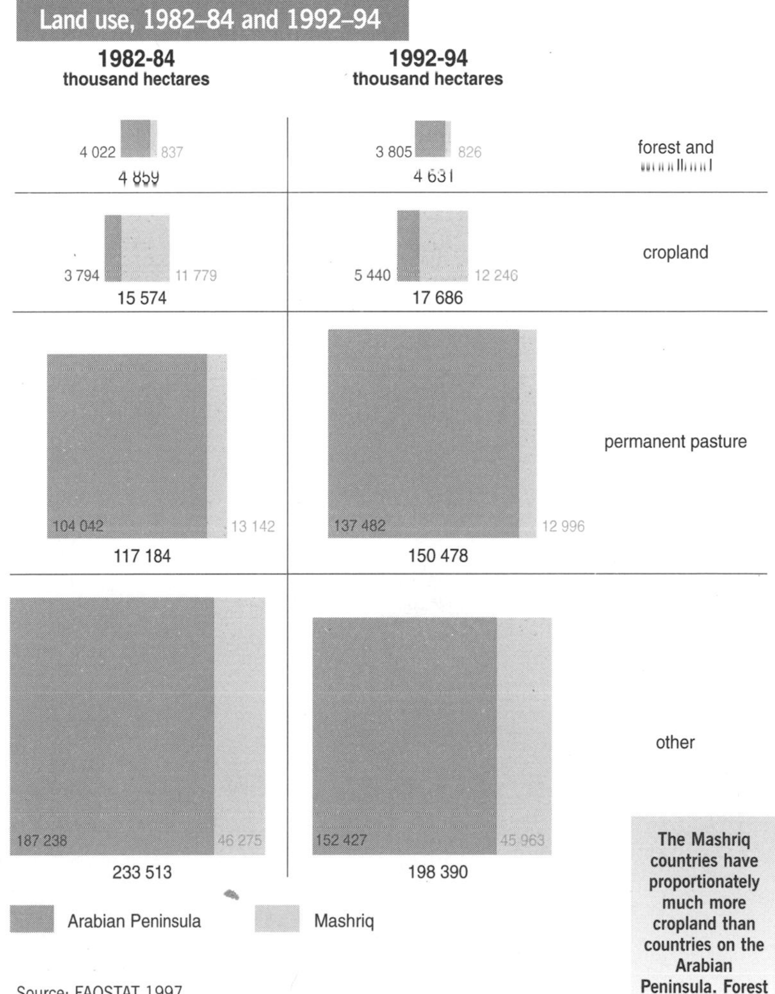

Source: FAOSTAT 1997

Note: data do not include Bahrain and Qatar (nor West Bank and Gaza). 'Other' is mainly desert but also includes built-up areas and wetlands.

The Mashriq countries have proportionately much more cropland than countries on the Arabian Peninsula. Forest cover has changed little

and Syria were felled to construct and operate the Hejaz railways. Lebanon alone lost nearly 60 per cent of its forest trees during the first three years of the war to provide fuel for the railways (Thirgood 1981).

Most forests are now classified as 'other wooded land' but there is still a small percentage of closed forests in the hills and mountains in the north of the region and in the southern part of the Arabian Peninsula. Total forest area is now estimated as 4.657 million hectares (FAOSTAT 1997), 1.25 per cent of the total land area.

Mangrove forests in the United Arab Emirates have been rapidly depleted by overcutting to feed camels and other livestock, but there have recently been intensive conservation and rehabilitation efforts. On the other hand, the forests in the Dhofar Mountains of southwestern Oman are being damaged by overgrazing, uncontrolled tourism and rapid development of rural communities.

The search for new agricultural land has also led to the clearing of forest areas on sloping terrain, causing severe soil erosion in the mountainous watersheds of Jordan, Lebanon, Syria and Yemen. Hostilities, road construction, quarrying and mining, and the construction of dams and irrigation canals have further reduced forest areas and destroyed forest habitats in several countries in the region.

Tourism has begun to affect forest areas in the past ten years. Unplanned activities and the haphazard development of forest sites for recreational purposes have reduced the capacity for regeneration, produced solid waste problems, polluted forest water resources and created new threats to forest ecosystems.

Increases in oil prices in Jordan, Lebanon and Yemen have forced nearby rural communities to depend heavily on forests for fuelwood supplies. The affluent communities of the Arabian Peninsula also use large amounts of fuelwood for cooking and water heating to satisfy their traditional preferences.

Forest areas of the northern part of the region were reduced by 5.8 per cent, from 863 000 to 813 000 hectares, during 1965–75 but reached 852 000 hectares again by 1994 (FAOSTAT 1997). Although the change seems minor, it is often the best stands that are felled. However, the problem is acute in some countries. For example, forest land in Yemen was reduced by nearly 50 per cent during 1980–85 (FAOSTAT 1997). Over the past ten years, however, forest areas have not changed significantly in most countries (see figure on page 161).

Wood forest productivity is rather low, in the range 0.02–0.5 m^3/ha/year, except on the coastal mountains of Lebanon and Syria (Nahal 1985). All countries depend on timber imports to satisfy their local needs.

West Asia's fragile land and marine environments are home to many threatened species

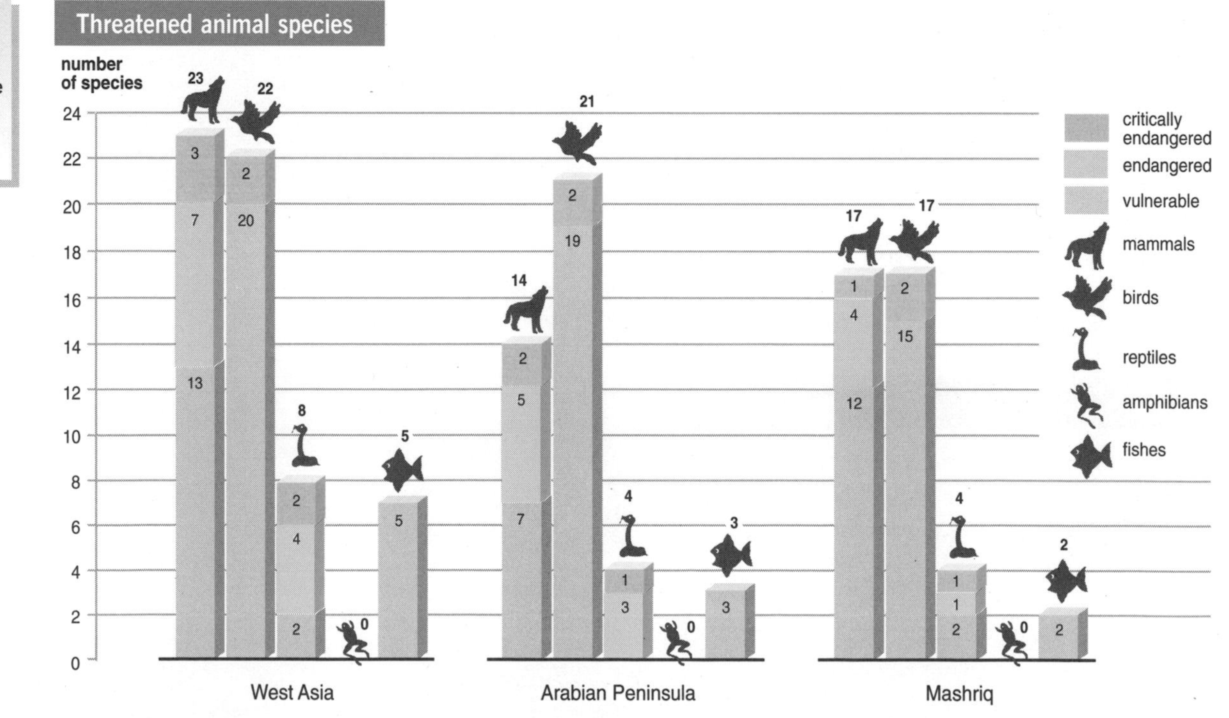

Source: WCMC/IUCN 1998

The value of imported forest products exceeded US$1 000 million in 1994 (FAOSTAT 1997).

Substantial afforestation and reforestation programmes have been launched to increase forest areas. Forest reserves have been declared in Jordan, Lebanon, Saudi Arabia and Syria. Work on sand dune fixation, green belts, roadside plantations and urban forests has been intensified. One result is a significant increase in afforested areas in some countries. For example, Lebanon's long-term programme aims at the afforestation of 200 000 hectares (some 20 per cent of its total area). The rate of afforestation in Syria has increased from 159 ha/year during 1953–70 to more than 24 000 ha/year during the 1980s (Ministry of Agriculture, Syria, 1996). These measures were enough to slow but not arrest deforestation.

Biodiversity

West Asian ecosystems are diverse. The terrestrial ones include Mediterranean forest in the north and sub-tropical mountainous vegetation in the south and southwest. Vast deserts with scant vegetation exist between the northern and southern parts of the region, particularly in the 'Empty Quarter' of Saudi Arabia. Marine ecosystems include extensive coastal areas bordering semi-closed water bodies such as the Persian Gulf, the Mediterranean and the Red Seas, and the open waters of the Arabian Sea. The main marine ecosystems include mudflats, mangrove swamps, sea grass and coral reefs. Large and small rivers in Iraq, Syria, Lebanon and Jordan are the focus of the freshwater ecosystems. Natural freshwater springs are found throughout the region.

The people of this region have traditionally made sustainable use of their natural habitats and conserved biodiversity – for example through the *Al Hema* system of rangeland protection and by prohibiting hunting during certain months of the year. Screening for genetic improvement was begun on cereals and sheep as long as 10 000 years ago (Ucko and Dimbleby 1969). However, more recently overgrazing, deforestation and hunting have contributed to desertification and the extinction of some native plants and animals. These include the Asian lion, *Panthera leo persicus,* which used to live in the northern parts of the region, but disappeared in 1918 (Kingdon 1990); the Syrian wild ass, *Equus hemionus hemippus*, which disappeared in 1928 (Balouet 1990); and the Arabian ostrich, *Struthio camelus syriacus*, which used to live in Syria and Arabia, but became extinct in the 1940s due to overhunting.

Size and number of protected areas

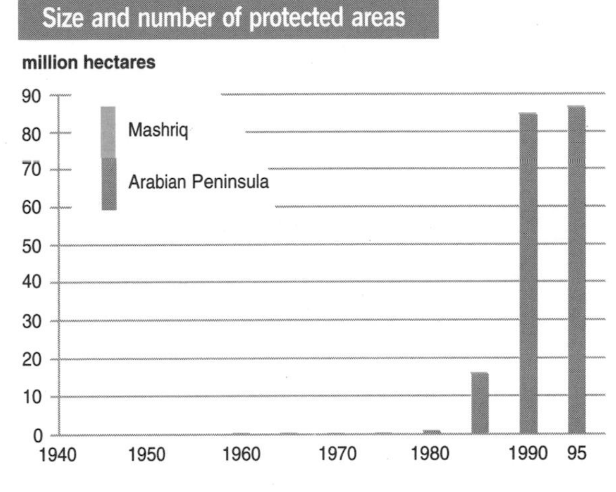

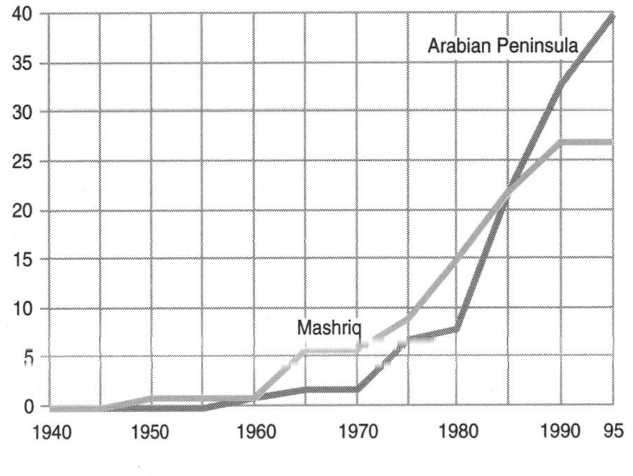

Source: WCMC 1998

The number and size of protected areas has greatly increased over the past two decades but the protected area in the Mashriq sub-region is still small

West Asia's ecosystems are inhabited by numerous species of flora and fauna. Numbers of recorded plant species range from 301 in Qatar (Batanouny 1981) to more than 3 000 in Syria (WRI, UNEP, UNDP and WB 1996). Marine algae range from 216 in the Persian Gulf to 481 in the Red Sea (Mohamed and others 1996); there are 21 species of mammals in Kuwait and 92 in West Bank and Gaza; numbers of birds range from 312 in Kuwait to 413 in Saudi Arabia; and reptiles range from 29 in Kuwait to 84 in Saudi Arabia (ACSAD 1997a, WRI, UNEP, UNDP and WB 1998).

The Red Sea and the Arabian Sea are known for the richness of their marine life. There are, for example, more than 330 species of corals, 500 species of molluscs, 200 species of crabs, 20 species of marine mammals and more than 1 200 species of fish (Fouda and others 1998). Marine biodiversity has been badly affected by overfishing, pollution and habitat destruction. As a result, the fish and shellfish harvest has declined in the Persian Gulf (ROPME/IMO 1996).

Many marine species, including Mediterranean monk seals, marine turtles and marine sponges, are threatened by the continuous deterioration of coastal water quality (Lakkis 1996, Tohmé 1996, Environmental Protection Council, Yemen, 1995). Seawater intrusion is also becoming a real threat to coastal ecosystems (AUB 1994, Youssef and others 1994). The extensive exploitation of sand for construction has aggravated the problem of seawater intrusion and destroyed the habitats of many coastal and marine biota, including marine turtles, along the Lebanese and Syrian coasts. Reclamation and infilling of intertidal areas in Bahrain and marshes in countries such as Iraq and Yemen are destroying habitats and jeopardizing their biological diversity (Environmental Protection Council, Yemen, 1995, UNDP 1998).

There are more than 800 endemic vascular plants in the region (Batanouny 1996), 7 endemic mammals and 10 endemic birds (WRI, UNEP, UNDP and WB 1998). The region has 20–23 endemic corals and 17 per cent of the fishes in the Red Sea are endemic (Sheppard and others 1991). More than 30 per cent of plant species are endemic and some 233 of them are threatened, including *Abies cilicica*, *Cedars libani* and *Juniperus excelsa* in Syria and Lebanon, which are threatened by deforestation. Thirty-two per cent of the plant species in Yemen's Socotra Island are endemic (Environmental Protection Council, Yemen, 1995). Endemic animal species such as the Arabian leopard *Panthera pardus nimr*, the striped hyena *Hyaena hyaena*, the Arabian tahr *Hemitragus jayakari* and the Arabian wolf *Canis lepus arabs* are also threatened (Kingdon 1990).

Protected areas and national parks have been established in all parts of the region. Examples include the Barouk Cedar Forest, Ehden Natural Reserve and Palm Island Marine Reserve in Lebanon, the Azraq Wetland Scientific Reserve in Jordan, the Umm Qusar Swamp Reserve in Iraq, the Harrat al Harra Reserve, Asir National Park and Al-Jubail Marine Sanctuary in Saudi Arabia, the Arabian Oryx Reserve at Jiddat al Harasis and the Sea Turtle Reserve at Ra's Al-Hadd in Oman, and the Cedar and Fir Reserve in Syria.

Formerly extensive plantations of the date palm have been substantially reduced, mainly through poor irrigation

The date palm is one of the most important crop plants in the region. The formerly extensive plantations have been drastically reduced over the past few decades as a result of poor irrigation systems which have lead to soil salinization. Urbanization and the introduction of plant pests have also affected the species. The depletion of underground water levels has led to the deterioration and loss of unique freshwater springs and wetlands with their associated flora and fauna.

Over the next decade, urbanization, industrialization, a growing population, abuse of agrochemicals, uncontrolled fishing and hunting, war chemicals and military manoeuvres in the desert are expected to increase pressures on the region's fragile ecosystems and their endemic species.

Freshwater

Water is the most precious and limited natural resource in West Asia. The Mashriq countries are potentially richer in surface water resources than the Arabian Peninsula, having two shared rivers (the Tigris and the Euphrates) which originate in the temperate zone outside the region. They also have a number of short seasonal and perennial rivers: Lebanon, for example, has 40 such rivers, draining more than 46 per cent of the country's precipitation (Government of Lebanon 1997) as well as a series of sizeable springs located in the mountains. The Arabian Peninsula, by contrast, is poor in surface water resources, which comprise only the erratic seasonal flow of *wadis* and a limited number of medium-quality springs.

Groundwater exists in both sub-regions, including both semi-confined and unconfined shallow aquifers, and deep confined aquifers. Recharge is faster in the Mashriq countries although the deep aquifers of the Arabian Peninsula contain much larger reserves. Data on surface water resources are better than those for groundwater resources where statistics on annual recharge rates, total reserves and safe yields are still unreliable (Al Alawi and Abdul Razzak 1993).

Until the end of World War II, it was believed that water resources were sufficient to meet demand. Population growth and economic development have since resulted in much increased demand. By the 1980s, it was clear there were heavy pressures on both the quality and the quantity of water resources.

During the past decade, the First and Second Gulf Wars seriously affected the economies of West Asia, and many water development schemes were cut back or delayed. Surface water resources in Syria, Iraq, and West Bank and Gaza were reduced by conflicts over water allocations from rivers and aquifers shared with neighbouring countries. This has led to the postponement of many planned agricultural schemes.

Annual renewable water resources in West Asia amount to 113 759 million m^3 (ACSAD 1997b, Zubari 1997). Although the annual per capita volume of renewable water in the region as a whole was 1 329 m^3 in 1995 (see figure right), which is relatively high, the distribution varies widely. Renewable water sources in the Arabian Peninsula are much lower, amounting to only 381 m^3/capita/year in 1995, much less than the benchmark level of 1 000 m^3/year often used to denote water scarcity (Johns Hopkins 1998). The per capita figure in the sub-region ranged from 199 m^3/year in Bahrain to 899 m^3/year in Oman.

In the Mashriq sub-region, renewable water resources are much higher, with an annual 2 181 m^3 per capita, ranging from 191 m^3 in Jordan to 3 089 m^3 in Iraq.

In 1995, about 92.0 per cent of the water in West Asia was used in the agricultural sector, while 7.0 per cent was used for domestic purposes and 1.0 per cent for industry (ACSAD 1997b). In the Mashriq countries, the share of the agricultural sector is 95 per cent compared to 85 per cent for the Arabian Peninsula. Countries of the Arabian Peninsula, by contrast, consume 13.7 per cent of their water in the domestic sector compared to only 4.0 per cent in the Mashriq (see figure below). The agriculture share varies from 25.2 per cent for Kuwait to 96.9 per cent for Iraq, and the industry share from 0.5 per cent for Oman to 71.6 per cent for Kuwait.

Groundwater resources in West Asia in general and on the Arabian Peninsula in particular are in a critical condition because the volumes withdrawn far exceed natural recharge rates. In the region as a whole, groundwater is being extracted much faster than its renewal rate of some 17 000 million m^3 a year (Zubari 1997). As a result, water levels in the shallow aquifers are continually declining.

This has many negative effects. The use of groundwater in Syria, for example, increased by 0.5 per cent annually during 1976–85 but by 7 per cent a year during 1989–93, largely because of the decrease in surface water availability (Ministry of State for Environmental Affairs, Syria, 1997). In the northeastern part of the country, some springs have dried up and the flow of permanent rivers such as the Khabur have been seriously reduced due to the overexploitation of groundwater. There is growing evidence of groundwater depletion in Syria (and other countries) and projections suggest that, if current rates of abstraction continue, overall demand will outstrip the supply in Syria by 2005 (Ministry of State for Environmental Affairs, Syria, 1997). In the West Bank and Gaza, the water table is dropping at a rate of 10–20 cm a year (UNEP 1996). The water table at Sana'a, capital of Yemen, has also dropped seriously as a result of heavy pumping (Environmental Protection Council, Yemen, 1995).

Falling groundwater levels also have a detrimental effect on the *Afalaj* system – a means of tapping

Renewable water resources, 1995

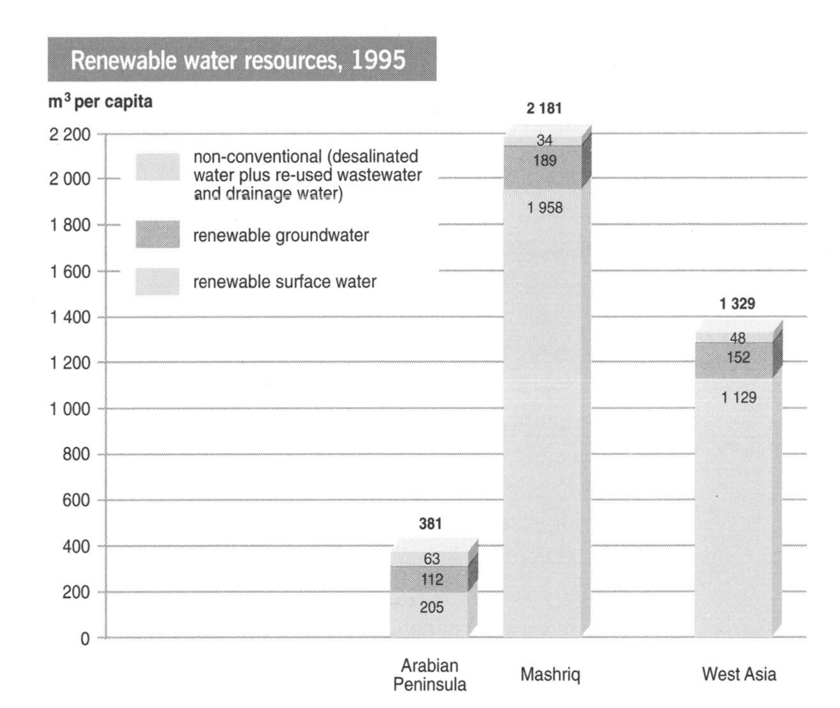

Sources: ACSAD 1997b, GCC 1996a, FAO 1997a, Al-Qasimi 1997, Durabi 1995, Al-Murad 1994, Al Alawi and Abdul Razzak 1993, Ismail 1995

Renewable water supplies in the Arabian Peninsula are well below the critical 1 000 m^3/capita value used to indicate chronic water shortage

Agriculture is by far the largest consumer of water; domestic and industrial consumption is low in comparison

Freshwater withdrawals by sector

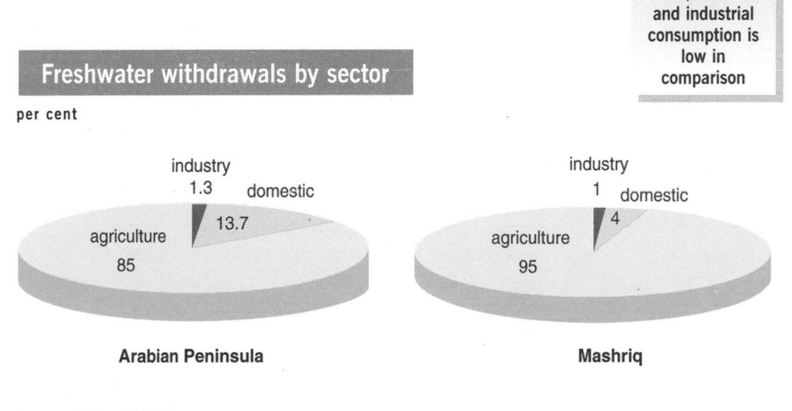

Source: ACSAD 1997b

Traditional Afalaj system of tapping shallow aquifers, which was common in West Asia and limited withdrawals to safe levels

aquifers using gravity-fed underground conduits – that flourished for thousands of years in countries such as the United Arab Emirates, Oman, Iraq and Syria. This effective system of using groundwater restricted withdrawals to within the safe yield of the shallow aquifers. Its use is now rapidly diminishing due to the over-exploitation of the shallow aquifers.

Over-abstraction has also affected the quality of groundwater. This has led to seawater intrusion along the shoreline, causing salinization of coastal agricultural lands. As a result, agricultural production has been reduced, and some arable land, such as the Batinah coastal plain of Oman, has been completely lost (UNEP/UNESCWA 1992). It is estimated that the saline interface between the sea and groundwater advances at an annual rate of 75–130 metres in Bahrain (UNEP/UNESCWA 1991).

If over-abstraction continues, many groundwater resources will eventually be lost as a result of quality degradation. This will result in a further reduction of the arable area because of salinization.

In the Mashriq sub-region, the discharge of raw and partially treated wastewater from agriculture, industry and municipalities into water courses has caused deep concern over health impacts, and has subjected agricultural land and water resources to severe pollution. Shallow aquifers have also been contaminated. It is reported that the nitrate concentration in some domestic wells in the West Bank and Gaza may reach 40 ppm (Zarour and others 1994) – four times the limit set by WHO (see box below). River basins in the Mashriq countries have shown similar symptoms (Hamad and others 1997, Ministry of State for Environmental Affairs, Syria, 1997).

Many efforts have been made to increase recharge rates and to reduce withdrawals by using non-conventional resources (such as desalinated water and recycled wastewater) and by water conservation techniques such as improved irrigation, reducing water subsidies, new legislation and public awareness campaigns.

In the GCC countries, only about 400 million m^3 of the annual 918 million m^3 of treated wastewater are tertiary-treated and used for irrigating non-edible and fodder crops and landscaped areas. About 60 per cent of the partially-treated wastewater is discharged to sea or to low-lying land. In the Mashriq countries, 200 million m^3 of wastewater are used annually for irrigation.

Since the 1950s, desalination plants have been built to increase the supply of freshwater to coastal cities. With nearly 50 plants now in operation, their output of some 1 700 million m^3 a year (about 50 per cent of world capacity) is still relatively small in the region as a whole (Zubari 1997): however, desalination plants operate mainly in the GCC countries where they supply about 40 m^3 out of the total renewable water supply of 381 m^3 per capita per year. Despite their high capital and running costs (desalinated water costs some US$1.0–1.5/m^3), desalination plants will continue to be built to meet the domestic water demands of GCC countries. Desalination capacity is expected to increase from 2 316 million m^3 a year in 1996 to more than 3 000 million m^3 a year in the year 2020 (GCC 1996a). All desalination plants produce some contamination, and the impact of heated brine on the marine environment needs further investigation.

The use of treated wastewater is expected to alleviate pressures on groundwater resources to some extent in some countries. Although wastewater recycling is still relatively little used, ambitious plans for expanding the use of this resource as a strategic alternative to meet future demand exist in many countries. Recycled treated wastewater volumes are expected to increase from about 392 million m^3 a year in 1996 to about 3 000 million m^3 by the year 2020,

Over-abstraction in the West Bank and Gaza

The shallow, sandy aquifer that underlies the West Bank and Gaza is heavily over-pumped and becoming polluted. The Strip currently supports 800 000 people, and a serious pollution risk is posed by the indiscriminate disposal of liquid and solid wastes. The aquifer is essentially the only source of water. The natural replenishment rate of the aquifer is estimated at 50-65 million m^3 a year. Abstraction rates are estimated at 80-130 million m^3 a year, of which most is used in inefficient forms of irrigation. Over-abstraction is causing saline intrusion, and irrigation with this water is causing soil salinization. Most of the population is not connected to main sewerage and uses latrines draining to cesspits, many of which overflow into the surface drains. Faecal contamination of groundwater is widespread and nitrate concentrations in some parts of the aquifer are reported to be ten times the WHO guidelines. Pesticide levels are also believed to be high. Groundwater is no longer potable in some central areas, and five million m^3 of drinking water are transported into the area every year.

Source: UNEP 1996

with the water used mainly to irrigate fodder crops, gardens, landscaped areas and parks (Zubari 1997).

The use of recycled drainage water from irrigation is currently limited to a few countries: Syria, for example, recycles 1 210 million m^3 a year of drainage water. This practice, however, has considerable potential for the future.

The region's population is growing much more rapidly than the pace of development of water resources. Consequently, per capita availability is decreasing. Of the 11 countries in the region, 8 already have a per capita water use of less than 1 000 m^3 a year, and four – Jordan, Kuwait, Lebanon and Yemen – use less than half of that. Only two countries, Iraq and Syria, exceed this benchmark in a sustainable way; two others, Saudi Arabia and the United Arab Emirates, do so by mining their groundwater reserves (ACSAD 1997b, United Nations Population Division 1997 and verified country reports).

Use of unconventional water resources

	desalination	wastewater	drainage
	(m^3 per capita per year)		
Arabian Peninsula	41	23	0
Mashriq	2	5	27
West Asia	20	14	14

Sources: ACSAD 1997b, GCC 1996a, FAO 1997a, Al-Qasimi 1997, Ismail 1995, Durabi 1995, Al-Murad 1994, Al Alawi and Abdul Razzak 1993, Ismail 1995

Unless improved water management plans are put in place, a series of water-related issues will interact to cause major environmental problems in the future. These issues include:

- escalating water demand;
- slow augmentation of water resources;
- the continuous deterioration of water quality and reduction in the yield of heavily-exploited aquifers;
- the modest programme for the treatment of water and sewage from developing urban communities;
- inefficient methods of wastewater treatment and solid waste disposal;
- escalating conflicts over shared surface and groundwater resources if agreements are not reached on equitable allocations;
- rapid population growth; and
- inadequate public awareness and public participation.

Research on the use of solar and nuclear energy for desalination and power generation, coupled with advances in agricultural research and techniques for saving irrigation water, could help ease the impact of these problems.

Marine and coastal areas

The coastlines of West Asia countries are short in Jordan (26 km) and Iraq (58 km) but reach 2 510 km in Saudi Arabia and 2 092 km in Oman. Marine resources have supported coastal populations for thousands of years, and nourished the development of a maritime and trading culture linking Arabia and Africa with Europe and Asia.

Until the turn of the century, the environmental impact of human development on coastal areas was limited to port areas. Fisheries were mainly artisanal, which left fish stocks nearly undisturbed. However, by the end of World War II the marine environment began to show symptoms of ecological imbalance caused by physical alteration of the coastline and coastal habitats by infilling and dredging, increased sewage output, the release of industrial effluents, dumping of oily wastes from tankers and oil-loading terminals, and dumping of litter from both land- and sea-based sources.

From the late 1970s to the early 1990s, the region was affected by the civil war in Lebanon and the two Gulf Wars, which had devastating effects on the environment in Lebanon, Iraq, Kuwait, Saudi Arabia and some other countries. Subsequent reconstruction resulted in substantial developments along the shores of the affected countries. The uncontrolled expansion

The fish catch is declining in the Mediterranean region but good catches are still being taken in the Arabian Peninsula region

Marine fish catch

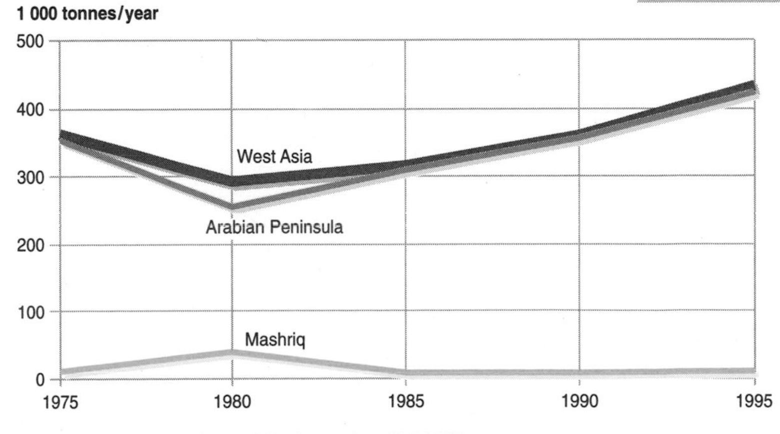

Source: data compiled by UNEP GRID Geneva from FAO 1997c

of coastal cities – in which much of the region's population lives – during the 1990s put even more pressure on the marine and coastal environment.

The population continues to encroach on coastal areas. In Syria, for example, 11 per cent of the population occupies the coastal areas which constitute only 2.2 per cent of total area (Grenon and Batisse 1989) while the figure for Lebanon is about 67 per cent (Government of Lebanon 1997). The situation is similar in the West Bank and Gaza. In some of the countries bordering the Persian Gulf, demand for coastal development is equally high, with coastal cities housing more than 90 per cent of the total population.

In the Mashriq countries and Yemen, discharges into the sea come mainly from domestic and agricultural sources and are dominated by sewage, organic pollutants such as pesticides, heavy metals and oils. Population growth and the concentration of population along the coasts do not match the pace of infrastructure development. Liquid wastes from coastal cities, villages and resort areas are often discharged directly or indirectly to the sea without treatment, causing eutrophication in coastal waters (AUB 1994, Environmental Protection Council, Yemen 1995). Tourism and recreational sites along the coasts contribute to the eutrophication problem along the eastern shore of the Mediterranean.

Many of the region's coral reefs are at risk from overfishing and the threat of oil spills

The threat to coral reefs

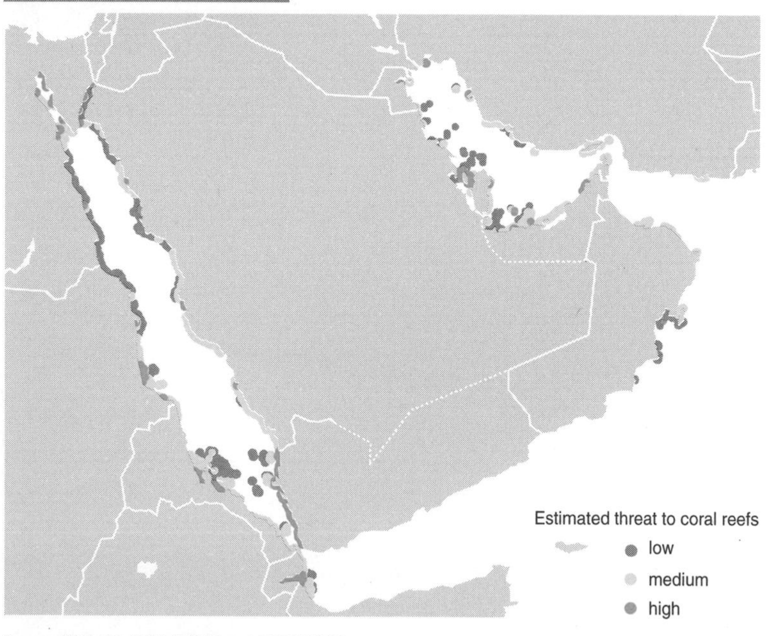

Source: WRI, ICLARM, WCMC and UNEP 1998

In the Arabian Peninsula, land-based pollution from industry dominates and includes:

- Petroleum hydrocarbons from refineries, petrochemical industries, oil terminals, oil spills from ships, pipeline accidents, disposal at sea of oil-contaminated ballast water and dirty bilge, and sludge and slop oil. Some 1.2 million barrels of oil are spilled into the Persian Gulf annually (ROPME/IMO 1996).
- Solid waste discharges include household refuse of 0.5–1.5 kg/person/day and food wastes of 1.4–2.4 kg/person/day (IMO 1995). However, this situation is being improved as a result of cooperation between ROPME, GCC and the European Union.
- Some 20–30 per cent of the sewage discharged into the sea is estimated to be untreated or only partially treated (ROPME 1996). This poses a potential threat of eutrophication in confined areas such as bays.
- Sand depositions from the atmosphere as high as 29 g/m^2/year have been reported (Gharib and others 1985).
- Levels of persistent organic pollutants (POPs) are still relatively low but screening of contaminants in marine sediments and biota have also revealed low levels of halogenated pesticides, PCBs and organic phosphorous compounds.
- Heavy metal concentrations are generally low but there are hot spots near the old outfalls of chemical plants where there are relatively high levels of mercury. Copper and nickel levels are also relatively high near the outfalls of desalination and power plants (Watanabe and others 1993).
- Discharges of concentrated and hot brines from desalination plants.

Oil pollution in the eastern part of the Mediterranean seems to be minor compared to the shores of the Persian Gulf and the Red Sea. However, the Mediterranean Sea, which constitutes 0.7 per cent of the global water surface, receives 17 per cent of global marine oil pollution (UNESCWA 1991).

Much previously uncultivated coastal land has now been reclaimed and converted to agricultural uses. The heavy use of fertilizers, pesticides and herbicides has caused water pollution problems in many countries.

The region contains only about 8 per cent of the

world's mapped coral reefs but almost two-thirds of those in the Persian Gulf are classified as at risk (see map opposite), mainly as a result of overfishing and because more than 30 per cent of the world's oil tankers move through this area every year (WRI, ICLARM, WCMC and UNEP 1998).

Fisheries are an important resource in the Mashriq countries. The fish harvest on the eastern coast of the Mediterranean is decreasing because of coastal pollution, overfishing, the use of destructive fishing techniques, and inadequate fisheries management. Good harvests are still obtained in the Red Sea, Arabian Sea and the Persian Gulf (FAOSTAT 1997).

War has caused extensive damage to the marine environment of the Persian Gulf. The Iran/Iraq war, which lasted eight years, targeted refineries, oil terminals, offshore oil fields and tankers. However, the war over Kuwait exceeded all other environmental disasters of the past four decades. Several million barrels of oil were released into the marine environment. Fallout from burning oil products produced a sea surface microlayer that was toxic to plankton and the larval stages of marine organisms. The long-term impacts of these wars on fisheries and the marine environment in general have yet to be assessed.

Over the next ten years, coastal areas will become more crowded and the pace of development, tourism, agricultural and industrial expansion will increase pressure on these areas.

Atmosphere

Until the middle of this century, the only source of air pollution was dust and sandstorms. Transportation was limited to a few cars, buses and trains and no efforts were made to identify or measure air pollutants.

After World War II, the development of the oil industry, coupled with rapid socio-economic development and high rates of industrial and population growth, led some countries to become high energy consumers: by 1990, Qatar, the United Arab Emirates and Bahrain were the leading per capita consumers of commercial energy in the world (WRI, UNEP and UNDP 1992).

There was an equally fast increase in the number of vehicles inside cities, which compounded the problem. Environmental and safety standards were exceeded in many cities, especially in the Mashriq sub-region, as a result of the growth of industries

Annual commercial energy consumption per capita

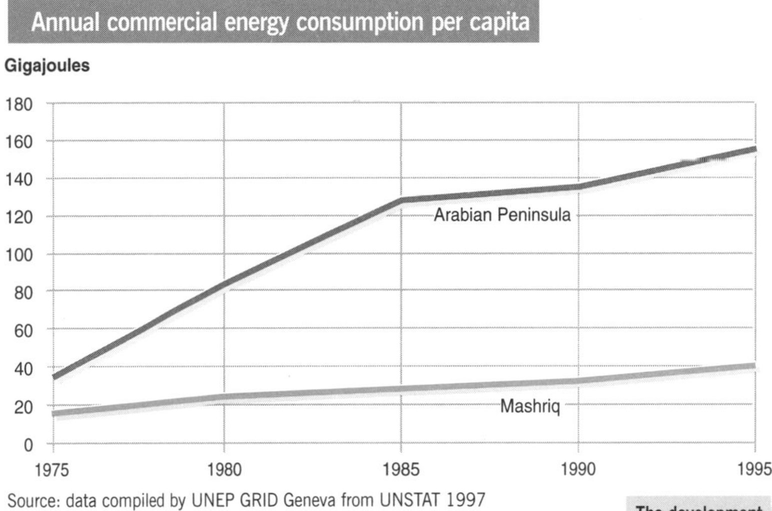

Source: data compiled by UNEP GRID Geneva from UNSTAT 1997

The development of the oil industry and rapid economic development turned some West Asian countries into high energy consumers. Comparable figure for North America is 340 GJ/capita

using heavy fuels, power stations and cement factories. For example, in 1995 Lebanon was estimated to be emitting an annual 3 million tonnes of CO_2, 100 000 tonnes of SO_2, 44 000 tonnes of NO_x and 3 000 tonnes of suspended particulates (Government of Lebanon 1997). In the countries bordering the Persian Gulf, air pollution occurs mainly during rush hours and under conditions of air stability and thermal inversion. Air contamination has risen to alarming levels, especially in cities with more than one million inhabitants such as Baghdad, Damascus and Beirut. SO_2 levels of more than 100 $\mu g/m^3$ are not unusual near industrial areas with refineries and power stations. Traffic also contributes to air pollution, emitting 5 per cent of total SO_2, 37 per cent of NO_x, 10 per cent of suspended particulates and more than 80 per cent of CO and hydrocarbons. It also contributes up to 90 per cent of lead emissions (World Bank 1994). The use of leaded gasoline in old and inefficient cars has made lead exposure a major health problem.

The climate plays a major role in increasing the intensity of pollution in urban areas. Sunshine and high temperatures prevail throughout most of the year. These two parameters play major roles in converting primary pollutants to secondary pollutants, such as ozone and sulphates, which can be more damaging to the environment and human health than the primary pollutants (Bahrain Environmental Protection Committee 1995). Concentrations of ozone higher than the WHO and USEPA accepted limits have been reported in cities such as Baghdad (Kanbour and others 1987), Bahrain (Bahrain Environmental

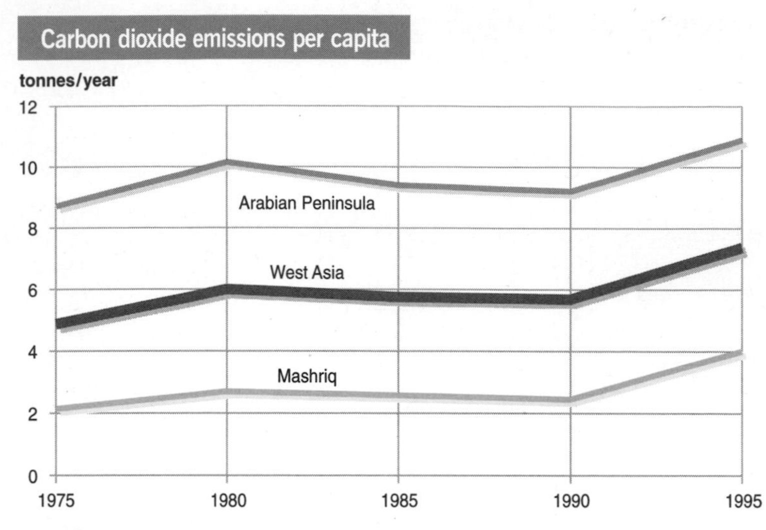

Source: Source: data compiled by UNEP GRID Geneva from CDIAC 1998

Per capita emissions of carbon dioxide on the Arabian Peninsula are well above the world average of 4.0 tonnes

Protection Committee 1995) and Dubai (Dubai Municipality Health Department 1993).

Seasonal dust storms also degrade the environment. The presence of suspended particulates in the air is a health risk, especially to people with asthmatic troubles (Al Awadi 1983). The risk is increased by the presence of other particulates emitted by industry and vehicles. Concentrations of total suspended particulates in several major cities have been found to be three times higher than the accepted WHO limit (Kanbour and others 1985, Environment Protection Department, Kuwait, 1984, Bahrain Environmental Protection Committee 1995 and Dubai Municipality Health Department 1994).

Vehicles are the major source of air pollution in urban areas. Lead additives are still being used in petrol throughout the region but most urban areas report the lead concentration to be within the WHO limit (Kanbour and others 1985, JMOH 1996, Dubai Municipality Health Department 1994, Vreeland and Raveendran 1989) except sometimes during heavy traffic congestion (Kanbour and others 1985).

Most West Asian countries are net energy exporters (except for Jordan, Lebanon and the National Palestinian Territories) and the petroleum and petrochemical industry is expected to grow further in the next decade. This need not necessarily increase air pollution alarmingly; in fact, it is feasible to increase industrial output up to threefold without increasing emission loads. A precedent has already been set by some heavy industries such as the Aluminum Bahrain Company (ALBA) which has reduced fluoride emissions from its factories by more than 98 per cent and suspended particulates by 95 per cent (Ameeri 1997). Refineries in Kuwait, Saudi Arabia and the United Arab Emirates have pledged to reduce sulphur emissions, gas flaring and other hydrocarbon releases as part of the drive towards efficiency and environmental protection.

Urban areas

The region is home to a number of pre-industrial urban centres including Damascus, Beirut and Baghdad, and seaports such as Basra, Aden and Jeddah. These areas supported bazaars, quarters for handicrafts and industries, and shipyards. Industries depended on man and animal power. Most of these urban areas were self-sufficient in food while others depended on shipping and ancient caravan routes.

Urbanization is a consequence of economic growth (World Bank 1997), which is also associated with high population growth rates as well as industrialization. Rapid urbanization can yield important social benefits but can also lead to negative environmental consequences.

There are sharp differences in the patterns of urban growth between the Mashriq and the Arabian Peninsula sub-regions. Urbanization in the Mashriq countries occurred as a result of the slow shift of population from agriculture to industry and services in such well-established urban centres as Damascus and Baghdad. Urbanization in the GCC countries was rapid and sudden, and occurred within the past four decades as GDP and revenues from oil increased. Modern urban infrastructures were created featuring new municipal and government buildings, new industries, and health and educational services. Nomadic communities and foreign workers flowed into the new urban centres. High rates of urban growth are expected to continue into the next century.

In 1950, 23.7 per cent of the region's population (4.7 million) lived in towns and cities. By 1980, the urban population had reached 27.5 million or more than 55 per cent of the total population. Average annual growth rates were 7.9 per cent (1960–65) and 6.8 per cent (1975–80), more than double the overall population growth rate. By 1995, 66.5 per cent of the total population lived in urban areas, and this was projected to reach 69.4 per cent by the year 2000.

Urbanization has been much more rapid in the GCC countries where the urban population reached 83.5 per cent by 1995 and is expected to exceed 86 per cent by the year 2000. Almost all the population in Kuwait (97 per cent) was urban in 1995; the figures for other countries were 90 per cent for Bahrain, 83 per cent for Saudi Arabia and 84 per cent for the United Arab Emirates (United Nations Population Division 1997). However, urbanization in Yemen is still very low; hence overall figures for the sub-region are similar to those for the Mashriq sub-region (see graph right).

In the Mashriq sub-region, the Lebanon urbanization growth rate was 8.14 per cent during 1950–55 but it decreased dramatically to 1.18 per cent during 1975–80 and to only 0.3 per cent during 1985–90, reflecting the effects of the protracted Lebanese war and political instability. In other countries, urbanization matched their slow but steady economic growth. The urban population in Iraq increased from 35.1 per cent in 1950 to 74.5 per cent in 1995. In Syria, the comparable figures are 30.6 per cent in 1950 and 52.2 per cent in 1995 (United Nations Population Division 1997).

In many countries, much urban growth has been concentrated in the one or two cities where new investment, employment opportunities, industries, government jobs, education and health services were concentrated. In 1960, only one city in the region had a population of more than 750 000 inhabitants. By 1990, seven cities had at least this number (United Nations Population Division 1997). In 1995, five cities had populations of more than one million (World Bank 1997).

In most cities, especially in the oil-producing countries, there has been strict land-use planning and zoning. However, this has not always prevented chaotic physical growth. It is now the norm to find residential zones next to industrial sites, and industries enveloped by housing estates, with all the potential risks this implies for human health and safety. Throughout the region, cities have encroached onto agricultural land, where the urban fringes and peripheries grow faster than the cities themselves. Spontaneous or squatter settlements tend to grow in the poorest parts of urban areas where local governments are short of the resources needed to provide basic services such as road networks, health care, sanitation and wastewater treatment plants.

Cities consume natural resources from both near and distant sources. In doing so they generate large amounts of waste that are disposed of within and outside the urban areas, causing widespread environmental problems. Estimated municipal waste generation increased from 4.5 million tonnes per year in 1970 to 25 million tonnes in 1995. In the GCC countries, urban waste generation ranges from 430 kg per capita per year in Qatar to 750 in Dubai (United Arab Emirates), while in the Mashriq sub-region these figures are 185 kg per capita per year in Syria and 285 in Iraq (Kanbour 1997).

In some countries up to 50 per cent of the waste

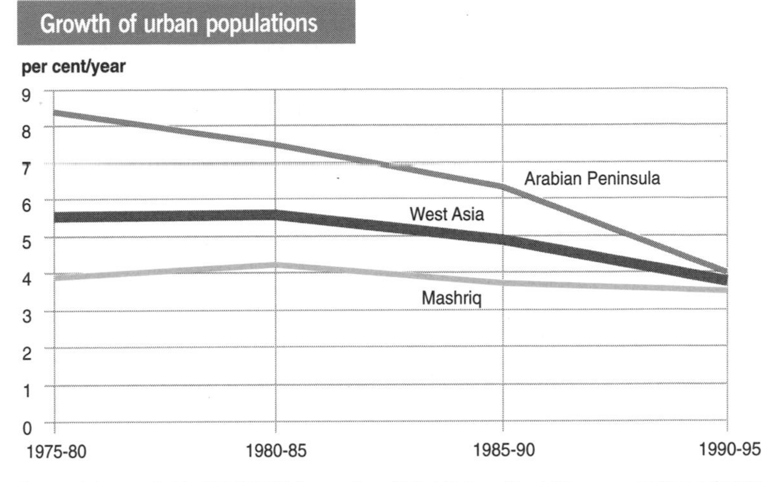

Source: data compiled by UNEP GRID Geneva from United Nations Population Division 1997

Extremely rapid urban growth in the 1970s and 1980s in the Arabian Peninsula has now slowed to almost the same level as in the Mashriq countries

Pollution loads by sector

	SO_2	NO_x	*suspended particulates*	*CO*	*hydrocarbons*
	(1 000 tonnes/year, with percentages in parentheses)				
Power generation	1 600 (39)	1 000 (34)	200 (17)	150 (<1)	50 (<1)
Industry	2 000 (49)	780 (26)	770 (65)	60 (<1)	330 (10)
Road transport	200 (5)	1 100 (37)	120 (10)	16 000 (>90)	3 000 (>80)
Residential	300 (7)	100 (<5)	100 (8)	20 (<1)	10 (<1)

Note: figures include the Islamic Republic of Iran Source: World Bank 1994

generated is left uncollected. With extremely warm temperatures, uncollected waste tends to decompose rapidly, causing serious health risks and an unpleasant nuisance. Another problem is the method of disposal. Several urban areas still dispose of their waste by open dumping and burning, causing potential water and air pollution. In some GCC countries, however, waste collection and disposal are highly efficient. Sanitary landfills are widely used. The high content of organic matter in waste has received some interest from municipal governments. Several composting plants are already in operation, producing organic manure and soil conditioners (Kanbour 1997).

Solid waste production, wastewater treated and garbage collection

	solid waste per capita (kg/year)	*wastewater treated (%)*	*households with garbage collection (%)*
Amman	220	51	100
Dubai	840	100	100
Sana'a	290	51	10
By comparison:			
Toronto	511	100	100

Source: Habitat 1997

While industrial growth is vital for economic development, it has also been a major cause of environmental problems. Nor are all of these problems due to modern industry. Many of the craft industries (such as tanneries and textiles) that are dispersed around and in cities use old and polluting technologies. Most industrial activities are still characterized by lack of pollution control and the absence of cleaner production technologies. In Syria, for example, industrial waste treatment plants are poorly managed and maintained, and are often not compatible with the production processes. In Lebanon, Syria and Jordan industrial activities suffer from inadequate infrastructure, especially for wastewater disposal.

To reduce their reliance on oil revenues, many countries have now embarked on programmes of industrial diversification. Multinational companies have invested in the development of petrochemical complexes, fertilizer plants, refineries and chemical plants. These industrial activities are major generators of hazardous waste. Other hazardous wastes are generated by small and medium-sized industries such as electroplating, tanneries, workshops and garages. A small amount of hazardous waste is also generated by hospitals, research laboratories and transport services.

There are no reliable data on the amount of hazardous waste produced in the region as a whole but some countries do report their hazardous waste inventories and it is possible to use this information to estimate the hazardous wastes of non-reporting countries using GDP as an indicator (World Bank 1989). The results obtained suggest that during 1990–95 there was an increase of 70 000 tonnes a year of hazardous waste in the region as a whole (Kanbour 1998). However, the most alarming result is obtained when per capita generation of hazardous waste is calculated using officially reported data. These results show that the per capita generation of hazardous waste in the non-oil producing country of Jordan is close to that of the United States, which is in the range of 16–28 kg/year. On the other hand, the oil-producing countries generate from 2–8 times more hazardous waste per capita than does the United States (Kanbour 1998). Only a few countries have built facilities for disposal of hazardous wastes, and these are generally inadequate to handle the large amount of waste produced. The problem requires urgent solutions.

A survey of obsolete and banned pesticides in the countries of Africa and the Middle East has recently been carried out by FAO. It found that more than 1 000 tonnes of deteriorated and obsolete pesticides had been disposed of in five countries of West Asia in abandoned areas or in an unsafe manner (FAO 1997b).

Conclusion

The region is facing a number of major environmental issues. The degradation of water and land resources is the most pressing. Degradation of the marine and coastal environment, loss of biodiversity, industrial pollution and management of hazardous wastes also threaten socio-economic development in the region.

Marine resources have been affected negatively by overfishing, pollution and habitat destruction. There is therefore an urgent need to protect coastal zones. Urbanization, industrialization, population growth, abuse of agrochemicals, uncontrolled fishing and hunting are expected to put more pressure on fragile ecosystems and threaten endemic species. Despite afforestation and reforestation activities, forest degradation is a continuing problem. Air pollution is a problem in urban areas and nearby industrial sites.

Water resources, particularly in the Arabian Peninsula, are in a critical condition because the volumes of groundwater withdrawn far exceed the recharge rate. Mismanagement of land resources, periodic droughts, overgrazing, desertification, intensification of agriculture, the wasteful use of irrigation water and uncontrolled urbanization have all contributed to land degradation.

The impending water crisis requires a new strategy to alleviate the impact of development activities on freshwater resources and to identify means of reconciling competing demands for water among stakeholders. These issues are dealt with in more detail on page 356.

References

ACSAD (1997a). *Proceedings of the Arab Experts Meeting on Biodiversity in the Arab World,* 1-5 October 1995. ACSAD, Cairo, Egypt

ACSAD (1997b). Water Resources and their Utilization in the Arab World, 2nd Water Resources Seminar, 8-10 March 1997, Kuwait

Al Alawi, Jamil and Mohammed Abdul Razzak (1993). Water in the Arabian Peninsula: Problems and Perspectives. In Rogers, Peter and Peter Lydon (eds.), *Water in the Arab World: Perspectives and Prognoses*. Harvard University Press, Cambridge, Massachussetts, United States

Al Awadi, A. A. (1983). Health Impacts of Urbanization and Development. Paper presented at the Seminar on Environmental Impact Assessment, 17-29 July 1983, WHO and PADC, University of Aberdeen, United Kingdom

Al-Murad, M. A. (1994). *Evaluation of Kuwait Aquifer System and Assessment of Future Well Fields Abstraction using a Numerical 3D Flow Model.* Arabian Gulf University, Bahrain

Al-Qasimi, H. A. (1997). Management of Wastewater in Qatar. Regional workshop on Technologies of Wastewater Treatment and Reuse, 2-4 June 1997, Bahrain

Ameeri, J.G. (1997). Environmental accomplishments of ALBA. Arab Environmental Day, 14 October 1997. UNEP/MHME, Bahrain

AOAD (1995). *Study on Deterioration of Rangelands and Proposed Development Projects* (in Arabic). AOAD, Khartoum, Sudan

AUB (1994). Position paper for Lebanon. Environmental Workshop in the Middle East: Education, Research Needs and Prospects for Cooperation. AUB/JUST, 28 April–1 May 1994, Irbid, Jordan

Bahrain Environmental Protection Committee (1995). *Report on Air Quality Monitoring.* Bahrain

Balouet, J. C. (1990). *Extinct Species of the World*. Letts, London, United Kingdom

Batanouny, K. (1981). *Flora of Qatar*. University of Qatar, Qatar

Batanouny, K. (1996). Biological Diversity in the Arab World. *Final Report and Proceedings of the UNEP Workshop on Biodiversity in West Asia,* 12–14 December 1995. UNEP/ROWA, Bahrain

CAMRE/UNEP/ACSAD (1996). *State of Desertification in the Arab Region and the Ways and Means to Deal with it* (in Arabic with English summary). Damascus, Syria

CDIAC (1998). *Revised Regional CO_2 Emissions from Fossil-Fuel Burning, Cement Manufacture, and Gas Flaring: 1751–1995.* Carbon Dioxide Information Analysis Center, Environmental Sciences Division, Oak Ridge, Tennessee, United States. http://cdiac.esd.ornl.gov/cdiac/home.html

Dubai Municipality Health Department (1993, 1994). *Air Pollution Bulletins.* Environmental Protection and Safety Section, the Emirate of Dubai, United Arab Emirates

Durabi, A. A. (1995). Water Resources Management in UAE, paper delivered at the Sixth Regional Meeting of Arab IHP Committees, 25–30 December 1995. Amman, Jordan

El-Khatib, A. B. (1974). *Seven Green Spikes. Water and Agricultural Development.* Ministry of Agriculture and Water, Riyadh, Saudi Arabia

Environment Protection Department, Kuwait (1984). *National Report on the State of Environment in Kuwait*

Environmental Protection Council, Yemen (1995). *The Status of the Environment in the Republic of Yemen* (in Arabic)

FAOSTAT (1997). *FAOSTAT Statistics Database.* FAO, Rome, Italy. http://www.fao.org

FAO (1997a). *Irrigation in the Near East in Figures*. Water Report No. 9, FAO, Rome, Italy

FAO (1997b). *Report on FAO project Prevention and Disposal of Unwanted Pesticide Stocks in Africa and the Middle East, Phase I 1994–96.* FAO, Rome, Italy

FAO (1997c). *FAO Fishstat-PC.* FAO, Rome, Italy

FAO/ESCWA (1994). *Analysis of Recent Developments in the Agricultural Sector of the ESCWA Region*; in Arabic. United Nations, New York, United States

FAO/RONE (1994). Desertification in Arab Countries, paper delivered at the JCEDAR Meeting of Arab Experts on Sustainable Rural and Agricultural Development in Arab Countries, Cairo, 25–29 September 1994

Fouda, M. M, Hermosa, G. and Al-Harthi, S. (1998). Status of Fish Biodiversity in the Sultanate of Oman. *Italian Journal of Zoology Speciale*, Vol. 65, Supplement 1

GCC (1996a). *Power Generation and Water Desalination Units in GCC Countries.* GCC General Secretariat, Riyadh, Saudi Arabia

GCC (1996b). *Economic Report No. 11.* GCC General Secretariat, Riyadh, Saudi Arabia

Gharib, I., Foda, M. A., Al-Hashash, M. and Marzouk, F. (1985). *A study of control measures of mobile sand problems in Kuwait Air Bases.* Kuwait Institute for Scientific Research. Report No. KISR 1696. Safat, Kuwait

Government of Lebanon (1997). *Report on the Regional Environmental Assessment: coastal zone of Lebanon.* ECODIT-IAURIF, Beirut, Lebanon

Grenon M. and M. Batisse, eds. (1989). *Futures for the Mediterranean basin: The Blue Plan*. UNEP/MAP, Oxford University Press, Oxford, United Kingdom

Hamad, I, G. Abdelgawad and F. Fares (1997). Barada River Water Quality and its Use in Irrigated Agriculture (Case Study), UNEP/ROWA/AGU Regional Workshop on the Technologies of Wastewater Treatment and Reuse, Bahrain 2–4 June 1997

IMO (1995). *Global Waste Survey.* IMO, Manila, Philippines

Ismail, N. (1995). Strategic Projection for Planning and Management of Water Resources in GCC Countries. *Attaawun*, Vol. 10, No. 38, pp. 47–62 (in Arabic)

JMOH (1996). *Report on Air Quality in the City of Amman.* Jordanian Ministry of Health, Amman, Jordan

Johns Hopkins (1998). Solutions for a Water-Short World, *Population Report,* Vol. XXVI, No. 1, Johns Hopkins Population Information Program, Baltimore, Maryland, United States

Kanbour, F. (1998). Generation of Hazardous Waste in West Asia, paper delivered at the Arab Meeting on the Implementation of the Basel Convention and the Establishment of a Regional Training Centre, Bahrain, 15–17 June 1998

Kanbour, F. (1997). General Status on Urban Waste Management in West Asia, paper delivered at the UNEP Regional Workshop on Urban Waste Management in West Asia, Bahrain, 23–27 November 1997

Kanbour, F., and others (1985). Elemental Analysis of Total Suspended Particulate Matter in the Ambient Air of Baghdad. In *Environ. Int.*, 11, 459

Kanbour, F., and others (1987). Variation of the Ozone Concentration in the City of Baghdad. In *Atmospheric Environment*, 21, 2673-2679

Kingdon, J. (1990). *Arabian Mammals: A Natural History*. Al-Areen Wildlife Park and Reserve, Bahrain

Lakkis, S. (1996). Biodiversity de la flore et la faune marines du liban, paper delivered at the National Seminar on Marine Sciences in Lebanon and the Region: Historical, Current and Future Prospects, Batroun, Lebanon, 25–26 November 1996

Le Houerou, H.N. (1955). Eco-climatic and bio-geographic comparison between the rangelands of the iso-climatic Mediterranean arid zone of northern Africa and the Near East. In Omar, A.S., and others (eds.), *Range Management in Arid Zones: proceedings of the second international conference on range management in the Arabian Gulf,* pp. 25-40. Kegan Paul International Ltd., London, United Kingdom

Ministry of Agriculture, Syria (1996). *Afforestation in Syria.* Damascus, Syria

Ministry of State for Environmental Affairs, Syria (1997). *The State of the Environment in Syria* (draft). Damascus, Syria

Mohamed, S. A, J. A. Abbas and P. W. Basson (1996). Biodiversity in the Arabian Gulf. *Final Report and Proceedings of the UNEP Workshop on Biodiversity in West Asia,* 12–14 December 1995. UNEP/ROWA, Bahrain

Nahal, I. (1985). *Fuelwood production in Syria* (FAO Mission Report). FAO, Rome, Italy

Nahal, I. (1995). Study on sustainable forest resources development in Syria. In University of Aleppo *Agricultural sciences series*, No. 23, pp. 29–67

ROPME (1996). *Review of the State of the Marine Environment, State of Kuwait.* Kuwait

ROPME/IMO (1996). The effect of oil on the marine environment – an overview. ROPME and IMO. Symposium on MARPOL 73/78, 29-29 February 1996, Kuwait

Sheppard, C., C. Price and C. Roberts (1991). *Marine Ecology of the Arabian Region*. Academic Press, London, United Kingdom

Thirgood, J. V. (1981). *Man and the Mediterranean Forest: a history of resource depletion.* Academic Press, London, United Kingdom

Tohmé, H. (1996). Les zones sensibles de la côte libanaise, leur préservation et les moyens de conservation, paper delivered at the National Seminar on Marine Sciences in Lebanon and the Region. Batroun, Lebanon, 25–26 November 1996

Ucko, P. .J and G. W. Dimbleby (1969). *The domestication and exploitation of plants and animals.* Duckworth, London, United Kingdom

UNDP (1998). *Achievements and Challenges of Human Development. Human Development Report, State of Bahrain.* UNDP, Bahrain

UNEP (1996). *Groundwater: a threatened resource.* UNEP Environment Library No. 15, UNEP, Nairobi, Kenya

UNEP/UNESCWA (1992). *The National Plan of Action to Combat Desertification in Oman.* UNEP, Oman

UNESCWA (1991). Discussion paper on general planning, marine and coastal resources, and urbanization and human settlements, delivered at the Arab Ministerial Conference on Environment and Development, 10–12 September 1991, Cairo, Egypt

UNESCWA (1997). *A Survey of Socioeconomic Development in ESCWA Region for 1995.* United Nations, New York, United States

United Nations Population Division (1996). *Annual Populations 1950-2050 (the 1996 Revision),* on diskette. United Nations, New York, United States

United Nations Population Division (1997). *World Urbanization Prospects: The 1996 Revision.* UN, New York, United States

UNSTAT (1997). *1995 Energy Statistics Yearbook.* United Nations Statistical Division, New York, United States

Vreeland, W. and Raveendran, E. (1989). *Lead in Air and Blood in the State of Bahrain,* Report to the Bahrain Environmental Protection Committee, Bahrain

Watanabe, Y., Y. Kanemoto, K. Takeda and H. Ohno. (1993). Removal of soluble and particulate organic material in municipal wastewater by a chemical flocculation and biofilm processes. In *Water Science Technology,* Vol. 27, 11, 201-209

WCMC (1998). WCMC Protected Areas Database http://www.wcmc.org.uk/protected_areas/data

WCMC/IUCN (1998). WCMC Species Database, data available at http://wcmc/org/uk, assessments from the 1996 IUCN Red List of Threatened Animals

WMO and others (1997). *Comprehensive Assessment of the Freshwater Resources of the World.* WMO, Geneva, Switzerland

World Bank (1989). *Safe Disposal of Hazardous Wastes: the special needs and problems of developing countries* (edited by Bastone R., Smith J.E., Wilson D.). World Bank Technical Paper No. 93, World Bank, Washington DC, United States

World Bank (1994). *Industrial Pollution Projection System.* World Bank, Washington DC, United States

World Bank (1997). *World Development Indicators.* World Bank, Washington DC, United States

WRI, ICLARM, WCMC and UNEP (1998). *Reefs at Risk: a map-based indicator of threats to the world's coral reefs.* Washington DC, United States

WRI, UNEP and UNDP (1992). *World Resources 1992–93.* Oxford University Press, New York, United States, and Oxford, United Kingdom

WRI, UNEP, UNDP and WB (1996). *World Resources 1996-97: A Guide to the Global Environment* (and the *World Resources Database* diskette). Oxford University Press, New York, United States, and Oxford, United Kingdom

WRI, UNEP, UNDP and WB (1998). *World Resources 1998-99: A Guide to the Global Environment* (and the *World Resources Database* diskette). Oxford University Press, New York, United States, and Oxford, United Kingdom

Youssef, A. K., A. Balleh and S. Noureddin (1994). Environmental Coastal Situation for Syria. The Environmental Workshop in the Middle East: Education, Research Needs and Prospects for Cooperation. AUB/JUST, 28 April–1 May 1994, Irbid, Jordan

Zarour, H., Jad, I., and Violet, Q. (1994). Hydrochemical Indicators of the Severe Water Crises in the Gaza Strip. In *Final Report on the Project Water Resources in the West Bank and Gaza Strip.* Applied Research Institute, Jerusalem

Zubari, W.K. (1997). Towards the Establishment of a Total Water Cycle Management and Re-use Program in the GCC Countries. The 7th Regional Meeting of the Arab IHP Committee, 8-12 September 1997, Rabat, Morocco

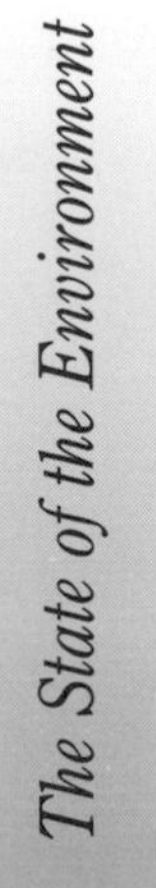

The Polar Regions

KEY FACTS

The once-pristine polar environments are becoming contaminated. They are also exposed to high levels of ultraviolet radiation, and their ice caps, shelves and glaciers are melting as a result of global warming.

- Polar regions act as sinks for contaminants such as persistent organic pollutants, heavy metals and radioactivity which threaten the health of Arctic inhabitants because of bioaccumulation in food chains.
- Some of the highest values of cadmium ever recorded in birds have been found in ptarmigan from northern Norway and the Yukon Territory in Canada.
- Norwegian domestic reindeer increased threefold between 1950 and 1989, during which time the lichen cover was grazed to exhaustion over an area of several thousand square kilometres.
- The Arctic ecosystem has a narrow window of growth determined by snow cover and availability of daylight. Changes to this window can have profound consequences.
- Levels of technetium 99 in some brown algae in Norway increased by a factor of five from 1996 to 1997 as a result of discharges from the Sellafield reprocessing plant.
- Many sub-Antarctic islands bear the distinctive imprint of human modification, particularly through the deliberate introduction of animal pests and predators.
- Conservative estimates put the annual albatross mortality from fishing in the Southern Ocean at 44 000; similar problems exist in the Arctic.
- The reported legal catch of Patagonian toothfish in the Antarctic was 10 245 tonnes whereas the illegal catch was estimated at more than 100 000 tonnes in the Indian Ocean sector of the Southern Ocean alone.

The Arctic and Antarctic are literally poles apart. While they share some characteristics, such as high latitude, cold and remoteness, they also exhibit significant differences. The Arctic is dominated by a large deep central ocean surrounded by land masses. The Antarctic is a large, partially ice-covered land mass surrounded by ocean. The two areas are covered separately in this section, except in discussions of ozone and polar sea ice (see pages 177 and 178).

The Arctic and Antarctic regions, as defined in this publication, are shown in the maps opposite. The Arctic corresponds to the Arctic area internationally accepted through the Arctic Council's Arctic Monitoring and Assessment Programme (AMAP). For the Antarctic, the Polar Front or Antarctic Convergence provides an oceanographically- and biologically-useful natural boundary. The Antarctic is thus defined as the area south of the Antarctic Convergence, unless otherwise specified.

The polar areas play a significant role in the dynamics that affect the global environment and are a good indicator of global change, particularly climate change, although more research is required to understand fully the processes involved and their effects (AMAP 1997). The consequences of an increase in global temperatures and local changes in precipitation and snow cover are not fully understood, but could be leading to the melting of polar ice caps, ice shelves and glaciers, the retreat of sea ice, sea-level rise, thawing of permafrost resulting in increases

The Arctic

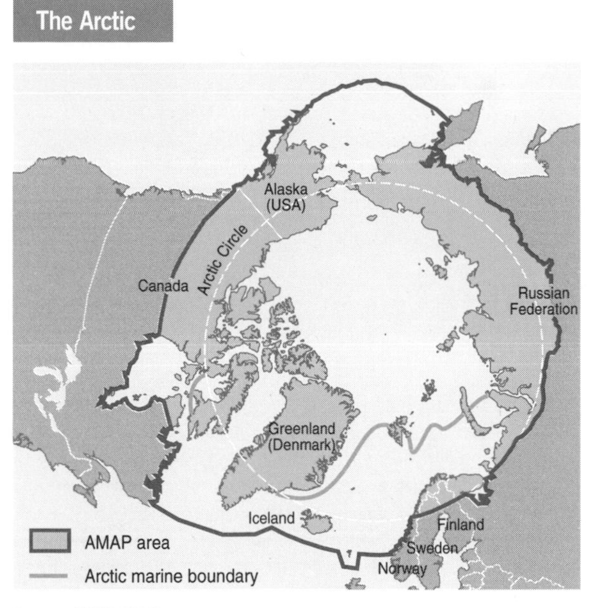

Source: AMAP 1998

The Antarctic

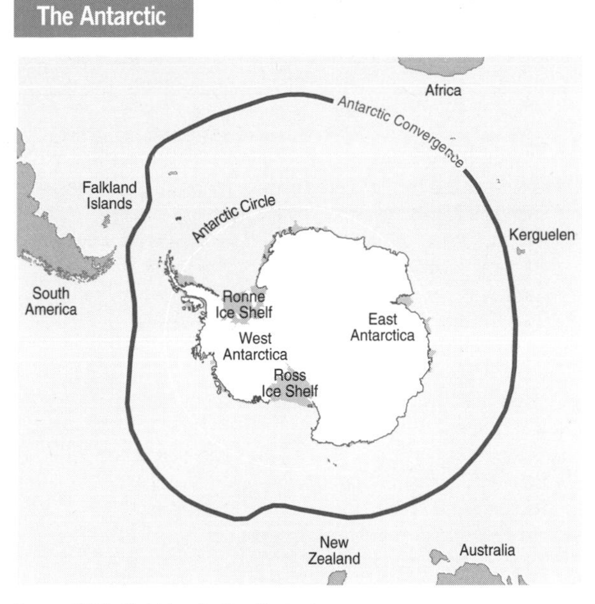

Source: ICAIR, Christchurch, New Zealand

The Arctic (left) as defined by the Arctic Council's Arctic Monitoring and Assessment Programme (AMAP). The Antarctic (right) is the area south of the Antarctic Convergence.

in the emissions of greenhouse gases such as methane and carbon dioxide to the atmosphere, and changes to the radiation balance. In the Arctic, while temperatures have increased in some areas (such as central Siberia and western Canada), in others (such as Greenland) they have decreased (Chapman and Walsh 1993).

Both the Arctic and Antarctic are valued for their relatively clean environments. Polar biota have adapted to the extreme conditions found there, characterized by large variations in temperature and light, and the effects of snow and ice. These adaptations have made some plants and animals more sensitive to human impacts on the environment. Both polar areas are affected by events that occur outside the region. In particular, they act as sinks for a variety of contaminants originating from more temperate latitudes, including persistent organic pollutants (POPs), heavy metals, radioactivity and acidifying substances. There is growing concern that some of

Stratospheric ozone depletion over polar areas

Ozone depletion is much more severe in polar areas than nearer the equator. Over the poles it is manifested by both a general lowering of total ozone amounts and the development of 'holes' in the stratospheric ozone layer.

Until now, the Arctic ozone reduction has been significantly weaker than that of the Antarctic. This may be due to the fact that mean winter temperatures in the Arctic are higher than in the Antarctic, the abundance of polar stratospheric clouds is lower, and the vortex is more variable and breaks down earlier in the winter than in the Southern Hemisphere. Whilst a large, distinct and persistent hole appears in the Antarctic ozone layer every spring, Arctic ozone depletion is characterized by the development of smaller holes, generally up to a few hundred kilometres in diameter, which last only a few days (AMAP 1997). As the latter are never as severe as those in the Antarctic, there is still disagreement on whether the Arctic version should be termed holes at all. The loss of ozone over the South Pole is due mainly to chemical reactions that take place inside the Antarctic polar vortex. Chemical destruction of ozone also occurs over the Arctic during winter and spring. In addition, Arctic ozone lows occur outside the polar vortex as a result of influxes of low-ozone air from middle latitudes.

As mentioned in *GEO-1*, the Arctic winter of 1994–95 was exceptionally cold, and ozone concentrations were 20–30 per cent below normal. Total ozone deficiencies deduced from observations by the Scientific Assessment Panel of the Montreal Protocol were 10–12 per cent lower over Europe than in the mid-1970s, about 5–10 per cent lower over North America, and as much as 35 per cent lower over Siberia. The 1995–96 winter was also extremely cold, with ozone losses of up to 40 per cent (WMO and others 1998).

The Antarctic ozone hole is formed when there is a sharp decline in total ozone over most of Antarctica during the Southern Hemisphere spring. A seasonal hole has developed every year since its advent in the late 1970s, with strong occurrences in 1992, 1993, 1996 and 1997. In 1998, the maximum area of the ozone hole was more than 26 million km^2 and it covered some populated areas of the Southern Hemisphere (WMO 1998).

Polar sea ice and climate change

Antarctic sea ice undergoes an annual change in area from around 4 million km^2 in late summer to 19 million km^2 in late winter (Allison 1997). The Arctic's sea ice varies from 9 million km^2 at its minimum around September to around 15 million km^2 between March and May (Gloersen and others 1992). This ice, and associated snow cover, play an important role in the global climate, with high ice albedo limiting surface absorption of solar radiation and extensive ice cover impeding ocean-atmosphere interaction.

Antarctic pack ice is significantly more mobile than ice in the central Arctic (Kottmeier and others 1992, Worby and others 1997). Consequently, there is generally more open water within the Antarctic pack, much of the ice is created by rapid growth and it is considerably thinner than in the Arctic (Allison 1997).

Climate change could have a considerable impact on sea ice in the Southern Ocean (Murphy and Mitchell 1995, Gordon and O'Farrell 1997). In turn, changes in the characteristics and extent of Antarctic sea ice will affect the vertical structure of the Southern Ocean. These oceanic variations are likely to be felt widely around the globe because the Southern Ocean, as the unifying link for exchanges of water masses at all depths between the world's major ocean basins, transmits climate anomalies around the globe (White and Peterson 1996).

Satellite observations of sea ice extent during 1978–95 suggest that, while there has been little change in the Antarctic, there has been a reduction in the extent of Arctic sea ice of about 4.5 per cent (Bjørgo and others 1997). This difference between Arctic and Antarctic sea ice change is consistent with global climate model experiments simulating future conditions under a gradual increase in atmospheric CO_2 (Stouffer and others 1989, Murphy and Mitchell 1995, Gordon and O'Farrell 1997, and Manabe and others 1992).

The extent of Antarctic sea ice over the past 60 or so years has recently been evaluated using Southern Ocean whaling records extending back to 1931 (de la Mare 1997). From October to April, the Antarctic sea ice edge moved southwards by 2.8° of latitude between the mid-1950s and the early 1970s. This suggests a decline in the sea ice cover of almost 25 per cent. Other observations during this period also support a decline.

General circulation models (GCMs) have been used to simulate the influence of a doubled atmospheric CO_2 level (considered possible in the next century) on polar sea ice and snow cover (Connolley and Cattle 1994, Tzeng and others 1994, and Krinner and others 1997). These models indicate no significant changes for the next two decades but significant reductions in both sea ice thickness and extent thereafter (Hunt and others 1995). Given doubled levels of CO_2, reductions in Antarctic sea ice cover are variously predicted to be around 25 per cent (Gordon and O'Farrell 1997) or almost 100 per cent (Murphy 1995, Murphy and Mitchell 1995). The greenhouse-induced temperature changes projected by GCMs are largest in the polar regions.

these contaminants pose a serious health hazard to some Arctic inhabitants, because of their bioaccumulation and biomagnification in terrestrial and aquatic food chains. Ecosystems may also be at risk from increased levels of UV-B resulting from stratospheric ozone depletion.

Overharvesting of commercially valuable marine species from relatively short food chains which have few species is a major ecological concern in both the Southern Ocean and the Arctic shelf seas. In the Arctic, these activities threaten the livelihood of a number of indigenous groups which traditionally support themselves from the sea. Several migratory bird species spend a significant period of each year in the Arctic, often using the region as a breeding and hatching ground. These species are particularly vulnerable to the effects of environmental contamination. Commercial forestry has fragmented and depleted boreal forests, especially in northern Russia and Fennoscandia (the term used to define an area including Scandinavia, Finland and adjacent areas of northwest Russia). Pressure is moving northwards, threatening the biodiversity of the timberline ecosystem.

Further environmental damage in the Arctic is attributable to natural resource extraction and processing. Industrial processing is causing severe local contamination, particularly in parts of the Russian Arctic. Local contamination is also caused by some mining activities. The Arctic contains some of the world's largest oil and gas reserves. Causes of existing and potential damage to the environment include localized leakage and blow-outs, tanker spills and pipeline leakages. The issue of Antarctic oil spills (for example, the *Bahia Paraiso* spill of 1989) is also of concern, since ships are used to bring fuel to Antarctic stations; 73 oil spills in excess of 200 litres were reported by 17 of the 29 National Antarctic Programmes for the period 1988–98 (COMNAP 1999). An additional threat to the Arctic coastal and marine environment is the development of shipping routes and, in particular, recent work to open up the northern sea route across the northern coasts of Norway and the Russian Federation.

The Arctic

Social and economic background

The Arctic is rich in natural resources. For several millennia, it has been home to people hunting off the land and ocean. The Inuit have lived and travelled throughout the Arctic for more than 5 000 years (Lynge 1993). Non-indigenous interest in the Arctic began in the 1500s with the search for shorter and faster shipping routes through the northeast and northwest passages. Although disputed by some, it is believed that the the American Robert Peary was the first non-indigenous person to reach the North Pole, in 1909.

During the 1950s and 1960s, the Cold War dominated the region. By the late 1970s, the Arctic Ocean was one of the major regions in the conflict between the superpowers. Security concerns dominated. Economic activities also had their geo-strategic component in oil, gas and mineral exploration and exploitation. In the late 1980s, the Cold War came to an end and a new openness and cooperation began (CIA 1978, Samson 1997).

Large numbers of immigrants have moved into the north. This has taken place over thousands of years in the European and Russian Arctic but only over the past 100 years in the North American Arctic, (Dallmann 1997, Samson 1997). There are now approximately 3.5 million people in the Arctic, with the indigenous population ranging from 80 per cent in Greenland to 15 per cent in Arctic Norway and as little as 3–4 per cent in Arctic Russia (AMAP 1998). The region is generally sparsely populated with large distances between population centres. Settlements range from a few large, industrialized cities to small nomadic communities following a traditional lifestyle. The extremes have different relationships with the environment, demonstrated by strong contrasts between their impacts on, and how they are affected by, the environment. Increasing activity in forestry, mineral resource extraction and tourism contribute significantly to the incomes of northern residents.

On a political level, the end to the Cold War brought a thaw in international relations which has had a profound effect on the Arctic (Østreng 1997). The region has emerged as an area of environmental cooperation involving all Arctic nations (see Chapter 3).

Sub-regional trends are important for the future. Alaska is rich in natural resources and the economic pressures to develop these resources put pressures on the environment. Oil and gas production is a large concern, with work in the Prudhoe Bay oil field being the most significant. The extension of this production area east into the Arctic National Wildlife Refuge is being debated at the highest political level with much pressure from local groups concerned with indigenous peoples issues and conserving the environment. The outcome will have major consequences for the local and national economy, the livelihood of local people, and local flora and fauna.

A significant issue affecting northern Canada is the April 1999 creation of the new territory of Nunavut, which means 'our land' in the Inuit language. This new territory includes the central and eastern portions of the existing Northwest Territories which are traditional lands of the Inuit, who comprise 80 per cent of the population in the region. The creation of Nunavut, with its new government, will give residents control over their education, health, social and other services.

There has also been a move towards regionalism and away from central 'southern' government rule in Fennoscandia. The Barents Council was created in

Total and indigenous populations in the Arctic. Some 3.5 million people now inhabit the region

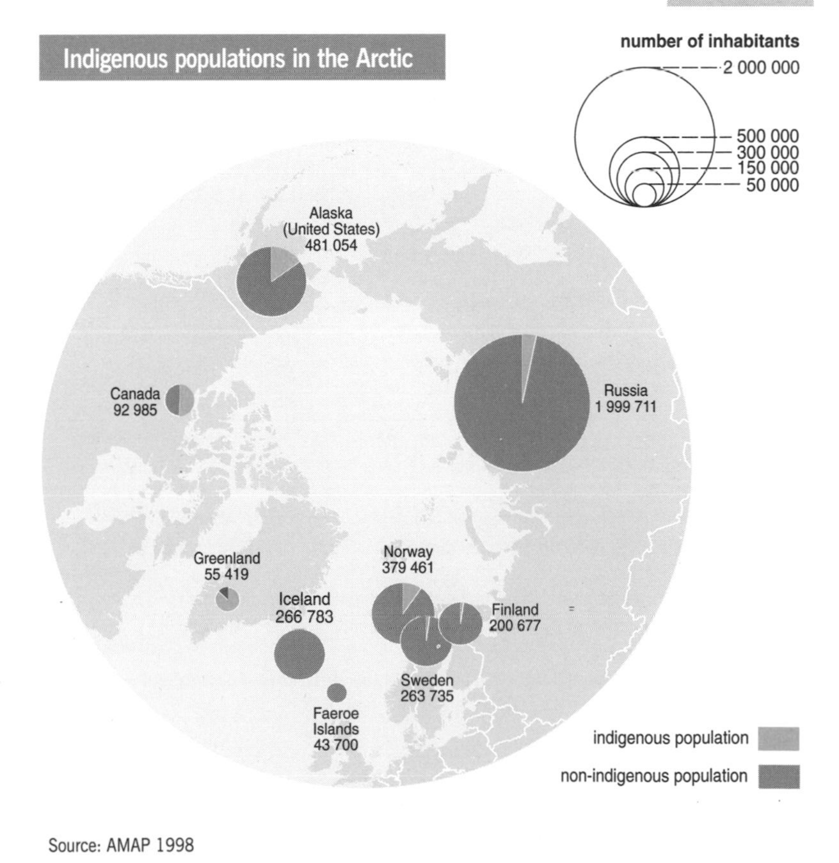

Source: AMAP 1998

1993 and given wide powers. A Saami Parliament has been created in Norway.

A major new trend in Greenland is the exploration for oil and gas. The Norwegian oil company Statoil is beginning exploration on the west coast of Greenland. The local Home Rule Government exercises control over mineral and petroleum activities. However, the extent to which these resources are exploited will have a significant impact on both the environment and the local economy.

The Russian economy has been in a state of transition since the late 1980s, with the northern areas particularly affected. Environmental conditions in the Russian Arctic are of concern. The greatest potential threats from radioactive contamination stem primarily from Russian sources. Accidental releases from power plants, stored nuclear waste and poorly maintained military facilities pose the worst threats. The Russian Federation also has large potential oil and gas reserves, particularly in the Barents Sea Region. The exploitation of these resources will be mainly driven by economic factors. The involvement of non-Russian companies is expected to increase once confidence grows in the political stability of the country but these activities will also increase the threat from accidental releases, particularly from pipelines and shipping transport.

Arctic assessment

The Arctic Monitoring and Assessment Programme (AMAP), established in 1991 to implement part of the Arctic Environmental Protection Strategy (1997), was responsible for preparing an assessment of the state of the Arctic environment with respect to pollution issues. Between 1991 and 1996, AMAP designed and implemented a monitoring programme, largely based on adaptation of ongoing national and international activities. Approximately 400 programmes and projects delivered data to the AMAP assessment. The assessment was produced by scientists and experts from the eight Arctic countries, observing countries and organizations, and representatives of indigenous peoples of the North.

The results of the AMAP assessment are published in two complementary reports. *Arctic Pollution Issues: A State of the Arctic Environment Report* (AMAP 1997) is a concise presentation of the results of the AMAP assessment, including recommendations for actions to be taken. Specifically addressed to the Ministers, this report was presented at a Ministerial Conference in Alta, Norway, in June 1997. *The AMAP Assessment Report: Arctic Pollution Issues* (AMAP 1998) is a more detailed and fully referenced documentation of the scientific background to the AMAP assessment.

The AMAP assessment is a compilation of current knowledge about the Arctic region, an evaluation of this information, and a statement of the prevailing conditions in the area. The assessment was prepared in a systematic and uniform manner to provide a means for inter-comparisons of regional environmental conditions and for assessing the nature and extent of anthropogenic influences on larger (global) scales.

Further information about AMAP and its assessments can be found via the AMAP web site (http://www.grida.no/amap).

Self-governance by Arctic indigenous peoples

Most indigenous peoples favour a move to self-governance. The current level of self-governance varies. Greenland, with its Home Rule Government, the formation of the new territory of Nunavut in Canada and the Norwegian Saami Parliament are the most advanced examples. In the Russian Federation, the Russian Association of Indigenous Peoples of the North, Far East and Siberia is working to link 30 indigenous minority groups and present a united voice to official, Moscow-led, governance. NGOs are connecting indigenous peoples across national boundaries. The Inuit Circumpolar Conference and the Saami Council are prime examples. The Arctic Council, the inter-governmental process towards sustainable development in the Arctic, has established an Indigenous Peoples Secretariat to support and coordinate the activities of the indigenous permanent participants to the process.

Source: AMAP 1997

Land and food

The Arctic terrestrial environment is affected by several factors, including direct impacts from development, pollution from local and distant sources, commercial forestry and grazing. This section focuses on three key issues: pollution, grazing and tourism.

Pollution

Radioactivity. Military activities and the testing of nuclear weapons have been a major source of radioactive contamination of the Arctic. Most atmospheric testing was carried out before 1962, with the Russian island of Novaya Zemlya being the major Arctic testing site. Fallout levels peaked in the 1960s and testing stopped in 1980.

Radionuclides on moss and lichen can reach humans through a simple three-member food chain with caribou in the middle. Radionuclides can also concentrate in mushrooms and berries. All these foods are part of the traditional diet. Their contamination not only affects people's nutrition but also their cultural identity. Radionuclide doses are generally higher for Arctic indigenous peoples living on traditional foods than for people further south. AMAP has calculated that radionuclide contamination from nuclear weapons testing has resulted in approximately 750 additional fatal cases of cancer in the Arctic (AMAP 1997).

The accident at the Chernobyl nuclear power

station in 1986 particularly affected Fennoscandia and northwestern Russia. The initial threat was through the contamination of milk by iodine 131. This was quickly replaced by the threat from caesium 137 with its longer-term contamination of berries, mushrooms and animals grazing on lichen and moss. After the accident, indigenous people in some parts of the Arctic had significantly increased radioactive levels (AMAP 1998).

Significant levels of the naturally-occurring radioactive isotopes polonium 210 and lead 210 exist in northern Canada and Alaska. These isotopes settle on vegetation, such as lichens, which are consumed by caribou. Levels in caribou are higher than in other mammals in the northern Canadian environment (Indian and Northern Affairs 1997a).

Persistent Organic Pollutants (POPs) are a group of chemicals that can travel long distances and resist degradation in the environment. They can be passed through the food web and thereby accumulate in animals. POPs have been in use since the 1950s when substances such as dichlorodiphenyl trichlorethane (DDT) appeared on the market. Circumpolar countries have banned the use of many of the more toxic pesticides. The appearance of these chemicals thus indicates transport by long-range pathways – pathways that concentrate POPs in particular areas, sometimes those of high biological productivity. Polychlorinated biphenyl (PCB) and DDT levels appear high around Svalbard, in the southern Barents Sea, and in eastern Greenland. Canada has higher levels of the pesticide lindane, and other forms of hexachlorocyclohexane (HCH).

Heavy metals. Some of the highest values of cadmium ever recorded in birds have been found in the livers of willow and rock ptarmigan from northern Norway and the Yukon Territory in Canada. These levels may reflect local geological conditions, although the reasons are not fully understood. Effects on the birds have not been studied but it is believed that concentrations can exceed values known to cause kidney damage. The same geographic variations are seen in kidney concentrations of reindeer and caribou. The main source of heavy metals for land mammals and birds is the food they eat (AMAP 1997).

Oil pollution has existed for the past 20 or 30 years and the threat is increasing for a number of reasons. First, exploration is on the increase and is taking oil companies into more remote areas. Secondly, the infrastructure in northwestern Russia is old. Large quantities of oil are transported through thousands of kilometres of pipeline over western Siberia. Many of these pipelines are in poor condition and leaks frequently occur. There were 103 major pipeline failures in the Russian Federation between 1991 and 1993 (AMAP 1997). In 1994 there was a major spill in the Komi Republic of Russia when thousands of cubic metres of crude oil reached watercourses due to the failure of dams being used to contain chronic pipeline leakages. Two major pipelines exist in Canada and Alaska, the Norman Wells pipeline and the Trans-Alaska pipeline. These pipelines are generally maintained to a higher standard.

Some of the highest values of cadmium ever recorded in birds have been found in the livers of willow and rock ptarmigan from northern Norway and the Yukon Territory in Canada

The Arctic environment is more vulnerable to oil pollution than more temperate regions because oil breaks down more slowly under cold and dark conditions. Ecosystem recovery from the effects of oil pollution also takes longer. Oil pollution is also a threat to the marine environment.

Human health concerns

Many Arctic species accummulate fatty tissue as an adaptation to the cold. This increased fat content enhances the biomagnification of contaminants. The traditional diets of some indigenous peoples are concentrated on such fat-accumulating species, making them more vulnerable than consumers from lower latitudes.

The nutritional benefits of these foods are still thought to outweigh the risks of consuming them. There is no conclusive evidence that current contaminant levels in these foods have effects on Arctic adults but effects on the unborn and new-born are of concern. Methyl mercury and several persistent organics are at levels in some mothers' blood which, if found in children, could be potentially detrimental to health through developmental effects. Some of these contaminants are able to cross the placental barrier. Levels of some of these contaminants are on average 2–10 times higher in the Arctic than in non-Arctic areas of Arctic nations. Health concerns also include reproductive impacts and effects on the immune system. In some areas, dietary advice is needed where levels of toxpahene, PCBs and chlordane exceed tolerable daily intakes (AMAP 1997).

These concerns apply to groups with dietary habits that concentrate on the consumption of predatory marine and terrestrial mammals. This differs from more southern regions of the world where diet is based more on agricultural products including animals grown for comparatively short periods, specifically for human consumption. Mammals found in the Arctic are often older and therefore more susceptible to the bioaccumulation of contaminants.

Acidification. Whilst acidification has affected parts of the Arctic for much of the 20th century, the problem did not receive adequate attention until the 1960s.

The most important substances are oxides of sulphur which form when sulphide ores are smelted or fossil fuels burnt. At present, acidification is mainly a local problem, notably around the nickel-copper smelting plants on the Kola Peninsula of northwestern Russia, and at Norilsk in central Siberia where trees, dwarf shrubs and lichens have been severely affected. Other areas of the Arctic are sensitive to acidification, and continue to receive low levels of acidifying substances as a result of long-range transport from sources to the south. However, no effects have yet been observed in these areas (AMAP 1998).

Tourism in the Arctic, early 1990s

area	*number of tourists*
Arctic Alaska	25 000
Yukon (Canada)	177 000
Northwest Territories (Canada)	48 000
Greenland	6 000
Iceland	129 000
Northern Scandinavia	500 000
Svalbard	35 000
Russian Federation (estimate)	a few tens of thousands

Source: http://www.ngo.grida.no/wwfap/tourism/touristmap.htn

Grazing

Reindeer husbandry is an important economic activity that is affecting the vegetation cover in northern Fennoscandia. In Sweden and Norway it is a central part of the Saami culture. In Finland, reindeer husbandry is also practised by more traditional southern farmers. Norwegian domestic reindeer increased almost threefold between 1950 and 1989 during which time the lichen cover in the north of Norway was grazed to exhaustion over an area of several thousand square kilometres. A similar problem exists in northern Finland (EEA 1996, Nordic Council of Ministers 1996). One of the main causes of this problem is that reindeer are individually owned but the grazing land is common property. This means that reindeer herders have no incentive to reduce numbers because if they do then their neighbours will benefit from the extra land to graze upon. One recent, successful change in policy has seen herders paid by weight rather than by head of animal.

Iceland's most serious environmental problem is probably erosion of its loose volcanic soils as a result of overgrazing (Nordic Council of Ministers 1996).

Tourism

Tourism is a recent development for the Arctic. There were nearly one million Arctic tourists annually in the early 1990s (see table). Tourism is most threatening in Scandinavia and Svalbard, where large numbers of people need travel only short distances to easily accessible places. Physical disturbance and noise effects from tourism are apparent in all Arctic countries. Garbage is a related problem. In the Canadian Arctic, the number of visitors to remote Arctic ecosystems is increasing rapidly and there is little understanding of environmental damage thresholds. Cruise ships are able to bring people into areas which were previously unspoilt by humans. On the positive side, ecotourism, when well planned, can help to preserve Arctic environments. An initiative led by the World Wide Fund for Nature has come some way in developing Arctic tourism guidelines (WWF 1997).

Forests

The development of infrastructure to support the extraction of natural resources has grown steadily over the past 100 years, leading to the fragmentation of vulnerable habitats. Commercial forestry is carried out in the boreal forests of northern Fennoscandia, northwest Russia, Siberia and Alaska. Pressures are increasing to harvest timber for pulp, paper and wood products. Once boreal forests are cut, they can regenerate only slowly because of the harsh climate. A considerable amount of commercial cutting occurs at the treeline, an important transition zone between the northern boreal forests and the treeless tundra. This zone varies in width and latitude around the globe.

Logging activities began as far back as the 1500s but it was not until the beginning of this century that the wood-processing industry, with its ability to use more of the available forest materials, started to devastate large tracts of forest. The technique of clear cutting was used, often with subsequent planting of foreign species. This has had a severe effect on the biodiversity of forest ecosystems. Few areas of virgin forest still exist, and those that do are amongst the few areas remaining globally. At the same time, infrastructure developed to meet the increase in forest felling and tourism over the past century has caused fragmentation of the forest ecosystem.

In addition to their use as a commercial resource, forests are important as wintering grounds to reindeer herders in northern Fennoscandia. In Iceland, large areas have been cleared and used as pasture land or for urban development. Some reforestation is now occurring in these areas though alien species are often introduced. Other local issues of concern include damage from smelting (see map) and overgrazing.

The prognosis for the future is mixed. In some commercial forestry areas, new and innovative management regimes are being implemented to allow sustainable exploitation of the natural forest systems. Large areas in the north of Sweden and Finland have been given national park or nature reserve status to protect against deforestation. As yet this has not been duplicated in Norway or the Russian Federation. In Fennoscandia, commercial forestry and infrastructure development are moving northwards and extending the area of habitat fragmentation. Timberline forests, known as shelter forests in Fennoscandia, need special management.

Biodiversity

Human activities have reduced the area of pristine landscape or wilderness in the Arctic. The eight Arctic countries of the Arctic Council have made a pledge, through the programme for the Conservation of Arctic Flora and Fauna (CAFF), to protect a minimum of 12 per cent of each Arctic ecozone (CAFF 1996). The recent status is shown in the table below. Although these figures show a positive trend, it is unclear how representative, or effective, the protected areas are of the variety of ecozones present in the Arctic.

Hundreds of species are endemic to the Arctic. Many of these show genetic uniqueness, a large proportion are migratory, and they are often found concentrated in restricted areas such as marginal ice edges and terrestrial migration corridors.

The limited numbers of plants, animals and micro-organisms living in the Arctic are subject to major climatic variations over very small distances; for example, raised beach ridges are subject to drought, frost heave and large temperature fluctuations. On a much larger scale, there have been major temperature variations over time; for example, ice-free areas with vegetation existed in isolated areas in Alaska, Norway and Novaya Zemlya 10 000–40 000 years ago.

There is evidence, at least for plants, that climatic variations have resulted in significant genetic variation within species. This is obvious in the coexistence of tufted and prostrate forms of purple saxifrage (*Saxifraga oppositifolia*) on Svalbard, distinct locally-adapted populations of mountain avens (*Dryas octopetela*) in fellfield and snow-bed communities, and the highly variable bitterworts (*Draba* species). These ecotypic variations have long-term survival value, equipping the species to withstand oscillating climatic conditions.

Recent application of molecular techniques has shown that high genetic diversity exists within some

Forest damage zones

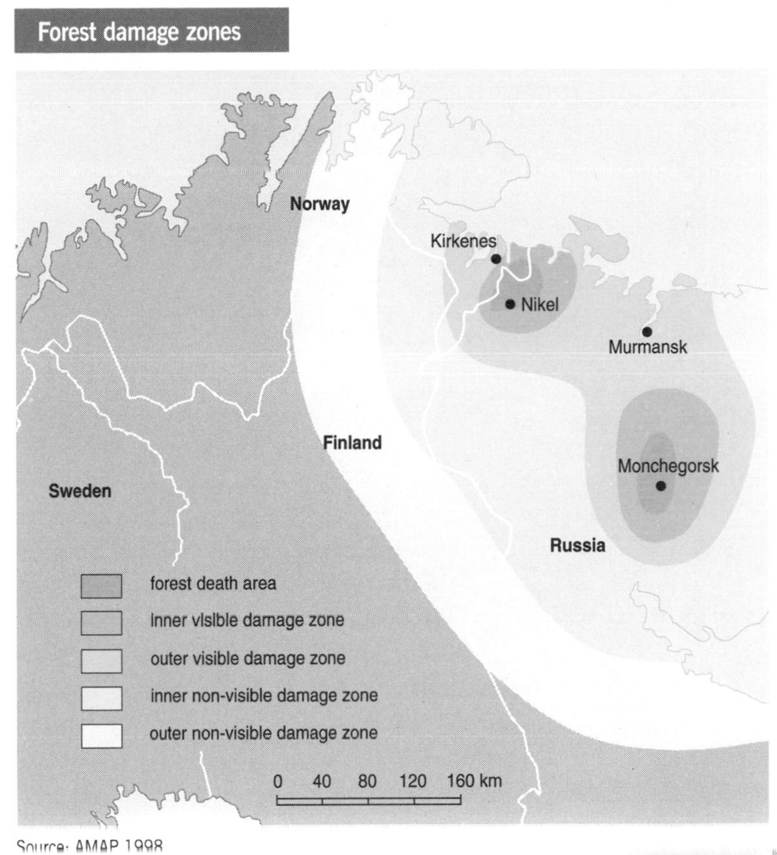

Source: AMAP 1998

Extensive damage to the boreal forest in northeast Russia has been caused by sulphur and heavy metal emissions from industrial sites

Protected areas, 1997

	number established in 1996–97	size (km²) established in 1996–97	total number	total size (km²)	% of Arctic protected
Canada	2	27 815	48	462 674	8.8
Finland			52	25 905	32.6
Greenland/ Denmark			14	993 023	45.7
Iceland	1	5	26	12 165	11.8
Norway			38	41 637	25.5
Russian Federation	5	76 157	31	313 818	4.9
Sweden	1	725	44	20 348	21.4
United States (Alaska)			41	331 425	56.1

Source: CAFF 1997a

Mercury concentrations in beluga whales, 1993–94

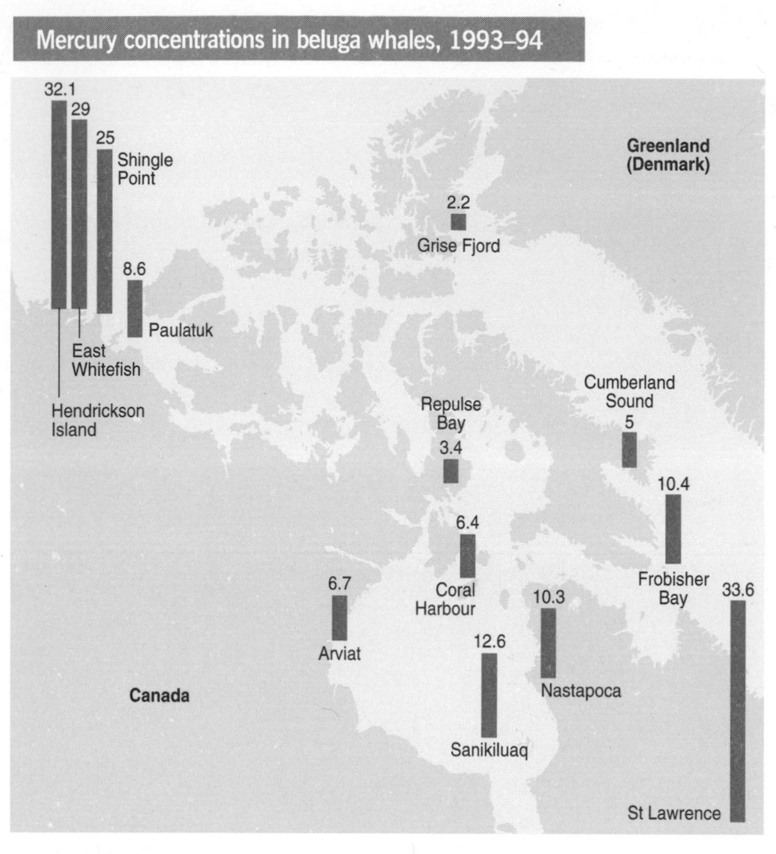

Source: Indian and Northern Affairs 1997a

Average mercury concentrations (µg/g) in the muscle of belugas (1993–94) was more than twice the Health Canada guideline in some areas of Canada

species and other evidence indicates that there has been significant gene flow between populations, both locally and over large distances. It is highly probable that the heterogeneity of sites and populations in the Arctic, coupled with the long history of climatic variation at high latitudes, means that the present day flora of the Arctic has the necessary resilience to accommodate substantial and even rapid changes without loss of species (Abbott and others 1995). Similar genetic diversity probably occurs within animal and microbial species, providing what might be considered as 'pre-adaptation' to climate change (Crawford 1995).

Whilst the levels of human impact on Arctic ecosystems and biological diversity appear to be relatively low compared with more temperate and tropical areas, these can have a greater effect on what are relatively simple systems. Arctic biodiversity is threatened from both direct and indirect human activities in and outside the region. Habitat fragmentation and disturbance of species and habitats are being caused by various forms of pollution, tourism, natural resource extraction, overharvesting of biological resources, introduction of alien species, and effects of climate change and increased UV-B radiation. Cumulative effects also need to be considered. Many activities do not threaten biodiversity on their own but do so in combination. For example, habitat fragmentation can become a problem due to the cumulative effects of forestry, tourism and mining (CAFF 1997b). Traditional human subsistence lifestyles could be particularly vulnerable to loss of biological diversity in the region.

Pollution

The effects of POPs are not fully understood but reproductive and developmental effects have been seen in Arctic birds. DDT is affecting reproduction of the Arctic Peregrine Falcon. DDE is causing egg shell thinning in some predatory birds. Although it has not yet been possible to link contaminant loads to effects, it seems likely that PCBs and dioxin-like compounds are causing reproductive effects in some marine mammals, in particular polar bears. Several Arctic species contain concentrations of POPs close to known thresholds associated with neurotoxic and immunosuppressive effects (AMAP 1997).

Trybutylin (TBT) has been detected in snails from the Norwegian, Icelandic and Alaskan coasts. Imposex (a disease in which a female organism develops male characteristics leading to sterilization) has been documented in snails in harbours of northern Norway, Svalbard, Iceland and Alaska, although not always accompanied by detectable TBT concentrations (AMAP 1998). Many of the Arctic countries now partially regulate the use of TBT. Regulations vary but generally only controlled release formulations are permitted. The relationships between cause and effect of TBT and imposex is as yet unclear so it is difficult to say what the recent bans on the use of TBT will achieve.

The take-up of mercury in Arctic biota has become a significant issue over the past two decades. The increase of mercury in the livers and kidneys of some marine mammals is possibly due to an increase in the global flux of mercury. The Arctic acts as a sink for mercury due to the cold climate. Marine mammals from northwestern Canada show the highest levels of mercury, with levels rising faster in the 1980s and 1990s than ever before (see map above).

There are high levels of cadmium in some terrestrial and marine mammals, and in marine birds. Levels might be high enough to cause kidney damage in marine birds and mammals in northeastern Canada and northwestern Greenland. Again it is unknown how much of this is attributable to local geology or to local sources.

Acidification has had some impact on Arctic biodiversity. The impact on Arctic vegetation of the nickel-copper smelters in northern Russia has already been mentioned. Other effects of acidification, evident in northern Fennoscandia and northwestern Russia, have been the disappearance of sensitive invertebrates from small lakes and streams. Some fish species are also being affected by acidification during the spring snow melt (AMAP 1997).

The situation over the next ten years is not expected to improve. The biomagnification effects on certain species may well get worse as some POPs and metals continue to accumulate in the Arctic environment. The projected increase in, and the effects of, methyl mercury will be of major interest over the next decade.

Fishing

Many species are at risk from the side effects of fishing. As a result of a major decline in capelin between 1987 and 1989 from overfishing, thousands of starving seals drowned in fishing nets, and large numbers of seabirds died of starvation and were washed up on the coast of northern Norway. Seabirds are also at serious risk of being caught in the bycatch. It is estimated that in 1996 200 000 birds were caught in this way in Russian waters and more than 11 000 off the Alaskan coast (CAFF 1998).

Selective fishing practices have adversely affected the populations of northern char and salmon. The selective removal of the largest fish may also affect interactions between species lower in the food chain.

Salmon farming has become an important industry along the coasts of the northern Atlantic. This may cause genetic loss in salmon as well as degeneration of local species populations due to competition from alien species. Fish have also been introduced to northern Fennoscandian lakes and streams to provide sport for tourists.

Climate change and UV-B radiation

The coming decade may help shed light on how the Arctic will respond to changes in climate, ozone and UV-B radiation. One thing is sure: the Arctic ecosystem has a narrow window of growth determined by snow cover and availability of daylight. Any changes to this window could have profound consequences. For the terrestrial environment, changes in permafrost, snow cover and ice caps will be important (AMAP 1997).

Little is known about the effects of UV-B radiation on Arctic terrestrial ecosystems but Arctic plants are more likely to be affected by increased UV-B radiation than plants at lower latitudes. Even less is known about effects on animals. Even though solar UV-B is relatively low in the Arctic compared to other regions, it is considered an important environmental concern for the future. Factors such as an increased albedo due to snow cover actually increase the effects of UV-B in the Arctic. The International Arctic Science Committee (IASC) is working with CAFF and AMAP to monitor effects of UV-B in the Arctic (IASC 1995).

Freshwater

The catchment area of the Arctic Ocean is extensive, with the watersheds of Arctic rivers penetrating far south (see map below). The quality of Arctic freshwater systems was close to pristine until the development of industrial activities and natural resource extraction in the region started to have a major impact on the environment after World War II.

The most affected areas include those on the Kola Peninsula and around Norilsk adjacent to metal smelters. Between 1991 and 1994 copper concentrations reached more than 2 500 times and nickel concentrations up to 130 times locally-set permissible limits. In some areas of Arctic Canada,

Watersheds of Arctic rivers

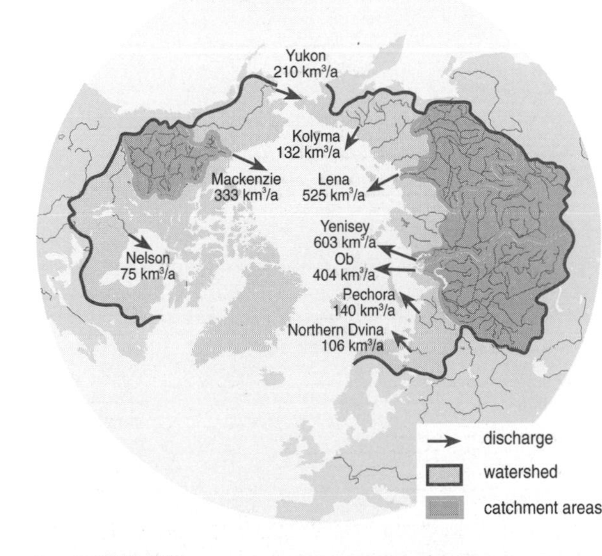

Source: AMAP 1998

The Arctic Ocean watershed and some of its catchment areas, with annual run-off of major rivers into the Arctic Ocean

Finland, the Russian Federation and Alaska, the levels of lead in rivers exceed water quality guidelines set out for more temperate latitudes. Metals are not the only problem. A mixture of metals, sewage, petroleum hydrocarbons, acidifying substances and other chemicals combine to attack ecosystems.

Lake sediments in both northern Canada and Fennoscandia contain increasing levels of mercury. The relative contributions from anthropogenic and naturally-occurring sources are unclear.

Several rivers and estuaries in northern Russia suffer from oil contamination. Hydrocarbon concentrations often reach several mg/litre in the lower part of the Ob River, for example, and most samples taken from rivers in northwestern Siberia have levels of hydrocarbons that exceed permissible limits. Oil and gas production poses a serious threat to wetlands where wastes are disposed of directly into wetland depressions which act as sinks for contaminated fluids. When these sinks spill their banks, they contaminate local rivers and lakes (AMAP 1997).

There are high levels of HCH in several Russian rivers, most notably the Ob, apparently due to the use of the pesticide lindane. DDT is present in Arctic rivers, ranging from 0.03 ng/litre in rivers flowing into Hudson Bay to 5 ng/litre in the Ob River. Russian rivers also have high levels of PCBs. The levels of POPs in some Russian rivers draining into the Arctic may be higher than those found in urban North America and Western Europe.

The rivers of the Kola Peninsula in the Russian Federation have suffered a significant decrease in alkalinity and some streams undergo rapid acidification during the spring snow melt. The general situation with respect to acidification is improving over northern Fennoscandia as a result of decreases in European emissions. The situation on the Kola Peninsula shows some sign of improvement but this reflects economic factors rather than emission controls.

Radioactive isotopes have been released into Russian rivers which flow into the Arctic. These releases have come from reprocessing plants and a nuclear weapons production facility, all south of the Russian Arctic. Levels in Russian rivers have dropped since the 1960s and it is believed that levels peaked in the late 1940s and 1950s (AMAP 1997). Today, the biggest threat is probably from ponds at, for example, the Mayak plant which are artificially dammed in order to contain radioactive waste. If these dams were to fail, then significant radioactive contamination could flow into the Arctic.

Contaminant pathways into the Arctic Ocean are from the North Atlantic, the Bering Sea and from major northward flowing rivers. Circulation is dominated by the Beaufort Gyre and the Transpolar Drift

Major currents in the Arctic Ocean

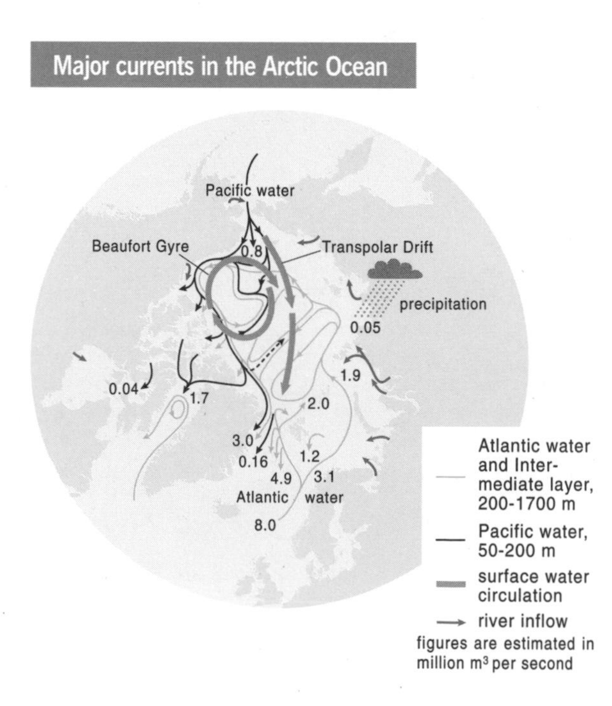

Source: AMAP 1998

Marine and coastal areas

Marine pollution

The Arctic marine area is dominated by a deep, ice-covered central ocean with surrounding shallow coastal seas. Contaminant pathways into the Arctic Ocean are from the North Atlantic, via the Norwegian coastal current, from the Bering Sea and from major northward flowing rivers. Within the Arctic Ocean, circulation is dominated by two major currents, the Beaufort Gyre and the Transpolar Drift (see map below). Ice forming in the shelf seas can be transported into the central part of the Arctic Ocean through these currents. The circulation and subsequent melting of this ice allows contaminants to be redistributed to deep ocean sediments and other shelf seas. Compared with atmospheric transport, movement is slow. Contaminants can take years to travel from temperate industrialized coasts to the Arctic.

The shelf seas, ice edges and polynyas – open-water areas in areas of sea ice – are seasonally some of the most biologically productive ecosystems in the world, providing an economic base for several large fishing fleets as well as a feeding ground for large populations of migratory birds. However, both Arctic marine sediments and biota are affected by contaminants. Levels of persistent organics are elevated around Svalbard, the southern Barents Sea and eastern Greenland. There is localized heavy metal contamination near mining activities, for example in Greenland. Mercury levels are increasing in marine sediments (AMAP 1997).

Radioactive isotopes are widely found in sediments as a result of fallout from atmospheric weapons testing, from military accidents – for example the crashing of a US B52, with nuclear weapons, at Thule in northwest Greenland in 1968 – and from discharges from European reprocessing plants. Wastewater containing radioactive isotopes has been carried north from plants at Sellafield and Dounreay in the United Kingdom and at Le Cap de la Hague near Cherbourg in France. Sea currents have taken isotopes up to the Norwegian and Barents Seas. Levels peaked in the 1970s and gradually returned to a relatively low level in the late 1980s (AMAP 1997). Concentrations of technetium 99 in brown algae collected in the outer Oslofjord increased by a factor of five from 1996 to 1997 as a result of discharges from the Sellafield reprocessing plant in the United Kingdom (Brown and others 1998).

Mining is contaminating some marine environments. Only a few examples are documented, such as the Black Angel lead and zinc mine in Greenland (AMAP 1997). Contamination is contained within a small radius of the mine (about 30 km in the Black Angel example).

Pollution from oil and gas activities can be devastating to the Arctic marine environment. Petroleum activities in the region are shown in the map opposite. Probably the main threat comes from tanker spills. Experience with the 1989 *Exxon Valdez* tanker spill off the southern Alaskan coast has shown that large spills can cause massive contamination over large areas. The *Valdez* spilled 35 000 tonnes of oil and was responsible for the death of approximately 250 000 birds (Platt and Ford 1996, AMAP 1998). The probability of a similar accident happening elsewhere in the Arctic will rise as production and therefore the need for transportation increases.

The development of the International Northern Sea Route across the northern coasts of Norway and the Russian Federation will provide a greater potential for accidental chemical releases and other damaging environmental impacts. However, this route is now much less used by domestic shipping than in the days of the former Soviet Union. While economic gain is driving the push for quicker sea routes, significant investment is also directed towards assessing environmental impacts (Østreng and others 1997).

Unsustainable fishing

Fishing provides an important source of income for all Arctic coastline countries. In Iceland, 70 per cent of the national income results from this industry (CAFF 1998). The Bering Sea is one of the world's largest fishing areas, and many indigenous people maintain sustainable traditional lifestyles from the sea. New techniques have increased the level of catches rapidly over the past 100 years. The Arctic is attractive for fishing as a few productive species dominate, thereby reducing wasted bycatch. The low number of species means that overfishing can have disastrous effects. After the herring industry declined in the 1970s, due to overharvesting, the capelin was the next to be overfished. The stock has collapsed twice since a peak catch of 3 million tonnes in 1977. At present there are signs of recovery (Gjosaeter 1995).

Atmosphere

The atmosphere contains relatively low amounts of contaminants compared with the other media. However, the atmosphere is the fastest transport mechanism for delivering contaminants to the Arctic.

Oil and gas exploration and exploitation

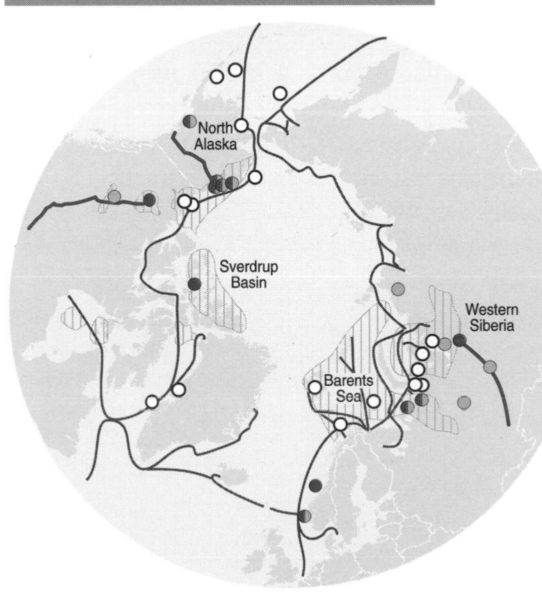

Source: AMAP 1998

The chances of major oil spills occurring in the Arctic will increase as hydrocarbon exploration and production expands

Pathways and sources for POP-contaminated air coming from outside the Arctic

Atmospheric pathways for POPs

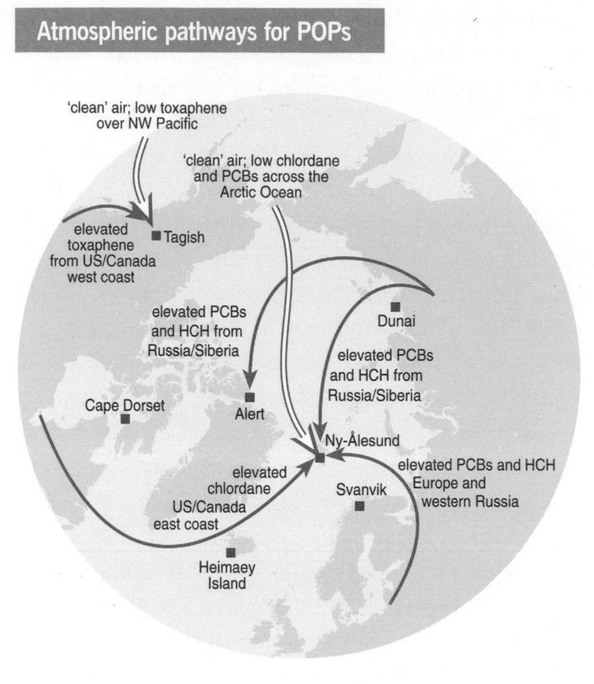

Source: AMAP 1997

Transport times can be days or weeks from more temperate agricultural and industrialized areas. The time of year and the prevailing weather systems determine the fate of contaminants in transport. Transport to the Arctic is more prevalent during winter and spring when an intense high pressure system over Siberia pushes the Arctic front far to the south. Large polluted areas of Eurasia are then within the Arctic air mass, the lower one to two kilometres of which can move contaminants across the pole. This activity is further amplified by the lack of clouds and precipitation during this time; thus the contaminants travel into the Arctic before they can be deposited in precipitation. These air flows transport a range of contaminants, including sulphur and nitrogen compounds, POPs, heavy metals and radionuclides, from parts of Eurasia, Japan and North America into the Arctic – see map above (AMAP 1997).

The phenomenon of Arctic haze was first identified by weather reconnaissance planes in the 1950s. The haze, which is densest in spring, consists mostly of sulphate with some soot and dust originating from anthropogenic sources outside the Arctic. Most of the particles originate in Eurasia from coal burning. Arctic haze has helped prove that emissions from Eurasia are transported into the Arctic and, in some cases, over into North America. Haze particles can also carry heavy metals and other contaminants, helping to explain how the long-range transport of pollutants into the Arctic is so efficient.

Urban areas

Arctic settlements range from a few large industrialized cities, with populations of several hundreds of thousands, to small nomadic herding communities of only a handful of people continuing to live a traditional indigenous lifestyle.

The three significant Arctic (Russian) cities are Murmansk, Arkhangelsk and Norilsk. Arkhangelsk was founded in 1584. Murmansk and Norilsk have grown to approximately 400 000 and 165 000 respectively this century. The populations of all three cities peaked at the beginning of the 1990s and have been in decline since then (Lappo 1994, State Committee for Statistics 1995).

In addition to urban settlements, ports and harbours and other coastal developments, as well as heavy industrial centres, are all present in parts of the Arctic, and the Russian Arctic in particular. Waste materials have also been dumped in the North American Arctic since the beginning of the Cold War period. A total of 1 246 sites have been identified as hazardous in the Yukon and Northwest Territories of Canada. By 1997, almost 500 sites had been cleaned up, with another similar number assessed as not hazardous. The remainder await assessment (Indian and Northern Affairs 1997b).

The Antarctic

Social and economic background

The Antarctic region was first penetrated by European explorers in the 18th century. Cook's circumnavigation in 1772–75, and the resulting awareness of the large populations of whales and seals there, ushered in the first era of marine mammal exploitation in the Antarctic waters. Fur seals were massively overexploited, and eliminated from some islands by the 1820s. Sealers and whalers became major explorers of the Antarctic and the sub-Antarctic in the 19th century while searching for new hunting grounds. National expeditions began to be sent to Antarctica in the middle of the 19th century. The first deliberate overwintering was in 1898 and the South Pole was reached by Amundsen in 1911, although the mapping of Antarctica was only completed in the late 1940s. Permanent human presence in Antarctica dates mainly from the establishment of year-round research stations in the 1940s – although the Argentine station 'Orcadas' has been continuously operated since 1904. In 1997 some 35 stations on the continent and islands south of 60° South, and a further 7 on sub-Antarctic islands operated year-round (SCAR 1998).

Seven states assert sovereignty claims to Antarctic sectors (three of these overlap and are mutually contested), both the United States and the Russian Federation have reserved the right to make claims, and most other states do not recognize any claims. The Antarctic Treaty seeks to 'freeze' the various positions on sovereignty, demilitarize the area, guarantee freedom of access and emphasize science as the primary currency of national activity in Antarctica. Subsequent agreements in what has been termed the Antarctic Treaty System (ATS) include the 1972 Convention for the Conservation of Antarctic Seals (CCAS), the 1980 Convention for the Conservation of Antarctic Marine Living Resources (CCAMLR), and the 1991 Protocol on Environmental Protection to the Antarctic Treaty. All three are in force. Parties to the Antarctic Treaty (increased from the 12 original signatories in 1959 to 44 by 1999) assert that the ATS is an effective and open system for the governance of Antarctica in the interest of the international community. The legitimacy of the ATS has been challenged by some states within the forum of the UN General Assembly, and the effectiveness of the system in relation to environmental protection has periodically been challenged by NGOs.

Scientific investigation has been the predominant human activity since the 1950s. Localized impacts from station operations are generally well recognized and seem unlikely to worsen. Operators are now legally required to conduct environmental impact assessments for all activities and to develop waste management plans.

Tourism is developing rapidly. Ship-borne tourists increased from 4 698 in the 1990/91 season (Enzenbacher 1992) to an anticipated 13 900 in 1999/2000 (IAATO 1999a). Although most tourists arrive in Antarctica by ship, a few travel by air and yacht. Most tourism cruises (96 of the 102 in 1998/99) are in the Antarctic Peninsula area, with the remainder generally in the Ross Sea region (IAATO 1999a). The sub-Antarctic islands are also visited on many Antarctic voyages (Cessford and Dingwall 1998).

Although the length of tourists' stays is much shorter than for personnel of national Antarctic programmes, tourists can cause adverse environmental impacts, particularly on the Antarctic Peninsula. Measures to assess, mitigate and prevent these impacts, as well as to repond to emergencies, are taken by industry (IAATO 1999b), by the Antarctic Treaty System (AT 1994) and by individual states (Prebble and Dingwall 1997). For the Antarctic Peninsula a guide to regular tourist sites has been produced which identifies environmental sensitivities (Naveen 1997).

The Protocol on Environmental Protection to the Antarctic Treaty specifically prohibits mineral resource activity other than for scientific purposes. The only extractive resource industry in the Antarctic is fishing

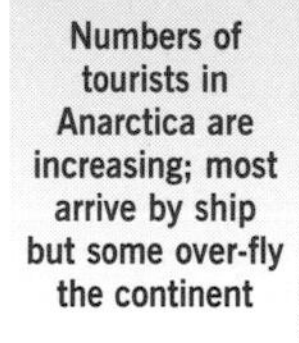

Numbers of tourists in Anarctica are increasing; most arrive by ship but some over-fly the continent

Antarctic tourists

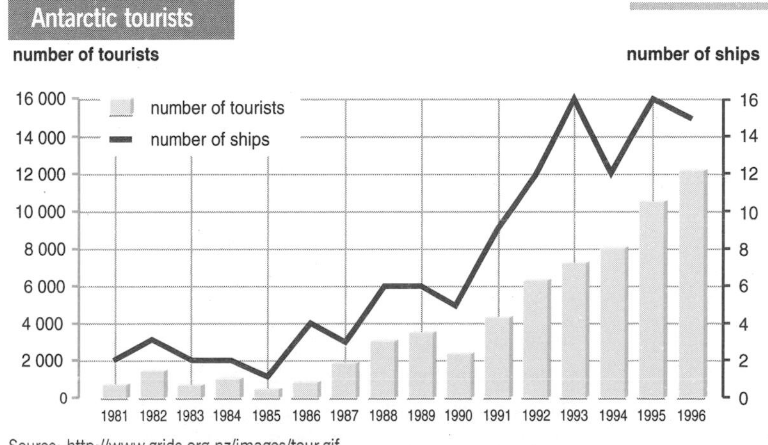

Source: http://www.gridc.org.nz/images/tour.gif

Area of jurisdiction of the 1980 Convention for the Conservation of Antarctic Marine Living Resources

CCAMLR extent

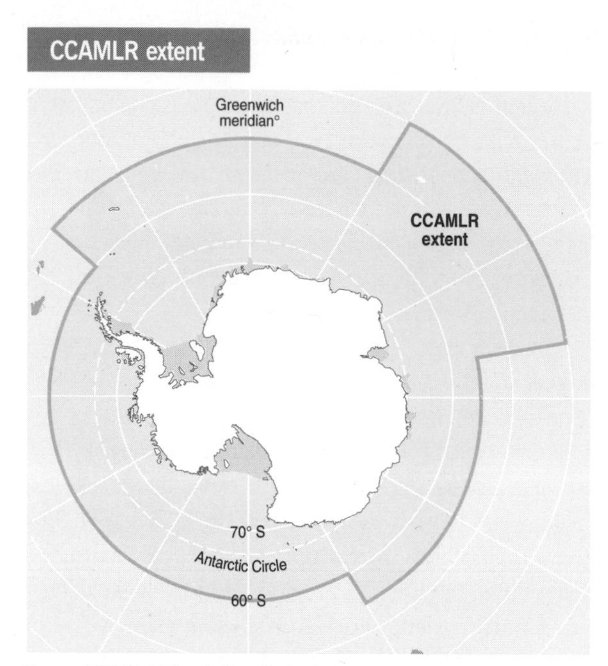

Source: GRID Christchurch, New Zealand

within the CCAMLR area, south of the Polar Front. The 1997/98 fisheries in the Convention area are Patagonian and Antarctic toothfish (*Dissostichus eleginoides* and *D. mawsoni),* mackerel icefish (*Champsocephalus gunnari*), krill (*Euphausia superba*) and squid (*Martialia hyadesi*). Illegal, unreported and unregulated fishing in the Convention area is a matter of major concern and poses a serious challenge to the Antarctic Treaty system (CCAMLR 1999).

Land

In the context of the Antarctic, land is used here to include all relatively stable ice surfaces attached to, or overlaying, the continent – ice shelves and the polar plateau.

The total ice-free area in a region dominated by a largely ice-covered continent and the surrounding Southern Ocean is some 2 per cent of the surface of the continent (about 280 000 km^2), and most of the approximately 26 000 km^2 area of the sub-Antarctic and cool temperate islands (Dingwall 1995). The Antarctic ice-free areas are largely found on the continental coastline (particularly in the Peninsula) and islands south of 60° South. These ice-free areas are the site of most biological activity (including stands of vegetation, and bird and seal colonies) and various sensitive periglacial and geological features. They are also the major scene of human activity and infrastructure, and accordingly much of the historic and current environmental impact is found at these sites – including oil spills (Cripps and Priddle 1991, Aislabie 1997), terrain modification (Campbell and others 1998), habitat loss (Thomas 1986), and the introduction of exotic (Gremmen 1997) and/or disease organisms (Gardner and others 1997). Human disturbance of biota, including Adelie penguins (Woehler and others 1994), also occurs although in some situations direct human influence may be negligible compared with other environmental changes (Fraser and Patterson 1997).

A number of stations have also been sited on the polar plateau, including at the pole itself. The plateau has been well traversed, and continues to be the scene of scientific and tourist activities. Although indications are that impacts may only be slight – and in most cases undetectable – this environment remains susceptible.

With ice shelves, the issue is not local impact but the consequences of climate change at regional and global levels, whether natural or human-induced. Recent ice shelf disintegration around the northern and western parts of the Antarctic Peninsula has been associated with regional atmospheric warming (Ward 1995, Vaughan and Doake 1996, Rott and others 1996, Lucchitta and Rosanova 1998, Rott and others 1998, and Skvarca and others 1998) which has been occurring for approximately the past 50 years. Retreat of other ice shelves, such as the Cook Ice Shelf (Frezzotti and others 1998) and the West Ice Shelf is probably also related to atmospheric warming (see map on page 193). The most vulnerable part of the Antarctic ice sheet is thought to be the West Antarctic Ice Sheet. Although climate-induced surface warming will take millennia to reach the base of the sheet, the presence of a lubricating water layer on a bed of deformable sediment suggests it may be unstable quite independently of climate forcing (MacAyeal 1992).

Over the next decade it seems certain that ice shelf retreat will continue in the Antarctic Peninsula, with parts of the Larsen, Wilkins and the northern ice front of the George VI ice shelf all being vulnerable. Although the processes of fracturing and rifting of the ice are not yet fully understood, the time scales over which they operate (a few years to a decade) probably means that weakening has already taken place and only the final impetus of a warm summer with increased surface melt water is necessary to cause further retreat. The processes of (re)formation of ice shelves are probably very different from those influencing their disintegration, and may require time scales of centuries. The only plausible scenario for ice shelf collapse which could result in appreciable global sea level rise would involve a non-climate induced collapse of the West Antarctic Ice Sheet from its own internal dynamics, which would probably cause a rapid rise in sea level of the order of a few millimetres a year over the next 50–100 years (Bentley 1997, Oppenheimer 1998). The likelihood of this occurring is extremely small.

Biodiversity

Terrestrial ecosystems

For the area south of 60° South, the terrestrial flora is confined to mosses, lichens, liverworts and two species of flowering plants. The sub-Antarctic islands have much greater diversity, including, in the case of the cool temperate islands, shrubs and trees. Sub-Antarctic islands may have 30–40 indigenous vascular plant species, while cool temperate islands are comparatively rich in plant life, some islands supporting more than 150 species of plants. The sub-Antarctic and cool temperate islands have high levels of endemism as a consequence of their long-standing geographical and ecological isolation from each other and from surrounding continental land masses. For example, the genetically distinctive vascular flora of the New Zealand southern islands, comprising about 250 taxa, include 35 taxa endemic to the region and several endemic to a single island group.

Ecological floras for the Antarctic are currently being compiled. Until this work has been completed, it is not possible to identify endangered species. Recent studies at the genetic level indicate that diversity within some moss species is considerably higher than was expected, suggesting either a high genetic mutation rate or a much greater frequency of introduction and establishment of exotic species than previously supposed. Glacial recession in the Antarctic Peninsula and its associated islands is one process providing opportunities for colonization by species that are new to Antarctica.

The terrestrial fauna of the Antarctic continent and nearby islands comprise mites, collembola (spring tails) and (in the Peninsula) two species of midge plus a limited range of microscopic soil invertebrates (protozoans, tardigrades, nematodes and rotifers). As with the flora, the fauna of the sub-Antarctic and cool-temperate islands display greater diversity and include

Winter and summer sea ice

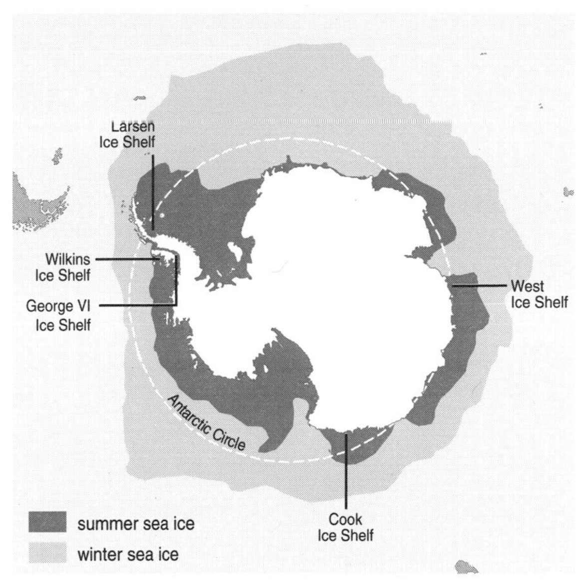

Source: GRID Christchurch, New Zealand

Map left shows extent of winter and summer ice, and location of vulnerable ice shelves

many species that are rare and/or endemic, particularly among the invertebrates. Some species also display specialized adaptation to their oceanic island setting, such as flightlessness among the insects.

Several of the islands are internationally important for science and conservation in that they harbour intact natural ecosystems that remain unmodified by human impacts. Many other islands, however, bear the distinctive imprint of human modification, particularly through the deliberate introduction of animal pests and predators. Of particular concern are the introduced mammals, notably rodents, cats, rabbits, sheep, cattle and reindeer. In recent years, there have been several cases of successful eradication of alien mammals from islands, and such efforts are continuing. The islands are also vital breeding and resting grounds for the seabirds and seals that feed in the Southern Ocean.

Increasing interest in biochemical compounds in Antarctic organisms may result in the identification of one of commercial significance. There have already been collections of micro-organisms for pharmaceutical purposes (SCAR 1999). As a biological prospecting interest is developing rapidly, there is the possibility of serious impacts on the target (and associated) species through harvesting, although compliance with the obligations of the Protocol should prevent this. A further problem likely to affect both the flora and fauna is the continuing possibility of inadvertent introductions of alien species, whether via the activities of national research programmes or tourists. Adelie and Emperor penguin populations have been affected by the incidental introduction of the Infectious Bursal Disease Virus (Australia 1997).

Marine ecosystems

There are relatively few modern pressures on biodiversity associated with the Southern Ocean. This was not so in the past, however, when the whaling industry had a significant impact on the populations of cetaceans, an impact from which recovery is slow.

Knowledge of Southern Ocean marine diversity, although not as complete as some other areas, is nevertheless quite good (Winston 1992, Arntz and others 1997). Many taxa have fewer species than might be expected on the basis of an even distribution of species throughout the seas. Good examples of these are gastropod and bivalve molluscs and benthic/demersal fish. The explanation for this low diversity is not clear. In some cases it may simply be part of a global pattern of reduced diversity towards polar regions (the latitudinal diversity cline), the explanation for which is not generally agreed amongst ecologists (Clarke 1992). In the case of fish, the low diversity may reflect the absence of habitats traditionally rich in fish species (Clarke and Johnston 1996). Some taxa are, however, well represented in the Southern Ocean. Examples of these include amphipod and isopod crustaceans, bryozoans and sponges (Arntz and others 1997). In common with most of the world's oceans, knowledge of Southern Ocean marine diversity is confined largely to the continental shelves and slopes. Almost nothing is known about the fauna of the deep sea around Antarctica.

Historically, the higher marine predators, especially southern fur seals (*Arctocephalus gazella*), were hunted to economic and almost biological extinction in the late 18th and early 19th century, and heavy exploitation of the great whales followed. The only exception to this trend of overexploitation came with the strict management of the take of southern elephant seals (*Mirounga leonina*) at South Georgia. There is no evidence that past exploitation of the Southern Ocean has resulted in a single biological extinction. The southern fur seal in particular has recovered fully. It is possible, however, that human activity has had important influences on Southern Ocean marine diversity. It is quite possible that overexploitation of the great whales may have switched the Southern Ocean to a new stable state where these whales may never again achieve the populations they once had. Documented increases in the southern fur seal and some penguin populations suggest this.

There are major concerns about the bycatch from fishing, particularly in relation to albatrosses and petrels. Conservative estimates put annual albatross mortality on longlines in the Southern Ocean at 44 000 (Brothers 1991). Populations at some sites are declining at a rate of up to 7 per cent per annum, a rate that populations cannot sustain (Alexander and others 1997). Longline fishing activity has been identified as the most serious threat facing albatross (SCAR 1996a).

Other fishing activities may have had important collateral impacts, notably the impact of bottom trawling on benthic communities with naturally slow turnover rates. There are currently no data on this topic. Overall, it is likely that the direct and indirect impacts of fishing will remain the main threats to Southern Ocean diversity. These impacts must, however, be assessed against the effects of natural variability in the Southern Ocean, and the impact of natural physical disturbance (such as that by ice).

Freshwater

Freshwater and saline lakes of Antarctica are found mainly in coastal regions, and in ice-free areas such as the Larsemann Hills, Schirmacher Oasis, Bunger Hills and Vestfold Hills. Such areas are rare in Antarctica and are often foci for human activities. These lakes are fed by glacier melt streams and many are particularly susceptible to contamination from human activities in the lake basins. In addition to those in ice-free areas, small lakes on ice are often associated with melt water from *nunataks* in inland Antarctica, and large lakes occur under the ice sheet in the central regions. Some, such as Lake Vostok are very large. Their global significance lies in the fact that they have not been exposed to the atmosphere for the past 500 000 years, and they have not yet been drilled into. International codes for the exploration of these lakes are being discussed by SCAR.

While the scale of threats facing Antarctic inland waters is not as great as those of the Arctic, significant threats do arise from local human activity, particularly for lakes in ice-free areas.

In the Larsemann Hills, at least one lake has been seriously contaminated by activities from an adjacent base which used the lake for cooling the base generators and as a dump for kitchen and other waste (Ellis-Evans and others 1997). Several other nearby lakes are adversely affected by road activities with a poor prognosis for the future (Lyons and others 1997).

In the McMurdo Dry Valleys, there have been diesel spills on some of the icecaps of the lakes and (small) losses of drilling material and fluid into the lakes from operations associated with the Dry Valley Drilling Programme (Parker and Holliman 1978). Small amounts of radioactive isotopes have been released into some of the lakes from accidental spillage and by deliberate introductions such as iodine 35 for tracing water movements (Vincent 1996). Small-scale contamination from camp sites adjacent to the lakes occurred during the first few years of exploration of the Valleys.

At least one of the lakes on the Schirmacher Oasis, Lake Glubokoye, has been recorded as receiving high amounts of phosphorus from wastewater inflow from the Russian Station Novolazarevskaya, and both this lake and Lake Stancionnoye show the effects of human influence from their elevated ammonium concentrations. The long-term threats to these lakes may have diminished in recent years due to a reduction in scientific activities in the oasis area (Borman and Fritsche 1995).

Lakes on the islands of the Antarctic Peninsula, particularly the South Orkney Islands, have undergone rapid eutrophication in recent years as a result of increasing populations of seals transferring marine nutrients onto their catchments. Other lakes in the more densely populated islands, such as King George Island, that are in close proximity to bases, may be used as drinking water, and supplies remain under threat of ongoing low-level pollution.

In the past decade, there has been a significant increase in awareness of the environmental fragility of the waters of the ice-free areas of Antarctica. An environmental code of conduct for science operations in the Dry Valleys had been agreed by New Zealand and the United States (Vincent 1996). This has been presented to SCAR for consideration for recommendation to all nations operating in ice-free areas (SCAR 1996b).

Marine and coastal areas

The Southern Ocean, defined as those waters south of the Polar Front or Antarctic Convergence, represents approximately 10 per cent of the world's oceans. In strong contrast to the Arctic basin, relatively little of this area is permanently covered by ice. Instead vast areas are subject to a strongly seasonal ice cover which forms in winter and melts the following spring. This seasonal ice zone includes all of the areas of continental shelf and slope around the Antarctic continent. The box on page 178 describes the status of sea ice in the polar areas.

As with previous eras of exploitation, fisheries have the potential to cause serious impacts on the marine ecosystem. The CCAMLR regime, with its ecosystem focus, should be able to prevent this. However, recent evidence of large-scale overfishing of the Patagonian toothfish, *Dissostichus eleginoides* – through illegal, unreported and unregulated activity – in the Convention area has caused serious doubts about the efficacy of the regime, the sustainability of the fishery, and the state of the ecosystem. For the period 1 July 1996 – 30 June 1997 the reported legal catch of Patagonian toothfish in the entire Convention area was 10 245 tonnes (97 per cent of the total finfish catch) whereas the illegal, unreported and unregulated catch was estimated at 107 000–115 000 tonnes from the Indian Ocean sector of the Convention area alone (CCAMLR 1998).

The Patagonian toothfish, *Dissostichus eleginoides*, is being severely overfished

Built-up areas

The Antarctic is uninhabited apart from wintering members of national scientific programmes, visiting summer scientists and support staff and tourists. In a few Antarctic locations, the clustering of human facilities is a potential environmental threat. These include the substantial US and New Zealand infrastructure at Hut Point on Ross Island; the concentration of Australian, Chinese and Russian stations in the relatively small ice-free area of the Larsemann Hills; and King George Island in the South Shetland group, where eight nations maintain year-round stations.

References

Abbott, R. J., Chapman H. M., Crawford, R. M. M. and Forbes D. G. (1995). Molecular diversity and derivations of populations of *Silene acaulis* and *Saxifraga oppositifolia*. *Molecular Ecology*, 4, 193-201

Aislabie, J. (1997). Hydrocarbon-degrading bacteria in oil-contaminated soils near Scott Base, Antarctica. In Lyons, W.B., Howard-Williams, C. and Hawes, I. (eds.). *Ecosystem Processes in Antarctic Ice-free Landscapes*. Balkema, Rotterdam, The Netherlands

Alexander, K., Robertson, G. and Gales, R. (1997). *The Incidental Mortality of Albatrosses in Longline Fisheries.* The First International Conference on the Biology and Conservation of Albatrosses, Hobart, Australia, September 1995

Allison, I. (1997). Physical processes determining the Antarctic sea ice environment. *Australian Journal of Physics*, 50, 759–771

Arctic Environmental Protection Strategy (1997). *Guidelines for Environmental Impact Assessment (EIA) in the Arctic*. Finnish Ministry of Environment, Helsinki, Finland

AMAP (1997). *Arctic Pollution Issues: A State of the Arctic Environment Report*. Arctic Monitoring and Assessment Programme, Oslo, Norway.

AMAP (1998). *AMAP Assessment Report: Arctic Pollution Issues*. Arctic Monitoring and Assessment Programme, Oslo, Norway

Arntz, W. E., Gutt, J. and Klages, M. (1997). Antarctic marine biodiversity: an overview. In Battaglia, B., Valencia, J. and Walton, D. W. H. (eds.). *Antarctic Communities: Species, Structure and Survival*. Cambridge University Press, Cambridge, United Kingdom

AT (1994). Recommendation XVIII-1, Tourism and non-governmental activities. In *Final Report of the Eighteenth Antarctic Treaty Consultative Meeting*, Kyoto 11–22 April 1994, 35–45

Australia (1997). *Introduction of Disease into Antarctic Birds*, Information Paper 51, XXI Antarctic Treaty Consultative Meeting, Christchurch, New Zealand

Bentley, C.R. (1997). Rapid sea-level rise soon from West Antarctic Ice Sheet collapse? *Science*, 275, 1077-8

Bjørgo, E., Johannessen, O. M., Miles, M. W. (1997). Analysis of merged SMMR-SSMI time series of Arctic and Antarctic sea ice parameters 1978–1995. *Geophysical Research Letters*, 24, 413-416

Borman, P., and Fritsche, D. (eds., 1995). *The Schirmacher Oasis, Queen Maude Land, East Antarctica, and its surroundings*. Justus Perthes Verlag, Gotha, Germany

Brothers, N. (1991). Albatross mortality and associated bait loss in the Japanese longline fishery in the Southern Ocean. *Biological Conservation*, 55, 255-268

Brown J., Kolstad, A. K., Lind, B., Rudjord, A. L. and Strand, P., (1998). *Technetium-99 Contamination in the North Sea and in Norwegian Coastal Areas 1996 and 1997.* Strålevern Rapport 1998, 3, Oslo, Norway

CAFF (1996). *Circumpolar Protected Areas Network (CPAN) Strategy and Action Plan*. CAFF Habitat Conservation Report No. 6, Directorate for Nature Management, Trondheim, Norway

CAFF (1997a). *Circumpolar Protected Areas Network (CPAN) Progress Report 1997.* CAFF Habitat Conservation Report No. 7, Conservation of Arctic Flora and Fauna, Iceland

CAFF (1997b): *Co-operative strategy for the conservation of biological diversity in the Arctic region*. Finnish Ministry of the Environment, Helsinki, Finland

CAFF (1998). *Incidental Take of Seabirds in Commercial Fisheries in the Arctic Countries. Technical Report No. 1 from the Circumpolar Seabird Working Group (CSWG)*. Conservation of Arctic Flora and Fauna, Vidar Bakken and Knud Falk (eds.)

Campbell, I. B., Claridge, G. G. C. and Balks, M. R. (1998). Short- and long-term impacts of human disturbances on snow-free surfaces in Antarctica. *Polar Record*, 34, 15-24

Cavalieri, D. J., Gloersen, P., Parkinson, C. L., Comiso, J. C. and Zwally, H. J. (1997). Observed hemispheric assymetry in global sea ice changes. *Science*, 278, 1104-1106

CCAMLR (1998). *Report of the CCAMLR observer to ATCM XXII*, Information Paper 21. XXII Antarctic Treaty Consultative Meeting, Tromsø, Norway

CCAMLR (1999). *Report of the CCAMLR Observer to ATCM XXIII*, Information Paper 64, XXIII Antarctic Treaty Consultative Meeting, Lima, Peru, 1999

Cessford, G. and Dingwall, P. R. (1998). Research on shipborne tourism to the Ross Sea region and the New Zealand sub-Antarctic islands. *Polar Record*, 34, 99-106

Chapman, W. L. and Walsh, J. E. (1993). Recent variations of sea ice and air temperature at high latitudes. *Bulletin American Meteorological Society,* 74, 34-47

CIA (1978). *Polar Regions Atlas*. Central Intelligence Agency, US Government Printing Office, Washington DC, United States

Clarke, A. (1992). Is there a diversity cline in the sea? *Trends in Ecology and Evolution*, 9, 286-287

Clarke A. and Johnston, I. A. (1996). Evolution and adaptive radiation of Antarctic fishes. *Trends in Ecology and Evolution,* 11, 212-218

COMNAP (1999). *An Assessment of Environmental Emergencies Arising from Activities in Antarctica*, Working Paper 16, XXIII Antarctic Treaty Consultative Meeting, Lima, Peru, 1999

Connolley, W .M. and Cattle, H. (1994). The Antarctic climate of the UKMO unified model. *Antarctic Science*, 6, 115-122

Cripps, G. C. and Priddle, J. (1991). Hydrocarbons in the Antarctic environment. *Antarctic Science*, 3, 233-250

Crawford, R. M. M. (1995). Plant Survival in the High Arctic. *Biologist,* 42-3, 101-105

Dallmann, W. K. (1997). Indigenous peoples of the northern part of the Russian Federation and their environment. INSROP Working Paper No. 90. International Northern Sea Route Programme, Oslo, Norway

de la Mare, W. K. (1997). Abrupt mid-twentieth-century decline in Antarctic sea ice extent from whaling records. *Nature*, 389, 57-60

Dingwall, P. R. (1995). Legal, institutional and management planning considerations in subantarctic island conservation. In Dingwall, P. R. (ed.). *Progress in Conservation of the Subantarctic Islands*, IUCN, Gland, Switzerland

Ellis-Evans, J. C., Laybourn-Parry, J., Bayliss, P. R. and Perriss, S. T. (1997). Human impact on an oligotrophic lake in the Larsemann Hills. pp. 396-404. In Battaglia, B., Valencia, J. and Walton, D. W. H. (eds.). *Antarctic Communities: Species, Structure and Survival*. Cambridge University Press, Cambridge, United Kingdom

Enzenbacher, D. J. (1992). Tourists in Antarctica: numbers and trends. *Polar Record*, 28, 17–22

EEA (1996). *The State of the European Arctic Environment*. European Environment Agency, Copenhagen, Denmark

Fraser, W. R. and Patterson, D. L. (1997). Human disturbance and long-term changes in Adelie penguin populations: a natural experiment at Palmer Station, Antarctic Peninsula. In Battaglia, B., Valencia, J. and Walton, D. W. H. (eds.). *Antarctic Communities: Species, Structure and Survival*. Cambridge University Press, Cambridge, United Kingdom

Frezzotti, M., Combelli, A., and Ferrigno, J. G. (1998). Ice-front change and iceberg behaviour along Oates and George V Coasts, Antarctica, 1912–96. *Annals of Glaciology*, 27, 643-650

Gardner, H., Kery, K. and Riddle, M. (1997). Poultry virus infection in Antarctic penguins. *Nature*, 387, 245

Gjosaeter H. (ed., 1995). *Ressuroversikt 1995*. Havforskningsinstituttet, Bergen, Norway

Gloersen, P., W. J. Campbell, D. J. Cavalieri, J. C. Comiso, C. L. Parkinson and H. J. Zwally (1992). *Arctic and Antarctic sea ice, 1978-1987: Satellite Passive Microwave Observations*. NASA, Greenbelt, Maryland, United States (NASA SP-511)

Gordon, H. B. and O'Farrell, S. P. (1997). Transient climate change in the CSIRO Coupled Model with dynamic sea ice. *Monthly Weather Review*, 125, 875-907

Gremmen, N. J. M. (1997). Changes in the vegetation of sub-Antarctic Marion Island resulting from introduced vascular plants. In Battaglia, B., Valencia, J. and Walton, D. W. H. (eds.). *Antarctic Communities: Species, Structure and Survival*. Cambridge University Press, Cambridge, United Kingdom

Hunt, B. G., Gordon, H. B., Davies, H. L. (1995). Impact of the greenhouse effect on sea ice characteristics and snow accumulation in the polar regions. *International Journal of Climatology*, 15, 3-23

IAATO (1998a). *Overview of Antarctic Tourism Activities,* Information Paper No. 86. XXII Antarctic Treaty Consultative Meeting, Tromsø, Norway

IAATO (1998b). *Education and training: a survey of IAATO member companies,* Information Paper No. 87. XXII Antarctic Treaty Consultative Meeting, Tromsø, Norway

IAATO (1999a). *Overview of Antarctic Activities*, Information Paper No. 98, XXIII Antarctic Treaty Consultative Meeting, Lima, Peru, 1999

IAATO (1999b). *Report of the International Association of Antarctica Tour Operators*, Information Paper No. 97, XXIII Antarctic Treaty Consultative Meeting, Lima, Peru, 1999

IASC (1995). *Effects of Increased Ultraviolet Radiation in the Arctic*. IASC Report No. 2. International Arctic Science Committee, Oslo. Norway

Indian and Northern Affairs (1997a). [illegible] (eds.), *Canadian Arctic Contaminants Assessment Report*. Indian and Northern Affairs, Ottawa, Canada

Indian and Northern Affairs (1997b). *Action, Arctic Environment Strategy: Progress report*. April 1996–March 1997. Indian and Northern Affairs, Ottawa, Canada

Kottmeier, C., Olf, J., Frieden, W., Roth, R. (1992). Wind forcing and ice motion in the Weddell Sea Region. *Journal of Geophysical Research*, 97, 20373

Krinner, G., Genthon, C., Li, Z, Le Van, P. (1997). Studies of the Antarctic climate with a stretched-grid general circulation model. *Journal of Geophysical Research*, 102, 13731-45

Lappo, G. M. (ed., 1994). *Goroda Rossii* (Russian cities). BRE, Moscow, Russia (in Russian)

Lucchitta, B. K. and Rosanova, C. E. (1998). Retreat of northern margins of George VI and Wilkins Ice Shelves, Antarctic Peninsula. *Annals of Glaciology*, 27, 41-46

Lynge, A. (1993). *Inuit–The story of the Inuit Circumpolar Conference*. ICC, Nuuk, Greenland

Lyons, W. B., Howard-Williams, C. and Hawes, I. (eds., 1997). *Ecosystem Processes in Antarctic Ice-free Landscapes*. Proceedings of an International Workshop on Polar Desert Ecosystems, Christchurch, New Zealand, 1-4 July 1996. Balkema, Rotterdam, The Netherlands

MacAyeal, D. R. (1992). Irregular oscillations of the West Antarctic Ice Sheet. *Nature*. 359, 29-32

Manabe, S., Spelman, M. J., Stouffer, R. J. (1992). Transient response of a coupled ocean-atmosphere model to gradual changes of atmospheric CO_2. *Journal of Climatology*, 5, 105

Murphy, J. M. (1995). Transient response of the Hadley Centre Coupled Ocean-Atmosphere Model to increasing carbon dioxide. Part I: control climate and flux adjustment. *Journal of Climate*, 8, 36-56

Murphy, J. M. and Mitchell, J. F. B. (1995). Transient response of the Hadley Centre Coupled Ocean-Atmosphere Model to increasing carbon dioxide. Part II: spatial and temporal structure of response. *Journal of Climate*, 8, 57-80

Naveen, R. (1997). *The Oceanites Site Guide to the Antarctic Peninsula*. Oceanites, Chevy Chase, Washington DC, United States

Nordic Council of Ministers (1996). *The Nordic Arctic Environment – Unspoilt, Exploited, Polluted?* Nord, Copenhagen, Denmark (1996, 26)

Oppenheimer, M. (1998). Global warming and the stability of the West Antarctic Ice Sheet. *Nature,* 393, 325-332

Østreng, W. (1997). *The post-Cold War Arctic: Challenges and transition during the 1990's*. In Vidas, D. (ed.). *Arctic Development and Environmental Challenges: information needs for decision-making and international cooperation*, pp. 33–49. Scandinavian Seminar College, Gentofte, Denmark

Østreng, W. (ed.), Griffiths, F. Vartanov, R. , Roginko, A. and Kolossov, V. (1997). *National Security and International Environmental Cooperation in the Arctic–the Case of the Northern Sea Route*. INSROP Working Paper No. 83. International Northern Sea Route Programme, Oslo, Norway

Parker, B. C., and Holliman, M. C. (eds., 1978). *Environmental Impact in Antarctica*. Virginia Polytechnic Institute and State University, Blacksburg, Virginia, United States

Platt, J. F. and Ford, R. G. (1996). How many seabirds were killed by the *Exxon Valdez* oil spill? In Prebble, M. and Dingwall, M. (1997). *Guidelines and procedures for visitors to the Ross Sea Region*. Ministry of Foreign Affairs and Trade, Wellington, New Zealand

Prebble, M. and Dingwall, M. (1997). *Guidelines and procedures for visitors to the Ross Sea Region*. Ministry of Foreign Affairs and Trade, Wellington, New Zealand

Rott, H., Rack, W., Nagler, T. and Skvarca, P. (1998). Climatically induced retreat and collapse of northern Larsen Ice Shelf, Antarctic Peninsula. *Annals of Glaciology*, 27, 86-92

Rott, H., Skvarca, P. and Nagler, T. (1996). Rapid collapse of Northern Larsen Ice Shelf, Antarctica. *Science*, 271, 788-792

Samson, P. (1997). *Thin Ice: International Environmental Cooperation in the Arctic*. Pacific Press, Wellington, New Zealand

SCAR (1996a). *Albatross populations: status and threats.* SC-CAMLR-XV/BG/21. Paper tabled by the Scientific Committee for Antarctic Research at the XVth Meeting of the Commission for the Conservation of Antarctic Marine Living Resources, Hobart, Tasmania, 16 October 1996

SCAR (1996b). Report of the XXIV SCAR Delegates Meeting, Cambridge, United Kingdom, 12-16 August 1996

SCAR (1998). Stations of SCAR nations operating in the Antarctic, Winter 1997. SCAR Bulletin 127 (1998) in *Polar Record,* 33, 361–374

SCAR (1999). *Scientific Research in the Antarctic*, Information Paper 123, XXIII Antarctic Treaty Consultative Meeting, Lima, Peru

Skvarca, P., Rack, W., Rott, H., and Ibarzábal y Donángelo, T. (1998). Evidence of recent climate warming on the eastern Antarctic Peninsula. *Annals of Glaciology*, 27, 628-632

State Committee for Statistics (1995). *Chislennost naseleniya.* Moscow, Russia

Stouffer, R. J., Manabe, S., Bryan, K. (1989). Interhemispheric asymmetry in climate response to a gradual increase in carbon dioxide. *Nature*, 342, 660

Thomas, T. (1986). L'effectif des oiseaux nicheurs de l'archipel de Pointe Géologie (Terre Adélie) et son évolution au cours des trente dernières années. *L'Oiseau* et la *Revue Francaise D'Ornithologie*, 56, 349-368

Tzeng, R. Y., Bromwich, D. H., Parish, T. R., Chen, B. (1994). NCAR CCM2 simulation of the modern Antarctic climate. *Journal of Geophysical Research*, 99, 23131–48

Vaughan, D. G. and Doake, C. S. M. (1996). Recent atmospheric warming and retreat of ice shelves on the Antarctic Peninsula. *Nature*, 379, 328-331

Vincent, W. F. (ed., 1996). *Environmental Management of a Cold Desert Ecosystem: the McMurdo Dry Valleys*. Desert Research Institute, University of Nevada, United States

Ward, C. G. (1995). Mapping ice front changes of Muller Ice Shelf, Antarctic Peninsula. *Antarctic Science*, 7, 197-8

White, W. B. and Peterson, R. G. (1996). An Antarctic circumpolar wave in surface pressure, wind, temperature and sea ice extent. *Nature*, 380, 699-702

Winston, J. E. (1992). Systematics and marine conservation. In Eldredge, N. (ed.). *Systematics, Ecology and Biodiversity Crisis*. Columbia University Press, New York, United States

WMO (1998). *Antarctic Ozone Bulletin,* various issues. WMO, Geneva, Switzerland. http://www.wmo.ch/web/arep/ozobull.html

WMO and others (1994). *Scientific Assessment of Ozone Depletion:1994*. WMO, Geneva, Switzerland, Global Ozone Research and Monitoring Project Report No. 37

WMO and others (1998). *Scientific Assessment of Ozone Depletion:1998*. WMO, Geneva, Switzerland, Global Ozone Research and Monitoring Project Report No. 44

Woehler, E. J., Penney, R. L., Creet, S. M. and Burton, R. H. (1994). Impacts of human visitors on breeding success and long-term population trends in Adelie penguins at Casey, Antarctica. *Polar Biology*, 14, 269-274

Worby, A. P., Massom, R. A., Allison, I., Lytle, V. I., Heil, P. (1997). East Antarctic sea ice: a review of its structure, properties and drift. *AGU Antarctic Research Series*

WWF (1997). Linking Tourism and Conservation in the Arctic. *WWF Arctic Bulletin* No. 4.97, WWF Arctic Programme, Oslo, Norway, 1997

Chapter Three

Policy Responses

Global and Regional Synthesis

KEY FACTS

- Policy assessment is made particularly difficult because of uneven monitoring, poor and missing data, and a lack of indicators. Continuous reporting and data on the environmental situation before and after implementation of policies are virtually non-existent.
- *Agenda 21* has had an impact on environmental governance and led to the creation or strengthening of multi-stakeholder organizations in many countries.
- Many new environmental institutions have to compete for staff and budgets with older and more powerful agencies, suffer from a serious lack of resources, and are easily overwhelmed by the ever-growing and more complex national and international legislation.
- Governments spend more than US$700 000 million a year subsidizing environmentally-unsound practices in the use of water, agriculture, energy and road transport.
- Governments in all regions have made substantial efforts to encourage industries to adopt cleaner production methods, with major successes in a number of countries.
- Harnessing private capital flows may be more important than increasing public capital flows but much more will need to be done to ensure that private investment is not used for unsustainable forms of development and that the poorest countries receive a much higher share.
- Public participation enables individual knowledge, skills and resources to be mobilized and fully employed, and the effectiveness of government initiatives to be increased.
- There is a need not just for more data on environmental issues but for standardizing data collection and storage, and making it accessible to technical and management levels, and the public.

This chapter describes the different types of policy response that are being used to address environmental issues, and tries to assess their success or failure. Quantitative assessment of the success or failure of policy initiatives and developments is not an easy task. Four questions need to be addressed:

- Have environmental problems been adequately addressed 'on paper'?
- Have these intentions been implemented?
- Has implementation had positive effects on the environment?
- Are these effects sufficient?

The last two questions are particularly hard to answer because of uneven monitoring, poor and missing data, and a lack of indicators, continuous reporting and data on the environmental situation before and after implementation. Furthermore, there are no proper mechanisms, methodologies or criteria to determine which policy contributes to which change in the state of the environment. It is usually impossible to single out a specific action or policy as having a particular impact; linkages between human actions and environmental outcomes are still poorly understood. In addition, political developments and bad governance can easily nullify the potential benefits of policy instruments.

Such problems often prevent valid comparisons between the current situation and what would have happened had no policy action been taken. The case of CFCs, where the sources are almost entirely man-made, regulations are so stringent that the impact on emissions is clear, and the impact of changes in emission levels on the natural environment is well known, is a rare exception. In general, a more complete and precise analysis will require the development of better mechanisms for monitoring and assessing the effects of environmental policies on environmental quality.

This chapter includes a general introduction followed by descriptions of policy responses in the seven GEO regions. The analysis is conducted in terms of eight policy clusters:

- multilateral environmental agreements (MEAs) and non-binding instruments, including implementation, compliance, effectiveness and reporting;
- laws and institutions;
- economic instruments;
- industry and new technologies, including cleaner production, eco-efficiency and ecodesign;
- the financing of environmental action;
- public participation;
- environmental information and education; and
- social policies.

MEAs and non-binding instruments

Multilateral environmental agreements

Although some international environmental treaties date back to early in the 20th century, it was not until the 1960s that concern about environmental pollution and the depletion of natural resources led to the kind of binding multilateral environmental agreements (MEAs) that we know today. The evolution of environmental agreements and legislation has fallen into two inter-related and overlapping generations. The first was one of single issue, use-oriented, mainly sectoral agreements and legislation, primarily addressing allocation and exploitation of natural resources such as wildlife, air and the marine environment; the second generation of agreements is more trans-sectoral, system-oriented and holistic. The second generation instruments are supplementing rather than replacing the first.

In this chapter, 10 major MEAs (see box below) are dealt with in much more detail than the others, and their current state of ratification is summarized both by region (see page 201) and by sub-region in the sections that follow. The graph on page 201 shows how ratification has grown over time.

In the early 1900s, environmental agreements, such as those covering fish or birds, were aimed more at regulating their exploitation and maintaining their

The ten conventions

The ten global conventions selected for more detailed analysis are:

CBD: Convention on Biological Diversity, Nairobi, 22 May 1992
www.biodiv.org/

CITES: Convention on International Trade in Endangered Species of Wild Fauna and Flora, Washington, 3 March 1973
www.wcmc.org.uk/cites/

CMS: Convention on the Conservation of Migratory Species of Wild Animals, Bonn, 23 June 1979
www.wcmc.org.uk/cms/

Basel: Basel Convention on the Transboundary Movements of Hazardous Wastes and their Disposal, Basel, 22 March 1989
www.unep.ch/basel/index.html

Ozone: Vienna Convention for the Protection of the Ozone Layer, Vienna, 22 March 1985 and Montreal Protocol on Substances that Deplete the Ozone Layer, Montreal, 16 September 1987
www.unep.org/ozone/

UNFCC: United Nations Framework Convention on Climate Change, New York, 9 May 1992
www.unfccc.de/

CCD: United Nations Convention to Combat Desertification in those Countries Experiencing Serious Drought and/or Desertification, Particularly in Africa, Paris, 17 June 1994
www.unccd.de/

Ramsar: Convention on Wetlands of International Importance especially as Waterfowl Habitat (Ramsar Convention), Ramsar, 2 February 1971
www.ramsar.org/

Heritage: Convention Concerning the Protection of the World Cultural and Natural Heritage, 23 November 1927
www.unesco.org/whc

UNCLOS: United Nations Convention on the Law of the Sea, Montego Bay, 10 December 1982
http://www.un.org/depts/los/losconv1.htm

economic usefulness than at protection for its own sake. One of the first international environmental agreements was the 1900 Convention for the Preservation of Animals, Birds and Fish in Africa, signed in London by the European colonial powers with the intention of preserving game in East Africa by limiting ivory exports from the region (Ruester and Simma 1975, Brenton 1994).

As knowledge about the environment increased, there was a gradual transition from such utilitarian approaches to a more general protection of endangered species. The prohibition, temporary or permanent, on taking and killing was supplemented by the protection of habitats, allowing the species to feed, rest and reproduce.

Many of the first-generation MEAs were concerned with the marine environment. Examples include the 1954 International Convention for the Prevention of Pollution of the Sea by Oil (OILPOL), 1972 Convention on the Prevention of Marine Pollution by Dumping of Wastes and other Matter, the 1973 International Convention for the Prevention of Pollution from Ships (MARPOL) and the 1982 UN Convention on the Law of the Sea (UNCLOS), which sets out the principles for dealing with threats to the marine environment and the biological resources of the sea. Many MEAs were also adopted for the protection of regional seas. Thirteen regional Action Plans and nine Regional Seas Conventions and their protocols today make up a web of detailed obligations for the majority of coastal states (Kiss and Shelton 1991, Sands 1995).

The UN Conference on the Human Environment held in Stockholm in 1972 was one of the first attempts to move away from a sectoral towards a more comprehensive approach including all aspects of environmental protection. This was reflected in the Declaration of the UN Conference on the Human Environment and the Action Plan for the Human Environment, which were adopted in Stockholm.

Environmental agreements drawn up in the lead-up to and aftermath of the Stockholm Conference, however, continued to emphasize conservation rather than addressing the totality of society's interaction with the environment. Examples include the 1971 Convention on Wetlands of International Importance

Structure of global conventions developed since 1972

The Conference of the Parties (COP) exists to:

- review implementation based on reports submitted by governments;
- consider new information from governments, NGOs and individuals to make recommendations to the Parties on implementation;
- make decisions necessary to promote effective implementation;
- revise the treaty if necessary;
- act as a forum for discussion on matters of importance.

The meetings of COPs are open to representatives of the Parties and others. This helps ensure transparency of operation and cooperation with other intergovernmental bodies and non-state actors. For example, more than 200 intergovernmental and non-governmental organizations were represented at the meetings of the COP of the UNFCCC in Bonn in 1995 and in Kyoto in 1997.

Several conventions invest the COP with the power to adopt amendments and additional protocols to help them adapt to new circumstances (the Montreal Protocol of the Vienna Convention and the Kyoto Protocol of the UNFCCC). Parties to the parent convention are not obliged to become parties to such protocols unless the convention requires the parties to do so. In some cases, non-parties can voluntarily comply with the requirements set forth in the protocols.

COPs can also create subsidiary bodies to ensure or oversee the functioning of the convention and assist in implementation of the convention between the meetings of the COP (such as the Implementation Committee of the Montreal Protocol).

A Secretariat to assist the COP. In some cases, such as for the CBD and CITES, administrative and policy support is provided to the secretariat by UNEP or other international bodies. The Secretariat of the Ramsar Convention, for example, is assisted by IUCN. Secretariats assist Parties in implementation by collecting reports on compliance and transmitting these to the COP, facilitating technology transfer, maintaining information on the development of projects relevant to the convention and, in certain cases, by enhancing compliance and implementation through a financial mechanism.

A Scientific Body, which generally includes members designated by the parties or the COP but is independent from governments. These bodies can be created by the convention itself or by the COP. Scientific Bodies are generally consulted before the discussion by the COPs on reports or information related to the implementation of the convention, as well as new or emerging scientific and technical issues related to the implementation of the convention or protocol. The Scientific Body of the Convention on Migratory Species (CMS), for example, meets immediately before a meeting of the COP, as well as once inter-sessionally. Scientific Bodies support the implementation of the convention by making proposals and providing advice to the Secretariat and the Parties.

Parties to major environment conventions (as at 1 March 1999)

	CBD (174)	CITES (145)	CMS (56)	Basel (121)	Ozone (168)	UNFCCC (176)	CCD (144)	Ramsar (114)	Heritage (156)	UNCLOS (130)
AFRICA (53)	50	48	19	23	44	47	51	28	39	38
ASIA AND PACIFIC (40)	36	25	5	21	33	37	25	18	26	28
EUROPE and CENTRAL ASIA (54)	47	35	25	39	48	48	29	41	51	28
LATIN AMERICA AND THE CARIBBEAN (33)	33	31	6	27	32	32	29	22	29	27
NORTH AMERICA (2)	1	2	0	1	2	2	1	2	2	0
WEST ASIA (11)	7	4	1	10	9	10	9	3	9	9

Key: percentage of countries party to a convention

0–25% 25–50% 50–75% 75–100%

Notes:

1. Numbers in brackets below the abbreviated names of the conventions are the total number of parties to that convention
2. Numbers in brackets after name of regions are the number of sovereign countries in each region
3. Only sovereign countries are counted. Territories of other countries and groups of countries are not considered in this table
4. The absolute number of countries that are parties to each convention in each region are shown in the coloured boxes
5. Parties to a convention are states that have ratified, acceded or accepted the convention. A signatory is not considered a party to a convention until the convention has also been ratified

Growth in numbers of Parties to selected MEAs

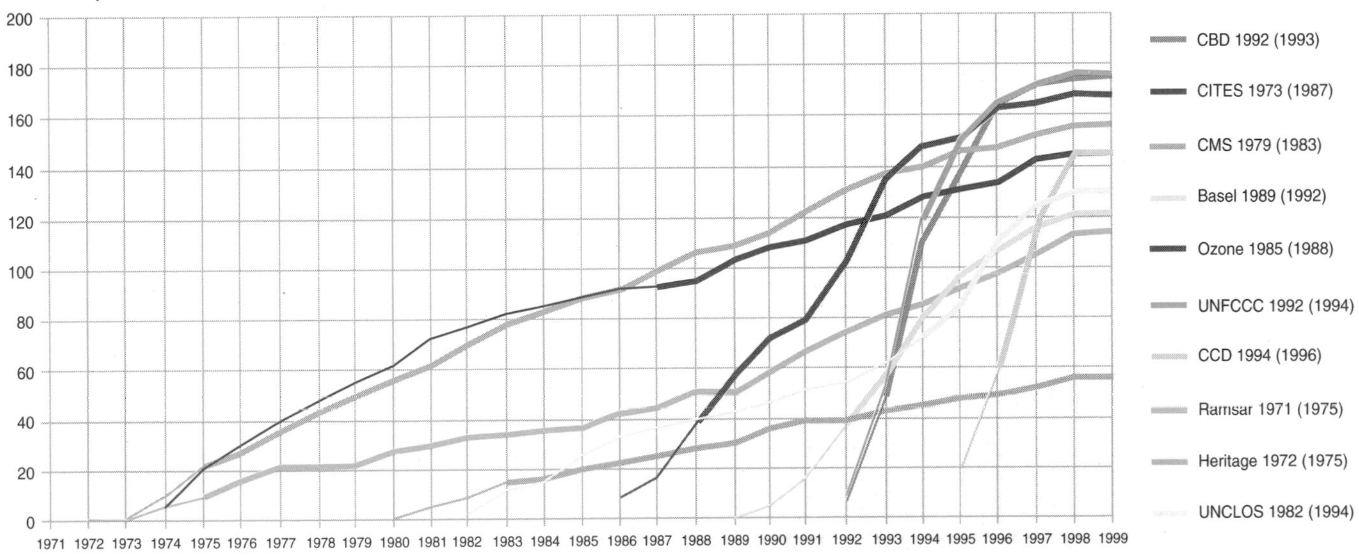

Note: years after Convention names are those of adoption followed by (in brackets) entry into force; lines are thin before entry into force of a Convention, thick after

Sources: Convention Secretariats via a GEO questionnaire, Convention Web sites, national governments and international organizations

Especially as Waterfowl Habitat (Ramsar Convention), the 1972 Convention concerning the Protection of the World Cultural and Natural Heritage (World Heritage Convention), the 1973 Convention on International Trade in Endangered Species of Wild Fauna and Flora (CITES), and the 1979 Convention on the Conservation of Migratory Species of Wild Animals (CMS) (Johnston 1997, UNEP 1997). These all address biological diversity, and the protection of wild fauna and flora has become one of the most developed areas of international environmental law covered by MEAs. Another sectoral MEA that emerged after Stockholm was the 1979 Convention on Long-range Transboundary Air Pollution (LRTAP), adopted by nearly all European States as well as Canada and the United States.

The second generation of multilateral environmental agreements and legislation is based on a holistic approach under which all species should be exploited either sustainably or not exploited at all, and their habitats protected, extended or improved to make this possible. This second generation was centred on the United Nations Conference on Environment and Development (UNCED), the Earth Summit, held in Rio de Janeiro in 1992. With all but six member states of the UN represented, the conference was a landmark in the history of environmental law, confirming the global character of environmental protection and its integration with development. Two new conventions were opened for signature: the UN Convention on Climate Change (UNFCCC), which is sectoral in that it deals with climate and the atmosphere but widespread in its effects, and the Convention on Biological Diversity (CBD) which seeks to bring together agriculture, forestry, fishery, land use and nature conservation in new ways. UNFCCC and CBD demonstrated the difficulty of pursuing agreements that affect multiple sectors. Both agreements were highly politicized with major diplomatic battles taking place during negotiations. Because holistic, multi-sector agreements involve so many different and cross-cutting areas of law, policy and politics, they can engender more conflict and problems than sectoral MEAs.

The Rio Conference also encouraged the development of other global MEAs (Burhenne and Robinson 1996), including those on Straddling and Highly Migratory Fish Stocks (1995), the Convention to Combat Desertification (CCD, 1994), the agreement on Prior Informed Consent (PIC) on Hazardous Chemicals (1998, see box on the Rotterdam Convention) and new regional MEAs including the Regional Agreement on the Transboundary Movement of Hazardous Wastes in Central America (1992), the Convention on Cooperation for the Protection and Sustainable Use of the Danube River (1994) and the Agreement on Cooperation for the Sustainable Development of the Mekong River Basin (1995).

Special mention must be also be made of a major new convention on Persistent Organic Pollutants (POPs) which is expected to be opened for signature early in the next century. Two meetings of the Negotiating Committee for an International Legally Binding Instrument for Implementing International Action on Certain Persistent Organic Pollutants have been held, and three more are planned before the treaty is opened for signature (POPs 1995).

Most of the global conventions developed since Stockholm are supported by a Conference of the Contracting Parties, a Secretariat and a Scientific Body (see box on page 200).

The Rotterdam Convention

The Rotterdam Convention was opened for signature on 11 September 1998, and by early 1999 the representatives of 63 countries and the European Community had signed it. The Convention incorporates a system called Prior Informed Consent (PIC) for Certain Hazardous Chemicals in International Trade. Operated by FAO and UNEP, PIC is a procedure that helps participating countries learn more about the characteristics of potentially hazardous chemicals that may be shipped to them, initiates a decision-making process on the future import of these chemicals and helps disseminate this decision to other countries (IRPTC 1999).

The Convention represents an important step towards ensuring the protection of citizens and the environment from the possible dangers resulting from trade in highly dangerous pesticides and chemicals. It will establish a first line of defence against future tragedies by preventing unwanted imports of dangerous chemicals, particularly in developing countries, and extend to all countries the ability to protect themselves against the risks of toxic substances.

The Convention is intended to enable the world to monitor and control the trade in very dangerous substances. It will give importing countries the power to decide which chemicals they want to receive and to exclude those they cannot manage safely. If trade does take place, requirements for labelling and provision of information on potential health and environmental effects will promote the safe use of these chemicals (Rotterdam 1999).

Non-binding instruments

Non-binding instruments are often forerunners of binding policy instruments and have at times had a more profound effect on environmental policy than binding ones. Non-binding instruments have also helped to bring about changes in attitudes and perceptions at all levels of society. While binding instruments receive the bulk of attention from policy makers and the public alike, non-binding instruments have played and are likely to continue to play a major role in the management of global and regional environmental problems.

Non-binding instruments, rather than creating commitments for action in the form of legally-binding targets and timetables, provide a looser framework. In some cases, the main aim is to set out important issues and priorities, foster discussion and attention, and stimulate new thinking and understanding about the relation of humans to the natural world. In other cases, such as the non-binding Prior Informed Consent (PIC) system for chemicals and pesticides, jointly administered by UNEP and FAO, non-binding instruments provide international procedures and arrangements that contributed to the development of a legally-binding regime (Victor and others 1998).

Examples of non-binding instruments include the Global Programme of Action for the Protection of the Marine Environment from Land-Based Activities and the Pan-European Strategy for Biological and Landscape Diversity.

Many non-binding instruments were developed during the run-up to the Rio Conference, some intended for discussion and definition by governments during the Conference. Two important non-binding instruments were adopted during the Conference: the *Rio Declaration* and *Agenda 21* (UN 1993).

The *Rio Declaration,* consisting of 27 principles, reaffirms the Stockholm Declaration on the Human Environment and builds upon it. Its central concept is sustainable development but it also includes several established and emerging principles, including common but differentiated responsibilities of states, the precautionary principle and the polluter pays principle. Many of the environmental agreements drawn up since the Rio Conference include principles contained in the Rio Declaration which are now becoming part of international law. These include the right of people to an appropriate environment, the right to information, and the right to participate in environmental management. The Rio Declaration has had an important impact on the development of national legislation in all regions, much of which now incorporates some or all of the principles contained in the Declaration, including the right of all people to an appropriate environment and the polluter pays principle.

Agenda 21 is essentially a programme of action covering 40 different sectors and topics. It pays particular attention to national legislation, measures, plans, programmes and standards, and to the use of legal and economic instruments for planning and management. It is probably the most prominent, significant and influential non-binding instrument in the environmental field and has become the guiding document for environmental management in most regions of the world. Its most important impact, through its core concept of sustainable development, has been to extend the environmental debate beyond

Implementation, compliance and effectiveness

Implementation of MEAs is usually through national legislation and regulatory measures. Failure to implement is often due to delays in setting up the necessary legislation (Brañes 1991). But legislation alone may not be enough – it usually needs to be followed up by specific programmes and activities and there may not be adequate resources to implement these.

Compliance with MEAs is often used as an indicator of implementation although compliance with the requirements of an MEA can occur without implementation, for example because a particular activity stops or is reduced for other reasons (such as economic crisis) or because compliance occurs as a result of some existing national legislation.

Compliance can be assessed in terms of the general spirit of an MEA as well as its specific requirements. Some MEAs cover many complex activities which make traditional means of measuring compliance impossible. For example, measuring compliance with the Basel Convention on the importation of hazardous wastes is a Herculean task and many Parties are unable to do so. For such MEAs, achieving compliance is a matter of developing self-implementing rules and incentives.

When comparing compliance for different agreements, it is essential to examine how demanding the agreement is – does it require large changes in behaviour (as in the Montreal Protocol) or little (as in the UNFCCC for developing countries)? The whaling accords of the early 20th century set allowable catch levels at such a high level that compliance was easy (Birnie 1985). The result was high compliance but, arguably, little implementation.

Effectiveness of an MEA is the extent to which the MEA achieves its objectives, for example by influencing the behaviour of target groups and, ultimately, the extent to which it actually improves the quality of the natural environment (Victor and others 1998, Weiss and Jacobsen 1998). Such improvement may, however, result from many factors, natural, social and political. For example, political changes in Central and Eastern Europe have recently led to major reductions in some emissions but these are not primarily due to the Convention on Long-Range Transboundary Air Pollution or its many protocols.

An MEA may be effective by one measure, for example, stemming the international trade of a particular endangered species, yet fail to protect that species because it is consumed domestically. Some MEAs have a wide scope and include several protocols whose effectiveness should really be evaluated independently. Some problems are simply easier to address than others. Substitutes for harmful substances may be more or less available. Some problems may require basic and costly changes in patterns of production, consumption or consumer behaviour and others may not. Some actions may need to cover a wide range of activities, others may be able to concentrate only on a limited number of 'hot spots'.

Ultimately, implementation and compliance, and thus the effectiveness of MEAs, depend primarily on the existence and effectiveness of the corresponding national legislation, institutions and policies, including those that ensure access to judicial and administrative fora, national capacity and political will (UNEP 1996a). While compliance is neither a necessary nor a sufficient condition for effectiveness, higher levels of compliance will usually produce greater environmental improvement (Werksman 1997).

Non-compliance: the choice between hard and soft approaches

The Montreal Protocol combines hard and soft approaches. For example, Protocol-related funding which flows through the GEF, rather than through the Protocol's Multilateral Fund, is withheld if recipients do not use the Non-Compliance Procedure. While this process is formally outside the terms of the Protocol, the lever of denial of GEF funding has enhanced the effectiveness of the Protocol's softer measures. Management of non-compliance seems to work well partly because it operates in the shadow of harder measures (Victor and others 1998).

The Non-compliance Procedure of the Montreal Protocol is one of the best examples of a formal, dedicated mechanism for addressing non-compliance. An Implementation Committee considers reports on implementation from other Parties, from the Secretariat or from a Party having difficulty complying. The Committee tries to find solutions rather than resolve disputes or punish offenders. A Technical and Economic Assessment Panel assists the Implementation Committee by analysing reasons for a Party's non-compliance and offering recommendations on how best to comply. Lessons learnt from the successful implementation of the Montreal Protocol have guided the development of similar procedures in more recent agreements such as the 1994 (Second) Sulphur Protocol to the 1979 LRTAP.

The 1994 Lusaka Agreement on Cooperative Enforcement Operations directed at Illegal Trade in Wild Fauna and Flora addresses non-compliance through an international Task Force which can investigate violations of national laws on illegal trade at the request or with the consent of concerned national authorities.

Many MEAs use less elaborate and formal mechanisms to promote and ensure compliance and full implementation. For example, the Convention on the Protection of the Marine Environment of the Baltic Sea Area (the Helsinki Convention) includes a complex system of implementation review and assistance.

Perhaps unsurprisingly, compliance tends to correlate with economic development but even some wealthy states face compliance problems. Japan, for instance, faced difficulties in implementing CITES because customs officials, who must identify regulated species at airports and ports, were until 1994 supplied with manuals of regulated species printed in English with only black-and-white photos (Victor and others 1998).

Innovative and effective means of limiting and addressing non-compliance will become increasingly important as MEAs develop and, as is likely, become more stringent. The most successful MEAs, such as the Montreal Protocol, have used a mix of formal and informal reviews and hard and soft measures; such an approach is likely to be the best way of preventing and managing non-compliance.

environmental departments and NGOs. It provides policy makers with a starting point for linking environmental and socio-economic issues. Although there is a long way to go, such linkages have become difficult to ignore. To fulfil the requirements of *Agenda 21*, most countries have prepared national environmental strategies or action plans and set up institutions for environmental management. *Agenda 21* has also had an impact on environmental governance and led to the creation or strengthening of multi-stakeholder organizations in many countries.

Implementation, compliance and effectiveness

While many environmental policy instruments have been put in place in recent years, there is growing concern about their implementation and effectiveness. Implementation, compliance, and effectiveness are separate but linked concepts (see box on page 203). Most policy instruments seek to change the behaviour of target groups – for example, traders of endangered species or emitters of greenhouse gases – leading to an improvement in the state of the environment. Implementation is an activity not only of government – international organizations, trade associations, local governments, NGOs and the general public may all be involved. In practice, society has to deal with a mixture of policy tools: a complex set of legislation that directs command-and-control measures, economic instruments, cleaner production processes, voluntary action by the public and business, etc.

Many procedures and mechanisms can be used to respond to non-compliance, including incentives and sanctions. Non-compliance can often be dealt with without resorting to the law: there may be many steps from noticing an infringement to initiating legal procedures. While some analysts argue for 'soft' measures such as persuasion, negotiation and assistance, others favour 'hard' sanctions of various types (Chayes and Chayes 1993, Downs and others 1996). The former claim that failure to comply is generally unintentional, and even accidental, rather than wilful. Better management of the causes of failure is preferable to punishment and is likely be more effective. The latter believe that the costs of non-compliance must be made high enough to act as a deterrent and ensure that significant resources and attention are devoted to compliance (see box left).

Reporting

Reporting on the implementation of MEAs generally improves effectiveness by making parties more accountable, diffusing information on successful strategies and methods, helping to direct assistance if needed, and providing information and assessments to guide any future development of the MEAs (Victor and others 1998). States are more likely to comply with the terms of an MEA if they are confident that others are also complying, and transparency provides some deterrence against non-compliance (Chayes and Chayes 1995).

Effective reporting depends on the willingness and capacity of parties to gather and report data accurately and objectively, and well-resourced Secretariats to process this information into accessible formats. The increase in the number of MEAs and non-binding instruments has led to more and more onerous reporting requirements and bureaucratic burdens.

Some of the most highly-developed and effective reporting and review systems are those in MEAs for wildlife conservation The implementation of the Ramsar Convention, for example, is enhanced by a Monitoring Procedure for the specific requirement of maintaining the ecological character of a designated wetland of international importance. A team, generally composed of one Secretariat representative and two international experts, undertakes field visits and office-based discussions with local experts and government representatives. A detailed report, with recommendations for action, is then compiled and submitted to the Government concerned.

Laws and institutions

Environmental legislation dates back at least to the middle of the 19th century – the first pollution control laws in the United Kingdom were introduced in 1863 to control acid fumes from alkali manufacture and smoke from coal-burning furnaces (Ashby 1981). Since then, there have been two periods of intense national activity in relation to environmental legislation and the setting up of institutions to oversee environmental issues.

The first dates from the 1960s when growing concerns for the environment, exacerbated by specific disasters such as the wreck of the oil tanker *Torrey Canyon* in 1967, releases of toxic chemicals such as those at Minamata in Japan and Bhopal in India, and concerns about persistent pesticides, led to many new environmental laws and regulations, mainly in the developed countries.

The second, which is still ongoing, has involved almost every country in the world. It dates from the 1992 Earth Summit. *GEO-1* reported on many of the efforts being made to set up the national legislation and institutions needed to promote development through environmental improvement. This activity continues, in both developed and developing countries. It is being implemented through national environment strategies, plans and action programmes, and is strongly influenced by international initiatives and agreements. This intense activity involves many, if not most, countries. It comprises:

- increasing national commitment to environmental issues and sustainable development, and a strengthening of institutions involved with environmental governance and protection;
- a shift from strictly sectoral laws to a more holistic approach seeking to integrate environmental and socio-economic issues such as the fight against poverty;
- attempts to make the environment a cross-cutting issue requiring the cooperation of many individual ministries and government organizations;
- the creation of National Environmental Action Plans and State of the Environment Reports;
- mandatory Environmental Impact Assessments (EIAs) for major projects;
- incorporation of citizens' rights to – and responsibilities for – a clean environment in national legislation;
- the close involvement and participation of local people and NGOs in EIAs and National Environmental Action Plans;
- a trend towards decentralization which is encouraging the birth of new grass-roots institutions to manage natural resources – such as local water user associations;
- local environmental activities, such as waste separation, which are citizen-led and many of which have followed the adoption of *Agenda 21* which explicitly emphasizes the importance of local action.

Not all this activity has been generated by international initiatives such as the Earth Summit. In some areas, trade and investment have led to rapid economic growth followed quickly by serious environmental degradation. This has led some governments to take action to reconcile trade and environmental interests, through the use of trade-environment related policies and agreements such as product standards, enforcement of the polluter pays principle, health and sanitary standards related to food exports, and ecolabelling.

Nor has all this activity necessarily led to environmental improvements. Many of the new institutions created face overwhelming tasks and responsibilities. They have to compete for staff and budgets with older and often more powerful sectoral agencies. They often suffer from a serious lack of human, technical and financial resources. And it is easy for the regulatory burden of ever-growing and more complex national and international legislation to overwhelm inadequately-resourced institutions.

One notable weakness is in implementation of legislation and enforcement of standards. In many countries, enforcement of legislative measures is far from satisfactory, mainly because of weak institutional capacity, shortage of trained staff, imported standards which are not always relevant or applicable, and government machinery unused to monitoring and enforcement. Uneasy relations with other government institutions – whose cooperation is essential in dealing with environmental issues – has also resulted in delays and failures to implement policies and enforce laws. As a result, the full and effective enforcement of environmental legislation and sanctions for non-compliance remains an elusive goal in many countries.

Another unfortunate result of the new popularity of environmental measures has been fragmentation and duplication of responsibilities. In one African country, 10 different ministries administer an estimated 20 environment-related laws and in another 8 ministries are responsible for applying 33 environmental laws (SADC 1998).

Activity is not restricted to countries where legislation is relatively weak. Both North American countries – where national legislation and institutions are highly developed – are engaged in major reviews of their environmental efforts. Canada is revising its major environmental legislation to emphasize pollution prevention, and is proposing more efficient assessment for an increased number of toxic substances. A new authority has been established to address Canadian sources of international water pollution. New enforcement tools will allow for negotiated settlements for offenders without the need for a court case, and opportunities for public participation have been expanded, including a right to sue for damage to the environment if the government fails to enforce its own legislation. In the United States, major reforms are also under way, and are expected to include new policy instruments and technologies, increasingly sophisticated and interlinked partner organizations in government and society in general, and new approaches that tackle complex issues holistically on the ecosystem scale.

Economic instruments

The Earth Summit placed great stress on economic incentives as a means of both making production and consumption patterns more sustainable and of generating the resources needed to finance sustainable development. This potential is reflected in Principle 16 of the Rio Declaration (see box above).

Rio Declaration on Environment and Development: Principle 16

Principle 16

'National authorities should endeavour to promote the internalization of environmental costs and the use of economic instruments, taking into account the approach that the polluter should, in principle, bear the cost of pollution, with due regard to the public interest and without distorting international trade and investment.'

Source: UN 1993

GEO-1 reported that economic instruments were increasingly being applied worldwide. They have great potential as powerful tools for stimulating sustainable development and the range of possibilities is wide (see box below).

Yet, five years after Rio, the fifth anniversary Special Session of the UN General Assembly concluded that reform had been inadequate and too slow. Despite pleas for more use of market-oriented instruments, increase in the use of such instruments has been limited. In many countries, in particular in the developed world, there have been many proposals for taxes on emissions, mineral oils and pesticides, for example. Some countries, including for example Sweden (IER 1997) and the United Kingdom (see box right), have recently introduced taxes of this kind.

Economic instruments for environmental protection and resource management

Fiscal instruments
taxes on inputs, exports, imports, pollution, resources, land use

Market creation
tradeable emission permits, catch quotas, water shares, land permits

Financial instruments
loans, grants, subsidies, revolving funds, green funds, low interest

Charge systems
charges for pollution, environmental impacts, access, road usage

Property rights
ownership, use and development rights

Bonds and deposit refund systems
bonds for forest management, land reclamation, waste delivery

Others plan to do so soon. Similar discussions have taken place at the international level, for instance concerning the imposition of a CO_2 tax, a tax on air tickets or on capital movements. As long ago as 1972, James Tobin suggested levying a tax on international currency transactions, and the idea has recently been revived as a means of financing development in the face of the 'growing need for international cooperation on problems such as the environment, poverty, peace and security ...' (ul Haq and others 1996). Suggestions have also been made for using a Tobin tax to solve pressing and more specific issues such as the global removal of land mines (Collins 1996). No form of international tax has yet been accepted, however.

One of the most promising of the many new economic instruments being developed, particularly in the context of air pollution, is emissions trading. Tradeable emissions permits enable pollution reduction measures to be applied where reductions are most cost-effective. A company that reduces emissions below the level required by law can receive emissions credits that can pay for higher emissions elsewhere. Companies can trade emissions among sources within a company, as long as combined emissions stay within a specified limit, or trade them with other companies.

In the case of electricity generation, for example, utilities can decide the most cost-effective way to use available resources to meet pollution regulations. They can employ energy conservation measures, increase reliance on renewable energy, reduce usage, employ pollution control technologies, switch to fuel with lower sulphur content or develop other strategies. Utilities that reduce emissions below the number of allowances they hold may trade allowances with other units in the system, sell them to other utilities on the open market or through auctions, or bank them to cover emissions in future years.

Emissions trading has been developed into a highly sophisticated system in the United States (see page 305). The same approach can be applied, in principle, on the international scale through trading between countries. The allocation of the emission reductions agreed in Kyoto will lead to a system of national allowances for greenhouse gas emissions which could be distributed via tradeable permits.

There is a growing awareness that, in addition to advocating new economic instruments to encourage sustainability, more attention should be focused on public policies that move society further away from sustainability. Government subsidies, for example, can often encourage wasteful behaviour and unsustainable practices (Myers 1998, OECD 1997 and 1998a).

Subsidies are widespread and pervasive in virtually every country, developing and industrialized; world-wide, governments are estimated to spend more than US$700 000 million a year subsidizing environmentally-unsound practices in the use of water, agriculture, energy and road transport. Many of these subsidies are economically inefficient, trade-distorting, ecologically destructive and socially inequitable, sometimes all at the same time (de Moor and Calamai 1997).

The largest subsidy is agricultural support in OECD countries. About US$335 000 million is spent annually on subsidizing farm production and farm incomes, a sum which is equivalent to about US$380 per capita or $16 000 per full-time farmer. Most of this support is linked to production or to supporting prices.

Tax initiatives to protect the environment: United Kingdom

The UK finance minister Gordon Brown announced several tax reforms to help protect the environment on 9 March 1999. The package included measures to tackle climate change through reducing emissions of greenhouse gases, to improve local air quality and support integrated transport, and to limit the environmental impacts of land use and water pollution. The reforms included:

- a 'climate change levy' on the business use of energy, with offsetting cuts in employers' national insurance contributions, and additional support for energy efficiency schemes and renewable sources of energy to be introduced from April 2001;
- reform of the company car taxation regime, on a revenue neutral basis, in April 2002;
- reduced road taxes for smaller cars;
- a package of seven tax measures to promote non-car commuting, the use of bicycles for business journeys and car-sharing;
- an increase in the rate of landfill tax intended to send a strong signal to waste producers and persuade waste managers and local authorities to consider alternatives to landfill; and
- an announcement that draft legislation will be published shortly for a tax on hard rock, sand and gravel used as aggregates – the forerunner of an aggregates tax – following a report on the environmental costs of quarrying, including noise, dust, visual intrusion, loss of amenity and damage to biodiversity.

Source: EDIE 1999

Reforming policy on subsidies: New Zealand

In 1984, New Zealand decided to pursue a drastic reform of its agricultural sector. By 1983 total farm assistance had risen to one-third of agricultural output, with huge costs to the public budget and taxpayers, and it was recognized that this level of support was unsustainable. Agricultural reform focused on removing all support such as concessionary loans, input subsidies and cost recovery for services as well as changing the tax and regulatory regime.

The results have been impressive: subsidies were almost completely eliminated and now apply to around 3 per cent of output, compared with 33 per cent in 1983. Total value of farm output initially declined in real terms but increased again in the late 1980s. The number of farms (80 000) is currently slightly higher than in 1983. It is also believed that farming has become more diversified and competitive in the international marketplace. When the subsidies went, land-clearing and over-exploitation of the most marginal lands stopped. Although the transition process was painful in some cases, the government accommodated the reform by suspending or reducing interest payments through write-offs; it helped more than 4 700 farmers to adjust. Some 300 farmers became eligible for buy-out and left the sector.

Source: Sandrey and Reynolds 1990

By raising prices received by farmers, these subsidies encourage the use of inputs and the over-intensive use of land. This kind of support is both costly and ineffective since only 20 per cent of it ends up as additional farm income, and three-quarters of that goes to the larger and richer farmers. The remaining 80 per cent leaks away, primarily to intermediate industries (OECD 1995).

Another example of perverse public policy is support for energy production. Global energy subsidies currently total US$200 000 million a year (de Moor, in press). OECD countries spend some US$82 000 million a year subsidizing energy production, the equivalent of about US$90 per person, mostly through tax breaks, cheap provision of public infrastructure and services, subsidized capital and price support (OECD 1997). A common feature in global energy policy is that more than 80 per cent of the subsidies concern fossil fuels, the most polluting energy sources. Nuclear energy, with its risks for human health and the environment, receives 8 per cent and gets more support than renewable forms of energy. Governments are thus actually subsidizing pollution. Removing all energy subsidies would reduce global CO_2 emissions by 10 per cent while at the same time stimulating economic efficiency and growth (OECD 1997).

There have been successes in removing damaging subsidies (see box above, for example). Bangladesh has abolished subsidies for the sale of fertilizers, Brazil has removed some subsidies for land conversion, China has removed price controls on coal, many countries in Central and Eastern Europe have reduced energy subsidies, Indonesia has reduced very high subsidies for pesticides, and many countries have raised tariffs for the public water supply (World Bank 1997).

An important economic key to sustainable development is to ensure that price and incentive structures reflect the true costs and benefits of production and consumption. In this, subsidies may play a role, despite the problems just discussed. The first step is to remove or reform current subsidies, or re-shape them to target their purposes more effectively, taking proper account of environmental concerns. Internationally-coordinated action may be the most effective strategy for subsidy reform. Institutional change is crucial: creating more transparency in support policies and regular monitoring increases the political costs of irresponsible policies and rewards responsible action by policy makers.

Industry and new technologies

Industry bears a major responsibility for environmental damage. The traditional government policy response used to be command-and-control legislation that set standards for emissions, monitored results, and imposed fines for infringements. As standards became increasingly expensive to meet, industry's traditional 'end-of-pipe' solutions, consisting mainly of adding an extra filter here or installing another settling tank there, also became increasingly inappropriate.

From this apparent impasse was born the concept of cleaner production (see Chapter 1) in which industrial products and processes are redesigned to minimize resource use, waste and emissions to the environment. Cleaner production has proved highly profitable (financially and environmentally) to industries that have embraced the concept, and governments have been provided with advice on how to set up and manage cleaner production strategies and policies (see flow chart above right).

In developed countries, many industries have not waited for government advice or assistance but have preferred instead to clean up their production methods as a voluntary response to a situation that was both environmentally untenable and financially wasteful (Rabobank 1998). Some results are shown in the box on the right. In addition, governments in all the GEO regions have made substantial efforts to encourage industries to adopt cleaner production methods (UNEP 1996b), with major successes in a number of individual countries, notably China (NEPA and others 1996). In 1998, UNEP launched a new International Declaration on Cleaner Production to foster widespread adoption of the concept. By 31 March 1999, the Declaration had collected 109 high-level signatories from national and local government, companies, business associations, NGOs and international agencies (UNEP 1999).

As mentioned in Chapter 1, meeting the goals of *Agenda 21* requires substantial reductions in the intensity of resource use in developed countries – typically by at least a factor of 10 (down to one-tenth of current levels). It is argued that such reductions are not, in fact, as difficult to achieve as might at first appear and could be attained within a generation (von Weizsäcker and others 1995). The Factor 10 approach has been endorsed by a number of governments and by the OECD (OECD 1998b). In the Netherlands, a five-year Sustainable Technology Development Programme, financed by five ministries, was completed in 1997 (IEEP-B 1994, DTO 1997). It addressed key sectors such as food production, housing, water management, transport and chemical industry, and met with widespread interest from science and industry.

Cleaner production has much in common with a similar approach which the World Business Council for Sustainable Development calls eco-efficiency, defined as the delivery of competitively-priced goods and services that satisfy human needs and improve quality of life while progressively reducing ecological impacts and resource intensity, throughout the life cycle of the product, to a level at least in line with the Earth's estimated carrying capacity (WBCSD 1995).

There is now ample precedent for the view that a

Overall strategy for cleaner production development

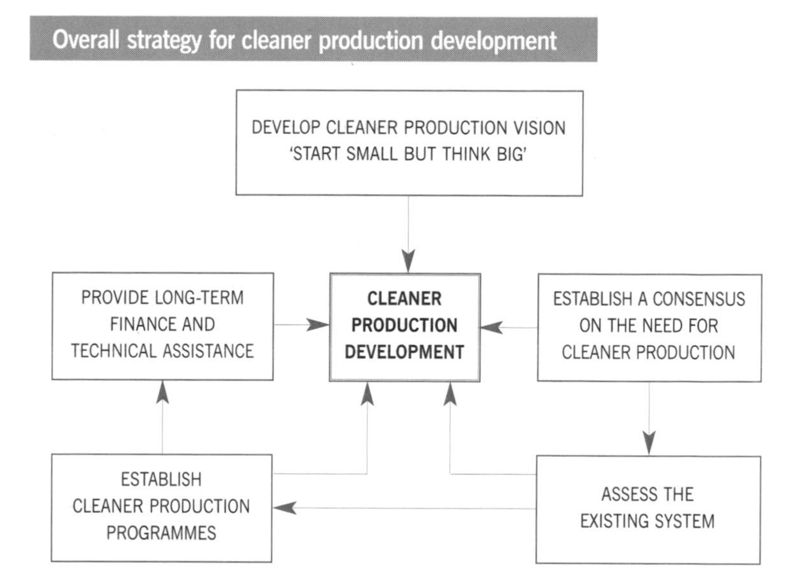

Source: UNEP 1994

Costs and benefits of cleaner production: worldwide examples

- a DuPont agricultural products team from LaPorte, Texas, reduced its toxic emissions by 99 per cent through closed-loop recycling, off-site reclamation, selling former wastes as products, and substituting raw materials – overall savings included US$2.5 million of capital and more than US$3 million in annual operating costs
- during 1987–94 IBM reduced its annual global emissions of hazardous wastes from 218 to less than 90 tonnes
- in North America, concentrated formulas for cleaners and washing up liquid have cut plastic packaging by 15-20 percent and corrugated cardboard containers by up to 30 per cent because concentrated products are sold in smaller packages requiring fewer raw materials;
- in the United Kingdom, a redesigned shipping container eliminated cartons that held bottles of Clearasil, saving 156 tonnes of packaging a year
- refill bags for powdered laundry detergents in North America use 80 percent less material than paper cartons, require less energy to ship and contain 25 percent recycled plastic
- during 1991–94, Dow Chemical reduced global emissions of the US EPA's 17 priority compounds by 65 per cent – further reductions of 75 per cent are planned for the year 2005
- in Denmark five plants on an industrial estate have cooperated with one another, with local authorities and with local farms to utilize each other's wastes, making savings in energy and water worth US$12–15 million a year
- in the Netherlands, the labels from Heineken's returned beer bottles are processed into light-weight bricks
- a new incinerator at Swissair's Zurich headquarters burns all non-recyclable paper, cardboard and wood from the company's offices and planes, heating the headquarters, nearby apartments and a hotel as well as generating electricity to cool computer equipment;
- a cleaner production audit of a Chinese chemical factory's penta-erythritol plant identified nine low-cost options which produced annual savings of US$30 000 for an investment of US$1200 – which was thus paid back in 15 days
- in California, wastewater from a Unilever tomato-processing plant is used as a soil conditioner to allow crop production on a site never previously farmed;
- in Brazil, liquid effluent per tonne of production from a Unilever factory is now less than 5 per cent of what it was in 1990 – a 20-fold improvement.

Sources: UNEP 1993 and Rabobank 1998

manufacturer must assume at least some responsibility for the environmental impacts of its products throughout their life cycles. Indeed, there is mounting public pressure for products that can be reused, recycled, returned to the manufacturer or better disposed of. Many industrialized countries are introducing regulations on these issues. This means that manufacturers need to find ways of acquiring data on life cycle impacts. A technique for doing so has been developed: life cycle assessment (LCA), a process for evaluating the 'cradle-to-grave' effects that a product has on the environment over its entire life cycle (UNEP and others 1996).

LCAs can be used for:

- evaluating the environmental performance of a new product against that of an old one;
- evaluating the environmental performance of a competitor's product;
- informing governments, pressure groups and other organizations about the environmental aspects of products;
- improving products and processes, and designing new ones;
- developing business strategies, including investment plans;
- setting ecolabelling criteria;
- developing product policies and strategies; and
- making purchasing decisions.

The business of redesigning products from scratch in order to minimize their environmental impact is called ecodesign. The basic idea behind ecodesign is to include environmental factors at the beginning of the design process. Ecodesign is a complex process and a manual is now available to guide industries through the process (Rathenau Institute and others 1997). The strategies available to reduce environmental impacts in product design include:

- developing completely new concepts;
- selecting low-impact materials;
- reducing materials usage;
- optimizing production techniques;
- optimizing the distribution system;
- reducing impacts during use;
- optimizing the lifetime of the product;
- optimizing the end-of-life system.

Overall, there has been encouraging progress towards more sustainable industry in many countries. Companies are committing themselves to sustainable development strategies through partnerships with customers, suppliers, government, NGOs and the general public, and they are becoming more open, for example through the publication of environmental reports. The development of voluntary self-regulating schemes in several countries is also a promising step forward (Rabobank 1998, Elkington 1997). However, while these policy responses are encouraging, they are insufficient: much faster action will be required in the next century if sustainable production and consumption is to become reality.

Financing environmental action

Participants at the Earth Summit agreed that implementing *Agenda 21* would need new and additional sources of funds. While much could be provided by a country's own public and private sectors, low-income countries would need substantial additional funding in the form of development aid or other foreign capital flows. The UNCED Secretariat estimated that implementing *Agenda 21* in low-income countries would cost an average of more than US$600 000 million annually between 1993 and 2000 (UN 1993), of which US$125 000 million would have to come in the form of international donations or concessions. At the Earth Summit, high-income countries reaffirmed their commitment to reaching the UN target of providing 0.7 per cent of their GNP for Official Development Assistance (ODA). Some agreed to reach this target by the year 2000.

Many low-income countries have increased their domestic investments in the social and environmental sectors but generally these increases have been small (See-Yan 1997). ODA finances nearly half of Gross Domestic Investment (GDI) in most low-income economies, except in India and the fast-growing economy of China – GDI is much larger than ODA in both these countries.

Governments of both high and low-income countries have a wide range of options for increasing domestic investment in sustainable development including tax reforms, environmental taxes, reducing perverse subsidies (see pages 207–208), and redirecting financial resources through macro-economic reforms.

The cost of implementing *Agenda 21* has to be met from domestic investment, aid and foreign investment. Sources and volumes of development finance in each

Overseas Development Assistance, 1997

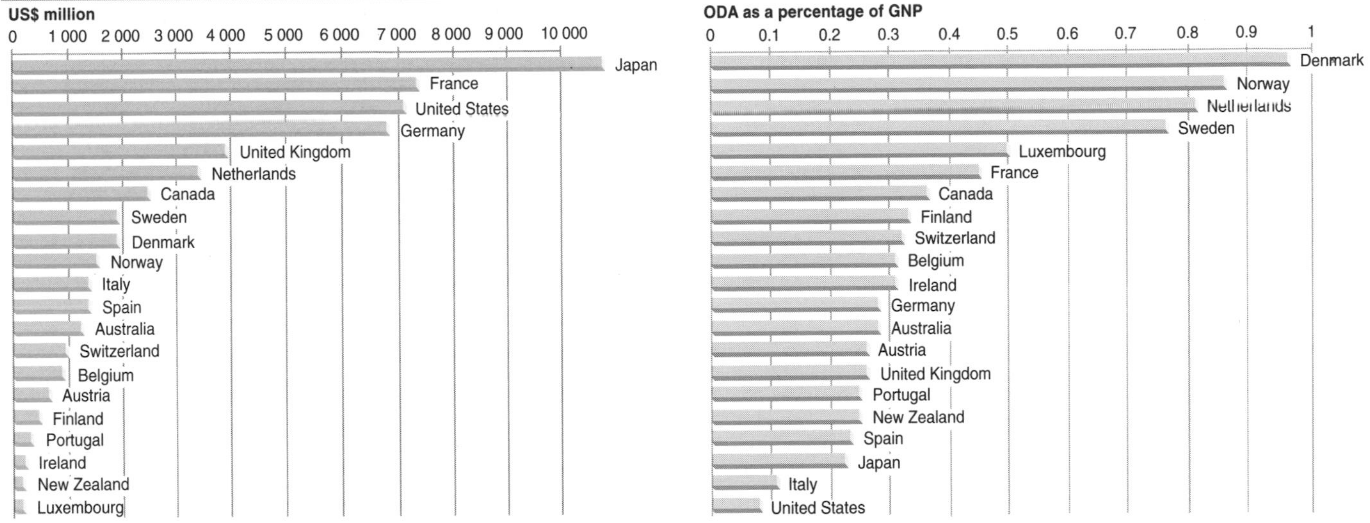

Source: OECD 1998c

region vary widely. For example, sub-Saharan Africa received in 1997 an average of some US$27 per capita of aid and US$3 per capita of foreign direct investment. By contrast, Latin America and the Caribbean received US$13 per capita of aid and US$62 per capita of foreign direct investment (OECD 1998c).

ODA was for many years the main source of external finance for low-income countries but this funding has been declining since 1991, both in absolute terms and as a percentage of Gross National Product. In 1992, industrialized nations were providing on average 0.35 per cent of their GNP as foreign aid – amounting to a total of a little less than US$60 000 million (less than half what was estimated as needed for *Agenda 21*). ODA dropped to an average of 0.22 per cent of GNP in 1997, the lowest rate of foreign aid ever. Between 1996 and 1997 alone, global development assistance fell by more than 14 per cent, to only US$47 580 million (OECD 1998c). Only four nations – Denmark, the Netherlands, Norway and Sweden – met the 0.7 per cent target (see diagrams above).

The decline has several causes which seem likely to persist. They include budgetary pressures in donor countries, the end of the Cold War justification for aid, a view that the poorest countries were becoming too dependent on aid, the decreasing need for aid among middle-income countries and scepticism about the historical effectiveness of aid. All this adds up to what the Commission on Global Governance has described as 'aid fatigue' (CGG 1995).

The most dramatic development in funding in recent years has been the rapid increase in private capital flows to low-income countries – by more than a factor of three since 1992 (though with a substantial fall since 1997 due mainly to the collapse of Asian economies). The increase has been concentrated on a few privileged countries, mainly the more dynamic economies in Asia, Europe, and Central and South America. In 1997, low-income countries as a group received a total of US$22 000 million of private flows, heavily concentrated in China and India. Countries in sub-Saharan Africa, including South Africa, received only US$2 000 million in foreign direct investment and roughly the same amount in bank flows. The difficulties that the poorest countries are experiencing in attracting resources for development point to a continuing need for aid to help establish conditions that will favour market investment, self-sustaining growth, and attainment of internationally-agreed development goals (OECD 1998c).

By 1997, total ODA had fallen to little more than US$4 500 million, and only four countries were achieving the *Agenda 21* goal of devoting 0.7 per cent of their GNP to ODA

Sources of funding in addition to ODA and private capital investment include the aid given to countries with economies in transition, transactions in which the main objective is not development, the Bretton Woods institutions (the World Bank family and the International Monetary Fund), various UN agencies and the Global Environment Facility (GEF).

Since the Earth Summit, official international financing for sustainable development has remained well below the level considered necessary to implement *Agenda 21*. Harnessing private capital flows may be more important than increasing or maintaining public capital flows but much more will need to be done to ensure that private investment is not used for unsustainable forms of development and that the poorest countries receive a much higher share (RIVM 1997).

Public participation

Broad public participation in decision-making is an important element of *Agenda 21* because, combined with greater accountability, it is basic to the concept of sustainable development. *Agenda 21* devotes separate chapters to involving many different groups including women, children and youth, indigenous people, NGOs, local authorities, workers and trade unions, business and industry, scientists and technologists, and farmers (UN 1993). In many cases, individuals and members of these groups are the best source of knowledge about the causes of and remedies for many environmental problems. Public participation enables such knowledge, skills and resources to be mobilized and fully employed, and the effectiveness of government initiatives to be increased (see box above).

In many countries, both developed and developing, this message has been taken to heart. In others, a start has hardly been made. Public access to environmental information has improved as governments have become more open and representative but in many regions more effort is required to achieve real public participation in environmental management (World Bank 1994).

Involving the public may require fundamental changes in social attitudes and individual behaviour. For example, following the political changes in 1989, the legal framework and institutions needed to secure public participation and access to justice have slowly begun to emerge in some Central European countries, but in others and in many Eastern European and Central Asian countries the absence of specific regulations or guidelines means that the real implementation of access to information and public participation has yet to take place. The efforts and resources needed to achieve active participation are considerable but the resulting people-motivated and collective actions are necessary if long-term sustainable growth is to be achieved.

Public participation is increasingly seen as a vital component of many environmental policy initiatives.

Rio Declaration on Environment and Development: Principle 10

Principle 10

'Environmental issues are best handled with the participation of all concerned citizens, at the relevant level. At the national level, each individual shall have appropriate access to information concerning the environment that is held by public authorities, including information on hazardous materials and activities in their communities, and the opportunity to participate in decision-making processes. States shall facilitate and encourage public awareness and participation by making information widely available. Effective access to judicial and administrative proceedings, including redress and remedy, shall be provided.'

Source: UN 1993

Many countries are encouraging public participation in environmental management, through local government and community-based groups, often as a result of the trend towards greater democracy. But institutional and legal participation is often restricted to only a few areas.

An increasingly important legislative tool for ensuring public participation is the EIA, which requires that the environmental impacts of major public and private projects are studied and published prior to decision making and often includes public hearings as a formal part of the process. A limitation is that EIAs are often required only for major projects. In some countries, institutions have polled the public by means of surveys dealing with specific questions on environmental degradation and possible courses of action. Citizens' participation has also materialized in a more direct way through formal representation of several social actors on various councils. The actors are entrepreneurs, environmental activists, municipal administrators, consumers and state institutions.

NGOs, in particular, have collaborated effectively with national and local governments on a wide range of issues and concerns. They have emerged as major players and partners at local, national, and regional levels in development and conservation activities, performing a multitude of roles including environmental education and awareness-raising among the public. NGOs have assisted in the design and implementation of environmental policies, programmes and action plans, as well as the setting out of specifications for EIAs. They also play a crucial advocacy role through their environmental campaigns.

They have established legal services to assist citizens, other NGOs and local communities in exercising public participation rights and gaining access to justice. Many studies have noted the importance of NGOs in monitoring state behaviour and promoting compliance of MEAs (Weiss and Jacobson 1998).

Citizens' demands for broader and institutionalized legal participatory channels are expected to increase. The changed perception of the role of civil society in achieving the objectives of the Earth Summit has resulted in the adoption of the principles of co-management of natural resources and a close collaboration between governments, NGOs, community organizations and the private sector in setting standards and preparing environmental policies or action plans. In some countries this collaboration is formalized in moves to decentralize governance to the community level. Decentralized actions to deal with environmental conflicts locally or at the provincial level may offer an effective way of channelling public participation.

Environmental information and education

Ideally, access to governmental information of any kind should be guaranteed through a 'Freedom of Information Act' as exists in the United States. This situation is rare at the national level and entirely absent at the international level. *Agenda 21* calls for greater public access to environmental information but, as yet, the level of access differs markedly in different regions and countries.

In some regions, public and government awareness of environmental issues has been high, as indicated by the early appearance of regional treaties, national environmental laws and policies, and the establishment of formal communication and awareness-raising strategies, targeting those directly affected and the general public. Examples include the UNECE Convention on Access to Environmental Information and Public Participation in Environmental Decision-making which was approved in June 1998 and opened for signature at the Århus Conference, and several European Union Directives which contain requirements for the disclosure and dissemination of data. NGOs are also involved in improving access to environmental information, for example by providing information on the location of pollution sources, the types of pollutants produced, and contact information to help catalyse local action.

Elsewhere, for example in the former Eastern Bloc countries until the political changes of 1989, information on the state of the environment has been difficult to come by, simply not collected or, when available, frequently altered to present a more favourable picture. In some countries, the provision of information is still characterized by strict regulations, multiple reasons for refusal, and long delays.

There is a growing recognition that national development plans and environmental policies have a better chance of being implemented effectively when supported by an informed, educated and involved public, and a general acceptance of the need for greater openness and transparency. A major problem, however, is the weakness of the information base in many countries. A shortage of reliable data and capability for data analysis undoubtedly hinders policy development, planning and programme implementation.

There is a need not just for more data on environmental issues but for standardizing data collection and storage, and making it accessible to technical and managerial levels. Reports may be located in different bodies between which there is little or no cooperation or exchange, resulting in gaps, duplication and limited utilization of data. This, in turn, hinders policy development, planning, implementation, and follow-up.

Even where adequate data are available, there may be incompatibility between different agencies and different countries. Networking and integration of data for environmental assessment are at a very early stage in some regions. Electronic information systems, networks and cooperation among relevant organizations all need to be strengthened to enable all users to benefit from data at the local, national, regional and international levels.

Attempts are being made to redress this situation. In many areas, national and regional databases have been set up, for example on climate, soils and biodiversity, through the efforts of the UN and other organizations. Better information-exchange networks are gradually evolving but are still constrained by problems of access to modern communication systems. Language barriers also frustrate networking and information exchange.

A number of international agencies are involved in strengthening institutional capacities to manage environmental information and assisting some countries to prepare national and regional state of the environment reports. Cooperative activities include standardizing, collecting, analysing and exchanging scientific data, improving national scientific and technical capabilities and research infrastructure, cooperating with research networks in other regions, providing scientific knowledge to the public and inputs to decision makers, and developing appropriate mechanisms for technology transfer.

While information on environmental issues has increased, particularly since the Earth Summit, the impact of this on decision-making cannot yet be assessed because many information programmes are still at an early phase of implementation.

Improving environmental education

Agenda 21 urges every country to prepare a national strategy for environmental education. Chapter 36, 'Education, Training and Public Awareness', recommends:

- making education on environment and development available to people of all ages;
- including environment and development concepts in all educational programmes;
- involving children in local and regional studies of environmental health;
- setting up training programmes to help school and university graduates attain sustainable livelihoods;
- encouraging all sectors of society to train people in environmental management;
- providing locally recruited and trained environmental technicians to give communities the services they require;
- working with the media, entertainment and advertising industries to stimulate public debate on the environment;
- bringing the understanding and experience of indigenous peoples into education and training programmes.

Source: UN 1993

A number of schemes have been launched, some voluntary and some compulsory, under which companies make public information on their environmental activities, including the degree to which they pollute the environment. For example, under the European Union's Pollutant Release and Transfer Register, information on potentially harmful releases to the air, water and soil, and on waste transport, is collected in a unified national reporting system, ensuring that the community, industry and governments gain greater access to relevant information on environmental pollution. The Indian Government has set up a network to deal with the collection, collation, storage, analysis, exchange and dissemination of environmental data and information (MOEF India 1995). Many governments regularly publish environmental indicator reports, for example on water and air quality, and issue regular environmental information bulletins and other books and pamphlets for wide dissemination.

Education, from elementary school up to the university, has an important role in providing environmental information and raising awareness (see box left) and there has been a substantial expansion in education programmes at all levels. A growing number of countries now include environmental education in school curricula, and specialised programmes at the technical and higher education levels lead to environmental masters degrees and doctorates. Many universities and institutions are running courses and organizing training programmes, seminars and postgraduate studies in various fields of the environment. In some countries, however, environmental education is voluntary and differs from school to school.

There has also been considerable interest in integrating environmental concepts into adult education and literacy programmes, for example using non-formal education centres as sources of environmental education. Consciousness-raising activities include educational campaigns on saving natural resources and reducing waste generation, ecolabelling, and publicity campaigns aimed at promoting recycling and consumption of non-polluting products.

In some areas, NGOs have played a key role in producing print and audio-visual materials for non-formal environmental education in schools and other institutions of learning.

Social policies

Improving health, education and living conditions remains the top policy imperative for many countries. Such improvement is vital for ensuring the political stability and social sustainability needed to move toward greater economic and environmental sustainability.

Nevertheless, social policies have had an unequivocal impact on the environment in many countries. Programmes devised to fight poverty and especially extreme poverty have often ignored environmental policies. Many projects that damage the environment have been considered valuable because of the employment they generate. Some housing programmes have fuelled urban growth and discouraged better use of existing urban areas. The regulatory approach characteristic of environmental management in most countries is dissociated from social policies, which have often clashed with environmental management. Practices and regulatory measures that benefit the industrial sector have often ignored environmental deterioration and its impacts on the quality of life.

Population growth in many countries continues to exceed the growth of agricultural production and the resulting food deficit is aggravated by the scarcity of land and water resources. A lack of education is often associated with a lack of environmental awareness.

New policies and strategies for environmental and natural resources management for sustainable development are beginning to emerge, in line with the innovative approach of *Agenda 21*. Many countries have adopted policies to stabilize or moderate population growth rates. Greater equity, however, is needed in the distribution of the opportunities and benefits of national economic development and international aid programmes. At present, too few national or international aid programmes reach or benefit the poor majority.

The success of efforts directly targeted at poverty alleviation has varied. For example, direct support programmes to provide subsidized food have sometimes resulted in food deliveries going largely to the better-off in urban areas, and subsidized credit

Policy goals for achieving sustainable development

	economic sustainability	*social sustainability*	*environmental sustainability*
Water	Ensure the adequate supply and efficient use of water for agricultural, industrial, urban and rural development	Ensure adequate access of the poor majority to clean water for domestic use and small scale agriculture	Ensure adequate protection of watersheds, aquifers and freshwater ecosystems and resources
Food	Increase agricultural productivity and production for regional food security and export	Improve productivity and profitability of small scale agriculture and ensure household food security	Ensure sustainable use and conservation of land, forest, wildlife, fisheries and water resources
Health	Increase productivity through preventative health care and improved health and safety at the work place	Enforce air, water and noise standards for protecting human health and ensure basic health care for the poor majority	Ensure adequate protection of biological resources, ecosystems and life support systems
Shelter and services	Ensure the adequate supply and efficient use of resources for buildings and transportation systems	Ensure adequate access to affordable housing, sanitation and transportation by the poor majority	Ensure sustainable or optimum use of land, forest, energy and mineral resources
Energy	Ensure the adequate supply and efficient use of energy for industrial development, transportation and household use	Ensure adequate access to affordable energy by the poor majority, especially alternatives for fuelwood	Reduce local, regional and global environment impacts of fossil fuels and expand the development and use of forest and other renewable alternatives
Education	Ensure the availability of trained people for all key economic sectors	Ensure adequate access for all to education for a healthy and productive life	Integrate environment in public information and education programmes
Income	Increase economic efficiency, growth and employment opportunities in the formal sector	Support small scale enterprises and job creation for the poor majority in the informal sector	Ensure sustainable use of natural resources needed for economic growth in the formal and informal sectors

Source: SADC 1996

programmes have resulted in loans failing to reach the poor, being used for consumption, and often not being repaid.

Some governments have developed social policies that have emphasized the promotion of sustainable human settlements. The priority targets have been the basic needs of the rural population, especially shelter and safe drinking water, and human resource development, for example education and training.

Some of the key policy concerns and main objectives for a new agenda and strategy on equity-led growth for sustainable development are set out in the table on page 215. The matrix links the critical basic needs (water, food health, shelter and services, energy, education and income) with the overall and interrelated goals of economic, social and environmental sustainability. Developed in the context of Southern Africa, the matrix applies equally to much of the developing world. The social sustainability and equity issues are the new and crucial link between economic and environmental sustainability issues and objectives.

References

Ashby, E. (1981). *Politics of Clean Air.* Clarendon Press, Oxford, United Kingdom

Birnie, P. (1985). *International Regulation of Whaling.* Oceana Publications, Dobbs Ferry, New York, United States

Brañes, R. (1991). *Aspectos institucionales y jurídicos del medio ambiente, incluida la participación de las organizaciones no gubernamentales en la gestión ambiental.* World Bank, Washington DC, United States

Brenton, T. (1994). *The Greening of Machiavelli: The Evolution of International Environmental Politics.* Earthscan, London, United Kingdom

Burhenne, W.E., and Robinson, N.A. (eds., 1996). *International Protection of the Environment: conservation in sustainable development.* Oceana Publications, Dobbs Ferry, New York, United States

Chayes, A., and Chayes, A. H. (1993). On Compliance. *International Organization*, Spring 1993

Chayes, A., and Chayes, A. H. (1995). *The New Sovereignty.* Harvard University Press, Cambridge, Massachusetts, United States

Collins, R. (1996). *Tobin Tax as a mechanism to finance de-mining, de-mining technology development and mine victim rehabilitation: An introduction to the idea.* United Nations Association in Canada http://www.ncrb.unac.org/landmines/UNACinfo/tobin-tax.htm

CGG (1995). *Our Global Neighbourhood: the report of the Commission on Global Governance.* http://www.cgg.ch/home.htm

de Moor, A. (in press). The perversity of government subsidies for energy and water. In *Greening the budget: proceedings of a workshop of the European Research Network on Market-based Instruments for Sustainable Development,* Edward Elgar, United Kingdom

de Moor, A., and Calamai, P. (1997). *Subsidising unsustainable development: undermining the Earth with public funds.* Earth Council Institute, Toronto, Canada http://www.ecouncil.ac.cr/econ/sud/

Downs, G.W., Rocke, D.M., and Barsoom, P.N. (1996). Is the Good News about Compliance Good News about Cooperation? *International Organization,* Vol. 50, No. 3, Summer 1996

DTO (1997). Interdepartementaal Onderzoeksprogramma Duurzame Technologische Ontwikkeling, *DTO Visie 2040–1998: technologie, sleutel tot een duurzame welvaart* (Sustainable Technological Development. *DTO Vision 2040-1998. Technology, key to sustainable prosperity*). Uitgeverij ten Hagen and Stam, Den Haag, The Netherlands

EDIE (1999). Environmental Data Interactive Exchange, 12 March 1999. http://www.edie.net/news/Archive/841.html

Elkington, J. (1997). *Cannibals with forks: the triple bottom line of 21st century business.* Capstone, London, United Kingdom

IER (1997). Green Taxes Proven Successful, Evaluation by Swedish EPA Says. *International Environment Reporter*, 16 April 1997

IEEP-B (1994). *The Dutch Governmental Programme for Sustainable Technology Development.* Institute for European Environmental Policy, Brussels, Belgium

IRPTC (1999). *Prior Informed Consent for Certain Hazardous Chemicals in International Trade* http://irptc.unep.ch/pic/

Johnston, S. (1997). The Convention on Biological Diversity: the Next Phase. *RECIEL,* Vol. 6, Issue 3, p. 219

Kiss, A., and Shelton, D. (1991). *International Environmental Law.* Transnational Publishers, Ardsley, New York, United States

MOEF India (1995). *Annual Report 1994-95.* Ministry of Environment and Forests, New Delhi, India

Myers, N. (1998). *Perverse Subsidies, Tax $s Undercutting Our Economies and Environment Alike.* International Institute for Sustainable Development, Winnipeg, Canada

NEPA, China NCPC, UNEP and World Bank (1996). *Cleaner Production in China: a story of successful cooperation.* UNEP IE, Paris, France

OECD (1995). *Adjustment in OECD Agriculture: issues and policy responses.* OECD, Paris, France

OECD (1997). *Reforming Energy and Transport Subsidies.* OECD, Paris, France

OECD (1998a). *Improving the Environment Through Reducing Subsidies.* OECD, Paris, France

OECD (1998b). OECD Environment Ministers share goals for action. News Release, 3 April 1998. OECD, Paris, France

OECD (1998c). *OECD 1998 DAC Report.* OECD, Paris, France http://www.oecd.org/dac/pdf/aid98a.pdf

POPs (1999). http://irptc.unep.ch/pops

Rabobank (1998). *Sustainability: choices and challenges for future development.* Rabobank International, Leiden, The Netherlands

Rathenau Institute, Delft University of Technology and UNEP (1997). *Ecodesign: a promising approach to sustainable production and consumption.* UNEP IE, Paris, France

RIVM (1997). *Agenda 21 Interim Balance, 1997.* Van Vuuren, D.P., and Bakkes, J.A. Globo Report Series No. 19, RIVM Report No. 402001008. RIVM, Bilthoven, The Netherlands

Rotterdam (1999). http://www.fao.org/WAICENT/FaoInfo/Agricult/AGP/AGPP/Pesticid/PIC/dipcon.htm

Ruester, B., and Simma, B. (eds., 1975). *International Protection of the Environment: Treaties and Related Documents, Vol. 4.* Oceana Publications, Dobbs Ferry, New York, United States

SADC (1996). *SADC Policy and Strategy for Environment and Sustainable Development: Toward Equity-Led Growth and Sustainable Development in Southern Africa.* SADC Environment and Land Management Unit, Maseru, Lesotho

SADC (1998). *Report of the SADC Regional Workshop on Integrating Economic, Environmental, and Equity Impact Assessments in Decision-Making*, Harare, Zimbabwe, October 1997. SADC Environment and Land Management Unit, Maseru, Lesotho

Sandrey, L., and Reynolds, R. (eds., 1990). *Farming without subsidies: New Zealand's recent experience.* MAF Policy Services, New Zealand

Sands, P. (1995). *Principles of International Environmental Law.* Manchester University Press, New York, United States

See-Yan, L. (1997). Chairman's Summary of the Fourth Expert Group Meeting on Financial Issues of Agenda 21. 8–10 January 1997, Santiago, Chile

Ul Haq, M., Kaul, I., and Grunberg, I. (eds., 1996). *The Tobin Tax, Coping with Financial Volatility.* Oxford University Press, New York, United States, and Oxford, United Kingdom

UN (1993). *Agenda 21: The United Nations Programme of Action from Rio.* Sales No. E.73.II.A.14 and corrigendum. UN Publications, New York, United States

UNEP (1993). *Cleaner Production Worldwide.* UNEP IE, Paris, France

UNEP (1994). *Government Strategies and Policies for Cleaner Production.* UNEP IE, Paris, France

UNEP (1996a). *Mid-term Review of the Programme for the Development and Periodic Review of Environmental Law for the 1990s and Further Development of International Environmental Law Aiming at Sustainable Development.* UNEP, Nairobi, Kenya

UNEP (1996b). Cleaner Production: Fourth High Level Seminar in Oxford. *Industry and Environment,* Vol. 19, No. 3, July-September 1996, UNEP IE, Paris, France

UNEP (1997). *Register of International Treaties and Other Agreements in the Field of the Environment - 1996.* UNEP, Nairobi, Kenya

UNEP (1999). http://www.unepie.org/Cp2/declaration/home.html

UNEP, CML, Novem and RIVM (1996). *Life Cycle Assessment: what it is and how to do it.* UNEP IE, Paris, France

Victor, D., Raustiala, K., and Skolnikoff, E. (eds., 1998). *The Implementation and Effectiveness of International Environmental Commitments: Theory and Practice.* MIT Press, Cambridge, Massachusetts, United States

von Weizsäcker, E., Lovins, A., and Lovins, H. (1995). *Faktor Vier*. Droemer Knaur, München, Germany

WBCSD (1995). *Eco-efficient Leadership – for Improved Economic and Environmental Performance.* World Business Council for Sustainable Development, Geneva, Switzerland

Weiss, E.B., and Jacobsen, H.K. (1998). *Engaging Countries: Strengthening Compliance with International Environmental Accords.* MIT Press, Cambridge, Massachusetts, United States

Werksman, J. (1997). *Five MEAs, Five Years Since Rio: recent lessons on the effectiveness of Multilateral Environmental Agreements.* Foundation for International Environmental Law and Development (FIELD), London, United Kingdom

World Bank (1994). *Forging a partnership for environmental action; an environmental strategy towards sustainable development in the Middle East and North Africa*. World Bank, Washington DC, United States

World Bank (1997). *Five Years after Rio: innovations in environmental policy*. World Bank, Washington DC, United States

Africa

KEY FACTS

- Many countries have recognized their need to move beyond *Agenda 21* in order to deal more effectively with the interlinked challenges of economic recovery, poverty reduction and sustainable development.
- Most countries have now developed National Action Plans to combat desertification and some have gone further: Tunisia, for example, has strengthened its 'Yellow Hand' environmental programme to combat desertification and promote the socio-economic development of rural areas.
- Some countries – for example Benin, Ethiopia, Eritrea, Ghana, Malawi, Mali, Mozambique, Seychelles and Uganda – have incorporated the environmental rights and responsibilities of their citizens in their constitutions.
- The move toward sustainable development will require major changes in many policies, programmes, laws and institutional arrangements, outside as well as within the environment field.
- Botswana and Namibia, for example, have recently developed natural resource accounts to enable a better assessment of their economic value and management options.
- Regional centres have been strengthened to provide expertise in environmental management but most are vulnerable because of their dependence on government and donor funding; their inability to compete with the salaries offered by the private sector and international organizations is leading to a brain drain within and outside Africa.
- The governments and people of Africa are increasingly setting their own agenda for change as a result of more democratic rule and improved governance in several countries and greater cooperation at the sub-regional, regional and global levels.

The policy background

In many traditional African societies, the management and conservation of nature and natural resources was largely a community responsibility until well into the 20th century. The survival of hunter-gatherer, pastoral and sedentary farming communities depended on extensive knowledge and sustainable use of the land, forests, plants and wildlife resources. That knowledge and related conservation practices were improved and passed on by successive generations, including social taboos and community sanctions for those who violated the norms. People and communities adapted to different ecological conditions, especially in the large dryland areas throughout the continent where people led pastoral lifestyles and migrated with their families and livestock in response to recurrent droughts and other environmental challenges. Culture was mainly linked with ecology in a balanced and dynamic manner (Achoka and others 1996).

During the 'scramble for Africa' in the late 19th century, European countries demarcated their new spheres of influence and territories in Africa by setting boundaries, which arbitrarily divided many peoples, cultures and ecosystems. In addition, during the first half of the 20th century, the European colonial powers imported and imposed new laws and regulations which undermined and replaced the traditional community-based approach to conservation. Local communities also lost access to many ecologically-rich areas set aside for European farming, mining and nature protection, especially in Southern and Eastern Africa (SARDC, IUCN and SADC 1994).

By the mid-20th century the responsibility for natural resource management had shifted from local communities to colonial administrations, with a centralized bureaucratic structure which many countries perpetuated and expanded after independence in the 1960s. However, in the closing decade of the 20th century some governments began changing their environmental policies and laws to give greater support once again to community-based approaches to conservation, especially in wildlife management (SARDC, IUCN and SADC 1994).

The colonial powers also established a development pattern focused mainly on economic growth, with the export of key commodities and natural resources as a major feature. In the 1960s, many newly-independent countries continued this growth strategy, an approach encouraged and reinforced over the next two decades by the aid programmes of industrialized countries and the lending policies of the World Bank and IMF. In addition to increased dependence on commodity prices in a world trade system dominated by the major industrialized countries, and vulnerability to fluctuations in these prices, African countries also paid a rising environmental cost (UNEP 1991).

Many African countries took an active but initially sceptical part in the negotiations for and at the United Nations Conference on the Human Environment held in 1972 in Stockholm. They considered some of the key problems on the Stockholm Conference agenda to be largely the concern of the developed countries while their own biggest problem was lack of development. African delegations also worried that stricter environmental measures would be used to constrain their future development by leading to non-tariff trade barriers and new conditions in aid programmes. They joined other developing countries in ensuring that the recommendations in the Stockholm Action Plan took account of their concerns (UNCHE 1971).

Over the following decade, many countries established new environmental agencies, policies and laws. However, the new environmental units were usually added on to the government decision-making structures, mainly as coordinating bodies, or combined with an existing natural resource or development department (for example in water, forestry or tourism). During the 1980s, some countries prepared new National Conservation Strategies as well as National Environmental Actions Plans (NEAPs) which were sometimes combined with National Action Plans to Combat Desertification. But in these and other countries the environment agencies often lacked the authority, budgets, staff, expertise and equipment to implement the new policies and action plans effectively or to enforce the laws. Among other constraints during the 'lost decade' of the 1980s were unfavourable terms of trade, economic stagnation, rising indebtedness, corruption and civil unrest which further undermined and marginalized environmental action in many countries (UNEP 1993).

A major new regional policy initiative was launched in 1985 with the convening of the first African Ministerial Conference on the Environment (AMCEN) in Cairo. The conference operates through close cooperation among three key organizations that provide the secretariat: the Organization of African Unity (OAU), the United Nations Economic Commission for Africa (UNECA) and UNEP (AMCEN 1992). Over the next decade, AMCEN provided political leadership and policy guidance through the Ministerial Conference and its Bureau. AMCEN established five committees focused on the main regional ecosystems (deserts and arid lands, forests and woodlands, regional seas, river and lake basins, and island ecosystems) and the eight technical cooperation networks on soils and fertilizers, climatology, water resources, education and training, science and technology, biodiversity, energy and environmental monitoring. AMCEN also played a notable role in catalysing village-level approaches towards sustainable development.

During its first decade, AMCEN played a crucial part in creating a distinctive and common position for Africa in the negotiations for, at and after the United Nations Conference on Environment and Development in 1992 in Rio de Janeiro. By the late 1980s, most countries were confronted by a large backlog of environmental degradation as well as many new environmental challenges caused by rapidly-growing populations and increasing urbanization and industrialization. But the top political imperative and policy challenge throughout the continent was – and remains – the poverty of the majority of the people. As repeatedly demonstrated by many of the priority issues highlighted in the African Common Position on Environment and Development, the poverty of the majority was increasingly recognized as a major cause and consequence of environmental degradation and resource depletion in the region (UNECA 1991).

In their joint report to the 1992 Earth Summit, the Southern African countries reflected the concerns of many others in concluding that:

> *'Throughout the negotiations before and during the Rio Conference in June 1992 we must never forget that the majority of people and countries in the SADCC region and the world are poor. If the poor sometimes behave in a way that degrades the environment it is not because they choose to do so. They only do so when they have no other choices. The Earth Charter and Agenda 21 must expand the development choices and opportunities for the majority of poor people, communities and countries ... No new political and economic arrangements within or among our countries can be called sustainable if they fail to change the present situation of a rich minority and poor majority by significantly reducing the gap between them. The Earth Charter and Agenda 21 must provide a new basis for a new deal for the majority of poor people and countries in order to secure and sustain our common future' (SADCC 1991).*

At the Earth Summit, however, poverty alleviation got far more attention in plenary speeches than in the action plan. After the Earth Summit, overall development aid largely stagnated or declined. No significant new or additional resources were provided for expanding environmental protection and improvement programmes in Africa or other developing regions (SADC 1996).

The 1992 Earth Summit thus produced a comprehensive but flawed action plan for application in Africa, especially with regard to practical policies and measures for reducing the poverty of the majority of people while combating environmental degradation. Although many countries used *Agenda 21* as an innovative guide for improving national policies and planning during 1993–97, they also recognized their need to move beyond it in order to deal more effectively with their triple and interlinked challenges of economic recovery, poverty reduction and environmental improvement (SADC 1996).

By the mid-1990s, the conclusion reached by the Southern African countries reflected the common challenges to all top policy-makers in all key development sectors throughout the continent (see box left). These countries then set a new agenda for equity-led growth and sustainable development focused on accelerating economic growth with greater equity and self reliance, improving the health, income and living conditions of the poor majority, and ensuring equitable and sustainable use of the environment and natural resources for the benefit of present and future generations. They especially emphasized in their report that 'These three goals constitute one agenda for action. None of the goals is achievable without the other two. In the SADC region economic growth is not sustainable without protecting the environment and resource base on which future development depends. Environment and social improvement programmes are not feasible without the financial resources generated by economic growth. Most importantly, economic and environmental sustainability are not achievable without significant improvements in the lives and livelihoods of the poor majority' (SADC 1996).

Policy challenges for Southern Africa

The countries of Southern Africa identified the focus and direction for needed policy change in their *Policy and Strategy for Environment and Sustainable Development*:

'Changes for greater equity and sustainable development are needed, for example, to shift the emphasis and priorities:

- In economic development policies focused largely on the formal sector toward policies supporting the much larger informal sector which is the main source of jobs, income and affordable goods and services for the poor majority.
- In agricultural policies promoting large scale production for export of food and horticultural crops toward policies focused on the food security of poor rural and urban households and fair returns for small scale farmers.
- In land tenure laws and policies excluding women toward policies recognising and expanding the rights of women to inherit and own land.
- In health policies which allocate a disproportionate share of the budget to specialized medical services and hospitals toward policies focused on primary health care services which are affordable and accessible for the poor majority.
- In human settlements policies which emphasize more planning, research and delivery of unaffordable housing toward policies giving top priority to the lethal shelter, water and sanitation problems of the poor majority in urban and rural settlements.
- In wildlife and parks policies where local people bear many of the costs toward policies which give local people and communities a greater voice in wildlife management and a significant share of the benefits of wildlife-based tourism.
- In international lending policies, especially structural adjustment policies which adversely affect the poor majority first and most, toward UN system-wide policies and programmes of economic reform for greater equity and sustainable development.'

Source: SADC 1996

MEAs and non-binding instruments

Global MEAs

Although some African countries signed MEAs as long ago as the 1930s, it was not until the 1992 Earth Summit that a number of global MEAs relevant to the African situation were signed by most countries. The situation regarding implementation is unclear and at times confusing. The enthusiasm generated at the Earth Summit, coupled with the support and encouragement of United Nations agencies, bilateral and multilateral donors and lending agencies, resulted in most countries initiating action to implement programmes and projects under the global conventions. Studies and training programmes recommended for implementation by the MEAs have taken place in most countries where external funding was provided.

The most favoured means of implementing MEAs, and environmental policy in general, at the national level has been the enactment of legislation and passing of decrees. Some countries, especially in Western and Central Africa, have favoured legislation; others, particularly in North Africa, have relied more heavily on Presidential and Ministerial decrees. Several countries have also adopted 'environmental codes' with provisions for the implementation of various environmental MEAs.

However, enforcement remains a problem as governments seldom have personnel with the skills needed to select and enforce policy packages. The causes of weakness, however, are not the same throughout Africa. Some relate to the choice of policies, some to the enforcement processes and some to low priority being assigned to the MEAs. Most countries have not yet adequately integrated the legislative, technical, administrative and other parameters into the fabric of their development process.

The importance attached to MEAs is reflected in the rate of their ratification (see figure below). Although there is some political commitment to most global MEAs, some are of more relevance than others. Desertification and drought, for example, is one of the most important environmental issues and the particular emphasis on the situation in Africa in the convention text has resulted in the CCD receiving a high degree of political commitment and extensive support; 51 African countries were parties to the CCD as of 1 March 1999.

Parties to major environment conventions (as at 1 March 1999)

	CBD (174)	CITES (145)	CMS (56)	Basel (121)	Ozone (168)	UNFCCC (176)	CCD (144)	Ramsar (114)	Heritage (156)	UNCLOS (130)
AFRICA (53)	50	48	19	23	44	47	51	28	39	38
Northern Africa (6)	5	5	4	4	6	5	6	5	6	4
Western and Central Africa (25)	24	23	13	9	20	23	24	14	18	18
Eastern Africa and (11) Indian Ocean Islands	10	11	1	3	8	9	10	4	7	8
Southern Africa (11)	11	9	1	7	10	10	11	5	8	8

Key: percentage of countries party to a convention

0–25% | 25–50% | 50–75% | 75–100%

Notes:

1. Numbers in brackets below the abbreviated names of the conventions are the total number of parties to that convention
2. Numbers in brackets after name of regions/sub-regions are the number of sovereign countries in each region/sub-region
3. Only sovereign countries are counted. Territories of other countries and groups of countries are not considered in this table
4. The absolute number of countries that are parties to each convention in each region/sub-region are shown in the coloured boxes
5. Parties to a convention are states that have ratified, acceded or accepted the convention. A signatory is not considered a party to a convention until the convention has also been ratified

While there have been many initiatives to combat desertification, most predate the CCD which came into force only in December 1996. Most countries have now developed National Action Plans (NAP) to combat desertification and some countries have gone even further. Tunisia, for example, has strengthened its 'Yellow Hand' environmental programme to combat desertification and promote the socio-economic development of rural areas. The Government has also updated the NAP (UNSO/UNDP 1995a), established a National Committee on Desertification and a national fund for natural resource conservation with the aim of financing projects to combat desertification (UNSO/UNDP 1995b). In Zimbabwe, the NAP process has been decentralized to the district level and District Environmental Action Plans have been produced. These are not tied only to the CCD – a complementary Desert Margins Initiative has also been implemented to enhance food security for rural populations, focusing on rain-fed crops, tree and livestock production systems in dryland areas receiving 100–600 mm of rainfall annually. In the Sahelian countries of Burkina Faso, Niger and Senegal, national structures have been put in place, and programmes to combat drought and desertification are being implemented.

Other initiatives relevant to the CCD include sub-regional action programmes emphasizing cooperation in sustainable management of shared natural resources, including rivers, lakes and aquifers, through streamlining, harmonizing and enforcing environmental laws, standards, mandates and responsibilities among member states. Notable examples include the Kalahari-Namib Action Plan and the extension of the Zambezi Action Plan to other shared water course systems (Maro 1995). The first phase of the Kalahari-Namib Action Plan was concluded in 1995 at a cost of US$11.9 million, and sought to improve the welfare of people living in the area through sustainable exploitation of natural resources and abatement of human-caused land degradation and the desertification process (SADCC 1991).

The UNFCCC also remains a high priority. Although African countries are not major contributors to the greenhouse gas emissions that lead to climate change, they will be severely affected by it – coastal areas and the drylands will be the most affected. As a result, 47 countries are parties to UNFCCC. Unlike the industrialized countries, there is no obligation to reduce emissions of greenhouse gases to a specific target and deadline since Africa's contribution to global CO_2 emissions is low – only 3.5 per cent (CDIAC 1998). However, the African countries should learn the lesson of the industrialized world and set targets for future emissions as their energy usage and industrialization increases. The region's preparedness to live in a warmer 21st century will depend on how actively the scientific community, governments and policy-makers respond to emerging global environmental issues (SARDC, IUCN and SADC 1994).

Projects being developed under UNFCCC include energy efficiency studies, alternative energy programmes, and inventories of greenhouse gases. Projects in Northern Africa include the use of non-fossil fuels and assessments of the impact of climate change on coastal zones and mitigation options. In Southern Africa, the focus has been on adapting to drought conditions resulting from climate change, for example through research on alternative crops and varieties, and strengthening the SADC Drought Monitoring Centre in Harare.

UNCLOS is of significant relevance to the life support systems of several countries, and many countries have ratified it since it was adopted in December 1982 (UNEP 1997). The Basel Convention on the Control of Transboundary Movements of Hazardous Wastes and their Disposal is also of importance as Africa has been the main destination of hazardous waste exports from industrialized countries. In spite of this, however, only 23 countries are parties to the convention. Substantial financial benefits can be derived from storing and disposing of hazardous wastes and, while some states oppose the trade in such wastes, clearly not all do.

The Ramsar Convention has 28 contracting parties in Africa. The Convention requires that parties designate at least one national wetland for inclusion in the List of Wetlands of International Importance (UNEP 1997). The floodplains of the Zambezi River and the Okavango delta are among Southern Africa's major wetlands, providing a wide range of functions such as water and nutrient retention and flood control. They are also important for tourism. The coral reefs of Tongaland and the St Lucia System (South Africa) and the Kafue Flats and Bangweulu Swamps (Zambia) have been designated as wetlands of international importance. In Eastern Africa, Ramsar sites include the Lake George ecosystem in western Uganda, and Lake Nakuru in Kenya.

The Convention on Biological Diversity (CBD) has led to the formulation of biodiversity plans and strategies, especially in countries where the depletion of tropical rain forests and the rapid disappearance of some animal species have attracted national and international attention. In Cameroon and Ghana, for example, forest policies and management plans, formulated specifically to address biological diversity, sustainable forestry and wildlife management, have been adopted as a result of the CBD (Cameroon 1996 and Ghana 1991), resulting in the implementation of community-based forestry management projects. National biodiversity strategies and action plans have been prepared following ratification by 50 countries, many with the support of the Global Environment Facility (see box).

CITES is closely related to the CBD. National programmes have been introduced in much of Africa to help in the sustainable utilization and trade in wildlife. Such programmes include the Communal Areas Management Programme for Indigenous Resources (CAMPFIRE) in Zimbabwe, the Peace Parks Concept in Mozambique and South Africa, and the Administrative Design Programme for Game Management Areas (ADMADE) in Zambia. In Northern Africa, the CITES Secretariat is running a series of workshops to introduce countries to all the CITES concepts and identify common needs. For Eastern and Southern Africa, the main significance of CITES has been in conservation of the African elephant through control of the ivory trade.

While some actions in the areas of drought, desertification and biodiversity have been triggered by global MEAs, it is too early to evaluate their overall impact. The Basel Convention and, to some extent, the Bamako Convention have resulted in the creation of 'Dump Watch', an advance warning agreement among West and Central African states with diplomatic representation in European capitals. According to this agreement, European governments, international NGOs and cooperating institutions and individuals provide information to African diplomatic missions on the movement of hazardous wastes from Europe to Africa. This information is then urgently transmitted to responsible institutions in Africa for necessary action. For example, a number of attempts to export toxic wastes to Côte d'Ivoire, Ghana, the Democratic Republic of the Congo and Nigeria have been aborted through 'Dump Watch' and other environmental organizations (Dorm-Adzobu 1995). However, this is a long-term process, and while some of the countries have started implementation, others are still securing the necessary funds.

The region is increasingly adopting production standards that aim to reduce ozone depletion. Most African countries (44 out of 53) have ratified the Montreal Protocol. Egypt has replaced ozone-depleting substances with cyclopentane in all of its nine refrigerator manufacturing companies. The new refrigerators also have lower energy consumption (UNIDO 1997). Funding was from the Protocol's Multilateral Fund, with most of the Egyptian companies making substantial contributions by complementing the conversion process with improvements of their plants, buildings and products. Sudan has also banned the use of CFCs in aerosols and refrigerants, and has begun a training programme

GEF and Africa

Most African countries are benefiting from GEF-funded projects focusing on biodiversity, climate change and international waters; by June 1998, a total of US$410 million or 22 percent of all GEF funding for approved projects had been allocated to the region, and considerable co-financing generated.

Africa has been the top regional recipient of GEF biodiversity funding, having been allocated one-third of all biodiversity funding totaling US$250 million by mid-1998. East Africa was one of the earliest recipients of GEF support in biodiversity through the East African capacity-building project implemented in Kenya, Uganda and Tanzania at a cost of US$10 million. One-third of the GEF funds devoted to biodiversity projects in coastal, marine and freshwater ecosystems was committed to sub-Saharan Africa, and significant funds are also addressing land degradation issues, particularly in Africa.

Other examples of regional and national biodiversity projects cover arid and semi-arid zones including the management of plant genetic resources, protected areas management, the conservation and management of habitats and species, island biodiversity, coastal, marine and freshwater ecosystems, forest ecosystems, mountain ecosystems, capacity building and institutional support as well as emergency responses.

In the climate change area, Kenya is part of a US$120 million Photovoltaic Market Transformation Initiative being implemented with US$30 million of GEF funds. Other climate projects include support to Mauritania for a project on wind power, to Morocco for re-powering of power plants, to Côte d'Ivoire and Senegal for control of GHG emissions, to Benin and Sudan for rangeland rehabilitation for carbon sequestration and greenhouse gas reduction, and support for renewable energy (Ghana, Mauritania, Mauritius, Tanzania, Tunisia, Uganda, Zimbabwe).

International waters projects include waterbody management programmes, such as the GEF-supported Lake Victoria Environment Management project being implemented at a total cost of US$78 million, pollution control in Lake Tanganyika, integrated land and water programmes, and containment-based programmes such as the oil pollution management project for the South-west Mediterranean Sea.

In recognition of the importance of land and water degradation, an interagency initiative to develop a coordinated action programme to address these issues through GEF has recently been agreed (GEF 1999b).

Sources: GEF 1998, 1999a and 1999b

CITES and the ivory trade

One of the high profile areas of CITES operations has been the regulation of the ivory trade. Parties to CITES put the African elephant on Appendix II of CITES in 1977, and in 1985 ivory trade controls were tightened through the introduction of a quota system. Despite reductions in volume, illegal trade continued. In mid-1989, against a backdrop of growing international pressure, some importing countries began introducing domestic legislation to stop the importation of raw ivory. In 1990, the Parties to CITES placed the African elephant in Appendix I, banning all international trade in elephant products with effect from January 1990 in the belief that only a complete end to the ivory trade would stop the continued losses to poachers. Since then, a number of African countries have attempted to reverse the total ban in the belief that their elephant populations were no longer threatened with extinction.

In 1997, the Parties to CITES decided to permit some highly controlled exports of elephant ivory for the first time since 1989, and agreed to transfer the elephant populations of Botswana, Namibia and Zimbabwe from Appendix I to Appendix II. The first, experimental sale of ivory took place in April 1999, when a total of 52 tonnes from these three countries were sold to Japanese buyers under conditions strictly controlled and monitored by the CITES Secretariat. Funds obtained from this operation must be re-invested in conservation programmes.

Another development that has been approved by the Parties to CITES is the exchange of ivory for conservation funds. Donor countries will symbolically buy the ivory which will then be destroyed. Again, all funds generated in this way must be used to support conservation activities (CITES 1999).

The history of CITES in Cameroon illustrates the obstacles to implementation and effectiveness (Weiss and Jacobson 1998). While Cameroon signed CITES in 1973, the report on the signing was lost for several years, and thus ratification occurred only in 1981. CITES received low levels of attention from the government. More recently, nearly two tonnes of ivory seized in Hong Kong was traced to sources in Cameroon.

to inform the public about alternatives to these ozone-depleting substances (Earth Council 1997). South Africa, the only CFC producer in the region, phased out its production by 1996 in line with the requirements of the Montreal Protocol (UNEP Ozone Secretariat 1998).

The effectiveness of CITES is unclear, despite its critical importance to this species-rich region. Besides the usual problems of tight budgets, low administrative capacity and corruption, many countries have large territories and national borders with few checkpoints. Much 'international trade' occurs without official notice. In some countries, CITES has resulted in little change in national law or practices. In Cameroon, for instance, while there have been several changes to Cameroon's wildlife laws during the past two decades, almost none could be attributed to the influence of CITES (Weiss and Jacobson 1998). The situation regarding the ivory trade is described in the box above.

Regional MEAs

Most regional and sub-regional MEAs seek to reinforce global MEAs by filling in gaps, facilitating joint action and mutual understanding in environmental policy and management, and enabling environmental issues to be treated on a regional rather than national basis.

Environmental cooperation is also enshrined in broad-based agreements such as Articles 56–59 of the OAU treaty establishing the African Economic Community relating to natural resources, energy, environment and control of hazardous waste. Other players in regional environmental issues include the biennial African Ministerial Conference on the Environment.

The boxes on pages 225 and 226 list the most important regional MEAs and provide some details on compliance and implementation. The major impact of these agreements has been the formulation of new national and sub-regional policies and action plans. Through such agreements member states have been able to correct imbalances in the sharing of resources. For example, under the Nile 2000 initiative, riparian states of the River Nile are negotiating a new framework for the development and equitable sharing of the river's water resources.

Obstacles to implementation

There are many obstacles to the effective implementation of MEAs in Africa, including:

- an inadequate policy framework for implementation;
- constraints on financial resources;
- lack of qualified personnel in the different disciplines related to MEAs at the national level;
- inadequate participation by national stakeholders, as well as the general public, in the negotiation of MEAs;
- failure of some MEAs to reflect national environmental priorities;
- absence of in-depth understanding of the contents of MEAs.

The policy actions needed for effective implementation include:

- increasing awareness of decision makers and the general public of the contents, required follow-up, and implications of MEAs – awareness raising should include information and analysis of available policy alternatives, particularly policies encouraging the private sector to help the implementation process;

- capacity building of national institutions, including securing finance from different sources, training and technology development and transfer;
- instituting an intra-regional mechanism for implementing programmes related to the MEAs;
- ensuring the active involvement of African experts in MEA negotiations;
- translation of MEAs into different languages and ensuring proper understanding of their texts.

Action Plans

During the past decade, many environment agencies in African countries, including Madagascar, Kenya, Uganda, Burkina Faso, Congo and Togo, have prepared new National Conservation Strategies, Forestry Action Plans, National Environmental Action Plans, and Plans of Action to Combat Desertification, often with the support of international organizations such as the FAO, IUCN, UNDP, UNEP, UNSO and the World Bank.

However, these new plans often run parallel to overall national development plans and are not linked or integrated with other economic and sectoral plans. Many of them therefore lack the full involvement and support of key ministries whose cooperation is needed to ensure effective implementation (UNEP 1993). Some have other inherent weaknesses, with recommendations that are too general and lacking such strategic details as the assignment of a specific implementing agency, time targets, detailed cost estimates and funding arrangements. Most also focus exclusively on national issues, without taking sufficiently into account the transboundary implications of the proposed actions.

Laws and institutions

As part of their follow-up to the 1992 Earth Summit, some countries – for example Benin, Ethiopia, Eritrea, Ghana, Malawi, Mali, Mozambique, Seychelles and Uganda – took the significant step of incorporating the environmental rights and responsibilities of their citizens in their constitutions. At the sub-regional level, the 1992 SADC Treaty included the following key objectives: 'to achieve development and economic growth, alleviate poverty, enhance the standard and quality of life of the peoples of southern Africa' and 'to achieve sustainable utilization of natural resources and effective protection of the environment' (SADC 1996). At the regional level, the OAU recognized a safe environment as a right by declaring that 'All people shall have the right to a generally satisfactory environment favourable to their development' (United Nations 1990).

Over the past two decades, many countries have put in place a wide range of new environmental laws and regulations. One unfortunate result of the commitment to strengthen environmental protection measures has been a significant fragmentation and duplication of authority and responsibilities. For example, in Zimbabwe 10 different ministries administer an estimated 20 environment-related laws while in Botswana 8 ministries are responsible for

Major regional MEAs

Treaty	*Place and date of adoption*
Convention on the African Migratory Locust	Kano 1962
Convention and Statute Relating to the Development of the Chad Basin	Fort-Lamy 1964
Phyto-Sanitary Convention for Africa	Kinshasa 1967
African Convention on the Conservation of Nature and Natural Resources	Algiers 1968
Convention Concerning the Status of the Senegal River	Nuakchott 1972
Convention Establishing a Permanent Inter-State Drought Control Committee for the Sahel	Ouagadougou 1973
Convention for the Protection of the Mediterranean Sea against Pollution	Barcelona 1976
Convention Creating the Niger Basin Authority	Faranah 1980
Convention for Cooperation in the Protection and Development of the Marine and Coastal Environment of the West and Central African Region	Abidjan 1981
Regional Convention for the Conservation of the Red Sea and Gulf of Aden Environment	Jeddah 1982
Convention for the Protection, Management and Development of the Marine and Coastal Environment of the Eastern African Region	Nairobi 1985
Protocol Concerning Protected Areas and Wild Fauna and Flora in the Eastern African Region	Nairobi 1985
Agreement on the Action Plan for the Environmentally Sound Management of the Common Zambezi River System	Harare 1987
Bamako Convention on the Ban of the Import into Africa and the Control of Transboundary Movement and Management of Hazardous Wastes Within Africa	Bamako 1991
Lusaka Agreement on Cooperative Enforcement Operations Directed at Illegal Trade in Wild Fauna and Flora	Lusaka 1994
Treaty Establishing the Lake Victoria Fishing Organization	Kisumu 1994
SADC Protocol on Shared Watercourse Systems	Johannesburg 1995

Implementation and compliance in selected regional MEAs

African Convention on the Conservation of Nature and Natural Resources
With 30 Parties, the Convention has resulted in the establishment of several national parks and natural reserves, providing economic returns through tourism. Zambia, for example, has 19 national parks covering 6.4 million ha (8.4 per cent of land area) and four bird sanctuaries. Game management areas cover an additional 16.6 million ha (22 per cent of land area), and the country also has extensive forest reserves (IUCN ROSA 1995). The Convention is reflected in national instruments such as the Trapping of Animals Control Act (1973) and the Parks and Wildlife Act (1975) in Zimbabwe.

Convention for the Protection of the Mediterranean against Pollution
All Mediterranean countries and the European Community are Parties to the Barcelona Convention which has established 123 protected areas.

Convention for Cooperation in the Protection and Development of the Marine and Coastal Environment of the West and Central African Region
Covers coasts from Mauritania to Namibia, a distance of some 8 000 km (Vernier 1995). Provides for the protection and preservation of rare and fragile ecosystems, for the development of guidelines for conducting environmental impact assessments, and for the formulation of procedures for the determination of liability and compensation (UNEP 1981). Ratified or acceded to by 10 countries.

Regional Convention for the Conservation of the Red Sea and Gulf of Aden Environment
Djibouti, Egypt, Somalia and Sudan are the African Parties (see page 315).

Convention for the Protection, Management and Development of the Marine and Coastal Environment of the Eastern African Region
Although adopted in 1985 with two Protocols (one on protected areas. fauna and flora, the other on combating marine pollution) and an Action Plan, ratification has been slow. The convention, which has been ratified or acceded to by Comoros, France (Réunion), Kenya, Seychelles, Somalia and Tanzania, is currently being revised to take into account developments since it was adopted.

Bamako Convention on the Ban of the Import into Africa and the Control of Transboundary Movement and Management of Hazardous Wastes Within Africa
Complementary to the Basel Convention, it caters for the special conditions of Africa not fully incorporated into that Convention. Places obligations on states to control and prevent transboundary movement or importation of hazardous wastes, and take precautionary measures against recurrence. Both conventions were conceived when three African countries became victims of dumping of toxic waste with disastrous consequences. Fewer than half of African nations have signed the Convention, which entered into force on 22 April 1998. Some countries may be stalling ratification in order to continue lucrative trade in hazardous wastes.

Lusaka Agreement on Cooperative Enforcement Operations Directed at Illegal Trade in Wild Fauna and Flora
Aims to reduce and ultimately eliminate illegal trade in wild fauna and flora. Six countries have ratified the agreement, which entered into force on 10 December 1996. The agreement works closely with CITES but is empowered to investigate violations of national laws.

Treaty Establishing the Lake Victoria Fisheries Organization
Open to all riparian states of Lake Victoria since 1994 but so far signed by Kenya, Uganda and Tanzania, and ratified only by Tanzania.

The Southern African Development Community (SADC) Protocol on Shared Watercourse Systems
Calls for equity and shared responsibility among riparian states in the utilization and management of watercourse systems. Member states are obliged to strive for a higher standard of living for their peoples, and conservation and enhancement of the environment to promote sustainable development. Signed by eight of the 12-member SADC, with Angola and Zambia requesting more time: no country has yet ratified.

Source: UNEP 1997

applying 33 environmental laws (SADC 1998). The situation is further complicated in countries such as South Africa where responsibilities are shared by central and regional authorities (DEAT 1996). As a consequence, some countries are now reviewing their legal framework with the aim of consolidating, streamlining and strengthening their environmental laws. Priority requirements include updating and bringing laws in line with current scientific knowledge, reducing overlap and conflicts, setting more realistic and higher penalties to encourage compliance, clarifying and harmonizing the responsibilities of different national ministries, and identifying and filling gaps in the legal framework for environmental protection (UNEP 1993).

The first environment Ministry in sub-Saharan Africa was created in 1975 in the former Zaire. By the early 1990s, most countries had established a wide variety of new institutional arrangements for environment protection and improvement. For example, many countries in Northern and Southern Africa created new environment ministries while most East African countries favoured separate environmental coordination bodies such as the National Environment Management Authority in Uganda and the National Environment Secretariat in Kenya. Countries in West and Central Africa have a mixture of environment ministries and coordination bodies. Environmental units were also combined with existing ministries. For example, in Cameroon the environment is combined with forestry, in Burkina Faso with water resources, in Ghana with science and technology, and in Zimbabwe with tourism.

Special institutional arrangements at the subregional level include the Environment Unit in the Secretariat of the Economic Community of West African States (ECOWAS), the Inter-Governmental Authority on Development (IGAD), and the

Environment and Land Management Sector Coordination Unit and the recently created SADC Water Sector Coordination Unit based in Lesotho.

At the regional level, the UNECA and OAU have long had separate natural resources departments. The African Development Bank has had a special environment unit since the early 1990s, and recently restructured its programme to put greater emphasis on the environment.

While providing new prominence and focus for environmental issues, these agencies and units still face overwhelming tasks and responsibilities. They have to compete for staff and budgets with the older and often more powerful sectoral agencies, whose activities often have more impact on the state of the environment and natural resources than those of the environmental agencies themselves. Since the 1980s, the adverse environmental impacts of the policies of key economic and sectoral agencies have continued to undermine public health as well as the resource base needed for future economic development. As repeatedly emphasized at and after the 1992 Earth Summit, a move toward sustainable development will require major changes in many present policies, programmes, laws and institutional arrangements, outside as well as within the environment field. In particular, it will require moving away from the tendency to sectoralize the environment by breaking through the many administrative and other constraints on inter-Ministerial coordination and cooperation. It will also require new approaches and methods for national development planning and the integration of the environment in decision-making throughout the public and private sectors.

Agenda 21 includes many recommendations calling for the wider use and integration of Environmental Impact Assessments (EIAs) in all major economic and sectoral policies and programmes. Since Rio, at least one-third of African countries have introduced new EIA policies, laws and procedures. However, although representing a significant step toward greater environmental sustainability, EIAs alone are not enough to secure sustainable development.

National and regional policies for securing economic, social and environmental sustainability in Africa need to be anchored and reinforced by incorporating impact assessments as an integral part of decision-making in at least three key respects: assessing the likely environmental impacts of economic policies and activities; assessing the likely economic impacts of environmental policies and measures; and assessing the likely equity impacts of both economic and environmental policies (SADC 1996).

As emphasized in the 1996 SADC report, 'The integration in all key policy sectors of simultaneous economic, environmental and equity impact assessments (EIA^3) will certainly not make decision-making easier though. EIA^3 will inevitably increase rather than reduce the number and complexity of the trade-offs involved in most major decisions. EIA^3 will, however, significantly improve the chances of making better decisions in support of sustainable development. By identifying and making those trade-offs more explicit, and preferably more public as well, EIA^3 will increasingly compel decision-makers to assess and defend their policy choices in terms of economic, social and environmental sustainability' (SADC 1996).

Land issues

Land degradation continues to be the main environmental challenge facing many African countries, especially in terms of deforestation, loss of biodiversity and declining soil fertility. Current national policies do not adequately respond to these challenges (World Bank 1995a), mainly because responsibilities and laws for protecting the soil, forests and biodiversity are fragmented among different government agencies. Those who have the most to lose and are the first to be affected by land degradation – the local people and communities – have seldom been adequately consulted or involved in the planning and implementation of the many different conservation and remedial programmes. Most of these people are poor and the existing agricultural, energy, forestry and wildlife policies do not provide them with the support and range of choices they need.

Land-use and access regimes are often complex; in Malawi, for example, the process of leasing land involves 33 steps (Okoth-Ogendo 1998). The regimes often do not allow a simple and direct integration of sound land-use principles with acceptable tenancy agreements and enforceable property rights.

Land policies (or lack thereof) and land laws are often based on feudal regimes originating in 19th century England. Their relevance to African cultures and property regimes has always been problematic and incommensurate. Most African governments have not responded to the need for participatory, appropriate and effective policies and legislation on land and natural resources.

However, some people are searching for a systematic articulation of visions, plans and institutional mechanisms to clarify questions of land ownership and develop cohesive regulations for just, sustainable and equitable use of land and land-based resources. Some governments, including those of Eritrea, Malawi, Mozambique, South Africa, Tanzania and Uganda, are responding through reviews, sectoral policies, Land Commissions and participatory or non-participatory task forces (Okoth-Ogendo 1998). The Uganda Land Act of 1998 provides for the tenure, ownership and management of land, giving security of tenure to people (such as squatters) who were originally at the mercy of the title holder (Land Act of Uganda 1998). This Act reinforces the provisions in the 1995 Constitution of Uganda which vests control of land in the citizens of Uganda, with the previously powerful Uganda Land Commission now only holding and managing land on behalf of the Government, having relinquished administrative issues regarding land to District Land Boards.

A further chronic problem for the environmental institutions is the lack of adequate staff, expertise, funds and equipment to implement and enforce many existing national laws and international conventions. Expanded capacity-building programmes and a stronger commitment by governments are needed in many countries. The largely regulatory approach to environmental protection and management in most countries needs to be supplemented by a wider use of economic instruments and legal incentives (UNEP 1993), with progressive empowerment of rural communities over natural resources on the land on which their lives depend (see also box on land issues on page 227).

Economic instruments

Confronted by tough challenges to accelerate economic growth and reduce poverty, African countries are constantly under pressure to pursue short-term growth policies, which shift ecological and economic costs to the next generation. But escalating environmental degradation and resource depletion are already recognized as both a consequence and a cause of poverty. Moreover, the costs of environmental neglect are substantial. In Nigeria, for example, the long-term cost of not acting to prevent environmental degradation has been estimated at more than US$5 100 million a year, more than 15 per cent of GDP (World Bank 1990a). Some of the degradation is irreversible. Groundwater polluted by industrial and agricultural chemicals cannot readily be cleaned. Topsoil washed or blown away in a few years takes centuries to replace. Extinct plant and animal species are lost forever, and with them disappear the many potential health, economic and other benefits they represent.

To avoid these escalating environmental and economic costs, some governments are starting to consider a much broader range and mix of regulatory measures and economic instruments to facilitate and accelerate their transition toward development that is economically, socially and ecologically sustainable. These economic instruments and incentives include changes in tax policies to allow accelerated depreciation allowances, tax write-offs and, as in Zambia, reduced or no import duties for pollution control equipment and environmentally-sound technologies (World Bank 1995a).

'Green' taxes, providing incentives for producers and consumers to act in ways that are environmentally friendly, are relatively new and many current tax structures and environment management systems are not compatible with such taxes (World Bank 1995a). Indirect taxes such as value-added tax and excise duty are the most likely tools for the future, for example as a result of regional economic integration through the SADC.

Under its Industrial Expansion Act of 1993 (GOM 1993), Mauritius provides manufacturing enterprises with duty exemptions, tax credits and other incentives on the importation of pollution abatement and environmental protection facilities, new machinery and equipment in order to promote economic, industrial and technological development. Disincentives, based on the polluter pays principle, in the form of fines and the recovery of costs for the clean-up of pollution are also provided for in the Environmental Protection Act (GOM 1991).

Policy changes that prevent environmental degradation can reduce government expenditure, for example by abolishing agricultural and other subsidies which encourage deforestation or the exploitation of marginal lands or, as in Egypt, reducing the hidden subsidies when user charges for public services such as water fall below actual costs. Other measures such as effluent charges, user charges, product charges and administrative fees can become a source of much-needed revenue which can then be used to finance additional environmental restoration and protection

The cost of inaction: Nigeria

'The conventional constraint on government and private sector action in our own and other countries has been concern about the costs of taking new environmental protection measures. This narrow preoccupation has overshadowed the equally important consideration of the mounting economic, social and ecological costs of not acting.

A recent World Bank study (World Bank 1990a) provides a stark assessment of the risks and enormous costs if no remedial action is taken on at least eight of our priority environmental problems.

In the absence of near-term remedial and mitigative measures, the long-term losses to Nigeria from environmental degradation in eight priority areas have been estimated [see table].

In sum, the long-term losses to our country of not acting on our growing environmental problems are estimated to be around US$5 000 million annually.'

Annual costs of inaction (US$million/year)

Soil degradation	3 000
Water contamination	1 000
Deforestation	750
Coastal erosion	150
Gully erosion	100
Fishery losses	50
Water hyacinth	50
Wildlife losses	10
Total	**5 110**

Source: FEPA 1991

measures. The use of tradeable permits is being examined by Ghana as a way of reducing and controlling industrial pollution in the Korle and Chemu lagoons (World Bank 1995a).

The Ministries of Finance and National Statistical Offices in some countries are considering how to adapt their system of national accounts to reflect better the extent to which economic development affects the quantity or quality of the natural resource base on which future development depends. Botswana and Namibia, for example, have developed natural resource accounts to enable a better assessment of their economic value and management options (Markandya and Perrings 1991). Madagascar has also made a similar first attempt, focusing on water and forest resources (ONE 1997). Other innovative approaches include special environment protection funds based on contributions by government, industry and other private sources, such as those already created in Benin, Côte d'Ivoire, Niger and Seychelles.

Industry and new technologies

Industries in Africa have only limited access to environmentally sound technology and historically have aimed at maximizing production, with little or no concern for the environment. Although National Cleaner Production Centres have been established in Tanzania, Tunisia and Zimbabwe (UNEP 1998), many industries are still unaware of the potential benefits of cleaner production and relevant legislation is usually non-existent. Countries are often more concerned with attracting foreign investment which could be deterred by making production standards too restrictive.

Some multinational corporations, local companies and large-scale mining companies are beginning to adopt precautionary environmental standards. For example, ISO 9000 production standards are now used in Mauritius, South Africa and Zimbabwe. Although the ISO and other standards are still voluntary, more companies are using them as a guide when targeting foreign markets which increasingly require environmentally-safe products. Mauritius is planning the introduction of industrial waste audit regulations to encourage industries to self-regulate and adopt cleaner technologies, as a precursor to the eventual adoption of ISO 14 000.

In some countries, cleaner production is being promoted through incentives and support services. Mauritius, for example, is using award schemes for environmental excellence, local accreditation and certifying bodies, as well as providing soft loans for investments in domestic solar heaters. The country is also participating in a UNEP pilot project to implement cleaner production with the industries involved – sugar production, food processing, knitwear manufacture and tourism.

Financing environmental action

The picture with regard to the trend in financing environmental action in Africa is fragmented and incomplete. *Agenda 21* recognized that developing countries, especially the least developed, would need substantial funds to implement sustainable development, and that ODA should be a main source of funds for these countries. Other sources include the public and private sectors, including debt reduction, and innovative financial mechanisms such as economic instruments, joint implementation programmes, national and international environment funds

For developing countries, particularly those in Africa, ODA is a main source of external funding, and essential for the effective implementation of *Agenda 21*, which cannot generally be replaced by private capital flows (Osborne and Bigg 1998). However, throughout the 1990s ODA has been declining both in real terms and as a percentage of GNP, and the outlook for a recovery remains poor. Nevertheless, sub-Saharan Africa's share of total aid has increased slightly from 37 per cent in 1990 to 39 per cent in 1998 (World Bank 1999).

There have been changes in the composition of ODA, with a sectoral reallocation of aid towards those with a greater share of public goods, and increasing aid to countries with sound policies (World Bank 1999). According to World Bank ratings, average policy performance has improved significantly in sub-Saharan Africa, and a number of countries – including Ethiopia, Mali and Uganda – have adopted a sound policy environment (World Bank 1999).

The data available, which are fraught with shortcomings, indicate that the shares of bilateral ODA commitments for the conservation and management of resources have risen over the period 1990 to 1996 in total and percentage terms, from about US$5 300 million to US$6 500 million, or from 18 per cent to 25 per cent of the total. Multilateral ODA commitments in this sector have fallen from

25 per cent to 16 per cent of the total over the same period (CSD 1998).

Although there have been unexpectedly large increases in private capital flows to developing countries, including foreign direct investment (FDI), these have been mostly concentrated in middle-income countries in Asia and Latin America. African countries are unlikely to have benefited significantly because of their high level of indebtedness, which has a negative effect on domestic investment, including the investment necessary to attract private capital (CSD 1997).

Some progress has been made towards alleviating the debt burden in Africa, under the Heavily Indebted Poor Countries (HIPC) Debt Initiative, which focuses resources on countries with a solid track record of performance. By January 1999, 10 of the 12 countries reviewed for eligibility under the HIPC Initiative were in Africa. Debt-relief packages, with an estimated total nominal debt service relief of US$4 800 million, have been agreed on for five of these countries (Burkina Faso, Côte d'Ivoire, Mali, Mozambique and Uganda), while packages with an estimated total nominal debt relief of a further US$2 450 million are being considered for Guinea-Bissau, Ethiopia and Mauritania. Six other African countries remain to be reviewed for eligibility under the HIPC framework – Chad, Guinea, Niger, Togo, Tanzania and Zambia (World Bank 1999). Nevertheless, further efforts to reduce indebtedness are still needed in sub-Saharan Africa.

With regard to financing by African countries from their own public resources, there has been an increase in national institutions established to deal with environmental issues in most African countries, demonstrating increased political and financial commitment to this sector. Public sector financing of environmental activities by African countries, including in-cash and in-kind counterpart funding of projects, is therefore likely to have increased significantly since UNCED, although also still likely to be inadequate.

Environment funds are innovative financing mechanisms that can pool revenues from various types of resources (such as earmarked taxes and charges, concessional grants or loans, debt-for-nature swaps, and interest on endowment funds) to provide long-term funding for environmental programmes. The success of environment funds is reflected in their growing number (CSD 1997). Environment funds based on contributions by government, GEF, industry and other private sector sources have been created in Benin, Congo, Côte d'Ivoire, Madagascar, Niger, Seychelles and Uganda, among others (World Bank 1990b and 1995b).

A few African countries have used debt-for-nature swaps to generate new additional financial resources for environmental conservation and management. This mechanism converts part of a country's external debt into a domestic obligation to support environmental activities and programmes. Madagascar, Sudan and Zambia have all used debt-for-nature swaps, and the funds generated have been used to support environmental education, sustainable development, conservation and ecosystem management of protected areas, and inventories of endangered species. They have also been used to establish endowment funds, with interest paid used to finance conservation activities.

International financial mechanisms such as the Global Environment Facility (GEF) and the Multilateral Fund for the Montreal Protocol have transferred financial resources to developing countries for investments related to protection of the global environment. By mid-1998, Africa had been allocated about 22 per cent of GEF funds, totalling US$419 million, including 33 per cent of funding for biodiversity, 38 per cent of funding for international waters, and 11 per cent of funding for climate change activities (GEF 1998).

There has been increased interest by the World Bank in the environmental and social effects of its projects in developing countries. The Bank's portfolio of environmental projects in Africa increased, from US$125 million approved in fiscal year (FY) 1990 to about US$282 million approved in FY1997 (including GEF funds). These projects generate considerable co-financing from a variety of sources including governments, bilateral and multilateral donors, NGOs and the private sector. There were 56 World Bank financed environment projects active in Africa in FY98, with a total cost of US$898 million (World Bank 1998).

Some progress has been made in increased financing for environmental action in Africa, although it is not easy to determine the magnitude from the information available. However, much more needs to be done by increasing public and private capital flows to this sector. With regard to external sources of finance, unresolved issues of particular relevance to African countries include falling ODA and the unfulfilled UNCED commitments on increasing ODA to the level considered necessary to implement *Agenda 21*, and reducing the debt burden. At the domestic level a wider range of instruments and mechanisms, reforms in public expenditure, greater private-sector participation and more innovative mechanisms need to be considered and further developed (CSD 1997).

Public participation

Public participation encourages people to take more responsibility for their actions and governments to address environmental issues more explicitly and more effectively. Public participation in decision-making is on the increase in Africa but this has yet to increase substantially the access of women and youth to decision-making processes. (SARDC/IUCN/SADC, 1994). Public participation in the state of the environment reporting process in such countries as Lesotho, Malawi, South Africa and Zimbabwe is a practical example of how all stakeholders can become involved in decision-making. Public consultation is now also common on new environmental legislation such as the Green Paper on the environment in South Africa (DEAT 1996) and the Environmental Management Bill in Zimbabwe.

Local people often know the causes and best remedies for such problems as deforestation or soil erosion, how to find and use plants with unique properties and how to prevent animals from damaging their crops. Public participation enables this knowledge and these skills and resources to be mobilized and increases the effectiveness of government initiatives. Equally, when people are allowed to take part in assessing problems, resources and opportunities, they acquire information and enhance their awareness of factors affecting their lives (FAO 1994).

Indigenous knowledge has been recognized and used for the benefit of wildlife management in Southern Africa through the Regional Network on Indigenous Knowledge Systems (SARNIKS). One example is the CAMPFIRE programme in Zimbabwe, which has enabled local communities to participate in the management of wildlife resources. While the programme initially targeted wildlife use, particularly large mammals, other natural resources such as forests are now also included. Some countries have adapted this concept to their own situation. Similar programmes include ADMADE in Zambia and the Community Based Natural Resource Management Programme in Namibia (SARDC, IUCN and SADC 1994).

There are also examples of weak community participation. In the Game Management Areas (GMA) in Zambia, the voice of the local people in relation to wildlife management appears to be weak, partly as a result of ignorance about the nature of conservation and partly because of the inability of institutions to define local needs effectively and translate them into development policy. Although local people are legally represented by their local leaders, among whom are chiefs and district councils, such institutions are often blamed for misrepresenting the views and interests of local people (Chenje 1997).

Environmental information and education

Many governments involved representatives of the private sector, the academic community, NGOs and community groups in their preparations for and follow-up to the 1992 Earth Summit. This led to a growing recognition that national development plans and environmental policies had a better chance of being implemented effectively when supported by an informed and involved public.

Environmental awareness and education programmes have expanded throughout the region (see, for example, the box below). Most countries now include environmental education in school curricula. In Kenya, the aims include 'to create new behavioural

Umgeni Valley Environmental Education Centre

Umgeni Valley is the name of the SADC Environmental Education Centre in South Africa. The centre is popular with environmental conservation clubs and has produced a number of environmental education tools including:

- A hands-on series of simple field guides, suitable for those with little environmental knowledge. They can be used both in the field and in the classroom.
- The beginners' guide series which provide a simple introduction to selected plants and animals.
- Water test kits, including an Action Starter Kit for the investigation of water quality and the Coliform Kit for assessing the coliform bacteria count in freshwater.
- Teachers' handbooks.
- Information fact books about different aspects of the environment.
- An action series on how to take action to solve environmental problems which focuses on remedial action such as how to propagate indigenous trees and how to eradicate invader plants.

Source: Share-Net Resources 1996

patterns of individuals, groups and communities towards the environment' and 'to provide every person with an opportunity to acquire knowledge and develop values, attitudes, commitments and skills needed to manage the environment' (Kenya 1994).

Non-formal environmental awareness and educational programmes are being promoted at the national and regional levels through special conservation demonstration projects, newsletters, posters, radio and television programmes, seminars and workshops. In Niger, many tree planting, soil conservation and restoration projects have been implemented on a voluntary basis by students and community groups (Niger 1998).

Regional centres and networks such as SARDC in Southern Africa, ACTS in Eastern Africa, CEDARE in North Africa and NESDA in West and Central Africa have all been strengthened to provide specialized services and expertise in environmental management. However, most of these are vulnerable because of their dependence on government and donor funding; their inability to compete with the higher salaries offered by the private sector and international organizations is leading to a brain drain within and outside Africa. The linkages among researchers in and among most countries are also weak and lead to unnecessary duplication. Networking needs to be expanded to make more effective, efficient and economic use of existing environmental expertise.

National and regional databases have been established on climate (IPCC), soils (FAO) and biodiversity (WCMC) through the efforts of the United Nations and other international and regional organizations. Better information-exchange networks are gradually evolving but are still constrained by problems of access to modern communication systems. Language barriers also frustrate networking and information exchange. The national reports for the 1992 Earth Summit provided a basis for more national state-of-the-environment reports which in turn provided new information for sub-regional assessments such as the *State of the Environment in Southern Africa* (SARDC, IUCN and SADC 1994).

African countries are addressing the issues of availability of and accessibility to environmental information through the creation of specialized environmental information units within government institutions, and of environmental information systems and networks at national and sub-regional levels, as well as participation in similar initiatives at regional and global levels. These units or information centres, such as Uganda's National Environment Information Centre (NEIC), produce and disseminate a range of information products in the form of statistics, issue-based, sectoral and State-of-the Environment Reports and public information products, both as documents and increasingly electronically through the Internet. Several African countries have begun adopting certain principles of the Århus Convention on Access to Information, Public Participation in Decision-making and Access to Justice in Environmental Matters.

Social policies

Poverty alleviation is the overriding goal and priority. Improving the health, income, education and living conditions of the poor majority remains the top policy imperative for ensuring the political stability and social sustainability needed to move toward greater economic and environmental sustainability. While adopting the innovative approach of *Agenda 21* for integrating environment and development, a third crucial element must be added to make *Agenda 21* more applicable and operational in most African sub-regions. This critical missing link is equity. Greater equity is needed throughout the region in the distribution of the opportunities and benefits of national economic development and international aid programmes. At present, too few national and international aid programmes reach or benefit the poor majority (SARDC, IUCN and SADC 1994).

SADC has developed a new policy and strategy for environmental and natural resources management for equitable and sustainable development. Its main aim is to support economically, socially and environmentally sustainable forms of development (SADC 1994). Although environmental sustainability is the starting point of the new policy, the equally important economic and social sustainability concerns are given high priority.

Conclusion

In 1983, a future perspectives study by the UNECA presented a number of forecasts for the year 2008 if no major changes were made to prevailing policies (UNECA 1983). The study projected figures on issues such as population and employment, urban poverty, food security, industrial growth, and donor aid dependency. The picture drawn was bleak (see box).

The future seen from 1983

In 1983, a future perspectives study by the UNECA suggested that, without policy changes, by the year 2008 Africa would have:

- a population of 1 100 million people, a workforce of 510 million and at least 44 million unemployed;
- 220 million people without adequate shelter, a poor urban population of 472 million and a growing gap between the rich and poor;
- serious food shortages and only marginal increases in the already low levels of per capita food consumption;
- industrial stagnation;
- an increased need for more foreign aid to avoid even greater poverty throughout the region.

Source: UNECA 1983

Africa's major problems have changed little since this study was made more than 15 years ago. High population growth and rising unemployment are still major concerns in many African countries, economic growth rates have not increased significantly, the number and proportion of poor people have increased, per capita food consumption has declined in two of the sub-regions (see page 56), industrial production has not increased significantly, the prospects for more foreign aid to tackle these issues has diminished, and environmental degradation and resource depletion have escalated.

However, there are encouraging signs. The governments and people of Africa are increasingly setting their own agenda for change as a result of more democratic rule and improved governance in several countries and greater cooperation at the sub-regional, regional and global levels. The new agenda is encouraging more anticipatory and preventive approaches to environmental problems.

Putting the poor at the centre of the new sustainable development agenda should give the new policies and plans a better chance of being economically, socially, ecologically and politically sustainable (Munro 1997).

Policy change should be reinforced and informed through regular assessments and reports on progress towards sustainable development. New analytical tools and methods are also needed such as EIA[3] measures, sustainable development audits, natural resource accounting and innovative macro-economic indicators such as sustainable net national product (UNEP 1993). All key ministries should report annually on the extent to which their activities are contributing to the degradation, protection or improvement of the environment and natural resource base. These reports should form an integral part of budget submissions and development plans. If any of their activities have had an adverse environmental impact, the plans should include specific proposals for correcting and avoiding such impacts in the future (SADC 1998).

Regular policy impact audits and reports will help achieve equitable and sustainable development. They will build into government policy and decision-making a timely mechanism for identifying and correcting remnants of unsustainable development. They will also provide a regular check for assessing progress made and needed (UNEP 1993). But most importantly they will reinforce the leadership role of governments as the pacesetters for the transition to sustainable development and rightfully earn them the votes of the present generation and the gratitude of future generations.

References

Achoka, A., Kapiyo and Karinge (1996). *OHAI, a model for sustainable livelihood and natural resources management in Africa*. KENGO, Nairobi, Kenya

AMCEN (1992). *The Cairo Programme for African Cooperation*. AMCEN Secretariat, UNEP Regional Office for Africa, Nairobi, Kenya

Cameroon (1996). *The National Environment Management Plan (Volume II)*. Ministry of Environment and Forestry, Yaoundé, Cameroon

CDIAC (1998). *Revised Regional CO_2 Emissions from Fossil-Fuel Burning, Cement Manufacture, and Gas Flaring: 1751–1995*. Carbon Dioxide Information Analysis Center, Environmental Sciences Division, Oak Ridge, Tennessee, United States.
http://cdiac.esd.ornl.gov/cdiac/home.html

Chenje, M. (1997). Community Participation Essential in Management of Environment. In *The Sunday Mail Magazine*, 30 March 1997, Harare, Zimbabwe

CITES (1999). www.wcmc.org.uk/cites/

CSD (1997). *Overall progress achieved since the United Nations Conference on Environment and Development*. Report of the Secretary-General, Addendum, Financial resources and mechanisms (Chapter 33 of Agenda 21). E/CN.17/2/Add.23, 22 January 1997. Commission on Sustainable Development, Fifth Session, 7-25 April 1997

CSD (1998). *Financial Flow Statistics*. Commission on Sustainable Development Background Paper No. 17. Sixth Session, 20 April-1 May 1998
http://www.un.org/esa/sustdev/finsd4.htm

DEAT (1996). *Green Paper: An Environmental Policy for South Africa*. Department of Environmental Affairs and Tourism, Pretoria, South Africa

Dorm-Adzobu, C. (1995). *New Roots: institutionalizing environmental management in Africa*. World Resources Institute, Washington DC, United States

Earth Council (1997). *Experiences and Recommendations from National and Regional Consultations for the Rio+5 Forum*. Earth Council, for the Fifth Session of the UN CSD, April 1997

FAO (1994). *Enhancing People's Participation*. Briefing Note for National Forestry Action Programmes, by G. Borrini. FAO, Rome, Italy

FEPA (1991). *Achieving Sustainable Development in Nigeria*. National Report for the 1992 United Nations Conference on Environment and Development. Federal Environment Protection Agency, Lagos, Nigeria

GEF (1998). *Project Implementation Review of the Global Environment Facility - 1997*. GEF, Washington DC, United States
http://www.gefweb.com/monitor/introme.htm#eval

GEF (1999a). Statement to African environmental leaders by Mohamed T. El-Ashry, Chief Executive Officer and Chairman, Global Environment Facility, Nairobi, Kenya, 4 February 1999
http://www.gefweb.com/GEFCEO/afrenvld.htm

GEF (1999b). *Conclusions of the GEF Heads of Agencies Meeting*. 11 March 1999. GEF, Washington DC, United States

Ghana (1991). *National Environmental Action Plan (Volume I)*. Environmental Protection Council, Accra, Ghana

GOM (1991). *Environmental Protection Act. Act No. 34 of 1991*. Ministry of Environment and Quality of Life, Port Louis, Mauritius

GOM (1993). *Industrial Expansion Act*. Ministry of Industry and Industrial Technology, Port Louis, Mauritius

IUCN ROSA (1995). *GEF Programme Preparation - Zambia: A Review of Global Environmental Conventions and the National Environmental Action Plan with Comments on Policy, Institutional and Capacity Issues*. IUCN ROSA, Harare, Zimbabwe

Kenya (1994). *National Environmental Action Plan*. Government of Kenya, Nairobi

Land Act of Uganda (1998). *Land Act No. 16 of 1988, Supplement 11*. UPCC, Entebbe, Uganda

Markandya, A. and Perrings, C. (1991). *Resource Accounts for Sustainable Development: A Review of Basic Concepts, Recent Debate and Future Needs*. LEEC Paper DP91-06. International Institute of Environment and Development, London, United Kingdom

Maro, P. S. (1995). *Report on the SADC Sub-Regional Planning and Programming Workshop on the Follow up of the International Convention to Combat Desertification and its Urgent Action for Africa*. SADC ELMS, Maseru, Lesotho

Munro, R. D. (1997). *Equity-led Growth for Sustainable Development in the SADC Region: Integrating EIA³ in Development Policies and Decisions*. Policy Paper for the SADC Regional Workshop on Integrating Economic, Environmental, and Equity Impact Assessments in Decision-Making, Harare, Zimbabwe, October 1997. SADC/ELMS, Maseru, Lesotho

Niger (1998). *Plan National de l'Environnement pour un Développement Durable*. Avant-projet, mars 1998. Niamey, Niger

Okoth-Ogendo, H. W. O. (1998). *A Comparative Analysis of Drivers, Processes and Outcomes*, unpublished

Osborne, D., and Bigg, T. (1998). *Earth Summit II: Outcomes and Analysis*. Earthscan, London, United Kingdom

ONE (1997). *Une Première Approche de la Comptabilité de l'Environnement à Madagascar*. IDA 2125/MAG, juillet 1997. Office National Pour l'Environnement, Cellule Système d'Information sur l'Environnement, Madagascar

SADC (1994). *Proposed SADC Policy and Strategy for Environment and Sustainable Development: Towards Equity-Led Growth and Sustainable Development in Southern Africa*. SADC Environment and Land Management Unit, Maseru, Lesotho

SADC (1996). *SADC Policy and Strategy for Environment and Sustainable Development: Toward Equity-Led Growth and Sustainable Development in Southern Africa*. SADC Environment and Land Management Unit, Maseru, Lesotho

SADC (1998). *Report of the SADC Regional Workshop on Integrating Economic, Environmental, and Equity Impact Assessments in Decision-Making*, Harare, Zimbabwe, October 1997. SADC Environment and Land Management Unit, Maseru, Lesotho

SADC (1999). http://www.sadc-usa.net/home.html

SADCC (1991). *Sustaining Our Common Future: Report to the 1992 Earth Summit*. SADC Environment and Land Management Unit, Maseru, Lesotho

SARDC, IUCN and SADC (1994). *State of the Environment in Southern Africa*. SADC Environment and Land Management Unit, Maseru, Lesotho, and IUCN ROSA, Harare, Zimbabwe

Share-Net Resources (1996). *Share Net Resources*. Share Net Resources, Howick, KwaZulu-Natal, South Africa

UNCHE (1971). *Development and Environment*, Report of the Expert Panel convened by the Secretary-General for the United Nations Conference on the Human Environment (UNCHE) in June 1971 at Founex, Switzerland. United Nations, Geneva, Switzerland

UNECA (1983). *ECA and Africa's Development, 1983–2008: A Preliminary Perspective Study*. UNECA, Addis Ababa, Ethiopia

UNECA (1991). *Cairo Common Position on the African Environment and Development Agenda*. Report of the African Regional Conference for the United Nations Conference on Environment and Development (UNCED). UNECA, Addis Ababa, Ethiopia

UNEP (1981). *Convention for Cooperation in the Protection and Development of the Marine and Coastal Environment of the West and Central African Region*. Protocol Concerning Cooperation in Combating Pollution in Cases of Emergency. United Nations Environment Programme, Nairobi, Kenya

UNEP (1991). *Regaining the Lost Decade: A Guide to Sustainable Development in Africa*. United Nations Environment Programme, Nairobi, Kenya

UNEP (1993). *Accelerating the Transition to Sustainable Development: Implications of Agenda 21 for West Africa*. United Nations Environment Programme, Nairobi, Kenya

UNEP (1997). *Register of International Treaties and Other Agreements: Relevance in the Field of the Environment - 1996*. United Nations Environment Programme, Nairobi, Kenya

UNEP (1998). *Cleaner Production: a guide to sources of information.* UNEP IE, Paris, France

UNEP Ozone Secretariat (1998). *Production and Consumption of Ozone Depleting Substances 1986–1996*. UNEP Ozone Secretariat, Nairobi, Kenya
http://www.unep.org/unep/secretar/ozone/pdf/Prod-Cons-Rep.pdf

UNIDO (1997). *Phasing out ODS in the Manufacture of Domestic Refrigerators in Egypt*. Success Stories from Africa: Special Session of the General Assembly to Review and Appraise the Implementation of Agenda 21. Rio +5 Forum. UNIDO, Vienna, Austria

United Nations (1990). *The African Charter on Human and People's Rights*, United Nations, Geneva, Switzerland

UNSO/UNDP (1995a). *Convention to Combat Desertification (CCD)*. Information Notes on Some of the Actions Being taken at Country Level. Information Note No. 2 (May). UNSO/UNDP, New York, United States

UNSO/UNDP (1995b). *Convention to Combat Desertification (CCD)*. Information Notes on Some of the Actions Being taken at Country Level. Information Note No. 3 (July). UNSO/UNDP, New York, United States

Vernier, N. (1995). A Framework for Integrated Coastal Zone Management in Sub-Saharan Africa. In *Building Blocks: Towards Environmentally Sustainable Development in Sub-Saharan Africa: A World Bank Perspective*, Paper No.4. World Bank, Washington DC, United States

Weiss, E.B., and Jacobsen, H.K. (1998). *Engaging Countries: Strengthening Compliance with International Environmental Accords.* MIT Press, Cambridge, Massachusetts, United States

World Bank (1990a). *Towards the Development of an Environmental Action Plan for Nigeria*, Report No. 9002-UNI. World Bank, Washington DC, United States

World Bank (1990b). *Taking Stock of National Environmental Strategies.* World Bank, Washington DC, United States

World Bank (1995a). *Towards Environmentally Sustainable Development in Africa.* World Bank, Washington DC, United States

World Bank (1995b). *Issues and Options in the Design of GEF Supported Trust Funds for Biodiversity Conservation*. K. Mikitin, Environment Department Papers, Biodiversity Series Paper No. 011. World Bank, Washington DC, United States

World Bank (1998). *Environmental Projects Portfolio Supplement in Environment Matters: Annual Review at the World Bank.* Fall 1998. World Bank, Washington DC, United States

World Bank (1999). *Global Development Finance 1999: Analysis and Summary Tables.* World Bank, Washington DC, United States

WWF (1997). *Conserving Africa's Elephants: Current Issues and Priorities for Action*. WWF, Gland, Switzerland

Asia and the Pacific

KEY FACTS

- Although the Convention to Combat Desertification is extremely important to the region, there is little interest from governments, civil society or NGOs in giving it priority.
- Due to coverage by local mass media, the Ganges Water Sharing Treaty is known even to poor farmers in the remotest parts of Bangladesh and West Bengal in India.
- China has some 2 900 environmental protection bureaus, more than 2 000 environmental monitoring stations and about 1 850 stations for monitoring and enforcing compliance. Nearly 100 000 people are directly employed in environmental protection.
- Economic instruments in Thailand have saved 295 MW of peak demand, 1 564 GWh of energy a year, reduced CO_2 emissions by more than 1 million tonnes and resulted in consumer savings of US$100 million a year.
- The Republic of Korea has sponsored potable water supply systems and wastewater system improvement in several countries and intends to expand environmental assistance in the future.
- Several Japanese companies have now taken voluntary actions on pollution control which include stricter standards than the national ones.
- NGOs have emerged as major players and partners in development and conservation activities, performing a multitude of roles including environmental education and awareness-raising among the public.
- The region's major strategy in attacking environmental problems should now be to combine command and control actions with economic incentives, with massive public consultation to gain widespread public acceptance for improved environmental policy actions.

The policy background

Many countries have made environmental conservation an integral feature of their development plans and have evolved the strategies, polices and legislation necessary for implementation. Policy relies mainly on command-and-control instruments, though economic incentives are becoming more widely used as economies mature and as the private sector demonstrates increasing sensitivity towards the environmental concerns of citizens. In most of Asia, business associations and private industry groups in high- and low-income countries have shown an increasing willingness to support environmental charters. In the Pacific sub-region, however, little progress has been made in bringing economic stakeholders into the dialogue on environment policy. Across the region, NGOs perform key roles in protecting the environment, especially through awareness-raising.

MEAs and non-binding agreements

Global MEAs

Global MEAs intended to promote sustainable development and reduce pollution have been signed by many countries but commitment to them varies widely: the CBD and the UNFCCC have been ratified by nearly 90 per cent of countries but the CMS by only 16 per cent (see diagram above right).

Parties to major environment conventions (as at 1 March 1999)

	CBD (174)	CITES (145)	CMS (56)	Basel (121)	Ozone (168)	UNFCCC (176)	CCD (144)	Ramsar (114)	Heritage (156)	UNCLOS (130)
ASIA AND PACIFIC (40)	36	25	5	21	33	37	25	18	26	28
South Asia (9)	8	7	3	7	7	8	7	6	8	4
Southeast Asia (5)	4	5	1	4	5	4	2	3	3	5
Greater Mekong (5)	4	4	0	2	4	5	4	2	5	3
Northwest Pacific and East Asia (5)	5	4	0	4	5	5	3	4	5	4
Australasia and the Pacific (16)	15	5	1	4	12	15	9	3	5	12

Key: percentage of countries party to a convention

0–25% 25–50% 50–75% 75–100%

Notes:

1. Numbers in brackets below the abbreviated names of the conventions are the total number of parties to that convention
2. Numbers in brackets after name of regions/sub-regions are the number of sovereign countries in each region/sub-region
3. Only sovereign countries are counted. Territories of other countries and groups of countries are not considered in this table
4. The absolute number of countries that are parties to each convention in each region/sub-region are shown in the coloured boxes
5. Parties to a convention are states that have ratified, acceded or accepted the convention. A signatory is not considered a party to a convention until the convention has also been ratified

National approaches and institutional capacity to address MEAs also vary widely. For example, Japan, China and India have played a key role in the global debate on climate change and have developed significant negotiating capability. Vulnerable countries have also played an active role, for example through the Alliance of Small Island States (AOSIS), and Bangladesh has developed significant scientific and analytical capability. Some countries are making a significant and growing input to climate change negotiations as members of the 'G-77 plus China' negotiating block.

Progress in implementing global MEAs has been slow as a result of the lack of institutional, administrative and financial capability, and a lack of integration of different MEAs. Although national plans and programmes exist in many countries, institutional arrangements for implementing MEAs are not well developed. Responsibility for environmental issues lies with specific ministries and their departments.

NGOs, regional networks and independent research institutes in several countries are helping to to get MEAs implemented by exerting pressure on governments and other concerned bodies. NGOs, in particular, are increasing public awareness and educating decision-makers and executives through training schemes, workshops, and newsletters and reports. They are playing a pivotal role in shifting from a command-and-control to a more participatory approach. Regional NGO networks often focus on a single MEA, for example the Climate Action Network in South Asia and Southeast Asia, and the Kiko Forum of Japanese NGOs, which was formed after the Kyoto conference and is now working with other regional and international NGOs. Similarly, in the context of the CBD and Ramsar conventions, regional and sub-regional NGO networks are working, with varying degrees of success, on developing awareness and providing policy support to national governments.

Some global MEAs have gained widespread public acceptance. Public pressure to reduce greenhouse gas emissions, for instance, is mounting, although the

UNFCCC has as yet had little impact. Most of the small island states and vulnerable coastal states such as Bangladesh will be seriously affected by sea level rise and look to the industrialized countries to discharge their obligation to prevent catastrophic consequences of climate change. In terms of meeting the objective of the UNFCCC, only Japan, the sole Annex 1 country in the region, is under an obligation to reduce emissions. Most other countries are developing their capabilities and have begun to make greenhouse gas inventories, while some have also developed abatement strategies and climate action plans.

Discussion on climate change has reflected the global debate but with differing national responses. While the Annex 1 countries have demanded voluntary commitments from non-Annex 1 countries such as China and India, most non-Annex 1 countries have opposed this vehemently and in their turn demanded higher reductions by Annex 1 countries. Similarly, vulnerable small islands, such as the Maldives and Fiji, and countries with major deltas threatened by sea-level rise, such as Bangladesh, have called for a greater emphasis on and financial support for adaptation measures. To some extent, this is reflected in the Kyoto Protocol of 1997 through the introduction of the Clean Development Mechanism. The concept of tradeable permits has been generally opposed by major developing countries in the region.

UNDP, the Asian Development Bank (ADB), the Global Environment Facility (GEF) and several bilateral donors have been funding projects under different global MEAs to improve national and regional implementation capabilities. The ADB, with funds from GEF through UNDP, has completed a 12-country regional study of an Asian Least-Cost Greenhouse Gas Abatement Strategy. The study addressed mainly the question of mitigation and several countries identified the need for a regional approach to adaptation to climate change. There has been a significant improvement in the capability of government agencies and research institutes in the countries involved.

Many initiatives have been taken to implement the Ozone Convention (the Montreal Protocol). Developing countries are required to begin their phase out of chlorofluorocarbons (CFCs) in July 1999 but several of them, notably Malaysia and Thailand, are well ahead of the Protocol's requirements. Many other countries have now prepared plans to reduce or phase out the use of ozone-depleting substances. On the other hand, CFC production in some of the larger developing countries has been increasing significantly (UNEP 1998).

In China, ozone depletion has received substantial attention. The programme to implement the Montreal Protocol involves more than a dozen central government agencies, including the State Environmental Protection Administration (SEPA), the Ministries of Finance and State Planning, and the Development Commission.

Japan implemented most of the provisions of the Montreal Protocol at an early date, and has provided assistance to other countries in the region to phase out the use of ozone-depleting substances (OAN 1997). Industry–government ties are close, and the Ministry of Trade and Industry (MITI) was placed in charge of implementation. While the consensus-oriented regulatory approach of MITI has been successful in addressing the early, production-related regulatory provisions of the Protocol, some analysts believe it may be less successful as the treaty evolves and addresses issues such as the retrieval, recycling and destruction of CFCs and other ozone-depleting substances (Weiss and Jacobson 1998).

Many countries have signed the Convention on Biological Diversity (CBD) but capacity to address biodiversity loss and appropriate scientific manpower is generally low. However, many grass-root initiatives have been undertaken, and NGOs and regional networks have succeeded in raising awareness about the issues. In several countries, indigenous peoples' organizations are also involved, and activities to implement biodiversity protection have been supported by GEF. There have also been some regional projects and some funding of small projects at the community level. However, overall awareness is still very low. The region contains areas of rich biodiversity including tropical and mangrove forests, wetlands and mountain ranges and needs far more vigorous efforts to implement the CBD. Key concerns, in addition to biodiversity loss, are the reduction of the genetic pool for rice – the key cereal in Asia – and questions relating to intellectual property rights, indigenous knowledge and rights of ownership of species.

The effectiveness of CITES in the region remains uncertain. For example, while seizures of CITES species in India have increased in value and volume, it is not clear whether this indicates better enforcement or increased smuggling. Elephant tusks fetch high prices in India and officials charged with implementing

CITES occasionally face serious threats; near Manas Park, one of India's most biodiverse regions, activity by the separatist Bodo movement has severely diminished CITES enforcement (Weiss and Jacobson 1998). Weapons were removed from the camps of anti-poaching forces, the camps were burned and officials attacked. Bodo guerrilla forces are, according to some sources, partly financed by the wild animal trade.

The situation is similar in China. Even the giant panda, China's most famous endangered species, is not immune. Dealers offer poachers US$3 000 for each panda pelt, and the pelts are later sold in Hong Kong or Japan for US$10 000 or more. These sums are enormous in the Chinese countryside, and this – coupled with the size of China and the poverty of many village dwellers – poses many problems.

The Convention to Combat Desertification (CCD) is extremely important. Several countries, such as Afghanistan, China, India and Pakistan, have vast desert areas and many others suffer from serious land degradation. However, there is little interest from governments, civil society or NGOs in giving the CCD priority. Unlike the UNFCCC and CBD, CCD has not succeeded in enthusing key research or activist groups at the country or regional level. One reason is that there are too many environmental conventions to be dealt with by the limited resources and skilled personnel that are available. The lack of a proper funding mechanism has also constrained the emergence of new initiatives and hindered national governments. Nevertheless, Asian Parties are launching three thematic programme networks at regional level: on monitoring and assessment (hosted in China); on agroforestry and soil conservation (hosted in India); and on rangeland and sand dune fixation (hosted in the Islamic Republic of Iran).

Regional MEAs

Regional and sub-regional MEAs are concerned mainly with shared facilities and the protection and proper management of the region's abundant but severely threatened resources. Several MEAs have resulted from negotiations on problems such as sharing river basins between different countries. The regional MEAs are shown in the table above.

Most attention is now being paid to the atmosphere (see box right), water, wildlife and natural disasters (see Chapter 1), to which the region is very prone and which appear to be increasing in frequency and severity. Many countries have developed their own strategies, independent of MEAs, for example to reduce air pollution and protect wildlife, but MEAs can help to reinforce the actions of planners and decision-

Major regional MEAs

Treaty	*Place and date of adoption*
Plant Protection Agreement for Asia and Pacific Region	Rome 1956
Interim Convention on Conservation of North Pacific Fur Seals	Washington DC 1957
Amendments to the International Convention for the Prevention of Pollution of the Sea by Oil 1954, Concerning the Protection of the Great Barrier Reef	London 1971
Convention on Conservation of Nature in the South Pacific	Apia 1976
South Pacific Nuclear Free Zone Treaty	Raratonga 1985
ASEAN Agreement on the Conservation of Nature and Natural Resources	Kuala Lumpur 1985
Convention for the Protection of the Natural Resources and Environment of the South Pacific Region	Noumea 1986
Protocol for the Prevention of Pollution of the South Pacific Region by Dumping	Noumea 1986
Protocol Concerning Co-operation in Combating Pollution Emergencies in the South Pacific Region	Noumea 1986
Agreement on the Network of Aquaculture Centres in Asia and the Pacific	Bangkok 1988
Convention for the Conservation of Southern Bluefin Tuna	Canberra 1993
Agreement on the Cooperation for the Sustainable Development of the Mekong River Basin	Chiang Rai 1995
Convention to Ban the Importation into the Forum Island Countries of Hazardous and Radioactive Wastes and to Control the Transboundary Movement and Management of Hazardous Wastes within the South Pacific	Waigani 1995

Regional MEAs on transboundary pollution

The ASEAN Cooperation Plan on Transboundary Pollution, adopted in June 1995, covers atmospheric and ship-borne pollution and the movement of hazardous waste (ASEAN 1995). ASEAN also adopted a Regional Haze Action Plan in 1997. Most recent is a South Asia declaration, known as the Malé Declaration on Control and Prevention of Air Pollution, which was adopted at SACEP's 7th Governing Council meeting, 22 April 1998, in Malé, Maldives. The Declaration encourages intergovernmental cooperation to address the increasing threat of transboundary air pollution and its impacts due to concentrations of pollutant gases and acid deposition. Besides laying down the general principles of intergovernmental cooperation for air pollution abatement, the Declaration includes plans for an institutional framework linking scientific research and policy formulation. It also calls for the continuation of this process in stages, with mutual consultation, to draw up and implement national and regional action plans and protocols based on a fuller understanding of transboundary air pollution issues. A follow-up action plan will be implemented at national, sub-regional and regional levels to strengthen the implementation of the Declaration (SACEP 1998).

makers. Because of the lack of resources, expertise and sometimes political will, regional MEAs are seldom fully translated into the national legislation that would be needed to ensure implementation.

Regional MEAs in which countries share and manage common resources result in relatively high levels of public information and awareness; generally, the smaller the membership of the agreement, the wider the coverage in local mass media. For example, the Ganges Water Sharing Treaty is known even to poor farmers in the remotest parts of Bangladesh and West Bengal in India. Similarly, the Indus Water Treaty between India and Pakistan has worked effectively despite many other unresolved issues between these two countries.

Few new national institutions have been created to adopt and implement regional MEAs. Legal adoption is slow and political acceptance is below the target levels set in the agreements. The pace of implementation depends on political will which, in turn, is controlled by the direct effect on the public of non-compliance. Secretariats are generally situated in the offices of international organizations such as FAO and ASEAN, or in the Foreign Ministries of individual countries, with national organizations responsible for implementation. For example, the Bangladesh Water Development Board is the implementing institution for the Ganges Water Sharing Treaty and the two countries involved, India and Bangladesh, have set up a Joint River Commission. In the case of plant protection, the responsible body in many countries is the Forest Department.

Incentives in the form of subsidies, tax reductions and penalties on organizations in breach of an agreement are now being considered as possible mechanisms for promoting compliance. Relevant national offices can be involved in such incentives. For instance, national bodies entrusted with the conservation of forests (under the Plant Protection Agreement) provide assistance for plantation and impose penalties for cutting wood in protected areas.

Reporting compliance is mandatory only in the case of bilateral projects and such reporting is carried out by designated agencies.

The lack of accepted indicators for assessing the impact of regional MEAs means that only a qualitative view can be given of their impacts. In general, it is hard to identify any positive effects; indeed the flora and fauna of the region are rapidly decreasing, vulnerability to flooding is increasing, drought and other extreme events are growing concerns and the traditional varieties of rice are disappearing. Government and non-government agencies responsible for implementation need immediate strengthening if the situation is to be improved.

Laws and institutions

Following the 1972 Stockholm Conference on the Human Environment, many countries developed a substantial body of environmental law and regulations dealing with the protection of the environment and the management of natural resources, partially as a result of obligations under MEAs. Widespread public concern over pollution led to legislation to curb emissions of effluents and airborne pollutants, while concerns over the depletion of natural resources led to legislation for resource conservation and the preservation of areas of special biological value.

In many Southeast Asian nations, this legislation was more comprehensive than the more sectorally-based and somewhat piecemeal earlier approach. Laws and regulations were revised, updated and expanded to cover such areas of concern as pollution control, nature conservation, protection of public health and the control of toxic substances and hazardous wastes. Comprehensive water protection measures are now in place, along with water quality standards, effluent standards, sanctions for violation, and plans to strengthen responsible bodies (ASEAN 1997). Similarly, increasing air pollution has prompted the definition of ambient and emission standards, especially in urban and industrialized areas. EIAs are now a common requirement.

Frameworks for implementing environmental legislation vary. In Thailand, in addition to being institutionalized, many environmental regulations were included in the Constitution of 1997 to make them more binding and easier to implement. (Government of Thailand 1997). In the Philippines, the administration of water resources and sewage management systems has been handed over to the private sector. Cambodia, the Lao People's Democratic Republic and Myanmar are all at the initial stages of strengthening institutional frameworks.

Challenges associated with this new legislation have arisen from conflicts between environmental and resource conservation and the need for rapid economic growth and development. The full and effective enforcement of environmental legislation and sanctions for non-compliance remain elusive goals, despite the strengthening of legislation in recent years, the wide availability of legal recourse, and judiciaries active in promoting environmental compliance and enforcement, and giving recognition to emerging principles of environmental law. This is due primarily to a lack of political will, the relative weakness of environmental institutions, and inadequate funds and technical expertise.

The situation in the Pacific is somewhat similar, although there are fewer constitutional safeguards and legal mechanisms on the books. In common with the rest of the region, enforcement is a problem, particularly in relation to illegal resource extraction, although some efforts have been made to establish codes of good practice. Enforcement through the sanction of traditional culture and community structures is tending to weaken throughout the Pacific as a result of continued migration to urban areas (either to the capital city on the main island or to a neighbouring developed country on the Pacific Rim) and the parallel pressure to increase cash income at the village level.

In South Asia, many institutions involved with environmental governance and protection are being strengthened. Many new public sector institutions have been established, including environmental ministries, while independent environment agencies and departments have been created to assist them.

Environmental impact assessments (EIAs) are becoming widely institutionalized while several countries are evolving National Environmental Action Plans (NEAPs), often implemented with the close involvement and participation of local people and NGOs. The trend towards decentralization of environmental decision-making and property rights will encourage new institutions to emerge at the grass-roots level to manage natural resources. The potential success of such developments is demonstrated by the participatory management of forest resources in Haryana, India (see box above).

Monitoring and enforcement of standards in East Asia have been generally weak. In Japan, legislative initiatives in the late 1960s, including the establishment of an Environment Agency, were compromised by rapid industrial growth and economic development. However, by the end of the 1980s Japan's growing international role and the generally poor state of the national environment forced a re-evaluation of environmental and development goals.

Forest resources in Haryana, India

Forests are the main source of fuelwood, fodder for livestock, building materials and medicinal plants for most of the tribal and rural poor who live in and around forests in India.

In June 1990, India's Ministry of Environment and Forestry issued a policy guideline supporting the greater participation of village communities and NGOs in the regeneration, management and protection of degraded forests. This policy was further elaborated by the Provincial Government of Haryana to enable the joint management of forest areas by the Haryana Forest Department and village groups known as Hill Resource Management Societies (HRMS).

The highlights of the programme are the:

- formation of nearly 40 HRMS to implement the programme;
- involvement of women in the decision-making process – at least 15 HRMS have two to three women members on their Management Committee;
- encouragement of bamboo basket makers by granting concessions in monthly felling permits and in reduced bamboo prices;
- rationalizing of royalty rates paid by the HRMS for *bhabbar* grass (*Eulaliopsis binata*, used to make rope) and fodder grass leases;
- raising of tree plantations by HRMS funded by the National Wastelands Development Board.

The programme's positive impact can be measured by the marked improvement in forest resources and the socio-economic status of the local people. Tree and grass cover and soil moisture have been enhanced and water run-off from the catchment area has been lessened, resulting in reduced silt loads and less downstream flooding. In addition, fewer forest offences, such as thefts, fires and illegal felling of trees, have been reported.

Source: original material supplied by the Tata Energy Research Institute, New Delhi, India

New legislation was enacted, for example to reduce vehicle emissions, and by 1993 the government had established a Basic Environment Plan (Environment Agency of Japan 1994) which outlined policies and policy instruments, and defined the roles of each sector of society. Businesses and factories are responsible for self-monitoring and evaluation, for example, whilst local government operates air and water monitoring networks. Similarly, after adopting the same fast-track development path as Japan, the Republic of Korea also encountered severe environmental degradation and has responded with comprehensive legislation and environment action (Government of Republic of Korea 1998).

Recent efforts of the Chinese government to implement environmental laws and regulations have culminated in a comprehensive Environmental Protection Law which focuses on implementation and enforcement, defines accountability and legal responsibility, and imposes sanctions for non-compliance. Standards constitute a major component of environmental policy and now embrace every aspect of environmental quality, pollution discharge, environmental management, and even monitoring methodology. Recent amendments to the criminal law have greatly strengthened this mandatory aspect of environmental protection. At the institutional level, significant progress is reported in implementing unified monitoring, inspection and management systems throughout the country, through a wide range of local and central environmental bodies. Growing numbers of environmental professionals are now also employed in both the state and the private industrial sectors. A total of 8 400 departments within the environmental protection network include some 2 900 environmental protection bureaus, more than 2 000 environmental monitoring stations and about 1 850 stations for monitoring and enforcing compliance. Nearly 100 000 people are directly employed in environmental protection (SEPA 1997a).

Environmental policy responses and law in Australia have attempted, particularly in recent years, to incorporate the guiding principles of ecologically sustainable development (CoAG 1992, Commonwealth of Australia 1996). Coordination for environmental management is effected mainly through the Council of Australian Governments and the relevant Ministerial Councils. In addition to existing strategies on major sectors of the environment, the recently implemented Natural Heritage Trust of Australia Act 1997 will allow for the spending of some US$800 million over five years and to keep US$193 million in perpetuity as a capital base for future environmental expenditure (Commonwealth of Australia 1999a).

Laws and institutions dealing with New Zealand's environment were reduced in number and made more coherent in the late 1980s and early 1990s. The philosophical centrepiece is the Resource Management Act 1991, which places most environmental decision-making in the hands of locally-elected authorities and requires them to develop policies and plans governing the use of air, land and water. Central government still has primary responsibility for environmental issues where there is a clear national interest, and it can also set national policies, standards or guidelines to ensure that local authorities manage environmental issues in ways that are nationally consistent (New Zealand Ministry for the Environment 1997).

At the regional level, several major cooperative mechanisms focusing on the environment have been established (see box left). Transboundary pollution has attracted significant regional cooperation. For instance, since 1993 the Environment Agency of Japan has been advocating the establishment of an Acid Deposition Monitoring Network in East Asia to establish uniform monitoring techniques, share data and information, create a common understanding of the state of acid deposition and provide inputs for decision-making at all levels. This monitoring network will bring together nine collaborating countries (China, Indonesia, Japan, the Republic of Korea, Malaysia, Mongolia, the Philippines, Thailand and the Russian Federation) (Environment Agency of Japan 1997a). Similarly, the Northwest Pacific Action Plan, adopted in 1994, deals with regional data collection, a survey of national legislation, marine pollution monitoring and

Cooperative projects: some examples

- The ASEAN Senior Officials on the Environment (ASOEN), of the Association of South East Asian Nations (ASEAN), has membership at the level of Permanent Secretaries. The July 1993 Meeting of ASOEN (representing Brunei Darussalam, Indonesia, Malaysia, the Philippines, Singapore and Thailand) agreed to develop an ASEAN Strategic Plan of Action on the Environment (1994–98). Recently ASEAN has been expanded to include the People's Democratic Republic of Lao, Myanmar and Viet Nam (ASOEN 1999).
- The South Asia Cooperative Environment Programme (SACEP), covering Afghanistan, Bangladesh, Bhutan, India, Maldives, Nepal, Pakistan, Sri Lanka and Iran, continues to implement an action plan known as the SACEP Strategy and Programme (1992–96). This covers a number of key areas including capacity building and awareness raising; systematic information exchange and intra-regional technology transfer; training on environmental management and institutional development; regional cooperation in the management of mountain ecosystems, watersheds and coastal resources; and wildlife and wildlife habitat conservation (SACEP 1992).
- The South Pacific Regional Environment Programme (SPREP), established in 1982, covers 22 Pacific Island countries and territories. It acts as the sub-region's interface with international agencies and in global environmental negotiations and by executing specific programmes to build national capacity. SPREP's *Action Plan for Managing the Environment of the South Pacific Region 1997-2000* addresses the diverse actions required 'to build national capacity to protect and improve the environment of the region for the benefit of Pacific island people now and in the future'. Current trends demonstrate that this goal, and the other objectives listed in the Action Plan, will be extremely difficult to achieve but there is significant forward movement in certain areas of environmental management, such as community-based nature conservation (SPREP 1997 and 1999).
- The Mekong River Commission (MRC), with representatives from Cambodia, the Lao People's Democratic Republic, Thailand and Viet Nam (with China as an observer), is an inter-governmental organization responsible for cooperation and coordination in the use and development of the water resources of the Lower Mekong Basin. In 1991, an Environment Unit was established within the Technical Support Division to deal with environmental issues in this sub-region (MRC 1999).
- The International Centre for Integrated Mountain Development (ICIMOD), representing Afghanistan, Bangladesh, Bhutan, China, India, Myanmar, Nepal and Pakistan, was established in Nepal in 1983, and continues to implement different programmes to attain environmental stability and sustainability in mountain ecosystems and the eradication of poverty in the Hindu Kush-Himalayas (ICIMOD 1999).

preparedness, and response strategies (O'Conner 1996). Some other examples are described in the box on the left.

One of the greatest policy challenges of the decade is to promote liberal trade while maintaining and strengthening the protection of the environment and natural resources. Trade and investment have been the principal engines of economic growth but they have resulted in serious environmental degradation. A number of governments are now taking action to reconcile trade and environmental interests, through the use of trade-environment related policies and agreements such as product standards, enforcement of the polluter pays principle, health and sanitary standards related to food exports, and ecolabelling. On this issue, ASEAN has recognized that any measures to promote better environmental management must be consistent with General Agreement on Tariffs and Trade (GATT) principles. It therefore calls for trade arrangements that support environment and development policies and seeks to improve capacity in trade-environment policy analysis, planning and evaluation (ASEAN 1997).

Economic Instruments

Many countries are beginning to make more use of economic instruments although often still in combination with command-and-control regulations. China is a typical example. Economic instruments such as pollution charges, pricing policy, favourable terms of investment for environmental technology, market creation, as well as ecological compensation fees, are being introduced and, in the coming decade, China aims to incorporate natural resource and environment values into the accounting system for its national economy and to establish a pricing system that reflects environmental cost (SEPA 1997b). Mongolia, in trying to move from a top-down, command-and-control approach to one with increased public participation, is relying on traditional patterns of resource use enhanced by economic incentives and user-pay principles (JEA 1994). Thailand has subsidized capital investment of the treatment of hazardous waste and toxic chemicals, implemented a service charge on community wastewater treatment, introduced a price differentiation between leaded and unleaded gasoline, and is considering granting community rights to conserve forest.

Economic incentives and disincentives are being employed to promote environmental conservation and efficient resource use. Incentives include preferential tax credits and accelerated depreciation allowances on pollution abatement and control equipment. For example, tax deductions stimulated the installation of industrial anti-pollution equipment in the Philippines and the Republic of Korea, while in India an investment allowance of 35 per cent, compared with the general rate of 25 per cent, is provided towards the cost of new machinery and plant for pollution control or environment protection (Government of India 1992). Another success story is the Demand-side Management Programme in the power sector of Thailand (see box below), partly funded by the GEF.

In a different sector, Malaysia has implemented tax exemptions for investment in timber plantations to complement efforts in sustainable timber production (Government of Malaysia 1994). For the most part, however, forest-related policies, which include the use of economic instruments, have failed to curb the degradation of Asia's forests (see box on page 244).

Demand-side management in the power sector of Thailand

Recognizing the severe impacts of accelerated energy demand, the Thai government has adopted a comprehensive Demand-side Management (DSM) Plan for the power sector. A five-year (1993–97) DSM Master Plan was formulated and implemented with a total budget of US$189 million. By the end of October 1997, the DSM programmes were saving 295 MW of peak demand and 1 564 GWh a year of electrical energy. The reduction in carbon dioxide emissions through implementing the DSM programmes was estimated at more than 1 million tonnes a year while investment requirement in power generation was reduced by US$295 million. The programmes also resulted in consumer savings of US$100 million a year in terms of electricity bills. The DSM programmes include:

- switching lamp production from fat tubes (40 W and 20 W) to slim tubes (36 W and 18 W) and promotion, by the Electricity Generating Authority of Thailand (EGAT), of compact fluorescent lamps instead of incandescent lamps through price differentials;
- the Green Building Program, through which commercial buildings can obtain CFLs at a subsidized price. For existing buildings, EGAT carries out an energy audit, design and retrofitting of electrical systems to comply with the energy efficiency requirements set by the government. EGAT also provides interest-free loans to building owners for energy-saving modifications;
- a programme to replace fluorescent lamps for rural street lighting with subsidized high-pressure sodium vapour lamps;
- a campaign to test refrigerators and air-conditioners for efficiency, and interest-free loans to purchase efficient air-conditioners;
- a programme under which EGAT encourages manufacturers and importers of electric motors to produce or import high-efficiency motors, and industrial entrepreneurs to utilize high-efficiency motors by providing interest-free loans to meet the additional cost.

Source: EGAT 1997

A number of deposit-refund schemes have been promoted to encourage recycling and re-use of products, especially packaging materials. For instance, manufacturers and importers of various goods in the Republic of Korea are required to deposit funds with the government to cover the costs of waste recovery and treatment (Government of Republic of Korea 1991).

Economic disincentives are often based on the polluter pays principle (PPP). Pollution fines are common; for example, in the Philippines fines are used to complement the enforcement of emission standards, and are based on the duration of the violation, and environmental conditions prevailing at the time, the quantity of effluent discharged, and the average deviation from the effluent or emission standards (Government of the Philippines 1992).

Degradation of Asia's forest cover – an example of market, policy and institutional failures

The degradation of Asia's forests clearly demonstrates market, policy and institutional failures. Explicit and implicit subsidies and volume-based taxes on timber removal encourage destructive logging, especially of marginal and fragile forest lands. When concessions are awarded, the goods and services a forest provides in addition to timber are rarely priced. This results in excessive deforestation and in conflicts between logging companies and local communities. Also, forest concessions are typically too short to provide incentives for conservation and replanting. The situation is further aggravated by the lack of secure property rights, both of agricultural land and often of forest resources. Without secure land tenure, farmers do not invest in soil conservation practices and, as it becomes impossible to maintain agricultural yields on existing land, people clear new land from the edges of forests.

Source: ADB 1997

Among the East Asian countries, Japan and the Republic of Korea have both adopted the PPP although, in Japan, it is yet to be applied comprehensively to pollution control because of existing systems of financial subsidies and tax credits (IDE 1995). In Malaysia, discharge fees have been in use since 1978 to complement a regulatory approach towards solving water pollution from palm oil mills (Panayotou 1994). With the gradual imposition of more stringent standards and higher discharge fees, biological oxygen demand in public water bodies dropped steadily from 222 tonnes per day in 1978 to 58 in 1980 and 5 in 1984 (Government of Malaysia 1994).

Singapore introduced road pricing in the early 1970s to reduce road congestion. Highly effective area licensing schemes were adopted which, by charging drivers to use the roads in the city centre during peak hours, reduced congestion significantly during these times. This system will be further improved by an automated Electronic Road Pricing System (Panayotou 1994). In 1990, to control the growth of private vehicles even further, Singapore introduced a vehicle quota system in which anybody wishing to own a car had to bid for a Certificate of Entitlement (O'Conner 1996).

South Asia and most of the Mekong Basin countries still rely more on regulatory mechanisms for achieving environment policy objectives than on market forces or economic instruments. However, there is a growing awareness of the importance of pricing resources, such as water, to reflect their real economic value and social cost, and there are some successful examples of price mechanisms influencing the more efficient use of water by the industrial sector, for example in India (World Bank/UNDP 1995). Property rights, specially in water and forests, remain ill-defined and insecure despite efforts to decentralize decision making to local levels and the need to consider the interests of the poor and take measures to prevent commercial interests from taking over.

In the Pacific Islands, almost no economic instruments are yet used as tools for environmental management. A lack of experience with such mechanisms, the important role of the informal economy and the traditional role of 'custom' in resource management at the local level, all weigh against market-based instruments. Nevertheless, the possibility of increased impacts stemming from globalization will make it essential for countries to consider the role that such mechanisms may need to play in future.

While economic and fiscal instruments are being promoted for many environmental uses in Australia, the opposite seems to be occurring in New Zealand, where the only fully-developed example of an economic instrument at present is a transferable quota system used to manage the major fisheries. The best-known economic instruments were the deposit-refund schemes that once operated for soft drink, beer and milk bottles. These disappeared in the 1980s as the growth of supermarkets and centralized distribution centres favoured plastic containers over glass ones (New Zealand Ministry for the Environment 1997).

Industry and new technologies

Industry is becoming increasingly sensitive to environmental concerns. Waste minimization, energy efficiency, waste recycling and substitute CFC programmes are among the many initiatives now being undertaken. While environmental auditing is not yet common, some countries have pioneered the practice. Major equipment manufacturers in Japan produced a package of environmental control and audit standards to prevent pollution as early as the 1970s (UNESCAP/ ADB 1995). In India, the Ministry of Environment and Forests issued a notification in 1992 for every industry to audit stocks and consumption of raw materials, outputs, wastes, methods of waste disposal, and the environmental impact of the industry on its surroundings (Government of India 1993). A number of companies have tried to develop a green image to increase market share, for instance by promoting environment-friendly products and allocating a proportion of their profits to environmental conservation activities.

Recognition of the importance of clean technology is reflected by regional interest in ISO 14 000 standards for manufacturing. National organizations to certify these standards have been established in Malaysia, Singapore and Thailand. The Philippines are adopting ISO 14 000 standards as part of their national standards (Philippine Council for Sustainable Development 1996). Industries in the Republic of Korea are preparing to adopt the ISO 14 000 environmental management system and some companies have already introduced an internal environmental audit (OECD 1997). Japanese companies have watched the ISO developments closely and many of them are planning to obtain the ISO 14001 registration which they see as essential to succeed in international markets (OECD 1994).

Environmental labelling is being promoted in a number of countries to encourage cleaner production and raise awareness among consumers of the environmental implications of consumption patterns. In Indonesia, for example, timber certification and eco-labelling are used as instruments to attain sustainable forest management (Government of Indonesia 1995). In Singapore, some 26 product categories are listed under the Green Labelling Scheme (Government of Singapore 1998) while the Indian government has prepared 'Ecomark' criteria for 14 product categories – soap and detergents, paper, paints, plastics, lubricating oil, aerosols, food items, packaging materials, wood substitutes, textiles, cosmetics, electrical and electronic goods, food additives and batteries (Government of India 1992). In New Zealand, the national ecolabel 'Environmental Choice' was launched in 1991 but six years later only three companies had earned the label (New Zealand Ministry for the Environment 1997).

Partnerships are emerging between governments and the private sector to provide environmental services and infrastructure. In Pakistan, the Federation of the Pakistan Chamber of Commerce and Industries has been working with the government to combat pollution (UNESCAP/ADB 1995), while in India the National Environmental Engineering Research Institute is developing a wide range of environmental technologies to improve pollutant monitoring, recycling and management of urban and industrial solid wastes, EIA analysis, water treatment and environmental support for rural development programmes (Government of India 1992). In Indonesia, the government, acting through the Environmental Impact Management Agency, is providing assistance for factories to develop cleaner and less polluting technology (Government of Indonesia 1995). In Thailand, the textile, pulp and paper, electroplating, chemical and food industries are all involved in promoting cleaner production initiatives. Reports by the Federation of Thai Industries and Thailand Environment Institute indicate that cleaner production is having a significant impact in terms of minimizing waste and pollution as well as promoting cooperation between government and industries, and among the industries themselves (TEI 1996). Other countries in this sub-region are expected to follow this trend.

Japan is leading the way in pursuing policies to encourage cleaner production and developing the required new technologies. The private sector finances some 60 per cent of all research and development into environmental technology and contributes heavily to a number of government research agencies (UNESCAP/ADB 1995). Japanese industry is particularly strong in certain clean energy fields such as photovoltaic cells and fuel cells, and in 'end-of-pipe' technology and clean motor vehicle technology. The country enforces the world's most stringent standards for automobile exhaust emissions, as well as strict standards to control smoke emissions from factories and other facilities. As a result, Japan has been successful in reducing atmospheric SO_2 and CO emission levels. Nine of Japan's largest steel makers are involved in a project to increase the use of scrap metal in steel manufacture, and the Japan Automobile Manufacturers Association (JAMA) has set standards for making vehicle parts in plastic for easy recycling. Consumer Cooperatives have become a powerful force in Japan to popularize green products (UNESCAP/ADB 1995) while local governments have progressively provided technological and financial support to small and medium-sized companies.

In the Republic of Korea, an Act for Promoting Environmentally Friendly Production Systems and the Environmentally Friendly Plant Certification System was passed in 1994 (Government of Republic of Korea 1994 and 1998).

In China, an elimination system in the chemical, metallurgical, machine tools, power generation and construction industries is getting rid of factories with high pollution costs and those based on old smokestack technology. By June 1997, some 64 000 enterprises with heavy pollutant emissions had either been closed for refurbishment or ceased production (SEPA 1997a). Heavy metal pollution from industrial workshops, which used to constitute a major water contamination problem, has been particularly targeted. For example, as part of the Three Rivers and Three Lakes water control project – covering the Huai He, Hai He and Liao He rivers and Lakes Tai Hu, Dian Chi and Chao Hu – an Interim Regulation for Controlling Water Pollution along the Huai He River was formulated. This was one of the seven largest water basin programmes in China. By 1997 when the programme ended, several thousand small enterprises that used to discharge heavy pollutants had closed down, upgraded their technology or transferred production to clean products, and water quality in the river had improved substantially (SEPA 1998).

Policies are being pursued to decrease atmospheric pollution, particularly of smoke and dust, and to expand smoke control areas. These policies include the levying of SO_2 emissions discharge fees and the introduction of clean-burning technology. The main obstacles are the lack of adequate capital and technology necessary for changing the present energy structure.

In Australia, the draft National Strategy for Cleaner Production examines activities to date in encouraging the implementation of cleaner production and recommends further measures, drawing on national and overseas examples (Commonwealth of Australia 1999b). The National Pollutant Inventory, established under the 1996 National Environment Protection Act, will produce a public database detailing the types and amounts of certain toxic chemicals entering different areas of the Australian environment (Commonwealth of Australia 1996c).

Cleaner production is also fostered in New Zealand by government agencies such as the Energy Efficiency and Conservation Authority and the Ministry of the Environment (New Zealand Ministry of the Environment 1997).

The inclusion in the Kyoto Protocol of a Clean Development Mechanism, together with other features of the continuing negotiations in the context of the UNFCCC are potentially important to all countries in the region. They open significant new prospects for the Pacific Islands in particular, since the small scale of their economic activity has not previously created scope for the transfer of clean technology outside a limited number of aid projects, and also because there is a need to build local capacity in applying the new technologies which are now available, for example in the management of solid waste and hazardous substances. This could have a crucial and beneficial impact on many Pacific Island communities whose remote situation invites the application of such technologies as solar photovoltaic cells and windpower.

Financing environmental action

National investment in the environment is increasing in most countries. A major thrust, particularly among developing countries, is on water supply, waste reduction and waste recycling. Environment funds, such as the US$200 million environment fund established by the Government of Thailand to clean-up cities and control industrial pollution (UNESCAP/ADB 1995), have been set up in many countries. In the Philippines, two mining companies acting on orders from the government have created an Environmental Guarantee Fund to rehabilitate and restore areas adversely affected by mining operations. A Reforestation Fund is also proposed as part of a scheme to counter deforestation (Government of the Philippines 1992).

Bilateral and multilateral aid is a significant source of environmental investment and expertise. A large portion of loans have been aimed at improving energy and industrial efficiency, water supply and sanitation facilities, afforestation, and marine and coastal resources management. The Asia Sustainable Growth Fund, sponsored by the Asian Development Bank, aims to raise US$150 million to invest as long-term capital for environmentally-sound companies in the developing countries of the Pacific Rim (UNCSD 1995). The OECD, ADB, World Bank and international financial markets have provided additional official development assistance (ODA) for environmental investments.

The financing of environmental action in the Pacific Islands has been problematic, partly because factors of scale and remoteness give rise to high transaction costs. Nevertheless, increased access has recently been obtained to new sources of funding such as the GEF. The Pacific Islands generally have relatively small economies and large distances to potential markets compared to their larger neighbours in the Asia and Pacific region at a similar stage of development (Commonwealth of Australia 1999d). Without corrective and anticipatory measures, the impacts of more open trade and investment regimes particularly for natural resources could intensify a number of environmental risks, such as those created by natural disasters and the consequences of climate change and sea-level rise for low-lying island ecosystems.

In ASEAN countries, the private sector has played an increasingly important role in stimulating economic development, especially with the gradual reduction in ODA. Its role in resource and environmental management is likely to expand with the introduction of economic instruments, third party monitoring and auditing, and the privatization of environmental management systems, but this requires huge investments. For example, the ADB estimated that, in the 1990s, Indonesia, Malaysia, the Philippines and Thailand alone must invest some US$5 400 million for environmentally-sound power generation systems over and above conventional power generation systems, while ASEAN countries require more than US$6 000 million additional investment to protect the environment from industrial pollution. Adopting clean technologies in these countries would consume a further US$72 000 million between 1991 and 2000 (ASEAN 1997). Mobilizing investment funds for environmental protection will be a major challenge for ASEAN countries in the next decade.

Japan is the largest source of project-related development assistance in the Northwest Pacific and East Asia. In 1992 Japan announced that environmental aid was to be increased to US$7 100 million over the next five years, and a Green Aid Plan of some US$2 650 million implemented over a 10-year period to transfer antipollution measures to developing countries and support joint research and development projects on the global environment (UNCSD 1995). Another notable example is the Japan Fund for Global Environment (see box below), an initiative of the Environment Agency of Japan.

The Japan Fund for Global Environment

The Japan Fund for Global Environment was established within the Japan Environment Cooperation (JEC) in 1993 to provide financial, information, educational and training assistance to NGOs inside and outside Japan. The fund is endowed by the national government as well as by citizens and corporations, and had a value of some US$62.5 million at the end of 1996. Projects fall within two categories and are funded by the interest earned on the fund:

- assistance to private organizations for environmental conservation activities in developing countries and in Japan;
- dissemination of the information necessary for promoting the activities of private organizations as well as education and training activities for the public and NGO staff members.

In addition, a Global Partnership Programme inaugurated in 1997 plays host to NGO fora held in conjunction with inter-governmental conferences, to help organize a worldwide network of NGOs and foster partnership among countries in the Asia-Pacific region for environmental conservation activities.

Source: Environment Agency of Japan 1997b

The Republic of Korea's ODA, which is provided by the Economic Development Cooperation Fund and the Korea International Cooperation Agency, exceeded US$520 million in 1996. The Republic of Korea also contributes to the Global Environment Facility. Although the environmental component of development aid remains limited, the Republic of Korea has also sponsored potable water supply systems and wastewater system improvement in several countries and intends to expand environmental assistance in the future (Government of Republic of Korea 1997).

Several countries receive funds from AusAID. Australia's ODA to GNP ratio in 1998–99 is expected to be 0.27 per cent, currently above the latest (1997) published average of 0.22 per cent for all donor nations. About 55 per cent is designated for Country Programmes (mainly country-specific poverty reduction development projects), and 32 per cent for Global Programmes, such as multilateral, humanitarian and NGO assistance (Commonwealth of Australia 1999d).

A number of national projects involve several financial partners. These can increase national experience in technology transfer as well as benefiting the environment. Examples include:

- the Alternative Energy Project in India, worth US$186 million, being funded by GEF, the World Bank, Switzerland and DANIDA;
- the Leyte-Luzon Geothermal Project in the Philippines, worth US$1 333 million, being funded by GEF, the World Bank, the Export Import Bank of Japan, and Sweden; and
- Promotion of Electricity Efficiency in Thailand, worth US$189 million, to be funded by GEF, the World Bank, Australia and Japan's Overseas Economic Cooperation Fund (OECF).

Public participation

NGOs have emerged as major partners in development and conservation activities, performing a multitude of roles including environmental education and awareness-raising among the public. NGOs have helped design and implement environment policies, programmes and action plans, and set out specifications for EIAs. They also play crucial advocacy roles through their environmental campaigns.

For example, in Sri Lanka NGOs have been active in preventing logging of the Singharaja Forest, setting up a Tiger-top lodge in Udawala National Park, stopping the construction of a thermal plant at Trincomalee, and questioning the blind implementation of the National Forestry Master Plan (Government of Sri Lanka 1994). In India, thousands of NGOs have helped raise awareness of environment-development issues and mobilized people to take action. Although not an NGO, the Narmada Bachao Andolan movement has brought together scattered voices of protest against the damming of the river Narmada and has raised awareness in India and among the international community (Government of India 1992). Another effective means of advocacy that NGOs in India have adopted is people's tribunals. For instance, the Permanent People's Tribunal (PPT) hears cases filed by individuals or communities affected by environmental degradation. Its judgements are widely publicized (South-South Solidarity 1992).

In the Pacific, local community and NGO inputs come mainly through programmes to build community awareness and expand environmental education. Some NGOs find trade opportunities which provide a return for local people wishing to manage resources such as timber on a sustainable basis. Others have linked with international NGOs to build ecotourism opportunities in partnership with landowners.

NGOs, while retaining their individual identity, have also collaborated effectively with national and local governments on a wide range of issues. In the Philippines, a consortium of 17 environmental NGOs (NGOs for Protected Areas, Inc.) received a US$27 million grant to implement a seven-year Comprehensive Priority Protected Areas Programme. The programme is a major component of the World Bank-GEF Sectoral Adjustment Loan initiative being managed by the Department of Environment and Natural Resources.

The Mongolian Government cooperates closely with NGOs, for example with the Mongolian Association for the Conservation of Nature and Environment, which coordinates the voluntary activities of local communities and individuals to protect nature and wildlife, and with the Green Movement which promotes public environmental education in support of traditional protection methods (MNE, UNDP and WWF 1996).

NGO networks are springing up throughout the region. Among the most prominent are the Asian NGO Coalition for Agrarian Reform and Rural Development and the Asia Pacific People's Environment Network (Government of Republic of Korea 1994). Networks also exist at national level. For example, the Korean Federation Environmental Movement in the Republic of Korea provides an umbrella organization for nearly 200 NGOs engaged in environment-related issues (Government of Republic of Korea 1994).

Many countries have encouraged public participation in environmental management, through local government and community-based groups. For example, in Thailand, Article 7 of the Environment Act of 1992 delegates the work on environmental management to provincial and local authorities, and encourages people's participation through environmental NGOs (Government of Thailand 1992). Article 56 of the Thailand Constitution (1997) recognizes the rights of people to participate in the protection of natural resources and environment (Government of Thailand 1997). In the Philippines, small fishing communities are given the right to manage their fishery resources (Panayotou 1994, Philippine Council for Sustainable Development 1996) and community-based forest resource management has helped protect and conserve forest resources. Similarly, many coastal community groups in Thailand protect mangroves and seagrass (OEPP 1997). Amongst SPREP members, a variant of the NEAP process – known as the NEMS (National Environmental Management Strategy) – was developed prior to the 1992 Earth Summit. The strength of this process was to involve all national stakeholders in a debate about environmental priorities and key actions, and then to present a national consensus to external counterparts, especially in the donor community (SPREP 1994).

Community participation is required by law under New Zealand's Resource Management Act. When developing their ten-year policies and plans, regional and district councils are required to consult widely with community stakeholders and interest groups, including the indigenous Maori people (New Zealand Ministry for the Environment 1997). The main constraints on citizen participation are the time and costs involved but motivation levels are reportedly high (Colmar Brunton 1990 and 1993, Gendall and others 1994).

The Chinese government has long been aware that public participation is an essential prerequisite for successful environmental protection and management (see box above).

Community-based approaches are also widely used in Australia. For example, the Landcare programme aims to address natural resource management problems, protect agricultural resources, and assist natural resource managers to improve their technical, management, communication and planning skills (Commonwealth of Australia 1999e). About one-third of Australian farmers are members of a Landcare group. The Coastcare programme provides opportunities for communities to work with local land managers to identify problems along their stretch of the coast and develop and implement solutions. Coastcare has formed 250 community groups since its inception in early 1996. The Endangered Species Programme administers public networks, such as the Threatened Species Network and Threatened Bird Network, to promote community involvement in recovery programmes for threatened species (Commonwealth of Australia 1999e).

Women's participation in environmental protection in China

The 'Law on the Protection of Women's Interest and Rights of the People's Republic of China', the Committee in charge of the Work of Women and Children established under the State Council in 1993, and the 'Development Program of Chinese Women' issued by the State Council in 1995 all aim to protect Chinese women's rights and ensure their participation in national management and environmental decision-making. Women are now playing more and more important roles. For instance, in most provincial and municipal environmental protection bureaus, at least one of the directors is a woman.

In rural areas, more than 60 000 'green bases' such as orchards, which have developed the economy and protected the environment, have been built by women farmers. Other environmentally-beneficial activities organized by women are publicizing fuel-efficient stoves, accumulating farm manure and reducing the use of chemical fertilizers. In poverty-stricken areas many women have played an important role in afforestation and ecological protection, including setting up and staffing demonstration sites and professional afforestation teams.

Source: SEPA 1997b

Many of these initiatives depend on voluntary action. Such actions have played a particularly important role in Japan where local communities, citizens' groups and government together take the initiative to negotiate with major polluters. Several Japanese companies have now taken voluntary actions on pollution control which include stricter standards than the national ones. In addition, the Japan Federation of Economic Organizations (Keidanren) adopted a Global Environmental Charter in 1991 which includes a provision that companies should carry out environmental impact assessments of their activities, use and develop low pollution technologies, and participate in local conservation programmes (OECD 1994).

Environmental information and education

With a few notable exceptions, such as Japan and the Republic of Korea, the information base in most countries is relatively weak. The shortage of reliable data and data analysis capabilities undoubtedly hinders policy development, planning and programme implementation. There is a need not just for more data on environmental issues but for standardizing data collection and storage, and making it accessible to technical and management levels. Data of poor quality are often recycled for studies and plans, which are later accepted without proper scrutiny. For example, the most recent (1987) estimate of forest cover in Vietnam (9.3 million ha or 28 per cent of the land area) has been repeatedly quoted and used unchanged since then, during which time the forests have been seriously exploited. Thus a serious environmental issue was not prioritized for action (MRC/UNEP 1997). The lack of baseline socio-economic data has also been identified as a serious constraint, and even where sufficient data exist there is no established mechanism for access or interchange.

Some attempts are being made to redress this situation. At national level, for example, the Indian Government Environmental Information System Network has been set up to deal with the collection, collation, storage, analysis, exchange and dissemination of environmental data and information (Government of India 1995). There have been some impressive achievements in ecosystem monitoring by satellite imaging. Remotely sensed data are increasingly being used by Commonwealth, State and Territory environmental departments and agencies in Australia (Commonwealth of Australia 1999f). Australia is also applying modelling techniques using environmental data to indicate areas of high environmental value as part of resource planning exercises, for example Comprehensive Regional Assessments for native forests (Commonwealth of Australia 1999f). Environmental indicators are being developed by several countries to assist with national state of the environment reporting.

Several sub-regional programmes have data and monitoring components and are addressing the standardization of databases and data sources to support environmental assessment, reporting, research and decision making. The ADB and UNEP, in collaboration with the Mekong River Commission, are running a Sub-regional Environmental Information and Monitoring System project to make environmental and natural resources data more easily available to national government agencies and regional organizations, and to allow such data to be quickly shared (ADB, UNEP and MRC 1996). In addition, a number of international agencies, including the ADB, UNDP, UNESCAP, the UN Statistical Institute for Asia-Pacific and UNEP are involved in strengthening institutional capacities to manage environmental information and helping some countries prepare national and regional state of environment reports. SPREP and the UN Economic and Social Commission for Asia and the Pacific (UNESCAP), with assistance from members of the Interagency Committee on Environment and Sustainable Development, prepare regional state of the environment reports every five years.

The intergovernmental Asia-Pacific Network for Global Change Research supports a range of regional cooperative activities relevant to the region. These include standardizing, collecting, analysing and exchanging scientific data; improving national scientific and technical capabilities and research infrastructure; cooperating with research networks in other regions; providing scientific knowledge to the public and input to decision makers; and developing appropriate mechanisms for technology transfer (APN 1997).

The Environment Agency of Japan initiated Eco Asia (Environment Congress for Asia and the Pacific) in 1991 as a forum for informal dialogue among environment ministers in Asia and the Pacific. Discussion topics have included the Long-Term Perspective Project on Environment and Development in Asia and the Pacific to identify policy options for sustainable development, the Asia Pacific Environmental Information Network Project utilizing the Internet, and Junior Eco-Club Activities to enhance environmental awareness among children, and to promote environmental conservation activities (Environment Agency of Japan 1997c).

Nearly all countries have public education and awareness programmes which seek to sensitize people to environmental issues and problems. China has set up an environmental information system in each of its 27 provinces and autonomous regions with a technical assistance loan from the World Bank (SEPA 1996). China has been publishing its annual report on the state of the environment for 10 years, and 46 Chinese cities issue a weekly report on urban air quality. Australia has adopted state of the environment reporting as a mechanism for public education on environmental matters and is increasingly using the Internet to promote information exchange among the scientific community, the Government and the public (Commonwealth of Australia 1996). Similarly, Pacific Island members of SPREP have participated in drawing up plans for individual and joint action to strengthen environmental education, training and information systems, the fifth main objective in the Action Plan for 1997-2000 (SPREP 1997). In Myanmar, the National Commission for Environmental Affairs has, since it was established in 1990, been instrumental in promoting public environmental education and awareness in the country.

The Republic of Korea is the first country in the region to have a Law for Public Information (1996) and the government regularly publishes several environmental indicators on water and air quality. The government also distributes a White Paper on the Environment to some 160 private organizations and issues a monthly *Environmental Information Bulletin* (OECD 1997). The Japanese Environmental Agency also publishes an annual White Paper on the Environment and many other books and pamphlets for wide dissemination. The Environment Agency and the Ministry of Education, in collaboration with the UNEP International Environmental Technology Centre (IETC) in Osaka and Shiga, promote popular environmental education by producing television programmes and films, holding seminars on environmental education, and distributing teaching material (Environment Agency of Japan 1997d).

The Asia-Pacific Network for Tertiary Level Environmental Training of UNEP focuses on enhancing the environmental expertise of decision makers, policy formulators and tertiary level trainers by establishing a self-sustaining network of trained individuals. It prepares and disseminates curriculum guidelines, resource materials, teaching aids and packages for environmental training. The network has grown rapidly over the past two years, and now covers 35 countries, with more than 200 tertiary institutional members and 2 000 individual members (UNEP/PROAP 1998). However, in general, relatively little effort has been directed specifically towards policy and decision makers in the region, particularly those concerned with resource use and resource allocation at senior government level.

In the formal educational system, the development of environmental education has concentrated on the primary and secondary levels, with pre-school and tertiary levels receiving less attention. Except in Nepal, environmental education became fully integrated into the school curricula of South Asian countries in the 1970s (UNESCO/ROAP 1992). Environmental protection has found its way into primary and secondary education and institutions of higher learning in China. The situation in Japanese schools is somewhat different as environmental education is voluntary and so differs from school to school.

There has also been considerable interest in integrating environmental concepts into adult education and literacy programmes. For example, the Asian-South Pacific Bureau of Adult Education established a network of environmental educators in 1992 (ASPBAE 1992). Many countries use informal education centres as sources of environmental education – Indonesia, for example, has promoted environmental consciousness through its 54 Environmental Study Centres (UNESCO/PROAP 1988).

NGOs have played a key role in producing print and audio-visual materials for non-formal environmental education in schools and other institutions of learning. For instance, in India, the Centre for Environmental Education has produced a wide range of books and audio-visual packages for the benefit of teachers and students (CEE 1995), while some television programmes have been successful in raising environmental awareness at all levels and drawing attention to the illegal poaching of tigers, rhinoceros and other endangered species.

Newspapers are increasingly addressing environmental issues. Until a few years ago, reporting on the environment was limited to reports of speeches on Environment Day or the coverage of tree planting campaigns. Today journalists, working closely with environment activists, are much more pro-active and are focusing on larger issues on a much wider scale. China's national level environment newspaper, *China Environmental News*, played a major role in improving public awareness of the environment. Broadcasting companies also play a major role. Chinese radio stations, for example, regularly broadcast programmes and conduct competitions on environment themes, one local radio station in Beijing in 1988 attracting more than 60 000 responses from more than one million listeners to a knowledge competition on environmental protection, a success that has since been repeated by other radio stations (Chaoran and Changhua 1993).

Social policies

While few environmental policies specifically target equity or poverty issues, there have been some policy initiatives in the social sector. Their thrust has been to address poverty directly through employment generation programmes and to improve equity through rural credit. At the same time, many countries have adopted policies to stabilize or moderate population growth rates. The success of efforts directly targeted at poverty alleviation has varied with notable progress in East Asia but less in South Asia (UNESCAP/ADB 1995). The direct support programmes set up by many Asian governments provide subsidized food or credit and introduce micro-finance programmes. Subsidy programmes have tended not to work well. Subsidized food deliveries are not easy to target, and have often gone largely to the better-off in urban areas. In subsidized credit programmes, loans typically fail to reach the poor, are often used for consumption, and are usually not repaid (ADB 1997).

The social policies of ASEAN countries have focused on sustainable human settlements. The priority targets have been the basic needs of the rural population, especially shelter and safe drinking water. Human resource development has also been emphasized, with a high priority on education and training (ASEAN 1997). Social policies have also supported resource decentralization and environmental management. The spread of HIV-AIDS, and increasingly severe air and water pollution, are emerging issues to be given priority in social development. ASEAN has also spelled out the need to support the development of a regional framework for integrating environment and development concerns in the decision-making process.

Over the past decade, the Chinese government has implemented a series of policies for science and education, population, women and social protection, in the interest of environmental protection. Both central and local governments have stepped up efforts to relieve the victims of natural disasters. In 1995, relief funds amounting to the equivalent of US$284 million benefited more than 31 million poor people. A further seven million households received relief funds from local government organizations while some two million households were able to rise above the poverty level (SEPA 1997b).

Conclusions

While the development of policy responses to environmental issues is very uneven across the region, it is possible to discern the major key priorities for action in the near future. These include:

- mobilization of environmentally-conscious forms of investment;
- the application of cleaner technologies much more widely;
- research on alternative forms of energy supply and the promotion of fuel switching and energy conservation;
- making mass transport systems more efficient and developing innovative alternatives;
- increasing public consultation;
- implementing efficient resource pricing;
- encouraging voluntary action on the environment;
- stimulating capacity building; and
- promoting regional cooperation.

The region's major strategy in attacking environmental problems should now be to combine command and control actions with economic incentives, with massive public consultation to gain widespread public acceptance for improved environmental policy actions.

References

ADB (1997). *Emerging Asia: Changes and Challenges.* Asian Development Bank, Manila, the Philippines

ADB, UNEP and MRC (1996). *Sub-regional Environmental Monitoring and Information System (SEMIS) – Project Implementation Document.* TA. No. 5562-REG. Asian Development Bank, Manilla, Philippines

APN (1997). Asia-Pacific Network for Global Change Research http://www.rim.or.jp/apn

ASEAN (1995). *ASEAN Co-operation Plan on Transboundary Pollution.* ASEAN Secretariat, Jakarta, Indonesia

ASEAN (1997). *First ASEAN State of the Environment Report.* ASEAN Secretariat, Jakarta, Indonesia

ASOEN (1999). ASEAN Senior Officials on the Environment http://www.brunet.bn/gov/modev/environment/asean.html

ASPBAE (1992). Asia-South Pacific Bureau of Adult Education, *Environmental Education Newsletter,* No.1, Quezon City, Philippines

CEE (1995). *Environmental Education in Asia: Regional Report for the UNESCO Inter-regional Workshop on Reorienting Environmental Education for Sustainable Development.* Centre for Environmental Education, Ahmedabad, India

Chaoran, Yu, and Changhua, Wu (1993). Environmental Education and the Media's Role in China. Paper presented at the SASEANEE Workshop in Ahmedabad, India, February 1993

CoAG (1992). *National Strategy for Ecologically Sustainable Development.* Council of Australian Governments, AGPS, Canberra, Australia.

Colmar Brunton (1990). *Project Green.* Report prepared for the New Zealand Ministry for the Environment. Colmar Brunton Research Ltd, Auckland, New Zealand

Colmar Brunton (1993). *Project Green.* Second report prepared for the New Zealand Ministry for the Environment. Colmar Brunton Research Ltd, Auckland, New Zealand

Commonwealth of Australia (1996). *Australia: State of the Environment 1996.* State of the Environment Advisory Council and Department of the Environment, Sport and Territories. CSIRO Publishing, Collingwood, Australia

Commonwealth of Australia (1998). *1998 Year Book Australia.* Australian Bureau of Statistics, Canberra, Australia

Commonwealth of Australia (1999a). National Heritage Trust of Australia Home Page http://www.nht.gov.au

Commonwealth of Australia (1999b). Ecoefficiency and Cleaner Production Home Page http://www.environment.gov.au/epg/environet/eecp

Commonwealth of Australia (1999c). National Pollutant Inventory Home Page http://www.environment.gov.au/epg/npc/home.html

Commonwealth of Australia (1999d). *1999 Year Book Australia.* Australian Bureau of Statistics, Canberra, Australia

Commonwealth of Australia (1999e). Environment Australia Home Page http://www.erin.gov.au

Commonwealth of Australia (1999f). Comprehensive Regional Assessments and Regional Forestry Agreements Home Page http://www.rfa.gov.au/index.html

Environment Agency of Japan (1994). *The Basic Environmental Plan.* Environment Agency, Tokyo, Japan

Environment Agency of Japan (1997a). *Acid Deposition Monitoring Network in East Asia – Achievements of Expert Meeting, March 1997.* Environment Agency, Tokyo, Japan

Environment Agency of Japan (1997b). *Japan's Environment Protection Policy.* Environment Agency, Tokyo, Japan

Environment Agency of Japan (1997c). *Environmental Cooperation Programme in Asia and the Pacific towards Sustainable Development (ECO-PAC).* Environment Agency, Tokyo, Japan

Environment Agency of Japan (1997d). *Quality of the Environment in Japan 1997,* in Japanese. Environment Agency, Tokyo, Japan

EGAT (1997). *Demand-side Management in Thailand: Experience and Perspectives.* Electricity Generating Authority of Thailand, Bangkok, Thailand

Gendall, P.J., Hosie, J.E. and Russell, D.R. (1994) *International Social Survey Programme: The Environment.* Department of Marketing, Massey University, Palmerston North, New Zealand http://www.massey.ac.nz/~wwmarket/issp.htm

Government of India (1992). *Environment and Development: Traditions, Concerns and Efforts in India: National Report to UNCED.* Ministry of Environment and Forests, New Delhi, India

Government of India (1993). Environment Statement, 5 June 1993 (part of Environmental Audit). Ministry of Environment and Forests, New Delhi, India

Government of India (1995). *Annual Report 1994-95.* Ministry of Environment and Forests, New Delhi, India

Government of Indonesia (1995). *Indonesian Country Report on Implementation of Agenda 21 1995.* The State Ministry of Environment, Jakarta, Indonesia

Government of Malaysia (1994). *Report to the United Nations Commission on Sustainable Development.* Ministry of Science, Technology and the Environment, Kuala Lumpur, Malaysia

Government of the Philippines (1992). *A Report on Philippine Environment and Development.* Department of Environment and Natural Resouces, Quezon City, Philippines

Government of Republic of Korea (1991). *National Report of the Republic of Korea to UNCED.* Ministry of Environment, Seoul, Republic of Korea

Government of Republic of Korea (1994). *Environmental Protection in Korea.* Ministry of Environment, Kwacheon, Republic of Korea

Government of Republic of Korea (1997). *Environmental Protection in Korea.* Ministry of Environment, Republic of Korea

Government of Republic of Korea (1998). *Environmental Protection in Korea 1997.* Ministry of Environment, Kwacheon, Republic of Korea

Government of Singapore (1998). *Make the Green Label Your Choice.* Ministry of the Environment, Singapore

Government of Sri Lanka (1994). *State of the Environment of Sri Lanka (*for submission to the SACEP). Ministry of Environment and Parliamentary Affairs, Colombo, Sri Lanka

Government of Thailand (1992). *The Enhancement and Conservation of National Environmental Quality Act, B.E. 2535*. Bangkok, Thailand

Government of Thailand (1997). *Constitution of the Royal Thai Kingdom 1997*. Bangkok, Thailand

ICIMOD (1999). International Centre for Integrated Mountain Development Home Page http://www.south-asia.com/icimod.htm

IDE (1995). *Development and the Environment: The Experiences of Japan and Industrializing Asia. Edited by* Kojima, Reeitsu, Nomura, Yoshihiro, Fujisaki, Shigeaki, and Sakumoto, Naoyuki. Development and the Environment Series No.1. Institute for Developing Economies, Tokyo, Japan

JEA (1994). *Environmental Governance in the Pacific Century,* edited by J.E. Nickum and J.R. Nishioka. Japan Environment Association, East-West Center

MNE, UNDP and WWF (1996). *Mongolia's Wild Heritage*. Ministry of Nature and Environment, Ulaanbaatar, Mongolia

MRC (1999). Mekong River Commission Home Page http://eco-web.com/register/02769.html

MRC/UNEP (1997). *Mekong River Basin Diagnostic Study: Final Report.* Mekong River Commission, Bangkok, Thailand

New Zealand Ministry for the Environment (1997). *The State of New Zealand's Environment 1997*. GP Publications, Wellington, New Zealand

OAN (1997). Thailand and Japan share phase-out success. *OzonAction News,* 21, January 1997

O'Conner, D. (1996). *Applying Economic Instruments in Developing Countries: From Theory to Implementation.* Paper prepared for EEPSEA, IDRC, 1996

OECD (1994). *OECD Environmental Performance Reviews: Japan.* OECD, Paris, France

OECD (1997). *OECD Environmental Performance Reviews: Republic of Korea.* OECD, Paris, France

OEPP (1997). *National Action Plan for Environmental Quality Promotion* (in Thai). Office of Environmental Policy and Planning, Ministry of Science, Technology and Environment, Bangkok, Thailand

Panayotou, T. (1994). *Economic Instruments for Environmental Management and Sustainable Development.* Environmental Economics Paper No. 16. UNEP, Geneva, Switzerland http://www.unep.ch/eteu/econ/e-pumenu.htm

Philippine Council for Sustainable Development (1996). *Onwards from Rio: Continuing Philippine Efforts in Sustainable Development.* Philippine Council for Sustainable Development, Manila, Philippines

SACEP (1992). *SACEP Strategy and Programme I (1992–96).* SACEP, Colombo, Sri Lanka

SACEP (1998). *Male Declaration on Control and Prevention of Air Pollution and its Likely Transboundary Effects for South Asia.* Report of 7th SACEP Governing Council Meeting, 22 April 1998, Male, Maldives

SEPA (1996). *China Environment Year Book, 1996.* State Environmental Protection Administration of China, Beijing, China

SEPA (1997a). *China Environment Year Book, 1997.* State Environmental Protection Administration of China, Beijing, China

SEPA (1997b). *National Report on Sustainable Development, 1997.* State Environmental Protection Administration of China, Beijing, China

SEPA (1998). *Report on the State of the Environment in China 1997.* State Environmental Protection Administration of China, China Environmental Science Press, Beijing, China

South-South Solidarity (1992). *South Link Newsletter,* Vol. II, Nos. ii-iii, July-October 1992, New Delhi, India

SPREP (1994). *Action Strategy for Nature Conservation in the South Pacific Region, 1994–98.* South Pacific Regional Environment Programme, Apia, Samoa

SPREP (1997). *Action Plan for Managing the Environment of the South Pacific Region 1997–2000.* SPREP, Apia, Samoa

SPREP (1999). South Pacific Regional Environment Programme http://www.sprep.org.ws/default.htm

TEI (1996). *Towards Environmental Sustainability – Annual Report 1996.* Thailand Environment Institute, Bangkok, Thailand

UNCSD (1995). *Financing the Transfer of Environmentally Sound Technology.* United Nations Commission on Sustainable Development, New York, United States, January 1995

UNEP (1998). *Production and Consumption of Ozone Depleting Substances 1986–1996*. Ozone Secretariat, United Nations Environment Programme, Nairobi, Kenya http://www.unep.org/unep/secretar/ozone/pdf/Prod-Cons-Rep.pdf

UNEP/ROAP (1998). *UNEP/ROAP Information Brochure 1998.* UNEP/ROAP, Bangkok, Thailand

UNESCAP/ADB (1995). *State of the Environment in Asia and the Pacific 1995.* United Nations Economic and Social Commission for Asia and the Pacific, and Asian Development Bank. United Nations, New York, United States

UNESCO/PROAP (1988). *Environmental Education at University Level: Report of a Seminar on the Strategy for Inclusion of Environmental Education at University Level.* Yogyakarta, Indonesia, 20 June – 4 July 1987. UNESCO Principal Regional Office for Asia and the Pacific, Bangkok, Thailand

UNESCO/PROAP (1992). *Final Report of Training Workshop in Environmental Education for Elementary Teacher Educators for South Asian Countries.* UNESCO/PROAP, Bangkok, Thailand

Weiss, E.B., and Jacobsen, H.K. (1998). *Engaging Countries: Strengthening Compliance with International Environmental Accords.* MIT Press, Cambridge, Massachusetts, United States

World Bank/UNDP (1995). *Water Conservation and Reallocation: Best Practices in Improving Economic Efficiency and Environmental Quality.* A World Bank–ODI Joint Study by Ramesh Bhatia, Rita Cessti and James Winpenny. World Bank/UNDP, Washington DC, United States

Policy Responses

Europe and Central Asia

KEY FACTS

- There is a danger that unsustainable lifestyle patterns in Western Europe will be adopted by countries in transition too indiscriminately. The accession countries need to find an acceptable balance between adapting to Western European policy and maintaining existing practices where these are environmentally beneficial.
- The overall level of ratification of global MEAs is relatively high, and reasonably balanced among sub-regions.
- Comprehensive National Environment Action Plans (NEAPs) have helped in the elaboration of new policy principles and the redesign of institutions. Sixteen transition countries have developed or are developing NEAPs.
- One of the main barriers to accession to the European Union is lack of financial resources – the cost of conforming to environmental regulations is estimated at ECU* 100 000–150 000 million for the 11 accession countries.
- In Hungary and Latvia, product charges are being imposed on manufacturers of goods, such as packaging or batteries, which can be recycled or alternately disposed of, encouraging private investment.
- Increasing VAT on resource and energy use is frequently recommended as an instrument with positive effects on both cleaner production and employment.
- In Western Europe, the European Commission estimates that during 1994–99 more than ECU 17 000 million was spent on environmental measures.
- Although subsidies are declining, they continue to have adverse impacts on the environment, especially in the energy, transport and agricultural sectors.

* The ECU was the forerunner of the European Euro, worth approximately US$1 in mid-1999.

The policy background

Until the late 1980s, there was a sharp division between east and west in Europe. In both parts, there was structured international cooperation on topics of environmental relevance. All Western European democracies participated in the OECD and in the Council of Europe, which was particularly active in human rights issues and the protection of the natural and cultural heritage. Most of these countries were also NATO members. Membership of the European Community (later the European Union) grew from the initial 6 to 15. In the socialist eastern part of the region, NATO was more or less mirrored by the Warsaw Pact and the European Union by COMECON.

Although some countries and environmental NGOs tried to build bridges between east and west, the only constant and respected bridge was the United Nations, and particularly the UN Economic Commission for Europe (UNECE). Since the UN Stockholm Conference of 1972, the UNECE focused considerable attention on environmental issues and, in spite of the difficult political situation and slow progress, its importance is hard to overestimate.

The disintegration of the socialist bloc triggered two important initiatives. In Western Europe, most of the international bodies just mentioned launched programmes to help the eastern countries in their transition to democratic systems with market economies. For example, the European Union complemented its ERASMUS programme, for the exchange of university students and teachers, with TEMPUS, established the PHARE and TACIS funds for financial assistance to Central and Eastern Europe, and participated in other international initiatives, such as the creation of the European Bank for Reconstruction and Development (EBRD) and the Regional Environmental Centre for Central and Eastern Europe. At the national level, bilateral east-west cooperation agreements were signed, and regions within countries and individual cities established reciprocal relations with counterparts in Central and Eastern Europe. Similar relations were established between private partners, such as business and agricultural organizations, trade unions and environmental NGOs.

The priorities of the eastern countries were to achieve a transition to a democratic market economy and join international organizations that were previously open only to Western European countries. Many countries have joined the Council for Europe, and the Czech Republic, Hungary and Poland have also joined OECD and NATO; the Russian Federation is at OECD's doorstep. Membership of the European Union is now the top priority, particularly for Central European countries,

There is a danger, however, that unsustainable lifestyle patterns in Western Europe will be adopted by countries in transition too indiscriminately. Not all Western European standards and policies are environmentally beneficial, and some management policies in eastern countries were environmentally beneficial. For example, forestry and farming systems in the Baltic states were comparatively sustainable throughout the communist era and maintained much higher levels of biodiversity than western systems; and the high level of private car ownership in Western Europe, with its accompanying high pollution levels in urban areas, is clearly not something to be emulated. The accession countries need to find an acceptable balance between adapting to Western European policy and maintaining existing practices where these are environmentally beneficial.

While the UNECE has continued and intensified its function as a link across the entire region, an important development was the Pan-European Conference of environment ministers in Dobříš Castle in 1991. This marked the beginning of the Environment for Europe (EfE) process and has led to many new initiatives and regular similar meetings in Lucerne, Sofia and Århus (see box above).

From Dobříš to Århus

The Dobříš Conference served as the formal beginning to the Environment for Europe (EfE) process – an international activity geared toward improving and rehabilitating the environment in countries in transition from centrally planned to market-based, democratic economies, creating a framework for expanding cooperation in Europe and converging European environmental policies over the long term.

An Environment Action Plan (EAP) for Central and Eastern Europe was endorsed at the Second Conference in Lucerne, Switzerland, in 1993 (World Bank 1994). This adopted a three-pillar approach: introducing policy reforms, strengthening institutional capacity, and developing cost-effective investments for environmental action. Following Lucerne, two mechanisms were created to help the transition countries implement the EAP and facilitate environmental investments. The first, the EAP Task Force, provides a forum for facilitating the development of National EAPs, exchanging information and experience, and assessing (and responding to) the needs of institutional development. The second, the Project Preparation Committee (PPC), provides a framework for the identification, preparation and financing of environmental projects. The EAP Task Force includes representatives from the transition countries and other UNECE countries, while the PPC is comprised of donors with an interest in supporting environmental finance in the region (OECD 1998a). An important, additional result of the Lucerne Conference was that the entire EfE process was put under the aegis of the UNECE.

The Third Conference, held in Sofia, Bulgaria, in 1995 called for increased business involvement in environmental protection and adopted the Pan-European Biological and Landscape Diversity Strategy. Four 'Sofia Initiatives' were introduced by Central European countries to accelerate the implementation of the EAP: Biodiversity, Economic Instruments, Local Air Pollution and Environmental Impact Assessment.

The Fourth Conference, in Århus, Denmark, in 1998 launched the first convention on citizens' environmental rights (the Århus Convention), two international protocols on limiting air pollution, a strategy to phase out lead in petrol, initiatives on energy efficiency, recommendations on financing environmental projects in Eastern Europe, and the European Biodiversity and Landscape Strategy.

MEAs and non-binding agreements

Global MEAs

Environmental factors are playing an increasing role in international relations within the region and between it and the rest of the world. Many countries, acting separately or as members of various political groupings, have played a major role in the development of global multilateral environment agreements (MEAs).

Parties to major environment conventions (as at 1 March 1999)

	CBD (174)	CITES (145)	CMS (56)	Basel (121)	Ozone (168)	UNFCCC (176)	CCD (144)	Ramsar (114)	Heritage (156)	UNCLOS (130)
EUROPE and CENTRAL ASIA (54)	47	35	25	39	48	48	29	41	51	28
Western Europe (25)	21	20	18	21	22	23	20	22	23	17
Central Europe (17)	15	10	6	13	16	15	2	15	17	8
Eastern Europe (7)	6	4	0	2	6	6	2	4	6	3
Central Asia (5)	5	1	1	3	4	4	5	0	5	0

Key: percentage of countries party to a convention

0–25% | 25–50% | 50–75% | 75–100%

Notes:

1. Numbers in brackets below the abbreviated names of the conventions are the total number of parties to that convention
2. Numbers in brackets after name of regions/sub-regions are the number of sovereign countries in each region/sub-region
3. Only sovereign countries are counted. Territories of other countries and groups of countries are not considered in this table
4. The absolute number of countries that are parties to each convention in each region/sub-region are shown in the coloured boxes
5. Parties to a convention are states that have ratified, acceded or accepted the convention. A signatory is not considered a party to a convention until the convention has also been ratified

The dynamic changes in recent years in the east and increasing moves towards integration in the west are of great relevance to global MEAs. As these are based in part on baseline pollution levels, the large economic and social shifts that are occurring in the transition countries may have unforeseen consequences. Disruption and economic collapse make it difficult to find resources to implement MEAs. On the other hand, transition presents opportunities for more flexible solutions, and young countries, such as those emerging from the break-up of the Soviet Union and Yugoslavia, which may lack national traditions or administrative experience, are tending to rely more on MEAs as a point of reference in international relations (UNEP 1998). In the west, economic integration is fuelling growth in important sectors, such as transport and tourism, which present severe challenges to the environment. But, in theory at least, increased integration can also lead to more effective transboundary cooperation in implementing MEAs.

The overall level of ratification of global MEAs is relatively high, and reasonably balanced among sub-regions (see table above). However, ratification, acceptance and implementation are affected by the particular environmental problems and priorities of the sub-regions. For example, the Convention on Migratory Species has been ratified by only a few of the countries undergoing economic transition.

In general, national legislation has been adopted for most global MEAs, including in Central Europe where much legislative drafting capacity has been committed to bringing environmental laws into line with European Union Directives prior to accession. However, technical difficulties have been encountered with respect to some MEAs. For example, under the Basel Convention several countries have reported discrepancies between their national lists of hazardous wastes and the lists in convention annexes (UNEP 1998). In some countries, particularly in Eastern Europe and Central Asia, international agreements become directly applicable upon ratification without the need to adopt national legislation. Many provisions of global MEAs had previously been covered by national legislation or regional agreements.

National strategies, plans and programmes have been adopted for most global MEAs, and national and regional institutions have been established to implement them, but the development of regulatory and enforcement measures has been uneven, with significant difficulties with the enforcement of some MEAs in the eastern parts of the region (UNEP 1998). This is especially true of MEAs that involve investigation and detection of violations, such as CITES and the Basel Convention. Special funds have been established to support transition countries with implementation. For example, Central European countries have received financial and technical aid through the Ramsar Small Grants Fund for Wetland Conservation and Wise Use, PHARE, GEF, EBRD and bilateral assistance programmes.

The Pan-European Biological and Landscape Diversity Strategy adopted at the Sofia Conference is an example of a regional contribution towards effective implementation of a global MEA. It is one of the instruments for the implementation of the Convention on Biological Diversity (CBD). The Strategy provides a framework for a consistent approach and common objectives for national and regional action to implement the CBD.

Traditional economic instruments such as taxes, subsidies, charges and fees are widely used to encourage implementation. MEAs may augment national regimes through the establishment of specific fees, such as waste disposal fees related to transboundary movements under the Basel Convention, which help to increase the effectiveness of hazardous waste disposal on a national level (UNEP 1997). Some newer types of economic instruments, such as emissions trading schemes, are still controversial and not yet well developed in connection with global MEAs.

While most countries abide by their general obligations for reporting, independent verification of reports is hampered by technical limitations and the legacy of totalitarian administrative systems in some countries. Nevertheless, some global MEAs have resulted in the establishment of effective monitoring systems. The Basel Convention has given rise to a well-established system for controlling transboundary movement of hazardous wastes, with good results in Western and Central Europe. Reporting under the Ramsar Convention is fairly good, with better reporting from Central than Western Europe (Ramsar Convention 1996).

While there have been substantial reductions in some emissions and improvements in some environmental conditions in areas covered by global MEAs, it is difficult to determine the extent to which these are the result of the MEAs themselves. In some cases national legislative regimes were in place before the MEAs were adopted; indeed national initiatives are often the first step towards environmental improvement, with further progress resulting from internationalization of goals through MEAs. In other cases, economic transition has had a direct and major impact, through reduced energy demand and more efficient production as well as economic collapse and the consequent reduction in the levels of industrial activity. Economic transformation in Eastern Europe has been a major contributor to a significant decrease in emissions of greenhouse gases (see page 114).

Transition to market economies with enhanced democracy and transparency can also have an indirect impact on the environment by creating conditions for more effective implementation of MEAs. One effect is the trend towards longer-term thinking and planning, driven in part by the interest of Central European countries in joining the European Union, which emphasizes the importance of implementation as well as legislation. International assistance from western donors aimed at integration into Euro-Atlantic structures helps to create the political will to carry out international obligations within the framework of civil society.

Some MEAs have clearly had positive effects. The Basel Convention is a good example, although the complexity of the situation means that much of the evidence is anecdotal. The reduction in the international market for the disposal of hazardous wastes and the political pressures against acceptance of waste from other countries have been credited to the successful working of the Basel Convention (Werksman 1997), partly through the availability of an array of enforcement tools.

Similarly, the Ramsar Convention has helped to protect particular wetlands from development. For example in Finland new habitats were established and vegetation returned to previous levels in identified wetlands. Water quality in estuaries has improved in areas where water treatment was enhanced as a result of Ramsar. On the other hand, Ramsar has been ineffective in preventing the loss of some listed wetlands in the face of intense pressures for development, especially from the transport and tourism sectors. In most countries the results are therefore mixed. In Germany for example, out of 14 sites studied, 6 showed substantial improvement, but

8 showed gradual negative impacts up to 1996 (Ramsar Convention 1996).

Regardless of the overall effectiveness of implementation, the improved monitoring that generally results from MEAs helps to focus international attention on problems of particular importance. The countries of Europe and Central Asia provide some of the best opportunities globally for the involvement of the public in independent and supplementary monitoring and reporting. CITES is one example where independent monitoring, an active secretariat, and the threat of sanctions by the conference of the parties have provoked positive legislative responses from slow-acting parties (ERM 1996).

Regional MEAs

On a regional level, concern for the environment has led to the establishment of new groupings based on shared natural resources, including the Baltic Sea, the Danube basin and the Rhine. Affected countries have used regional MEAs to create protection regimes, with a varying degree of success in terms of implementation and effectiveness. Differences result in part from the re-emergence of divergent forces shaping national priorities among the states that were formerly part of the Eastern Bloc.

On the whole, overall trends are similar to those for global MEAs: acceptance and compliance, including the establishment of formal arrangements and bodies, are generally good. For example, regional MEAs pertaining to the North Sea, the Rhine and the Baltic are highly developed, well implemented and are expanding in scope (UNEP 1998). The effectiveness of some other regional MEAs, however, is limited by the difficulties that some countries have in implementing their obligations. The protection of the Black Sea ranks low among national priorities in the sub-region and is not well funded. NGOs and the public have, however, taken up complex initiatives in place of the authorities, for example establishing information, education and resource centres (Black Sea Environmental Programme 1996 and 1997).

The implementation of the Sofia Convention on Protection of the Danube, not yet in force, will provide an indication of the extent to which countries in transition may have improved the execution and enforcement of environmental protection measures. This regional agreement, as well as the Espoo Convention on Environmental Impact Assessments in a Transboundary Context, may lead to new forms of cooperation in Europe.

On the regional level, the impacts of economic and social transformation in the transition countries may become more apparent. European integration should increase the influence of Western European environmental protection norms throughout the region but countries emerging from decades of authoritarianism may also be able to contribute their own ideas and solutions.

The more specific the scope of regional MEAs, the greater the need for national legislative and regulatory measures. The Barcelona Convention for the Protection of the Marine Environment and the Coastal Region of the Mediterranean involves many parties from several regions (Europe, Africa and West Asia) and has spawned several protocols. Because of the complex and varied nature of the Mediterranean area and its environmental problems, the convention operates primarily through programmes, action plans and initiatives, with great flexibility in terms of prescriptive measures. While the overall degradation of the Mediterranean Sea appears to have been halted (UNEP-MAP 1996), recovery has proved harder to achieve, due in part to a low level of commitment on the part of the few Central European parties to the Convention. Mixed results can also be found in coastal areas faced by intense pressures from building and tourism industries. Only where these industries take a longer view are the interests of conservation sufficient to hold off development pressures. Implementation and enforcement of legislation is mixed due to a lack of coordination of environmental management and legislation across economic sectors and sub-sectors. The main obstacles to effective implementation of the Barcelona Convention are thus institutional (WWF 1997).

This regime can be compared to the Helsinki Convention on the Baltic Sea, in which the member states are fewer in number and closer to each other in terms of regulatory tradition, social organization and state administration. This convention is typical of some regional MEAs which are characterized by relatively strong secretariats which supervise implementation and review, and take decisions on programmes and institutional matters. Under the Baltic Sea Convention, a strong commission (HELCOM) has been vested with power to make recommendations for specific legislative measures to be adopted by the parties. The confidence shown in

Regional MEAs

Treaty	*Place and date of adoption*
Agreement Concerning Measures for Protection of the Stocks of Deep-sea Prawns (*Pandalus borealis*), European Lobsters (*Homarus vulgaris*), Norway Lobsters (*Nephrops norvegicus*) and Crabs (*Cancerpagurus*)	Oslo 1952
Convention Concerning Fishing in the Waters of the Danube	Bucharest 1958
Convention Concerning Fishing in the Black Sea	Varna 1959
Protocol Concerning the Constitution of an International Commission for the Protection of the Mosel Against Pollution	Paris 1961
Agreement Concerning the International Commission for the Protection of the Rhine Against Pollution	Berne 1963
European Agreement on the Restriction of the Use of Certain Detergents in Washing and Cleaning Products	Strasbourg 1968
European Convention for the Protection of Animals During International Transport	Paris 1968
European Convention on the Protection of the Archaeological Heritage	London 1969
Agreement for Cooperation in dealing with Pollution of the North Sea by Oil	Bonn 1969
Benelux Convention on the Hunting and Protection of Birds	Brussels 1970
Convention on Fishing and Conservation of the Living Resources in the Baltic Sea and Belts	Gdansk 1973
Convention on the Protection of the Environment Between Denmark, Finland, Norway and Sweden	Stockholm 1974
Convention on the Protection of the Marine Environment of the Baltic Sea Area	Helsinki 1974
Convention for the Protection of the Mediterranean Sea Against Pollution	Barcelona 1976
European Convention for the Protection of Animals Kept for Farming Purposes	Strasbourg 1976
Agreement Concerning the Protection of the Waters of the Mediterranean Shores	Monaco 1976
Convention on the Protection of the Rhine Against Chemical Pollution	Bonn 1976
Convention Concerning the Protection of the Rhine Against Pollution by Chlorides	Bonn 1976
European Convention for the Protection of Animals for Slaughter	Strasbourg 1979
Convention on the Conservation of European Wildlife and Natural Habitats	Berne 1979
Convention on Long-range Transboundary Air Pollution	Geneva 1979
European Outline Convention on Transfrontier Cooperation Between Territorial Communities or Authorities	Madrid 1980
Benelux Convention on Nature Conservation and Landscape Protection	Brussels 1982
Agreement for Cooperation in Dealing with Pollution of the North Sea by Oil and Other Harmful Substances	Bonn 1983
European Convention for the Protection of Vertebrate Animals Used for Experimental and Other Scientific Purposes	Strasbourg 1986
European Convention for the Protection of Pet Animals	Strasbourg 1987
Convention on Environmental Impact Assessment in a Transboundary Context	Espoo 1991
Convention Concerning the Protection of the Alps	Salzburg 1991
Agreement on the Conservation of Bats in Europe	London 1991
Convention for the Protection of the Marine Environment of the North-east Atlantic	Paris 1992
Convention for the Conservation of Anadromous Stocks in the North Pacific Ocean	Moscow 1992
Agreement on the Conservation of Small Cetaceans of the Baltic and North Seas	New York 1992
Convention on the Protection and Use of Transboundary Watercourses and International Lakes	Helsinki 1992
Convention on the Protection of the Marine Environment of the Baltic Sea Area	Helsinki 1992
Convention on the Protection of the Black Sea Against Pollution	Bucharest 1992
Agreement on the Protection of the Meuse	Charleville Mezières 1994
Agreement on the Protection of the Scheldt	Charleville Mezières 1994
Convention on Cooperation for the Protection and Sustainable Use of the Danube River	Sofia 1994
The Energy Charter Treaty	Lisbon 1994

Source: UNEP 1997

Implementation of the 1979 Convention on Long-range Transboundary Air Pollution

The Convention on Long-Range Transboundary Air Pollution (LRTAP) in Europe is a classic example of regional environmental management. The first protocols organized finance and addressed acidification and photochemical pollution. Acidification was addressed again in 1994 since it was found that the first sulphur protocol did not provide sufficient protection. Recently, attention has been focused on the problems caused by persistent organic pollutants and heavy metals. Future priorities include development of an innovative, multi-effect, multi-pollutant protocol aimed at nitrogen oxides and related substances, which will include protection of the environment as well as human health.

Participating countries commit themselves to periodic reporting on emissions, national strategies and programmes. Many participating countries have developed action plans or long-term strategies based on a system of cost-effective, differentiated obligations.

Clear financing, the involvement of national scientific bodies and joint implementation have contributed to LRTAP's status as one of the most successful regional MEAs. Emissions of acidifying substances have decreased in all areas since the first protocols came into force. The decrease is greatest for sulphur dioxide, the pollutant causing the major problem, with expected national reductions in 2000 relative to 1980 of about one-third in Central and Eastern Europe, and two-thirds or three-quarters in Western Europe. However, reductions of the emissions of nitrogen oxides, ammonia and hydrocarbons are more difficult to achieve. This is the subject of a new protocol, due to be signed in autumn 1999. To exploit the advantages of the multi-effect-multi-pollutants approach, sulphur emissions will also need to be further reduced. This should increase the cost-effectiveness of controlling air pollution for all participating countries.

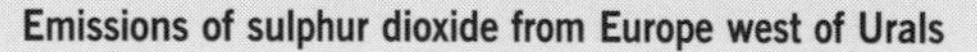

Emissions of sulphur dioxide from Europe west of Urals

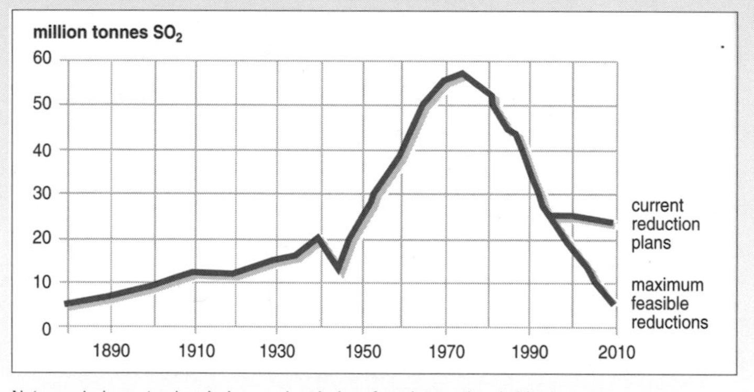

Note: excludes natural emissions and emissions from international shipping

Source: compiled by RIVM, the Netherlands, based on data from Mylona 1996 and EMEP/MSC-W 1998

HELCOM by the parties leads to a high level of co-ordination of programmes (Baltic Marine Environment Protection Commission 1996). Weaker institutional arrangements in other conventions may contribute to a mixed level of implementation, resulting only in isolated and localized improvements.

Regional MEAs often give rise to bilateral agreements for coordination and assistance with implementation. The Espoo Convention, which came into force in 1997, gave rise to several bilateral agreements following its signature in 1991, including cooperation on transboundary environmental impact assessment between Germany and Poland. The Espoo convention has also had an impact on other international instruments, including the Convention on the Transboundary Effects of Industrial Accidents, the Convention on the Marine Environment of the Baltic Sea Area, and several ministerial declarations. The Helsinki Convention on the Protection and Use of Transboundary Watercourses and International Lakes has provided a basis for international cooperation with respect to certain water bodies (Oder River cooperation between Poland and Germany and cooperation on water bodies between Finland and Estonia). Finland and Estonia have entered into a specific agreement that implements both the Baltic Sea Convention and the Helsinki Convention.

Traditional economic incentives, including taxes, subsidies, charges and fees, have been used in the context of transboundary watercourses but their effectiveness has been limited by economic disruption in transition countries, where rapid currency fluctuations earlier in the decade affected the system of incentives. Where charges failed to keep pace with inflation, polluters preferred to pay fines rather than change their behaviour (UNEP 1998). More recently, in most countries, inflation has come under control or economic measures have been indexed.

The North Sea regime, composed of the Oslo Convention for the Prevention of Marine Pollution by Dumping from Ships and Aircraft (1972) and the Paris Convention for the Prevention of Marine Pollution from Land-Based Sources (1974), has relied heavily in recent years on non-binding instruments created during conferences of the environment ministers of the participating states (Victor and others 1998). These International North Sea Conferences resulted in ambitious goals for pollution reduction, which were acceptable to the parties because they were non-binding. The goals adopted at the conferences were effective but became more so when they were later codified into legally-binding measures through the Oslo and Paris (OSPAR) Commissions and the European Union.

The level and effectiveness of monitoring and compliance of regional MEAs varies greatly, a major factor being the effectiveness of the institutions and arrangements for collecting and processing data and information. The Baltic Sea Convention, for example, is adequately supported by the parties and effective in information gathering, enabling the commission to

Selected actions plans and programmes

Baltic Sea Joint Comprehensive Environmental Action Programme
Twenty-year (1993–2012), ECU 18 000 million programme to prevent and eliminate pollution, mainly from municipal and industrial sources, and promote the ecological restoration of the Baltic Sea. Programme includes policies, laws and regulations, institutional strengthening, investment activities, management programmes for coastal lagoons and wetlands, applied research, public awareness and environmental education.

Baltic 21 Programme
Also known as *Agenda 21 for the Baltic Sea Region*, a regional action programme for sustainable development, focusing on agriculture, energy, fisheries, forests, industry, tourism and transport.

Baltic Sea Experiment (BALTEX)
Covers meteorological, hydrological and oceanographic aspects related to the energy and water balance of the Baltic Sea and its catchment region. The programme includes numerical modelling, data assimilation, experimental and numerical process studies, re-analysis of existing data, and application of remote sensing. It aims to improve weather forecasting and climate models, and to support flood forecasting systems.

Strategic Action Plan for the Rehabilitation and Protection of the Black Sea
Aims at sustainable development in the Black Sea region and a biologically diverse Black Sea ecosystem. Policy actions include reducing pollution from land-based sources, vessels, and dumping, waste management, assessment and monitoring of pollutants, biodiversity, habitat and landscape protection, environmental impact assessment, integrated coastal zone management, sustainable aquaculture and tourism, and public involvement.

WHO Programmes in Environment and Health
Mediterranean Action Plan
The WHO/Europe project office for the Mediterranean Action Plan provides information on health-related aspects of pollution in the Mediterranean Sea, develops common regional guidelines, criteria and standards for coastal recreational areas, shellfish areas and land-based sources of pollution and related activities, and helps countries in monitoring for the prevention and control of marine pollution.

National Environmental Health Action Plans (NEHAPs)
A series of national plans under WHO auspices, an *Environmental Health Action Plan for Europe* and the *Health for All by the Year 2000* programme, covering accidents, environmental health policy, environmental health service, water, air, food, soil and waste, human settlements and the work environment.

Danube
The *Environmental Programme for the Danube River Basin* includes the creation of an operational emergency warning system for sudden pollution, a water quality monitoring strategy, a strategic action plan, enhancing public awareness and participation, and supporting actions to rehabilitate wetlands. The *Danube Pollution Reduction Programme* aims to reduce pollution in the basin and Black Sea in support of the *Danube River Convention*. Activities include an information system for the Danube basin, a Danube water quality model, a wetlands rehabilitation study, a social and economic impact analysis and the development of financing mechanisms.

National Plans
While there is no international obligation in Western Europe for national environmental plans or programmes, some countries, such as the Netherlands, have been developing sophisticated planning systems, including long-term planning, since the 1970s and their system of intermediate planning and evaluation has been used as a model for other countries and for the 5th EAP of the European Union.

In Central and Eastern Europe, the EfE process has resulted in the development of National Environmental Action Plans in 16 transition countries (OECD 1998b). They include new environmental framework laws, implementation of new policies (the polluter pays principle, the right to know), medium- and long-term plans, mobilization of financial resources to tackle priority problems, improved public participation, strengthening of institutions, and enhancement of environmental management skills.

Local Initiatives
The EfE process and *Agenda 21* have been an important stimulus, both for local action and for national and international cooperation. Coordinating bodies such as the Earth Council, the Global Action Plan and the International Council for Local Environmental Initiatives have been involved in launching the European Sustainable Cities and Towns Campaign, drafting charters, organizing meetings and training courses, and creating a website (http:/www.iclei.org) with information on initiatives of individual communes and cities. In Western Europe, these developments have been encouraged by a European Union strategy stimulating urban and regional research.

make well-founded recommendations for incorporation into national legislation. The level of monitoring and compliance of other MEAs, including the Barcelona and Black Sea Conventions, have been affected by technical limitations and economic disruption. Even where reporting obligations are met, however, independent verification may be difficult. HELCOM and the OSPAR Commission have sufficient resources for gathering basic data but it is difficult to monitor the extent to which member states follow recommendations, since reporting is done by each country on a mandatory basis but without provisions for enforcement. Implementation of the 1979 Convention on Long-range Transboundary Air Pollution is discussed in the box on the left.

Bilateral agreements among neighbouring states

are used to build capacity and transfer technology from Western European states to countries in transition. Examples include agreements between Germany and Poland, and Finland and Estonia. In Poland, debt-for-nature swaps implemented as part of these schemes have been used to improve water protection. Regional arrangements based on particular shared natural resources provide scope for capacity building and technology transfer. Funds for such agreements have been established through regional MEAs developed under UNECE auspices.

Regional action plans

The European Union's Fifth Environmental Action Programme (5th EAP) was approved in 1993 and runs until the year 2000 (CEC 1993). Despite progress in some areas, emissions of several pollutants need to be further reduced so that targets already agreed – and new ones in prospect – can be met. Progress has been least evident in the agriculture, tourism and transport sectors (EEA 1998). Fundamental changes and improvements are also required for the Common Agricultural Policy, particularly in light of the European Union harmonization process of Central European countries. The 5th EAP has been criticized for being too broad and without clear mandates for action, budgets and deadlines. Nevertheless, the European Union is the most advanced form of international integration and cooperation in the world and it has achieved unprecedented progress

Examples of other action plans at the sub-regional, national and local level are given in the box on page 263.

Laws and institutions

There is no international obligation for national governments to draft environmental plans or programmes. Some countries, such as the Netherlands, have been developing sophisticated planning systems, including long-term planning, since the 1970s. Their system of intermediate planning and evaluation has been used as a model for other countries, such as Austria and the Flanders region in Belgium, and for the 5th EAP of the European Union. For other countries, the need to draft national environmental reports for the UNCED conference was an important incentive to environmental planning.

However, plans and strategies have to be implemented and this is not always simple, particularly in sectors where the easier measures have already been implemented but further improvement is still needed.

In Central and Eastern Europe and in Central Asia, the EfE process has been an important driver of national plans. Countries which had environmental management capacity, such as Belarus, the Russian Federation and Ukraine, have welded some EAP elements into their traditional environmental policies. Comprehensive National Environment Action Plans (NEAPs) have been produced, for example in Armenia, Georgia and the Republic of Moldova, which served many purposes, including elaborating new policy principles and redesigning institutions. Sixteen transition countries have developed or are developing NEAPs (OECD 1998b).

NEAPs have resulted in new environmental framework laws, implementation of new policies (such as the polluter pays principle and the right to know), medium and long-term plans, mobilization of financial resources to tackle priority problems, improved public participation, strengthening of institutions, and enhancement of environmental management skills. However, there is still a need for more effective private sector involvement in the NEAP process.

Continued progress is limited by the lack of financial resources and weak capacity, especially in economic and financial analysis, priority setting and monitoring. Most NEAPs lack an adequate financing strategy (OECD 1998b). With the exception of Lithuania, there has been no systematic effort to identify and estimate costs and revenue sources, and to balance them.

Since the early 1970s, citizens in many countries have initiated local environmental activities such as separate waste collection, energy conservation measures, avoiding pesticide use and local education projects. Gradually, these activities have become more structured and sometimes integrated into 'clean' city plans. The adoption of *Agenda 21,* which explicitly emphasizes the importance of local action, has been a major stimulus. Outside Western Europe, the EfE process has also been a positive factor.

The substantial body of European Union environmental legislation that exists today has developed mainly over the past 10 years, and now provides a common frame-work for the development of the national policies of all member states. The current trend is to integrate this diverse body of legislation into framework laws, to update existing legislation, and to adopt legislation on entirely new subjects. Implementation is, however, proving more difficult. Examples include the Nitrates Directive, the Habitat Directive and the Natura 2000 plan for a European Ecological Network, and the amendment to the Directive for the Protection of Wild Birds which has been under discussion since 1994. The European Union has also begun to implement legal measures aimed at the recovery of resources and waste reduction.

A series of new measures have been introduced to combat air pollution. These include addressing the emissions of cars, trucks and non-road vehicles, emissions of incineration plants for different types of waste, and emissions of volatile organic compounds and petrol and diesel fuel. A proposal for a revised Directive on sulphur dioxide, nitrogen oxides, particulates and lead has recently been published.

European Union Member States have to adhere to the common European Union environmental policy developed over many years (see box right). Central European countries are required to adapt their national policies and legislation to the European Union's body of legislation if they wish to qualify for accession.

Eleven countries have begun the process of adapting their domestic legislation, institutions and structures to European Union standards. In July 1997, the European Commission published its opinion on the progress of the accession countries in *Agenda 2000*. It held open the possibility of accession to all candidates but considered five countries (Czech Republic, Estonia, Hungary, Poland and Slovenia) eligible for accession discussions. Negotiations with these countries (and with Cyprus) began in March 1998 with a view to full membership during 2002–2005. 'Screening' is being carried out with a further five countries: Bulgaria, Latvia, Lithuania, Romania and Slovakia (ED 1999). Five countries in southeast Europe (Albania, Bosnia and Herzegovina, Croatia, the Former Yugoslav Republic of Macedonia and Yugoslavia) are not included in the European Union accession process but are in need of assistance and have expressed the will to adapt their laws and policies accordingly. Most other Central European countries do not view European Union membership as a near-term possibility.

One of the main barriers to accession is the lack of financial resources – the cost of conforming to environmental requirements is estimated at ECU 100 000–150 000 million for the 11 accession countries. In order to fund the necessary investments 3–5 per cent of these countries' GDP would have to be re-allocated to environmental expenditures over a period of 20 years. In contrast, the average environmental expenditures of OECD member countries is only 1–2 per cent of GDP. Some help will be forthcoming from the European Union, with a total of ECU 53 800 million earmarked for the period 2000–2006 (ED 1997a). This amount covers all sectors rather than just the environment.

Drafting and implementing the legislation needed for accession to the European Union is a somewhat distant objective for many Eastern European and Central Asian countries, although the Republic of Moldova and Ukraine have already indicated their desire to begin the process (OECD 1998b). For the remaining Eastern European and Central Asian countries, current legislative development relates mainly to implementing new media-specific laws with attainable standards and regulations, coupled with stricter implementation and enforcement.

Environmental policy in the European Union

The European Union is largely responsible for environmental policy in Western Europe. However, the OECD Environmental Policy Committee, the UN Economic Commission for Europe (UNECE), the Council of Europe and other organizations have also initiated important initiatives.

Before 1987, European environmental policy was determined mainly by developments in the individual member states, which were 'harmonized' or paralleled by similar, but often weaker, legislation at the European Union level. This created policy gaps and inconsistencies.

The Single European Act of 1987 was devoted primarily to the completion of the internal market by simplifying decision-making procedures but also introduced environmental protection as a formal goal.

The Maastricht Treaty of 1992 expanded the common activities with two new 'pillars': a Common Foreign and Security Policy, and a Home Affairs and Justice Policy. The notion of sustainable development was also reflected in the treaty.

The 1997 Amsterdam Treaty was largely determined by the need to change both policies and decision-making procedures before opening the European Union doors to countries in Central and Eastern Europe. The treaty emphasizes the need to integrate environmental and sustainable development concerns in all policy areas.

Economic instruments

Direct instruments

The aim of direct economic measures in policy making is to encourage responsible use of natural resources, avoid pollution and waste, and make prices include environmental costs. The application of market-based instruments that affect costs directly is encouraged by the 5th EAP but is still rare at the European Union level. Progress in increasing civil liability has been slow. A draft Directive concerning waste has been stalled, while a more general *White Book* on civil liability is still forthcoming. Energy or carbon taxes which may help decrease energy or carbon use have been discussed for several years without much progress. In some countries, eco-taxes are being added to fuel and other energy products but their effects can be offset by increasing energy consumption if adequate policy responses are not in place. There is a reluctance to apply the tax instrument in other areas. An additional complicating factor is the general need for unanimity in the area of taxation. However, tax instruments appear to be effective in, for example, Belgium, Denmark, the Netherlands and Sweden.

Fees and fines are applied across the region, levied on polluters and users of the environment and through the stricter implementation of standards and regulations, primarily targeting the private sector. In general, they succeed when they address specific environmental issues and do not threaten the competitiveness of key sectors.

The European Union does not impose fees or fines directly in the environment area but they are used in some member states. In spite of the positive approach in the 5th EAP, the European Union is more a barrier than a help in this area. It is extremely difficult for member states to focus national measures on products since the European Union is adamant in defending the open market. If states do apply discouraging economic instruments, their scope is usually limited because stringent national measures would distort international competitiveness. In 1997, the European Commission formulated its view in *Environment taxes and charges in the single market* (EC 1997).

The implementation of fees and fines in Central and Eastern Europe and in Central Asian countries is limited by weak enforcement, large differences in levels of modernization and profitability between enterprises, and the rapidly growing number of small and medium-sized enterprises. In Hungary and Latvia, product charges are being imposed on manufacturers of goods, such as packaging or batteries, which can be recycled or alternately disposed of, encouraging private investment. The striking differences between imposed charges and fines and the amount actually paid suggest an inability and in some cases an unwillingness on the part of enterprises to pay, a lack of adequate sanctions against non-payment, and a lack of institutional capacity to collect fees (REC 1998). Furthermore, inflation – relatively high until 1995 – had the effect of eroding the value of fixed charge and tax levels (REC 1994a).

Indirect instruments

Several indirect initiatives combine elements of awareness-raising with improving market position through better environmental performance, such as the Eco-Management and Audit Scheme (EMAS) and the (later) ISO 14 000 standard. Participation is voluntary although both have the status of formal legislation. Participating industries undertake to install environmental management systems and produce environmental performance reports that must be verified by a certified environmental auditor. If standards are met, the industry can market an approved label in its advertisements and other promotion schemes.

The European Union's eco-label (92/880/EEC) is another instrument that can be awarded to products that meet certain European Union standards, in anticipation that consumers will prefer those products. Unfortunately, there is disagreement about the form the label should take and several member states have maintained or threatened to develop their own labels.

In some countries, businesses have negotiated voluntary agreements with governments to meet certain environmental targets within a given time span. In the Netherlands, such agreements, or 'covenants', have been widely used and they are being given increasing attention in France, Germany and other countries (EEA 1997). Because the application of this instrument may interfere with competition requirements and anti-cartel legislation and with the duty of member states to implement European Union Directives by means of formal, national legislation, the European Commission has entered the debate by publishing basic rules in a Communication on Environmental Agreements (97/C321/02).

Fair market conditions and competition are also promoted by the strict implementation and enforcement of legislation. As a rule, this is the task of the individual member states. However, the Commission has also played an active role as 'prosecutor', taking member states to the European Court for non-compliance. Until recently, these legal actions bore no financial consequences. However, on the basis of Article 171 of the Maastricht Treaty, penalties for non-conformity with European Union environmental legislation may be imposed.

During the 1992 Dutch presidency of the European Union, the Environment Inspectorate of the Netherlands ordered a report on the state of enforcement in the European Union (ERM 1991) and invited their European Union colleagues to discuss it. Since then there have been regular meetings between the national enforcement agencies and the European Commission.

Industry and new technologies

Cleaner production

During the 1995 Environmental Ministerial conference in Sofia, the Environment Ministers of the countries involved in the EfE process endorsed the cleaner production (CP) approach as a preferred strategy and called for a basic capacity level to be reached for cleaner production activities, requiring an active core of advisors, the establishment of a number of centres, training materials in local languages, case studies, demonstration projects and business plans, the introduction of the principles of CP in university curricula and a monitoring framework.

In Western Europe, the European Union has endorsed the cleaner production strategy through preventative measures such as the IPPC (Integrated Pollution and Prevention Control) Directive (96/61/EC). This entails the use of the best available technologies and takes into account all the environmental consequences and the entire life cycle of the product.

Acceptance of EMAS, initially slow, is growing rapidly with 1 502 sites registered by April 1998. Austria, Germany, Sweden and the United Kingdom are leading countries in terms of registration. With considerable funds being made available in the European Union for the development of cleaner production processes and technologies, this number is likely to increase further, although EMAS still remains voluntary. Demonstration projects, such as those funded by the Community Financial Instrument for the Environment (LIFE), and various sub-projects under the umbrella of the Framework Programmes for Research and Development of the European Union are also contributing to CP.

Economic incentives can also help to promote CP. In their June 1998 meeting, the European Union environment ministers coupled the introduction of greenhouse gas emission reductions to the introduction of an energy tax but the decision about when to introduce it has still to be made. Increasing VAT on resource and energy use, while simultaneously decreasing taxes on labour, is also frequently recommended as an instrument with positive effects on both CP and employment.

In Central Europe, with European Union harmonization approaching, some countries are already preparing for EMAS registration and have begun introducing the similar ISO 14001 environmental management standard; a bridging document highlighting the differences between the two as well as facilitating conversion has recently

Cleaner production in the Czech Republic

The Cleaner Production Centre (CPC) in the Czech Republic, founded with the support of the Norwegian Government, is a professional non-governmental and non-profit organization dealing with prevention practice in waste and waste production. The CPC was established in 1994 under the auspices of the Czech-Norwegian project on cleaner production aimed at establishing domestic professional capacities. Since 1995, the CPC has been a member of a UN-organized international network of national centres for cleaner production and has been awarded a long-term grant under the cleaner production scheme of the United Nations Industrial Development Organization (UNIDO) and the United Nations Environment Programme (UNEP).

The CPC organizes long-term interactive CP courses open to employees of industrial companies and consulting firms. A course includes 8–15 days of theoretical work spread over six months and the preparation of a case study for a selected site. The study is prepared by the participants under the professional assistance of a consultant from the CPC. Some of the proposed CP measures are implemented at the site immediately, further measures being taken according to the decisions of its management.

The Czech-Norwegian project on cleaner production has held three interactive CP courses since its inception. The first two courses were tutored by Norwegian specialists while the third was administered by Czech specialists. Thirty-four case studies have been prepared and 122 specialists have been trained.

These studies have produced savings of a total [illegible] million Czech crowns annually and prevented annual discharges of:

- 2 100 tonnes of emissions of volatile organic substances (from fuels and solvents);
- 12 000 m^3 of sewage wastewater;
- 12 000 tonnes of waste classified as 'special and hazardous' – that is, waste that can be stored only at secured dumps or which demands other methods of neutralization, such as incineration.

Source: REC 1995a

been published (REC 1997/1998). In addition, the EAP Task Force has played an active role in encouraging the adoption of CP activities outside Western Europe. The Czech Republic, Hungary and Poland, for example, have already achieved the necessary basic capacity level with a further nine countries, including the Russian Federation and Ukraine, close behind.

Experience in policy formulation in the field is limited, however, with government support ranging from active to non-existent. While the Czech Republic, Poland and Slovenia have been active supporters from the outset, other countries have made more limited progress. This is partly because CP is still not a top priority among key government and industry decision-makers. The current regulatory framework, often outdated and poorly enforced, concentrates on end-of-pipe measures and thus does not provide an incentive to adopt a preventive approach among enterprises. The slow response of enterprises in Central and Eastern European countries to adopting CP can also be attributed to limited investment capital, the crowding out of CP investments by other projects, lack of awareness among enterprises, unwillingness to take risks, and the slow privatization process which delays transfer of ownership and management (OECD 1998a). An EBRD survey, however, indicated that 85 per cent of financial sources were willing to support CP investment, particularly since some observers say that pollution reductions of as much as 50 per cent can be achieved by investing in energy efficiency and waste minimization.

Further work is needed to strengthen support for CP among policy-makers, promote the use of a wider range of policy instruments, implement further environmental management systems and standards, and improve financing mechanisms. While the EAP Task Force supports these objectives, critical gaps in knowledge and information are still common. Overcoming these problems and raising awareness among industrial company managers about the available systems, technologies, and the costs and benefits of CP programmes also remains a priority of the Task Force. An international declaration supporting CP and sound environmental management practices is under preparation under the aegis of the Task Force, and was endorsed in Denmark at the Århus Ministerial Conference in June 1998 (OECD 1997).

Making technology more sustainable

In Western Europe the awareness that production processes and products have to be made more environment-friendly is by now widespread, both in science and industry (Schmidheiny 1992). In the scientific world, several institutes such as the DOMUS Academy in Italy and the Wuppertal Institute in Germany are now devoting attention to the design of 'eco-products' and services (von Weizsäcker and others 1995).

In practice, government activities in this area are rare. An exception is the Netherlands, where a five-year Sustainable Technology Development Programme was financed by five ministries (IEEP-B 1994). It addressed key sectors such as food production, housing, water management, transport and the chemical industry, and has found widespread interest in science and industry.

The introduction of concepts such as ecological space and carrying capacity has led to a growing emphasis on eco-efficiency – developing or choosing between alternative products and services on the basis of their total environmental impact. An important tool for this is Life Cycle Analysis (LCA) in which all environmental impacts (such as energy and resource consumption, pollution, and impacts on bio-diversity) are determined from 'cradle-to-grave' (or, in agriculture, from 'stable-to-table'). For a product or device, this implies that the impacts of mining, production of raw materials, the production process, product use and final disposal (including reuse or recycling) are all taken into account. In Western Europe, the European Union has been active in developing LCA (Udo de Haes 1996) and promoting its application. For example, the application of LCA is demanded in the EMAS regulation and the directives on IPPC and eco-labelling.

The concept of tradeable permits (see page 207) can in principle provide an important incentive for increasing eco-efficiency – companies that can do better than their permit demands can benefit by selling all or part of their emission rights. The concept has been developed into a highly sophisticated system in the United States (see page 305) but there is as yet no counterpart in the European Union and it is applied only rarely in individual member states. This may change, however, as a result of the joint implementation agreed to in the Kyoto Protocol. The allocation of the emission reductions agreed in Kyoto and presently under discussion in the European Union will inevitably lead to national emission allowances for greenhouse gas emissions that could be further distributed via tradeable permits.

Financing environmental action

Financing by international bodies

The European Commission estimates that during 1994–99 more than ECU 17 000 million was spent in Western Europe on environmental measures. The money was invested in a wide range of activities including agro-environmental initiatives in Spain, wastewater treatment in Greece, protection of biotopes in Ireland, institutional capacity-building in Central and Eastern Europe, and toxic waste treatment along the Baltic coast.

There are currently three main sources of European Union finance with provisions for environmental expenditure: Structural Funds, the Cohesion Fund and the Community Financial Instrument for the Environment (LIFE). In addition, the European Investment Bank (EIB) provides loans under favourable conditions. Another important source of finance are the many European Union research budgets which are available for policy support or research. Assistance programmes are also available for the transition countries – PHARE covers Central Europe and TACIS covers Eastern Europe and Central Asia.

The Structural Funds are used for agriculture, and social and regional development. For the period 1994–99, their budget is about ECU 150 000 million.

The Cohesion Fund concentrates on environmental protection and the development of infrastructure in those member states where GDP is below 90 per cent of the European Union average. For 1994–99 the total budget is about ECU 14 500 million, of which 50 per cent is spent on environmental projects. Priority areas include water supply infrastructure, wastewater and urban waste treatment and nature conservation. The development of the infrastructure of long-distance Trans-European transport Networks has its own financing scheme.

LIFE is devoted to practical demonstration models of sustainable measures and activities, and for strengthening administrative structures. Its budget for 1996–2000 is about ECU 450 million.

These financial sources are a powerful force for development but they have also had negative impacts on the environment – for example, when used to finance the drainage of valuable wetlands for

Donor environmental commitments to CE, EE[a] and CA[a] countries, 1994–97

	Technical cooperation		*Investments*	*Total[b]*	*Total per capita*
	Policy development	*Investment preparation*			
	(million ECU)		*(million ECU)*		*(ECU)*
Central Europe					
Albania	20.7	0.9	24.9	60.5	17.9
Bosnia-Herzegovina	0.3	0.9	32.2	33.5	9.3
Bulgaria	20.2	3.0	90.1	136.3	16.0
Croatia	0.9	1.2	88.8	90.9	20.2
Czech Republic	39.8	5.0	313.5	397.3	38.5
Estonia	7.5	7.1	73.5	132.1	88.8
FYR Macedonia[c]	1.3	0.0	5.4	10.3	4.8
Hungary	16.1	0.5	172.9	208.4	20.4
Latvia	9.5	7.0	96.5	123.9	48.8
Lithuania	15.7	10.7	86.9	138.5	37.1
Poland	34.6	18.2	339.3	603.5	15.6
Romania	12.4	25.1	169.3	249.4	11.0
Slovakia	9.6	2.2	132.1	145.2	27.2
Slovenia	19.1	0.3	20.2	43.9	22.8
Region-wide	12.6	16.6	23.0	107.2	–
Total	**220.3**	**97.6**	**1668.6**	**2486.1**	**20.9**
Eastern Europe and Central Asia					
Armenia	0.1	0.3	0.0	0.4	0.1
Azerbaijan	0.4	0.3	63.4	64.0	8.5
Belarus	3.2	3.2	1.0	7.4	0.7
Georgia	42.0	0.4	18.0	60.4	11.1
Kazakhstan	14.5	1.1	0.0	15.6	0.9
Kyrgyzstan	3.0	0.0	0.0	3.0	0.7
Moldova Republic	4.8	1.3	1.4	7.5	1.7
Russian Federation	103.0	17.7	94.6	375.2	2.5
Ukraine	22.2	11.7	22.8	56.7	1.1
Uzbekistan	11.6	8.4	67.5	87.4	3.8
Region-wide	36.5	0.0	0.0	36.5	–
Total	**240.3**	**44.5**	**268.8**	**714.2**	**2.6**
CE-, EE- and CA-wide	11.6	2.2	0.0	13.7	–
TOTAL	**472.2**	**144.3**	**1937.4**	**3305.2**	**8.4**

a. preliminary data

b. totals are larger than the sums of technical cooperation and investment assistance, as some donors did not classify commitments.

c. Former Yugoslav Republic of Macedonia

Source: OECD 1998a

agriculture, the development of tourist capacity in unspoiled areas and the construction of transport infrastructure. With the exception of LIFE, the shares of the member states in these Funds are fixed and the Commission cannot force member states to allocate them to particular projects or in particular ways.

PHARE and TACIS are aimed at assisting transition to democratic free-market economies and supporting the reintegration of economies and societies with the European Union and the world. PHARE's budget for the environment during 1990–95 was ECU 483 million, or 9 per cent of its total budget. TACIS's budget for the environment for 1991–95 was ECU 429 million, or 19 per cent of its total budget (PHARE 1997).

In the past, PHARE has catalysed funds for important projects from other donors through studies, capital grants, guarantee schemes and credit lines. It has also invested directly in infrastructure projects, such as wastewater treatment, pollution monitoring, hazardous waste disposal, nature conservation, education and training, and air pollution abatement. Such investment spending is likely to continue since PHARE has recently been restructured to focus on compliance with European Union standards, with up to 70 per cent of funding now supporting investments for compliance with European Union law (ED 1997a).

TACIS helps Eastern European and Central Asian countries by funding know-how transfer through advice and training, advising on the reform of legal and regulatory frameworks, institutions and organizations, setting up partnerships and networks, and promoting political and economic links between countries. The programme is due to be renewed in 1999 (OECD 1998a).

In addition to supporting PHARE and TACIS, the EIB offers finance for transport, energy, telecommunication and environment projects. In January 1997 it was given a mandate to lend ECU 3 500 million to the ten European Union candidate countries, with maximum priority to the environment. The Bank, however, cannot finance 100 per cent of projects, and other financial institutions, such as the European Bank for Reconstruction and Development and the World Bank, are also expected to co-finance, along with government and industry (ED 1997b).

The Project Preparation Committee (PPC) is a network of donors and funding institutions established to facilitate environmental investment in Central and Eastern European and Central Asian countries. The financial support offered is largely catalytic, with the PPC matching donors to priority projects. By 1995, 45 projects had been thus matched (OECD 1998b).

The table on page 269 summarizes the foreign assistance offered to Central and Eastern European and Central Asian countries. Clearly, Central European countries benefit the most, with per capita environmental assistance being more than eight times that of countries in Eastern Europe and Central Asia.

Foreign direct investment: the main recipients

	total 1996 (US$million)	*per capita (US$)*	*cumulative 1989–96 (US$million)*
Czech Republic	1 264	123	7 120
Hungary	1 986	195	13 260
Poland	2 741	71	5 398
Kazakhstan	1 100	67	3 067
Russian Federation	2 040	14	5 843
Total five countries	**9 131**	**41**	**34 688**
Total all CE/EE/CA	**12 330**	**31**	**43 888**

Source: EBRD 1997

Private foreign investment

In mid-1997, the private sector accounted for 50 per cent or more of GDP in all Central European and in about half of the countries of Eastern Europe and Central Asia. Although data are difficult to obtain, foreign direct investment (FDI) appears to be roughly at the same level as official aid and finance flows. In global terms, FDI is low compared to Latin America, the Caribbean and Asia and the Pacific. FDI is unevenly spread, with almost 75 per cent going to the Czech Republic, Hungary, Poland, Kazakhstan and the Russian Federation (see table left).

National measures

A distinction must be made between financial measures aimed at individuals and those aimed at companies. Subsidies for citizens who install energy and water-saving devices have been quite successful in many countries, as have been tax incentives to encourage the use of unleaded petrol and cars with low fuel consumption and, in the Netherlands and some other countries, tax freedom for dividends from private investment in 'green funds'.

Financial incentives for companies, however, are a delicate issue because of their potential to distort competition. At the least, they have to be equally accessible to all enterprises within the European Union. For these reasons, the European Commission has forbidden several measures, imposing high fines and ordering the refund of unjustified subsidies. Subsidies and other financial incentives may, however, be acceptable for 'pre-competitive' industrial research, investments in technology that goes beyond environmental standards, and favourable depreciation conditions for environmental technology.

In Central Europe, the primary source of finance for environmental investment is enterprises; other sources include state, regional and local budgets, commercial banks and extra-budgetary environmental funds. In Eastern Europe and Central Asia, the state remains the major source of finance but the share of the state budget allocated to the environment is 0.5 per cent or less.

Many environment ministries have established funds for specific environmental investments such as municipal infrastructure (water, waste, heating conversion), industrial pollution control, prevention technologies, education and the establishment of monitoring systems. These funds usually redistribute environmental fees, fines, eco-taxes and other resources, disbursing them in the form of grants and soft loans.

Many funds have helped to create a domestic source of financing for environmental services. For example, environmental funds account for 40 per cent of total environmental expenditures in Poland and about 20 per cent in Hungary, Lithuania, and Slovenia. With the exception of a few countries in southeast Europe, all Central European and some Eastern European countries now have funds at the national level and some have also been established at the regional and municipal level. Eastern European and Central Asian countries have expressed interest in strengthening their funds and have received some assistance. In Poland and Bulgaria, special ecofunds have been created, capitalized by debt-for-environment swaps. The Polish Ecofund, in particular, is considered highly successful (OECD/PHARE 1998).

Commercial banks are becoming more active in offering loans for commercially-viable projects in the environmental field and freeing up resources for the private sector. There is still a tendency, however, for funds to be offered for investments which yield immediate profits (not typical of many environmental projects) and an aversion to projects involving net costs or only long-term profits (typical of many environmental projects). Improved macro-economic conditions should enable banks to play a stronger role.

Those countries that have progressed the furthest in their economic transitions are beginning to make use of mechanisms commonly used in western countries for funding environmental investments. These include loans, bonds, equity investments, public-private partnerships and user fees. Commercial banks in Poland, Hungary, the Czech Republic and Slovenia are becoming more active by offering loans for commercially-viable projects – sometimes in cooperation with institutions such as the EBRD, World Bank, European Investment Bank and the Nordic Investment Bank. Municipal bonds have also been issued by some municipalities (in Poland, for example) to finance environmental investments. Initial steps are also being taken with the use of 'green equity' to generate financing for environmentally-beneficial investments at the enterprise and municipal level, although with mixed results.

Domestic and international efforts to strengthen environmental financing still face a number of serious obstacles, many of which are related to profound economic, political and social problems. Economies are still in transition and financing conditions and mechanisms typical of more mature market economies are only beginning to emerge. Traditional sources of financing for environmental protection, particularly state budgets, have dwindled or disappeared entirely. In many cases, the finance needed by companies and municipalities for environmental protection exceeds their ability to repay loans. Subsidized financing from institutions such as central governments, environmental funds, bilateral donors and international development banks is therefore still necessary for some projects. Household budgets will also continue to be affected, as they became increasingly obliged to pay the full costs of many goods and services previously subsidized by the state, or simply not provided at all.

Furthermore, an effective financing system requires environmental strategies with clear goals and priorities, another area where progress is needed. There is a need for training and education in environmental management and financing, especially at the local level where local governments are becoming increasingly responsible for providing major investments for public services such as water and waste management. In many countries, the capacity for preparing financially- and environmentally-sound projects should be increased.

Public participation

Ideally, citizens and their organizations should have a right of appeal and recourse in all administrative procedures and a full right of standing in procedures under civil law. The Netherlands is the only country in which this situation has almost been achieved – without distinction between nationals and foreigners – and most other countries lag far behind.

In the European Union, there is no specific legislation on public participation. European Union citizens and organizations have some access to the European Court (EEB 1994) but a recent denial to Greenpeace from taking the European Council to court shows that these rights have their limitations. This implies that the citizens of Western Europe have to rely on national legislation, although the Århus convention may change this in time (see box on page 257).

The most important tool for ensuring public participation is Environmental Impact Analysis (EIA) which was introduced at the European Union level only in 1985; only the Netherlands has developed an EIA system that approaches that of the United States in effectiveness. EIA requires that the environmental impacts of major public and private projects are studied and published prior to decision making. However, while this offers the public an opportunity to make its voice heard, and grants rights to appeal against investments, decision makers are entitled to proceed in the face of public opposition. In addition, the strict and enforceable demand of the US Environmental Protection Authority that the EIA must be complete and correct is lacking in the European Union Directive. Another limitation is that an EIA is required only for major projects. In 1997 legislation was drafted to extend this requirement to plans and programmes.

In Central Europe, the legal framework and institutions needed to secure public participation and access to justice have slowly begun to emerge since 1989. The first countries to take initiatives were Poland, Hungary, the Czech Republic, Slovenia and to some extent the Slovak Republic, countries where the NGO movement had always been strong and democratic traditions had previously existed. Public participation in these countries has now become normal practice, with good access to information systems. The major tools for public participation are EIAs and local referenda. In terms of access to justice (in the form of access to constitutional courts, and ombudsmen), appeals, administrative and civil court cases are part of normal practice.

In other non-western countries, however, the absence of specific regulations or guidelines means that the real implementation of access to information and public participation has yet to take place. This is most pronounced in such countries as Albania, Bulgaria, Romania, some of the countries of former Yugoslavia, and the countries of the former Soviet Union. Citizens in many of these countries experience difficulties due to administrative barriers to court access, lack of liberal standing rules, high court fees, lack of interim and permanent injunctive relief, and slow court procedures (REC 1994b).

In Eastern Europe and Central Asia, while legislation may have been implemented, the constitutional rights of people and legislation pertaining to the decision-making process are often violated. In the Balkan area, appeals and court cases are still exceptional and legal service is weak because of severe social and economic crises. For similar reasons, public participation has been even less feasible in Croatia, Bosnia-Herzegovina and Yugoslavia. However, recent positive political changes may encourage openness and participation in such countries as Romania and Bulgaria.

The role of NGOs

NGOs can play a major role in ensuring public participation in environmental issues.

In Western Europe, several national and regional governments support NGOs financially, either continuously or by the creation of funds for specific NGO projects. At the European Union level, following the publication of the 1st Environmental Action Programme of 1973, NGOs from the member states created an European Union umbrella – the European Environment Bureau (EEB), which now has more than 150 member organizations, including NGOs from Central European countries as associate members. Following the 1987 decision to complete the internal market by 1992, several international NGOs (including WWF, Birdlife and Friends of the Earth) established European Union offices. From the beginning, the European Union Environmental Directorate-General has welcomed NGO participation and provided financial support. Consultation with NGOs is common both at the European Union level and in most member states, sometimes through official advisory structures (Vonkeman and others 1996).

In Central Europe, many leading NGOs have established legal services to assist citizens, other NGOs and local communities in exercising public participation rights and gaining access to justice. Organizations such as the Regional Environmental Centre for Central and Eastern Europe (REC) have a mandate to provide project grants and capacity-building programmes for NGOs and to support environmental protection through the participation of all stakeholders. Plans are now being implemented to establish new RECs in Eastern Europe and Central Asia.

There is still a lack of understanding of the positive role that NGOs can play in the decision-making process. In some countries, initial public support for environmental [illegible] when governments began formulating environmental programmes without consultation or dialogue with NGOs. This lack of a cooperative attitude, however, is now giving way to more constructive cooperation as the environmental movement becomes more capable of proposing alternative solutions. In Poland, the Environment Ministry has begun to organize periodical meetings with NGO representatives to exchange opinions on current issues, while in Hungary an environmental advisory committee was established in 1996 to review government policies, plans, draft laws and development proposals related to the environment.

Positive results are also being found in the NEAPs. In the former Yugoslav Republic of Macedonia, for example, NGOs have participated in NEAP working groups and are now even jointly responsible with the Ministry of Environment for their implementation. Similarly in the Russian Federation, the Caucasus countries, the Republic of Moldova and Central Asian countries, consultation processes with stakeholders including NGOs were successfully created for NEAP elaboration (OECD 1998b). In other cases, NGOs are still seen to lack the expertise to become effective partners, for example, in using EIA as a tool to participate in environmental decision-making.

Environmental information and education

Availability of information

Until the political changes of 1989, information on the state of the environment in the former Eastern Bloc countries was difficult to come by and, where available, was frequently altered to present a more favourable picture. In many other cases, information was simply not collected.

There have also been data availability problems in Western Europe (CEC 1993). The European Environment Agency (EEA) and its national topic centres are now playing a key role in improving the situation. In the rest of the region, the EfE process has been a powerful catalyst, with help from the EEA (EEA 1998).

Access to information

Ideally, access to governmental information of any kind should be guaranteed through a 'Freedom of Information Act' as in the United States. This is rare in European nations and absent at regional and sub-regional level. However, some provisions have been made regarding access to environmental information.

In Western Europe, the European Union has been relatively slow to address this issue, compared with some of its member states, because of large differences in openness between them. The European Union Directive on Access to Environmental Information (90/313/EEC), currently under review, entitles the public to request environmental information and a judicial review, should a request be refused; however, the Directive specifies reasons for non-disclosure.

First steps towards better access to information at the pan-European level were taken by the adoption of the *Sofia Guidelines* at the 1995 Sofia conference. In response to the need to establish a suitable legal framework with basic constitutional and other citizen rights, and environmental laws that guarantee access to information, public participation and access to justice, the UNECE prepared the *Convention on Access to Environmental Information and Public Participation in Environmental Decision-making*. This 'public participation' convention was drafted with the strong involvement of a European environmental NGO coalition, and was signed by most of the UNECE country representatives in June 1998 at the intergovernmental ministerial meeting in Århus, Denmark. The Convention should greatly help to facilitate the adoption of the necessary regulations and guidelines, and to harmonize and improve practices.

Over the coming years, policy and legislation are also likely to be significantly influenced by the European Union harmonization and approximation processes (REC 1995b). Several European Union Directives and Regulations contain requirements for the disclosure and dissemination of data.

On a national level, the first Central European countries to show initiative in providing access to information were the Czech Republic, Hungary, Poland and Slovakia, where access to information systems functioned well in the past and information is now readily available upon request. The Ministry of Environment in Hungary, for example, launched its information office in 1997 with the sole aim of answering requests for environmental information from the general public. Developments are not limited to Central European countries, however, and information centres have also been established in some Central Asian countries; Kazakhstan, for example, has an office to disseminate environmental knowledge to NGOs (OECD 1998b).

In some Central European countries, however, the practice of providing information is still characterized by strict regulations, multiple reasons for refusal, and long delays (REC 1995b). With official publications being the only available information, there is a need for greater openness and transparency. The drafting of laws for access to environmental information has already started in many Central European countries, and this should help to harmonize the practice across the pan-European region.

In the European Union, the Directive on Access to Information (90/313/EEC), introduced in 1990, has helped to encourage the collection and wider dissemination of environmental information. Through it, the public became entitled to request and receive information on the state of the environment. Access to environmental information is also provided by the Seveso Directive (82/501/EEC) and the adoption of *Agenda 21* which calls for greater public access to environmental information such as industry emissions.

The voluntary EMAS Regulation (EEC/1836/93) was launched in 1993, under which registered companies make public information on their environmental activities, including the degree to which they pollute the environment. A further scheme, supported by the OECD and now being integrated within the European Union's IPPC Directive, is the Pollutant Release and Transfer Register. Information on potentially-harmful releases to the air, water and soil as well as wastes transported to treatment and disposal sites is collected in a unified national reporting system, ensuring that the community, industry and governments gain greater access to relevant information on environmental pollution. From 1999, implementation of pollutant register systems will be compulsory within all European Union member states on an initially limited scale. Encouraged by strong NGO movements, countries such as the Czech Republic and Hungary have also started to implement pilot projects (REC 1997).

Public awareness

Public awareness of environmental issues is improving but is much more limited in Central and Eastern Europe and in Central Asia than in Western Europe. In Central and Eastern Europe and in Central Asia, public support for the environment declined following the political changes of 1989–90, even though this issue spearheaded many of these changes.

Some regional MEAs, particularly recent ones such as the Espoo Convention, have provisions for awareness building and public information. The public's expectations concerning Espoo's provisions are high, creating a short-term implementation problem for parties. Regional MEAs often make provisions for observer status for NGOs or representatives of intergovernmental bodies. The Baltic Convention, for which public awareness and environmental education is a key requirement, is an example of the successful involvement of NGOs in meeting such a requirement.

Education

Environmental education is mainly seen as a national responsibility. At the secondary and university level, the European Union has developed a series of activities, however, mainly based either on the conviction that Europeans 'should get to know each other' or on the desire to harmonize higher education levels and diplomas.

Although not specifically addressing the environmental area, the exchange programme for university teachers and students, ERASMUS, and the Marie Curie programme for scholarships and the funds for post-academic courses available in the budget of DG XII, are playing an important role. Most of these programmes have been opened to participation from Central Europe and Eastern Europe, and ERASMUS has a counterpart in TEMPUS. Member states also sometimes support international environmental education programmes. Thus the one-year European Post-academic Course in Environmental Management (EPCEM), supported by the Dutch government, usually has a majority of students originating from Central Europe and the European Union.

Sectoral policy integration

Integration was one of the key themes of the 4th Environment Action Programme of the European Union (Council Resolution 86/485) and has also ranked high on the OECD agenda (Haigh and Irwin 1990). Integration requires that environmental concepts, intentions, principles, plans, commitments and policy goals should be 'internalized' and treated by decision-makers in other sectors as equal in importance to their own concepts and intentions. Although integration is still high on the European Union's 'unfinished agenda', there has been little progress, but the Commission has identified the Kyoto Protocol and *Agenda 2000* as being among the most urgent policy packages where integration should play a key role.

A similar situation exists in Central and Eastern Europe and Central Asia. Countries there are faced with the same problems, and also with the reduction in international competitivity of sectors such as agriculture, transport, industry and tourism that could result from the environmental constraints associated with integration. Further problems are the low political priority assigned to the environment, the failure to separate the role of the state as a source and regulator of economic activity, the weakness of environmental ministries and the failure to demonstrate the economic benefits of environmental measures.

Conclusion

The main positive development in Western and Central Europe is continued progress in harmonization with European Union environmental legislation, prompting countries to adhere to relatively high standards. The chief problem is that environmental issues are not integrated into sectoral issues or, in the transition countries, into the economic restructuring process (see box above). Enforcement of environmental regulations also poses a difficult problem in the transition countries.

Regional action plans have proved effective mechanisms for environmental improvement and, in some cases, closer compliance with European Union standards. The European Union's Fifth EAP has been successful in forging a common environmental policy based on a philosophy consistent with the principle of sustainable development. However, a number of key targets may not be met, particularly in the agriculture,

Priorities for policy action

	Western Europe	*Central Europe*	*Eastern Europe*	*Central Asia*
Agriculture	Subsidy reform	Land privatization	Land privatization	Land privatization
	Land-use planning		Subsidy reform	Diversification of agricultural production
Energy	Subsidy reform	Renewal of energy production facilities	Renewal of energy production facilities	Renewal of energy production facilities
	Energy efficiency	Shift to gas		
	Energy taxation	Replacement of industrial technology	Replacement of industrial technology	
			Enforcement of proper standards in oil and gas extraction	Enforcement of proper standards in oil and gas extraction
Transport	Urban planning and traffic management	Urban planning and traffic management	Urban planning and traffic management	Urban planning and traffic management
	Subsidy reform	Modernization of vehicle technology	Modernization of vehicle technology	Modernization of vehicle technology
	Green taxation	Keeping/making public transport attractive		

tourism and transport sectors. Barriers to policy implementation include the decision-making structure of the European Union, conflicting national interests, and an emphasis on economic priorities.

The EAP for Central and Eastern Europe has served as a catalyst for the development of NEAPs. Although most environmental improvement in the transition countries is attributable to industrial decline, NEAPs in Central Europe have resulted in better policy making, stronger institutional frameworks, increased environmental investments and cost-effective accession strategies. In Eastern Europe and Central Asia, action plans are less advanced, mainly because of weak institutional capacity and the slower pace of political reform and economic restructuring.

Since the main drivers of environmental problems in Europe and Central Asia are agriculture, energy and transport, an attempt has been made in the table on page 275 to highlight the priorities for policy action in relation to these driving forces for each sub-region.

The most important action items for improving the implementation and impacts of the major MEAs are:

- to complete ratification of agreements such as the Vienna Convention;
- to assure sufficient financing for weaker conventions, such as the Convention on Biological Diversity;
- to establish independent verification systems for monitoring and enforcement;
- to enhance non-compliance mechanisms with individual complaint bodies;
- to establish sanctions under criminal law in relation to MEAs;
- to extend the scope of selected agreements beyond first-step solutions, as part of a continuous process.

The implementation of economic instruments has ranged from fair to poor. Although the polluter pays principle is widely recognized, economic interests often take precedence over attempts to internalize external environmental costs. Environmental funds in the transition countries remain an important domestic financing source, though greater transparency and freedom from political manipulation are necessary. Although subsidies are declining, they continue to have adverse impacts on the environment, especially in the energy, transport and agricultural sectors.

Initiatives to implement cleaner production practices and environmental management systems have been successful in Western Europe and, to a lesser extent, in Central Europe. Progress is almost non-existent in Eastern Europe and Central Asia.

Public participation in environmental issues is satisfactory in Western Europe, with positive trends in Central and Eastern Europe. Greater awareness of EIA as a tool for public participation has played a positive role, although many countries still lack a proper legislative framework for public participation. This situation should improve with the expected ratification and implementation of the public participation convention. Access to environmental information has increased significantly with the formation of the European Environment Agency and other information resource centres.

Priority action areas include:

- integration of environmental issues into economic processes and other policy areas;
- assessment of the costs and benefits of European Union harmonization;
- building up capacity in Central Europe to implement and enforce European Union standards; and
- building up capacity in the transition countries to implement action plans.

In Western Europe, it is important to strengthen green taxation and reduce the adverse impacts of subsidies from the Cohesion and Structural Funds. In the transition countries, additional priorities are to strengthen enforcement of fees and fines, increase the transparency of environmental funds, increase the capacity of local governments and businesses to prepare projects for financing, establish incentives for promoting cleaner production, and build up the capacity of enterprises to introduce environmental management systems. All European countries should be encouraged to ratify the public participation convention, while the transition countries should be encouraged to eliminate institutional barriers to public participation.

References

Baltic Marine Environment Protection Commission (1996). *Protection of the Baltic Sea – results and experiences.* Baltic Marine Environment Protection Commission, Helsinki, Finland

Black Sea Environmental Programme (1996 and 1997). *Annual Reports 1996 and 1997.* BSEP, Istanbul, Turkey http://www.dominet.com.tr/blacksea/

CEC (1993). *Towards sustainability A European Community programme of policy and action in relation to the environment and sustainable development* (COM 1992/150 27.03,1992 EN 103), *Vol. III State of the Environment.* Commission of the European Communities, Luxembourg http://europa.eu.int/comm/dg11/actionpr.htm

EBRD (1997). *Transition Report 1997.* European Bank for Reconstruction and Development, London, United Kingdom

EC (1997). *Environment taxes and charges in the single market.* EC Communication COM-97-9, Brussels, Belgium

ED (1997a). PHARE Programme. *European Dialogue*, issue 6, November-December 1997, European Commission (Directorate-General for Information), Brussels, Belgium http://europa.eu.int/en/comm/dg10/infcom/eur_dial/index.html

ED (1997b). Agenda 2000: enhanced role for EIB. *European Dialogue*, issue 6, November-December 1997, European Commission (Directorate-General for Information), Brussels, Belgium http://europa.eu.int/en/comm/dg10/infcom/eur_dial/index.html

ED (1999). Enlargement: screening process continues. *European Dialogue*, issue 2, March-April 1999, European Commission (Directorate-General for Information), Brussels, Belgium http://europa.eu.int/en/comm/dg10/infcom/eur_dial/index.html

EEA (1997). *Environmental Agreements.* European Environment Agency, Copenhagen, Denmark

EEA (1998). *Europe's Environment. The Second Assessment.* European Environment Agency, Copenhagen, Denmark

EEB (1994). *Your rights according to the environmental legislation of the European Union* (in all European Union languages). European Environment Bureau, Brussels, Belgium

EMEP/MSC-W (1998). *Transboundary acidifying air pollution in Europe. Part 1: calculation of acidifying and eutrophying compounds and comparison with observations.* Research Report 66, EMEP/MSC-W Status Report 1998. Det Norske Meteorologiske Institutt (Norwegian Meteorological Institute), Blindern, Norway

ERM (1991). *The Structure and Functions of Environmental Enforcement Organisations in EC Member States.* ERM for the Ministry of Housing, Physical Planning and Environment of the Netherlands, London, United Kingdom

ERM (1996). *How to Improve the Effectiveness of the Convention on International Trade in Endangered Species of Wild Fauna and Flora (CITES).* ERM, London, United Kingdom

Haigh, N. and Irwin, F. (eds., 1990). *Integrated Pollution Control in Europe and North America*

IEEP-B (1994). *The Dutch Governmental Programme for Sustainable Technology Development.* Institute for European Environmental Policy, Brussels, Belgium

Mylona, S. (1996). Sulphur dioxide emissions in Europe 1880-1991 and their effect on sulphur concentrations and depositions. *Tellus*, 48B, 662-689

OECD (1997). Oslo Roundtable on CP: presentation on *Towards Sustainable Environmental Programmes.* OECD, Paris, France

OECD (1998a). Workshop Report *Environmental Financing in CEE/NIS.* OECD EAP Task Force, Paris, France

OECD (1998b). *Evaluation of Progress in Developing and Implementing National Environmental Action Programmes in Central and Eastern Europe and the New Independent States.* OECD, Paris, France

OECD/PHARE (1998). *Sourcebook on Environmental Funds.* OECD, Paris, France

PHARE (1997). *An Interim Report*, European Commission, Brussels, Belgium

Ramsar Convention (1996). *Overview of the Implementation of the Convention in the Western European Region* and *Overview of the Implementation of the Convention in the Central and East European Region.* Ramsar, Gland, Switzerland

REC (1994a). *Use of Economic Instruments in Environmental Policy in Central and Eastern Europe.* Regional Environmental Center for Central and Eastern Europe, Szentendre, Hungary

REC (1994b). *Manual on Public Participation in Environmental Decision-making: current practice and future possibilities in Central and Eastern Europe.* Regional Environmental Center for Central and Eastern Europe, Szentendre, Hungary

REC (1995a). *Competing in the New Environmental Marketplace.* Proceedings of Workshops for Environmental Professionals, Regional Environmental Center for Central and Eastern Europe, Szentendre, Hungary

REC (1995b). *Report on Status of Public Participation Practices in Environmental Decisionmaking in CEE*, Regional Environmental Center for Central and Eastern Europe, Szentendre, Hungary

REC (1997). *The Bulletin*, vol. 7, no. 1. Regional Environmental Center for Central and Eastern Europe, Szentendre, Hungary

REC (1997/1998). EMAS-ISO Bridge Building. *The Bulletin*, vol. 7, no. 3. Regional Environmental Center for Central and Eastern Europe, Szentendre, Hungary

REC (1998). *Sourcebook on Economic Instruments in the CEE.* Regional Environmental Center for Central and Easten Europe, Szentendre, Hungary

Schmidheiny, W. (1992). *Changing Course.* MIT Press, Cambridge, Massachusetts, United States

Udo de Haes, H.A. (1996). *Towards a Methodology for Life Cycle Impact Assessment.* SETAC Europe, Brussels, Belgium

UNEP (1997). *Register of International Treaties and Other Agreements in the Field of the Environment 1996.* United Nations Environment Programme, Nairobi, Kenya

UNEP (1998). *Report on the Status of Multilateral Environmental Agreements in the European Region.* UNECE/ARH.CONF/BD.12, Background document prepared for the fourth Ministerial Conference Environment for Europe, Århus, Denmark, 23–25 June 1998. UNEP, Geneva, Switzerland

UNEP-MAP (1996). *The State of the Marine and Coastal Environment in the Mediterranean Region.* MAP Technical Report Series No. 100. UNEP, Athens, Greece

Victor, D., Raustiala, K., and Skolnikoff, E. (eds., 1998). *The Implementation and Effectiveness of International Environmental Commitments: Theory and Practice.* MIT Press, Cambridge, Massachusetts, United States

von Weizsäcker, E., Lovins, A., and Lovins, H. (1995). *Faktor Vier.* Droemer Knaur, München, Germany

Vonkeman, G. H. and Stielstra, H. B. C. (1996). *Structured Expert Opinion on Environment – a review and analysis of scientific advisory bodies in the area of environment policy.* Study for the Austrian Ministry of Environment (Nr. GZ: 01 3145/1-I/7/95), Vienna. Institute for European Environmental Policy, Brussels, Belgium

Werksman, J. (1997). *Five MEAs, Five Years Since Rio: recent lessons on the effectiveness of Multilateral Environmental Agreements.* Foundation for International Environmental Law and Development (FIELD), London, United Kingdom

World Bank (1994). *Environmental Action Plan for Central and Eastern Europe.* Abridged version of the document endorsed by the Ministerial Conference in Lucerne, Switzerland. World Bank, Washington DC, United States

WWF (1997). Transregional Project 9E0106: Barcelona Convention. www.panda.org/resources/countryprofiles/mediterranean/six.htm

Latin America and the Caribbean

KEY FACTS

- Public environmental agencies, with their limited and unfocused mandate, have had little impact on industrial and other productive activities.
- Global MEAs and non-binding instruments have increased awareness of environmental issues and contributed to an environmental conscience which would have been unimaginable a quarter of a century ago.
- Rules and regulations are hard to enforce because many institutions cannot monitor compliance and systematic enforcement can have negative economic effects.
- In Brazil, charges are levied for the use of natural resources ... and revenues are distributed to the Federal Government and the states where the exploitation took place.
- In Costa Rica, a series of forestry laws has established the principle that people involved in reforestation or forest conservation should be rewarded for the environmental and social services provided by forests.
- The 1996 Declaration of Santa Cruz commits the signatories to supporting and encouraging broad participation by civil society in designing, implementing, and evaluating policies and programmes in all countries of the Americas.
- Chile's National System of Environmental Information was launched in 1994. Based on a decentralized, low-maintenance platform, it has its own web site.
- In Argentina, legal initiatives empowering the National Secretary of Natural Resources and Human Environment to publish a list of violators of environmental regulations are causing negative publicity for the offending industries.
- Programmes devised to fight poverty are usually unrelated to environmental policies.

The policy background

Over the past decade, domestic and international pressures to fight environmental degradation have increasingly resulted in environmental issues being dealt with in the context of overall development. Preparations for the 1992 United Nations Conference on the Environment and Development led to the establishment of fora to examine environmental and natural resource issues, and a new approach to North-South differences. Gradual economic globalization has led to new international trading practices with significant environmental implications. Governments have sought to strengthen environmental policies through institutional change and legal, technical and economic initiatives, both at the domestic level and through sub-regional cooperation agreements. The open debate that followed the return to democracies in the region has increased pressures for the development of environmental policies and planning systems.

However, these changes have not, on the whole, led to more efficient management or significant environmental progress. In spite of institutional strengthening, public environmental agencies, with their limited and unfocused mandate, have had little impact on industrial and other productive activities and have been involved in clashes with other public agencies and NGOs. The environmental consequences of public policy decisions and private sector initiatives are not being adequately assessed (Brzovic 1993).

The fundamental economic objective remains the implementation and expansion of a liberal approach which relies on export growth and foreign capital inflows, regardless of the consequences for the environment and the preservation of natural resources, and with no internalization of environmental costs (Gligo 1997). Economic policies continue to be drafted according to criteria that imply unsustainability and, in some cases, sheer indifference to environmental impacts (CEPAL/PNUMA 1997). Economic development programmes to fight poverty continue to be unrelated to environmental policy, and poor inter-agency coordination and the lack of focus on the broader picture have limited progress under *Agenda 21*. Examples of the lack of sectoral integration in environmental policy are given in the box below.

Implementation of environmental policies is often difficult because of inadequate control, monitoring and enforcement mechanisms. In some cases, the legal framework for environmental management is dispersed among many legal texts in diverse institutions, and one environmental issue may be delegated to several public institutions at various political levels. New policies and institutions have not always included the revision of the old legislation. Environmental regulations include complex and sophisticated instruments and norms that are difficult to enforce because of financial restrictions and lack of human and operational resources (IDB 1996).

Environmental policies: the lack of sectoral integration

Soil and land use

Explicit policies for soil conservation and better land management have mostly failed because of ineffective legislation, institutional weakness, lack of information, inadequate public awareness and emphasis on short-term productivity goals.

The indiscriminate use of pesticides and other agrochemicals is still a problem, and a major challenge in terms of technological change (Gligo 1997). Land management problems are exacerbated by high rural population growth rates, the absence of land-use planning and persistent land tenure difficulties.

Forests

Policies to protect forests have failed, mainly because they do not impact on the underlying factors that cause deforestation. Industrial-scale agriculture and agricultural settlement programmes continue; for example agriculture in Bolivia, the midwest region of Brazil and Paraguay is still expanding at the cost of forests (Paraguay 1995). Firewood is still a major source of cheap energy.

There is, however, increasing recognition of the environmental and social value of forests and their ecosystems, including the roles they play in water management, conserving biodiversity, absorbing greenhouse gases, and creating landscapes and scenic value. There is a growing interest in the use of second-growth forests to reduce pressure on native forests, particularly in countries with humid forests.

Biodiversity

Virtually all countries have drafted national strategies to preserve biodiversity but only a few have done so on a comprehensive basis. Policies have been successful only through the implementation of laws aimed at regulating and improving the management of wildlife in conservation areas. Governments have generally failed to strengthen agencies in charge of biodiversity preservation.

Water

Many national policies fail to provide for sustainability or involve groups with a specific interest in water issues. In Chile, for instance, water rights are traded on a free market basis, with no restrictions related to conservation or to strategic or social values (IDB 1996). In many countries laws to control water use have failed to take a comprehensive view because of the lack of coordination between the different regulatory agencies involved (CCAD 1997).

In 1997, however, Brazil approved a National Law of Hydraulic Resources which includes water charges and which assigns management to Watershed Committees and Water Agencies which are required to execute integrated policies with public participation.

Marine areas

Policies have been limited to countering the threats of social upheaval caused by reduced fishing activities, and temporary enforcement of regulatory measures in response to warnings from scientific institutions and international programmes such as the Regional Seas Programmes for the South Pacific and the Caribbean. On the Atlantic coast, Argentina and Uruguay have a joint programme for the La Plata estuary with a bi-national executive secretariat.

A few countries are beginning to experiment with the more integrated approach which is vital for coastal areas. Panama, for example, created the Maritime Authority of Panama in 1997 to integrate all issues related to fishing, coastal management and ports (Government of Panama 1998)

Atmosphere

Measures to regulate atmospheric pollution have had some positive impact but continue to be insufficient. Mexico City, Santiago de Chile and São Paulo have adopted stringent measures to restrict the circulation of vehicles to improve atmospheric quality. However, little consideration is given to controlling the growth of vehicle numbers, improving urban management or upgrading public transport in areas where emissions fail to meet WHO standards.

MEAs and non-binding instruments

Global MEAs

Global MEAs and non-binding instruments, especially those generated by the Stockholm (1972) and Rio de Janeiro (1992) Conferences, have had an important influence on the development of national legislation to protect the environment and support sustainable development during the 1990s, a period of growing macro-economic stability for almost all the countries of the region.

The level of participation in global MEAs is now high (see table opposite). This contrasts with the situation in the early 1990s, when a UNEP study found that only 26 per cent of the countries of the region participated in some or all of 53 multilateral global instruments considered (PNUMA-ORPALC 1993). One reason for the improvement is greater harmonization between national and international environmental priorities.

Global MEAs and non-binding instruments have increased public awareness of environmental issues and have contributed to an environmental conscience, in the public as well as the private sectors, which would have been unimaginable a quarter of a century ago when environmental problems were identified with specific pollutants and were considered to be problems only of rich countries.

Certain MEAs are, of course, highly relevant to specific problems in the region – such as protection of the ozone layer in the southernmost countries and the effects of climate change on small Caribbean island states.

Few national institutional structures have been created specifically for the implementation of global MEAs (PNUMA-ORPALC 1996), with most countries absorbing the new functions associated with implementation into existing national structures. The creation of National Committees of Biological Diversity in Meso-american and other countries is an exception.

In some instances, there has been good communication between those responsible for implementation in different countries – for example, in the Regional Networks of ODS Officers for the Montreal Protocol that exist for South America, Central America and the Caribbean (OAN 1998).

Implementation of global MEAs at the national level has been carried out through several instruments, especially a number of specific programmes and funds which have been recently developed (see examples in table below).

Other economic instruments can also be used for that purpose, even though they were not created specifically for the implementation of MEAs. For example, user tariffs which take account of the environmental cost of producing certain goods or services are becoming more common. In Chile, the Law on General Bases of the Environment foresees tariffs to cover the costs of pollution prevention or decontamination and, in Panama, the Law on Forest Incentives exempts owners of forest plantations from paying income tax and makes investments in this sector 100 per cent deductible (PNUD/PNUMA 1996).

A system of tradeable emission permits is also being introduced, for example through the Chilean Law on General Bases of the Environment and the 1996 amendments to Mexico's General Law on Ecological Balance and Protection of the Environment. However, the specific regulations required for the application of these instruments do not yet exist (González 1997).

A problem with the implementation of global MEAs is the lack of adequate international financing to

Special financial funds related to the implementation of the Convention on Biodiversity

	Fund	*Relevant legislation*
Brazil	Brazilian Biodiversity Fund	Law on National Environment Fund, 1998
Costa Rica	Fund for Wild Life	Law on the Conservation of Wild Life, 1992
Ecuador	National Fund for Forestation and Reforestation	Regulation of 1993
Panama	National Fund for Wild Life	Law on Wildlife, 1995
Paraguay	Special Fund for the Conservation of Wild Life	Law on Wildlife, 1992
Paraguay	Special Fund for Wild Areas	Law on Protected Wild Areas, 1994

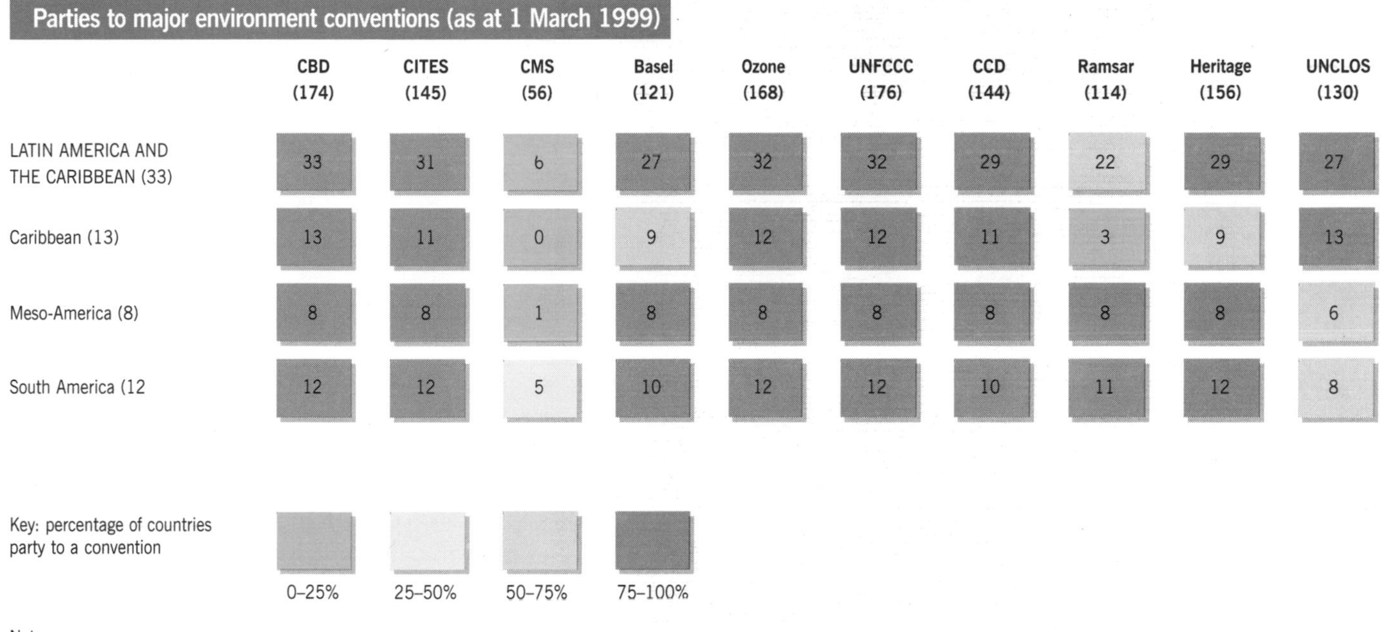

Parties to major environment conventions (as at 1 March 1999)

	CBD (174)	CITES (145)	CMS (56)	Basel (121)	Ozone (168)	UNFCCC (176)	CCD (144)	Ramsar (114)	Heritage (156)	UNCLOS (130)
LATIN AMERICA AND THE CARIBBEAN (33)	33	31	6	27	32	32	29	22	29	27
Caribbean (13)	13	11	0	9	12	12	11	3	9	13
Meso-America (8)	8	8	1	8	8	8	8	8	8	6
South America (12	12	12	5	10	12	12	10	11	12	8

Key: percentage of countries party to a convention

0–25% | 25–50% | 50–75% | 75–100%

Notes:

1. Numbers in brackets below the abbreviated names of the conventions are the total number of parties to that convention
2. Numbers in brackets after name of regions/sub-regions are the number of sovereign countries in each region/sub-region
3. Only sovereign countries are counted. Territories of other countries and groups of countries are not considered in this table
4. The absolute number of countries that are parties to each convention in each region/sub-region are shown in the coloured boxes
5. Parties to a convention are states that have ratified, acceded or accepted the convention. A signatory is not considered a party to a convention until the convention has also been ratified

ensure compliance and complement national financial efforts. Advances have been made when such financing has been provided. For example, the Ramsar Convention and its Small Grants Funds for conservation and rational use of the wetlands financed 25 projects in 13 countries between 1992 and 1995 to a total of US$800 000 (Ramsar 1998). Overall, however, such mechanisms have not yet been very effective. The Capacity 21 programme does not have sufficient resources to meet the demands generated by MEA implementation.

Many MEAs include regulations on monitoring and the preparation of compliance reports. This should result in adequate information on the development of the MEAs but these regulations are not always fully implemented. An exception is the Montreal Protocol which insists on strict compliance with data-reporting procedures through an Implementation Committee (UNEP Ozone Secretariat 1997).

Adopting national legislation to meet the requirements of MEAs usually takes some years, depending on the country, the MEA in question, and the matter being regulated.

Implementation of the United Nations Convention on the Law of the Sea (UNCLOS) is a particularly successful case. Originating in the Declaration of Santiago on the Maritime Zone (subscribed in 1952 by Chile, Ecuador and Peru), UNCLOS rapidly gained consensus in Latin America and worldwide, with the obvious exception of the major fishing powers.

In contrast, implementation of the Convention on Biological Diversity (CBD) has differed from country to country. Brazil, for example, established a National Programme on Biological Diversity in 1994. The Brazilian Institute of the Environment and Renewable Natural Resources (IBAMA) has approved and is responsible for the management of 165 conservation units in different ecosystems, comprising 39 national parks, 24 biological reserves, 21 ecological stations, 11 extractive reserves, 46 national forests and 24 environmental protection areas (MOE Brazil 1998).

In Peru, a Law on the Conservation and Sustainable Use of Biological Diversity, covering the majority of CBD commitments, was put into effect in 1997, and Costa Rica approved a Law on Biodiversity at the beginning of 1998 (ALDA 1997). Some countries apply CBD, or its objectives, through the inclusion of regulations in general or sectoral law. These include Costa Rica, Cuba, Honduras, Mexico, Nicaragua and Panama.

Major regional MEAs

Treaty	*Place and date of adoption*
Convention on Nature Protection and Wildlife Preservation in the Western Hemisphere	Washington DC 1940
Treaty for the Proscription of Nuclear Weapons in Latin America and the Caribbean (Tlatelolco Treaty)	Mexico City 1967
River Plate Basin Treaty	Brasilia 1969
Convention on the Defence of the Archaeological, Historical and Artistic Heritage of the American Nations (Convention of San Salvador)	Santiago 1976
Treaty for Amazonian Cooperation	Brasilia 1978
Convention for the Conservation and Management of the Vicuña	Lima 1979
Convention for the Protection of the Marine Environment and Coastal Area of the South-East Pacific	Lima 1981
Agreement on Regional Cooperation in Combating Pollution of the South-East Pacific by Hydrocarbons and Other Harmful Substances in Cases of Emergency	Lima 1981
Supplementary Protocol to the Agreement on Regional Cooperation in Combating Pollution of the South-East Pacific by Hydrocarbons and Other Harmful Substances in Cases of Emergency	Quito 1983
Protocol for the Protection of the South-East Pacific against Pollution from Land-Based Sources	Quito 1983
Convention for the Protection and Development of the Marine Environment in the Wider Caribbean Region	Cartagena de Indias 1983
Protocol Concerning Cooperation in Combating Oil Spills in the Wider Caribbean Region	Cartagena de Indias 1983
Central American Convention for the Protection of the Environment	San José 1989
Protocol Concerning Specially Protected Areas and Wildlife to the Convention for the Protection and Development of the Marine Environment of the Wider Caribbean Region	Kingston 1990
Convention for the Conservation of the Biological Diversity and the Protection of Priority Wilderness Areas in Central America	Managua 1992
Regional Agreement on the Transboundary Movement of Hazardous Wastes	Panama 1992
Regional Convention for the Management and Conservation of Natural Forest Ecosystems and the Development of Forest Plantations	Guatemala City 1993

Source: UNEP 1997

It is not yet possible to make an overall assessment of the impact of global MEAs on the region's environment, mainly because the degree of development and hence impact varies with the instrument and the country in question. Nevertheless, some MEAs seem to have developed more successfully than others. The Montreal Protocol, for example, has produced significant decreases in the consumption and production of ozone-depleting substances, although not at the same rate in all countries. Annual production of these substances fell in Argentina from 5 574 to 1 050 tonnes and in Brazil from 10 218 to 9 434 tonnes during 1986–96 (UNEP Ozone Secretariat 1998). Consumption of CFCs in Mexico had fallen to 52 per cent of the base year (1989) level by 1996, a reduction of more than 4 000 tonnes (UNEP Ozone Secretariat 1998). The Mexican strategy included agreements with industry, regulation of imports and exports of controlled substances, development of technical training programmes and implementation of clean technologies. Projects have covered commercial and household refrigeration, solvents, foams, and central and automotive air-conditioning, many with support from international agencies such as the World Bank, the United States Environmental Protection Agency, UNDP and UNEP (PNUMA-ORPALC/ALDA 1998).

CITES is also critically important to Latin America, which contains an extraordinary range of species. Brazil, perhaps the most species-rich nation in the world, has had difficulty implementing and enforcing CITES (Weiss and Jacobson 1998) although IBAMA now has more than 400 employees helping to control wild animal traffic, deforestation and other environmental crimes in the Amazon. The vastness of the Amazon combined with tight government budgets has also made implementation of the International Tropical Timber Agreement difficult. However, the Pilot Programme to Conserve the Brazilian Rain Forest has been set up as a joint undertaking between the government, civil society, NGOs and the international community, and is now entering its second phase (World Bank 1997).

Regional MEAs

None of the regional MEAs cover the entire region, all being sub-regional and limited to a group of countries (Central America), certain regional seas (the Southeast Pacific and the Wider Caribbean) or a group of ecosystems (the Amazon or the La Plata River basin). One relates to the protection of a specific species, the vicuña. There are also some important inter-American agreements, such as the Washington Convention of 1940 and the San Salvador Convention of 1976.

These MEAs are not only relevant to specific problems not covered by global MEAs but also help to make such international agreements more effective through stronger participation of the countries involved and a more realistic vision of their ability to implement and comply with those agreements. The agreements are listed in the table opposite.

The degree of participation in regional MEAs is high but not the same for all countries or instruments. In general, the regulations contained in the regional MEAs have been implemented through existing legislation rather than developed into new national legislation.

Few regional MEAs have resulted in substantial modifications to national institutional structures. At best, they have led to the establishment of administrative units within existing structures, specializing in the topics covered by the MEAs. The National Council for the Conservation of the Vicuña, established in Peru, is an exception.

Some sub-regional MEAs, such as the Treaty for Amazon Cooperation and the Convention of the South-East Pacific, have Secretariats which rotate amongst the signatory countries. Others have permanent secretariats.

Economic instruments have not generally been used in the implementation of these MEAs but there has been some use of national funds, not necessarily created for the implementation of the MEAs, such as the Amazonian Environmental Fund in Colombia and the Fund for Amazonian Regional Eco-development in Ecuador (PNUD/PNUMA 1996), and of 'debt-for-nature' swaps.

Most of these MEAs do not have their own financing. An exception is the La Plata River Basin Treaty with its Financial Fund for the Development of the La Plata River Basin, created to finance pre-feasibility and feasibility studies, engineering designs, and projects in member countries. This fund is also intended to attract co-financing from international bodies.

Some regional MEAs, such as the Treaty for Amazon Cooperation (ACT) and the Central American Convention for the Protection of the Environment, receive international funding for certain projects. The ACT's Special Commission receives financing from the GEF for its Regional Strategic Project for the Conservation and Sustainable Management of Natural Resources in the Amazon. The Amazon Zonification Support Program receives financing from the Inter-American Development Bank (IDB) and FAO, and the Regional Project for Planning and Management of Protected Areas of the Amazonian Region is supported financially by the European Union and some resources come from FAO (Treaty for Amazonian Cooperation 1997).

The effectiveness of these sub-regional MEAs is difficult to assess because of the difficulty of establishing direct relationships between MEAs and changes occurring in the environment. The ACT, for example, has led to some important political results following meetings of the Ministers of Foreign Affairs of the signatory countries dealing with the sovereignty of the Amazonian States over this territory. However, the quantity and quality of the projects launched through this Treaty do not reflect the environmental importance of the Amazon region.

The Convention of the Vicuña, one of the most modest of the regional MEAs in terms of its objectives, is a good example of successful implementation. After 25 years of operation, there has been a satisfactory re-population of the species in the participating countries. In Bolivia, for example, a 1996 census showed that the number of vicuñas had doubled since 1986 (National Census of the Vicuña 1996).

Despite the difficulties in assessing the MEAs, their existence has clearly contributed to placing important issues more firmly on national environmental agendas and strengthening the consciousness of the public and private sectors, both of environmental problems in general, and of the specific problems dealt with in the MEAs.

Regional MEAs in the Caribbean

The Convention for the Protection and Development of the Marine Environment in the Wider Caribbean Region has three protocols covering oil spills, specially protected areas and wildlife, and land-based sources of marine pollution. Several articles of the Cartagena Convention refer to the need for measures to prevent, reduce and control pollution caused by discharges from land-based sources, ships, dumping and seabed activities.

High priority has been accorded to waste management in the Caribbean and most countries have ratified the relevant international conventions. National and regional institutions are involved in the preparation of plans, and a number of initiatives are being taken including the Solid Waste Management Project of the Organization of Eastern Caribbean States which aims at comprehensive improvements in the management of waste over the medium term.

A Regional Response Mechanism has been established with telecommunication networks that link all Natural Disaster Coordinators, the Eastern Caribbean Donor Disaster Coordination Group, the Caribbean Disaster Relief Unit and the Caribbean Disaster Emergency Response Agency. These agencies can make quick assessments of damage, establish needs and mobilize resources to provide initial relief to affected communities.

Action plans

Several regional meetings at the highest level have been held during the past few years, mainly to review development topics, including environmental issues, from a sustainable development perspective. The most relevant was the Summit of the Americas on Sustainable Development, held in Santa Cruz de la Sierra, Bolivia, in December 1996 as a follow-up to the First Summit of the Americas held in Miami in 1994. In the Declaration of Santa Cruz de la Sierra, the signatories stated: 'Development strategies need to include sustainability as an essential requirement for the balanced, interdependent, and integral attainment of economic, social, and environmental goals'. The Action Plan they adopted is ambitious and includes 65 initiatives on health, education, agriculture, forests, biodiversity, water resources, coastal areas, cities, energy and mining (Summit of the Americas 1997)

The Santa Cruz Summit led to the formation of the Interagency Task Force for the Bolivia Summit Follow-Up. Its main objective is to improve coordination among technical assistance organizations, international financing institutions and OAS member countries in implementing the initiatives contained in the Action Plan. It includes the participation of more than ten international agencies, with the OAS functioning as chair and technical secretariat.

The Second Summit of the Americas, held in Santiago, Chile, in April 1998, reinforced the mandates of Santa Cruz and particularly the continued role of the OAS, as well as the Interagency Task Force for the Bolivia Summit Follow-Up and the Inter-American Strategy for Public Participation (see also box on page 290). Implementation is still at an early stage.

Advances at the sub-regional level have been more significant. In Central America, for instance, although environmental progress by individual countries has been uneven, there has been better harmonization and coordination of national activities. The environment became a significant issue in 1989, following the signature of the Central American Convention for the Protection of the Environment (CPC), and the subsequent creation of the Central American Commission for the Environment and Development (CCAD). The signature of the Alliance for Sustainable Development (ALIDES) in 1994 was even more significant in that it generated a conceptual and operational framework for sub-regional and national goals and strategies (see box left).

Though there is little cooperation on biodiversity issues between research institutions and other academic bodies, the introduction of biological corridors such as the Central American Biological Corridor is significant. It implies the integration of conservation into land-use planning (CCAD 1998).

MERCOSUR (the Southern Common Market) is basically a commercial agreement in which environmental issues do not play a leading role but it has contributed to the discussion of important changes in environmental policy. The MERCOSUR legislation relating to environmental protection includes rules to regulate the levels of pesticide residues acceptable in food products, levels of certain contaminants in food packaging, eco-labelling and regional transportation of dangerous goods (IDB 1996). Most progress is being made in the area of the environmental impacts of new physical infrastructures for which an environment protocol is being negotiated (Gligo 1997).

An initiative to establish the Caribbean Sea as a Special Area was discussed by the Caribbean Ministers during their Meeting on the Implementation of the SIDS Programme of Action, held in November 1997 (UNEP/UWICED/EU 1999). In the energy field, a regional energy information network for the

Central American Alliance for Sustainable Development (ALIDES)

Objectives:

- Create the isthmus as a region of peace, liberty, democracy and development, by changing attitudes and promoting a sustainable development model in the political, economic, social, cultural and environmental areas, in the framework of *Agenda 21*;
- promote sustainable integral management of the territories to guarantee the conservation of biodiversity;
- inform the international community about the achievements of the Alliance and the potential benefits of its development model;
- promote the capacity and participation of society to improve the quality of life.

Instruments:

- National Council for Sustainable Development, with representation from the public sector and civil society, to maintain the coherence and consistency of national policies, programmes and projects within the strategy of sustainable development;
- Central American Council for Sustainable Development involving the Central American Presidents and the Prime Minister of Belize;
- Foreign Affairs Ministers Council and the Foreign Secretary of Belize to co-ordinate the presidential decisions, supported by the General Secretary of the Central American Integration system;
- Central American Fund for Environment and Development to fulfil the environmental objectives of ALIDES.

Source: ALIDES 1999

Caribbean has been established as part of a Regional Energy Action Plan and a renewable energy centre has been established in St Vincent and the Grenadines. A Caribbean Action Plan in support of the International Coral Reef Initiative has also been established.

Laws and institutions

Countries throughout the region have begun to adapt their legal and institutional frameworks to the new paradigm of sustainable development. A major feature is that the principle of environmental protection has been enshrined in constitutional law by all nations, albeit to varied levels. In 14 countries, the new constitutions promulgated during the past 25 years contain regulations of an environmental nature that were often inspired by ideas shaped in world fora. Others have amended existing constitutions to incorporate environmental regulations.

Most countries have established a general environmental legal framework together with one dealing with specific areas such as water resources, mineral resources, marine and land areas, hunting and fishing, forestry resources, tourism, chemical products, pesticides and air pollution. Many countries have also developed national environmental plans and strategies.

During the 1980s and 1990s, many Latin America and Caribbean countries created new environmental institutions in the form of ministries, commissions and councils, and others combined or reorganized existing institutions. Mexico (see box above), Honduras and Nicaragua are good examples of countries that work at the Ministerial level (IDB 1996). Countries such as Chile, Ecuador, Guatemala and later Peru have chosen to set up Coordinating Commissions (Gligo 1997).

There is a large body of rules and regulations on specific environmental issues, such as EIAs, hazardous wastes, environmental crime, protection of natural resources, regulations for the production of, use of, and access to natural resources, and protection of human health from harmful environmental effects (PNUMA 1993).

Laws intended to regulate the use of natural resources often include provisions to punish non-compliance (Orozco and Acuña 1997). However, the rules and regulations often include no criminal or administrative sanctions. An exception is the Brazilian Environmental Crimes Law, passed on March 1998, perhaps the most modern legal text focusing on environmental crime. Rules and regulations are hard to enforce because many institutions cannot monitor compliance and systematic enforcement can have negative economic effects. For example, pollution by domestic, industrial and agricultural sewage in Nicaragua is regulated by many rules but none can be properly enforced (Dourojeanni 1991).

A recent significant development has been the introduction of mandatory EIAs in many countries (see box on page 286). Most of the EIAs conducted so far, however, have been for specific projects previously approved in some economic area unrelated to the environment, rather than on the basis of general environmental policies or programmes. They have focused mainly on reducing negative environmental impacts and have seldom altered a proposal significantly, let alone led to its rejection.

Progress in implementing environmental laws, policies and regulations has been adversely affected by institutional hurdles created by weak coordination with other public, social and economic agencies, overlapping responsibilities between sectoral and the environmental institutions, budgetary restrictions, a lack of technical training and qualified human resources to carry out environmental management, and lack of political will (Figueroa 1994).

National environmental laws and institutions will probably be strengthened over the next few years as a result of international demands and further environmental deterioration. Such laws face a variety

Mexico's environmental policy

The Mexican federal government established its Environmental Programme 1995–2000, based on guidelines from the National Development Plan 1995–2000 and the National Strategy for Sustainable Development, the Agriculture and Rural Development Plan 1995–2000 and the Sectoral Agriculture Programme 1995–2000. National environmental institutions were reorganized to implement these programmes under a new Ministry of the Environment, Natural Resources and Fishery Management. The new ministry includes the Office of the Under-Secretary of Fishery Management, the National Water Commission and the Office of the Under-Secretary of Natural Resources, which manages forest and soil programmes.

The ministry's main jobs are the reversal of environmental degradation, the encouragement of sustainable production, the contribution to social wealth and the fight against poverty, with management strategies based on social participation, decentralization, inter-sectoral coordination and ecological planning.

Other programmes and initiatives include the 1995–2000 Hydraulic Programme, the 1995–2000 Fishing and Fish Farming Programme, the Protected Areas Programme 1995–2000, the Forestry and Soil Programme 1995–2000, the Programme of Wildlife Conservation and Productive Diversification in the Rural Sector 1997–2000, and the Sustainable Regional Development Programme.

Source: Mexico 1996 and SEMARNAP 1996

Environmental Impact Assessments in MERCOSUR

Environmental Impact Assessment systems in MERCOSUR member-states (Argentina, Brazil, Paraguay and Uruguay), and in its associate members (Bolivia and Chile), have been examined in a CIPMA (1997) study to establish whether and to what extent their different approaches could affect the dynamics and the objective of the MERCOSUR Treaty.

The study showed that all countries had adopted EIAs as a preventive environmental management tool but that they were applied in different ways. Brazil, which instituted EIAs 15 years ago, was the only nation with significant experience in this area. But the effectiveness of Brazilian EIA procedures was questioned in most of the studies reviewed. The other member countries were just beginning to use EIAs, and only Chile seemed determined to strengthen its EIA system.

Recurrent statements made at MERCOSUR's Special Meetings on the Environment (REMA) suggested that MERCOSUR was confident that fully enforceable EIAs were an important contribution to the Treaty's principles on environmental quality. However, Argentina, Paraguay and Uruguay stated that they opposed discrimination against their products on the basis of the need for additional environmental requirements such as EIAs. This suggests that countries equipped with less developed EIA systems will seek to lower standards, thus placing countries that take EIAs more seriously at a disadvantage.

Source: CIPMA 1997

of challenges. Brazil, for example, has to deal with management problems generated by the existence of federal as well as state environmental jurisdictions. Mexico is facing the new challenges of NAFTA, and Colombia needs to start implementing environmental policy through its Autonomous Regional Corporations. Argentina has to face the complexities of its own federal system, which features provincial legislative bodies and provincial property rights over natural resources. Conversely, Chile, where a statement of an overall environmental policy has recently been adopted (Chile CONAMA 1998), Ecuador, Paraguay, Uruguay and Central American nations have to deal with an excessively centralized institutional situation. Some developments in the Caribbean in relation to land-use planning are summarized in the box below.

Economic instruments

The use of economic incentives is still limited, and has been directed mainly at controlling pollution and access to certain natural resources. Subsidies and tax exemptions are the most common instrument used. Fiscal tools successfully deployed include reforestation subsidies, first implemented in Brazil and Chile more than 20 years ago. In Argentina, the waste management law allows a reduction of the annual charge to waste generators and operators of treatment and disposal plants if they apply waste recycling and improve their plants. Similar tax exemptions are offered to industries to encourage the use of natural gas (IDB 1996).

Environmental taxes for gasoline are levied in Mexico and Costa Rica, on the basis of lead content, to reduce polluting emissions and limit consumption. Green taxes to support environmental policies are beginning to be introduced in the region but they are unlikely to become a major priority due to serious arguments raised by revenue agencies. Green taxes on gasoline have been argued against in Chile on the

Caribbean initiatives on the environment and planning

Almost all Caribbean countries have strengthened their environmental institutions and administrative capacities to integrate environmental considerations into physical planning. Current initiatives in sustainable tourism, protected area management, integrated management of coastal and marine resources, and biodiversity conservation are expected to reduce pressures on natural resources.

Land-use plans or strategies have been prepared for a number of countries and territories including Anguilla and the British Virgin Islands and are under preparation in Antigua and Barbuda, Dominica, St Kitts and Nevis, and St Vincent and the Grenadines. National planning for Jamaica has, since 1978, included plans to reverse migration from the rural areas to Kingston and implement measures to control urban sprawl. More recently many of the other states, including Barbados, Cuba, Dominican Republic, Grenada, St Lucia, and Trinidad and Tobago, have included decentralization strategies in land-use planning exercises.

Measures to upgrade the quality of the built environment include giving priority to the improvement of basic services including shelter, potable water, and environmentally-sound sewage treatment and disposal.

There are efforts at the national level, particularly in Guyana, Jamaica and St Vincent and the Grenadines, to upgrade settlements and improve infrastructures through housing and land-management policies and strategies.

The Caribbean Planning for Adaptation to Global Climate Change is one of the most important initiatives in preparing to cope with the adverse effects of global climate change.

Biodiversity strategies are being developed in 10 countries and biodiversity conservation is also supported through activities in support of conventions such as Ramsar and CITES.

Several governments are seeking to increase the effectiveness and efficiency of preparations for and responses to natural disasters. Early warning systems for hurricanes and tropical storms have been put in place and national disaster response coordinating agencies have been established.

Source: UNEP/UWICED/EU 1999

grounds that they would affect overall tax structure, could not easily be used exclusively to finance environmental management, might put gasoline prices out of reach of many people and would have limited environmental impact.

The implementation of market instruments is often difficult. A system of transferable fishing permits was first adopted in Chile in 1991 to regulate access to some fishing grounds but this initiative faces a variety of hurdles due to the opposition of fishing enterprises and the lack of adequate control (O'Ryan 1996). Nevertheless, its implementation has allowed the recovery of resources of high commercial value which were overexploited under free-access regimes and inadequate traditional command-and-control measures (Borregaard and others 1997). Meanwhile, implementation of tradeable emission permits is still pending although the private sector backs the use of such tools and progress has been made towards designing a technically consistent system and defining the necessary legal framework. Practical difficulties are hard to overcome since environmental management schemes and capacities often lack a consolidated basic regulatory framework (Chile CONAMA 1998).

Systems based on fees and tariffs are more widely used and some have been in place for some time, although many such instruments were conceived to support overall economic policies rather than environmental management (Borregaard 1997). In Brazil, for instance, charges are levied for the use of natural resources (petrol, minerals and water) in a federal regulation dating from 1991: companies pay a tax proportional to the economic value of the exploited resources, and revenues are distributed to the Federal Government and the states where the exploitation took place (IDB 1996).

Mexico first established a fee on effluent discharge in 1991 in order to reduce pollution and encourage enterprises to take quality control measures. The charges per cubic metre of effluent vary with location; a similar initiative is being implemented in Uruguay and Colombia (CEPAL/PNUMA 1997). Charges for solid waste collection are also common.

Drinking water pricing, including charges for sewage collection and treatment, where applicable, is a common practice, for example in most Caribbean countries, although the environmental effectiveness of these charges has been minimal due to limited coverage of the users and a pricing policy that does not cover the capital costs. Peru and Central American countries are seeking to assess the value of their water resources, in order to reflect decreasing water availability in their tariffs and foster a more rational use of the resources (CEPAL/PNUMA 1997).

Charges to exploit certain resources, especially in the mining sector and activities involving the extraction of construction materials, are also collected in several countries (CEPAL/PNUMA 1997). A glass bottle deposit-refund system in Trinidad and credit subsidies to promote solar energy in Barbados have been successful (UNEP/UWICED/EU 1999).

The box below describes progress towards a new economic instrument which seeks to recognize and value the environmental services rendered by forests, and provides mechanisms for appropriate payment to forest owners.

In spite of some successes, implementation of economic instruments is often impaired by the weakness of public institutions, the lack of integrated legal frameworks, poor technical and administrative capacities, and political and ideological conflicts which, together with equity issues, result in a lack of political will to act (Borregaard 1997). In some cases, a failure to make economic instruments and direct regulations

Forestry incentives in Costa Rica

The concept that the environmental services provided by natural forests and plantations should be payed for is endorsed by several of Costa Rica's Forest Laws, of which the latest was passed in 1996. The results have helped implement the national policy of reducing deforestation rates which have remained at an average of 14 000 hectares a year over the past four years compared to 50 000 hectares a year in the 1980s and early 1990s.

Forest Law No. 7575 proposes paying owners of forested property, or property in the process of reforestation, to compensate for the services that their activities provide society in general. This Law established the framework for the development of cooperative projects, and strengthened the National Forestry Finance Fund (FONAFIFO). The projects of the Joint Coordination Office and the gasoline tax, passed within the framework of the UNFCCC, are the principle sources of finance for FONAFIFO. The latter tax raised close to US$7 million in 1997, and is expected to provide similar amounts over the next few years.

A series of Transferable Compensation Certificates (TCC) was issued on the stock market as a financial mechanism to facilitate the international commercialization of credits for carbon fixation from cooperative projects. The proceeds of this stock market issue will be used to pay small and medium-sized property owners for the environmental services of reforestation and voluntary forest conservation. In addition, a number of studies have been conducted to promote the incorporation of the cost of protecting watersheds in monthly charges for potable water. Finally, the Forest Law and the Wildlife Conservation Act also include fiscal and administrative incentives as compensation mechanisms for forest conservation and the management of national wildlife sites. These incentives include tax exemption, protection from squatters and technical assistance.

Source: Costa Rica 1998

complementary has had negative environmental effects. Command-and-control economic instruments are likely to remain in use over the next decade but the use of penalties, taxes, fees, tax deductions and subsidies will probably be intensified as more prevention-oriented environmental policies come into being. However, there is a tendency in some sectors to expect too much from regulatory programmes based on economic instruments. Direct regulation is still necessary for several environmental problems and a mix of direct regulations and economic instruments is likely to be needed.

There is a growing recognition of the need to adopt a system of environmental accounting and to appraise the value of natural resources but no significant practical progress has been made. Mexico is the only country in the region to use a 'satellite environmental accounting system' – a second accounting system which complements conventional economic accounting which captures changes in natural resources. A panel charged with making progress towards an environmental accounting system has recently been established in Brazil. In Chile, trials with satellite accounts for the forestry, mining and fishing sectors are continuing (Aguilar 1996). The concept of environmental accounting has not been dropped entirely but governments that are committed to strict free market policies, especially in South America, often question the usefulness of a tool that requires a substantial investment in resources and information (Sejenovich and Panario 1996).

Cleaner production in Chile

Chile's Cleaner Production Policy, dating from September 1997, was designed by the Ministry of Economy to encourage competitivity and good environmental management in business, support environmental preventive actions and develop cleaner production processes, including a more efficient use of energy and water. The action programme for 1997–2000 is designed to:

- promote cleaner production through technology transfer, open up markets for technological services and foster research and development;
- promote cooperation through voluntary programmes;
- strengthen the country's technological and information infrastructure;
- strengthen management and coordination in cleaner production, and integrate cleaner production with other national programmes;
- encourage regulatory and enforcement agencies to distribute cleaner production information, particularly to small and medium-size enterprises.

The plan is being implemented through existing machinery and funds such as the Metropolitan Region Prevention and Decontamination Plan, which includes rules to improve fuel composition, distribution and storage, and a tax to encourage the use of cleaner sources of energy. The new policy will be coordinated with other plans, such as the Energy Efficiency Plan of 1992 and the Frame Agreement of 1997 to optimize energy management by small and medium-size enterprises. Funds will include the National Environmental Commission's Technological Conversion Subsidy, set up for conversion of refrigeration and plastic foam manufacturing equipment to non-ozone depleting substances, which has an endowment of US$5 million for 1997-2000, and the 1992 Technological Innovation Programme.

Source: Chile Ministerio de Economía 1998

Industry and new technologies

Economic globalization and the development of markets sensitive to environmental issues are creating pressures to improve the environmental quality of products and promote clean industrial processes. Environmental demands are seen as challenges rather than limitations. Producers in Argentina, Brazil and Mexico are vigorously adapting productive processes to ISO 14 000 as a means of demonstrating compliance with international norms. In some countries, the lead has been taken by the most competitive sectors. In Chile, environmental and sectoral public agencies are seeking to transform the nation's productive structure through a series of economic incentives (see box on the left).

In other countries, some progress has been made as a result of voluntary agreements, for example with coffee entrepreneurs in Costa Rica and the programme of alcohol addition to gasoline in Brazil, demonstrating that strict regulation of resource use may not be the most efficient way of fostering technological change.

Research and technological development initiatives cover agriculture (genetic engineering), fisheries, forestry, waste management and the pharmaceutical sector (taking advantage of biological diversity to manufacture medicinal products). New methods of exploiting biotechnology and genetic engineering, new machinery, computerized drip irrigation and radioactive isotopes are gradually being introduced in areas where intensive agriculture is practised.

Some developments in the Caribbean are summarized in the box on page 289.

Financing environmental action

In most countries, the environmental sector is subsidized mainly by government funds, with resources coming from national budgets, donations, grants, transferable compensatory certificates, fees, loans, contribution legacies, fines, indemnification, auction sales of confiscated products and other resources determined by legislation (IDB 1996). In recent years there has also been important support from international aid and bilateral technical cooperation programmes, aimed mainly at setting up and strengthening environmental institutions.

The creation of special funds is a recent development. An initiative contained in ALIDES seeks to establish a Central American Fund for Environment and Development that would fill the funding gap in national and regional conservation projects. Similar initiatives of this kind in South America are Colombia's National Fund for the Environment, the Amazon Fund and ECOFONDO, which works with NGOs. Brazil has a National Environmental Fund, created in 1989 to finance projects related to the sustainable use of natural resources and the management and improvement of environmental quality; government agencies and environmental NGOs can request funds for activities that meet environmental policy aims. Some countries are also creating specific funds to finance forestry activities, offering better interest conditions than the financial market (Acuña and Orozco 1997).

There are other funding initiatives, such as those that take advantage of the restructuring of bilateral debts with the United States (Fund for the Americas), with examples in Argentina, Bolivia, Brazil, Chile, Costa Rica, Ecuador and Uruguay (CEPAL/PNUMA 1997, IDB 1996); these are oriented particularly to NGOs. Other initiatives relate to specific environmental issues. Examples include:

- Bolivia's National Environmental Fund (FONAMA) established in 1990, which aims to capture and manage funds oriented towards biodiversity;
- Paraguay's Protected Wilderness Areas, Wildlife and Forest Fund;
- Chile's Environmental Protection Fund;
- Brazil's Federal Fund for Forest Replacement, supported since 1973 by payments for the exploitation of forest resources; and
- the Rain Forest Trust Fund administered by the World Bank.

Brazil's National Environment Programme, with 70 per cent financing from the World Bank, was set up to strengthen environmental bodies, implement the National System of Conservation Units, protect endangered ecosystems and help reconcile economic interests with environmental protection.

Cleaner production in the Caribbean

The adoption of cleaner technologies was initiated in the early 1990s. Several countries, including Jamaica, Trinidad and Tobago, and Guyana, have undertaken clean technology initiatives through public and private sector partnerships, with research contributions from major universities.

The sectors covered include agriculture, tourism and mining. For example, JAMALCO, a joint venture between the Government of Jamaica and Alcoa Minerals of Jamaica Ltd., has pioneered two types of bauxite residue disposal technology.

Cleaner energy technologies are being promoted in a number of countries including Barbados, Cuba, Dominica, Jamaica and St Lucia, in the form of energy efficiency and alternative energy sources including solar, hydroelectric, biomass and biogas. Wind energy is being exploited in Curacao, Jamaica and Barbados. Since 1993, Curacao has been operating a 3-MW wind farm, while Jamaica plans to install an 18–20 MW wind farm by 2000. The first Ocean Thermal Energy Conversion plant, which uses heat energy from warm surface areas to generate power, has been constructed in Cuba, followed by the development of a 2-MW demonstration plant in Jamaica. Biomass has been used as an energy source in the sugar cane industry in Cuba.

Source: UNEP/UWICED/EU 1999

Public participation

Public participation has grown over the past few years, mainly in tune with a greater public awareness of threats to the quality of life and with the restoration of democracy in some countries. In most countries, however, institutional and legal participation is restricted to a few areas, such as EIA procedures, where public hearings are part of a formal process. Bolivia's Popular Participation Law grants important rights and roles to all citizens, unions and community organizations, the most important role being the watch kept by Vigilant Committees on the use of public funds by municipalities. In contrast, in Argentina environmental rule-making at the national level does not require formal consultative mechanisms; however, newly-created sectoral regulatory entities (electricity, natural gas and water) have institutionalized mechanisms for consultation on draft regulations or permits, and public hearings are called regularly as part of their decision-making processes (IDB 1996). In Chile, Supreme Court rulings have backed environmental groups that object to projects promoted by the government, thus creating important legal precedents (IDB 1996). Environment-related controversies in Argentina, particularly at the provincial level, have also confirmed the validity of resorting to the courts (IDB 1996).

Citizens' participation has also materialized in a more direct way through formal representation of individual and bodies with environmental interests on

Inter-American Strategy for Public Participation

The Plan of Action of the Second Summit of the Americas in Santiago, Chile, 1998, calls for dialogue and partnership between the public sector and civil society, and entrusts the Organization of American States (OAS) with encouraging support among governments and civil society and promoting appropriate programmes to carry out this initiative.

OAS, in compliance with the Bolivia Summit of the Americas (1996) mandate, is formulating the Inter-American Strategy for Public Participation (ISP) to identify mechanisms for securing transparent, accountable and effective participation by individuals, civil society and governments, and promoting participatory decision-making in environmental and sustainable development issues. The strategy is being formulated by conducting demonstration studies, analysing relevant legal and institutional frameworks and mechanisms, sharing information and experience, and establishing a basis for long-term financial support for public-private alliances. The work is being supported by the Global Environment Facility in collaboration with UNEP, OAS, USAID, UNESCO, IDB, and other donors and institutions. Several consultations and meetings have been held and technical studies are being conducted to review lessons learned and identify best practices for public participation mechanisms. The first draft of the ISP is available on the Internet at http://www.ispnet.org/strategy.htm.

Source: ISP 1998

numerous councils. In Mexico, for example, these councils include the National and Regional Consultative Councils for Sustainable Development, Basin Councils, the Consultative Council on Environmental Normalisation, Metropolitan Areas Councils on Air Quality Management, the National Council on Protected Wilderness Areas and the National Technical Consultative Councils on Forestry and Soil Rehabilitation and Conservation (Chacón 1997).

Bolivia's Council on Sustainable Development (created in 1996) has a consultative role and includes representatives of government, NGOs, the private sector, the press, the academic sector, indigenous people and trade unions. The National Council of Environment in Brazil and CONAMA's Consultative Council in Chile advise their governments on the policy-making process at national and regional level.

Citizens' demands for broader and institutionalized legal participatory channels are expected to increase. Decentralized actions to deal with environmental conflicts locally or at the provincial level may offer an effective way of channelling public participation.

At recent regional summits, governments have recognized that the strong engagement of civil society in decision-making is fundamental for enhancing democracy, promoting sustainable development, achieving economic integration and free trade, improving the lives of people, and conserving the natural environment for future generations. In the areas of sustainable development, the 1996 Declaration of Santa Cruz de la Sierra specifically endorses this principle, and commits the signatories to supporting and encouraging broad participation by civil society in designing, implementing, and evaluating policies and programmes. The Bolivia Summit Plan of Action entrusted the Organization of American States with the formulation of a strategy to promote public participation in decision-making for sustainable development (see box left).

Some public participation initiatives in the Caribbean are described in the box below.

Environmental information and education

The information generally available on environmental issues has increased, particularly since the Rio Conference. Programmes to develop information systems and data management in support of environmental policies have been created in several countries but their impact on decision-making cannot yet be assessed because they are still at an early phase. In Chile, for example, a National System of Environmental Information was launched in 1994, based on a decentralized, low-maintenance, open and flexible platform. It has a pilot web site with

Public participation initiatives in the Caribbean

In many Caribbean countries, the changed perception of the role of civil society in achieving the objectives of the Earth Summit has resulted in close collaboration between governments, NGOs, community organizations and the private sector in setting standards and preparing environmental policies and action plans. In some countries, this collaboration is formalized in moves to decentralize governance to the community level.

NGOs have made significant contributions in the creation and management of protected areas such as the Kingshill Forest Reserve in St Vincent and the Grenadines, the Montego Bay Marine Park in Jamaica, and in Trinidad and Tobago where concerned individuals and NGOs, with government support, introduced a 'co-management' arrangement entrusting responsibility for the management of beaches where turtles nest to villagers.

Public participation is also becoming institutionalized through legislative requirements. St Lucia, for example, has included special provisions for public participation in amendments to its National Trust Act and the St Lucia National Trust, along with the Caribbean Natural Resources Institute, is spearheading nation-wide participation in protected area management.

Source: UNEP/UWICED/EU 1999

information organized in environmental modules and topics (CHIPER 1999). Information policies have concentrated mainly on natural resources, with little information on the dynamics of ecosystems.

The most common problem in collecting and organizing environmental information is the incompatibility of data between different agencies and different countries. Some attempts have been made in Brazil since 1984 to maintain a National Environmental Information System, despite the difficulties of coordinating federal and state environmental agencies. A National Information Centre connected with national and international scientific organizations is being built and its implementation has begun (IDB 1996).

In several countries, there are different information systems working on specific aspects, managed by sectoral institutions, such as the Information System of Protected Areas of the National Office for Biodiversity Conservation in Bolivia. On a sub-regional level, in Central America, the UNDP Programme of Sustainable Development Networks (SDN) began in Honduras in 1994, with the aim of improving mechanisms for processing and exchanging information in support of sustainable development, involving the government and all actors of civil society at a national and regional level (SDN 1998). In Bolivia, the Council on Sustainable Development disseminates regular reports (Bolivia 1996 and 1997). The box on the right describes progress in the Caribbean.

Consciousness-raising activities include an increasing number of major educational campaigns on saving natural resources and reducing waste generation, and publicity campaigns aimed at promoting recycling and consumption of non-polluting products. In Argentina, legal initiatives empowering the National Secretary of Natural Resources and Human Environment to publish a list of violators of environmental regulations are causing negative publicity for the offending industries. In Chile and Uruguay, eco-labelling has been introduced for products not containing ozone-depleting substances.

Formal educational institutions have made progress at the technical and higher education levels thanks to the establishment of specialized post-graduate programmes, especially in Colombia, Brazil and Mexico. Bolivia has a specific law that creates a Ministry of Education and Culture as well as a National Secretariat and Departmental Councils, which are responsible for defining policies and strategies to plan and develop formal and informal environmental education in coordination with other public and private institutions, while the Ministry also contributes at national level to promoting seminars and short courses (IDB 1996). Some institutions devoted to technical and scientific research have added environmental issues to their programmes, responding to the incipient demand of the private sector, and the establishment of private universities is encouraging emerging subjects such as the environment.

Modest progress is being made at the elementary and secondary levels of formal education, where some environment-related courses and programmes are being offered on an experimental basis. In Peru, for instance, some progress has been made through the development of a strategy on education for sustainable development and by changing the educational system at the curriculum level, extending the scope of 'natural sciences' to 'sciences and environment'.

Social policies

Social policies have a major impact on environmental issues in the region. The needs of the very poor, who have to struggle to survive with virtually no education or environmental awareness, contribute substantially to the degradation of the environment.

In most countries, environmental management is dissociated from social policies – hence the controversies unleashed by projects that are detrimental to the environment but are nevertheless

Information initiatives in the Caribbean

Environmental information is increasingly being used in decision-making for sustainable development in Caribbean countries. Government policy is to establish environmental management institutions that are also responsible for information management. Jamaica, Trinidad and Tobago, and Guyana have established institutions that will also be responsible for the development of National Environmental Information Systems.

NGOs are increasingly involved in data collection, public education and capacity building. The Caribbean Conservation Association in Barbados runs an information management programme and participates in information dissemination and public education. The Guyana Environmental Management Conservation Organization conducts ecological research and, in St Lucia, the National Trust is moving from the collection of scientific data to the incorporation of these data into a management system for its National Parks. Regional networks for information exchange include the Fisheries Newsnet of CARICOM, Caribbean Community. Others such as AMBIONET, CARISPLAN, CEIS, INFONET, and UNEPNet focus on the creation and maintenance of regional databases on socio-economic and environmental data and information.

Source: UNEP/UWICED/EU 1999

considered valuable because of the employment they generate. Another example is housing programmes that fuel urban growth and discourage better use of existing urban areas. Inequality persists as a result of practices and regulatory measures that benefit the industrial sector or high-income social groups which ignore environmental deterioration and its impact on their quality of life.

The challenge posed by sustainable development clearly goes beyond the boundaries of environmental quality. It also relates to social factors that are highly critical for development. For instance, high population growth, inadequate food supply, polluting energy sources and threats to ecosystems all have a critical impact on poverty and inequity in urban and rural areas alike. In practice, social and environmental policies barely relate to one other. Populations are confronted by serious gaps in terms of basic social needs, particularly in the less developed countries. Programmes devised to fight poverty and especially extreme poverty are usually unrelated to environmental policies and no effective actions have yet been undertaken to complement those two policy areas to their mutual benefit. Along with the imperative of overcoming poverty, the lack of consistent and consolidated environmental policies is one of the main challenges for the future.

Conclusions

During the past decade, there has been a strong increase in concern for environmental issues. Almost all countries have created environmental institutions and developed new environmental laws and regulations. While it is too early to judge the effectiveness of the measures that have been taken, preliminary analyses indicate that environmental management continues to concentrate on sectoral perspectives, without coherent and explicit integration with economic and social strategies. The lack of financing, technology, personnel and training and, in some cases, overly large and complex legal frameworks are the most common problems.

The level of ratification and adoption of global and regional MEAs is high. While the impact of MEAs on the environment is difficult to assess, they have succeeded in raising awareness of environmental matters among decision-makers and the public. This, however, has seldom led to the prioritization of environmental issues in the political agenda or national budgets. Policy action should focus on overcoming the numerous barriers that exist to the successful implementation of MEAs. These include:

- lack of funding;
- the need to transfer appropriate technology to implement MEAs;
- lack of national legislation to enforce obligations;
- lack of institutions to implement MEAs; and
- poor or non-existent systems to monitor MEAs.

Priorities for policy action in key sectors are:

Forests and land use
- halting forest and land degradation;
- implementing and enforcing national land-use planning policies;
- developing and implementing economic instruments to promote the sustainable management of forest resources and agriculture, especially in fragile ecosystems;
- developing and implementing legal instruments to overcome land tenure problems and thus reduce pressure on forest and soil resources.

The urban environment
- establish economic incentives for the introduction of clean technologies, especially for small and medium-sized industries;
- enforce urban planning to avoid further uncontrolled expansion of cities, develop more efficient public transport and discourage the use of the private car;
- establish economic instruments to reduce the generation of waste from domestic and industrial sources, and improve sanitation infrastructure; and
- develop education/information strategies to promote sustainable consumption patterns.

The coastal and marine environment
- develop and implement coastal zoning to reduce pressure on coastal areas from intensive land use (from tourism, aquaculture, artisanal fisheries, and other industries).

References

Acuña, M. and Orozco, J. (1997). *Fortaleciendo las perspectivas para el desarrollo sostenible en Costa Rica*. De. E5, San José, Costa Rica

Aguilar, X. (1996). Cuentas Ambientales en Chile. In E. Claro and others, *Valoración económica de la diversidad biológica en America Latina y el Caribe*. CONAMA/Environment Canada, Santiago, Chile

ALDA (1997). Asociación Latinoamericana de Derecho Ambiental. *Boletín semestral de ALDA,* No. 1, January-June 1997

ALIDES (1999). http://www.sicanet.org.sv/alides/

Bolivia (1996). Ministerio de Desarrollo Sostenible y Medio Ambiente, Secretaría Nacional de Planificación, and Capacidad 21/PNUD. *Consejo Boliviano de Desarrollo Sostenible,* various issues

Bolivia (1997). *Río más cinco. De la Agenda a la Acción*. Ministerio de Desarrollo Sostenible y Medio Ambiente, Sercretaría Nacional de Planificación, Santa Cruz, Bolivia

Borregaard, N. (1997). Instrumentos Económicos en la Política Ambiental. Oportunidades y Obstáculos para su implementación en Chile. *Ambiente y Desarrollo,* Vol. XIII, No. 3, September 1997, CIPMA, Santiago, Chile

Borregaard, N., C. Sepúlveda, P. Bernal and E. Claro (1997). Instrumentos Económicos al Servicio de la Política Ambiental en Chile. *Ambiente y Desarrollo,* Vol. XIII, No. 1, March 1997, CIPMA, Santiago

Brzovic, F. (1993). *Crisis económica y medio ambiente en América Latina y el Caribe*. CEPAL, LC/R.818, Santiago, Chile

CCAD (1997). *Estado del ambiente y los recursos naturales en Centroamérica*. Comisión Centroamericano de Ambiente y Desarollo, Taller de Trabajo, Guatemala City, Guatemala

CCAD (1998). *Estado del ambiente y los recursos naturales de Centroamérica*. Comisión Centroamericano de Ambiente y Desarollo, San José, Costa Rica

CEPAL/PNUMA (1997). *Instrumentos Económicos para la Gestión Ambiental en América Latina y el Caribe*. CEPAL/PNUMA, Mexico

Chacón, C. M. (1997). *Desarrollo Sostenible en Centroamérica: políticas públicas, marco legal e institucional*. INCAE, San José, Costa Rica

Chile CONAMA (1996). *Permisos de emisión transables en Chile. Propuesta de sistema para los recursos aire y agua*. Comisión Nacional de Medio Ambiente, Santiago, Chile

Chile CONAMA (1998). *Una Política Ambiental para el desarrollo sustentable*. Comisión Nacional de Medio Ambiente, Santiago, Chile

Chile, Ministerio de Economía (1998). *Política de Fomento a la Producción Limpia.* Versión Final aprobada por el Comité Interministerial de Desarrollo Productivo, 30 de septiembre de por el Consejo Directivo de Ministros, 9 de enero de 1998. Ministerio de Economía, Santiago, Chile

CHIPER (1999). Chile Information Project Environmental Report http://www.chiper.cl/index.asp

CIPMA (1997). *Los Sistemas de Evaluación de Impacto Ambiental en los Países del MERCOSUR.* CIPMA, Santiago, Chile

Costa Rica (1998). *Informe de Pais a la Convención sobre Diversidad Biologica.* MINAE, San José, Costa Rica

Dourojeanni, A. (1991). *Procedimiento de gestión para el desarrollo sustentable, aplicados a municipios, microregiones y cuencas*. CEPAL, LC/R. 1002/Rev. 1., Santiago, Chile

Figueroa, E. (ed., 1994). *Políticas Económicas para el Desarrollo Sustentable de Chile*. Centro de Economía de los Recursos Naturales, Universidad de Chile, Santiago, Chile

Gligo, N. (1997). Institucionalidad Pública y Políticas Ambientales Explícitas e Implícitas. *Revista de la CEPAL 63*, December 1997

González, J. J. (1997). *Nuevo derecho ambiental mexicano (Instrumentos de política)*. Universidad Autónoma Metropolitana, Mexico City, Mexico

Government of Panama (1998). Acuerdo Ley 7 de 10 de febrero de 1998. *Gaceta oficial de la República de Panamá*, No. 23484

IDB (1996). *Environmental Management in the Southern Cone: A study on the legal and institutional framework.* Background Studies. Interamerican Development Bank (ATN/II-5109-96), Washington DC, United States

ISP (1998). *Inter-American Strategy for Public Participation in Environment and Sustainable Development Decision Making in the Americas.* Organization of American States, Washington DC, United States; http://www.ispnet.org/strategy.htm

Mexico (1996). *Ley Orgánica de la Administración Pública Federal.* Colección Porrúa, Mexico City, Mexico

MOE Brazil (1998). Ministry of the Environment, Brazil http:/www.mma.gov.br

National Census of the Vicuña (1996). *National Census of the Vicuña in Bolivia 1996.* http://coord.rds.org.bo/vicuna/censo/conclusi.htm

OAN (1998). Map of ODS Officers Networks. *OzonAction News*, 28, October 1998

Orozco, J. and Acuña, M. (1997). *Cambio estructural y ambiente en los últimos veinte años*. CINPE, San José, Costa Rica

O'Ryan, R., and Ulloa, A. (1996). Instrumentos de regulación ambiental en Chile. In O. Sunke (ed.), *Sustentabilidad ambiental del crecimiento económico chileno.* CAPP, Universidad de Chile, Santiago, Chile

Paraguay (1995). *Diagnóstico del sector forestal paraguayo. Proyecto Estrategia Nacional de los Recursos Naturales.* Subsecretaría de Recursos Naturales y Medio Ambiente, Ministerio de Agricultura y Ganadería, and GTZ, Asunción, Paraguay

PNUD/PNUMA (1996). *La recepción en los sistemas jurídicos de los países de América Latina y el Cariba de los compromisos asumidos en la Conferencia de las Naciones Unidas sobre el Medio Ambiente y el Desarrollo (1992)*. Propuestas para la cooperación hemisférica. PNUMA-ORPALC, Mexico City, Mexico

PNUMA (1993). *Legislación ambiental general en América Latina y el Caribe*. Serie de Legislación Ambiental No. 1. PNUMA-ORPALC, Mexico City, Mexico

PNUMA-ORPALC (1993). *Situación actual del derecho internacional ambiental en América Latina y el Caribe*, Serie de documentos de derecho ambiental No. 2. PNUMA-ORPALC, Mexico City, Mexico

PNUMA-ORPALC (1996). *Estudio camparativo de los diseños institucionales para la gestión ambiental en los países de América Latina y el Caribe*. Documento UNEP/LAC-IC-2/7. PNUMA-ORPALC, Mexico City, Mexico

[illegible]. *Respuesta al cuestionario de los acuerdos ambientales multilaterales*. Unpublished

Ramsar (1998). *Report of Allocations 1992–97, The Ramsar Convention Small Grants Fund.* Ramsar, Gland, Switzerland http://www.iucn.org/themes/Ramsar

SDN (1998). www.sdnp.undp.org

Sejenovich, H., and D. Panario (1996). *Hacia otro desarrollo. Una perspectiva ambiental*. Editorial Nordan, Montevideo, Uruguay

SEMARNAP (1996). *Programa de Areas Naturales Protegidas de México 1995-2000; Programa de Conservación de Vida Silvestre y de Diversificación Productiva en el Sector Rural, 1997-2000; Síntesis Ejecutiva del Programa Forestal y de Suelos 1995-2000; Poder Ejecutivo Federal, 1996. Programa de Medio Ambiente 1995-2000; Programa Hidráulico 1995-2000; Programa de Pesca y Acuacultura 1995-2000; Programa Sectoral Agrario 1995-2000; Plan Nacional de Desarrollo 1995-2000*. SEMARNAP, Mexico City, Mexico

Summit of the Americas (1997). http://environment.harvard.edu/cumbre/eng/docs.html

Treaty for Amazonian Cooperation (1997). *Boletín Informativo*. Treaty for Amazonian Cooperation Temporary Secretariat, Caracas, Venezuela

UNEP (1997). *Register of International Treaties and Other Agreements in the Field of the Environment 1996.* UNEP, Nairobi, Kenya

UNEP Ozone Secretariat (1997). *The Handbook for the International Treatment for the Protection of the Ozone Layer* (1997 update). UNEP Ozone Secretariat, Nairobi, Kenya

UNEP Ozone Secretariat (1998). *Production and Consumption of Ozone Depleting Substances 1986–1996*. UNEP Ozone Secretariat, Nairobi, Kenya http://www.unep.org/unep/secretar/ozone/pdf/Prod-Cons-Rep.pdf

UNEP/UWICED/EU (1999). *Caribbean Environment Outlook.* UNEP, Nairobi, Kenya

Weiss, E.B., and Jacobsen, H.K. (1998). *Engaging Countries: Strengthening Compliance with International Environmental Accords.* MIT Press, Cambridge, Massachusetts, United States

World Bank (1997). *Pilot Programme to Conserve the Brazilian Rain Forest.* World Bank, Brasilia, Brazil

North America

KEY FACTS

- North America has pioneered environmental policy development, first through command-and-control measures, and later through voluntary and market-based approaches.
- The United States and Canada are among the most active countries in developing and complying with global MEAs.
- The importance of the North American Agreement on Environmental Cooperation may go beyond the region since successes and failures in dealing with cross-border environmental impacts, the migration of industries seeking cheaper labour and more permissive environmental standards, and the sale of products with high environmental risks can serve as important examples for the entire global community.
- The goal of the Accelerated Reduction/Elimination of Toxics (ARET) programme in Canada is to reduce the emission of persistent, bio-accumulative and toxic substances by 90 per cent, and the emission of all other toxic substances by 50 per cent, by the year 2000.
- The tradeable permits system in the United States for sulphur dioxide could save as much as US$3 000 million a year compared with a traditional command-and-control approach.
- Erosion reduction credited to the Conservation Reserve Program in the United States may be as high as 630 million tonnes of soil annually, or 42.75 tonnes/hectare/year.
- Some provinces in Canada provide funding for citizens making legal interventions on issues of public concern.
- Providing information to the public has been a powerful incentive for encouraging action by industry to improve the management of toxic chemicals through reduced use and decreases in releases and transfers.

The policy background

The United States and Canada have extensive experience with environmental policies. Not all have been successful but, compared with most other countries, many have been. The region also has a well-developed set of institutions for implementing environmental policies. Finding a pattern that defines the success or failure of policy initiatives is challenging, given the multiplicity and complexity of environmental issues. There is a common thread, however, indicating that the successful policies are those based on approaching issues in their full socio-economic and ecological context, and understanding their dynamic changes over time, their interactions across geographic regions, and their importance to a variety of stakeholders, from communities through government to business. Since *GEO-1* reported on comprehensive management and creative partnerships in environmental policy-making in the region, interest in ecosystem management, stakeholder participation and consultative processes has been increasing, with sustainable development as an overriding objective. This is perhaps most evident in areas such as regional fisheries and climate where earlier policy measures failed to bring the expected results.

Discussion of Mexico in this section is limited to cross-border issues such as conservation of biodiversity and migratory species, transportation management, watershed management, air pollution and the North American Free Trade Agreement. For other issues, Mexico is included under Latin America and the Caribbean.

During most of this century, government regulation was the strategy of choice to deal with environmental concerns. Since the early 1990s, the growing need for cost-effectiveness, voluntary action, flexibility and consensus-building has led to a shift from command-and-control regulation towards a mixed set of policies, with an increasingly important role for market-based mechanisms, public-private partnerships and voluntary initiatives. When combined with essential regulatory measures, these mechanisms are compatible with the overall framework of sustainable development.

The concept of sustainable development, now widely recognized by government agencies, has helped to extend the debate about environmental issues beyond environmental agencies and interest groups. Fora such as the national, provincial and community round tables on the economy and environment in Canada, and the President's Council for Sustainable Development (PCSD) in the United States, have provided good opportunities to examine environment and development issues. They have also created an opportunity for dialogue between the public and private sectors and civil society. Sustainable development principles are being translated into sector-specific initiatives to create tangible objectives, targets and strategies for government agencies and private corporations. However, by the second half of the 1990s some of these organizations, including many provincial round tables in Canada, had disbanded as a result of either weak political support or unreasonable expectations that they would find quick and easy solutions to sustainable development issues.

A number of major changes have affected environmental policy-making in the 1990s.

- The business community increasingly accepts the need for environmental protection, and calls for policy changes that would make reaching environmental objectives more efficient and reward innovators. Environmental management standards, such as ISO 14 000 and the CERES Principles, are accepted by many corporations.
- As part of overall government efforts to reduce financial deficits, environmental departments in Canada are experiencing steep budget cuts. This has significantly reduced the capacity of these agencies to fulfil their mandates and meet their responsibilities.
- The costs of environmental protection and the struggle to reduce government budget deficits has highlighted the issue of accountability and cost-effectiveness, leading to a search for alternative policy instruments.
- Particularly in the United States, federal officials are being increasingly used as facilitators to find the most effective and efficient solutions. Environmental policy instruments are increasingly developed in consultation with the public and the business community.
- Participation by NGOs and community residents is increasingly viewed as a valuable part of any environmental protection programme.

Perhaps the most significant regional policy initiative has been the North American Free Trade Agreement (NAFTA) between Canada, Mexico and the United States. NAFTA is designed primarily to liberalize trade among the parties and, along with the environmental and labour side agreements, the North American Agreement on Environmental Cooperation (NAAEC) and the North American Agreement on Labor Cooperation (NAALC), to regulate economic, environmental and labour cooperation. The environmental impacts of NAFTA are yet to be understood. The potential for relocation of polluting industries to regions with more lenient environmental standards and enforcement, the increase in the intensity of agricultural production and its impact on land resources, and the impact of increased transportation are some of the trade-environment problems that the agreements are intended to help resolve.

The environmental policy scene is changing in response to changing conditions and social expectations. In Canada most of the emphasis is on regulatory reform, federal/provincial policy harmonization and voluntary initiatives. In the United States, the need for new types of environmental policy has increased and the country is moving faster and further on market-based policies. Examples include the use of tax incentives to phase out ozone-depleting substances, the use of tradeable emissions permits to help reduce the costs of air pollution controls, the requirement that corporations disclose to the public their releases of toxic and hazardous pollutants, government-initiated challenge programmes and voluntary action of corporations to reduce pollution, agricultural subsidy reform to create incentives for farmers to take highly erodible cropland out of production, and increased emphasis on performance reporting.

MEAs and non-binding instruments

Global MEAs

The United States and Canada have been among the most active countries in developing and complying with global MEAs. The requirements of many of the conventions are built into federal, state and provincial legislation. In several cases, awareness of environmental issues, legislation, and national and bilateral policies preceded the ratification of particular MEAs.

Neither Canada nor the United States has put into place any additional regulations specifically aimed at implementing the UNFCCC. Instead, they have relied on voluntary measures which so far appear insufficient to fulfil the UNFCCC goal of stabilizing emissions at 1990 levels by the year 2000. These measures are described in detail in the next section, 'Voluntary action', on page 301. While there is no dedicated legislation to address the UNFCCC in the United States, sufficient legal authority does exist to regulate emissions from sources such as cars and power plants under national environmental statutes such as the Clean Air Act and constitutional clauses on commerce and taxation.

MEA implementation and compliance are of critical importance. The region is a leading economic powerhouse, with the highest per capita rates of material production and consumption in the world. As a result there has been a rapid and extensive loss or transformation of native ecosystems and intensive use of natural resources. In per capita terms, the United States and Canada are also among the most intensive users of the global environmental commons such as the atmosphere and oceans. Compliance with MEAs without sacrificing quality of life is a major long-term challenge for North America and one that has the potential to set a powerful example to other countries.

Monitoring and reporting are essential to ensure the accountability and effectiveness of MEAs. Compliance is highest where the agreements include clear targets, performance measures and reporting mechanisms. While both Canada and the United States report as required to the conferences of the parties of the treaties that they have ratified, the public picture of compliance at the overall national level is sometimes unclear.

NGOs play an important role in monitoring compliance and issue ratings on overall performance in some key areas. For example, the Worldwide Fund for Nature monitors the progress of biodiversity conservation efforts through its Endangered Spaces Programme, by keeping representative samples of Canada's marine and terrestrial eco-regions under protection and assigning grades to provinces based on their performance (World Wildlife Fund 1998).

Involvement in global MEAs is high, although there are several such as CMS that await ratification

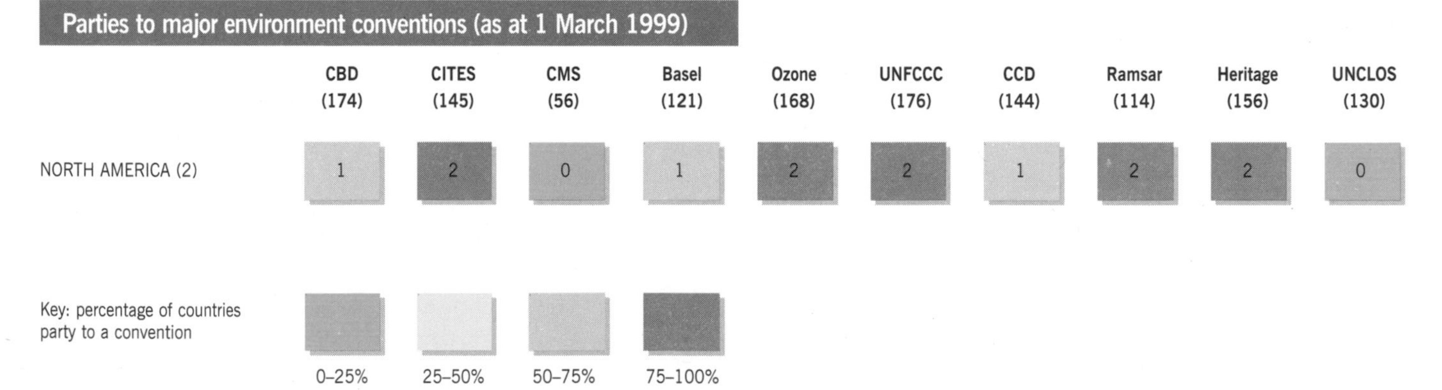

Parties to major environment conventions (as at 1 March 1999)

	CBD (174)	CITES (145)	CMS (56)	Basel (121)	Ozone (168)	UNFCCC (176)	CCD (144)	Ramsar (114)	Heritage (156)	UNCLOS (130)
NORTH AMERICA (2)	1	2	0	1	2	2	1	2	2	0

Notes:

1. Numbers in brackets below the abbreviated names of the conventions are the total number of parties to that convention
2. The number in brackets after the name of the region is the number of sovereign countries in the region
3. Only sovereign countries are counted. Territories of other countries and groups of countries are not considered in this table
4. The absolute number of countries that are parties to each convention in the region are shown in the coloured boxes
5. Parties to a convention are states that have ratified, acceded or accepted the convention. A signatory is not considered a party to a convention until the convention has also been ratified

(see figure left). In some cases – for example, Canada's role in shaping the CMS – the region took a leading role in negotiations and made a major contribution to shaping specific agreements. Since in most cases national and regional agreements preceded MEAs, a lack of ratification does not necessarily mean that the issue is not on the political agenda.

In Canada the ratification of MEAs by the national government may require implementing legislation by provinces. This has important implications for negotiations, the political consensus necessary to implement conventions, and evaluation of compliance on the national level. For example, in preparing for the ratification of the Kyoto Protocol and the drafting of detailed implementation strategies, provinces have agreed that no region should be asked to bear an unreasonable burden as Canada seeks to reduce its greenhouse gas emissions. Without appropriate assurances and consensus, provinces can delay national ratification or make it more difficult.

Implementation of CITES

CITES monitors and controls the international trade of more than 30 000 species of animals and plants. The control of illegal trade is enforced in the United States through measures that include interception at border entries, spot-checks of wildlife-related businesses, monitoring of hunting, and prosecution under criminal law. For example, recent cases in the United States involved the prosecution of a smuggling ring dealing with neo-tropical parrots, and the fining of a West Coast fishing company for falsifying fishing licences to hide excessive catches. In addition to a US$100 000 fine, the company was required to make an announcement on television urging others to comply with the law.

During 1995–96, Environment Canada conducted more than 3 000 inspections and more than 200 investigations into claims of illegal wildlife trafficking. Offenders face penalties of up to CAN$150 000 for individuals or up to CAN$300 000 for corporations if found to be guilty of illegally importing or exporting or simply possessing endangered species or products made from them (Environment Canada 1996).

Both countries apply a wide range of laws and policy measures to meet environmental objectives, whether or not they have ratified the related MEAs. In the United States implementing legislation is usually required to cover the obligations, unlike in many other countries where the MEA essentially takes on the status of legislation once it is adopted by the legislature. For instance, the Resource

Enforcement activities in Canada for endangered species

	inspections	*investigations*	*prosecutions*	*convictions*
1994–95	1 083	93	20*	43**
1995–96	3 369	207	46*	17**

* Some prosecutions are handled by other agencies and not reported to Environment Canada
** Includes some convictions obtained from prosecutions initiated in previous years

Source: CEC 1996

Conservation and Recovery Act regulates hazardous wastes in the United States, even though the country is not yet party to the Basel Convention. Canada has ratified the Basel Convention, and implements it through the Export and Import of Hazardous Waste regulations of the Canada Environment Protection Act (CEPA). Commitments under CITES are met mainly through the Wild Animal and Plant Protection and Regulation of International and Inter-provincial Trade Act in Canada, while this is achieved through several separate regulations dealing with specific flora and fauna in the United States (see box left and table above).

The legislation includes provisions for enforcement and penalties for non-compliance. In Canada, for example, offences are established in CEPA, which has an associated enforcement policy (CEPA 1998). In the United States, the general principles of enforcement are laid out in the Operating Principles for an Integrated EPA Enforcement and Compliance Assurance Program. Enforcement usually involves multi-agency task forces from environment, customs, agriculture, fisheries and other departments. Illegal activities, such as smuggling of ozone-depleting substances or endangered species, may lead to fines and/or prison sentences.

Many organizations, including a strong and mature NGO community, the media and multi-stakeholder organizations, exert pressures to comply with MEAs. Canada has national networks in some areas, such as the Canadian Ramsar Network and other provincial and national organizations that deal with Ramsar. Two of the most important regional coordinating bodies are the North American Wetlands Conservation Council and the North American Waterfowl Management Plan (NAWCC 1999).

The Commission for Environmental Cooperation (CEC), established in 1994, is an important forum for regional dialogue on compliance. The annual reports of

The development and implementation of Canada's Biodiversity Strategy

Canada was the first industrialized country to sign and ratify the Biodiversity Convention. Following ratification in 1992, Canada began to develop a Biodiversity Strategy to provide guidance for national implementation. Recognizing that successful implementation would require the effort of Canadian society at large, the strategy was drafted by a Federal-Provincial-Territorial Working Group, in consultation with a Biodiversity Advisory Group made up of representatives of industry, the scientific community, conservation groups, academia and indigenous organizations.

The development of the strategy began with an assessment of what additional work was needed for full implementation of the Convention. The assessment revealed that many elements of a successful national response were in place but that in some cases it would be more effective to enhance current efforts rather than develop entirely new initiatives.

The Canadian Biodiversity Strategy was published in 1995. Its main elements include a biodiversity vision, five goals and a series of mechanisms to help implementation. The five goals are:

- conserve biodiversity and use biological resources in a sustainable manner;
- improve understanding of ecosystems and increase resource management capability;
- promote understanding of the need to conserve biodiversity and use biological resources in a sustainable manner;
- maintain or develop incentives and legislation that support the conservation of biodiversity and the sustainable use of biological resources; and
- work with other countries to conserve biodiversity, use biological resources in a sustainable manner and share equitably the benefits that arise from the utilization of genetic resources.

Implementation mechanisms include reporting by all jurisdictions on any plans or action to implement the strategy, coordination of national and international implementation efforts, encouragement of NGO participation, and reporting on the status of biodiversity.

A 1998 report on the status of compliance with the Convention (Commissioner of the Environment and Sustainable Development 1998) points out that although the Canadian Biodiversity Strategy was a good first step, its implementation has been slow and there are knowledge, capacity and resource gaps that need to be addressed to meet national commitments. Some of the key recommendations include: develop biodiversity targets, identify the activities necessary to meet those targets, assign responsibilities for activities, clarify timelines and provide a budget for these activities and the monitoring of biodiversity indicators.

the CEC review measures that Parties to the NAAEC have undertaken to comply with their obligation under the agreement to enforce their domestic laws and regulations. These reviews include occasional discussion of the domestic enforcement measures each Party has undertaken in relation to selected MEAs (CEC 1996). At the national level, most conventions ratified by Canada and the United States have offices that serve as national focal points and are responsible for reporting to the international convention secretariats. Canada's Commissioner of the Environment and Sustainable Development has, as part of his sustainable development mandate, chosen to report on the implementation of international environmental treaties. His office has begun a series of reports on MEA compliance with a report on the Montreal Protocol (Auditor General of Canada 1997). In the United States, the Bureau of Oceans and International Environmental and Scientific Affairs of the State Department is the lead agency for foreign policy formulation and implementation in global environment, science and technology issues.

Financing of MEAs is an issue to the extent that it affects the capacity of public agencies to help meet commitments. Changes are planned to increase flexibility in service delivery, decrease the overall costs of environmental protection, and promote subsidiarity by assigning responsibility for environmental measures to those – usually lower – levels of government believed to be the most effective at implementation.

It is not easy to assess the impact of all global MEAs. Despite the leadership role that Canada and the United States have taken in negotiating some MEAs, there is insufficient information to evaluate fully the effectiveness of national efforts.

For the Vienna Convention, results are demonstrable and can be clearly attributed to the Montreal Protocol. Canada was one of the first countries to ratify the Vienna Convention on the Protection of the Ozone Layer. Both Canada and the United States took active measures to reduce CFC emissions, facilitate the introduction of alternatives, and enact and enforce legislation to meet and even exceed Convention limits. By 1996 the production of CFCs, halons and HCFCs and methyl bromide were all below Montreal Protocol targets (UNEP Ozone Secretariat 1998).

The situation is more complex with MEAs such as the CBD. Although North American environmental monitoring capacities are among the best in the world, how to measure biodiversity is an unresolved issue and there are significant data gaps. Tracking progress is difficult without clear indicators of success. In addition, many human activities and policies can influence biodiversity, including land-use changes, industrial development, consumption habits and recreation. It is seldom possible to separate the impact of an MEA from that of other policies.

Regional MEAs

For nearly a century, Canada and the United States have used bilateral agreements to protect the environment. The first was the Boundary Waters

Treaty, signed in 1909, which was developed to provide a mechanism to help prevent and resolve disputes, primarily those concerning water quantity and quality along the boundary between Canada and the United States. The treaty created the International Joint Commission (IJC), an impartial binational organization which oversees the implementation of the Treaty. The Great Lakes Water Quality Agreement (GLWQA) was signed by both countries in 1972 and reaffirms the commitment of each country to restore and maintain the chemical, physical, and biological integrity of the Great Lakes basin ecosystem. The IJC monitors and assesses progress under the GLWQA and advises governments on matters related to the quality of the boundary waters of the Great Lakes system. The 1972 Agreement was revised in 1978 and amended by a Protocol in 1987. Among other things, the Protocol calls for the federal governments, as well as local, state, and provincial governments, to designate areas of high degradation in the Great Lakes region and cooperate in the formation and implementation of remedial action plans. Officials also cooperate in inspection and enforcement actions.

The United States and Canada also have a 1986 accord on hazardous waste: the Canada-USA Agreement on the Transboundary Movement of Hazardous Waste. This accord requires each party to ensure that as far as possible their domestic environmental laws on hazardous waste are adequately enforced and to cooperate in monitoring transboundary movements of waste. The US Environmental Protection Agency (US EPA), for example, coordinates spot checks at the border with Environment Canada officials and customs officials (Fulton and Sperling 1996).

The first major MEA between the United States and Mexico was the Treaty on the Utilization of Waters of the Colorado and Tijuana Rivers, and of the Rio Grande. This treaty extended the authority of the International Boundary and Water Commission, which has the authority to initiate water quality and conservation projects, to aspects of US-Mexico boundary waters. The Commission also monitors water quality and collects data.

The 1983 Agreement on Cooperation for Protection and Improvement of the Environment in the Border Area, the La Paz Agreement, provides a more institutionalized framework for cooperation in the border area, defined as 100 kilometres on either side of the international boundary. Many businesses and production plants, known as *maquiladoras*, operate in the border area. The number of *maquiladoras* has grown in the wake of NAFTA. Concern over the practices of *maquiladoras* and their impact on the fragile border ecology is one of the factors that has led to periodic updates of the La Paz Agreement. Annexes to the Agreement address pollution accidents, hazardous waste transport, air pollution, and sanitation. For example, the Hazardous Waste Tracking System, known as HAZTRAKS, was developed as part of the La Paz regime. The system monitors waste movements across the border, matching information from both countries. Officials on both sides of the border also coordinate compliance

Major regional MEAs

Treaty	*Place and date of adoption*
Boundary Waters Treaty	Washington DC 1909
Migratory Bird Convention	Washington DC 1916
Convention on Nature Protection and Wildlife Preservation in the Western Hemisphere	Washington DC 1940
Treaty on the Utilization of Waters of the Colorado and Tijuana Rivers, and of the Rio Grande	Washington DC 1944
Convention for the Establishment of an Inter-American Tropical Tuna Commission	Washington DC 1949
Treaty Concerning the Diversion of the Niagara River	Washington DC 1950
The Great Lakes Water Quality Agreement	Ottawa 1972/78/87
Convention on Future Multilateral Cooperation in the North-West Atlantic Fisheries	Ottawa 1978
Agreement on Cooperation for Protection and Improvement of the Environment in the Border Area (La Paz Agreement)	La Paz 1983
The Canada-US Agreement on the Transboundary Movement of Hazardous Waste	Ottawa 1986
Agreement on the Cooperative Management of the Porcupine Caribou Herd	Ottawa 1987
Canada-US Agreement on Arctic Cooperation	Ottawa 1988
Canada-Mexico Agreement on Environmental Cooperation	Mexico City 1990
The Canada-US Air Quality Accord	Ottawa 1991
Convention for the Conservation of Anadromous Stocks in the North Pacific Ocean	Moscow 1992
North American Agreement on Environmental Cooperation (NAAEC)	Ottawa and Mexico City 1993
US-Mexico Agreement concerning the Establishment of a Border Environment Cooperation Commission and a North American Development Bank (BECC-NADBank Agreement)	1994

North American Agreement on Environmental Cooperation

The North American Agreement on Environmental Cooperation (NAAEC) entered into force for all three countries on 1 January 1994. In Canada, three provinces – Alberta, Manitoba and Quebec – have so far ratified the agreement. The agreement was negotiated on a 'parallel track' with NAFTA and was designed to promote trilateral environmental cooperation and the effective enforcement of environmental laws in the context of trade liberalization under NAFTA.

NAAEC is unique in its broad mandate, its intention being to strengthen cooperation in the development and harmonization of environmental laws and standards, with emphasis on transparency and public participation. A trilateral Commission for Environmental Cooperation (CEC) includes a Council, its governing body made up of the federal environmental ministers of the three countries, a Secretariat in Montreal to provide technical and administrative help and information to the Council and committees, and a Joint Public Advisory Committee of citizens from the party countries.

The parties to NAAEC are obliged to enforce their own domestic laws effectively, while seeking to raise environmental protection standards. This unusual emphasis on national standards stemmed from the unique context of NAFTA. The inclusion of Mexico in what had been a US-Canada free trade area raised the possibility that many firms would relocate production to Mexico so as to evade stricter US and Canadian environmental control. Some environmental groups feared that this process could trigger a competitive 'race to the bottom' in which the NAFTA parties competed to attract businesses and jobs by having the lowest regulatory standards. While Mexican environmental laws are strict, their implementation and enforcement are relatively weak. The primary aims of the NAAEC were to ensure that Mexico and the other NAFTA parties enforced their laws, and that citizens and NGOs had the ability to raise awareness of problems and initiate governmental actions through submissions to the CEC.

The CEC's general mandate includes regular publication of state of the environment reports, development of environmental emergency preparedness measures, promotion of environmental education, furthering scientific research with respect to the environment, appropriate use of environmental impact assessments, and promotion of the use of economic instruments in meeting environmental policy objectives. One of the core functions is to provide a forum for negotiated prevention and, where necessary, resolution of conflicts. The process for dispute settlement and prevention is expected to play an increasingly important role.

Source: CEC 1999

workshops with the *maquiladora* plants (Fulton and Sperling 1996). The La Paz agreement and its protocols thus establish an extensive cooperative process between the United States and Mexico, aimed at improving the border environment and its resources.

North American MEAs have often been precursors to addressing issues on a global scale. Examples include the Great Lakes Water Quality Agreement, the United States-Canada Air Quality Agreement, the Migratory Bird Convention and the Canada-USA Agreement on the Transboundary Movement of Hazardous Waste.

The broadest framework yet created for environmental cooperation between Canada, Mexico and the United States is the North American Agreement on Environmental Cooperation (NAAEC) (see box left). The importance of NAAEC may go beyond the region, since successes and failures in dealing with issues such as cross-border environmental impacts, the migration of industries seeking cheaper labour and more permissive environmental standards, and the sale of products with higher environmental risks can serve as important examples for the entire global community.

The negotiation of NAFTA and NAAEC was a high profile event that received significant public attention. NAAEC contains provisions to ensure that environmental laws and regulations are well publicized and citizens have access to legal remedies in relation to environmental matters. Specifically, the agreement requires that the Parties make it possible for individual citizens to request the investigation of alleged violations and that public authorities respond to such requests. Twenty such submissions had been received by May 1999 (CEC 1999). Submissions have addressed such matters as new timber-related legislation in the United States and the construction of a dock for cruise ships near a fragile coral reef in the Mexican resort of Cozumel. The timber submission was eventually rejected on the grounds that NAFTA parties have the right to modify their environmental laws, though they must enforce the laws they do have. An assertion that Canada is failing to enforce, with respect to certain hydropower projects in British Columbia, portions of its Fisheries Act designed to protect fish habitat was under investigation in mid-1999.

The CEC has the power to initiate Secretariat reports to Council. These can address any matter within the scope of the CEC's Annual Program. These reports may not address any alleged failure by a party to enforce its environmental laws. One example of this kind of document is a report concerning the death of large numbers of birds at the Silva reservoir in Mexico in 1994. The Secretariat is currently preparing a report on the use of groundwater in the San Pedro riparian area in Arizona.

The first evaluation of NAAEC was prepared by an Independent Review Committee and released in June 1998. The Committee commented that 'The citizen submission process is unique among international organizations, but is reflective of a trend toward increased citizen involvement in international mechanisms to address environmental issues. The purpose of the process is to provide some 350 million pairs of eyes to alert the Council of any "race to the bottom" through lax environmental enforcement'.

The review recommended the development of a medium-term, three-year programme that should focus on regional issues, build relationships between elements of different projects, promote sustainable development, trade and environment factors, exploit the comparative advantage of the CEC, and ensure appropriate resources for the mandatory work programme elements (CEC 1998).

Voluntary action

Voluntary policies and private sector initiatives, often in combination with civil society, are gaining in importance as policy tools. In Canada alone there are some 90 voluntary initiatives involving industry, including the Accelerated Reduction/Elimination of Toxics, the National Packaging Protocol, the North American Waterfowl Management Plan, the Major Industrial Accidents Council of Canada and Responsible Care to ensure responsible management of chemical products and others (Industry Canada 1998).

This section deals first with reductions in greenhouse gas emissions, where voluntary initiatives have met with little success, and secondly with reductions of other forms of chemical pollution, where voluntary and non-regulatory actions have been successful.

Reducing greenhouse gas emissions

Until now, the primary policy approach to meeting the UNFCCC commitments in Canada and the United States to stabilize greenhouse gas emissions around 1990 levels by the year 2000 has been to rely on voluntary efforts to increase energy efficiency or otherwise reduce emissions, although there have been no significant changes to the incentive system. In both countries, emissions have continued to increase and in the year 2000 will significantly exceed 1990 levels. For example, Canada's Energy Outlook 1996–2020 estimated that GHG emissions in 2000 would be 8.2 per cent higher than in 1990 (Natural Resources Canada 1996). The efficiency improvements that have been achieved have been overwhelmed by economic growth and increased travel, both of which have increased demand for energy and consequently greenhouse gas emissions.

Upon ratification, both Canada and the United States will have negotiated new, more stringent provisions under the Kyoto Protocol which would require more direct measures. Although the United States will not meet its 2000 commitment, it did have a strategy called the Climate Change Action Plan (CCAP) to initiate responses. CCAP included more than 50 federally-supported programmes, including incentives for voluntary measures. A range of new initiatives has been proposed since Kyoto. The new federal proposals consist of three main stages: immediate action to stimulate the development and use of technologies that can minimize the cost of meeting national emission reduction targets; options created through ongoing technology development, leading to detailed plans for establishing a domestic emission trading system; and implementation of the emission trading system. The plan, known as the Climate Change Technology Initiative, comes with a proposed budget of US$6 300 million to be spent on research and development and on tax credits for energy efficiency improvement. The research and development component is to fund research on buildings, industry, transportation and electricity. The tax credit programme will be used by the federal government to support the purchase of energy-efficient products such as vehicles, solar electricity and hot-water systems, energy-efficient new homes, and an extension of the wind and biomass tax credit (US EPA 1998a).

The emphasis in the United States is thus on making the transition as free from economic disruption as possible through emission trading and technology development. In addition scientific research is being supported to reduce uncertainties in climate scenarios. The barriers to success are significant if North Americans continue to lead high-consumption, high-mobility lifestyles and the energy sector continues its attempts to maintain the status quo. As in other developed countries, much depends on what states, municipalities and civil society will do.

There are many state and local initiatives that will contribute substantially to overall national performance (US EPA 1998b).

At Kyoto, Canada negotiated to reduce its GHG emissions to 6 per cent below 1990 levels by 2008–2012. Since emissions are actually growing rapidly, this translates into a reduction of 21–25 per cent (IISD 1998). Canada's strategy for meeting the Kyoto commitments builds on the lessons of the 1995 National Action Program on Climate Change (NAPCC) which focused on mitigation, science and adaptation (Environment Canada 1995). Stakeholder participation involving federal and provincial government, industry and municipalities has been identified as a key to implementation. At the federal level, responsibility for climate change policy is now shared mainly by Natural Resources Canada (for domestic implementation) and Environment Canada (for international commitments, outreach and education). A Federal Climate Change Secretariat has been established by the Cabinet to facilitate overall coordination. In April 1998, federal, provincial and territorial ministers of energy and environment agreed to move ahead on a process for developing a national implementation strategy on climate change, establishing credit for early action to reduce GHG emissions and strengthening voluntary action (Natural Resources Canada 1998). In June 1998, Canada launched the Greenhouse Gas Emission Trading Pilot (GERT) in British Columbia, its first project for GHG emission trading. The detailed strategy should be available by the year 2000.

Emission reductions through ARET

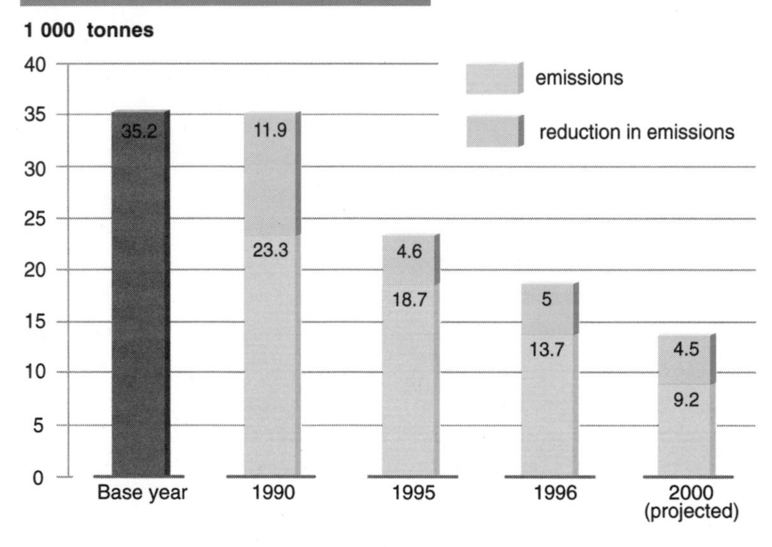

Source: ARET 1999

The voluntary Canadian Accelerated Reduction/ Elimination of Toxics (ARET) programme has substantially reduced emissions of toxic substances

Pollution reduction

Reduction in the production and release of a group of pollutants does not necessarily mean immediate declines in risk of exposure, and the effectiveness of these measures needs to be assessed in the context of contaminants in the environment and food, and impacts on humans and biota. Cross-border transport through the atmosphere, rivers and seas, and the persistence of some pollutants, can cause large lag times in achieving improvements, particularly in 'sink regions' such as Canada's Arctic (Environment Canada 1998a). Regardless of such complexities, voluntary action is becoming one of the essential tools in the environmental policy toolbox.

The US 33/50 Program is a US EPA voluntary pollution reduction initiative that targets 17 high-priority chemicals listed in the Toxics Release Inventory (TRI). The US EPA proposed to reduce releases and transfers of these chemicals by 33 per cent below the 1988 baseline by 1992 and 50 per cent by 1995. The effort has been remarkably successful, enlisting the support of more than 6 000 facilities. Both goals were achieved a year ahead of time. By 1995, releases and transfers were 55.6 per cent below and by 1996 60 per cent below the baseline (US EPA 1998c). The reductions were achieved through process/equipment change, waste recovery and recycling, chemical substitutions, modernization/ equipment upgrade and waste treatment (US EPA 1998d).

In Canada, the Accelerated Reduction/Elimination of Toxics (ARET) programme grew out of a proposal to the federal government by a coalition of leading business organizations and environmental groups to identify and then reduce or eliminate the use and emission of toxic substances. The goal of ARET is to reduce the emission of persistent, bio-accumulative and toxic substances by 90 per cent, and the emission of all other toxic substances by 50 per cent (compared to the base year of 1988) by the year 2000 (see figure). ARET is based on a collaboration of stakeholders from industry, government and civil society that identifies and lists toxic substances, challenges the main producers to commit themselves publicly to their reduction or elimination and then turn commitments into action plans. ARET participants also agree to track and report progress in meeting emission targets.

Chemical companies are increasingly realizing that pollution prevention can have 'win-win' outcomes, with the potential to reduce both pollution and costs. In some cases, pollution prevention is also built into legal settlements between regulators and companies. As part of a legal settlement with the US EPA, a DuPont facility in New Jersey agreed to implement pollution prevention programmes for 15 manufacturing processes. DuPont expects that waste from all 15 processes will be cut roughly in half. The total up-front investment is expected to be about US$6 million but the company anticipates annual savings of about US$15 million.

Increased emphasis on voluntary initiatives means more flexibility in setting and meeting pollution reduction targets, and increased responsibility for the private sector in meeting overall environmental quality objectives. Given that most voluntary measures would in effect allow corporations to self-regulate, there is a need for a code of practice to ensure that emerging self-regulatory or voluntary mechanisms are able to deliver the expected results. In recognition of this challenge, the New Direction Group (NDG), a coalition of major Canadian corporations and environmental NGOs, published *Criteria and principles for the use of voluntary or non-regulatory initiatives to achieve environmental policy objectives*. The objective is to ensure the quality, effectiveness and credibility of voluntary or non-regulatory initiatives, realizing that public trust is essential for their acceptance and success (see box). They are intended to apply to voluntary or non-regulatory initiatives that are employed instead of, or to complement, regulation. These could include government-industry negotiated agreements, industry-community and industry-NGO agreements, challenge programmes and regulatory exemption programmes (New Directions Group 1997).

Principles for the use of voluntary or non-regulatory initiatives

Credible and effective voluntary and non-regulatory initiatives should:

- be developed and implemented in a participatory manner that enables the interested and affected parties to contribute equitably;
- be transparent in their design and operation;
- be performance-based with specified goals, measurable objectives and milestones;
- clearly specify the rewards for good performance and the consequences of not meeting performance objectives;
- encourage flexibility and innovation in meeting specified goals and objectives;
- have prescribed monitoring and reporting requirements, including timetables;
- include mechanisms for verifying the performance of all participants;
- encourage continual improvement of both participants and the programmes themselves.

Source: New Directions Group 1997

Laws and institutions

The region has an extensive set of environmental laws and governmental institutions at all levels to enforce them. Compliance is relatively high. However, in some areas governments are now making changes in the legal and institutional framework for environmental protection and natural resource management, reviewing existing policies and introducing new ones. The clarification of responsibilities for environmental protection among various layers of government is particularly important.

One of the most important regulatory reforms involves renewal of the Canadian Environment Protection Act (CEPA). Since 1988 CEPA has given the federal government powers to control the use of toxic substances, pollution from the use of fuels and nutrients, ocean dumping and Canadian sources of international air pollution. In 1994 the government began a review of the Act and in 1998 it introduced legislation to renew the CEPA to reflect the increased complexity of environmental problems, advances in science and experience with existing regulatory policies.

The revised Act emphasizes pollution prevention, and proposes a more efficient screening and assessment process to identify toxic substances, with deadlines for preventive or control actions. Users or producers of new substances will be required to demonstrate that the material will not pose an unacceptable risk to the environment or human health. It will provide increased powers to control the transboundary movement of hazardous waste, hazardous recyclable material and non-hazardous waste destined for final disposal. A precautionary approach for the disposal of wastes at sea reflects recent amendments to the 1972 London Convention on Ocean Dumping and a new authority is included to address Canadian sources of international water pollution. CEPA will also be used to set emission standards for new vehicles and other engines. New enforcement tools include the Environmental

Alternative Measures Program which allows for negotiated settlements for offenders without the need for a court case. Opportunities for public participation have been expanded, including a right to sue for damage to the environment if the government fails to enforce CEPA (Environment Canada 1998a).

Harmonization of environmental protection responsibilities where federal and provincial governments share jurisdiction over the environment has been the top priority of Ministers of the federal government for some time. On 29 January 1998, Environment Ministers from all Canadian jurisdictions, with the exception of Quebec, signed the Canada-wide Accord on Environmental Harmonization. The Ministers of Canada, nine provinces and the territories also signed sub-agreements dealing with environmental assessment, inspection activities and nation-wide standards in areas such as air, water and soil quality. The plan is that governments will work in partnership to achieve the highest level of environmental quality for all Canadians. Governments will perform complementary tasks to achieve agreed, well-defined environmental results (Environment Canada 1998b).

Both countries in North America reduced CFC production to Montreal Protocol requirements using new policy measures

CFC production

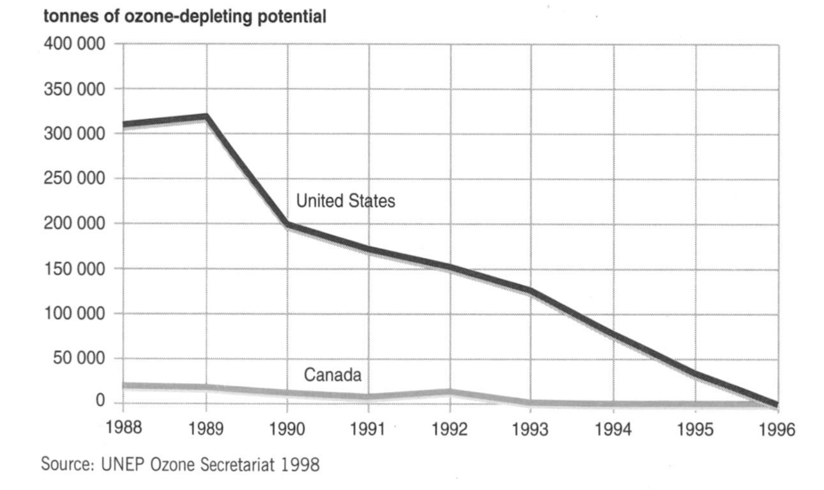

Source: UNEP Ozone Secretariat 1998

In the United States, the US EPA is leading an agency-wide, multi-year effort to reform the way environmental policy is set and delivered by the federal government. The main drivers of the initiative are cost-cutting, rationalization and improving organizational effectiveness. Although many past environmental protection measures achieve good results, there are new reasons to shift policies and programmes and review how the agency works with other stakeholders. Developments are expected to include new policy instruments and technologies, increasingly sophisticated and interlinked partner organizations in government and society in general, and new approaches that tackle complex issues holistically on the ecosystem scale.

While strengthening its programmes on environmental media, the US EPA is aiming to be more effective in dealing with multi-media problems by reducing the regulatory burden, increasing emphasis on inter-agency and public partnerships and testing new policy instruments that promise to provide better protection at reduced cost. Better monitoring and compliance reporting are among the mechanisms that would ensure that the agency keeps a close watch on its environmental performance (US EPA 1998a).

Economic instruments

Economic instruments are of growing importance and are used extensively. Taxes, subsidies, tradeable permits and other economic measures are established and enforced by legislation. The following examples show how such instruments can provide significant cost savings in managing environmental problems.

Phasing out ozone-depleting substances

In response to the Montreal Protocol, the US government introduced tradeable permits, excise taxes and other market-based instruments, and efforts to remove regulatory barriers and encourage the development of alternatives. Its programme involved business leaders, environmental advocacy groups and the scientific community.

The tax on ozone-depleting chemicals was a central element of the strategy. Legislation enacted in 1989 created an excise tax on manufacturers' or importers' sale or use of ozone-depleting chemicals, taxable imported products, and stocks. The base tax was set at US$3.01/kg in 1990 and 1991, with incremental increases of US$0.99 a year after 1994. In 1998, the tax was US$6.70/kg.

The combination of the tax with other regulatory legislation lowered production more quickly than required by the Montreal Protocol. US production of the five CFCs covered by the Montreal Protocol fell significantly since the tax and regulatory caps were imposed (see figure left), and the incentive-based approach has considerably reduced the costs of monitoring and enforcement. Subsequently, however, a black market in CFCs developed, and substantial quantities of CFCs were smuggled into the country (Brack 1997, Gale and others 1996).

In Canada, the control of ozone-depleting substances was achieved through the use of codes of practice for industry, recovery, recycling, emission reduction regulations and the training of refrigeration and recovery technicians. Although the strategy did not involve tax incentives, it did achieve the required reductions (see figure left).

Tradeable emission permits

National tradeable emissions permit systems are one of the most promising of the many new approaches being developed to reduce air pollution. They enable pollution reduction measures to be applied where reductions are most cost-effective. Such systems enable a company that reduces emissions below the level required by law to receive emissions credits that can 'pay for' higher emissions elsewhere. Companies can trade emissions among sources within a company, as long as combined emissions stay within a specified limit, or trade them with other companies.

Tradeable emission permits are an integral part of the US federal effort to reduce acid rain. The 1990 amendments to the Clean Air Act require annual SO_2 emissions in the year 2010 to be 9 million tonnes below the 1980 level (about 16.2 million tonnes). Emissions of nitrogen oxides by the year 2000 will be below the 1980 level of about 16.2 million tonnes. The law requires a two-phase tightening of restrictions on fossil fuel-fired power plants. Phase I affects 445 industrial units, mostly coal-burning electric utility plants in 21 eastern and midwestern states. Phase II, which begins in the year 2000, tightens the annual emissions limits on these larger plants and also sets restrictions on about 2 000 smaller, cleaner plants.

Under the allowance trading system, utilities were allocated allowances based on their historic fuel consumption and a specified emission: each allowance permits a unit to emit 1.0 tonnes of SO_2 during or after a specified year. For each tonne of SO_2 discharged in a given year, one allowance is retired. During Phase II, the law limits the number of allowances issued each year, effectively 'capping' annual emissions at 8.05 million tonnes (US EPA 1997).

Allowances may be bought, sold or banked. Utilities, with assistance from brokers, trade hundreds of thousands of allowances between each other each year. In addition, the US EPA holds an annual allowance auction, which helps to send a price signal to the market. After 1994, when allowances were initially US$150, the auction price declined substantially, reaching a low of US$70 in March 1996. It then rose again, reaching more than US$180 by mid-1998 (see figure below).

Sulphur dioxide allowance prices

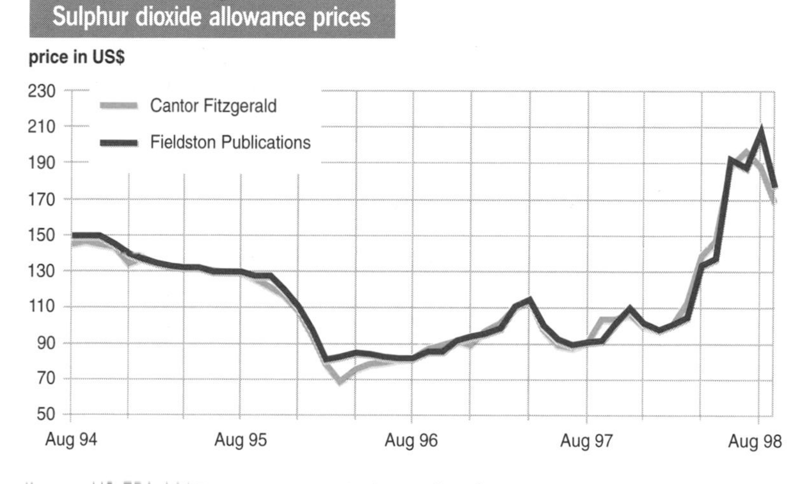

Source: US EPA 1997, prices from the brokerage firm Cantor Fitzgerald and a market survey by Fieldston Publications

In the United States, prices for tradeable allowance emissions for sulphur dioxide have varied widely since the scheme began in 1994

Under this system, utilities can decide the most cost-effective way to comply with the law. They can increase reliance on renewable energy, reduce usage, use pollution control technologies, switch to fuel with lower sulphur content, or develop other strategies. Units that reduce emissions below the number of allowances they hold may trade allowances with other units in the system, sell them to other utilities on the open market or through US EPA auctions, or bank them to cover emissions in future years.

The programme has resulted in a faster-than-expected reduction in emissions and considerable cost savings. In 1995, SO_2 emissions from the 445 Phase I units were 4.7 million tonnes, 40 per cent below the required level and less than half the 1980 level. The cost of reducing 0.9 tonne of SO_2 from the utility sector continues to decline: scrubber costs have dropped about 40 per cent below 1989 levels, removal efficiencies have improved from about 90 per cent in 1988 to about 95 per cent or more in new retrofits, and expected increases in cost associated with the increased use of low-sulphur coal have not materialized. Overall, it is estimated that the system could save as much as US$3 000 million a year compared with a traditional command-and-control approach.

Canada's first tradeable permits are being tested in the Windsor-Quebec corridor in Southern Ontario. The region has a recurrent smog problem due to traffic and industry-related emissions and transboundary pollution from the United States. The Pilot Emissions Reduction Trading Project (PERT) is focused on nitrogen oxides, volatile organic compounds, carbon monoxide and carbon dioxide. An environmental benefit is guaranteed by allowing only 90 per cent of the earned credits sold to be used internally, and permanently cancelling the remaining 10 per cent.

PERT is monitored by a range of regional stakeholders, including federal and provincial agencies, industry and public health bodies, and environmental groups. It is only a pilot scheme and there are challenging questions still to be answered. PERT's principles have already been used to design another pilot scheme in British Columbia to reduce greenhouse gas emissions (PERT 1998).

Agricultural subsidy reform

As in most other regions, governments in North America spend large amounts of tax revenue on subsidizing a variety of economic activities. Many of these are perverse, in that they exert adverse effects on both the environment and the economy in the long run. In the United States alone, the direct cost of perverse subsidies to the average taxpayer is US$2 000 a year, with another US$2 000 reflecting increased spending on goods, services and environmental damage. Road transportation, the energy sector and agriculture are some of the largest recipients of perverse subsidies (Myers 1998).

The pressure to reduce perverse subsidies comes from taxpayer groups and environmental organizations. Twenty-two NGOs in the United States publish a periodic *Green Scissors* report on perverse subsidies that should be cut (Friends of the Earth 1998).

Government incentives for agriculture have been used since the European-led settlement of the Great Plains. Free land in the early days, and government subsidies later, created a situation where the choice of farming practices failed to reflect market realities, and agriculture was extended to ecologically-sensitive, marginal land. Although encouraging settlement was desirable from the perspective of government social policy, by the late 1980s it had become clear that the practice was economically wasteful and environmentally unsustainable.

Many US programmes and policies developed in the 1980s and 1990s were designed to encourage more sustainable agriculture. One of the most successful is the Conservation Reserve Program (CRP), first established in the 1985 Food Security Act.

CRP offers farmers subsidies to retire highly erodible and other environmentally-sensitive croplands temporarily from production and plant them with grasses or trees. The programme had a double purpose – to reduce soil erosion and deal with commodity surpluses.

Farmers with land that qualifies can bid it into the programme by offering it at an annual rental rate for 10–15 years. The first five CRP sign-ups focused on highly erodible cropland. In 1988, the programme expanded to include vegetative filter strips along water bodies to trap sediment, nutrients and pesticides. From 1989, CRP temporarily accepted the restoration of previously cropped wetlands.

Since the first sign-up in 1986, farmers have enrolled more than 15 million hectares in CRP. Although farmers and ranchers have planted most of the land involved to grass, 0.8 million hectares have been planted to trees, and another 0.8 million hectares are for special wildlife needs. Nearly 0.2 million hectares are restored wetlands, and more than 14 000 kilometres of filter strips help protect waterways.

Erosion reduction credited to the CRP may be as high as 630 million tonnes of soil annually, or 42.75 tonnes per hectare per year. CRP also provides benefits in terms of wildlife habitat and populations, water quality, and restored wetlands and forest lands. The programme has reduced federal outlays for farm deficiency payments, strengthened farm income, and helped balance supply and demand for agricultural commodities.

The success of the CRP has also contributed to increasing political support for agricultural conservation generally. The 1996 farm law extended the CRP through the year 2002 and created a variety of new programmes to address high-priority environmental protection goals.

In Canada, the best example of agricultural subsidy reform was related to the Western Grain Transportation Act. Originally created in 1897, the Act aimed to support the movement of grain to export markets. During the late 1980s and early 1990s the federal government spent US$340–570 million per year on transport subsidies. The subsidies went directly to national railway companies and targeted the movement of grains and oilseeds as a raw product. Through this policy the government created an economic redistribution in favour of these products and discouraged farmers from diversifying into products not covered by the subsidy. Given that part of the transport costs were paid by the subsidy, the reduced cost made it economically sensible to keep a larger area of available land in production, including some low-yielding, marginal farmland (Wilson and Tyrchniewicz 1995).

The federal government recognized that the policy itself could encourage unsustainable agriculture on marginal land, at significant cost to the federal budget, and phased out transport subsidies between 1993 and 1995. The consequences were not easy to predict. Marginal farmland has been converted into lower-impact pasture to support an emerging cattle industry, but not to the levels predicted. There has been a strong move towards feedgrain production in support of diversification of livestock production and processing. This suggests that leaving the agricultural sector to find efficient alternative economic and agricultural solutions on its own may not produce results that benefit the environment.

Industry and new technologies

Cleaner and more efficient processes have been widely adopted as a result of regulatory actions, market-based incentives and a growing environmental consciousness on the part of industry. Many sectors of the economy are extremely innovative, developing and bringing to the market new technologies with reduced environmental impacts.

Environmental technologies are developed and introduced by a wide range of academic and industry centres and a network of federal laboratories supported by government through policy measures such as strategic investment, tax credits and regulatory reform. The Technology Partnership of Industry Canada, for example, provides repayable investment funding in selected areas such as pollution prevention, water treatment, recycling and clean car technologies.

The region is a world leader in the development and implementation of many environmental technologies. California's leadership in renewable energy is an example of the use of a mix of fiscal and regulatory incentives to help the diffusion of environmentally-sound technology, with an investment tax credit as a key element. Commercial and residential developments can reduce their income tax liability by 10 per cent if they invest in eligible projects. The tax credit applies to both federal and state taxes and, although they can be applied only if tax is actually owed, they can be carried back or forward to reduce past or future taxes. At the production end, a similar 10 per cent tax credit can be received for wind, biomass, solar and geothermal energy projects.

Canada and the United States are among the largest producers of waste, in per capita terms, both through production and consumption. Although technology development alone – without changes in public attitude – cannot resolve waste problems, it is crucial in finding solutions. Waste reduction is promoted through regulatory, voluntary and market-based measures but when the incentives are compared with the subsidies that go into promoting unsustainable resource use, it is clear that there is much more to be done to make waste production economically inefficient. However, there are many examples of successful waste reduction through cleaner production and closed-loop manufacturing projects, often in large eco-industrial parks (Smart Growth Network 1997 and UNEP 1997).

Public participation

Environmental justice and public participation in environmental decision-making has been a strong priority for North American governments, NGOs and communities.

The basis of environmental justice concerns is that minority and low-income groups are often more exposed to adverse environmental conditions than the general population. In 1994, the President of the United States issued Executive Order 12898 to focus the attention of federal agencies on these issues (US EPA 1994). The US EPA Environmental Justice Strategy, based on the Executive Order, aims to ensure that no segment of the population suffers disproportionately adverse environmental effects as a result of US EPA policies, and that those who have to live with the consequences of decisions can take part in reaching those decisions. Action items cover health and health research, public access to information, enforcement and compliance with international environmental laws, partnerships with communities, other levels of government, tribes, business and NGOs, and the integration of environmental justice into all departmental activities (US EPA 1995).

Public participation in environmental assessments has been a regular practice for many years. Some provinces in Canada provide funding for citizens making legal interventions on issues of public concern but this was discontinued in the late 1990s in some provinces, such as Ontario, as part of a general drive for environmental deregulation. The new CEPA confirms the rights of citizens to take issues of concern to Environment Canada and request investigation. The Commission for Environmental Cooperation established under NAFTA, as already mentioned, has the responsibility to receive, investigate and report on submissions from citizens about environmental concerns.

Public participation has been at the heart of many new management initiatives connected with local resources, with federal, state and provincial programmes based on collaboration with local stakeholders. The US Clean Water Action Plan of 1998 builds on the participation of local communities in water and land resource issues on a watershed scale. Most elements of the programme, such as the restoration of wetlands, protection of coastal waters and land-based pollution control, focus on providing resources to local organizations (US EPA 1998e).

Conservation Districts in Canada are based on the partnership of local communities, landowners, NGOs, industry and government. The most successful and innovative of these organizations, such as Manitoba's Conservation Districts, receive baseline support from the provincial government. The Districts are governed by a board of local people who decide about priority actions on a broad range of natural resource management issues, from water and soil conservation to public education and outreach. Combining modest but stable funding from government with a clear institutional structure, long-term thinking, a mandate for conservation and local participation is proving to be a successful model for other regions (MCDA 1998).

Public participation is an important component of two ecosystem management institutions concerned with the watersheds along the northern and southern boundaries of the United States: the International Joint Commission (IJC) created by Canada and the United States on the basis of the Boundary Waters Treaty of 1909, which has played a major role in reversing environmental degradation of the Great Lakes system; and the Border XXI Program recently set up to address environmental issues in a 200-km wide strip along the US/Mexico boundary.

The success of the IJC is based on a number of key elements, including consultation and consensus building, providing a forum for public participation, engagement of local governments, joint fact-finding, objectivity and independence, and flexibility. Despite the obvious progress made in reducing emissions, improving water and air quality and other ecosystem variables, the Great Lakes ecosystem faces continuing challenges. Population and economic growth, climate change, technological development and environmental awareness are some of the key counteracting forces of change (International Joint Commission 1998).

The Agreement for the Protection and Improvement of the Environment in the Border Area (the La Paz Agreement) of 1983 is the overarching legal framework for the Border XXI Program. This Program is also related to the Integrated Environmental Plan for the Mexico-US Border Area (IBEP), released in 1992. Whereas the IJC has had a long history that provides a basis for its evaluation, Border XXI is new. The geographic area covered is defined primarily in jurisdictional terms – a 100-km wide zone north and south of the US/Mexico border – some of which coincides with the watersheds of two important rivers: the Rio Grande and the Colorado.

The mission of Border XXI is to achieve a clean environment, protect public health and natural resources, and encourage sustainable regional development. The initiative includes five-year objectives and outlines mechanisms to meet those objectives. Its key strategies include public involvement, capacity building and the decentralization of environmental management, and ensuring interagency collaboration. A Strategic Planning and Evaluation Team will identify performance measures that can be linked to budget processes and management activities. These will allow participating governments and other stakeholders to focus planning efforts on meeting identified targets, assessing programme effectiveness and reporting progress to the public.

Border XXI builds on participation by a range of regional stakeholders, both in Mexico and the United States, including federal, state and local governments, Indian tribes, international institutions, educational centres, NGO and industry organizations, and grass-roots, community based associations. Binational Working Groups in five geographic regions along the border area will implement the initiative. Biennial public meetings and biennial progress reports further the connection to local stakeholders and communities (Border XXI 1999).

Environmental information and education

Public and government awareness of environmental issues is high in North America. The signature of treaties is usually followed or, in a number of cases, preceded by the establishment of formal communication and awareness-raising strategies, targeting those directly affected and the general public. For example, following the signature of CITES, Canada launched public awareness campaigns, targeted mailings to affected interest groups, held public displays and information sessions, professional training, conferences and trade shows and used other communication methods to help increase the agreement's effectiveness.

In the United States, the legal basis for public information disclosure is the Emergency Planning and Community Right-to-Know Act (EPCRA) which created the Toxics Release Inventory (TRI). The Canadian equivalent of the TRI is the National Pollutant Release Inventory.

TRI is a database on releases, off-site transfers, and other waste-management activities for more than 650 chemicals and chemical categories from manufacturing and other industry sectors. Companies releasing or transporting TRI chemicals in volumes above the threshold set out in the inventory are required to report it. Providing information to the public has been a powerful incentive for encouraging action by industry to improve the management of TRI chemicals through reduced use and decreases in releases and transfers.

In 1997, the United States took steps to provide the public with improved information about toxic chemical releases and transfers. Seven industry sectors, including electricity-generating facilities and solvent recyclers, were added to the categories subject to TRI reporting. The United States is presently considering adding persistent bio-accumulative and toxic pollutants (PBTs) and lowering the reporting threshold for chemicals subject to TRI requirements so that even smaller amounts of transported or released chemicals can be tracked.

In addition to establishing TRI, the EPCRA provides for States to establish state and local emergency planning groups to develop emergency response plans for each community. More than 3 400 local emergency planning committees have now been established. Industrial facilities are required to provide information to these local committees about the hazardous chemicals present at their facility.

The National Pollutant Release Inventory (NPRI) was developed under Canada's Green Plan initiative in 1992; the first reporting year was 1993. The requirements for the NPRI were set using a consultative process that included a working group with representatives from governments, industrial management and labour force, and environmental groups. Their consensus recommendations form the basis for NPRI. The programme has many similarities to the US TRI, which means that a large portion of the reported releases is comparable between the two countries.

Growth in NPRI activity

	number of reporting facilities	*number of substance reports*
1993	1 504	5 339
1994	1 713	5 928
1995	1 758	6 294

Source: Environment Canada 1997

The NPRI requires corporations to report the off-site transfer and on-site release of 176 toxic substances (Environment Canada 1997). Any company in Canada that manufactures, processes or uses the NPRI-listed substances in quantities of 10 tonnes or more per year, and which employs 10 or more people, is required by law to prepare a report for the NPRI. Since 1993, the number of facilities and the number of reports have been increasing (see table above). However, the amount of reported releases shows a steady downward trend. Environment Canada is continuing its consultations on amending the programme and reviewing the list of substances.

Since 1997 the NPRI has also collected qualitative information on pollution prevention initiatives. The information is available through the NPRI website on a facility and toxic substance basis (Environment Canada 1997). The simple and user-friendly structure of the

NPRI has received international recognition, and served as an example for Mexico's national information base on toxics.

Some NGOs are also involved with reporting on pollution. For example, the Environmental Defense Fund's Chemical Scorecard provides information on the location of pollution sources in every community in the United States, the types of pollutants produced, the government's response, and contact information to help catalyse local action (EDF 1998).

Conclusions

Environmental policies in the region have evolved significantly since the late 1960s. The region has pioneered environmental policy development, first through command-and-control measures, and later through voluntary and market-based approaches. The United States has used a more direct approach with strong enforcement and increasing reliance on market-based measures. Regulatory reforms have incorporated decentralization strategies and have emphasized multi-stakeholder processes and consensus building.

Implementation, enforcement and compliance of MEAs have generally been at a high level. Even when MEAs have not been ratified, national laws or regional MEAs often accomplish the same purpose: for example, the United States has laws and extensive measures to control hazardous wastes and preserve biodiversity, although it has only signed the Basel Convention and neither signed nor ratified the Convention on Biological Diversity.

Major barriers to more successful implementation of MEAs include fiscal incentives which encourage resource use, the absence of economic instruments (such as cap-and-trade or taxation) to encourage appropriate resource use, and difficulties in translating environmental cooperation goals into specific local policies or adapting them to local conditions.

It is more difficult to assess the impact of global MEAs than those of national or regional agreements although some, such as the Montreal Protocol, have clearly been effective. There is a growing interest in evaluating progress under MEAs, particularly in Canada.

Dominating the agenda for policy action under MEAs is the UNFCCC, which will require countries to begin reshaping their energy economies and their industrial structures in this, the most energy-intensive region in the world. Successful implementation will not be easy without raising energy prices. Yet demonstrating an ability to comply with this MEA without sacrificing key aspects of the quality of life would send an important message to other regions; equally, given the economic and political importance of North America, failure to meet international commitments under the Kyoto Protocol would send a powerful negative message.

Many environmental problems persist because technological improvements and gains in ecological efficiency have been overtaken by growth, as in the case of transport-related air pollution. In other cases, improvements have been inadequate, for example for some non-point source emissions and long-term exposure to cumulative toxic chemicals. New concerns, such as climate, biological diversity and persistent organic pollutants, have emerged and require coordinated and effective responses. The recognition of multimedia and cross-sectoral, ecosystem-based linkages calls for coordination, consensus-building and effective stakeholder involvement in policy design and implementation.

Institutional change has also been important: budget and staff reductions have led public agencies to rethink the design and enforcement of policies. Market-based instruments, voluntary measures and co-management are becoming more important. While some of these measures promise more cost-effective solutions, they will contribute to reaching environmental objectives only if public agencies continue their essential functions of standard setting, monitoring and enforcement. Devolving environmental responsibilities with the notion that local governments are more effective in dealing with them will work only if there is sufficient capacity and resources to make the transition.

Better accountability at all levels and better measurement of the performance of environmental policies are widely needed. Target setting, monitoring, scientific analysis and public reporting are essential to keeping stakeholders involved and policies under control. The US EPA's revision programme on environmental protection and the requirement of Canadian federal departments to report on their sustainable development strategies to the Auditor General's office are particularly relevant. Better accountability is expected to become important not only in federal agencies but also increasingly in lower levels of government and private corporations, as they become involved in sharing responsibilities through voluntary action or market-based initiatives.

References

ARET (1999). http://www.ec.gc.ca/aret/el2u/2u_c2.html

Auditor General of Canada (1997). *Report of the Auditor General of Canada to the House of Commons.* Chapter 27, Ozone Layer Protection: The Unfinished Journey. Minister of Public Works and Government Services, Ottawa, Canada

Border XXI (1999). http://www.epa.gov/usmexicoborder/

Brack, D. (1997). *The Growth and Control of Illegal Trade in Ozone-Depleting Substances*. Paper presented at the Taipei International Conference on Ozone Layer Protection, 9–10 December 1997

CEC (1996). *Annual Report*. Annex I. North American Report on Environmental Enforcement. CEC, Montreal, Canada

CEC (1998). *Four-Year Review of the North American Agreement on Environmental Cooperation: Report of the Independent Review Committee*. CEC, Montreal, Canada http://www.cec.org/english/procurement/cfp3.cfm?format=2

CEC (1999). http://www.cec.org

CEPA (1998). *Enforcement and Compliance Policy of Environment Canada.* http://www.doe.ca/enforce/policy/english/ content.htm

Commissioner of the Environment and Sustainable Development (1998). *Global Challenges. Report of the Commissioner of the Environment and Sustainable Development to the House of Commons*. Minister of Public Works and Government Services, Ottawa, Canada

EDF (1998). *The Chemical Scorecard.* Environmental Defense Fund, New York, United States ; http://www.scorecard.org/

Environment Canada (1995). *National Action Program on Climate Change*. Environment Canada, Hull, Quebec, Canada http://www.doe.ca/climate/resource/cnapcc

Environment Canada (1996). Environment Canada News Release, 6 June 1996. http://www.ec.gc.ca/cws-scf/es/wappa/proccong.htm

Environment Canada (1997). National Pollutant Release Inventory (NPRI). http://www.ec.gc.ca/pdb/npri/index.html

Environment Canada (1998a). *Strengthening Environmental Protection in Canada. A Guide to the New Legislation.* Environment Canada, Hull, Quebec, Canada http://www.ec.gc.ca/cepa/guide_e.html

Environment Canada (1998b). *A Canada-Wide Accord on Environmental Harmonization.* Environment Canada, Hull, Quebec, Canada http://www.ccme.ca/ccme/accord draft.html

Fulton, S., and Sperling, L. (1996). The Network of Environmental Enforcement and Compliance Cooperation in North America and the Western Hemisphere. In *The International Lawyer* 30, 1

Friends of the Earth (1998). *Green Scissors '98* http://www.foe.org/eco/scissors98/

Gale, R., Barg, S., and Gillies, A. (1996). *Green budget reform: an international case book of leading practices*. IISD, Winnipeg, Canada

IISD (1998). *A Guide to Kyoto. Climate Change and What it Means to Canadians.* International Institute for Sustainable Development, Winnipeg, Canada

Industry Canada (1998). *The Power of Partnerships*. Industry Canada, Ottawa, Canada http://strategis.ic.gc.ca/SSG/ea01494e.html

International Joint Commission (1998.) *The IJC and the 21st Century*. International Joint Commission, Washington DC, United States, and Ottawa, Canada

MCDA (1998). Conservation Districts in Manitoba http://www.cici.mb.ca/pvcd/progsumm.htm

Myers, N. (1998). *Perverse Subsidies, Tax $s Undercutting Our Economies and Environment Alike*. International Institute for Sustainable Development, Winnipeg, Canada

NAWCC (1999). North American Wetlands Conservation Council http://www.wetlands.ca/whoswet/nawcc.html; and the North American Waterfowl Management Plan http://www.wetland.sk.ca/nawmp/nawmpint.htm

Natural Resources Canada (1996). *Canada's Energy Outlook 1990–2020.*Natural Resources Canada, Ottawa, Canada http://www.es.nrcan.gc.ca

Natural Resources Canada (1998). Joint Meeting of Federal, Provincial and Territorial Ministers of Energy and Environment. News Release 24 April 1998 http://www.nrcan. gc.ca/css/imb/hqlib/jmm.htm

New Directions Group (1997). *Criteria and Principles for the Use of Voluntary or Non-regulatory Initiatives to Achieve Environmental Policy Objectives.* New Directions Group, Canmore, Alberta, Canada http://www.expertcanmore.net/pgriss/English.htm

PERT (1998). *Backgrounder* http://www.pert.org/backgrounder.htm

Smart Growth Network (1997). Eco-*Industrial Case Studies* http://www.smartgrowth.org/library/eco_ind_case_intro.html

UNEP (1997). *The Environmental Management of Industrial Estates.* UNEP IE Technical Report No. 39. UNEP, Paris, France

UNEP Ozone Secretariat (1998). *Production and Consumption of Ozone Depleting Substances 1986–1996*. UNEP Ozone Secretariat, Nairobi, Kenya http://www.unep.org/unep/secretar/ozone/pdf/[illegible].pdf

US EPA (1994). *Federal Actions to Address Environmental Justice in Minority and Low-Income Populations* http:// http://www.epa.gov/swerosps/ej/html-doc/execordr.htm

US EPA (1995). *Draft Environmental Justice Strategy for Executive Order 12898* http://es.epa.gov/program/iniative/justice/ej-strtg.html

US EPA (1997). *Acid Rain Program* http://www.epa.gov/acidrain/ats/prices.html

US EPA (1998a). *Global Warming – National Initiatives* http://www.epa.gov/globalwarming/actions/national/index.html

US EPA (1998b). *Global Warming – Actions Being Taken to Prevent Global Warming* http://www.epa.gov/globalwarming/actions/index.html

US EPA (1998c). *Toxics Release Inventory – Related National and International Programs* http://www.epa.gov/opptintr/tri/national.html

US EPA (1998d). *Pollution Reduction Method – Success Stories* http://cyber22.dcoirm.epa.gov/oppt/sstories.nsf

US EPA (1998e). *The Clean Water Action Plan* http://www.epa.gov/cleanwater/2pg.html

Wilson, A., and Tyrchniewicz, A. (1995). *Agriculture and Sustainable Development: Policy Analysis on the Great Plains.* International Institute for Sustainable Development, Winnipeg, Canada

World Wildlife Fund (1998). *Endangered Spaces Progress Report* http://www.wwfcanada.org/reportcard/index.htm

West Asia

KEY FACTS

The key issues for policy action in West Asia are the development of environmentally-sound water resource management, land-use planning, combating various forms of land degradation and desertification, management of hazardous and toxic wastes, integrated management of the marine and coastal environment, biodiversity conservation and the management of urban air quality.

- Of the MEAs, the Montreal Protocol has received most attention because international interest is high and funding, capacity building and systems for monitoring and reporting are all available.
- All eight signatories to the Kuwait convention on protecting the marine environment have prepared national action plans.
- The Gulf Cooperation Council (GCC) summit in Kuwait in December 1997 adopted a by-law to protect and develop wildlife in the GCC countries.
- Most state environmental institutions suffer from shortages of qualified manpower, inadequate funding and uneasy relations with other government institutions whose cooperation is essential in dealing with environmental issues. This has resulted in delays and failures to implement policies and enforce laws.
- In one country in the Arabian Peninsula, many apartment blocks collect wastewater, recycle it *in situ* and recirculate it in separate networks. It is claimed that this results in more than 40 per cent saving in total water consumption.
- The deficit in food production is growing and is aggravated by the scarcity of resources (land and water) which are nearly fully utilized. Water security and increased food production have been the dominant strategies behind most development policies during the past two decades.

The policy background

Most West Asian countries have begun to formulate and implement environmental policies over the past two decades. The initial approach was sectoral, consisting mainly of developing methods of managing individual environmental resources without giving due consideration to the environment as a whole. However, governments are now reformulating laws and regulations, and establishing cross-cutting approaches to deal with the real complexity of environmental issues. Environmental Impact Assessment (EIA) is receiving more attention and its use is spreading. Environmental considerations are slowly becoming part of the planning process.

Key issues for policy action are the development of environmentally-sound water resource management, land-use planning, combating various forms of land degradation and desertification, management of hazardous and toxic wastes, integrated management of the marine and coastal environment, biodiversity conservation, and the management of urban air quality.

The main obstacles to formulating and implementing environmental policies are the potential for political and administrative conflicts, limitations in financial and skilled human resources, inadequate planning of industrial and urban development, and conflicting interests in the use of water and land (see Chapter 2, pages 164–167).

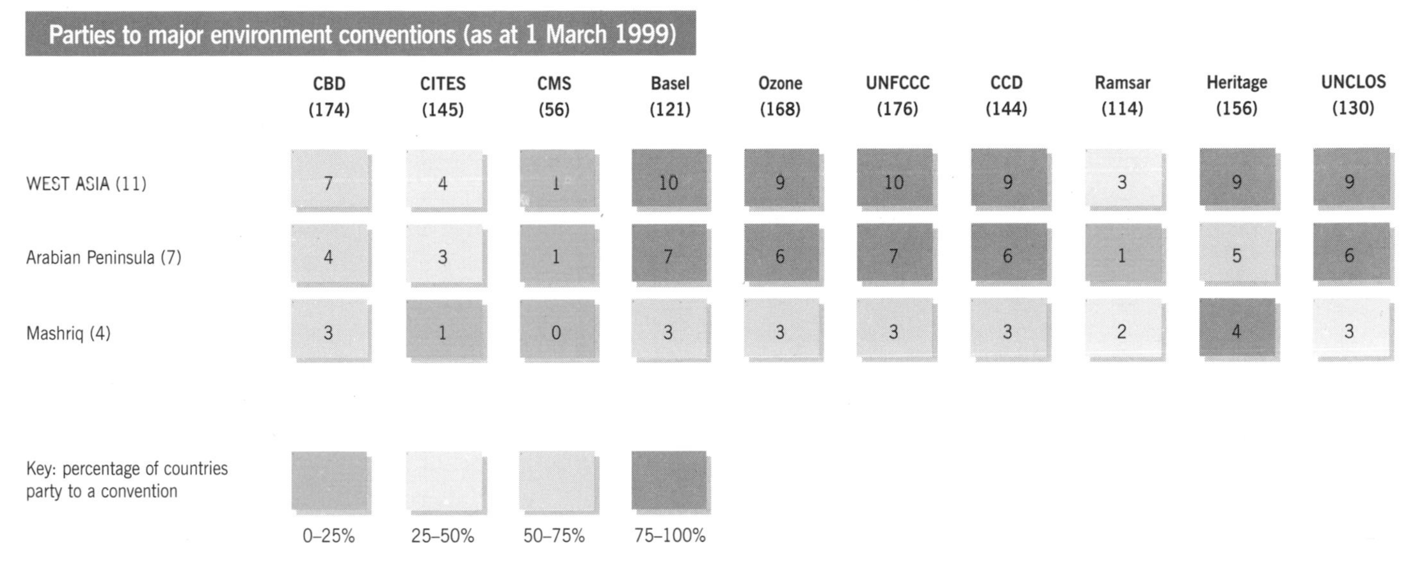

Parties to major environment conventions (as at 1 March 1999)

	CBD (174)	CITES (145)	CMS (56)	Basel (121)	Ozone (168)	UNFCCC (176)	CCD (144)	Ramsar (114)	Heritage (156)	UNCLOS (130)
WEST ASIA (11)	7	4	1	10	9	10	9	3	9	9
Arabian Peninsula (7)	4	3	1	7	6	7	6	1	5	6
Mashriq (4)	3	1	0	3	3	3	3	2	4	3

Key: percentage of countries party to a convention

0–25% 25–50% 50–75% 75–100%

Notes:

1. Numbers in brackets below the abbreviated names of the conventions are the total number of parties to that convention
2. Numbers in brackets after name of regions/sub-regions are the number of sovereign countries in each region/sub-region
3. Only sovereign countries are counted. Territories of other countries and groups of countries are not considered in this table
4. The absolute number of countries that are parties to each convention in each region/sub-region are shown in the coloured boxes
5. Parties to a convention are states that have ratified, acceded or accepted the convention. A signatory is not considered a party to a convention until the convention has also been ratified

MEAs and non-binding instruments

Global MEAs

West Asian countries have ratified or acceded to some 64 international and regional environmental conventions and agreements (UNEP 1997). The extent to which countries are parties to the ten main global MEAs is shown in the table above.

While ratification is at a fairly high level, compliance has been limited, mainly due to lack of funding. Awareness campaigns have not targeted decision-makers, stakeholders or the national figures who are the key to successful public participation. In addition, few pressure groups have been formed to influence business sectors.

While some countries are in the process of formulating by-laws relating to specific conventions, most rely on existing legal instruments for dealing with all major environmental issues. Economic instruments to improve the implementation of MEAs are still not well developed. Instruments such as incentives, taxes and charges, pricing strategies and other indirect measures are, however, starting to be applied to some MEAs.

The Montreal Protocol has received most attention, despite the 10-year grace period for compliance, because international interest is high and funding, capacity building and systems for monitoring and reporting are all available. Laws and decrees have been issued in several countries, including Bahrain, Jordan and Kuwait, to regulate the import, export and use of ozone-depleting substances and encourage the use of alternatives. Saudi Arabia is developing comprehensive by-laws on the implementation of the convention, and Syria is amending existing laws and regulations, and drafting new ones. Responsibility for implementation, follow-up and reporting lies mainly with existing national environmental bodies. However, special commissions or ozone units have been established within existing institutions, including ozone units in Bahrain, Jordan, Kuwait, Lebanon, Saudi Arabia, Syria and Yemen.

The programmes in Bahrain, Jordan, Lebanon and Syria have received financial support from the Multilateral Fund, and a number of projects are being assisted by other international organizations. In Syria, some aerosol manufacturers have substituted hydrocarbons for CFC-12. The refrigeration industry has started to use HCFC-134a instead of CFCs and has installed equipment to recover and re-use refrigerants (Government of Syria 1997).

CBD implementation is due to start at the end of 1999. However, there have been pilot projects in several countries. National committees have been set up in all countries to conduct national studies and formulate strategies and action plans. Most countries have prepared national studies on biodiversity and a number of protected sites have been declared. Studies have been completed in Jordan, Lebanon, Oman and Syria, and are under way in other states. Saudi Arabia's 6th development plan (1995–2000) includes measures to develop and preserve biodiversity. The most significant element of the Saudi strategy is the principle of addressing biodiversity in EIAs (ACSAD 1997). The League of Arab States called for a meeting of Arab Experts on Biodiversity in 1995. In addition, the Arab League Education, Scientific and Cultural Organization (ALECSO) adopted a comprehensive programme to encourage Arab States to participate fully in CBD activities.

Capacity building is needed to formulate national strategies and action plans for biodiversity. In a joint effort to promote the implementation of CBD, the GEF and its implementing agencies are providing technical assistance and have sponsored a regional workshop in Bahrain.

Responsibility for formulating strategies and implementing action plans to combat desertification lies mainly with agriculture ministries. Two important meetings were held after the adoption of the CCD in June 1994: the Bahrain regional meeting on implementation of the CCD in 1995, and the Sub-regional Meeting for Consultation on Implementation of the CCD in West Asia, held in Damascus in 1997. All West Asian countries participated. In addition, countries have formulated or are preparing national plans of action to combat desertification. UNESCWA and UNEP assisted Bahrain, Oman, the United Arab Emirates and Yemen to prepare their national plans (UNEP/UNESCWA 1991, 1992a and b, and 1994). Implementation of Yemen's national strategy (1991–2010), which aims to halt desertification by the year 2010, has started but full implementation is dependent on international financial support (Faras 1996).

The Council of Arab Ministers Responsible for Environment (CAMRE) has a committee for combating desertification and increasing green areas in the Arab region. The committee is preparing a study on the state of desertification in the region with the help of the Arab Centre for Studies of Arid Zones and Drylands (ACSAD) and UNEP.

With the exception of the Montreal Protocol, the monitoring of compliance with MEAs is not well developed or enforced. In all cases, lack of resources is a major obstacle and capacity building is badly needed. As to the UNFCCC, the contribution of West Asian countries to climate change is minimal compared to other regions but increases in efficiency in power generation and industry, and a switch to natural gas (already effected in some locations), can reduce greenhouse gas emissions. The 10 countries in the region that are signatories to the UNFCCC are preparing their national communications, with technical and financial support from United Nations agencies in capacity building. Most countries have prepared first drafts of their greenhouse gas (GHG) inventories. A few, including Jordan and Lebanon, are working on GHG mitigation analysis and turning to the more difficult task of analysing vulnerability and adaptation possibilities.

Public awareness of MEAs varies. The Montreal Protocol, CBD, CCD and UNFCCC have received considerable attention. The mass media have played a major role in enhancing public awareness by drawing attention to the seriousness of the environmental issues involved. Television and radio programmes on land degradation and desertification, loss of biodiversity and marine pollution have been broadcast regularly in all countries. These issues have also been covered at national and international events (such as biodiversity day, ozone day, world environment day and Arabian environment day). Newspapers and magazines in some countries allocate special pages for environmental affairs.

Governmental organizations and NGOs also help raise public awareness. The role of NGOs is widely recognized, especially in countries such as Lebanon where NGOs are well developed and vocal. However, in most countries the role of NGOs is still weak.

Overall, most countries have concentrated their efforts on developing institutions or expanding the mandates of existing institutions to implement MEAs. The focus is on priority studies, the preparation of strategies and the development of action plans and programmes, with little progress with implementation at the field level. Although it is too early to assess the impact of the conventions, public awareness is growing and positive attitudes towards most MEAs are developing rapidly.

Regional MEAs

The most important regional agreements are shown in the table on the right. Of these:

- the Kuwait Regional Convention for Cooperation

on the Protection of the Marine Environment from Pollution includes all six Gulf Cooperation Council (GCC) countries, and Iraq and Iran;

- The Regional Convention for the Conservation of the Red Sea and Gulf of Aden Environment (PERSGA) includes three countries in West Asia: Jordan, Saudi Arabia and Yemen;
- The Convention for the Protection of the Mediterranean Sea Against Pollution (Barcelona 1976) and its five protocols (including the Mediterranean Action Plan, MAP) includes Lebanon and Syria from West Asia.

The level of response and compliance for regional MEAs is generally higher than for global MEAs. For example, national action plans have been developed by all eight signatories to the Kuwait Regional Convention and its protocols.

No new national institutions have been initiated to implement the conventions dealing with the marine environment, responsibility lying with the national institutions responsible for managing coastal areas or marine resources. Similarly, no new national laws have been issued in relation to these conventions, the protection of the marine and coastal zone being covered by existing national laws.

The contracting parties to the Barcelona Convention, Lebanon and Syria, have complied with the convention and are implementing its requirements through various ministries and existing legislative frameworks. The Blue Plan Regional Activity Centre has carried out a number of studies within the MAP framework, related to past, present and future interactions between the environment and development in the Mediterranean basin, taking account of the impacts of population growth, urbanization, industry, agriculture, trade, energy, tourism and transport. The Priority Action Programme Regional Activity Centre carries out pilot projects concentrating on integrated coastal zone management in Lebanon and Syria. These countries also participated in the recent UNEP/GEF initiative on land-based sources of pollution under MAP. This was a multi-faceted exercise covering the identification of hot spots and sensitive areas, a strategy and action plan and estimates of the costs of remedial action (Government of Lebanon 1995, Government of Syria 1997). The Regional Marine Pollution Emergency Response Centre for the Mediterranean is developing a regional information system for preparedness and response to accidental pollution.

Under a Euro-Mediterranean initiative, the Specially Protected Areas Regional Activity Centre, in collaboration with the contracting parties, has identified 123 areas in the Mediterranean area in need of special protection. Capacity-building for national institutions is also planned. Revision of the Barcelona Convention in 1995 led to the development in 1997 of a plan to control and eliminate most land-based pollutants by 2025.

After almost 15 years, a Strategic Action Plan developed by PERSGA for the Red Sea and the Gulf of Aden has been approved for funding by GEF for US$19 million (Al-Sambouk 1998).

Major regional MEAs

Treaty	*Place and date of adoption*
Agreement for the Establishment of a Commission for Controlling the Desert Locust in the Near East	Rome 1965
Convention for the Protection of the Mediterranean Sea against Pollution	Barcelona 1976
Kuwait Regional Convention for Cooperation on the Protection of the Marine Environment from Pollution	Kuwait 1978
Regional Convention for the Conservation of the Red Sea and Gulf of Aden Environment	Jeddah 1982
Protocol Concerning Regional Cooperation in Combating Pollution by Oil and Other Harmful Substances in Cases of Emergency	Jeddah 1982
Protocol Concerning Marine Pollution Resulting from Exploration and Exploitation of the Continental Shelf	Kuwait 1989
Protocol for the Protection of the Marine Environment Against Pollution from Land-Based Sources	Kuwait 1990

The Regional Organization for the Protection of the Marine Environment (ROPME), with a head office in Kuwait, was established in 1979 by the parties to the Kuwait Regional Sea Convention to serve as its Secretariat (under the supervision of UNEP). It has a network of collaborating focal points in all eight member countries. A Marine Emergency Mutual Aid Centre was established in Bahrain in 1982 to assist ROPME in matters related to information, capacity building, reporting and cooperation with other organizations. A major survey on critical marine habitats, initiated by ROPME and carried out by the member states, identified all sensitive marine habitats in the region.

Regional organizations with environmental interests

- Arab Centre for Studies of Arid Zones and Drylands (ACSAD)
- Arab Industrial Development and Mining Organization (AIDMO)
- Arab League Educational, Cultural and Scientific Organization (ALECSO)
- Arab Organization for Agricultural Development (AOAD)
- Centre of Environment and Development for Arab Region and Europe (CEDARE)
- Council of Arab Ministers Responsible for Environment (CAMRE)
- Gulf Cooperation Council secretariat (GCC)
- International Centre for Agricultural Research in the Dry Areas (ICARDA)
- Islamic Educational Scientific and Cultural Organization (ISESCO)
- Joint Committee on Environment and Development in the Arab Region (JCEDAR)
- Mediterranean European Technical Assistance Programme (METAP)
- Regional Organization for Conservation of Environment of the Red Sea and Gulf of Aden (PERSGA)
- Regional Organization for the Protection of the Marine Environment (ROPME)

Although most countries are fulfilling their commitments as signatories, it is difficult to evaluate the impacts of regional MEAs on the protection of the environment. These depend on the agreement itself, the level of economic development of the country and the barriers to implementation in each sub-region and between and within countries. Limited access to adequate information also makes assessment of the impact of the MEAs difficult. In many cases, it is too early to identify impacts as countries are still at the stage of building up institutional and technical capacities and have barely started on implementation.

Regional action

Countries have made substantial joint efforts at the regional level to protect natural resources and the environment. Many regional institutions (see table) contribute to this task. Most cover the whole of the Arab Region but three relate specially to parts of West Asia (ROPME, PERSGA and METAP). The transboundary nature of environmental problems means that they also cover countries outside the region. Some (such as AOAD, ALECSO, AIDMO and ISECSO) do not deal specifically with environmental issues but address them insofar as they relate to their areas of concern (agriculture, industry, education and science). Even ACSAD covers concerns other than purely environmental ones in its studies of arid zones and drylands.

The organizations that are more sharply focused on environmental issues are CAMRE, JCEDAR and CEDARE. CAMRE is concerned mainly with policy issues at the national and regional levels. Only JCEDAR and CEDARE specifically address sustainable development and the interactions of the environment and development. However, even some of the more general organizations can make quite specific contributions to the environment: a recent example is the decision taken during the GCC summit held in Kuwait in December 1997 to adopt a by-law to protect and develop wildlife in the GCC countries. An example of the successful protection of an endangered species is given in the box below.

In addition, many UN organizations and their regional offices operate in the region. They help raise funds for technical assistance and support environmental management programmes,

Saving the Arabian oryx from extinction

For a long time, the Arabian oryx (*Oryx leucoryx*) was abundant all over the Arabian Peninsula. This endemic mammal is well adapted to living in the harsh environment of the region. Since the beginning of the 20th century, there has been a continuous decline in numbers, mainly as a result of hunting. By 1950, the population in the Great Nufud desert became extinct. The decline in the southern part of the region continued, with the final stronghold of the species in an area in Oman called Jiddat al Harasis. At the end of 1972 the last wild herd of six oryx was eliminated (Ghandour 1987). The main causes of the extinction were hunting for meat, skin, horns and internal organs, and the accessibility of the main habitats to outsiders.

Regional and international cooperation to save the Arabian oryx from extinction was started in 1962 and a World Herd was established in 1963, when nine animals were transferred to Phoenix Zoo in the United States to start a breeding programme. By the end of 1976, the World Herd had increased to 105 animals, and a programme was started to send back some animals to the states of the region where oryx used to live (Stanley Price 1989).

Most countries have succeeded in conserving and establishing separate herds. At present there are around 1000 animals in reserves and parks in Bahrain, Jordan, Kuwait, Oman, Qatar, Saudi Arabia and the United Arab Emirates. The successful re-introduction of oryx to the wild at Jiddat al Harasis has been recognized internationally.

Water policy in West Asia

There is an urgent need to review policy on water resources throughout the region. Current water resources cannot satisfy water demand much past 2005 unless steps are taken to rationalize demand management, increase and augment supplies, and impose realistic controls on use. Countries need to address three main issues.

Legal and institutional reform

There is an urgent need to review legislation and how it relates to policy options. The areas that require amendment include water rights, water abstraction, water quality and environmental standards, charges, pollution and environmental protection, protection of groundwater, wastewater treatment and solid waste disposal.

Amended legislation will not be effective unless water administrations are reorganized, with decentralization of the power of central government bodies responsible for water resources. Institutional weakness constitutes a major constraint on the management of water resources in most countries. This is a direct consequence of ill-defined responsibilities of the institutions involved and the absence of legislation to enforce coordination between authorities at local, regional and national levels. Coordination is also needed between government bodies responsible for water, agriculture, housing, industry and planning. Capacity building among the technical staff of research institutes and other water bodies is also badly needed.

Economic considerations

National development strategies directly influence water allocation and use, while policies to promote exports and foreign exchange earnings from highly-priced cash crops call for increased investment in irrigation schemes.

Because a lack of funds prevents the implementation of effective water resource policies, sustainable water policies should have positive impacts on central government finances from new tax revenues, charges and the reduction of subsidies.

Economic incentives could provide effective means of rationalizing water use. Possible incentives include tariffs for domestic and industry supply, charges for abstraction, irrigation, wastewater and pollution, and soft loans for modernizing equipment. Making charges for polluting water, proportional to the volume and the quality of effluent, may be the best way of discouraging industrial water pollution. Irrigation charges could be based on metering, area irrigated, type of crop, or length of irrigation time. Groundwater pricing can be based on quantity or on transferable pumping entitlements.

Water conservation

Water is wasted in all sectors. Huge losses (at least 45 per cent) in agriculture arise from inefficient irrigation systems, while there is a 20 per cent leakage from supply networks and general 10 per cent losses in industrial use. To reduce water losses, all countries must incorporate conservation programmes.

In the agricultural sector this can be achieved through:

- reviewing the economics of irrigation and agricultural production, and reappraising agricultural policies;
- improving traditional irrigation systems, introducing modern technology and promoting conservation techniques;
- reviewing irrigation incentives and tariffs;
- improving programmes to raise awareness of water as a scarce resource;
- providing subsidies and soft loans for modern irrigation systems.

The domestic and industrial sectors may require:

- water prices that reflect true costs, including wastewater treatment;
- applying escalating tariffs for increasing consumption;
- installing modern water-saving technology for distribution systems and households;
- improving leakage detection in supply networks;
- modifying building codes to promote efficient use of wastewater for landscaping;
- applying heavy pollution charges against industrial units violating regulations;
- obliging industrial units to treat water before discharging it.

environmental policies and institutional capacity building.

National initiatives

The policy aim of most governments is to limit further environmental degradation and achieve sustainable use of environmental resources. West Asian countries have made substantial efforts to integrate environmental aspects into their development schemes and strategies. Most have formulated national environmental action plans (NEAPs), which include identification and prioritization of key issues, and have set targets and timetables for implementation. For example, Lebanon, Oman, Saudi Arabia and Syria have initiated coastal management programmes, and WHO has prepared a plan of action in the Eastern Mediterranean Region in which priorities have been set for different countries with regard to the environment and related health problems (WHO/EMRO 1997). However, in most cases, NEAPs are mainly checklists of desirable actions, based on rather limited and doubtful information. They are generally short on reliable cost estimates, time schedules, division of responsibilities for implementation, and identification of sources of funding.

Water being a priority in the region, most countries have formulated national strategies but much more needs to be done (see box on page 317). For example, Bahrain's water strategy was initiated in the early 1970s. However, because it failed to reduce dependence on the country's fast-dwindling groundwater reserves, a new national water management strategy was endorsed to cover the period 1990 to 2010. The new strategy includes increasing production of desalinated water, a leak detection/reduction and system renewal programme, and reform of the agricultural sector by expanding the use of treated wastewater for irrigation. Similarly, the national water strategy in Lebanon seeks to reduce water losses through the use of more modern irrigation techniques, and promote methods for wastewater treatment and reuse for irrigation. The national water strategy in Saudi Arabia was formulated in the 1980s and covers national water policy up to the year 2020. Like the programmes in Bahrain and Lebanon, the Saudi strategy concentrates on the modernization of irrigation techniques and better use of treated wastewater. In addition, the Saudi plan includes the establishment of dams for surface water conservation and groundwater recharge, protection of groundwater resources from deterioration, capacity building and public awareness.

Laws and institutions

Laws

The command-and-control approach, through legislation, is still the main environmental management tool in almost all countries. Other approaches are being investigated and introduced, including technical assistance, advisory services, training, tax exemptions, cheap credit and fiscal disincentives.

Some environmental laws date back to the 1930s. However, legislation dealing with a wide range of environmental issues, including desertification, scarcity of freshwater, pollution, management of hazardous and toxic wastes, and conservation of biodiversity, has been developed more recently; many national laws and decrees dealing with environmental protection have been adopted within the past two decades.

Several new initiatives are being taken. In Bahrain, for example, ten legal instruments now address issues related to the protection of environmental resources (Fakhro 1997, Government of Bahrain 1998). In Saudi Arabia, several laws dealing with various aspects of the environment are being effectively enforced (see box above). Lebanon has undertaken an overall review of its environmental legislation and has drafted an environmental code, laws for the protection of natural sites and monuments, a law on integrated pollution control and a framework law on protected areas. Environmental impact assessment decrees and procedural guidelines have also been developed in Lebanon. These draft laws and regulations have been discussed in national consultative fora (Government of Saudi Arabia 1992). Oman is in the process of preparing a new set of regulations on environmental impact assessments.

Implementation of legislation and enforcement of standards varies. In many countries, as in other developing regions, enforcement of legislative measures is far from satisfactory. This can be

The Meteorology and Environmental Protection Administration (MEPA) of Saudi Arabia

The Meteorology and Environmental Protection Administration, established in 1981, is the central agency responsible for the environment in Saudi Arabia. It coordinates efforts towards the enforcement of Article 32 of the basic Rules of the Kingdom, which state: 'The government shall endeavour to conserve, protect and develop the environment as well as to prevent pollution'. Successes include:

- Combating desertification and land degradation through planting greenbelts, establishing parks, conserving forests, rangelands and animal resources, protecting water resources and providing potable water through large desalination plants. These efforts were supported through:
  - the forests and rangeland rule;
  - the uncultivated land regulation;
  - the regulation for conservation of water sources;
  - the regulation for fishing and protection of living water life in Saudi Arabia's territorial waters;
  - the regulation for veterinary quarantine.
- Collection and treatment of natural gas accompanying the production of crude oil, reducing emissions from the combustion of these gases;
- Application of strict environmental standards by local refineries to reduce lead content in gasoline as a first step towards lead-free gasoline;
- Incorporation of environmental considerations in major developmental plans, such as those for industrial development in Jubail and Yanbu, which later received the UNEP Sasakawa Environmental Prize for outstanding environmental achievements;
- Integrating local industries to recycle wastes, minimize pollution and use wastewater for cooling and landscape irrigation.

attributed to weak institutional capacity for environmental management, shortage of human and technical capabilities, adoption of imported standards which are not always relevant or applicable, the sectoral nature of environmental laws, unimpressive record of government machinery in monitoring and enforcing regulations and standards, political and economic constraints and lack of public and NGO participation (UNEP 1995).

There is a need to review, rationalize, update and integrate laws, modify norms and standards, and enforce monitoring procedures to address the inadequacies revealed in current legislation. An example of successful adaptation of laws and management strategies is the reforestation programme in Syria (see box below).

Institutions

During the past decade, there has been significant interest and improvement in environmental institutions to implement environmental policies, enforce laws, and set standards and norms. Some countries have environmental ministries (Lebanon, Jordan, Oman and Syria), others have general directorates and/or environmental councils (Bahrain, Iraq, Kuwait, Qatar, Saudi Arabia, United Arab Emirates and Yemen). A Palestinian Environmental Authority was established in December 1996. Committees and commissions have also been created to deal with specific environmental concerns such as ozone-depleting substances, pollution, wildlife conservation and biodiversity.

Continuous amendments in institutional structures and responsibilities reflect the changing attitude of states to developments in environmental policies. Cross-cutting policy institutions are difficult to establish in a system of government based on line-management structures. Most state environmental institutions suffer from shortages of qualified manpower, inadequate funding and uneasy relations with other government institutions whose cooperation is essential in dealing with environmental issues. This has resulted in delays and failures to implement policies and enforce laws. Both governmental and non-governmental organizations need to strengthen their institutional structures and increase their financial

Forest activities in Syria

Forest coverage was reduced from 32 per cent to 2.6 per cent (484 000 ha) of Syria's total area during the period 1900–95 (Government of Syria 1996, FAO 1997). The country's different forest ecosystems have suffered from deforestation, overgrazing, over-cutting, and man-made fires affecting forest biodiversity. Nearly 2 440 ha of forest were cleared for agriculture during 1985–93. During the past 15 years, more than 20 000 ha of coastal forest have been affected by fire. The *Pistacia atlantica* forests, which used to cover some 3 000 ha, are now reduced to a few hundred hectares. Large tracts of original forests have degenerated into secondary plant communities of low economic and environmental value.

Recognizing the importance of forest and trees in combating desertification and arresting land degradation, a Higher Commission for Afforestation was established in 1977. The aim of the commission was to afforest progressively 15 per cent of the country through planting of forest and fruit trees in different ecoregions. Cooperation between the Ministry of Agriculture and the Commission has led to:

- revision of the 1953 forest law;
- demarcation of forest lands;
- establishment of a management plan for major forest ecosystems;
- protection of forest ecosystems from fires;
- increase of numbers and capabilities of forest nurseries;
- distribution of forest tree seedlings to farmers at a nominal price; and
- the launching of extensive university education and forestry training programmes.

The outcome of these activities has been a substantial increase in afforestation and conservation of natural forests (see table).

Source: Government of Syria 1996

year	*afforested area (ha)*
1953–70	2 779
1971–76	5 273
1977–84	90 105
1985	23 459
1986	24 621
1987	25 586
1988	24 639
1989	24 988
1990	22 900
1991	21 027
1992	21 007
1993	24 177
1994	22 578
1995	27 026
1996	24 000

resources if they are to participate effectively in the formulation and implementation of environmental policies and action plans.

Economic instruments

West Asian countries generally rely far more on regulatory mechanisms than market forces. However, most states levy some charges and fees for environment-related services. Because of other socio-economic priorities, only a portion of the revenue raised is used to finance environmental protection schemes. The use of economic and financial instruments to control and prevent pollution through incentives and disincentives is rare. Soft loans are used in Bahrain, Jordan, Kuwait, Oman, Syria and the United Arab Emirates to encourage water-saving irrigation techniques and the use of tunnels and greenhouses to improve water productivity.

Where the polluter pays principle is applied, it is usually in the form of rather trivial fees levied for emitting polluting discharges of specified levels. These are difficult to collect because of the rather weak institutional framework for monitoring and enforcement. Other economic instruments, such as taxation for air pollution and levies on municipal services operations, are also widely applied. While the levies are relatively easy to collect, air pollution taxes require frequent and continuous sophisticated monitoring, which is seldom in place. Some countries (such as Oman) have initiated self-monitoring schemes but these are still at an early stage.

Resource conservation is practised through the pricing of some scarce resources, particularly water. Previous policies of state subsidy for water prices are changing. Pricing structures are generally applied to domestic and industrial water use as a rational instrument to control wasteful use. However, the pricing policy adopted still involves heavy subsidies in most countries. Furthermore, water used for irrigation, by far the largest category of water use, is either free or heavily subsidized (AOAD 1995).

The general trend over the past ten years has been towards liberalization and privatization of the economy. The growing role of the private sector and the reduction or removal of subsidies on a number of commodities, including pesticides and fertilizers, have helped reduce soil and water contamination. Relaxing strict price controls, including prices of agricultural products, can have a positive effect on agricultural output and food security. Future economic development, and increasing the role of the private sector, will undoubtedly have environmental effects but these may or may not be beneficial. The full impact of adopting market economics has yet to be clarified. The region is beginning to look into these issues but there are no clear indicators of the way policies may develop to deal with the myriad problems of a region in which the public sector is prominent in rich as well as less rich states.

Industry and new technologies

Industrial development is still dominated by public enterprises, which benefit from a range of protective policies such as subsidies, credits, discrimination in governmental procurement and preference over competing imports. Large-scale manufacturing facilities are located mostly in congested cities to benefit from urban infrastructure.

The pattern of current industrial development falls into two distinct categories. The first comprises the comparatively modern industries in the GCC countries which depend primarily on oil as a raw material; typical activities include petrochemicals, fertilizers, aluminium, iron and steel, and cement, with some diversification into the engineering and construction industries. In the past, the abundance of capital resources enabled these industries to finance the cost of cleaner production and pollution control techniques (see box right). Their strong economic base permits incentives for attracting domestic and foreign investment in state-of-the-art technology.

In Saudi Arabia, for example, cleaner production practices have been incorporated in major development projects at the Jubail and Yanbu industrial complexes, including intensive programmes for resource saving, minimizing waste generation, and recycling of resources and by-products (Government of Saudi Arabia 1992).

Another success story is the introduction of cleaner production concepts in the old aluminium smelter and a new extension at the Aluminum Bahrain Company. The new technologies reduced fluoride emissions by 98 per cent, total suspended particulates (including polyaromatic hydrocarbons) by 95 per cent and energy consumption by 15 per cent (Ameeri 1997). Refineries in Bahrain, Kuwait, Saudi Arabia and the United Arab Emirates use new technologies to reduce sulphur emissions, gas flaring and other

Cleaner production in West Asia

The major sources of industrial pollution in West Asia are the resource-based industries which include petroleum, chemical and petrochemical, mining, and agro-industries. Other sources include small and medium-size industries, such as metal finishing, tanneries and textile mills.

The command-and-control approach in dealing with industrial pollution has its limitations. Government agencies responsible for the environment have gradually been introducing more friendly working relationships with industrial management. This has succeeded in convincing major industries to comply with regulations and prevent pollution through the application of cleaner production procedures, cleaner technologies and pollution control.

Cleaner production procedures have been applied successfully by Dubai Cable in the United Arab Emirates, by Meshal International, BLAXECO and Al Zamil in Bahrain (Kanbour 1996), and by the National Titanium Dioxide company in Saudi Arabia (Harrison 1998).

Environmental agencies are requiring new industrial enterprises to use cleaner technologies and cleaner production procedures. Environmental laws in Bahrain, Iraq and Jordan encourage all industrial enterprises to apply pollution prevention procedures and reduce waste generation.

Waste recycling has increased, for example with many small and medium-size metal recycling plants. Metal recovery plants from waste generated by metal smelters are thriving. Dross from aluminium smelters and aluminium from other sources are being recovered in Bahrain. In Saudi Arabia, metal catalysts are collected and regenerated for reuse. Lead from lead storage batteries is recovered in Iraq, Jordan and Saudi Arabia for reuse. In other Gulf countries, used car batteries are exported to India and Indonesia. Used engine oil is collected and recycled. Other materials recycled include plastics, paper and cardboard.

Several industrial establishments are in the process of registering for certification under ISO 14 000, and evaluating the benefit of such certification to their general activities and product marketing. Enterprises such as refineries, petrochemical complexes and metal smelters in Bahrain, Kuwait, Oman, Qatar, Saudi Arabia and the United Arab Emirates have already begun procedures for obtaining certification. Implementation of such standards will achieve commitment from top executives to protecting the environment, and help to implement pollution prevention techniques and environmental awareness and training. Environmental agencies are encouraging enterprises to obtain such certification. In the Emirate of Dubai, United Arab Emirates, all new enterprises must submit a commitment to seek ISO 14 000 certification within two years of their establishment (Kanbour 1996).

In countries with agro-industries, waste minimization procedures have been in use for some time. The solid wastes from these industries are converted to animal feed or compost to be re-used as soil conditioners. In the dairy industry, the liquid waste from the manufacture of cheese is being bottled and sold in Iraq and Jordan as a soft drink. On-site recycling of water in these industries is becoming the norm rather than the exception.

Sources: Kanbour 1996 and Harrison 1998

hydrocarbon releases as part of their efforts towards environmentally-friendly production. Dubai (United Arab Emirates) has introduced unleaded gasoline for motor vehicles. It is expected that the other GCC countries will do the same by the year 2000.

The second category of industrial development involves countries with less affluent economies such as Jordan, Lebanon, Syria and Yemen. Many industries in these countries employ labour-intensive and heavily polluting technologies. Typical activities include mining, textiles, metal finishing and food processing. Due to the inadequacy of their infrastructure, and serious debt problems, these countries are seldom able to provide sufficient investment for industrial modernization and pollution control.

Changes in prices, taxes and subsidies normally fail to evoke the desired response in state-owned public enterprises. In countries where subsidies for inputs such as energy, water and material inputs are substantial, industrial pollution is usually aggravated. Controls over the prices of manufactured goods, on the other hand, discourage waste recovery and recycling, and often have negative environmental consequences. However, the emergence of new economic liberalization policies may force industry to pay a price that reflects environmental costs.

Privatization, promotion of energy conservation and removal of subsidies should ultimately result in an overhaul of industry. Structural adjustment policies are being pursued in countries such as Jordan, Lebanon and Syria but what effect this will have on environmental impacts is unclear. If manufacturers have to cut corners to keep their competitive edge, the environment may suffer. But where lowering tariffs results in cheaper imports of cleaner technologies and pollution abatement equipment, environmental benefits could follow. Given appropriate measures, structural adjustment policies can yield both economic and environmental gains.

Despite increased interest in cleaner technology, West Asian countries have yet to benefit significantly from the experience of the industrialized nations. This is mainly because of a lack of information on waste minimization technologies, management resistance to employing what they view as disruptive changes, and lack of policy measures conducive to investment in such technologies. There is a need to influence environmental actions for new industrial activities in the region and identify likely problems and proper mitigation measures for pollution control, particularly in industries that emit hazardous wastes. The EIA regulations being instituted in most countries may encourage proper environmental planning for future industrial development. A regional system is also needed to provide information concerning emission standards, cleaner production and waste minimization technologies and other relevant issues that may influence decisions on environmental management in industry.

Development of human resources for effective environmental management in industry is badly needed. There is a need to link enforcement with evolving environmental jurisdiction and competitiveness on both international and local markets under the new global trade regime. The institution of effective monitoring systems will support enforcement and reflect a commitment to develop environmentally-sustainable industry.

Generally, the trend is towards reducing emissions to comply with ambient air quality standards and control water pollution. For example, pollution control measures being implemented in industrial establishments and refineries in Syria include closed water circulation systems and treatment before discharge, implementation of safety procedures, and tightening standards for the addition of lead to gasoline (Government of Syria 1997). Many countries are placing more emphasis on Integrated Pest Management and organic farming to reduce the harmful impacts of agrochemicals discharged to the environment.

An important technical approach to resource conservation has been the growing interest in recycling of scarce resources, particularly water. In many states on the Arabian Peninsula, municipal wastewater is subjected at least to secondary treatment, and is widely used in irrigation of trees planted in determined efforts to green the landscape. Tertiary treatment is also practised in some countries. In Saudi Arabia, more and more homes and blocks of flats collect wastewater from toilet flushing, wash basins and baths, recycle it in situ and recirculate it in separate networks. It is claimed that this results in more than 40 per cent saving in total water consumption (Faheih Research and Development Centre 1997). Solid waste sorting and recycling is gathering momentum and trade in reclaimed resources is currently practised across national borders. The

Public participation in Oman: the case of Coastal Zone Management

The coastal zone of Oman is complex, dynamic and vulnerable, heavily used and under great pressure from offshore and inshore development activities. Its development involves many different authorities with overlapping interests and jurisdiction. This created the need for integrated action by all concerned authorities to safeguard the coastal zone.

A Coastal Zone Management Plan (CZMP) was started in the early 1980s and is now being implemented. It is cross-sectoral in its approach to wildlife, habitats, human use and management of the coastal area. Main objectives of the plan are to:

- establish a comprehensive policy to guide coastal development;
- legalize the control of development activities;
- identify a leading agency to coordinate planning, development and resource management activities;
- safeguard natural and cultural resources, including conservation areas;
- safeguard and restore scenic areas for enjoyment;
- improve monitoring, field studies and enforcement activities;
- identify and protect sensitive habitats including mangroves and coral reefs;
- ensure sustainability of the resources;
- control coastal erosion.

The plan started with an official communication to all concerned authorities to inform them of the project (its goals, objectives and anticipated benefits) and to ask them to identify a focal person for liaison and collaboration. A series of meetings was held to ascertain the interests and activities of each authority. The plan evolved through a flexible process of dialogue, trial, error and adaptation. Later meetings focused on specific issues and solutions to individual problems.

Responsibility for the execution of issue-specific actions was shared among concerned parties. A single plan for each target stretch of coast was produced through a participatory process. Each concerned authority was required to endorse the plan and its recommendations, and to accept responsibility for implementing agreed actions. In several instances, this process led to actions being implemented and issues developed before the plan was printed, so implementation even overtook the planning process.

Source: Government of Oman 1998

Islamic Relief Fund, which operates worldwide, has successfully promoted recycling schemes for aluminium cans in Saudi Arabia; the cans are exported to Bahrain. Waste paper is also being recycled, thus providing a sizeable income to the Fund.

Public participation

Public participation is a complex process that requires fundamental changes in rooted social attitudes and individual behaviour. The efforts and resources needed to achieve active participation are considerable but are essential in the long term.

Public access to environmental information has improved as governments have become more open but much more effort is required to achieve real public participation in environmental management (World Bank 1994). Public awareness of pressing environmental issues, particularly water scarcity, desertification, and pollution of air and marine resources, has increased rapidly during the past decade but in most countries public participation is still in its early stages. The role of NGOs, which now exist in most countries, is becoming more important (there are more than 50 NGOs in Lebanon – Government of Lebanon 1995).

Despite the proliferation of NGOs, few are truly viable or effective. Many continue to rely on state subsidies and are therefore neither self-sustaining nor truly independent. Many command little public respect or credibility. Consequently, assistance is needed to develop the capabilities of NGOs in the design, implementation and evaluation of actions to promote equitable access to resources and environmental services, particularly for the disadvantaged sectors of society. In addition, decentralizing decision-making seems vital for enhancing community participation in the formulation, execution and evaluation of local development projects.

Projects aimed at strengthening national capacities for environmental affairs have been initiated in several countries under *Agenda 21* and World Bank programmes. The UNDP Regional Bureau for Arab States and the GEF have initiated capacity-building projects in biodiversity, climate change and international waters.

Environmental technology transfer needs to expand in parallel with the development of improved capacity, human and financial resources. Dissemination of the results of successful examples of cleaner technologies, through newsletters, manuals, and the environmental press, will greatly strengthen public awareness and promote capacity-building strategies.

Environmental information and education

There is a general lack of reliable, up-to-date information and data on the state of the environment. This relates to the lack of standardization of data formats and consistent environmental monitoring, data collection and reporting. Reports are often located in different public and private organizations, between which there is little or no cooperation. This results in gaps and duplication of data, and limited utilization of the information. This, in turn, hinders policy development, planning, implementation and follow-up.

Networking and integration of data for environmental assessment are at an early stage. Electronic information systems, networks and cooperation and coordination among relevant organizations all need to be strengthened to enable all users to benefit from data at the local, national, regional and international levels

Many universities and institutions are running courses and organizing training programmes, seminars and postgraduate studies in different fields of the environment. Courses on the environment have also been introduced in school curricula (UNEP/ROWA 1994). However, environmental training has yet to be institutionalized throughout the region.

Social policies

In the past, social traditions, an improving economic climate and policies to encourage population growth have all resulted in increasing population during the past three decades. This growth has recently become difficult to control, and policies for controlling population growth are generally failing to produce significant results.

With nearly 92 million people in 1998, and an annual growth rate of more than 3.1 per cent, population pressure is now the core problem for economic development policies. The growth rate exceeds the anticipated growth rates of the region's economies, particularly in agriculture. The deficit in food production is growing and is aggravated by the scarcity of resources (land and water) which are nearly fully utilized. Water security and increased food production have been the dominant strategies behind most development policies during the past two decades. These have been only partially successful, mainly as a result of poor or ill-defined strategies to combat pollution, degradation and over-exploitation of resources, institutional weakness and lack of coordination, inadequate technical and financial resources and lack of public participation. A further major problem is that of refugees and dislocated people, especially in Jordan, Lebanon and Syria, where nearly one million refugees live in poor conditions in camps around major cities, putting pressure on the already over-stretched infrastructure in these countries.

Conclusions

There has been a significant increase in national commitments to environmental issues and sustainable development. Environmental institutions have been given higher priority and status, and the level of policy commitment has increased.

Many countries have introduced legislation covering a wide range of environmental issues. Economic and financial instruments are still used only to a limited extent. Cleaner production is being promoted, and environmental education and training have increased.

Water and land management problems are sufficiently severe to merit much stronger action and the consideration of alternative policies. Alleviating water depletion, land degradation and desertification, and achieving a sustainable use of these resources, requires the formulation and implementation of integrated national land and water plans incorporating improved planning and analysis, legal and institutional reforms, and new water and land resource projects and programmes.

Economic development needs to go hand in hand with sound environmental policies within the framework of sustainable development. Policy initiatives should concentrate on institutional strengthening, information management, attracting investment and effective incentives.

Success in implementing MEAs is mixed and economic instruments to improve implementation are still not well developed. Countries depend mainly on command-and-control measures, but instruments such as incentives, taxes and charges, pricing strategies, cleaner production and other indirect measures are being applied to a few MEAs, particularly the Montreal Protocol. National reporting is weak and compliance is not well monitored, except for the Montreal Protocol.

References

ACSAD (1997). *Proceedings of Expert Meetings on Biodiversity in Arab Countries*. ACSAD, Cairo, Egypt (in Arabic)

Al-Sambouk (1998). *Newsletter of the Regional Organisation for the Conservation of the Environment of the Red Sea and the Gulf of Aden*, No. 7

Ameeri, J. G. (1997). *Environmental accomplishment of ALBA*. Arab Environmental Day. UNEP-ROWA/MHME, Bahrain, 14 October 1997

AOAD (1995). *Study of Efficient Water Use in Agriculture in Arab Countries and Project Proposals for Development*. AOAD, Khartoum, Sudan (in Arabic)

Faheih Research and Development Centre (1997). *Water Reuse: an effective method to overcome water shortages in the Kingdom of Saudi Arabia.* Faheih Research and Development Centre, Riyadh, Saudi Arabia

Fakhro, R. M. (1997). *The Advent of Environmental Policy in Bahrain*. WHO/EA National Seminar on the Role of Economic Policies in Health and Environment, Bahrain, 27–30 September 1997

FAO (1997). Time Series for SOFA'97. *Country Time Series*. FAOSTAT TS software, FAO, Rome, Italy

Faras, Ahmad (1996). *Forestry National Report*. FAO/UNEP Expert Meeting on Criteria and Indicators for Sustainable Forest Management in the Near East. Cairo, Egypt, 15-17 October 1996

Ghandour, A. M. (1987). *The Oryx: from captivity to re-introduction*. NCWCD Publication, Riyadh, Saudi Arabia

Government of Bahrain (1988). *Environmental Management in Bahrain: an action plan.* Environmental Protection Committee, Bahrain

Government of Lebanon (1995). *Lebanon: Assessment of the State of the Environment*. Final Report. Ministry of the Environment, Beirut, Lebanon

Government of Oman ([illegible]). *Coastal Zone Management Plan*. Ministry of Regional Municipalities and Environment, Oman

Government of Saudi Arabia (1992). *National Report on Environment and Development.* Ministerial Committee on Environment, Riyadh, Saudi Arabia

Government of Syria (1996). *Productive and Protective Afforestation in the Syrian Arab Republic*. Ministry of Agriculture, Damascus, Syria (in Arabic)

Government of Syria (1997). *The State of the Environment in Syria.* Ministry of State for Environmental Affairs, Damascus, Syria

Harrison, I. (1998). *Benefits of an ISO 14 000 system.* Third Annual Conference on Environmental Management Systems and ISO 14 000, 14–18 February 1998, Dubai, United Arab Emirates

Kanbour, F. (1996). The Regional Seminar on Cleaner Production in Abu Dhabi. *Industry and Environment,* 19, 3, 52

Stanley Price, M. R. (1989). *Animal re-introductions: the Arabian Oryx in Oman*. Cambridge University Press, Cambridge, United Kingdom

UNEP (1995). *UNEP's New Way Forward: Environmental Law and Sustainable Development*. United Nations Environment Programme, Environmental Law Unit, Nairobi, Kenya

UNEP (1997). *Register of International Treaties and Other Agreements in the Field of the Environment - 1996*. United Nations Environment Programme, Nairobi, Kenya

UNEP/UNESCWA (1991). *The National Plan of Action to Combat Desertification in the Republic of Yemen.* UNEP, Yemen

UNEP/UNESCWA (1992a). *The National Plan of Action to Combat Desertification in Bahrain.* UNEP, Bahrain

UNEP/UNESCWA (1992b). *The National Plan of Action to Combat Desertification in the Sultanate of Oman.* UNEP, Oman

UNEP/UNESCWA (1994). *The National Plan of Action to Combat Desertification in the United Arab Emirates.* UNEP, Dubai, United Arab Emirates

UNEP/ROWA (1994). *Regional Directory for Tertiary Level Environmental Training Institutions in West Asia*. UNEP/ROWA, Bahrain

WHO/EMRO (1997). *Plan of Action for Health and Environment in the Eastern Mediterranean Region.* Ministerial conference on Health, Environment and Development, Damascus, Syria, 18–19 December 1997

World Bank (1994). *Forging a partnership for environmental action; an environmental strategy towards sustainable development in the Middle East and North Africa*. World Bank, Washington DC, United States

The Polar Regions

KEY FACTS

- Except for a few countries, polar issues are largely peripheral to domestic politics and the economy. At the global level, this has resulted in an inability or unwillingness, or both, to pay much attention to the problems of the polar environment, not least in the allocation of funds.
- The eight Arctic countries have, in the decade since the end of the Cold War, developed a dynamic regional cooperation focusing on Arctic environmental issues.
- An International Code of Safety for Ships Navigating in Polar Waters, setting specific safety and anti-pollution standards, is being drafted under the auspices of the International Maritime Organization.
- The Antarctic is unique in two ways: it is the only part of the world that is primarily managed cooperatively by interested countries on the basis of international agreements; and its policies are mainly proactive, seeking to address potential problems before they arise.
- The annual Antarctic Treaty Consultative Meetings have now adopted more than 100 measures, decisions or resolutions as well as several international agreements that relate specifically to the environment.
- Growth in tourism is likely to continue. 'Mass tourism' could challenge the existing policy structures or the International Association of Antarctica Tour Operators.
- The most serious policy challenge is in relation to Antarctic fisheries. The Convention for the Conservation of Antarctic Marine Living Resources (CCAMLR) needs to develop enforcement mechanisms to ensure that conservation measures are complied with by the activities of an overcapitalized global fishing fleet, operating under a complicated system.

The policy background

Cooperation has become the key to development of polar environmental policies. In the Arctic, there has been a transformation from the military secrecy of the Cold War era to pan-Arctic cooperation on sustainable development and the environment, with a keen focus on traditional lifestyles of indigenous peoples. The Antarctic environment and, in particular, the human activities that take place in it are regulated and managed on a cooperative basis by the parties to the complex of multilateral agreements of the Antarctic Treaty System.

Rather than describing policy responses under different policy instruments, as is done for the six other GEO regions, this section first describes the few major policy developments that are common to both polar regions, and then considers the Arctic and Antarctic in turn.

Common policy instruments

Several global international instruments make special provision for polar areas. For example, the 1992 Earth Summit and Agenda 21 led to the adoption of a Global Programme of Action for the Protection of the Marine Environment from Land-based Activities in 1995. This has been given an Arctic focus through the Regional Programme of Action for the Protection of the Arctic Marine Environment from Land-Based Activities, endorsed by Arctic Council Ministers in the Iqaluit declaration (Arctic Council 1998).

Similarly, the 1982 UN Convention on the Law of the Sea includes a special provision for ice-covered areas, applicable to pollution from vessels. An International Code of Safety for Ships Navigating in Polar Waters, setting specific safety and anti-pollution standards, is being drafted under the auspices of the International Maritime Organization (IMO) (Brigham 1999).

A Global Plan of Action for the Conservation, Management and Utilization of Marine Mammals (MMAP) has been developed by UNEP and FAO together with the International Whaling Commission and the World Conservation Union (IUCN). UNEP serves as the secretariat for MMAP and continues to support activities that promote and assist countries to achieve sound conservation and management of marine mammals.

Neither the Arctic nor the Antarctic is covered by the UNEP Regional Seas Programme but the UNEP-led Intergovernmental Negotiating Committee is preparing an MEA on persistent organic pollutants (POPs) – negotiations are expected to be completed by the year 2000. POPs are found in both polar regions as a result of long-range transport.

Policy overview – Arctic

Current situation

Cooperation among the eight Arctic countries (Canada, Denmark/Greenland, Finland, Iceland, Norway, Russia, Sweden and the United States) was initiated by Finland in 1989 and adopted as the Arctic Environmental Protection Strategy (AEPS) by a declaration of the First Ministerial Conference on the Protection of the Arctic Environment, held in Rovaniemi, Finland, in 1991. The declaration set out a joint Action Plan in which the Arctic Eight undertook to cooperate in scientific research to specify sources, pathways, sinks and effects of pollution, as well as to share data. Priority was given to pollution by POPs, oil, heavy metals, radioactivity and acidification. Moreover, the Arctic countries agreed to assess potential environmental impacts of development activities and to implement measures to control pollutants and reduce their adverse effects on the Arctic environment. The AEPS included special reference to accommodating the traditional and cultural needs, values and practices of local populations and indigenous peoples (AEPS 1991).

The AEPS also established a number of cooperative programmes:

- the Arctic Monitoring and Assessment Programme, which monitors the levels and assesses the effects of anthropogenic pollutants in all compartments of the environment;
- an Emergency Preparedness, Prevention and Response programme, which provides a framework to address the threat of environmental emergencies;
- Conservation of Arctic Flora and Fauna, which facilitates the exchange of information and coordination of research into species and habitats;
- Protection of the Arctic Marine Environment, which takes measures to prevent marine pollution.

These programmes reported to the Ministers of Environment of the Arctic countries approximately every two years. The Ministers then identified priority areas for further action. Four Ministerial conferences were held under the AEPS framework with the final conference in Alta, Norway, in June 1997.

MEAs affecting the Arctic

Many MEAs are as relevant to the Arctic as they are to more temperate regions. In addition, there are unique problems that need to be addressed through specific provisions of international instruments or through international organizations. MEAs of particular importance to the Arctic are:

- The 1979 UN Economic Commission for Europe's Convention on Long-Range Transboundary Air Pollution (LRTAP) and its protocols, ratified by all Arctic countries. The objective is to prevent, reduce and control transboundary air pollution from existing and new sources. The new protocols on heavy metals and POPs, signed in Århus in June 1998, are of special significance.
- The 1991 Convention on Environmental Impact Assessment in a Transboundary Context (the Espoo Convention). Aimed at preventing, reducing and controlling significant adverse transboundary environmental impacts, it obliges parties to conduct EIAs on proposed activities that may have such impacts. The Convention covers all land-based sources (except for POPs) but is limited to transboundary effects. It provides for notification but only through self-assessment. All the Arctic countries are eligible to become parties to Espoo but only two of them, Norway and Sweden, have ratified it.
- The 1992 Convention for the Protection of the Marine Environment of the North-East Atlantic up-dates and combines two previous agreements: the 1972 Oslo Convention (dumping from ships and aircraft) and the 1974 Paris Convention (land-based sources). This agreement introduced the precautionary and polluter pays principles into environmental protection in the region but it covers only some Arctic areas and it does not specifically address coastal development.

The only Arctic-specific MEA is the 1973 Agreement on Conservation of Polar Bears. However, the 1957 Interim Convention on Conservation of North Pacific Fur Seals and its protocols also cover the Arctic, and the 1995 Waterbirds Agreement under the Bonn Convention covers a large Arctic segment.

The programmes of the AEPS have now been subsumed under the Arctic Council – a high-level forum established by the Arctic Eight under a declaration signed in Ottawa in September 1996. The Arctic Council has provided a wider means for promoting cooperation, coordination and interaction among the Arctic states, involving indigenous communities and other inhabitants, particularly on sustainable development and environmental protection issues. In the Arctic Council, the category of Permanent Participants provides for active participation and full consultation with indigenous representatives. The Council has a complementary role in nurturing regional identity. Indeed, the Arctic Council Declaration takes a wider view on regional cooperation than was possible under the AEPS.

Sub-regional cooperation also began in the 1990s, building on and reinforcing earlier bilateral contacts. For example, the Barents Euro-Arctic Region (BEAR), established under the 1993 Kirkenes Declaration, focuses on the environment and operates at two levels – between the governments of Finland, Norway, Russia and Sweden, as well as between the eight northernmost counties and/or provinces in these countries (Kirkenes Declaration 1993). The BEAR Regional Council is comprised of representatives from local government and includes a Saami representative. A similar concept of institutionalizing bilateral dialogue originated in 1991 when the Northern Forum, involving regional authorities and also operating with an environmental agenda, was established in Anchorage (Nordic Council of Ministers 1995).

Cooperation amongst groups of indigenous peoples is now organized at sub-regional level – for example, through the Inuit Circumpolar Conference and the Saami Council – and at regional level through the Arctic Leaders' Summit.

There is also international cooperation between professional groups. For example, the International Arctic Science Committee, composed of scientific organizations in the countries that conduct research in the Arctic, was founded in 1990, with the primary goal of coordinating research (Nordic Council of Ministers 1995). Organizations such as the Nordic Council and the Standing Committee of Parliamentarians of the Arctic Region, established in 1993, have also shown an interest in Arctic environmental cooperation (Haarde 1997).

Policy trends

The end of the 1990s may signal the start of consolidation leading to even more intense environmental cooperation. Several governments, of Nordic countries in particular, have recently started a thorough re-examination of their northern policies. The new initiatives are aimed at:

- broadening the context – putting environmental protection on the wider agenda of the Arctic Council;
- intensifying sub-regional cooperation – particularly in the Barents region; and
- enlarging the picture – the European Union's 'northern dimension'.

Unlike AEPS, which focused on 'threats to the Arctic environment and the impact of pollution on fragile Arctic ecosystems', the Arctic Council Declaration considers the environment in a much wider context. Formulating a proper relationship between sustainable development and environmental protection has emerged as a key policy requirement.

Although there has already been some cooperation in relation to the Barents area (Schram Stokke and Tunander 1994), realizing the economic potential of this sub-region requires the integration of environmental concerns into energy production, forestry, transport, industry, natural resource exploitation and land-use planning. The scale of the environmental actions needed and the severe economic problems facing the Russian Federation, however, mean that financial resources need to be found to implement projects (Ojala 1997).

The closer involvement of the European Union in Arctic cooperation, creating the Union's 'northern dimension' (Heininin and Langlais 1997), will be taken into consideration in the preparation of the European Union's Sixth Environmental Programme. This cooperation will extend the provision of 2000–06 funding through the next round of the TACIS, PHARE and Interreg programmes (Lipponen 1997).

Barriers to progress

For some, especially the major states, Arctic issues are largely peripheral to domestic politics and the economy. This has resulted in an inability or unwillingness, or both, to pay much attention to the problems of the Arctic and its environment, not least in the allocation of funds. There is no agreement on funding at the international level. Cooperative activities, and especially the hosting of Arctic programme secretariats, depend on voluntary contributions of the participating countries.

There are also more specific problems related to particular aspects of environmental cooperation. For example, the Arctic Monitoring and Assessment Programme had problems in accessing sources of information on pollution because a number of national institutions were reluctant to provide the necessary raw data, often on grounds of security. Some agencies preferred to provide already interpreted data, and others failed to respond at all. Despite a political commitment within AEPS to provide the data, it was impossible to enforce this. Nevertheless, the AEPS produced valuable basic studies of ecosystem functioning, and provided a starting point for Arctic decision-makers to create further policy measures for environmental protection, now within the ambit of the Arctic Council.

Within its wide policy embrace, the Arctic Council has defined certain 'no-go' issues. For instance, although the United States, in a mid-1994 inter-agency review of Arctic policy, listed environmental protection as the top priority and simultaneously downgraded national security and defence considerations, freedom of navigation remains the strategic military interest of the US Navy in Arctic waters, particularly for submarine operations (Griffiths 1999). Environmental protection related to military activity in the Arctic is dealt with under separate arrangements among individual states, such as the trilateral military environmental cooperation between Norway, the Russian Federation and the United States, entered into in September 1996 (AMEX 1996). The separation of security from other issues in the Arctic is a characteristic of Arctic collaboration. Whilst this may be seen by some as an important precondition for successful cooperation, others may see it as a valid cause for stalemate on particular issues.

Policy overview – Antarctic

Current situation

The environmental policy situation in the Antarctic is unique in two ways:

- it is the only continent that is primarily managed cooperatively by interested countries on the basis of international agreements;
- policies are mainly proactive, seeking to address potential problems before they arise, in contrast to other parts of the world where they tend to be reactive and remedial.

Antarctica is uninhabited apart from the wintering members of national scientific programmes, summer visitors and tourists. The surrounding Southern Ocean has for a long time been exploited for whaling, sealing and fishing. The legal status of the Antarctic is quite unlike that of the Arctic. Seven states assert claims on the continent (Argentina, Australia, Chile, France, New Zealand, Norway and the United Kingdom) – three of which overlap (those of Argentina, Chile and the United Kingdom). The United States and the Russian Federation do not recognize these claims but reserve the right to make their own claims, and the majority of other states do not recognize any claims.

The sub-Antarctic islands surrounding the continent and north of latitude 60° S are, with two exceptions, subject to uncontentious national sovereignty. The exceptions are the South Sandwich Islands and South Georgia, where Argentina contests the current jurisdiction of the United Kingdom. This situation is linked to the Falklands (Malvinas) dispute between these states. Territorial seas and exclusive economic zones (EEZ) are asserted around a number of sub-Antarctic islands. Between these and the waters that are subject to regulation under the Antarctic Treaty System, there are large areas of the high seas.

The absence of agreed national sovereignty has shaped the international regime. The area south of 60° S is subject to a form of international governance involving 44 states (US 1999) under the Antarctic Treaty System, although other states have contested the propriety of this subset of the global community regulating, outside the UN system, what they assert is a global commons.

The Antarctic Treaty (concluded in Washington DC, United States, in 1959) has the primary objective of ensuring 'in the interests of all mankind that Antarctica shall continue for ever to be used exclusively for peaceful purposes and shall not become the scene or object of international discord' (Antarctic Treaty 1959). It prohibits any measures of a military nature, promotes international cooperation in scientific research, prohibits nuclear explosions and the disposal of radioactive wastes, and removes the potential for sovereignty disputes between Parties.

The treaty does not itself contain provisions relating directly to the environment, apart from the prohibition of nuclear explosions and waste disposal. However, it is now complemented by three other agreements that do: the Convention for the

Conservation of Antarctic Seals, the Convention for the Conservation of Antarctic Marine Living Resources (CCAMLR), and the Madrid Protocol on Environmental Protection to the Antarctic Treaty, all of which have entered into force. Together with the Antarctic Treaty itself, these are known collectively as the Antarctic Treaty System (see box). A Handbook provides the texts of all these agreements (US DOS 1994).

The annual Antarctic Treaty Consultative Meetings have now adopted more than 100 measures, decisions or resolutions as well as several international agreements that relate specifically to the environment. Particularly significant are recent measures on tourism, and designation and management plans for different categories of protected area, and resolutions on fuel storage and handling, inspection checklists for current stations, abandoned stations, vessels, waste disposal sites, emergency response action and contingency plans.

Independent bodies such as the Scientific Committee for Antarctic Research (SCAR) and the Council of Managers of National Antarctic Programmes (COMNAP) have developed year-round operation, with specialist working groups and groups of experts addressing different environmental issues. Specialized workshops are increasingly used to address particular issues, such as by SCAR and COMNAP on Monitoring of Environmental Impacts (SCAR/COMNAP 1996), by IUCN on Cumulative Impacts (IUCN 1996), by the United Kingdom (Norway/UK 1998) and by Peru (Peru 1999) on Protected Areas, by Chile during the combined XXV SCAR/X COMNAP meetings on the concept of 'dependent and associated ecosystems' in 1998, and by Australia on Diseases of Antarctic Wildlife (Australia 1999). In each case, these specialist meetings and their reports feed back into policy discussions at the ATCM.

During the 1980s, in response to pressure from the environmental movement and non-Antarctic Treaty states at the United Nations, the ATS became more open and accessible. Expert organizations were admitted to the ATCMs and CCAMLR meetings, meeting documents became publicly available and Parties paid increasing attention to aspirations and perceptions outside the meeting.

There has been some reduction in this openness during the 1990s. Expert organizations external to the ATS were denied access to the 11 meetings of the Group of Legal Experts on liability for damage to the environment between 1992 and 1998. Set against this, however, is the creation of a web site for the XXII and XXIII ATCMs that allows access to documents following the meeting (ATCM 1998).

Organizations invited to attend the ATCMs include United Nations organizations (IMO, IOC, UNEP and WMO) and international organizations such as IAATO, IUCN, Pacific Asia Travel Association, World Tourism Organization, and the Antarctic and Southern Ocean Coalition.

The region is also covered by a number of global MEAs which have been the subject of recent discussion within the Antarctic Treaty System (Chile 1996 and UK 1996): the UNFCCC, the Vienna Convention and its Montreal Protocol on Substances that Deplete the Ozone Layer, CITES, CBD and the 1972 Convention on the Prevention of Marine Pollution by Dumping of Wastes and Other Matter (London Convention). The UN Convention on the Law of the Sea (UNCLOS) is also relevant, especially Part XII on the protection and preservation of the marine environment.

In 1990, the region south of latitude 60° S was designated a Special Area under Annexes I (oil) and V

The Antarctic Treaty System

The Antarctic Treaty System comprises, in addition to the Antarctic Treaty of 1959, three other MEAs:

- The 1972 Convention for the Conservation of Antarctic Seals which provides international regulations for commercial sealing. Four species of seals are totally protected and catch limits are set for others. Since there has been no commercial sealing since 1964, the Convention has only collated annual kills or captures for scientific purposes.

- The Convention on the Conservation of Antarctic Marine Living Resources (CCAMLR) was adopted in Canberra, Australia, in 1980. It aims at protecting the ecosystem of the seas surrounding Antarctica by regulating the exploitation of living marine resources. It identifies protected species, sets catch limits, identifies fishing regions, regulates fishing period and methods, and establishes fisheries inspection procedures. Parties meet in an annual Commission, which is supported by a permanent Secretariat (the only one across the entire Antarctic Treaty System) based in Hobart, Tasmania. Catch and other data (such as seabird by-catch) are collated by the Secretariat. A Scientific Committee is established to provide technical advice on which catch levels may be based. The CCAMLR area is divided into a number of statistical areas which are, in a sense, individually managed. An Observer and Inspection system provides a mechanism for ensuring compliance. CCAMLR has managed fisheries for, among others, finfish, krill and squid.

- The Antarctic Treaty's Madrid Protocol on Environmental Protection was adopted in 1991. Under the protocol, mineral resource activities, apart from scientific research, are prohibited for a minimum of 50 years and Environmental Impact Assessments (EIAs) are required for all activities. A Committee for Environmental Protection (CEP) advises Parties on implementation. Technical annexes establish Standards and Procedures for EIAs, Conservation of Antarctic Fauna and Flora, Waste Disposal and Waste Management, Prevention of Marine Pollution, and Area Protection and Management.

(garbage) of MARPOL 73/78, banning the disposal, at sea or on shore, of oily residues and garbage from ships.

Whaling is not covered by the Antarctic Treaty System, as it is dealt with under the International Convention for the Regulation of Whaling. The International Whaling Commission (IWC), concerned with the negative impact of whaling, established a Southern Ocean Whale Sanctuary in 1994. Japan voted against the sanctuary and entered an objection to it with respect to Antarctic Minke whale stocks. In 1998, the Commission passed a resolution requesting Japan to refrain from issuing a special permit for the take of southern hemisphere Minke whales, particularly within the sanctuary (IWC 1998).

A number of the albatrosses breeding on the sub-Antarctic islands have been placed on the list of species which have an unfavourable conservation status and require international agreement for their conservation and management (Appendix II of the 1979 Convention on the Conservation of Migratory Species of Wild Animals, the Bonn Convention). One species (the Amsterdam albatross, *Diomedea amsterdamensis*) has been placed on the Appendix I list of species which are endangered (CMS 1997).

The sub-Antarctic Macquarie and Heard islands have been nominated by the Australian Government for inscription on the World Heritage List (Australia 1996) under the Convention for the Protection of the World Cultural and Natural Heritage. The New Zealand Government has nominated the Antipodes Island, Auckland Islands, Bounty Islands, Campbell Island and the Snares (NZ 1997).

Commercial tourism in Antarctica has accelerated in the past decade, both in numbers of passengers on ships and, more recently, in overflying aircraft. To prevent or mitigate the possible environmental impacts of this growing demand, the International Association of Antarctica Tour Operators (IAATO) has taken a number of measures in conjunction with the parties to the Antarctic Treaty. These measures include evaluation of the environmental impact of activities proposed by the company members of IAATO, and the introduction of Ship-board Oil Pollution Emergency Plans on all IAATO member vessels. Furthermore, Parties to the Antarctic Treaty have recommended the use of a standard form for post-visit reporting in order to obtain consistent information that will facilitate analysis of the scope, frequency and intensity of tourism and other non-governmental activities (ATCM 1997).

Science and the Antarctic

Science is given a high status in all four components of the Antarctic Treaty System. Indeed, with the possible exception of CCAMLR, where recent commercial pressures may compete more evenly with science, scientific advice has frequently been pivotal in setting the direction of policy.

Antarctica offers outstanding scientific opportunities to improve understanding and monitor global changes and major interactions in the biosphere. The Madrid Protocol designates Antarctica as a 'natural reserve, devoted to peace and science' and stresses the value of Antarctica as 'an area for the conduct of scientific research, in particular research essential to understanding the global environment'. The number of people participating in scientific programmes grew steadily up to 1989/90. Research activities are changing as global issues increasingly drive the directions of science and as technological development extends the range of science and make the region more accessible. There is now a balance between research into global change and local human impacts. Research projects are also becoming more integrated, gathering specialist information from different areas and countries to address major scientific issues such as past glaciations, sea levels and atmospheric composition, ozone depletion, ice-sheet changes, sea-ice dynamics and oceanic and atmospheric circulation.

To encourage more effective research, the Parties to the Antarctic Treaty are promoting international and multi-disciplinary science and efficient operational management. SCAR provides a forum for the development of cooperation, through its various Working Groups and Groups of Specialists. COMNAP provides opportunities for logistic cooperation and exchange of ideas and information. SCAR and COMNAP work closely together, in particular on Antarctic Data Management.

Policy trends

Two major global problems pose particular risks for the Antarctic – increased UV-B resulting from the depleted ozone layer, and climate change. Whilst overall policies to address these issues fall under appropriate global MEAs, regional initiatives should contribute to the collection of the necessary scientific data, and help galvanize international action.

Growth in tourism and fisheries activities is likely to continue. Current policies for management of these industries are inadequate. While tourism does not currently present a serious problem, existing policy is based on limited, largely ship-based touring, where most operators are members of an environmentally-sensitive industry association, IAATO. Any emerging mass tourism industry in Antarctica could become a challenge for existing policy structures or IAATO.

However, by far the most serious policy challenge is in relation to fisheries. CCAMLR needs to develop enforcement mechanisms to ensure that conservation measures are complied with by an overcapitalized global fishing fleet, operating under complicated ownership, operation, control and flagging arrangements.

References

AEPS (1991). *Arctic Environmental Protection Strategy*, First Ministerial Conference on the Protection of the Arctic Environment, Rovaniemi, Finland, 14 June 1991

AMEX (1996). *Declaration among the Department of Defense of the United States of America, the Royal Ministry of Defence of the Kingdom of Norway and the Ministry of Defence of the Russian Federation, on Arctic Military Environmental Cooperation.* Bergen, Norway, 26 September 1996

Arctic Council (1998). *Iqaluit declaration*. Ministerial meeting of the Arctic Council, Iqaluit, Canada, 17 to 18th September 1998

ATCM (1997). *Final Report of the XXI Antarctic Treaty Consultative Meeting*. New Zealand Ministry of Foreign Affairs and Trade, Wellington, New Zealand

ATCM (1998). http://www.antartica-rcta.com.de

Australia (1996). *Heard Island and McDonald Islands: Nomination by the Government of Australia for Inscription on the World Heritage List;* and *Nomination of Macquarie Island by the Government of Australia for Inscription on the World Heritage List.* Canberra, Australia

Australia (1999). *Report to ATCM XXIII on Outcomes from the Workshop on Diseases of Antarctic Wildlife*, Working Paper 32, XXIII Antarctic Treaty Consultative Meeting, Lima, Peru

Brigham, L. (1999). The Emerging International Polar Navigation Code: a bi-polar relevance? In Vidas, D. (ed.), *Protecting the Polar Marine Environment: law and policy of pollution prevention*. Cambridge University Press, Cambridge, United Kingdom

Chile (1996). *Relationship between the Protocol on Environmental Protection to the Antarctic Treaty and Other International Environmental Protection Treaties*. Working Paper 30, XX Antarctic Treaty Consultative Meeting, Utrecht, The Netherlands

CMS (1997). Convention for the Conservation of Migratory Species of Wild Animals. *Proceedings of the Fifth Meeting of the Conference of Parties*, Geneva, Switzerland, 10-16 April 1997

Griffiths, F. (1999). Environment in the US Discourse on Security: the case of the missing Arctic waters. In Østreng, W. (ed.), *National Security and International Cooperation in the Arctic – the case of the Northern Sea Route.* Kluwer Academic Publishers, Dordrecht, The Netherlands

Haarde, G. H. (1997). International cooperation and action for the Arctic environment and development: an overview of parliamentarian efforts. In Vidas, D. (ed.), *Arctic Development and Environmental Challenges*. Scandinavian Seminar College, Copenhagen, Denmark

Heininen, L. and Langlais, R. (eds., 1997). *Europe's Northern Dimension: The BEAR meets the south*. University of Lapland, Rovaniemi, Finland

IUCN (1996). De Poorter, M., and Dalziell, J. C. (eds.). *Cumulative Environmental Impacts in Antarctica: Minimisation and Management*. IUCN, Gland, Switzerland

IWC (1998). http://ourworld.compuserve.com/homepages/iwcoffice/

Kirkenes Declaration (1993). *Declaration on Cooperation in the Barents Euro-Arctic Region*, Conference of Foreign Ministers, Kirkenes, Norway, 11 January 1993

Lipponen, P. (1997). The European Union needs a policy for the Northern Dimension. In Heininen, L., and Langlais, R. (eds.), *Europe's Northern Dimension: The BEAR meets the south.* University of Lapland, Rovaniemi, Finland

Nordic Council of Ministers (1995). *Cooperation in the Arctic Region.* Nordic Council of Ministers, Copenhagen, Denmark

Norway/UK (1998). *Report of the Antarctic Protected Areas Workshop*. XXII ATCM/WP 26. Tromso, Norway

NZ (1997). *Subantarctic Islands Heritage: Nomination of the New Zealand Subantarctic Islands by the Government of New Zealand for inclusion in the World Heritage List*. Department of Conservation, Wellington, New Zealand

Ojala, O. (1997). Environmental actions in the Barents Region. In: Heininen, L. and Langlais, R. (eds.), *Europe's Northern Dimension: The BEAR meets the south.* University of Lapland, Rovaniemi, Finland

Peru (1999). *Report of the Second Workshop on Antarctic Protected Areas*, Working Paper 37, XXIII Antarctic Treaty Consultative Meeting, Lima, Peru

SCAR/COMNAP (1996). *Monitoring of Environmental Impacts from Science and Operations in Antarctica*. SCAR/COMNAP, Oslo, Norway

Schram Stokke, O. and Tunander, O. (eds., 1994). *The Barents Region: Cooperation in Arctic Europe.* Sage, London, United Kingdom

UK (1996). *The Relationship between the Protocol on Environmental Protection to the Antarctic Treaty and Other International Agreements of a Global or Regional Scope*. XX ATCM/WP 10 (Rev. 1). Utrecht, The Netherlands

UNEP (1997). *Register of International Treaties and Other Agreements in the Field of the Environment - 1996*. United Nations Environment Programme, Nairobi, Kenya

US (1998). *Report of the Depository Government of the Antarctic Treaty and its Protocol*. Information Paper 74, XXII Antarctic Treaty Consultative Meeting, Tromso, Norway

US (1999). *Report of the Depository Government of the Antarctic Treaty and its Protocol (USA) in accordance with Recommendation XIII-2*, Information Paper 104, XXIII Antarctic Treaty Consultative Meeting, Lima, Peru

US DOS (1994). *Handbook of the Antarctic Treaty System*, 8th edition, April 1994. US Department of State, Washington DC, United States

Chapter Four

Future Perspectives

Future Perspectives

KEY FACTS

- The key environmental issues in the 21st century may result from unforeseen events and scientific discoveries, sudden, unexpected transformations of old issues, and well-known issues that currently do not receive enough policy attention.
- A survey of emerging issues carried out among scientists for *GEO-2000* cited pollution and scarcity of water resources (57 per cent) and climate change (51 per cent) as major issues. Then came deforestation/ desertification (28 per cent) and problems arising from poor governance at national and international levels (27 per cent).
- The survey identified six issues cited with similar frequency by respondents from all regions: freshwater scarcity, environmental pollution (mainly chemical), invasive species, reduction in human immunity and resistance to disease, fisheries collapse and food insecurity.

The region-specific alternative policy studies carried out for *GEO-2000* show that:

- Business-as-usual will not lead to sustainability.
- There is a clear need for integrated policies.
- Market-based incentives, particularly subsidy reforms, are important in all regions.
- Environmental institutions in most regions are weak and plagued with limited mandates and power, small financial resources and few human resources.
- A main obstacle to successful policy implementation is the low priority generally afforded to the environment.

The importance of the future

For most of history, the capacity of human beings to affect the environment was limited and local, though the massive projects of ancient civilizations and the denuding of lands around the Mediterranean show what was possible even in pre-industrial times. Today, human activities have grown to the point where they affect many of the large-scale physical systems of the planet. Present day actions also have consequences that reach far into the future. The impact of present policies, for example on energy and infrastructure, will extend well beyond the lifetimes of those who initiated and implemented these projects. Despite their inherent uncertainties, projections and forecasts are becoming increasingly popular as a basis for decision making. Conversely, the 'future' is playing an increasing role in the present. The future impacts of today's decisions are becoming more and more prominent in current-day policy making.

This chapter looks at the environmental issues that will require priority attention in the 21st century. It assesses which old problems are rapidly becoming worse or for which solutions are becoming more and more difficult. It looks at issues that currently do not receive enough policy attention. And it identifies a number of key issues for the future through a series of alternative policy studies, carried out region by region, for the *GEO-2000* report.

Issues for the 21st century

The environmental issues that may become priorities in the 21st century can be clustered in the following groups:

- unforeseen events and scientific discoveries;
- sudden, unexpected transformations of old issues; and
- well-known issues to which the present response is inadequate – although their long-term environmental consequences are well known.

Unforeseen events and scientific discoveries

The huge increase in environmental research over recent decades has made the possibility of sudden and unexpected surprises about the environment less probable. Many hypotheses on possible future problems have already been analysed in detail or are under continuous investigation. However, the northern bias of this research means that an environmental issue that no one has predicted, foreseen or studied could emerge in the less developed regions. The best guarantee against unforeseen events is the stimulation of scientific research and the application of current knowledge through policy-oriented assessments.

In the past, several unforeseen environmental issues have been brought to light by the scientific community. The best known recent example is probably stratospheric ozone depletion caused by emissions of CFCs and other ozone-depleting substances. The phenomenon was not discovered until 1974 and it took until 1985 – when the presence of the Antarctic ozone hole was discovered – for it to be accepted as a major international issue. Similarly, acid rain was not foreseen as one of the results of industrial expansion, and its discovery in the 1960s – including the first observations of its impact – was one of the critical events that led to the Stockholm Conference on the Human Environment in 1972.

The oceanic flip-flop theory suggested by several scientists since the early 1960s (see, for example, Broecker 1987) is another example. Global warming, it is argued, could interrupt the system by which cold, salty water in the North Atlantic periodically sinks to the ocean floor, a mechanism that is vital to the general circulation of the oceans and particularly to the Gulf Stream that warms much of Europe. If global warming caused increased rainfall or reduced wind speeds over the North Atlantic or led to the melting of freshwater glaciers in Greenland, salt concentrations in surface waters could fall, leading to less mixing of surface and deep waters. This would interrupt the flow of the Gulf stream, bringing a cooler climate to northern Europe. More recently, Broecker (1997) has suggested that these effects could turn off the deep-ocean conveyor belt completely, triggering an ice age. Evidence for such flip-flops has been found in geological records obtained from ice cores and deep-sea sediments. Of particular concern is the fact that these events have occurred over time periods as short as four years. Broecker refers to the oceans as the Achilles' heel of the climate system.

Current research may well bring to light other unexpected consequences of the increasing human manipulation of nature and biological processes. The possible effects of accidentally or intentionally introducing genetically-modified organisms (GMOs) on the gene pools, survival and overall health of wild populations of cultured species is an active area of research. While GMOs are expected to be widely used in the United States by the year 2000 in crops such as soya and maize, in other regions there is serious concern about the risks involved, and their commercialization has been postponed until more is known about possible impacts.

The rapid evolutionary nature of microbes, viruses and some insects is another area where surprises could be in store. Similarly, the enormous disruption that chemicals can have on ecosystems and human health is now well known. But every year many new chemicals are brought into circulation. The fact that many are introduced without sufficient research into their impacts is a major worry, as is the potential impact of mixtures of chemicals of which we currently have little understanding. More research into the whole range of chemical issues, including recently recognized topics such as endocrine disrupters, is needed.

Unexpected transformations of old issues

Many of the issues that will require priority attention in the next century will be aggravated forms of today's issues. Many of these continue to evolve and broaden in response to changing socio-economic, cultural and environmental conditions, although they are becoming better understood through increasing scientific and technical knowledge.

A classic example is the chemical time bomb

Environmental surprises since 1950

Environmental surprises can result from many causes including unforeseen issues, unexpected events, new developments, changes in trends and shifts in environmental perception.

Unforeseen issues

New issues rarely appear without warning; however, their lack of frequency does not undermine their importance.

- *CFC-induced ozone depletion.* Stratospheric ozone depletion was unknown at the time of Stockholm. Chlorofluorocarbons (CFCs), which were thought to be chemically inert and harmless to the environment, are now recognized as the primary cause of stratospheric ozone depletion. CFC-induced ozone depletion, first hypothesized in 1974, gained some public attention as a result of an article in the *New York Times* the same year but became an alarming issue worldwide following the discovery of a large stratospheric ozone 'hole' over Antarctica in 1985.

Unexpected events

Some events may be well known but their severity, timing and location unanticipated. As populations expand and industrial activity intensifies, these events may become more common or more serious.

- *Oil spills* such as the *Torrey Canyon* (1967), *Amoco Cadiz* (1978) and *Exxon-Valdez* (1989) and through war, as in the Persian Gulf in 1991
- *Accidental poisonings and toxic chemical events*: methylmercury poisoning, Minimata, Japan (1959); PCB poisoning (Itai-Itai), Kyushu, Japan (1960s); dioxin leak, Seveso, Italy (1976); methyl isocyanate leakage, Bhopal, India (1984); chemical warehouse fire, Basel, Switzerland (1986)
- *Severe smog and air pollution events*: London (1952), Indonesia forest fires (1997)
- *Nuclear accidents*: Urals (1958); Three Mile Island (1979); Chernobyl (1986)
- *Biological Invasions*: zebra mussel, Great Lakes (1980s); *Mesquite* trees, intentional introduction with unexpected agricultural effects, Sudan (about 1950)

New developments

Existing issues may be highlighted by new developments or findings or by media involvement.

- *Pollution*: *Silent Spring*, the 1962 publication about effects of pesticides and herbicides; discovery of toxic substances in Love Canal, New York, United States; discovery of contaminants in the Arctic, far from areas of use (continuing)
- *Acid rain*: the Swedish Case Study presented at Stockholm in 1972 stimulated international concern and action
- *Climate change*: media sensationalized hot summer of 1988 in North America as evidence of global warming which stimulated scientific research and increased public awareness; the first IPCC Scientific Assessment in 1990 introduced new perspectives on climate change
- *Tropical deforestation*: satellite images of tropical deforestation in the 1980s vividly demonstrated the extent of the loss of biodiversity
- *Consequences of resource management*: ecological, social, economic and health impacts of the Aswan Dam and the shrinking of the Aral Sea from water diversion for irrigation

Changes in trends

Deviations (real or perceived) from the expected course of events can be ecologically and economically harmful.

- *Climate change*: increased occurrence of severe weather events and *El Niño*
- *Resource depletion*: oil crisis from perceived oil depletion (1970s); Atlantic cod fishery collapse (1990s)

Shifts in environmental perception

The way in which we view environmental issues changes, often with remarkable rapidity.

- 'Earthrise' photo from Apollo II (1969) instilled perception of the Earth as a fragile, unified ecosystem
- Focus of issues moved from local to national to international due to increased awareness of transboundary nature, or spatial extent of issues such as marine and air pollution, habitat loss
- Focus on connections: population, pollution and resource depletion, once viewed as independent issues, now seen as related and in the broader context of sustainable development, climate change and biodiversity loss.

Prepared by Sarah Kalhok and Glynn Gomes,
University of Toronto, Canada

(Stigliani 1991). Chemicals, either produced naturally or as a result of industrial and agricultural activities, tend to accumulate slowly and harmlessly over many years in soils, sediments, lakes and other environmental reservoirs. However, when the carrying capacity of the receiving ecosystems is finally exceeded, there can be a sudden release of the chemical. Alternatively, the chemical may be released because of changed environmental conditions, as happens when harbours are dredged and wetlands drained. The environmental consequences of chemical time bombs can be severe, as the following examples show:

- Acid rain falling into a lake may have no effect on the pH of the water over a decade or longer. Quite suddenly, however, the buffering capacity of the lake may be exhausted, the pH may drop precipitously, heavy metals may be mobilized, and aquatic life may be seriously impaired. The first observed case of rapid acidification after a long time delay was in Big Moose Lake in the Adirondack Mountains, United States (NRC 1984).
- Acid rain may also lower the buffering capacity of soils. If soil pH drops below 4.2, naturally-occurring aluminium may be mobilized, posing a threat to forests and water courses.

- If a wetland dries up, for whatever reason, the area may become a source of, rather than a sink for, toxic substances. Anaerobic conditions will be replaced by aerobic ones, and immobile sulphide compounds will be oxidized to sulphates, lowering the pH and mobilizing metals such as iron and aluminium. This phenomenon was first documented by Renberg (1986) in a wetland in Sweden but acid sulphate soils of this type occur mainly in tropical coastal wetlands, particularly in Southeast Asia.
- Both active and abandoned coal mines sometimes fill with water, which becomes highly acidic, thus mobilizing heavy metals. During heavy rains, the water may overflow, and downstream water courses may become seriously polluted (Robb 1994).
- Heavy metals may be released suddenly into the environment by a breach in toxic waste stores. In 1998, a waste storage lagoon burst at the Los Frailes zinc mine in Spain, sending a toxic wave down the Guadiamar river towards the Doñana National Park. Coastal toxic dump sites could become hazardous to surrounding communities during a storm surge or if sea level were to rise.

Another recent example of a 'transformation in action' is the current surge in number and severity of forest fires and natural disasters. Forest fires have raged periodically throughout history. However, in the last couple of years, due to a variety of factors, including human activities, the prevailing weather and severe degradation of natural resources, their frequency and intensity seems to have increased (see page 31), particularly in the Amazon and Southeast Asia. Addressing these and other disasters, such as flooding, may well become an environmental priority in the decades to come.

A further example of a contemporary issue becoming more serious is coral bleaching. First described nearly 80 years ago, coral bleaching occurs when corals expel the symbiotic algae that live in their tissues. This is a response to environmental stress, in particular high sea temperatures but also high solar radiation, fluctuating salinities, extremely low tides and often a combination of these factors (ISRS 1999, Pomerance 1999 and ITMEMS 1998).

In the mid-1980s, coral bleaching began to occur on a large scale. In 1998, coral bleaching was more severe than ever before and occurred in at least 60 countries (ISRS 1999 and ITMEMS 1998). Although the links between global climate change, *El Niño* phenomena and extensive coral bleaching are still subject to debate (ISRS 1999), it has been suggested that only global warming could have induced such extensive bleaching simultaneously throughout the disparate reef regions of the world (Pomerance 1999). What permanent effect the alarming 1998 event will have remains to be seen.

Similarly, the incidence of biological invasions by non-indigenous species seems to be on the increase. First identified as a problem in the 1950s when the *mesquite* tree was introduced into Sudan with quite unexpected agricultural results, more recent and even more costly effects have resulted from the inadvertent introduction of the zebra mussel into the Great Lakes area of North America (see page 145) and the water hyacinth which now clogs many of Africa's waterways (see page 61), and for which no solution is yet in sight. Invasive species are the second leading cause of biodiversity loss after habitat destruction (UNEP 1995). Continued globalization, and increases in travel and trade, might well make such invasions more common in the future and in need of even greater international attention.

Neglected issues

Most issues that will require policy attention in the next century are, however, issues that are currently existing and well known. As time goes on they will become more severe and pose major local and global challenges. If these challenges are not addressed, they will give rise to major environmental crises in the 21st century. As such, they are emerging due to lack of avoiding actions. More effort is also needed to understand the mechanisms through which emerging issues become issues for policy. Social and political processes, science and the trend towards public involvement all have a role to play, at least in some countries.

There are numerous examples from the past. Increased and accelerating emissions of carbon dioxide have led to the climate change issue; the continued intensification of fishing activities has led to the collapse of fisheries in many seas, and the relentless pace of urbanization has creating a series of problems for local authorities in developing and developed countries alike. A classic example of an environmental disaster caused by lack of action is the fate of the Aral Sea. Policy makers were well aware that continued and uncontrolled water abstraction for irrigation would lead

Global and regional trends likely to worsen in the next century

Nitrogen overload
We are fertilizing the Earth on a massive scale. Specific impacts are being studied but we are still largely in the dark about the overall effects of this huge disruption of the nitrogen cycle. Additional synergistic effects between major biogeochemical cycles (nitrogen and carbon, for example) and human activities are still very uncertain.

Environment-related disasters
Some natural, some exacerbated by human activity, disasters are becoming more frequent and severe, killing and injuring millions of people every year and causing severe economic loss.

Degradation of coastal areas and their resources
Fisheries have been grossly mismanaged; coastal land suffers from poorly planned and regulated urbanization, industrialization, aquaculture, tourism, port development and flood control; and nearshore coastal waters continue to deteriorate. Resource exploitation, changes to habitats and disruption of ecosystem functions probably pose more serious threats to many marine and coastal areas than pollution.

Chemicals
Chemical compounds that persist in the environment and which can affect the health and reproduction of organisms at the molecular or the reproductive level are now considered to be a much greater global problem than many of the 'old' poisons, such as lead, where adverse effects and the measures needed to reduce them are now reasonably well understood.

Species invasions
The deliberate and accidental introduction of non-indigenous species is increasing. Through competition, pathogenic impacts and other mechanisms, many indigenous organisms have been severely threatened by exotic species.

Climate extremes
The year 1998 was the warmest year on record, and the 20th consecutive year with an above-normal global surface air temperature. The 1997/98 *El Niño* was the most powerful on record. Is this setting the scene for things to come?

A looming global water crisis
Increasing water stress, especially in low-income populations.

Land degradation
Increased vulnerability of land to water-induced erosion, especially where marginal land conversion is widespread.

Urbanization
Soon half the world population will be urbanized. Where urbanization is uncontrolled or badly managed, it creates many environment-related problems, including waste disposal and a range of chronic health impacts.

Environmental importance of refugees
Refugees are forced to make unrestricted assaults on the natural environment for their survival. Refugee numbers reached an all-time high of 27.4 million in 1995 (UNHCR 1998).

Vulnerability of Small Island Developing States
Characterized by their remoteness, insularity, fragile ecosystems, lack of natural resources and high dependency on imports, high coast-to-inland ratio, and small physical and economic size, these states are particularly vulnerable to forces beyond their control – including global warming, natural hazards, shortage of fresh water, coastal threats and the vagaries of energy supplies.

to the death of the Aral Sea. They could, however, find no other way of meeting the economic imperatives of the time than through ignoring the problem.

In *GEO-1*, a list of fundamental global issues that threaten long-term sustainability was derived from the analysis of regional and global trends. These issues still stand and will, if not urgently addressed, become even bigger problems in the 21st century:

- The use of renewable resources – land, forest, fresh water, coastal areas, fisheries and urban air – is beyond their natural regeneration capacity and therefore is unsustainable.
- Greenhouse gases are still being emitted at levels higher than the stabilization target internationally agreed under the United Nations Framework Convention on Climate Change.
- Natural areas and the biodiversity they contain are diminishing due to the expansion of agricultural land and human settlements.
- The increasing, pervasive use and spread of chemicals to fuel economic development is causing major health risks, environmental contamination, and disposal problems.
- Global developments in the energy sector are unsustainable.
- Rapid, unplanned urbanization, particularly in coastal areas, is putting major stress on adjacent ecosystems.
- The complex and often little understood interactions among global biogeochemical cycles are leading to widespread acidification, climate variability, changes in the hydrological cycles, and the loss of biodiversity, biomass, and bioproductivity.

GEO-2000 draws attention to additional global and regional trends that are likely to get worse in the next century. These issues are summarized in the box above. Many of these environmental problems result from the continuous unfolding of socio-economic processes that have yet to be properly controlled or managed. The continued poverty of the majority of the planet's inhabitants and the excessive consumption by the minority are the two major underlying causes of continued environmental degradation. These causes, combined with rapidly changing political, social, institutional, financial and technological developments, present policy makers with intractable problems without easy or obvious solutions. The complexity and magnitude of the problems at hand should, however, not be a reason for complacency. We must not baulk at designing and implementing preventive policies today. Even if the costs seem high, they will be small in relation to the enormous risks and irreversible damage in the future associated with inaction or 'business-as-usual' paths.

Pointers for the 21st century

A survey of emerging issues

During the preparation of *GEO-2000*, a global survey on emerging environmental issues was conducted by ICSU's Scientific Committee on Problems of the Environment (SCOPE) as part of the GEO programme (UNEP/SCOPE 1999). The survey involved 200 environmental experts, including many research scientists, in more than 50 countries. The survey suggests that many of the major environmental problems expected in the next century are problems that exist now but which are not receiving enough policy attention.

The issues that were cited most frequently (see bar chart) were climate change, freshwater scarcity, deforestation/desertification and freshwater pollution. Then came problems arising from poor governance at national and international levels. The two other social issues most often mentioned related to population growth and population movements (including environmental refugees), and changing social values (principally increased consumerism and accumulation of economic wealth).

The reference to issues that are largely social in nature is particularly significant because the individuals targeted by the survey were selected for their environmental expertise. This indicates that views of environmental issues have expanded to include human dimensions in ways that would not have been typical a few years ago. At the same time, issues such as trade and environment, financing and accounting practices within the public and private sectors (including environmental accounting and valuation of natural resources) were not mentioned although experts from business, finance and

Major emerging issues identified in the SCOPE survey

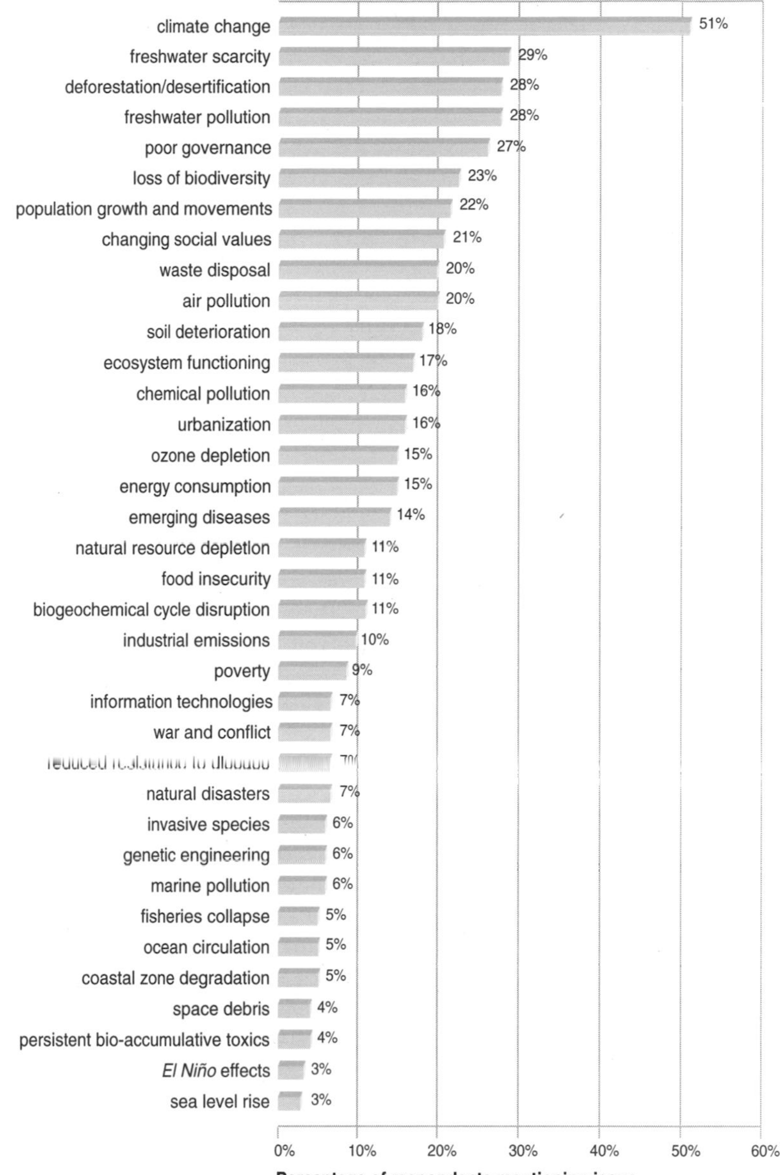

economics would almost certainly have done so. Similarly, issues such as environmental security and environmental justice were not mentioned. They probably would have been if the survey had included more environmental NGOs. The ability to link science, social and environmental data is itself a major emerging environmental challenge resulting from a broader recognition of the multi-faceted nature of environmental problems.

Chemical pollution appears relatively far down the list although it is the focus of major international negotiations. Only a few people seemed concerned

Climate change was the most cited issue in the SCOPE survey although, taken together, water scarcity and pollution ranked higher

about emerging diseases, the dangers of genetically-altered species (including their entering the wild gene pools of plants, microbes, fish and other animals), the synergistic effects of chemicals and endocrine disrupters, and the potential hazards of space debris. However, these issues should not necessarily be dismissed because they were rarely cited in the survey. Many of the environmental issues which now confront governments began with similar warnings from a few 'lone voices' in the past.

Many people emphasized that 'old' issues were being given a new significance with the recognition that they were components in a larger set of interactions among mega-issues. These complex systems demanded new ways both of studying them and of dealing with them. It was this recognition which led participants to describe some well-known problems as emerging issues. For example, climate change, which is already high on the public policy agenda, was described as an emerging issue by more than half of those consulted.

The issues cited differed from region to region. Only a few issues were mentioned with the same frequency in all regions: freshwater scarcity, environmental pollution (mainly chemical), invasive species, reduction in human immunity and resistance to disease, fisheries collapse and food insecurity. It seems that the more closely linked an issue is to social problems and processes, the more likely it is to be viewed differently in different regions, although food insecurity and freshwater scarcity had similar frequencies of citation in all regions and are closely related to social resource use.

Of the issues defined as environmental rather than social, three were mentioned more frequently as emerging problems in Africa, Asia and West Asia than elsewhere. These were air pollution (particularly urban air pollution), industrial emissions, and contamination from waste disposal. Concerns were raised about toxic waste and non-biodegradable waste, particularly plastics, which are a rapidly increasing environmental risk throughout many developing countries without appropriate landfill sites.

In contrast, experts in North America mainly described biodiversity, ocean system change, emerging diseases (including both new infectious diseases, and changing disease patterns brought about by global change), sea-level rise and space debris as key emerging issues. Concerns about biodiversity included potential interactions with climate change and with genetic engineering.

There were also strong regional patterns in the frequencies with which socially-related issues or causal factors were cited as part of the emerging environmental issues. For example, poverty was mentioned in Africa, Asia and West Asia but not in Europe, Latin America or North America. Similarly, urbanization as a driving force for environmental problems was mentioned mainly in Africa, Asia and West Asia. However, population growth and population movements around the world were most often cited in North America.

Poor governance at national and international levels was cited most often in Latin America, and relatively frequently in Africa and Asia. Changing social values towards consumerism and materialism were most frequently given as an emerging environmental issue in Europe and Asia whereas war and conflict were often cited as emerging issues in Asia. In North America, the region with the highest energy consumption per capita, experts did not point to energy use as an emerging environmental issue. This may either be because they thought that energy consumption was no longer emerging, or because they were less sensitive to the global implications of high energy use than they might be. These may also be the same reasons why nuclear power and the threat of a nuclear winter were rarely cited as an emerging environmental issue, even in North America.

Downstream environmental effects of new information and communication technologies (ICTs) were cited almost only in North America, where they are today most common. They were not seen as an emerging issue elsewhere. However, although there are currently major constraints to their adoption and use in large parts of the world, ICTs will almost certainly become as global a technology tomorrow as radio and television are today. The current rapid growth of mobile phones in many developing countries is leapfrogging the need for rural areas to be 'wired'. The establishment of Internet service providers and community 'telecentres' in poor urban and rural areas is also an emerging global phenomenon with unknown environmental implications.

A backdrop scenario for the future

Scenario analysis is another technique for exploring the future. By investigating and comparing the outcomes to which different scenarios lead, it is possible to assess current and alternative policies.

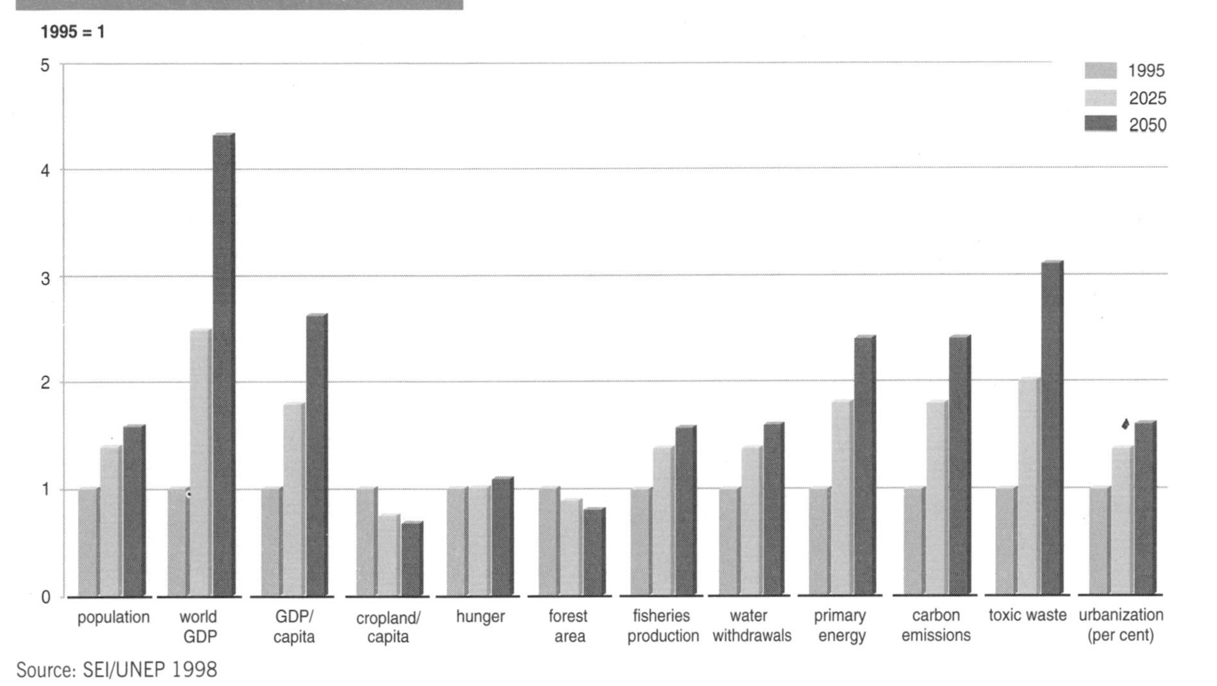

Source: SEI/UNEP 1998

The backdrop scenario suggests that the world could prosper by 2050 but hunger would remain roughly constant, and forest area and cropland/capita decrease. Fisheries production is increased entirely through aquaculture. Such a world is essentially unsustainable and would 'be a risky bequest to our 21st century descendants'

Since *GEO-1*, a new, quantitative analysis of global scenarios has explored the implications of the conventional development or 'business-as-usual' scenario in more detail (SEI/UNEP 1998).

Under this backdrop scenario, world population increases 65 per cent and economic output more than quadruples by 2050. At the same time, income per capita, expressed in purchasing power parity, would grow 2.6 times. Under these conditions, energy and water requirements are expected to increase by factors of 2.4 and 1.6 respectively, and food requirements to almost double, driven by growth in population and income. It is anticipated that sufficient food would be available globally to feed all the growing population but that hunger would remain, due to inequalities of access. Some 6 700 million people, 72 per cent of the predicted world population of 9 300 million, would live in urban areas. Despite rising average incomes, the number of people still in poverty and unable to feed themselves adequately would rise slightly, rather than decline, over the next 50 years as populations grew and traditional sources of material support eroded. Persistent poverty has negative implications both for sustainable development and for more sustainable management of natural resources.

The OECD share in world economic output, expressed in purchasing power parity, would decrease from 55 per cent in 1995 to 40 per cent in 2050. In terms of the ratio of average GDP per capita between OECD countries and non-OECD countries, a convergence is anticipated from the current [illegible] to [illegible]. However, in absolute figures the difference in purchasing power parity between OECD and non-OECD countries would increase from US$17 000 per capita in 1995 to US$47 000 in 2050.

With such a continuation of present trends in population growth, economic growth and consumption patterns, the natural environment would be increasingly stressed. Many environmental gains and improvements would be offset by the pace and scale of global economic growth, increased global environmental pollution and accelerated degradation of the renewable resource base.

Resource use, although growing less rapidly than the economy due to improved efficiencies, would put tremendous stress on non-renewable and renewable natural resources. For instance, making the optimistic assumption that the harvest of marine resources, already overexploited, remains stable at about 88 million tonnes a year, the projected increase in demand from 100 million tonnes now to 170 million tonnes by 2050 would require a substantial expansion of aquaculture, with its associated impact on mangroves and coastal zones.

Water, in particular, could prove a limiting factor for development in a number of regions; by 2050, more than 2 000 million people would live under conditions of high water stress. Land use under agriculture would rise from the current 37 per cent to 42 per cent and a further 17 per cent of forest area would be lost. Global CO_2 emissions from fossil fuel combustion would increase 2.4 times. Also, with the expanding industrial activity projected in this scenario, toxic emissions could triple by 2050 and increase nearly five-fold in developing regions, posing potential ecological and human health threats.

This backdrop scenario:

- Assumes that there will be no surprising economic, political, technological or environmental developments or crises, which would itself be surprising over a 50-year period.
- Does not consider adjustments that may be necessary to achieve long-term sustainability, such as reducing fossil fuel consumption because of climate change. The complex interplay between different pressures and trends has not been worked out, and would be difficult to predict.
- Reflects the value set of the dominant Western development paradigm in which income is the important parameter.

Such a scenario reflects what many people think the future will be like although it is not meant to be a realistic prediction. While scenarios need to be considered with caution, they do provide a framework for studies into the longer-term future and for understanding what is or is not possible.

For example, widespread policy reform could make a significant difference to these outcomes (SEI/UNEP 1998). Analysis of a policy reform scenario explored whether widespread adoption of policies that have already been proposed could shift global patterns toward a more sustainable pattern and specifically whether it was possible, with such policy changes, to reach ambitious social and environmental targets in the year 2050.

The analysis suggests that more equitable growth patterns – both between regions and within countries – are crucial to reducing poverty and achieving other social goals. This will require a range of policies and structural changes that help increase the incomes of the poor. Under such a scenario, population growth also moderates. Policies that could lower greenhouse gas emissions and achieve other environmental goals are also well known, if not yet widely implemented: for example, energy policies that favour rapid introduction of cleaner technology and fuel switching, regulatory standards and removal of institutional barriers. Meeting global food needs in ways that respect both social and environmental goals requires a range of measures to encourage sustainable farming practices, significant increases in water-use efficiency, and measures to prevent additional land degradation and forest loss. The analysis concludes that appropriate policies could make a major difference. The real question is whether the necessary awareness and change of political attitudes will come through an act of voluntary will or only after some environmental catastrophe.

The alternative policy studies

The global scenarios just described form the backdrop for a number of regional studies conducted for *GEO-2000*. Although varied in scope and detail, these studies deepen our insight into some of the apparent shortfalls and pitfalls of current policy, and what can be done to address the key issues on a regional basis. The GEO Collaborating Centres in each region selected one or a few of the major environmental problems for analysis. The methodology used (see box below) encompassed the economic, institutional, social and environmental aspects of each problem.

All the studies concluded that a continuation of business as usual would not lead to the required results – environmental protection and the quality of

Methodology used in the alternative policy studies

Each regional study comprised six steps.

- Define scope of the study and the primary policy question to be answered (for example 'what can be achieved by moderate additional measures and will the achievement be enough?').
- Define a reference scenario to describe likely social and economic developments up to 2010, including consistent projections of the key driving forces under current policies; the purpose of the reference scenario is to describe what could happen without alternative or additional policies.
- Estimate impacts of the reference scenario in terms of selected environmental issues.
- Define alternative policy packages, focusing on physical measures (such as fuel switching), the policy instruments needed to achieve them (such as taxation), or both.
- Estimate changes in the impacts caused by the alternative policies and compare them to those of the reference scenario.
- Draw conclusions about the effectiveness of the alternative policy packages.

Environmental focus of the region-specific alternative policy studies

Asia and the Pacific	Air pollution
Africa	Land and water resource management
Europe and Central Asia	Energy-related issues
Latin America	Use and conservation of forests
North America	Resource use, greenhouse gas emissions
West Asia	Land and water resource management

life needed for a sustainable future. The studies identified sets of alternative policies which – if implemented immediately and pursued with vigour – could indeed adjust trends in the regions towards a more sustainable course. However, even some of the more positive scenarios produce results which fall short of acceptable limits.

The table on the left lists the environmental issues on which these studies focused. The table below summarizes the policies on which the studies focused. All the studies included some analysis of the role of market-based incentives. Institutional aspects and the promotion of new technologies formed part of most studies. The studies themselves are briefly summarized below and described in more detail at the end of this chapter (see page 346).

Africa

The studies for Africa focus on different sub-regions – but all indicate the negative environmental consequences of a business-as-usual scenario driven by demographic change and slow economic growth: loss of biodiversity, land shortages for agriculture and greater susceptibility to natural disasters. Proposed policy options differ between the different sub-regions – but can be generally defined as integrated land and water management in combination with an effective population strategy. Alternative policies regarding land

Policy focus of the region-specific alternative policy studies

	Africa	Asia and the Pacific	Europe and Central Asia	Latin America and Caribbean	North America	West Asia
Laws and institutions						
Constitutions, institutions	■	■	□	■	■	■
Planning	■	■	□	■	□	■
Policy instruments						
Command and control	□	□	□	□	□	■
Promotion of new technologies	■	■	■	□	□	■
Env. information and education	■	□	■	□	□	■
Voluntary action by private sector	□	□	■	□	■	□
Widespread public participation	■	□	□	□	□	□
Enforcement monitoring	□	□	□	□	□	■
Other sectors						
Market-based incentives	■	■	■	■	■	■
Capital flows	■	□	■	■	□	□
Trade policies	■	□	□	■	□	■
Social policies	■	□	□	□	□	□
Regional cooperation	■	□	■	□	□	□

resources hinge on reform of tenurial rights; above all, at the political level, the controversial character of land tenure reform calls for transparency and broad participation. Intra-regional cooperation – for instance with respect to water resources – is also important.

Asia and the Pacific

Air pollution is expected to increase considerably in most countries in the region. In addition, acid deposition is becoming increasingly problematic. Under business-as-usual conditions, regional emissions of sulphur dioxide are expected to increase fourfold by 2030 over those of 1990; emissions of nitrogen oxides are expected to increase threefold. The alternatives focus on clean technology, increasing energy efficiency, and fuel switching. Fuel switching needs to be carefully adapted to the situation in each country. When combined with the other options, fuel switching could reduce the 2030 emission of sulphur oxides to below the 1990 level, and limit the increase of nitrogen oxides to 40 per cent. The study shows that the technology to reduce environmental pressures in the region to sustainable levels is actually available; however, capital will be required to make the necessary changes politically and financially feasible.

Europe and Central Asia

The study for Europe and Central Asia focuses on transport and electricity use as important drivers for environmental problems across the region. Notwithstanding cleaner vehicles, air pollution in cities and transport corridors will increasingly be dominated by mobile sources, making urban air pollution and ground-level ozone persistent problems. In the western parts of the region, the situation is likely to improve with regard to acid deposition and transboundary air pollution but not quite enough to achieve policy targets with respect to the protection of ecosystems. The toughest problem is the emission of greenhouse gases. The study concludes that the technical potential is available to meet the region's Kyoto commitments. The most cost-effective way is obviously by trading emission rights across the region. Integrating policies for different environmental problems could also reduce total costs.

Latin America and the Caribbean

The Latin American study shows that under current conditions deforestation is likely to continue to be driven by expansion of the agricultural sector, demographic pressure, logging and inequitable land distribution. So far, forestry policies in the region have not been effective, mainly because they have failed to take account of the differing needs of different forest users. Many more promising policy options are available, including direct control of government-owned forests and indirect control using fiscal incentives in the form of taxation, subsidies and forest credits, and other incentives such as the granting of private property rights, market reforms, the introduction of community forestry schemes, and improvements in extension, research and education. Packages of these policies could reduce deforestation rates, forest fires, numbers of threatened animal and plant species, and regional carbon dioxide emissions; slow down agricultural expansion onto forest land; improve forest ecosystem health, the quality of urban and rural life, and regional and local economies; and provide appropriate technologies to forest dwellers as a tool for sustainable development.

North America

The North American study deals with policies that alter fiscal incentives – reducing or eliminating environmentally-perverse incentives and increasing incentives for constructive change. Subsidies on resource use in the United States in 1996 were estimated to be about US$30 000 million for energy and probably more than US$90 000 million a year for transport. Agricultural producer support in 1998 amounted to about US$47 000 million. Reducing or eliminating direct and indirect subsidies for road transport, energy use, grazing and timber production could play a significant role in environmental improvement. The potential benefits of such reforms include reducing traffic congestion, improving urban air quality, increasing competitiveness, and slowing the increase of carbon dioxide emissions to meet climate goals.

West Asia

The study for West Asia looks at the management of water and land resources. It shows that a business-as-usual development will leave the region with serious water shortages – in particular in the Arabian Peninsula where the annual water deficit could increase to as much as 67 per cent of demand by 2015. In fact, it is clear that current water resources cannot satisfy future water demand much past 2005 without alternative policies. Two alternative policies are examined: in one, water supplies are increased, and in the other water supplies are increased and policies altered. The scenarios show that the water deficits could be reduced (though not eliminated) and that substantial savings could be made by giving priority to domestic and industrial water uses over the need for irrigation in agriculture.

Conclusions from the regional studies

A number of key conclusions emerge from the alternative policy studies:

- There is a clear need for integrated policies. For example, in Latin America a broad inter-sectoral approach is needed to achieve sustainable forest development. In Europe and Central Asia, combined strategies to deal with acidification, urban air pollution and climate change could lead to an optimal use of opportunities for energy efficiency and fuel switching. However, even integrated policies come up against such well-known and difficult issues as resource pricing, land tenure and financial rigour;
- Market-based incentives, particularly subsidy reforms, play a role in all regions. Although subsidies have many purposes and can act as a lifeline for the poor and small-scale enterprises in developing countries, they can become perverse, encouraging uneconomic practices, straining public budgets and leading to severe environmental degradation. Reform of perverse subsidies can encourage the more efficient use of resources such as energy, and thus help reduce pollution and degradation. Careful implementation is essential. For example, the first steps towards reform may simply involve greater public transparency on subsidies. Other studies have highlighted the possibility of leaving subsidies in place but not linking them to levels of resource use;
- Institutional aspects are stressed in most regions. While progress has been made in working out environmental legislation on paper and setting up environmental institutions, implementation lags far behind intention. Most of the institutions are weak and plagued with limited mandates and power, small financial resources and few human resources;
- The scope and content of the alternative policies correlate with level of economic development. For example, the African studies place major emphasis on the social aspects of environmental problems, including poverty alleviation, birth control, property rights, education, public participation and application of indigenous knowledge systems. At the other end of the spectrum, the developed regions focus more on the economic and fiscal aspects of environmental issues;
- A main obstacle to successful policy implementation is the low priority generally afforded to the environment. In Africa as well as in Eastern Europe and Central Asia, attention is repeatedly drawn to the simple but crucial point that environmental management usually needs financing. But money is always scarce. In many countries, issues such as housing, education and health care take priority over the environment when allocations are made.

References

Broecker, W. S. (1987). Unpleasant surprises in the greenhouse? *Nature* 328, 123-126

Broecker, W. S. (1997). Thermohaline circulation, the Achilles heel of our climate system: will man-made CO_2 upset the current balance? *Science* 278, 1582-1588

ISRS (1998). *Statement on Global Coral Bleaching in 1997–98.* International Society for Reef Studies, Florida Institute of Oceanography, Florida, United States

ITMEMS (1998). *Statement on Coral Bleaching.* International Tropical Marine Ecosystems Management Symposium, 24 October 1998, Townsville, Australia

NRC (1984). *Acid Deposition: Processes of Lake Acidification.* National Academy Press, Washington DC, United States

Pomerance, R. (1999). *Coral Bleaching, Coral Mortality and Global Climate Change.* Report to the US Coral Reef Task Force Meeting in Hawaii, 5-6 March 1999

Renberg, I. (1986). Diatoms and Lake Acidity: reconstructing pH from siliceous algal remains in lake sediments. *Developments in Hydrology* 29, Junk Publications, Dordrecht, The Netherlands

Robb, G. A. (1994). Environmental consequences of coal mine closure, *Geograph. Journal* 160, 33-40

SEI/UNEP (1998). Raskin, P., Gallopin, G., Gutman, P., Hammond, A., and Swart, R. *Bending the Curve: Toward Global Sustainability*. PoleStar Series Report No. 8, Stockholm Environment Institute, Stockholm, Sweden, and UNEP/DEIA/TR.98-3, UNEP, Nairobi, Kenya http://www.gsg.org/

Stigliani, W. M., (1991). *Chemical time bombs: definitions, concepts, and examples.* International Institute for Applied Systems Analysis, Laxenburg, Austria

UNEP (1995). *Global Biodiversity Assessment.* Edited by R. T. Watson, V. H. Heywood, I.Baste, B. Dias, R. Gamez, T. Janetos, W. Reid and R. Ruark. Cambridge University Press, Cambridge, United Kingdom

UNEP/SCOPE (1999). *Emerging Environmental Issues for the 21st Century: a study for GEO-2000.* Scientific Committee on Problems of the Environment, Paris, France, and UNEP, Nairobi, Kenya (UNEP/DEIA&EW/TR.99-5)

UNHCR (1998). *State of the World's Refugees, 1997-1998: A Humanitarian Agenda.* United Nations High Commissioner for Refugees, Geneva, Switzerland http://www.unhcr.ch/sowr97/statsum.htm

The alternative policy studies

Africa

Scenarios for the future in Africa are available from six existing studies, published between 1989 and 1997. These scenarios have been reviewed for *GEO-2000* in a technical report and they are summarized in the table opposite. None of these scenarios deals with the whole continent and each focuses on sub-regions.

All these studies include a 'business-as-usual' scenario driven by demographic change, particularly population growth and migration, and lacklustre economic development. These scenarios suggest a bleak future for human and natural environments in Africa. The indicators are similar in all the scenarios: slow or stagnant economic growth, shortages of agricultural land, ineffective governance and institutions, loss of biodiversity and susceptibility to natural disasters. Most reports also include a sustainable scenario, a future that is largely dependent on a resurgence of African culture, human resource development (especially with regard to education), outreach programmes (especially those focused on land management and sustainable agricultural practices), and public participation in the overall development process.

A key variable distinguishing African success scenarios is the extent of economic autonomy of African countries and global markets. Some of the scenarios see African 'delinkage' from these markets as essential to building strong and self-reliant institutions. Others, such as the World Bank scenarios, see African take-off as closely related to the capacity of the continent to enter the global economy through macro-economic adjustment, trade liberalization and better governance.

Another key variable is the level of intra-regional cooperation. Some scenarios suggest that water- and food-sharing arrangements between nations are critical for a desirable future; others place greater emphasis and confidence on scientific and technological progress to meet basic human needs.

Policy options regarding land and water in Africa have been re-analysed for *GEO-2000*, sub-region by sub-region, in a qualitative fashion (UNEP 1999a).

The *GEO-2000* analysis for Northern Africa focuses on whether the current policy frameworks for water and land use are likely to lead to the attainment of food security objectives by 2015. Findings indicate that, under current policies, water stress levels will rise to critical levels, at best straining regional food security and some national balances of payments. Continuing with business as usual implies that conflicting claims for water will increase, not only across boundaries but also between sectors within countries. Further projected impacts include accelerated land degradation and negative impacts on employment and population health, especially of the poor. Alternative policies are generally defined as integrated land and water management in combination with an effective population growth strategy. More specifically, alternative land and water policies include: regulated access to resources; further reform of subsidies on agricultural inputs; and economic diversification away from dominant resource-intensive agriculture. In particular, accepting intra-regional trade as a means to achieve regional food security is identified as a key policy to avoid the most extreme stresses as projected in the business-as-usual scenario.

For West and Central, as well as East and Southern Africa, the analyses focus on the demand to reform land and water tenure systems. This is a controversial issue, particularly with regard to land.

For West and Central Africa, key pressures on land under a business-as-usual scenario relate to: the difficulties of rain-fed agriculture and forestry, and especially the timber industry, to cope with climate variability; increasing deforestation; increasing land fragmentation and land insecurity affecting land use practices; demographic pressures, with a present population growth rate of nearly 3 per cent, and one-half of the population below 15 years of age. The resulting land degradation results in deteriorating input-to-output levels in agriculture, shortage of fuelwood, landlessness and increased migration to urban areas and other farming areas.

For East and Southern Africa, key pressures increase as a result of growth in export-oriented agriculture and tourism; population growth; migration to urban areas and into marginal land. In Southern Africa, migration to rural areas, following the relaxation of labour regulations, is a further trend. It is anticipated that current pressures will lead to more degradation and fragmentation of land resources, resulting in lower agricultural productivity and worse food security. In addition, food production for an increasingly urbanized population may demand three times as much water for irrigation by 2020 in this sub-region – an indication of increasing inter-sectoral competition.

Alternative policies on land resources hinge on reform of tenurial rights. At the same time, the available material emphasizes a number of points regarding the design and implementation of reform. Above all, at the political level, the controversial character of land tenure reform calls for transparency and broad participation (as emphasized in the Arusha

African scenario studies

Study	*Region*	*Horizon*	*Scenarios*
IIED (1997)		2015	**Doomsday Scenario:** rapid population growth, economic stagnation, conflict, and ineffective governance lead to deteriorating environmental conditions. **Building a Sustainable Future:** new development approach based on sustainability principles and cooperation leads to improved socio-economic conditions, peace, and a clean environment. **Real Future:** a compromise containing elements of both of the above.
World Bank (1996)		2025	**Looking Thirty Years Ahead:** Africa joins the information society but land and food shortages cause population migration and shift environmental burdens to urban centres.
SARDC (1996)		2020	**Current Trends:** environmental conditions deteriorate in the context of economic stagnation. **Desired Future:** policies mobilize region's physical and human resources through vigorous research, education and institution building.
Club du Sahel (1995)		2020	**Laissez-faire:** trade continues to lead to cheap imports, not economic diversification, as social inequity rises, international support wanes, and social breakdown is a threat. **Orthodox Growth:** good governance guides market towards development of new competitive sectors, international investments, and support. **Regional integration:** supports local development of small enterprises, modest economic growth, but stronger regional ties, and less tension and conflict.
Beyond Hunger (1989)		2057	**Current Perspective:** slow economic growth with increasing population leads to uncertain environmental conditions. **Big Lift:** a distinctively Afro-centric development process leads to economic independence and a clean environment.
Blue Plan (1989)		2035	**Trend scenarios:** focus on macro-economic success and a laissez-faire policy towards population growth jeopardizes economic and social development and leads to environmental deterioration. **Alternative scenarios:** goal-oriented development policies focus on domestic objectives leading to a cleaner environment.

Sources: Achebe and others 1995, Dalal-Clayton 1997, OECD 1995, SARDC 1996, UNEP 1989, World Bank 1989 and 1996

and Manila declarations). How far reform should go towards privatization of resources is by no means settled. Secondly, experiences in East Africa indicate that land registration by itself may create, rather than reduce, uncertainty and conflicts over land, may work out differently for different categories of the population and therefore leave unanswered important questions about the eventual effect on land management.

The use of a range of instruments is proposed, depending on the situation. Conceivably, this could even include a land tax to stimulate land productivity in large-scale extensive farming. Intensification of agricultural production systems would be a key element of alternative policies. Although no less ambitious and demanding than reform of land tenure, this is by itself thought to be less controversial. Further key elements of alternative policies go beyond the land/agriculture domain: rational population policies, development of employment in, for example, wood and food-processing industries in West and Central Africa, as an alternative to agriculture. Last but not least, as underlined in other analyses, all reforms depend on legal, institutional and political will as enabling factors.

An important possibility with respect to water resources in East and Southern Africa is to manage water within the boundaries of a river basin rather than within administrative boundaries (UNEP 1999b). International and transboundary sources of water will play a more important role in future national water supplies. Trust, and a modest amount of institution building, are obvious conditions for river basin management, as are the capacity to carry out systematic assessments and to share information between the partners involved. Water resource legislation needs to be reviewed to bring it in line with current prospects. Where appropriate, customary laws should be considered as these are usually easy to enforce. Another key element for sustainable water management in East and Southern Africa is the use of economic instruments in allocating water to competing uses.

Asia and the Pacific

For all its heterogeneity, the Asia-Pacific region does have an increasing air pollution problem in most of its countries. Twelve of the world's 15 cities with the highest levels of particulate matter are in Asia, as well as 6 of the 15 with the highest concentrations of sulphur dioxide (United Nations 1995). In many countries, ambient concentrations of these pollutants exceed WHO standards. Health losses in the form of premature death, chronic bronchitis and other respiratory symptoms are high or very high in at least 16 metropolitan centres in Southeast and South Asia (World Bank 1997). Air quality has improved in Japan and is improving in some other parts of the region, such as the Republic of Korea, although it does not meet health standards there. Health losses experienced in Japan in the 1970s, before it managed to improve ambient air quality, underline the health risk of current and projected air pollution elsewhere in the region.

In addition to the large and increasing urban air pollution, acid deposition is becoming increasingly problematic. Large sections of southern and eastern China, northern and eastern India, the Korean peninsula, and northern and central Thailand are expected to receive high levels of acid deposition by the year 2020 (Downing and others 1997).

For *GEO-2000*, a region-specific study (UNEP 1999c) analysed policy options to reduce the emissions of air pollutants, with a focus on urban air pollution in continental Asia (defined as the Asia and Pacific region less Australasia and the Pacific; the text that follows refers to this area as 'the region').

For comparison, the study considered a business-as-usual scenario. This follows the World Bank projection on global population (World Bank 1994), and assumes a partial convergence of per capita levels of GDP in OECD and non-OECD countries. In 2030, the GDP per capita of OECD countries in Asia and the Pacific is projected to exceed US$41 000 a year while that of non-OECD countries in Asia and the Pacific is projected to be about US$4 000. Non-OECD GDP growth rates decrease gradually from their mid-1990s levels. A frozen pollution reduction and control technology case is considered under the business-as-usual scenario. 'Frozen clean technology' means that, in addition to the baseline assumptions, diffusion rates of pollution reduction and control technologies are fixed at 1990 levels and technology existing in OECD countries is not transferred or introduced to developing countries. Thus, no emission mitigation from clean technology is envisaged in this case. Also, no special legislative measures are introduced to encourage new clean technologies.

Under business-as-usual conditions, the regional

Effectiveness of policy packages in reducing sulphur and nitrogen oxide emissions in continental Asia up to 2030

Policy package	*assumptions*	*ranking of SO_2 mitigation*	*ranking of NO_x mitigation*
A: business-as-usual	frozen clean technology	8th	8th
B: diffusion of cleaner technologies	B1: income-dependent introduction of clean technology	6th	7th
	B2: accelerated introduction of clean technology	4th	4th
C: promotion of non-motorized and public/mass transport	C1: transportation efficiency increase with fixed clean technological scenario	7th	6th
	C2: transportation efficiency increase with accelerated introduction of clean technologies	3rd	3rd
D: fuel switching and increasing efficiency	D1: fuel switching with a fixed clean technological option	5th	5th
	D2: fuel switching with accelerated introduction of clean technologies	2nd	2nd
E: mixture of B, C and D	accelerated introduction of clean technology, transportation efficiency increase and fuel switching	1st	1st

emissions of sulphur oxides in 2030 are projected to be four times those of 1990 and those of nitrogen oxides, three times. Concentrations of suspended particulate matter in China and India are already quite high and are projected to increase in most areas of the region.

The options explored for alternative policies include diffusion of clean technologies; promotion of non-motorized and public transport; fuel switching and increasing energy efficiency; and a combination of all these measures (clean technology, transportation efficiency increase, and fuel switching). The options are listed in the table above.

As one variant for the introduction of cleaner technology, the study analysed what would happen if clean technologies were introduced in developing countries only after a certain income level has been reached, and special legislation supporting the introduction of such technologies has been passed (case B1 in the table above). For example, the threshold income level for abatement of sulphur dioxide emissions would be US$3 500 per capita. In contrast, the analysis also considered a variant where emission control technologies would be introduced in an accelerated fashion, starting in 2005 (case B2 in the table above).

Regarding the promotion of non-motorized and public/mass transport, the study analysed two cases where measures would result in a 30 per cent increase of overall energy efficiency in transport between now and 2030. The cases differ in how this is achieved. Transportation efficiency improvement with fixed technology (case C1) would aim at a modal shift to public/mass transport. This would affect the emissions of sulphur and nitrogen oxides although the pollution control technologies would remain at 1990 levels. In contrast, transportation efficiency improvement with accelerated clean technologies (case C2) would reduce energy demand of the transportation sector through the introduction of pollution control technologies from 2005 onwards. Possibilities in the transport sector include replacing existing vehicles/technologies with more efficient ones (for example, four-stroke engine motor cycles in place of two-stroke ones, investment in public transport), fuel switching (such as the use of compressed natural gas in place of gasoline), phasing out leaded gasoline, and the adoption of strict emission standards for vehicles.

Fuel switching is assumedly brought about by a carbon tax, aimed at replacing coal, either with or without additional application of cleaner technologies from 2005 onwards. Options for fuel switching vary throughout the region.

The figure above shows the estimated values of sulphur dioxide emissions under the various scenarios. In the business-as-usual scenario, emissions in 2030 would be more than three times the 1990 level.

The table above ranks the various policy options according to their emissions reduction potential. From this table, it is clear that a combination of instruments is the most effective. The combination analysed assumes the accelerated introduction of clean technology, improvement in transportation energy

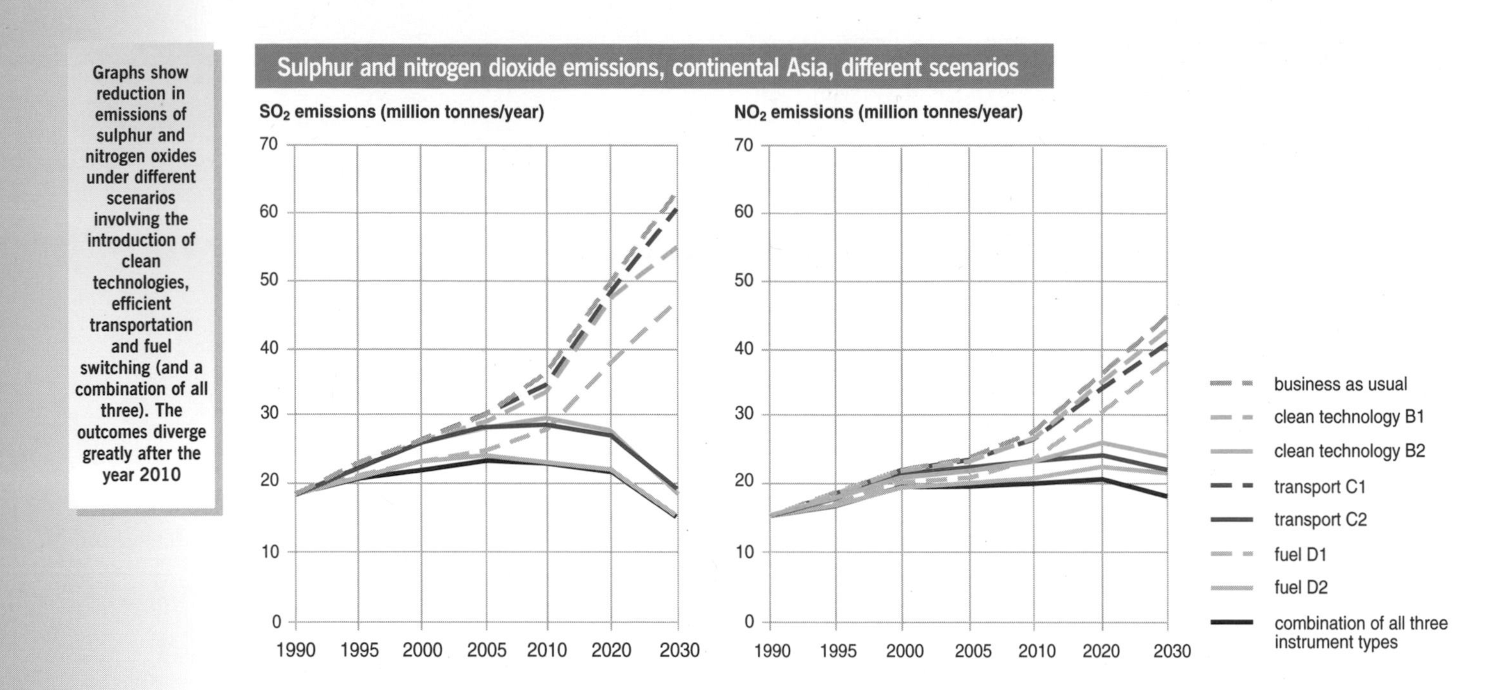

Graphs show reduction in emissions of sulphur and nitrogen oxides under different scenarios involving the introduction of clean technologies, efficient transportation and fuel switching (and a combination of all three). The outcomes diverge greatly after the year 2010

demand through efficiency improvement, and fuel switching.

From the various other policy packages, accelerated introduction of clean technologies as of 2005 could reduce the increase in emission levels from 200 per cent back to 6 and 60 per cent, for sulphur and nitrogen oxides respectively (regional totals relative to 1990). The promotion of public transport systems could reduce the increase even further, to 1.5 and 46 per cent.

Fuel switching requires careful consideration of energy resources and policies as they vary within the region. When combined with accelerated introduction of clean technologies, fuel switching could reduce the emission of sulphur oxides in 2030 by 17 per cent as compared to 1990 level (instead of tripling). For nitrogen oxides, it could limit the increase to 40 per cent (also instead of tripling).

The study emphasizes that increased income levels throughout the region will be required to make the necessary changes politically and financially feasible. Conversely, this connection can also be interpreted as the prospect that rising income levels will increase the demand for action on urban air pollution (World Bank 1997).

Europe and Central Asia

Energy plays a central role in many environmental problems of Europe and Central Asia. The alternative policies study for this region explores what can be achieved by 2010 with full implementation of accepted environmental policies or, alternatively, with additional moderate energy and environmental policies (UNEP/RIVM 1999). It considers five environmental issues directly affected by energy: climate change, acidification, summer smog, urban air pollution and the accident risks of nuclear power generation.

The study assumes population growth in line with the median UN population estimate and swift economic recovery, first in Central Europe, followed by Eastern Europe and Central Asia, with a strong increase in regional linkages and trade. Energy consumption is projected to increase between 1 per cent (Western Europe) and about 2.5 per cent per year (Central Asia) in the 1995–2010 period.

With full implementation of accepted environmental policies, transport and electricity use are expected to become important drivers for environmental problems. Notwithstanding cleaner vehicles, air pollution in cities will increasingly be dominated by mobile sources, making urban air pollution and summer smog persistent problems in all four sub-regions (see figure on page 352). Assuming that accepted environmental policies are fully implemented, acid deposition in the western parts of the region will be reduced considerably. However, 6–8 per cent of the area will continue to be exposed to excessive acid deposition and the targets of the EU Acidification Strategy will not be met. Acid deposition in some parts of Siberia will become an increasing problem (Stevenson and others 1998, Bouwman and van Vuuren 1999). Likewise, if current policies are fully implemented, by 2010 the occurrence of summer smog in Europe could be reduced by one-third but, even then, WHO guidelines will be exceeded, especially in Western and Central Europe – and increasingly in Central Asia as well. Accident risks from reactors for nuclear power generation will not diminish much under present policies and will still be dominated in the region by the relatively few reactors in Eastern Europe.

The toughest problem for much of the region is emission of greenhouse gases. For Western and Central Europe, the assumed trends result in 6 and 3 per cent more emissions of the three most important greenhouse gases in 2010 than in 1990, respectively. This means that the Kyoto commitments of these two sub-regions (8 and 5.5 per cent below the 1990 level, respectively) will not be met. For Eastern Europe, projected 2010 emissions are still almost 10 per cent below the 1990 level – complying with the commitment to keep emissions below this level. In Central Asia (no Kyoto target), developments without additional policies result in a 2010 emission level that is 3 per cent above the 1990 level.

Thus, if existing policies are fully implemented and fully effective, the environmental situation in the region will improve compared to 1990 – except for climate change. However, the projected improvement is generally insufficient to meet policy targets. Moreover, full implementation of currently existing policies will certainly not come about without policy effort. Experience has shown that often a 'policy failure gap' exists between expected and actual results (see, for example, Hoek and others 1998). Two specific policy gaps relate to the fact that transport may grow much faster than assumed and that environmental policies must compete with other priorities which may affect their implementation rate.

The contrasting scenario assumes the same demographic and economic trends but adds measures that are either necessary to meet existing policy commitments (such as Kyoto) or that feature very low costs and considerable impacts. Since the same driving forces act on many of the five energy-related environmental problems, policy measures for one problem can also help to reduce others. Most importantly, energy-saving measures to mitigate climate change will potentially also reduce emissions of acidifying substances and summer smog precursors, and hence reduce acidification, summer smog and urban air pollution.

On climate change, several analyses indicate that sufficient technical potential exists to meet Kyoto commitments in all relevant sub-regions (WEC 1995, OECD/IEA 1996, OECD/IEA 1997, Capros and Kokkolakis 1996, Gielen and others 1998, and Phylipsen and Blok 1998). Some studies also looked at how technical potential can be implemented through policy instruments (Blok and others 1996, OECD/IEA 1997). They indicate that financial stimuli should be an important element of environmental policies – including energy and carbon taxation or removal of subsidies. In particular in Central and Eastern Europe and Central Asia, the high subsidies on energy use and lack of consequences of the non-payment of energy bills have provided little incentive for energy efficiency (UNEP 1998). Although subsidies in most transition countries have been reduced since 1990, they are typically still around 25 per cent of average world prices. In all four sub-regions, further reform of the large variety of energy subsidies offers ways to induce more energy savings and increase the market share of cleaner fuels.

The figure on page 352 shows two variants for implementing additional measures to target climate change. In the 'no-trade' variant, it is assumed that each of the European sub-regions will seek to meet its Kyoto commitment by measures within the sub-region. The 'trade' variant assumes that Western Europe uses the flexible instruments introduced under the Kyoto agreement (emission trading and joint implementation) in order to reduce the costs of emission reduction. In fact, it seems macro-economically attractive for Western Europe to realize somewhat more than half of the required reductions through trading (Bollen and others, in press).

The potential to decrease further the emissions of acidifying compound and summer smog precursors, in order to meet ecosystem and health targets, has also been identified in various studies. Most of these scenarios try to meet existing targets by typical, but costly, end-of-pipe measures. However, if combined with measures taken to decrease carbon dioxide emissions, the costs of combating acidification and summer smog could be decreased considerably (Amman and others 1998). Meeting Kyoto targets by emission trading obviously also transfers the concomitant gains in decreasing pollution elsewhere.

Development of energy-related environmental problems in Europe and Central Asia

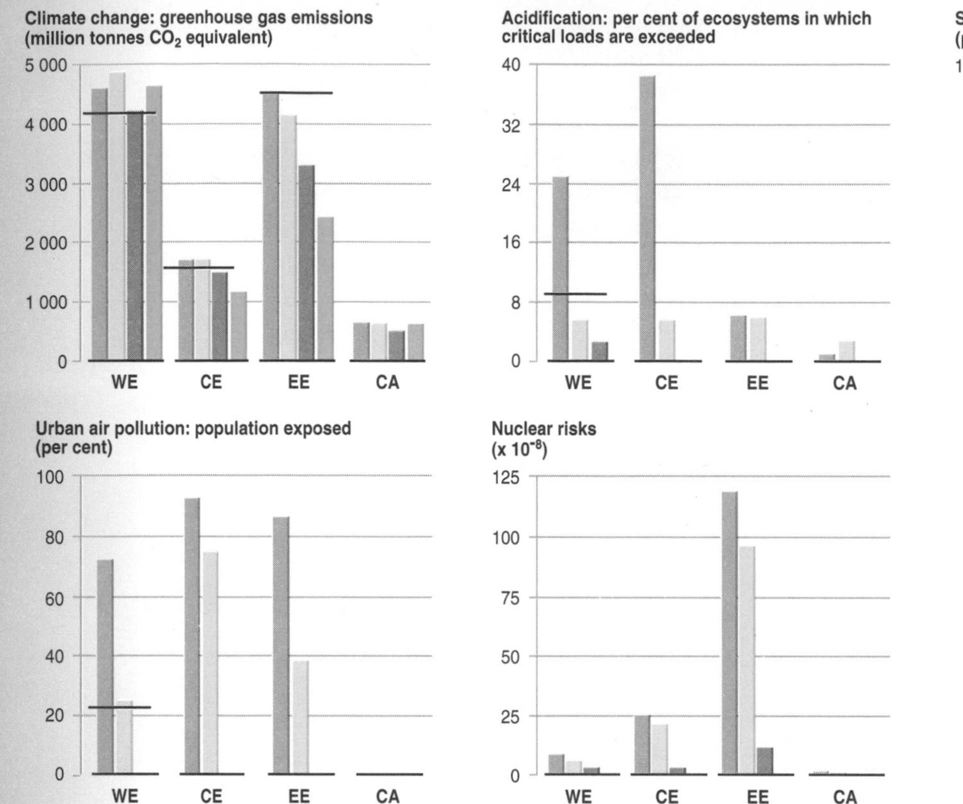

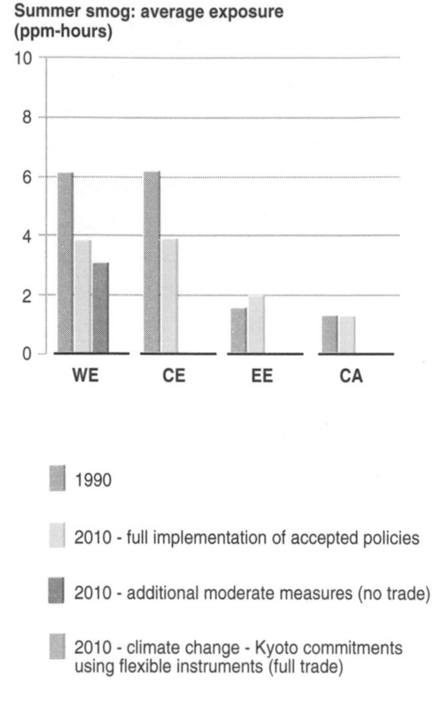

Indicators:

Climate change: greenhouse gas emissions

Acidification: ecosystems exposed to deposition of acidity in excess of critical loads

Summer smog: average time-weighted exceedence of ground-level ozone background concentration over 60 ppb eight-hour average

Urban air pollution: urban population exposed to levels of NO_2, particulate matter (pm10), benzene, SO_2 and benzo(a)pyrene in excess of of EU Air Quality Guidelines

Nuclear risks: excess risks of cancer fatalities due to accidents in nuclear power plants (deaths per 100 million per year)

Note: Horizontal line indicates relevant target or reference level; Kyoto commitments for climate change; EU acidification target for acidification; EU Air Quality Guidelines for urban air pollution.

Sources: RIVM/EFTEC/NTUA/IIASA 1999, UNEP/RIVM 1999

Climate change: Hendriks and others 1998, BP 1997

Acidification and summer smog: Amann and others 1998, Cofala and Klimont 1999, Bouwman and van Vuuren 1999, Stevenson and others 1998

Urban air pollution: Eerens and others 1999

Nuclear risks: Stoop and others 1998

Full implementation of existing policies will decrease the severity of energy-related environmental problems but not enough to meet most policy targets

For urban air pollution, most of the moderate end-of-pipe measures have already been included in the existing policies scenario. Further improvements may come from energy efficiency or specific measures to reduce the distance travelled by cars in urban areas. However, as their effect is hard to quantify at the sub-regional level, this has not been shown in the figure. On the whole, moderate policies to reduce the effect of the growth in transport are bound to be eroded by the rapidly growing volume. This has happened in Western Europe and is probable in Central Europe in the near future, especially where there is accession to the European Union.

Nuclear risks throughout the region can be reduced by measures to improve management and accident procedures for the 19 most unsafe nuclear plants in Central and Eastern Europe. Costs, however, are highly uncertain and vary from plant to plant (Stoop and others 1998). In the oldest cases, replacement – although costlier – might be more appropriate to reduce risks.

The reduction of risks in Western Europe hinges on the safety of the nuclear plants in Eastern Europe. In fact, the Europe and Central Asia alternative policy study observes that for all five environmental problems, the four sub-regions are connected. This is even true for urban air pollution, which is increasingly determined by transboundary background air pollution as well as by regional-scale driving forces such as EU enlargement. For acidification, in particular for Western and Central Europe and part of Eastern Europe, a significant share of the total deposition comes from emissions in the other sub-regions. For summer smog, the increasing level of background ground-level ozone in the northern hemisphere contributes to summer smog in all four sub-regions. Climate change is a global problem by its very nature but an additional inter-regional link is now provided by the potential for emissions trading. These linkages argue for environmental cooperation at the regional level, something that is already increasing.

Conclusions

- Full implementation of existing policies could improve the environmental situation in Europe and Central Asia. However, improvement will be insufficient to protect all ecosystems against acidification and climate change, and to prevent exceedence of WHO and EU health guidelines for summer smog and urban air pollution.
- A combined strategy for reduction in the framework of the Kyoto Protocol (greenhouse gases) and other environmental problems (transboundary and urban air pollution) could optimize use of the opportunities for energy efficiency and fuel-switching.
- The electricity and transport sectors deserve special attention as they are becoming increasingly important as driving forces behind environmental pressures in Europe and Central Asia.
- The potential to reform environmentally-damaging subsidies has not yet been fully used, in any of the four sub-regions.

Latin America and the Caribbean

Natural forests cover 47 per cent of the land area of Latin America, and the Amazon basin accounts for one-third of the world's tropical forest area. These forests are an important source of products, fuelwood and employment for local people, a major source of foreign exchange for governments, serve important functions in protecting watersheds and freshwater resources, act as a storehouse for carbon and support a significant portion of the world's biodiversity (Daily 1997). As reported in Chapter 2, Latin America is still losing around 58 million hectares of natural forest per year, although the rate of deforestation in, for instance, the Amazon has slowed considerably since the mid-1990s.

A recent study by CIAT/PNUMA (1998) shows that 84 per cent of deforestation in Latin America is due to expansion of the agricultural sector, 12.5 per cent is due to logging and 3.5 per cent is due to the construction of infrastructure. In addition, demographic pressure, unemployment and inequitable land distribution are important drivers for the further degradation of forests. An increased and unsustainable intensification of land use is projected. At the same time, there is a gross imbalance between destruction and reforestation, with only 1 hectare planted for every 25 hectares destroyed. The combination of these trends leads to the prospect of increased soil degradation, more frequent flooding and the degradation of freshwater resources.

Forestry policies in the region have been focused largely on the tropical rain forests, concentrating on protection and sectoral perspectives, and have not been well integrated with economic and social strategies. Lack of finance, technology, personnel and training have all hindered environmental management; furthermore, in some countries, large and complex legal frameworks are coupled with unclear definitions of the responsibilities of the environmental institutions involved. Clearly, forestry policy in the region has not, overall, been effective. One important reason is that policies have failed to take account of the diverse functions of the forest and the differing needs of forest users and inhabitants.

Comprehensive forest policy frameworks at national level are fundamental to sustainable forest management in Latin America. National forest programmes demand a broad inter-sectoral approach at all stages (formulation, implementation and monitoring). Moreover, they should be tailored to each country's social, economic, cultural, political and environmental situation. Appropriate policies could include direct control of government-owned forests and indirect control using fiscal incentives in the form of taxation, subsidies and forest credits; other incentives such as the granting of private property rights; market reforms; the introduction of community forestry schemes; and improvements in extension, research and education.

The alternative policy study carried out for *GEO-2000* (UNEP 1999d) indicates that these policy tools could be used to stimulate programmes in ten different areas:

- Rehabilitation of degraded areas, particularly those damaged by slash-and-burn agriculture.
- Development of agroforestry, which experience in Rondonia (World Bank/MMA 1998) has shown to be a viable economic alternative to crude logging, providing that areas and species are carefully selected and that forest 'mining' is addressed by severe taxes.
- Development of sustained forestry management through zoning, the establishment of forest corridors and complete bans on the exploitation of floodable forest areas.
- Establishment of networks of protected areas, buffer zones and ecological corridors in which traditional forest uses by indigenous peoples would be encouraged – such areas would need to be established in all the major ecological zones from dry scrublands to the important wetlands.
- Finding mechanisms to allow more of the wealth of forest regions to be returned to those who live there. In the Amazon, for example, mahogany is extracted from Indian reserves for US$50/m^3, fetches US$800/m^3 in the port of Belém City, Pará State, in Brazil and is then sold in European markets for US$1 500/m^3 (IBAMA 1998). The primary forest producers are poorly rewarded for what ends up as an expensive product. These price distortions indicate the necessity of improving regulation in the mahogany timber market, coupled with an inspection system and an inventory of mahogany resources to reduce pressure on mahogany species. Local forest products can also be encouraged, such as the vegetable leather handbags and clothes now being produced from latex in the Brazilian state of Amazonas.
- Construction of environmentally-sustainable cities in tropical forest regions which would stimulate economic development while preserving the forest resources on which all forest settlements are based.
- Disciplined exploitation of natural resources through national policies to control pollution from mining and the activities of small-scale illegal mining activities. The mineral wealth of the Amazon should be controlled by integrated activities of all Amazon countries to provide effective pollution control.
- Development of ecotourism to catalyse economic development and protect the environment. This has to be properly set up and managed, in order to prevent negative impacts. Good high-quality examples with large turnover do exist in the region, in particular in Costa Rica (Costa Rica 1996).
- Development of effective forestry institutions capable of licensing and monitoring activities.
- Improvements in forest-related training, information and educational activities.

Packages of these policies adapted to national conditions could achieve a number of goals simultaneously. These include:

- reducing deforestation rates, forest fires, numbers of threatened animal and plant species, and regional carbon dioxide emissions;
- a slowing down of agricultural expansion onto forest land;
- improvements in forest ecosystem health, the quality of urban and rural life, and regional and local economies; and
- the provision of appropriate technologies to forest dwellers as a tool for sustainable development.

A broad range of alternative policy tools could be used, and individual countries could select from this mix a policy package suited to their own social, cultural and economic conditions, thus making for better integrated forest policies.

North America

Over the past three decades, North American environmental policies have been reasonably successful at dealing with conventional, mostly local, environmental problems such as air and water pollution. The scale of the region's economic activity has brought increased well-being and opportunities for most of the North American population. The downside of this robust economy, however, is increased stress on environmental quality, with major impacts regionally and worldwide. Perhaps the most obvious example of such environmental stress is the fact that North America remains, on a per capita basis, the world's largest source of greenhouse gas emissions that threaten to alter the Earth's climate. Contributing to these emissions are heavy dependence on automotive and air transport and sprawling suburban residential patterns that put heavy demands on energy resources. The region's robust economy also draws heavily on other natural resources – water, forest products, agricultural goods, fisheries, minerals – sometimes contributing to environmental degradation. Policies favouring low-cost energy and subsidies for natural resource extraction may encourage high levels of production and use, and thus make it more difficult to attain environmental goals.

Given that the scale of North American economic activity has both beneficial as well as harmful effects, alternative policies that change the pattern of economic activity in ways that reduce environmental harm without retarding overall economic growth are worth careful consideration. Policies that alter fiscal incentives – both reducing or eliminating perverse incentives and increasing incentives for constructive change – seem to offer particular promise. Such alternative policies are the focus here (UNEP 1999e).

Subsidies for natural resource extraction are one widely-used though difficult-to-measure fiscal incentive. The United States, for example, indirectly subsidizes logging in national forests by providing logging roads built with public funds, and grazing of livestock on federal lands, by charging less than market rates for grazing permits (US Congress 1995). Similar subsidies support use of water for irrigation in the arid, western portions of the country, and mineral extraction from and recreational use of public lands. These subsidized activities have placed heavy burdens on the country's natural resources (US Government 1997). According to the OECD, the United States provided about US$46 960 million in agricultural producer support in 1998; the corresponding figure for Canada was US$3 184 million (OECD 1999). Farmers using irrigation water from federally-supported projects pay on average only about 17 per cent of the actual cost, and the total water subsidy in the western United States is estimated at about US$4 400 million (Repetto 1986, Pimentel and others 1997). Energy subsidies in the United States, according to one recent estimate, amount to perhaps US$30 000 million per year (Myers 1998) although a 1992 study by the Energy Information Administration estimated that approximately 30 per cent of energy subsidies were spent on renewable energy sources and improving energy efficiencies.

Subsidies for road transportation are also large. Road transportation in the United States, with 220 million vehicles (International Road Federation 1997) and more than 6 million kilometres of roads, has the world's largest system, and accounts for 80 per cent of the energy used in transport and 25 per cent of the country's carbon dioxide emissions. Subsidies for road transport include those for road construction, oil extraction, production and use, automotive research and safety programmes, highway patrols, and related government programmes; they have been estimated at US$91 000 million a year (Roodman 1996). Other related subsidies include those for parking, estimated at US$50 000 million (Myers 1998).

Reducing or eliminating direct and indirect subsidies for road transport could play a significant role in reducing congestion, improving urban air quality, and slowing the growth of carbon dioxide emissions to meet climate goals. Without parking subsidies, for example, many people might choose mass transit alternatives. Subsidized investments in road construction can have enormous leverage, since they appear to have a significant influence on per capita automobile travel rates (Litman 1996).

Similarly, US policy could eliminate subsidies for grazing livestock on public lands. The US Forest Service and the Bureau of Land Management administer the country's livestock grazing programme. The two agencies charge ranchers a fee based on an 'animal unit month', which is the amount of forage that one adult cow with calf or five adult sheep require for one month. Studies show that the fees do not cover the costs for administering the programme, and that they are below the market rate for grazing on private lands (Maxwell 1995). Such subsidies are opposed by those who argue that low grazing fees and lax supervision have encouraged overgrazing and have led to soil erosion, watershed destruction, loss of native grasses and other vegetation needed as food for wildlife and livestock, and elimination of forage reserves needed to withstand periodic drought (Hess and Holechek 1995). The most direct reform policy would be to raise grazing fees to cover the administrative costs, or to match estimates of market value (Maxwell 1995), thus eliminating subsidies.

In short, the North American region constitutes a good example of the potential to reduce environmental pressures by reform of various forms of subsidies. Proposals for such reforms are already being made. In relation to the Western States, for example, the maturation of economies with reduced reliance on resource extraction, the increased value placed on the environment by the public, and a desire to reduce the federal deficit are motivating new objectives for federal natural resource policy which will place more emphasis on market mechanisms while reducing subsidies (US Government 1997). Ironically, this form of alternative policy ultimately results in the reduction of government involvement – contrary to the widely-held belief that environmental policy always requires more government involvement.

The impact of taxes in the marketplace makes taxes another potentially powerful policy tool. In current industrial economies, there are visible costs that are paid by economic actors and external, hidden costs from environmental degradation that are paid by society as a whole or by other economic actors. In the case of air pollution from energy production, for example, these external costs may be borne by the population at large in the form of health costs and by other economic sectors such as agriculture in the form of reduced productivity. If energy producers had to bear these costs, then alternative, less-polluting energy sources might become more competitive. One way to internalize these hidden costs is through taxation of pollution, resource depletion or ecosystem degradation.

Environmental taxes have already been used in North America by individual states and provinces, and by national governments. Ontario, for example, has implemented a Tax for Fuel Conservation that taxes new car purchases based on their fuel efficiency and offers rebates for the most efficient vehicles. For more than a decade, California has offered state tax credits for renewable energy producers that have helped to stimulate the industry within the state. At a national scale, the US tax on ozone-depleting chemicals is credited with helping to phase out production of these chemicals in a rapid manner, coupled with an accelerated phase-out schedule imposed by the United States, supporting US commitments under the Montreal Protocol (see page 304).

Taxes on greenhouse gas emissions such as carbon dioxide have been proposed as one policy measure to help the United States and Canada reduce these emissions and thus lower the high burden that North America places on the global climate (Dower and Zimmerman 1992). Such taxes, known as carbon taxes, are controversial, even though their ability to reduce emissions is not in doubt. There is strong resistance in North America to higher energy taxes, and some economic studies argue that taxes high enough to reduce emissions markedly would also slow economic growth. However, Scandinavian countries including Finland, Sweden and Denmark have introduced such taxes, providing a strong incentive for greater energy efficiency and for fuel-shifting to renewable energy sources or to less carbon-intensive fuels. The United States intends to meet its commitments under the UNFCCC primarily through the use of another form of market mechanism: emission caps and emissions trading, which better fits the US domestic context and may, in fact, provide greater assurances of compliance than tax approaches.

West Asia

Unless improved water management plans are implemented in the West Asian region, a series of water-related issues will interact to cause major environmental problems (see Chapter 2, page 164). The immediate issues include overdrawing of aquifers and shallow groundwater, which are causing saltwater intrusion and a breakdown of traditional water supply systems, and the indiscriminate discharge of wastewater causing contamination of shallow groundwater and health hazards. Eventually, many groundwater sources will be lost by quality degradation; this will result in a further reduction of the arable area because of salinization. Many efforts have been made to increase recharge rates and reduce withdrawals. Prospects of improving the situation are hampered by high population growth and the intensive agricultural use of water for irrigation. This situation is aggravated by a policy of food self-sufficiency and by a general weakness of the institutions dealing with water affairs.

One of two studies on West Asia prepared for *GEO-2000* (UNEP 1999f) focuses on water demand and availability. The study examines three scenarios: business as usual, increasing supplies, and increasing supplies and rationalizing consumption (see diagram opposite).

Records and studies indicate the excessive and wasteful use of water in all sectors (agricultural, domestic and industrial) throughout West Asia. Huge water losses of at least 45 per cent in agriculture arise from inefficient irrigation systems, while there is a 20 per cent leakage from water supply networks and general 10 per cent losses during industrial use. Because agricultural use accounts for the lion's share of water use in both sub-regions (85 per cent in the Arabian Peninsula and 92 per cent in the Mashriq in 1995), water conservation is dominated by what can be achieved in the agricultural sector.

West Asia: water balance under three scenarios

Arabian Peninsula | Mashriq | West Asia

Scenario 1 assumptions: business as usual

- no new water sources developed after 1995
- constant domestic and industrial water demand
- research cuts agricultural water use by 17% by 2015
- settlement of shared water resources disputes

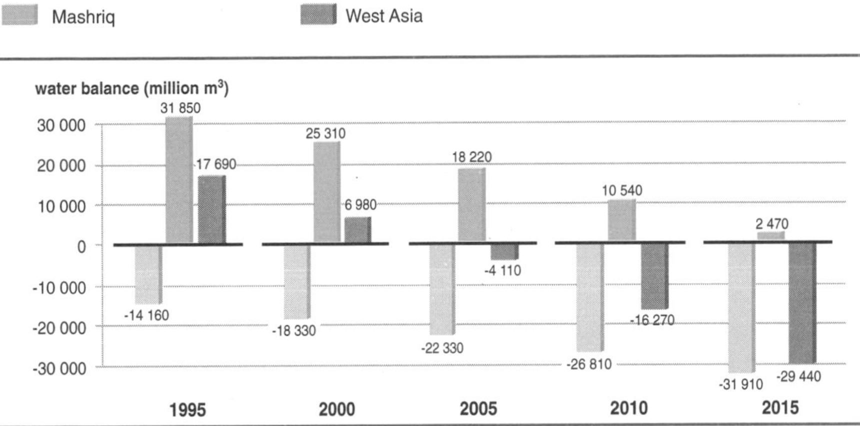

Scenario 2 assumptions: increasing supplies

as Scenario 1 (except first bullet) plus:

- groundwater resources developed by 100 million m^3 a year – 2 000 million m^3 for each sub-region by 2015
- desalination plants capacity is increased gradually to reach 3 000 million m^3 in the Arabian Peninsula by 2015.
- wastewater plant capacity is increased to reach 3 000 million m^3 in the Arabian Peninsula and 2 000 million m^3 in the Mashriq by 2015

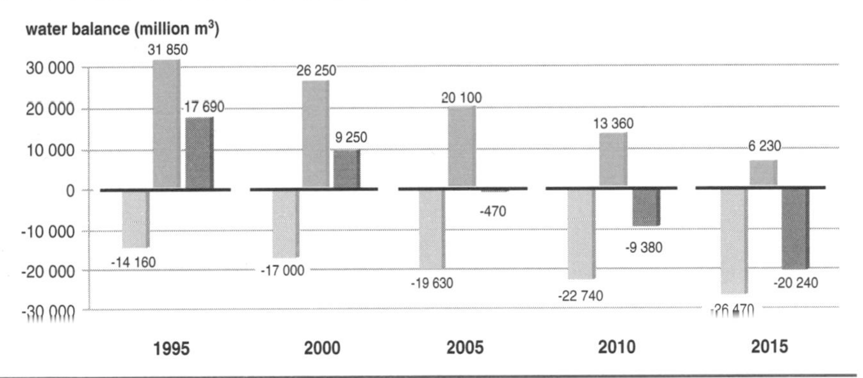

Scenario 3 assumptions: increasing supplies and rationalizing consumption

as Scenario 2 plus:

- more rational water use results in a saving in water demand of 5 600 million m^3 a year in the Arabian Peninsula and 6 000 million m^3 a year in the Mashriq by 2015

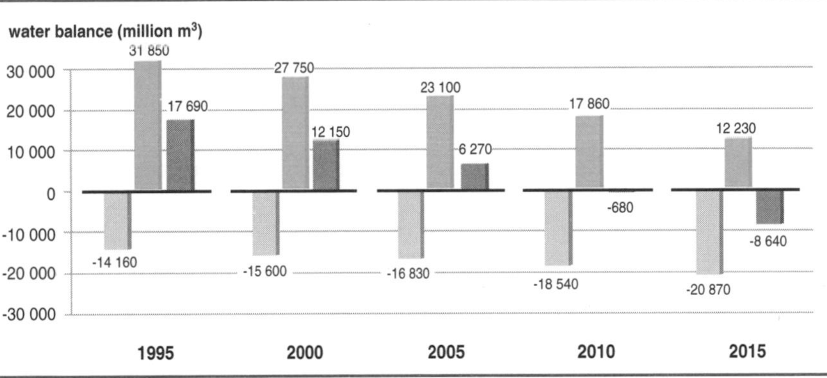

Bar charts left show water balance in West Asia until the year 2015, following three different scenarios. The situation deteriorates in both sub-regions, regardless of scenario, though the water balance remains positive in the Mashriq

The business-as-usual scenario will leave the Arabian Peninsula sub-region suffering from acute water shortages. The sub-region currently has a negative water balance, with 48 per cent of demand in 1995 not covered by available water resources (surface water, rechargeable groundwater and non-conventional resources, such as desalinated water). Water basins in which less than 70 per cent of demand can be met from available resources are considered water stressed. The deficit is being met unsustainably by overdrawing and depleting aquifers and further installation of very costly desalinization plants. Business as usual increases the annual deficit to an unrealistic 67 per cent of demand by 2015. In fact, it is clear that current water resources cannot satisfy future water demand much past 2005 without alternative policies.

The Mashriq sub-region is comparatively better off under this scenario, with no overall water deficit. But this sub-region is also clearly on an unsustainable course, with water stress becoming more and more extreme. Even this scenario requires saving 17 per cent in agricultural water demand by 2015 as compared to 1995, and it requires solving transboundary disputes over shared water resources.

This pessimistic picture for West Asia is compounded when the related issue of land resources management is factored in. If current policies continue until 2015, the continued pressures on groundwater resources, with extraction rates far beyond annual recharge limits, will increase groundwater salinity, leading to greater salinization of land resources. A recent study (CAMRE/UNEP/ACSAD 1996) has shown how the area of salinized irrigated land in both sub-regions would increase (see Chapter 2 for details). A business-as-usual scenario will result in the final abandonment of salinized land in certain areas. Furthermore, eradication of the natural plant cover of rangelands through overgrazing or cultivation of marginal rainfall areas will lead to enhanced wind erosion and severe loss of soil materials in many West Asian countries. Pollution of land resources from the disposal of urban, sewage and industrial wastes, as well as by agricultural chemicals, will increase.

Scenario results for water resource management

Scenario	*Water balance in 2015 (available resources as % of demand)*	
	Arabian Peninsula	*Mashriq*
Business as usual	–67	2.6
Increasing supplies	–56	6.5
Increasing supplies and rationalizing consumption	–50	14

The projected growing rate of land degradation in West Asia under the business-as-usual scenario will lead to job losses in the agricultural sector and increase migration to urban areas, adding to the pressures on urban infrastructure. In addition, the projected food trade balances indicate a widening divergence from food self-sufficiency.

Alternative policies targeting water problems could increase the supply of water, in particular through more research on available resources, increased desalinization capacity and wastewater recycling. By 2015, this could reduce the water deficit for the Arabian Peninsula to 56 per cent of demand while allowing the Mashriq a small positive water balance of 6.5 per cent (see table above).

Policies that in addition gradually rationalize water consumption further through more efficient irrigation, price reviews and improved waste water management could reduce the projected water deficit of the Arabian Peninsula by a further 6 per cent. For the Mashriq, such policies could produce a positive water balance of 14 per cent of demand in 2015, which is a less dramatic deterioration than under current policies which would produce a positive water balance of only 2.6 per cent of demand in 2015 (see table left).

One important component of scenario 3 is water conservation. This can be approached by:

- Reviewing the economics of irrigation and agricultural production, and reappraising agricultural policies.
- Improving the efficiency of traditional irrigation systems, introducing appropriate modern irrigation technology, and promoting water conservation techniques among farmers and water users.
- Reviewing current irrigation incentives and tariffs, implementing the necessary legislation to enforce and update regulations on water use, and strict enforcement of regulations.
- Improving extension services and programmes to raise awareness among the public and farmers of the economic value of water as a precious and scarce resource.
- Providing subsidies or soft loans to encourage application of modern irrigation systems.

Related alternative policies improving management of land resources would work in the same direction. Conceivably, such policies would combine two strategic lines:

- optimization by agricultural systems of the limited land and water resources that are available; and
- combating and alleviating the negative environmental impacts that result from agricultural activities on fragile ecosystems and lead to land degradation and desertification.

Specifically in the Mashriq, such policies would be geared towards the optimization of production from the available land resources and expansion in the unused arable area. In the Arabian Peninsula, alternative policies could be geared towards the rational use of land and water resources, maximization of productivity per unit area, and adoption of innovative agricultural techniques.

Implementing these alternative policies would require well designed detailed resource planning, legal and institutional reforms, application of water tariffs, pollution charges and groundwater pricing, and cutting water losses, especially in the agricultural sector. Even so, it is clear that the Arabian Peninsula will continue to experience a deficit in its water resources if the regional targets of food security are adhered to. In fact, the *GEO-2000* scenario analyses for both water and land confirm that this goal needs to be rethought. In the Mashriq, the settlement of potential conflicts over shared water resources remains a fundamental and pressing issue.

References

Achebe, C., A. Okeyo, G. Hyden, and C. Magadza (1990). *Beyond Hunger in Africa.* Heinemann Kenya Limited, Nairobi, Kenya

Amann, M., Bertok, I., Cofala, J., Gyarfas, F., Heyes, C., Klimont, Z., Schöpp, W., Makowski, M., and Syri, S. (1998). *Cost-effective Control of Acidification and Ground-level Ozone. Sixth Interim Report–Part B: Emission Control Scenarios*. International Institute for Applied Systems Analysis, Laxenburg, Austria

Blok, K., van Vuuren, D.P., van Wijk, A., and Hein, L. (1996). *Policies and measures to reduce CO_2 emissions by efficiency and renewables, a preliminary survey for the period 2005*. Department of Science, Technology and Society, Utrecht University, The Netherlands

Bollen, J. C., Gielen, A. M. and Timmer, H. (in press). Clubs, Ceilings, and CDM, Macro-economics of compliance with the Kyoto protocol. *Energy Journal* (in press)

Bouwman, A. F., and van Vuuren, D. P. (1999). *Global assessment of eutrophication and acidification*. National Institute of Public Health and the Environment (RIVM), Bilthoven, The Netherlands

BP (1997). *BP Statistical Review of World Energy*. British Petroleum Co., London, United Kingdom

CAMRE/UNEP/ACSAD (1996). *State of Desertification in the Arab Region and the Ways and Means to Deal with it* (in Arabic with English summary). Damascus, Syria

Capros, P., and Kokkolakis, E. (1996). *'CO_2 - 10% Target' Scenario 1990–2010 for the European Union: Results from the MIDAS Model*. National Technical University of Athens, Athens, Greece

CIAT/PNUMA (1998). *Atlas de Indicadores Ambientales y de Sustentabilidad para América Latina y el Caribe.* Version 1. CIAT/PNUMA, Cali, Colombia

Cofala, J., and Klimont, Z. (1999). *Economic assessment of priorities for a European Environmental Policy Plan. Update of the Baseline and the MFR scenarios for Acidification, Eutrophication and Tropospheric Ozone*. International Institute for Applied Systems Analysis, Laxenburg, Austria

Costa Rica (1996). *Plan de Politica Ambiental – ECO 2005.* Ministerio de Planificación Nacional y Política Economica, Ministerio del Ambiente y Energia y Gobierno de Costa Rica, San José, Costa Rica

Daily, G. (1997). *Nature's Services: Societal Dependence on Natural Ecosystems.* Island Press, Washington DC, United States

Dalal-Clayton, B. (1997). *Southern Africa Beyond the Millennium: Environmental Trends and Scenarios to 2015.* International Institute for Environment and Development, London, United Kingdom

Dower, R.C., and Zimmerman, M.B. (1992). *The Right Climate for Carbon Taxes.* World Resources Institute, Washington DC, United States

Downing, R.J., Ramankutty , R., and Shah, J. (1997). *RAINS ASIA: An Assessment Model for Acid Deposition in Asia*. The World Bank, Washington DC, United States

Eerens, H. C., Sluyter, R., Kroon, I. C., van Oss, R. F., Hootsen, H., Claesens, E., Smeets, W., van Pul, A., Hammingh, P., and de Waal, L. (1999). *Urban Air Quality in Europe: 1990–2010*. National Institute of Public Health and the Environment (RIVM), Bilthoven, The Netherlands

Gielen, D.J., P.R. Koutstaal, T. Kram and S.N.M. van Rooijen (1998). *Post-Kyoto: Effects on the Climate Policy of the European Union.* Energieonderzoek Centrum Nederland, Petten, The Netherlands

Hendriks, C. A., de Jager, D., and Blok, K. (1998). *Emission Reduction Potential and Costs for Methane and Nitrous Oxide in the EU-15*. Interim report. Ecofys, Utrecht, The Netherlands

Hess, K., and Holechek, J. L. (1995). *Beyond the Grazing Fee: An Agenda For Rangeland Reform*. Policy Analysis 234 http://www.cato.org/pubs/pas/pa-234.html.

Hoek D., C.H.A. Quarles van Ufford, J.A. Hoekstra, G. Duvoort, P. Glasbergen, P.P.J. Driessen, M.C. Das, J.P. de Poorter, N. Habermehl, P.J.Klok, R.A. van de Peppel and J. van de Ploeg (1998). *Milieubeleidsanalyse, de balans opgemaakt. Methodiek en toepassing in milieubalans 1995, 1996 en 1997* (Environmental policy analysis, the balance made up). National Institute of Public Health and the Environment (RIVM), Bilthoven, The Netherlands

IBAMA (!998). *Mahogany Operation (Operação Mogno)*. IBAMA Departamento de Fiscalização, Brasilia, Brazil

International Road Federation (1997). *World Road Statistics* 1997 Edition. IRF, Geneva, Switzerland, and Washington DC, United States

Litman, T. (1996). *Transportation Cost Analysis: Techniques, Estimates and Implications*. Victoria Transport Policy Institute, Victoria, Canada

Maxwell, G. (1995). Grazing on the Public Range. In Munson, R. (ed). *Reforming Natural Resource Subsidies*. Northeast-Midwest Institute, Washington DC, United States

Myers, N. (1998). *Perverse Subsidies, Tax $s Undercutting Our Economies and Environment Alike.* International Institute for Sustainable Development, Winnipeg, Canada

OECD (1995). *Preparing for the Future: A Vision of West Africa in the Year 2020.* Organization of Economic Coordination and Development and the African Development Bank, Paris, France

OECD (1999). *Agricultural Policies in OECD Countries: monitoring and evaluation 1999.* OECD, Paris, France

OECD/IEA (1996). *World Energy Outlook–1996 Edition*. International Energy Agency, Paris, France

OECD/IEA (1997). *Energy Efficiency Initiative–Volume 1: Energy Policy Analysis*. International Energy Agency, Paris, France

Phylipsen, D. and Blok, K. (1998). *A Review of the Stage of Implementation of European Union Policies and Measures for CO_2 Emission Reduction*. Department of Science, Technology and Society, Utrecht University, Utrecht, The Netherlands

Pimentel, D., Huang, X., Cordova, A., Pimentel, M. (1997). Impact of Population Growth on Food Supplies and Environment. *Population and Environment*, Vol. 19, No. 1

Repetto, R. (1986). *Skimming the Water: Rent Seeking and the Performance of Public Irrigation Systems*. World Resources Institute, Washington DC, United States

RIVM/EFTEC/NTUA/IIASA (1999). *Economic assessment of priorities for European environmental policies*. National Institute of Public Health and the Environment (RIVM), Bilthoven, The Netherlands

Roodman, D. M. (1996). *Paying the Piper: Subsidies, Politics and the Environment*. Worldwatch Institute, Washington DC, United States

SARDC (1996). *The State of the Environment in Southern Africa.* Penrose Press, for Southern African Research and Documentation Centre (SARDC), IUCN and SADC, Johannesburg, South Africa

SEI/UNEP (1998). Raskin, P., Gallopin, G., Gutman, P., Hammond, A., and Swart, R. *Bending the Curve: Toward Global Sustainability*. PoleStar Series Report No. 8, Stockholm Environment Institute, Stockholm, Sweden, and UNEP/DEIA/TR.98-3, UNEP, Nairobi, Kenya http://www.gsg.org/

Stevenson, D.S., Johnson, C.E., Collins, W.J., and Derwent, R.G. (1998). *Three-dimensional model studies of the coupling between the regional and global scale formation of tropospheric oxidants*. UK Meteorological Office, Bracknell, United Kingdom

Stoop, P., Blaauboer, R., Matthijsen, A., Slaper, H. (1998). *Risk assessments for accidental releases from nuclear power plants in Europe.* National Institute of Public Health and the Environment (RIVM), Bilthoven, The Netherlands

UNEP (1998). *Scanning subsidies and policy trends in Europe and Central Asia*. van Beers, C., and de Moor, A. UNEP/DEIA/TR.98-2. UNEP, Nairobi, Kenya

UNEP (1989). *Futures for the Mediterranean Basin – The Blue Plan.* Prepared by Grenon, M. and M. Batisse. Oxford University Press, New York, United States,and London, United Kingdom

UNEP (1999a). *GEO-20000 Alternative Policy Studies for Africa.* htttp://wwww.unep.org/geo20000/aps-africa

UNEP (1999b). *Early Warning of Selected Emerging Environmental Issues in Africa: change and correlation from a geographic perspective.* Singh, A., Dieye, A., Finco, M., Chenoweth, M.S., Fosnight, E.A., and Allotey, A. UNEP, Nairobi, Kenya (UNEP/DEIA&EW/TR.99-2)

UNEP (1999c). *GEO-20000 Alternative Policy Studies for Asia and the Pacific* htttp://wwww.unep.org/geo20000/aps-asiapacific

UNEP (1999d). *GEO-20000 Alternative Policy Studies for Latin America and the Caribbean* htttp://wwww.unep.org/geo20000/aps-lac

UNEP (1999e). *GEO-20000 Alternative Policy Studies for North America and the Caribbean* htttp://wwww.unep.org/geo20000/aps-namerica

UNEP (1999f). *GEO-20000 Alternative Policy Studies for West Asia* htttp://wwww.unep.org/geo20000/aps-wasia

UNEP/RIVM (1999). *Energy related environmental impacts of scenarios with and without additional policies, GEO-2000 Alternative Policy Study for Europe and Central Asia 1990–2010*. National Institute of Public Health and the Environment (RIVM), Bilthoven, the Netherlands, and UNEP, Nairobi, Kenya (UNEP/DEIA&EW/TR.99-4)

United Nations (1995). *Energy, Environment and Sustainable Development.* Energy Resources and Development Series No. 35. United Nations, New York, United States

US Congress (1995). *Taking from the Taxpayer: Public Subsidies for Natural Resource Development.* US Congress Committee on Natural Resources, Subcommittee on Oversight and Investigations, Washington DC, United States

US Government (1997). *Economic Report of the President*. Transmitted to the US Congress, February 1997, together with the *Annual Report of the Council of Economic Advisors*. US Government Printing Office, Washington DC, United States

WEC (1995). *Efficient Use of Energy Utilizing High Technology*. World Energy Council, London, United Kingdom

World Bank (1989). *Sub-Saharan Africa – From Crisis to Sustainable Growth.* World Bank, Washington DC, United States

World Bank (1994). *World Population Projections,* 1994/95. World Bank, Washington DC, United States

World Bank (1995). *Toward Environmentally Sustainable Development in Sub-Saharan Africa.* World Bank, Washington DC, United States

World Bank (1997). *Can the Environment Wait? Priorities for East Asia.* World Bank, Washington DC, United States

World Bank/MMA (1998). *Agroforestry Experiences in the Brazilian Amazon: Constraints and Opportunities.* Pilot Programme to Conserve the Brazilian Rain Forest. MMA/World Bank, Brasília, Brazil

Chapter Five

Outlook and Recommendations

Outlook and Recommendations

Outlook for the 21st century

The beginning of a new millennium finds the planet Earth poised between two conflicting trends. A wasteful and invasive consumer society, coupled with continued population growth, is threatening to destroy the resources on which human life is based. At the same time, society is locked in a struggle against time to reverse these trends and introduce sustainable practices that will ensure the welfare of future generations.

Whilst the overall outcome is still unclear, the *GEO-2000* assessment shows both that time is running out on some issues and also that new problems are surfacing to compound an already difficult situation. However, there have also been some remarkable achievements in halting environmental degradation.

Time is running out

There used to be a long time horizon for undertaking major environmental policy initiatives. Now time for a rational, well-planned transition to a sustainable system is running out fast. In some areas, it has already run out: there is no doubt that it is too late to make an easy transition to sustainability for many of the issues discussed in Chapter 2 of this publication. Full-scale emergencies now exist on a number of issues. For example:

- The world water cycle seems unlikely to be able to cope with the demands that will be made of it in the coming decades. Severe water shortages already hamper development in many parts of the world, and the situation is deteriorating.
- Land degradation has reduced fertility and agricultural potential. Replacing lost top soil takes centuries or even millennia. These losses have negated many of the advances made through expanding agricultural areas and increasing productivity.
- Tropical forest destruction has gone too far to prevent irreversible damage. Even if current trends were reversed, it would take many generations to replace the lost forests; the cultures that have been lost with them can never be replaced.
- Many of the planet's species have already been lost or condemned to extinction because of the slow response times of both the environment and policy makers; with one-quarter of the world's mammal species now at significant risk of total extinction, it is too late to preserve all the biodiversity that our planet once had.
- Many marine fisheries have been grossly over-exploited, and their recovery will be slow. Future

growth in demand for fish will have to be satisfied by aquaculture – itself a practice fraught with environmental dangers.
- More than half of the world's coral reefs are threatened by human activities, with up to 80 per cent at risk in the most populated areas. While some may yet be saved, it is too late for many others.
- Urban air pollution problems are reaching crisis dimensions in many of the megacities of the developing world, and the health of many urban dwellers has been impaired.
- Finally, the indications are that it is too late to prevent global warming as a result of increased greenhouse gas emissions; in addition, many of the targets agreed on in the Kyoto Protocol may not be met.

New problems ...

Since *GEO-1* was published in 1997, further dimensions to the major environmental issues facing the planet have been recognized, and the situation differs from what it was even two years ago. New events or insights include the following:

- Emerging recognition of a global nitrogen problem, with some areas receiving nitrogen compounds in quantities that lead to unwanted ecosystem changes, such as excessive plant growth.
- More and larger forest fires, caused by a combination of unfavourable weather conditions and land use that made susceptible areas more prone to burning; both forests and the health of inhabitants have been threatened over areas of millions of hectares.
- An increased frequency and severity of natural disasters, now killing and injuring many millions of people every year, and causing severe economic losses.
- With 1998 the warmest year on record, climate change problems coupled with the most severe *El Niño* to date have caused major losses of life and economic damage.
- The economic and ecological importance of species invasions, an inevitable result of increasing globalization, appears to have become more significant.
- Declines in some countries in the quality of governance which have weakened capabilities to solve national and regional problems and to manage the environment.
- Declining government and media focus on urgent environmental issues as attention has been diverted by political and economic upheavals.
- New wars which, like all wars, threaten not only the environment of those directly involved but that of neighbouring states, and those downstream on major rivers.
- The environmental importance of refugees, who are forced to make unrestricted assaults on the natural environment for their survival.

... and continued successes

This record of planetary negligence must be tempered, however, by a series of remarkable achievements, many of which would have been unthinkable even two decades ago. These gains promise major benefits for the future. Examples include:

- The increase in public concern over environmental issues. Until recently, few individuals cared about or even knew of the environmental issues facing the planet. Today, popular movements in many countries are forcing authorities to make changes.
- Voluntary action taken by many of the world's major industries to reduce resource use and eliminate waste. The fact that these actions are in industry's own economic self-interest does not detract from their environmental significance. On the contrary, the happy discovery that what is good for the environment can also be good for business may do much to reverse trends for which industry itself was originally largely responsible. This 'win-win' situation bodes well for the planet.
- Remarkable successes by governments in developed regions in reducing levels of air pollution in many major cities. Innovative legislation has been introduced, and the goal of zero emissions in several important areas is no longer considered utopian.
- The halt and reversal of deforestation in parts of both Europe and the North America. In other regions, eco-certification of forest products is on the increase.
- Local *Agenda 21* initiatives, which have proved an effective way of developing and implementing sustainable development policies that involve communities and political agencies alike.

At the international level:

- The ozone layer is now expected to have largely recovered within half a century as a result of the Montreal Protocol. While most other environmental problems are not as straightforward, the fact that the international community is well on the way to resolving one major issue should certainly give the critics of international organizations pause for thought.
- Since 1992, the first international steps – the United Nations Framework Convention on Climate Change and its Kyoto Protocol – have been taken to tackle the issue of global climate change. In addition, the world's scientists, meteorologists and climatologists are gaining major new insights into climate variation. Prediction of climatic variations of all kinds, human induced or not, are likely to become a regular feature of life in the 21st century.

The overall trend

GEO-2000 confirms the overall assessment of *GEO-1:* the global system of environmental policy and management is moving in the right direction but much too slowly. Dramatic and instructive variations can be found between regions, between economic sectors and between environmental issues. But, on balance, gains by better management and technology are still being outpaced by the environmental impacts of population and economic growth. As a result, policy actions that result in substantial environmental improvements are rare.

The present course is unsustainable and postponing action is no longer an option. Inspired political leadership and intense cooperation across all regions and sectors will be needed to put both existing and new policy instruments to work. Global issues have been recognized and effectively tackled for the first time in the latter part of the 20th century. It is important to make sure that these global lessons filter down to regional, national and local levels – concerted action always achieves more than individual initiatives, at every level. Similarly, global efforts could benefit substantially from the wealth of collaborative experience gained over the years at regional and sub-regional levels.

Recommendations for action

One of GEO's tasks is to recommend measures and actions that could effectively reverse unwelcome trends and reduce threats to the environment. Specific conclusions and recommendations have been formulated throughout this report. In addition, on the basis of the *GEO-2000* assessment, UNEP recommends that future action be focused on four key areas:

- filling the knowledge gaps;
- tackling root causes;
- taking an integrated approach; and
- mobilizing action.

None of these areas is discrete – the actions required to address one area are likely to benefit other areas as well. And each of the areas is open-ended – the suggestions for action described below are only a sample of those that could be included under each heading. In some areas, a good start has already been made. However, UNEP believes that increased and concerted policy development and action in these four 'cross-cutting' areas would do much to break the stalemate which currently prevails on too many pressing environmental issues. The rationale for selecting these key areas is described in the text that follows; the boxes at the end of each of the four sections summarize suggestions for action that have arisen since *GEO-1.*

Filling the knowledge gaps

GEO-2000 shows that we still lack a comprehensive view of the interactions and impacts of global and inter-regional processes. Information on the current state of the environment is riddled with weakness. There are few tools to assess how developments in one region affect other regions, and whether the dreams and aspirations of one region are compatible with the sustainability of the global commons.

Another serious omission is the lack of effort to find out whether new environmental policies and expenditures have the desired results. These knowledge gaps act as a collective blindfold that hides both the road to environmental sustainability and the direction in which we are travelling. However, whilst it is imperative to address these gaps, they should not be used as an excuse for delaying action on environmental issues that are known to be a problem.

Knowledge gaps need to be addressed in the following areas.

Environmental data and information

Integrated environmental assessment needs to be firmly based on reliable technical data and information. In attempting to obtain a strong database for the *GEO-2000* assessment, a number of critical problems with existing datasets became very apparent:

- Many datasets, including those on air pollution and water quality, are incomplete; others are non-existent, especially in developing countries. In addition there is a lack of reliable socio-economic data that can be related to the environment.
- The quality of much existing data is of equal concern – there are problems of reliability and of consistency between subject areas and countries. There are, for example, no comparable urban air quality data for the cities of the world.
- The kinds of data that are collected are often not useful for answering the key questions about the environment.
- Most of the available data apply to quantitative aspects of the environment. Little attempt has been made to measure the qualitative parameters that are equally important indicators of sustainability.
- Trend detection requires time-series data. Many datasets are once-off collections of figures that have never been repeated.
- Environmental assessments need data that are geo-referenced and information that is broken down by spatial areas other than administrative units. These data are still relatively scarce.
- Assessment at regional and global scale requires the aggregation of data relating to smaller areas. Data can be aggregated only if they measure the same thing in the same way and with the same precision. Available data frequently do not measure up to these specifications.

These shortcomings make integrated, cross-sectoral global assessment and trend analysis always difficult and sometimes impossible.

Environmental progress and policy effectiveness can be assessed only if quality data are routinely collected through monitoring systems. However, environmental monitoring infrastructure is poorly developed in most countries, making regular production of policy-relevant environmental data and indicators impossible. The situation is being exacerbated by the decline of some existing monitoring systems due to shrinking resources.

Earth observations by satellite now provide a means of collecting data for large areas relatively cheaply and uniformly. While this technology will certainly reduce the need for ground measurements, it does not make all direct observations or ground truthing redundant. More importantly, many of the data categories that are needed to draw up policy-relevant assessments – on resource efficiency, impacts on human well-being and suchlike – cannot be detected from space.

Much effort has gone into the search for suitable indicators for reporting purposes and hundreds have been suggested. Which are the really useful ones is still not known, nor has an aggregated group yet been selected to form the environmental equivalent of the Human Development Index.

The monitoring and reporting of data and indicators require a coordinated approach and the strengthening of many existing local, national and international initiatives. An analysis of data issues undertaken within the GEO framework highlights the need for a mechanism that brings together the compilers of global assessments on sustainable development (as data users) and key actors in the production and dissemination of the required data. The mechanism would identify and take initiatives to address critical and common data gaps for global assessment, paying particular attention to data access and data sharing. Arrangements are also needed for sharing data with the secretariats of multilateral environmental agreements (MEAs) in order to provide a consistent basis for assessment and reporting, while at the same time reducing the response burden on governments. In parallel, institutional, technical and other resources need to be provided for monitoring, and data collection standards improved. Potential environmental indicators must be tested worldwide, and a set of indicators identified that can be used to report on environmental progress.

Policy performance

An element of uncertainty is associated with most environmental policy measures. Yet indicators of policy effectiveness and underlying observing mechanisms are lacking everywhere, from local level initiatives to multilateral agreements. These deficiencies prevent the monitoring and assessment of policy performance.

This situation includes most MEAs, where a lack of uniformity in monitoring data, regularly-updated indicators and continuous reporting prevents comparisons being made between the current situation and what would have happened if no agreement had been concluded. A start has been made for some MEAs. Performance monitoring is already possible for stratospheric ozone and for greenhouse gas emissions.

Monitoring the impacts of current policies should precede and pave the way for the formulation of alternative or additional policies. The key is to consider policy instruments as tools for learning and adaptation, and to treat them with flexibility.

Routine assessment of the performance of environmental policies, including international agreements, is therefore urgently needed to fill this gap in the policy process. To carry out such assessments, suitable indicators need to be agreed and capacities developed to handle statistical and geographical data. Assessment results must obviously be made easily accessible to policy makers and the general public.

Trade and environment linkages

The purpose of the World Trade Organization (WTO), and regional trade accords such as the North American Free Trade Association and MERCOSUR, is to build neutral trade policies to avoid border or internal restrictions, thus promoting the free flow of goods and capital. Despite the huge volume and rapid growth of global trade, the environmental implications of the new regime are far from fully known. Trade liberalization should lead to the more rational use of resources between countries, thereby leading to greater efficiencies (economies of scale) and increases in global economic growth. However, there is a risk of eroding high environmental standards if lowest common denominators are adopted amongst trading partners. At the same time, trade liberalization could lead to inappropriate resource use and a shifting of environmental pressures from one region to another, not necessarily towards the region where they can best be handled.

Not only is the effect of trade liberalization on environmental quality unknown but, conversely, the extent to which the 200 or so international environmental agreements impact on trade is largely undocumented. Whether environmental agreements create, obstruct or divert international trade is still unknown.

Measures are needed to address these unknowns. Resources are needed for research on trade–environment linkages, setting up an international mechanism to monitor the impacts of these linkages, and enhancing the capabilities of countries to assess the environmental, social and economic implications of trade liberalization. In addition, the active engagement of countries, particularly developing countries, in environment–trade related negotiations and agreements will improve understanding of the issues involved. Better information and insight could lead to the development of policies that promote sustainable trade.

International finance and the environment

Interaction between the financial world and the environment is another crucial area where comprehensive global knowledge and action are currently limited. A start has been made – studies conducted within the framework of the Commission on Sustainable Development have led to several new proposals, including Tobin-type taxes (see page 207) which would raise money for the environment through an international tax on financial transactions. Canada took this idea a step further when its House of Commons voted, in March 1999, to authorize the federal government to promote the Tobin tax internationally. In addition, many banks and lending organizations, including the World Bank, have incorporated environmental considerations into their operations.

Despite these positive signs, a comprehensive global overview is urgently needed, particularly because the volume of overseas development aid continues to decline, amounting in 1996 to only about one-fifth the volume of foreign direct investment (World Bank 1997).

It has been estimated that 3 per cent of gross domestic product is the minimum amount needed for environmental protection and restoration. In addition, industry and the public currently allocate more than US$450 000 million a year towards environmental

Filling the knowledge gaps suggestions for action

- Identify a set of indicators that could be used to report on environmental progress.
- Establish a mechanism that brings together the compilers of global assessments on sustainable development and key actors in the production and dissemination of the required data.
- Provide the institutional, technical and other resources needed to improve monitoring and data collection standards.
- Channel resources into research on trade–environment linkages and developing proposals for an international mechanism to monitor the impacts of these linkages.
- Assess the extent to which international financial markets meet *Agenda 21* targets, with the eventual aim of channelling investment flows into areas that provide a sound basis for sustainable development.
- Find innovative ways of meeting the financial shortfall on sustainable development – such as a small tax on tourism.
- Implement policy performance monitoring by identifying suitable indicators, developing capacities to handle statistical and geographical data, and ensuring that assessment results are easily accessible to policy makers and the general public.

protection. Yet there are still no global-level tools to assess – and, if required, improve – the ways in which such huge amounts of money are spent.

Assessing the extent to which international financial markets address *Agenda 21* targets, with the eventual aim of chanelling investment flows into areas that provide a sound basis for sustainable development, is thus a top priority. Finding innovative ways to meet the financial shortfall on sustainable development – such as a small tax on tourism, which is now responsible for 8 per cent of GDP and exerts heavy pressure on the environment – is another high priority need.

Tackling root causes

Means must be found to tackle the root causes of environmental problems, many of which are unaffected by strictly environmental policies. Resource consumption, for example, is a key driver of environmental degradation. Policy measures to attack this issue must reduce population growth, reorient consumption patterns, increase resource use efficiency and make structural changes to the economy. Ideally, such measures must simultaneously maintain the living standards of the wealthy, upgrade the living standards of the disadvantaged, and increase sustainability. Inevitably, this will require a shift in values away from material consumption. Without such a shift, environmental policies can effect only marginal improvements.

Some policy measures are better than others in dealing with root causes. Taxing resource use rather than employment is one possible measure introduced with success in some countries of the European Union. Reforming subsidies for resource-intensive, polluting sectors is another. Use of the best available technology and production processes – incorporating the principles of cleaner production and eco-efficiency – could reduce environmental pressures by a factor of two to five.

Both consumption and production drivers are targeted in *GEO-2000* for reform through policy action. Three specific areas of reform are recommended.

Subsidies

Subsidies for natural resources are widely used to stimulate economic development. All have the effect that the user pays less than the market price for commodities such as energy, land, water and wood. While some subsidies are useful for stimulating economic or social development, protecting dependent communities or reducing dependence on imported resources, they can also encourage uneconomic practices and lead to severe environmental degradation. Without subsidies for irrigated water, for example, farmers in the western United States would be less likely to grow rice and other water-intensive crops in arid regions. Without crop supports, farmers would be less likely to overuse fertilizers and pesticides, a major source of water pollution. Without road transport subsidies, traffic congestion, urban air pollution and carbon dioxide emissions could be significantly reduced worldwide. And without energy subsidies, energy prices would rise, encouraging the use of more efficient vehicles and industrial equipment, and reducing pollutant emissions. Some subsidies established long ago for sound economic or social reasons no longer serve their original purpose. Subsidies can take many forms and are often hidden so that even the beneficiaries may be unaware of the adverse environmental impacts they are having.

Policy packages that reduce distorting subsidies without causing hardship, particularly to poor sectors of the population and small-scale industry, are badly needed. Cutting the link between support measures and resource use, leaving the support intact but removing the perverse incentive, is a first step. It is also important to raise awareness, amongst the general public and others, of the linkages between subsidies and environmental degradation, and of the impact of subsidy size.

Energy consumption

In *GEO-1*, energy demand was projected to grow 80 per cent between 1990 and 2015, even with substantial increases in energy efficiency. This prospect has not changed much. Most growth in energy use will occur in developing regions, especially in Asia. Without major policies changes, the projected increase in energy use will result in a large increase in the emissions of greenhouse gases. Two global developments will affect future energy use and emissions of greenhouse gases: energy prices and the Kyoto Protocol.

Excess production capacity has recently led to low oil prices while new insights into economically-recoverable fossil fuel reserves suggest lower energy prices for at least the next few decades, particularly for oil and natural gas. These low fossil fuel prices make it unlikely that the market share of renewable energy sources will grow significantly in the next few decades without major policy interventions favouring non-fossil energy resources combined with taxes on fossil fuel use to reduce urban air pollution, acidification and climate change.

The Kyoto Protocol is only an early step in this direction. The Protocol itself will not be sufficient to stop growth in global greenhouse gas emissions and must be followed by other major steps by both developed and developing countries. This poses a major policy challenge because developing countries have legitimate demands for economic growth.

What is needed are policies that promote economic development while simultaneously limiting greenhouse gas emissions. Policies that favour a transition to an energy system that is less dependent on fossil fuels and that build on the principle of common but differentiated responsibilities for developed and developing countries in relation to the equitable use of the global atmosphere should be vigorously pursued. Efforts should also be made to speed up the transfer of efficient technologies, in view of the long lead times needed to make technical changes, and to develop international strategies for de-carbonization.

Production technology

GEO-2000 reinforces the conclusions of *GEO-1* that 'worldwide application of the best available technology and production processes has yet to be ensured through the exchange and dissemination of know-how, skills and technology'. It clearly demonstrates that adoption is hindered by ignorance of cleaner production potentials and lack of dissemination of improved technology to target groups.

New financial mechanisms, particularly loan systems, are needed to bring about the rapid dissemination of cleaner and more efficient production techniques. Greater efforts have to be made to expose industrialists, especially in developing countries and countries with economies in transition, to the potential advantages of investing in cleaner and more efficient production techniques, especially to the 'win-win' outcomes that are possible in the medium and longer term. Small levies on polluting emissions can

Tackling the root causes suggestions for action

- Design new policy packages that reduce the role of subsidies without causing hardship, particularly to poor populations and small-scale industry.
- Take steps to raise awareness of the linkages between subsidies and environmental degradation.
- Design policies that favour alternative energy use, with differentiated responsibilities for developed and developing countries in relation to the equitable use of the global atmosphere.
- Take early action to catalyse the adoption of energy-efficient technologies.
- Develop international strategies for de-carbonization.
- Develop new financial mechanisms, particularly loan systems, to bring about the rapid dissemination of cleaner and more efficient production techniques.
- Find ways of disseminating the advantages of cleaner and more efficient production techniques to more industrialists, particularly in developing countries and countries with economies in transition.
- Introduce small levies on emissions to shift market conditions in favour of cleaner technology in the energy and other sectors.

be used to shift market conditions in favour of cleaner technology in the energy and other sectors

Taking an integrated approach

Agenda 21 championed the concept of environmental integration. There are two aspects to this – the way the environment is thought of and the way it is dealt with. *GEO-2000* shows there is inadequate integration on both. Further efforts are called for in three areas.

Mainstream thinking

The environment remains largely outside the mainstream of everyday human consciousness and is still considered as an 'add-on' to, rather than an integral part of, the social, economic and institutional fabric of life. And the environment is rarely taken as seriously as the social, economic and other components of national and regional planning. There is a critical gap between macroeconomic policy-making and environmental considerations. Although there are positive exceptions, many macroeconomic institutions – treasury, budget office, central banks, planning departments – still ignore sustainability questions and the long-term benefits of environmental choices against short-term economic options. The state of natural resources is often ignored when national macro-economic policies are evaluated.

Options for add-on environmental policies have been exhausted in many sub-regions. Better integration of environmental thinking into the mainstream of decision-making relating to agriculture, trade, investment, research and development, infrastructure and finance is now the best chance for effective action. This will require innovative policy, social, institutional and economic changes, and considerable perseverance at the political level backed up by convincing and forceful arguments. Environmental economics can be put to good use, for instance, to stress the high economic value of environmental goods and services, and the high costs of poor environmental management or inaction.

Integrated management

Sectoral policies conceived in isolation from related sectors do not always yield the desired results – and, indeed, can even have negative impacts, particularly when viewed over a longer time frame. Environmental policies that encompass broad social considerations are the most likely to make a positive and lasting impact. This holds good across the gamut of environmental issues – for example, water, land and other forms of natural resource management, forest conservation, air quality control, and urban and coastal area management.

Integrated management requires an understanding of the interlinkages involved, and an assessment of the results and risks that actions may have. Furthermore, management policies must always take into account the realities of the situation. For example, it may make no sense to try to improve land and water management if secure property rights are not in place.

Clean water and food security: putting integrated policies into practice

Many are still denied the clean water and food security which are basic human rights. Assessments confirm a dramatically-rising pressure on land resources, particularly in mainland Asia and Africa. Future freshwater problems look even more severe than they did two years ago. *GEO-2000* emphasizes how land issues are inextricably interwoven with water management at both national and regional levels.

A holistic approach to the management of water and food requires:

- Making full use of economic instruments that treat land and water as scarce economic resources that are part of the Earth's natural capital.
- Coordinating the management of land and water resources as closely as possible
- Establishing secure land and water property rights where these do not exist.
- Reorganizing land and water management policies on a river basin level.
- Introducing the concept of shared and equitable water use to resource allocation strategies.
- Reformulating regional and national agricultural and food security strategies to bring them into line with the principles of sustainable development.
- Providing people with alternatives to the use of marginal land.
- Reducing water wastage in urban areas.

Further research is needed on the socio-economic causes of environmental deterioration and the interlinkages within and among environmental and sustainability issues in order to define the priority issues and suggest ways of addressing them. Multisectoral approaches are needed at national level, with planning carefully tailored to local or regional circumstances as appropriate. Stakeholders need to be involved from the start when formulating and introducing integrated policies.

International coordination

Improved international coordination on environmental issues is a third prerequisite of the trend towards a more integrated approach.

Taking an integrated approach — suggestions for action

- Promote sustainable development as the central theme in policies relating to agriculture, trade, investment, research and development, infrastructure and finance by stressing the high economic and social value of environmental goods and services, and the high costs of poor environmental management.
- Conduct more research on the socio-economic causes of environmental deterioration and the interlinkages within and among environmental and sustainability issues in order to define the priority issues and suggest ways of addressing them.
- Work towards integrated multisectoral policies at national level, involving all stakeholders from the start.
- Improve coordination between MEAs at several levels – secretariat-level management, national-level implementation and regional or global-level performance monitoring.
- Establish a multi-agency, multi-stakeholder task force to develop proposals for strengthening global coordination and governance structures to protect the global commons.

Bilateral and multilateral environmental agreements have proven powerful instruments of change. Understanding of the key factors governing the success of agreements has evolved considerably. The ultimate and combined effect of the many global and regional agreements remains uncertain but it is clear that all multilateral agreements can make positive contributions to environmental policy.

There is a trend towards agreements with a wider scope, not only at the global but also at the regional and sub-regional levels. At the same time, the common ground between many global conventions is becoming increasingly apparent. This provides room for synergy and avoiding duplication of effort.

Coordination between MEAs and regional agreements needs strengthening at several levels, including cooperation between secretariats, national implementation, and regional and global performance monitoring.

There is also scope for the improvement of global environmental governance. Global environmental problems require strengthened global coordination structures that protect the global commons, ensure the long-term sustainability of planet Earth, encourage governments to take actions, and provide agreed frameworks to do so. These structures will need to be reinforced by environmental observing systems, scientific research programmes, policy advice and

Linking science, policy, environment and basic human needs

UNEP, the World Bank and the National Aeronautics and Space Administration (NASA) of the United States have collaborated to identify the key scientific and policy linkages amongst environmental issues (climate change, loss of biodiversity, fresh and marine water degradation and others) and linkages between these issues and meeting basic human needs for adequate food, clean water, energy and a healthy environment. The interlinkages assessment report (UNEP, World Bank and NASA 1998) shows that there are several facets to these linkages:

- The Earth's physical and biological systems provide humans with essential goods and services;
- A set of physical, chemical and biological processes link global environmental problems so that changes in one have repercussions for others;
- Actions taken to meet human needs have local, regional and global consequences;
- The same driving forces – population size, consumption levels and choice of technologies – underlie all global environmental problems; and
- All people affect the environment, and *vice versa,* but the rich have a disproportionately higher impact and the poor tend to be most vulnerable to the effects of environmental degradation.

Three important recommendations emerge from the report:

- To respond effectively to global environmental problems and, in turn, meet human needs more effectively, global environmental issues must be addressed in a holistic, integrated manner, building on the same technologies and policy instruments that are currently used to contend with these issues in a sectoral manner.
- New institutional partnerships involving governments, the private sector, academia, NGOs and civil society are needed at the global, regional and national levels.
- Taking into account that most changes to the global environment cannot be reversed quickly, and despite scientific uncertainties, decision makers need to adopt wise, cost-effective and adaptive management approaches that can be implemented now.

Source: UNEP, World Bank and NASA 1998

assessment panels, legislative bodies and international policy action mechanisms – some of which are already at an embryonic stage of development.

Mobilizing action

Solutions to environmental issues must come from cooperative action between all those involved –individuals, NGOs, industry, local and national governments, and international organizations. The need to involve all the parties concerned underlies all the regional analyses of current and alternative policies in *GEO-2000*. Specific examples include the increasing role of NGOs in multilateral agreements, the involvement of stakeholders in property rights issues, and instances where manufacturing and resource industries are taking a leading role on some issues. Decentralization has been decisively instrumental in involving a wider range of groups in issues of concern in some countries.

People

Of all the groups just mentioned, individuals are vitally important –they experience the deteriorating environment at first hand and they often know the best solutions. Their cumulative lifestyles make a huge impact– a small adaptation made millions of times over can add up to a significant change. Although public awareness and concern about the environment has continued to grow, public participation in many decision-making processes is still limited and environmental regulation is often viewed as a burden rather than a promoter of sustainable growth.

The *GEO-2000* policy analyses confirm that public participation is a key element in improved environmental management. Policies implemented without the full participation of stakeholders, particularly the poor and socially-deprived groups, have proven largely unsustainable. However, many citizens still lack a feeling of ownership with regard to national environmental legislation and management.

The general public's knowledge of the environment is the foundation on which environmental policies build. This knowledge is often seriously deficient. Education and public awareness programmes can change attitudes and produce more environment-friendly and sustainable life styles, at the same time encouraging public participation and action on environmental issues. Formal and non-formal education on the environment is therefore critically important. Public participation in environmental management could be improved by:

- Making environmental education, like mathematics, an integral part of the standard educational curriculum.
- Including regional and global issues and perspectives in environmental education syllabi.
- Expanding public awareness and education programmes to target more groups in society, especially engineers and economists.
- Improving access to environmental information.
- Encouraging the media to devote as much attention to environmental issues as they do to crime, politics, sport and finance.

Community groups and NGOs

NGOs and civil society groups have become increasingly well established and organized in many countries during the past decade. By addressing issues that matter to the individual, they now form influential lobby groups in many national and international arenas as well as spearheading a wide range of environment-related activities at grassroots level. These groups have much to offer, particularly in an intermediary role.

NGOs and community groups will become more important and more influential in the years to come. They need to be given more and specific responsibilities for environmental management. Their involvement should also be enlisted more widely in, for example, environmental monitoring and assessment.

The private sector

One of the indirect conclusions that can be drawn from most regional outlook studies is the diminishing power of local and national governments while the private sector becomes more influential. Business and industry now shoulder many of the responsibilities formerly taken on by governments. Multinational corporations, for long powerful forces in the global economy, have led the action to establish and implement voluntary actions such as codes of conduct, responsible care programmes and voluntary reporting on environmental performance (environmental auditing). This has occurred usually where a conducive national framework was in place in the country where the multinational headquarters was located. Although it is impossible to assess the exact contribution of these actions to overall environmental protection and global environmental stewardship, there is no doubt of their general utility. As yet, however, the improved environmental performance of large-scale industry has not been echoed by small and medium-sized firms, which need both help and encouragement.

Global coordination and information exchange amongst industries and between industrial sectors could lead to more widespread action, and large industries can be encouraged to help small and medium-sized industries with voluntary action and implementing the 'triple bottom line' – social, economic and environmental accountability, or 'people, profit and planet'. The overall impact of industrial initiatives should be assessed and recommendations made on how they can become more effective.

National governments

National governments have many obligations on the environmental front. Whilst they may work in close collaboration with others, they have the ultimate responsibility for national policy development and implementation, enforcing national environmental legislation, ensuring national level compliance with international agreements, public education and awareness building, and so on. A climate of stability is required before national decision makers will turn their attention to the environment and make real headway on environmental problems. Likewise, good governance and security are prerequisites for sustainable development, and it is up to citizens, governments and multilateral organizations to bring these about, both within and across national boundaries.

Many public environment agencies have experienced severe cutbacks over recent years. These cuts have affected the ability of these agencies to fulfil their core responsibilities. While sustainable development requires that sectors and agencies integrate the environment into their decision-making, this does not make environment agencies redundant nor is it a reason to reduce their budgets or to cut back on environmental spending. On the contrary, environmental agencies should take the lead in integrating environmental considerations in other policy domains from a position of strength. This will require the establishment of cross-cutting institutions whilst maintaining strong environmental agencies able to assess the overall state and trends of the environment, implement environmental policies and enforce environmental laws.

Other measures that could assist governments with their environmental mandates include setting up a dedicated system for mediation on environment-related disputes – by establishing an environmental ombudsman, for example – to supplement the increasing role of the judiciary for achieving environmental protection and sound management. In addition, multi-stakeholder involvement could be promoted by opening up decision-making processes, encouraging dialogue and information exchange between stakeholders to develop the mutual trust and goodwill that underlies all successful partnerships, and

Mobilizing action suggestions for action

- Improve public access to environmental information.
- Make environmental education, like mathematics, part of the standard educational curriculum.
- Encourage the media to devote as much attention to environmental issues as they do to crime, politics, sport and finance.
- Open up decision-making processes to all stakeholders and ensure that they have a share of benefits – particularly from profits accrued by the exploitation and subsequent export of natural resources.
- Provide more opportunities for NGOs and community groups to participate in environmental action.
- Encourage large industries to help small and medium-sized industries with voluntary action and implementing the 'triple bottom line' of social, economic and environmental accountability.
- Establish processes to evaluate the impacts of industry-led initiatives on the environment.
- Strengthen national cross-cutting institutions while maintaining strong environment agencies able to implement environmental policies, enforce environmental laws and assess the overall state of the environment.
- Accompany government decentralization with local capacity building and a redistribution of financial powers and accountability.
- Ensure that adequate systems are in place to deal efficiently with environment-related disputes.
- Increase support for international environmental organizations to enable them to improve their advisory, coordinating, mediating, implementing and assessment functions.

ensuring that all stakeholders have a share of benefits – particularly from profits accrued by the exploitation and subsequent export of natural resources.

International organizations

International organizations within and without the United Nations system have a range of environmentally-related mandates. These include providing local and national level assistance in environmental and natural resources management, coordinating regional and sub-regional development programmes, guiding intergovernmental negotiations on convention protocols, financing arrangements, conflict resolution and global-level environmental assessment. Establishing clear responsibilities, eliminating overlap and duplication, and improving information exchange remain major challenges.

Increased support to international environmental organizations will enable them to improve international cooperation at regional and global levels, strengthen conflict-resolution mechanisms, and implement environmental programmes and projects more effectively.

References

UNEP, World Bank and NASA (1998). *Protecting Our Planet – Securing Our Future: Linkages Among Global Environmental Issues and Human Needs.* Watson, R.T., Dixon, J.A., Hamburg, S.P., Janetos, A.C., and Moss, R.H. (eds.). UNEP, Nairobi, Kenya

World Bank (1997). *Global Development Finance 1997*. The World Bank, Washington DC, United States

Acronyms and Abbreviations

AAFC	Agriculture and Agri-food Canada
ACSAD	Arab Centre for Studies of Arid Zones and Drylands
ACTS	African Centre for Technology Studies
ADB	African Development Bank
ADB	Asian Development Bank
AEC	African Economic Community
AEPS	Arctic Environmental Protection Strategy
AGROSTAT-PC	FAO Agricultural Statistics database for PC
AGU	Arabian Gulf University, Bahrain
AIDMO	Arab Industrial Development and Mining Organization
AIDS	Acquired Immunodeficiency Syndrome
AIM	The Asia Pacific Integrated Model
AIT	Asian Institute of Technology, Thailand
ALDA	Asociación Latinoamericana de Derecho Ambiental, Mexico
ALECSO	Arab League Educational, Cultural and Scientific Organization
ALIDES	Central American Alliance for Sustainable Development
AMAP	Arctic Monitoring and Assessment Programme
AMCEN	African Ministerial Conference on the Environment
AOAD	Arab Organization for Agricultural Development
ARET	Accelerated Reduction/Elimination of Toxics
ASEAN	Association of South East Asian Nations
ASOC	Antarctic and Southern Ocean Coalition
ASOEN	ASEAN Senior Officials on the Environment
ATCM	Antarctic Treaty Consultative Meeting
ATS	Antarctic Treaty System
BAT	Best available technology
BCAS	Bangladesh Centre for Advanced Studies
BEAR	Barents Euro-Arctic Region
BOD	Biological oxygen demand
CAFF	Conservation of Arctic Flora and Fauna
CAMRE	Council of Arab Ministers Responsible for the Environment
CAP	The Common Agricultural Policy of the European Union
CBD	Convention on Biological Diversity
CCAD	Central American Commission on Environment and Development
CCAMLR	Convention on the Conservation of Antarctic Marine Living Resources
CCAP	Climate Change Action Plan
CCAS	Convention for the Conservation of Antarctic Seals
CCD	Convention to Combat Desertification
CCIAD	Central American Inter-Parliamentary Commission on the Environment
CDIAC	Carbon Dioxide Information Analysis Center of ORNL
CD-ROM	Compact disk - read only memory
CEC	Commission of the European Communities
CEC	Commission for Environmental Cooperation of NAAEC
CEDARE	Centre for Environment and Development for the Arab Region and Europe
CEPA	Canadian Environmental Protection Act
CEU	Central European University, Hungary
CFC	Chlorofluorocarbon
CGIAR	Consultative Group on International Agricultural Research
CIAT	Centro Internacional de Agricultura Tropical
CIESIN	Consortium for International Earth Science Information Network
CILSS	Permanent Inter-state Committee on Drought Control
CIS	Commonwealth of Independent States
CITES	Convention on International Trade in Endangered Species
CMS	Convention on Migratory Species
CO	Carbon monoxide
CO_2	Carbon dioxide
COMECOM	Council for Mutual Economic Assistance
COMEMIS	Coastal and Marine Environment Management Information System
COMNAP	Council of Managers of National Antarctic Programmes
CONABIO	National Commission for the Knowledge and Use of Biodiversity, Mexico
COP	Conference of the Parties
CP	Cleaner production
CSD	Commission on Sustainable Development of the UN
CSI	Common Sense Initiative
CZMP	Coastal Zone Management Plan
DALE	Disability-adjusted life expectancy
DDT	Dichloro Diphenyl Trichlorethane
DEIA&EW	Division of Environmental Information, Assessment and Early Warning of UNEP
DPCSD	Department for Policy Coordination and Sustainable Development of the UN
EAS	East-Asian Seas (Action Plan)
EATR	Environment Assessment Technical Report of DEIA
EBRD	European Bank for Reconstruction and Development
EC	European Commission
ECOWAS	Economic Community of West African States
ECU	New currency of the European Union
EDP	Eco-domestic product
eds	Editors
EEA	European Environment Agency
EEPSEA	Economy and Environment Programme for Southeast Asia
EfE	Environment for Europe, a European Union initiative
EFTA	European Free Trade Association
EGAT	Electricity Generating Authority, Thailand
EIA	Environmental impact assessment
EIB	European Investment Bank
EIS	Environmental information systems
EMAP-E	USEPA Monitoring and Assessment Programme for Estuaries
EMAS	Eco-Management and Audit Scheme
EMEP	European Monitoring and Evaluation Programme
ENRIN	Environment and Natural Resources Information Network of UNEP
EPPR	Emergency Prevention, Preparedness and Response
EQA	Environmental Quality Act, Malaysia

ERASMUS	European Action Scheme for the Mobility of University Students
FADINAP	Fertilizer Advisory, Development and Information Network for Asia and the Pacific
FALCAP	Framework for Action on Land Conservation in Asia and the Pacific
FAO	Food and Agriculture Organization of the United Nations
FASE	Foundation for Advancements in Science and Education
FDI	Foreign direct investment
FFI	Forest Frontiers Initiative
FSU	Former Soviet Union
GAP	Ganga River Action Plan
GATT	General Agreement on Tariffs and Trade
GCC	Gulf Cooperation Council
GCOS	Global Climate Observing System
GEF	Global Environment Facility
GEMS	Global Environment Monitoring System
GEO	Global Environment Outlook
GESAMP	Joint Group of Experts on the Scientific Aspects of Marine Environment Protection
GDP	Gross domestic product
GHGs	Greenhouse gases
GIEWS	Global Information and Early Warning System on Food and Agriculture
GIWA	Global International Waters Assessment
GNP	Gross national product
GLASOD	Global Assessment of Soil Degradation
GLWQA	Great Lakes Water Quality Agreements
GOOS	Global Ocean Observing System
GRID	Global Resource Information Database
GTOS	Global Terrestrial Observing System
GTZ	German development assistance agency
GWS	Global Waste Survey
Habitat	United Nations Centre for Human Settlements (UNCHS)
IAATO	International Association of Antarctic Tour Operators
IAEA	International Atomic Energy Authority
IBAMA	Instituto Brasileiro do Meio Ambiente e dos Recursos Naturais Renováveis (Brazilian Institute of Environment and Renewable Natural Resources)
ICARDA	International Center for Agricultural Research in the Dry Areas
ICCROM	International Centre for Conservation in Rome
ICIMOD	International Centre for Integrated Mountain Development
ICLEI	International Council for Local Environmental Initiatives
ICOMOS	International Council of Monuments and Sites
ICRISAT	International Crops Research Institute for the Semi-Arid Tropics
ICSU	International Council for Science
ICWE	International Conference on Water and the Environment
ICZM	Integrated Coastal Zone Management
IE	Industry and Environment Office of UNEP
IEEA	Integrated System of Environmental and Economic Accounting
IFF	International Forum on Forests
IFPRI	International Food Policy Research Institute
IIASA	International Institute of Applied Systems Analysis
IIED	International Institute for Environment and Development
IISD	International Institute for Sustainable Development, Canada
IIUE	International Institute for the Urban Environment
IFFWS	Inland Fisheries, Forestry and Wildlife Sectors in SADC
IGAD	Inter-Governmental Authority on Development
ILO	International Labour Organization
IMAGE	Integrated Model to Assess the Greenhouse Effects
IMERCSA	The India Musokotwane Environment Resource Centre for Southern Africa
IMO	International Maritime Organization of the UN
INBIO	National Biodiversity Institute, Costa Rica
INC	Intergovernmental Negotiating Committee
INFOTERRA	Global Environmental Information Exchange Network of UNEP
IOC	Intergovernmental Oceanographic Commission of UNESCO
IOC	Indian Ocean Commission, Mauritius
IPCC	Inter-Governmental Panel on Climate Change
IPM	Integrated Pest Management
IRPTC	International Register of Potentially Toxic Chemicals
ISESCO	Islamic Educational Scientific and Cultural Organization
ISO	International Standardization Organization
ISRIC	International Soil Reference and Information Centre, Netherlands
IUCN	World Conservation Union
IWC	International Whaling Commission
JCEDAR	Joint Committee on Environment and Development in the Arab Region
LAC	Latin America and the Caribbean
LCA	Life cycle assessment
LIFE	Community Financial Instrument for the Environment
LRTAP	Long-range Transboundary Air Pollution
MAP	Mediterranean Action Plan of UNEP
MARPOL	International Convention for the Prevention of Pollution from Ships
MEA	Multilateral Environmental Agreement
MEPA	Meteorology and Environmental Protection Administration of Saudi Arabia
MERCOSUR	Mercado Comun del Sur; Southern Common Market of Latin America
METAP	Mediterranean European Technical Assistance Programme
MOSTE	Ministry of Science, Technology and Environment, Malaysia
MRC	Mekong River Commission
MSU	Moscow State University, Russia
NAAEC	North American Agreement on Environmental Cooperation
NAFTA	North American Free Trade Association
NAPAP	National Acid Precipitation Assessment Programme
NATO	North Atlantic Treaty Organization
NBS	National Biological Service, United States
NCP	Northern Contaminants Programme, Canada
NCS	National Conservation Strategy
NEAPs	National Environment Action Plans
NEMA	National Environment Management Authority, Uganda
NEPA	National Environment Protection Agency, China
NESDA	Network for Environment and Sustainable Development in Africa
NFAP	National Forestry Action Plan
NIE	Newly Industrialized Economies
NIES	National Institute for Environmental Studies, Japan
NIP	National Implementation Plan
NIS	Newly Independent States
NO_x	Nitrogen oxides
NOAA	National Oceanic and Atmospheric Administration, United States
NOWPAP	North-West Pacific Action Plan

NPACD	National Plan of Action to Combat Desertification
NRAP	National River Action Plan
NRC	National Register of Chemicals
NSF	National Science Foundation
NSTC	National Science and Technology Council, United States
NGO	Non-governmental organization
OAS	Organization of American States
OAU	Organization of African Unity
OCHA	Office for the Coordination of Humanitarian Affairs of the UN
ODA	Overseas Development Assistance
ODS	Ozone-depleting substance
OECD	Organization for Economic Co-operation and Development
OEPP	Office of Environmental Policy and Planning, Thailand
ORNL	Oak Ridge National Laboratory, United States
OSPAR	Convention for the Protection of the Marine Environment of the North-East Atlantic
OTA	Office of Technology Assessment
PAHO	Pan American Health Organization
PAME	Protection of the Arctic Marine Environment
PC	Personal computer
PCBs	Polychlorinated biphenyls
PCSD	The President's Council on Sustainable Development, United States
PERSGA	Regional Organization for Conservation of Environment of the Red Sea and Gulf of Aden
PHARE	Poland and Hungary Assistance for Economic Restructuring, of the European Union
PIC	Prior informed consent
PLEC	People, Land Management and Environmental Change
[illegible]	[illegible]
PM-10	Fine particulate matter
PPP	Purchasing power parity
RAPA	Regional Office for Asia and the Pacific of FAO
RAINS	Regional Acidification Information and Simulation Model of IIASA
REC	Regional Environmental Center for Central and Eastern Europe, Hungary
RIOD	Réseau International des Organisations Non-gouvernemental sur Désertification
RIVM	National Institute of Public Health and the Environment, the Netherlands
ROPME	Regional Organization for the Protection of the Marine Environment
RSS	Royal Scientific Society, Jordan
SACEP	South Asia Cooperative Environment Programme
SADC	Southern African Development Community
SADC-ELMS	Environment and Land Management Sector of SADC
SAF	Society of American Foresters
SARDC	Southern African Research and Documentation Centre
SCAR	Scientific Committee on Antarctic Research
SCE	State of the Canadian Environment
SCOPE	Scientific Committee on Problems of the Environment of ICSU
SEI	Stockholm Environment Institute
SEPA	State Environment Protection Administration, China
SME	Small and medium-sized enterprises
SNA	System of national accounts
SO_2	Sulphur dioxide
SOE	State of the environment
SPREP	South Pacific Regional Environment Programme
SSO	Sahara and Sahel Observatory
TACIS	Technical Assistance to the Commonwealth of Independent States
TARGETS	Tool to Assess Regional and Global Environmental and Health Targets for Sustainability, model developed by RIVM
TEI	Thailand Environment Institute
TERI	Tata Energy Research Institute, India
TRI	Toxic Release Inventory Programme
UN	United Nations
UNCED	United Nations Conference on Environment and Development
UNCHE	United Nations Conference on the Human Environment
UNCHS	United Nations Centre for Human Settlements (Habitat)
UNCLOS	United Nations Convention on the Law of the Sea
UNCTAD	United Nations Conference on Trade and Development
UNDAF	United Nations Development Assistance Framework
UNDP	United Nations Development Programme
UNECA	United Nations Economic Commission for Africa
UNECE	United Nations Economic Commission for Europe
UNECLAC	United Nations Economic Commission for Latin America and the Caribbean
UNEP	United Nations Environment Programme
UNESCAP	United Nations Economic and Social Council for Asia and the Pacific
UNESCO	United Nations Educational, Scientific and Cultural [illegible]
UNESCWA	United Nations Economic and Social Commission for West Asia
UNFCCC	UN Framework Convention on Climate Change
UNFPA	United Nations Populations Fund
UNICEF	United Nations Children's Fund
UNITAR	United Nations Institute for Training and Research
UNSO	United Nations Sudano-Sahelian Office
UNU	United Nations University
US	United States
USA	United States of America
USAID	US Agency for International Development
USDA	US Department of Agriculture
USEPA	US Environmental Protection Agency
UV	Ultraviolet
UV-B	Ultraviolet-B radiation
VOC	Volatile organic compounds
WB	World Bank
WBCSD	World Business Council for Sustainable Development
WCMC	World Conservation Monitoring Centre
WCN	*World Climate News*
WHO	World Health Organization
WMO	World Meteorological Organization
WRI	World Resources Institute, United States
WTO	World Trade Organization
WWF	World Wide Fund For Nature

Collaborating and Associated Centres

Collaborating Centres

Asian Institute of Technology (AIT)
P.O. Box 4, Klong Luang
Pathumthani 12120, Thailand
Tel: +66-2-516 0110
Fax: +66-2-516 2126

Arab Centre for the Studies of Arid Zones and Drylands (ACSAD)
P.O. Box 2440
Damascus, Syria
Tel: +963-11-532 3039 / 532 3087
Fax: +963-11-532 3063

Arabian Gulf University (AGU)
P. O. Box 26671
Manama, Bahrain
Tel: +973-265 227 / 277 209
Fax: +973-272 555 / 274 028

Bangladesh Centre for Advanced Studies (BCAS)
House 23 (620 Old), Road 10 A (New) Dhammondi
Dhaka 1209, Bangladesh
Tel: +880-2-815 829 / 911 3682
Fax: +880-2-811 344

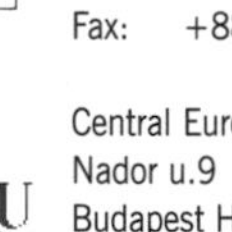

Central European University (CEU)
Nador u.9
Budapest H-1051, Hungary
Tel: +36-1-327 3000
Fax: +36-1-327 3001
http://www.ceu.hu/departs/envsci

Centre for Environment and Development for the Arab Region & Europe (CEDARE)
21/23 Giza Street, Nile Tower Building, 13th Floor
P.O. Box 52, Orman
Giza, Egypt
Tel: +20-2-570 1859/570 0979
Fax: +20-2-570 3242
http://www.cedare.org.eg

European Environment Agency (EEA)
Kongens Nytorv 6
DK-1050 Copenhagen, Denmark
Tel: +45-3336 7100
Fax: +45-3336 7199
http://www.eea.eu.int

Instituto Brasileiro do Meio Ambiente e dos Recursos Naturais Renováveis (IBAMA)
Sain Av. L4 Norte
Ed. Sede do IBAMA
CEP: 70 800 200
Brasilia DF, Brazil
Tel: +55-61-316 1005
Fax: +55-61-316 1025

International Institute for Sustainable Development (IISD)
161 Portage Avenue, East, 6th Floor
Winnipeg, Manitoba, Canada R3B 0Y4
Tel: +1-204-958 7700
Fax: +1-204-958 7710

Moscow State University (MSU)
119899 Moscow, Russia
Tel: +7-095-939 3962
Fax: +7-095-932 8836

National Institute of Public Health and the Environment (RIVM)
Antonie van Leeuwenhoeklaan 9
P. O. Box 1
3720 BA Bilthoven, The Netherlands
Tel: +31-30-274 9111
Fax: +31-30-274 2971

State Environmental Protection Administration (SEPA)
No 115 Xizhimen Nei Nanxiaojie
Beijing 100035, P.R. China
Tel: +86-10-6615 1937
Fax: +86-10-6615 1762

National Institute for Environmental Studies (NIES)
Environment Agency of Japan
16-Onogawa
Tsukuba, Ibaraki 305-0053, Japan
Tel: +81-298-502 347
Fax: +81-298-582 645

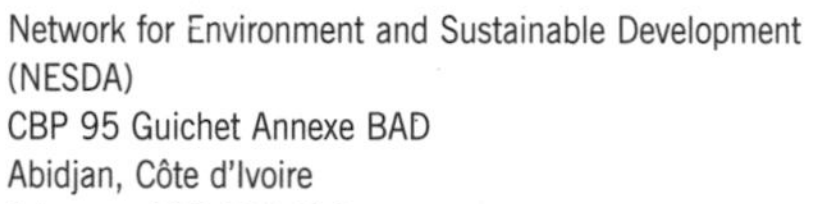

Network for Environment and Sustainable Development (NESDA)
CBP 95 Guichet Annexe BAD
Abidjan, Côte d'Ivoire
Tel: +225-205 419
Fax: +225-205 922
http://www.rri.org/nesda

Southern African Research and Documentation Centre (SARDC)
India Musokotwane Environment Resource Centre for Southern Africa (IMERCSA)
15 Downie Avenue, Belgravia
P. O. Box 5690
Harare, Zimbabwe
Tel: +263-4-738 894/5
Fax: +263-4-738 693

Stockholm Environment Institute (SEI)
Lilla Nygatan 1
Box 2142
S-10314 Stockholm, Sweden
Tel: +46-8-412 1400
Fax: +46-8-723 0348

Tata Energy Research Institute (TERI)
Darbari Seth Block, Habitat Place
Lodhi Road
New Delhi 110 003, India
Tel: +91-11-460 1550
Fax: +91-11-462 1770

Thailand Environment Institute (TEI)
210 Sukhumvit 64
Bangchak Refinery Building 4, 2nd floor
Prakhanong
Bangkok 10260, Thailand
Tel: +66-2-331 0047/331 0060
Fax: +66-2-332 4873

The Regional Environmental Center for Central and Eastern Europe (REC)
Ady Endre ut 9-11
2000 Szentendre, Hungary
Tel: +36-26-311 199
Fax: +36-26-311 294

University of Chile
Diagonal Paraguay 265, Torre 15 Of.1303
Santiago, Chile
Tel: +56-2-678 2077
Fax: +56-2-678 2006

University of Costa Rica
San José 2060, Costa Rica
Tel: +506-283 7619 / 283 7483
Fax: +506-283 7563

World Resources Institute (WRI)
10 G St., NE
Washington DC 20002 United States
Tel: +1-202-729 7600
Fax: +1-202-729 7610

Associated Centres

Asociación Latinoamericana de Derecho Ambiental (ALDA)
José María Velasco No. 74-701
03900 - México, D.F., Mexico
Fax: +52-5-651 2888

African Centre for Technology Studies (ACTS)
P.O. Box 45917
Nairobi, Kenya
Tel: +254-2-521 450
Fax: +254-2-521 001

Centro Internacional de Agricultura Tropical (CIAT)
Apartado Aereo 6713
Cali, Colombia
Tel: +57-2-445 0000
Fax: +57-2-445 0273
http://www.ciat.cgiar.org/indicators/project.html

Commission for Environmental Cooperation (CEC) of the North American Agreement on Environmental Cooperation (NAAEC)
393 rue St-Jacques W.
Montreal, Quebec, Canada, H2Y 1N9
Tel: +1-514-350 4300
Fax: +1-514-350 4314

Earth Council
Apartado 2323-1002
San Jose, Costa Rica
Tel: +506-256 1611
Fax: +506-255 2197

National Environment Management Authority (NEMA)
6th Floor Communications House
1 Colville Street
P.O. Box 22255,
Kampala, Uganda
Tel: +256-41-251 064/251 065
Fax: +256-41-257 521
http: www.uganda.co.ug\environ\

Indian Ocean Commission - Regional Environment Programme (IOC- REP)
Q4, Avenue Sir Guy Forget
Quatre Bornes, Republic of Mauritius
Tel: +230-425 9564
Fax: +230-425 2709

Scientific Committee on Problems of the Environment (SCOPE)
51 Bd de Montmorency
75016 Paris, France
Tel: +33-1-4525 0498
Fax: +33-1-4288 1466

South Pacific Regional Environmental Programme (SPREP)
P.O. Box 240
Apia, Western Samoa
Tel: +685-21 929
Fax: +685-20 231
Website: www.sprep.org.ws

University of West Indies
Centre for Environment and Development (UWICED)
3 Gibraltar Camp Road
U.W.I, Mona
Kingston 7, Jamaica
Tel: +1-876-922 9267
Fax: +1-876-922 9292

World Conservation Monitoring Centre (WCMC)
219 Huntingdon Road
Cambridge CB3 0DL, United Kingdom
Tel: +44-1223-277 314
Fax: +44-1223-277 136

Contributors

Those listed below have contributed to *GEO-2000* in a variety of ways, as authors, reviewers, participants in GEO consultations and survey respondents.

Africa

Asmaa Abdalla, Islamic Educational, Scientific and Cultural Organization, Morocco; Ahmed Abdel Rehim, Centre for Environment and Development for the Arab Region and Europe, Egypt; Adel Farid Abdel-Kader, Centre for Environment and Development for the Arab Region and Europe, Egypt; Ettajani Abdelkebir, Sécretariat d'Etat chargé de l'Environnement, Morocco; Khaled M. Abu-Zeid, Centre for Environment and Development for the Arab Region and Europe, Egypt; Laqualy Ada, Direction de l'Environnement, Niger; Sylvain Adokpo-Migan, Agence Béninoise pour l'Environnement, Benin; James Oppong Afrani, Ministry of Environment, Science and Technology, Ghana; Laurent Agossa Ogowa, Université Nationale du Bénin, Benin; Samir Anwar Al-Gamal, Egyptian Atomic Energy, Egypt; Fatema Al-Mallah, Technical Secretariat, Council of Arab Ministers Responsible for the Environment, League of Arab States, Egypt; Rose Sirali Antipa, Ministry of Environment Conservation, Kenya; Shawki Ibrahim Asaad, Egypt; Nadir Mohamed Awad, HCENR, Sudan; Marcel Ayité Baglo, Agence Béninoise pour l'Environnement, Ministère de l'Environnement de l'Habitat et de l'Urbanisme, Benin; Abou Bamba, Network for Environment and Sustainable Development in Africa, Côte d'Ivoire; Vania nee Assenoua Bamisso, Institut National d'Economie, Université National du Benin, Benin; Kamal H. Batanouny, Faculty of Science, Cairo University, Egypt; Bothwell Batidzirayi, Southern Centre for Energy and Environment, Zimbabwe; Abdelkrim Ben Mohamed, Institute for Radio-Isotopes, Niger; Zida Bertrand, Gestion des Ressources Naturelles et Lutte Contre la Désertification, Burkina Faso; Innocent Bizimana, Ministry of Agriculture, Livestock, Environment and Rural Development, Rwanda; Abdel-Rehani Boucham, Department of Environment, Ministry of Environment, Morocco; Bouazza Bouchra, Department of Environment, Ministry of Environment, Morocco; Munyaradzi Chenje, Southern African Research and Documentation Centre, India Musokotwane Environment Resource Centre for Southern Africa, Zimbabwe; Brian Chirwa, Environmental Council of Zambia, Zambia; R.J.M. Crawford, Sea Fisheries Research Institute, South Africa; A. Bram De Villiers, Potchefstroom University for Christian Higher Education, South Africa; Berhe Debalkew, Inter-Governmental Authority on Development, Djibouti; Abdi Mohamed Dirieh, Ministry of Environment, Tourism and Craft, Djibouti; Clement Dorm-Adzobu, Sustainable Development Consultancy Services Ltd, Ghana; Francois Ekoko, Center for International Forestry Research, International Institute for Tropical Agriculture, Cameroon; M. El-Raey, Institute of Graduate Studies and Research, Alexandria University, Egypt; Mona M. El-Agizy, Egypt; Nadine El-Hakim, Centre for Environment and Development for the Arab Region and Europe, Egypt; Hassan Salim El-Hassan, Arab Organization for Agricultural Development, Sudan; Mohamed El-Kassas, Botany Department, Cairo University, Egypt; Osama Amin El-Kholy, Technology Management Programme, Egyptian Environmental Affairs Agency, Egypt; Mai El-Remeisy, Centre for Environment and Development for the Arab Region and Europe, Egypt; Fahima El-Shahed, General Organization for Physical Planning, Egypt; David Everard, Division of Water, Environment and Forestry Technology, Council for Scientific and Industrial Research, South Africa; Mohamed Beshir Fares, Technical Center for Environment Protection, Libyan Arab Jamahiriya; Samir Gabbour, Department of Natural Resources, Institute of African Research and Studies, Cairo University, Egypt; M. Waleed Gamaleldin, Terra Incorporated, Egypt; Michael N.K. Gichobi, Office of the President, Kenya; Salwa Gomaa, American University in Cairo, Egypt; Ahmed Hamza, Ministry of Environment, Egypt; Ahmed Farghally Hassan, Center for Environmental Research Studies, Cairo University, Egypt; Baccar Hedia, Ministry of Environment and Land Use Planning, Tunisia; Ahmed Hegazy, Centre for Environment and Development for the Arab Region and Europe, Egypt; Mahmoud Hewehy, Ain Shams University, Egypt; Piet Heyns, Department of Water Affairs, Namibia; Edward G. Howard-Clinton, Environment Division, Organization of African Unity, Ethiopia; J.M. Hutton, Africa Resources Trust, Zimbabwe; Abdulrahman S. Issa, Eastern African Regional Office, IUCN - The World Conservation Union, Kenya; Seewoobaduth Jogeeswar, Ministry of Local Government and Environment, Mauritius; Maurice Kamto, Université de Yaoundé II, Cameroon; Micah Katuruza, Central Statistics Offices, Zimbabwe; François-Corneille Kedowide, Departement Planification Environnementale et Suivi-Evaluation, Agence Béninoise pour l'Environnement, Benin; Abbas Kesseba, International Fund for Agricultural Development, Egypt; Shova Khatry, Kenya; Michael K. Koech, Ministry of Environment and Natural Resources, Kenya; Joseph Mannaw Koroma, Network for Environment and Sustainable Development in Africa, Côte d'Ivoire; Tendayi Kureya, Southern African Research and Documentation Centre, India Musokotwane Environment Resource Centre for Southern Africa, Zimbabwe; Fatou Kuyateh, Department of State for Presidential Affairs, Fisheries and Natural Resources, Gambia; Elton Laisi, Southern African Research and Documentation Centre, India Musokotwane Environment Resource Centre for Southern Africa, Zimbabwe; Kalle Mohamed Lamine, Ministry of Environment, Mali; Mpouel Bala Lazare, Ministry of Environment and Forestry, Cameroon; Adama Ly, Ministère de l'Environnement, Senegal; Izekiel Machingambi, Department of National Parks and Wildlife Management, Zimbabwe; Saada Madjid, Secretariat d'Etat à l'Environnement, Algeria; Clever Mafuta, Southern African Research and Documentation Centre, India Musokotwane Environment Resource Centre for Southern Africa, Zimbabwe; Rosemary M. Makano, Ministry of Environment and Natural Resources, Zambia; Nelson Marongwe, ZERO-A Regional Environmental Organization, Zimbabwe; Solami Sipho Mavimbela, Ministry of Natural Resources, Swaziland; Pierre Mbouegnong, Ministry of Environment and Forestry, Cameroon; Esther Joyce Mede, Ministry of Research and Environmental Affairs, Malawi; Mbassi Menye, Ministry of Environment and Forestry, Cameroon; Yaqoub Abdalla Mohamed, Higher Council for Environment and Natural Resources, Sudan; Santaram Mooloo, Department of Environment, Ministry of Local Government and Environment, Mauritius; Ahmed Taher Moustafa, Soil, Water and Environment Research Institute, Egypt; John Mugabe, African Centre for Technology Studies, Kenya; Simon K. Mugera, Ministry of Environment Conservation, Kenya; Mary Mukinda, African Centre for Technology Studies, Kenya; Sheila Mwanundu, African Development Bank, Côte d'Ivoire; Robert T. N'Daw, c/o FAO Representative, Bamako, Mali; Godwell Nhamo, Solusi University, Zimbabwe; Mutasim Bashir Nimir, Sudanese Environmental Conservation Society, Sudan; Harouna Oumarou, Conseil National de l'Environnement pour un Développement Durable, Niger; Mahmoud Hellmy Moustafa Ousef, Egypt; Emmanuel Pouna, Cameroon; Adham Ramadan, Egyptian Environmental Affairs Agency, Egypt; Kamal A. Sabet, Centre for Environment and Development for the Arab Region and Europe, Egypt; Hamed Saleh, Agriculture Research Center, Egypt; Omar M. Salem, General Water Authority, Libyan Arab Jamahiriya; Prosper Sawadogo, Conseil National pour la Gestion de l'Environnement, Burkina Faso; Dina Ahmed Sayed, TERRA Incorporated, Egypt; Charles Sebukeera, Division of Environmental Information and Monitoring, National Environment Management Authority, Uganda; Wahida Patwa Shah, Earth Council for East and Southern Africa, Kenya; Henry M. Sichingabula, University of Zambia, Zambia; Fondo Sikod, Université de Yaoundé II, Cameroon; Lovemore Sola, Southern African Research and Documentation Centre, India Musokotwane Environment Resource Centre for Southern Africa, Zimbabwe; Thomas Fofung Tata, Network for Environment and Sustainable Development in Africa, Cameroon; Mostafa Kamal Tolba, International Center for Environment and Development, Egypt; Martine Tahoux Touao, Centre de Recherche en Ecologie, Côte d'Ivoire; Mohamed Lankan Traore, Direction Nationale des Eaux et Forêts, Guinea; Godber Tumushabe, African Centre for Technology Studies, Kenya; Jacqueline Van Staalduinen, Environment and Land Management Sector, Southern African

Development Community, Lesotho; Ahmed Wagdy, Faculty of Engineering, Cairo University, Egypt; Shem O. Wandiga, Kenya National Academy of Sciences, Kenya; Stephen M. Zuke, Swaziland Environment Authority, Swaziland.

Asia and the Pacific

Prakash Chandra Adhikari, Ministry of Population and Environment, Nepal; Mahshid Agir, Department of the Environment, Environmental Impact Assessment Bureau, Islamic Republic of Iran; Julian Amador, Environmental Management Bureau, Department of Environment and Natural Resources, Philippines; Raja M. Ashfaque, Pakistan Forest Institute, Pakistan; Lilita Bacareza-Pacudan, European Commission and the Association of South-East Asian Nations, COGEN Programme, Asian Institute of Technology, Thailand; J.T. Baker, Commissioner for the Environment A.C.T., Australia; Mahesh Banskota, International Centre for Integrated Mountain Development, Nepal; Tom Beer, Division of Atmospheric Research, Commonwealth Scientific and Industrial Research Organization, Australia; Wby Beiying, Institute of Atmospheric Physics, Chinese Academy of Sciences, China; Preety Bhandari, Tata Energy Research Institute, India; Gaurav Bhatiani, Tata Energy Research Institute, India; PG. Shamhary Bin, Ministry of Development, Brunei Darussalam; Jarupong Boon-Long, Pollution Control Department, Ministry of Science, Technology and Environment, Thailand; Phongsri Boonyasirikool, National Research Council of Thailand, Thailand; Damrong Boonyoen, Ministry of Public Health, Thailand; R.K. Bose, Tata Energy Research Institute, India; Surasit Chaiyaphum, Natural Resources and Environment Management Division, Office for Environmental Policy and Planning, Thailand; Duongchan Apavatjrut Charoenmuang, Social Research Institute, Chiang Mai University, Thailand; Chaninthorn Charuchandra, Agricultural Land Reform Office, Thailand; Kalipada Chatterjee, Development Alternatives, India; Qing Cheng, Beijing Medical University, China; Weixue Cheng, State Environmental Protection Administration, China; Surapong Chirarattananon, Asian Institute of Technology, Thailand; Yong-Seung Chung, Korea National University of Education, Republic of Korea; Joy Cohen, Earthglass, Taiwan Province of China; John Cole, Environment Management Industry Association, Australia; Bismark Crawley, South Pacific Regional Environment Programme, Western Samoa; Diwakar Dahal, Environment Assessment Program for Asia-Pacific, Asian Institute of Technology, Thailand; Pham Ngoc Dang, Center for Environmental Engineering of Towns and Industrial Areas, Hanoi University of Civil Engineering, Viet Nam; Aditi Dass,* Tata Energy Research Institute, India; G. Dembereldorj, Ministry of Nature and the Environment, Mongolia; Sawat Dulyapatch, Royal Forest Department, Thailand; Muhammad Eusuf, Bangladesh Centre for Advanced Studies, Bangladesh; Kiyoshi Fukuwatari, formerly Center for Global Environment Research, National Institute for Environmental Studies, Japan; E. Gumbira-Said, Indonesian Business Council for Sustainable Development, Indonesia; Lalith A. Gunaratne, Lalith Gunaratne and Associates, Sri Lanka; Xiaomin Guo, State Environmental Protection Administration, China; Allan Haines, Environment Australia, Australia; Hironori Hamanaka, Global Environment Department, Environment Agency of Japan; Colin Harris,* Global Resource Information Database, Christchurch, New Zealand; Shiro Hatakeyama, Center for Global Environmental Research, National Institute for Environmental Studies, Environment Agency of Japan, Japan; Ian Hawes, National Institute of Water and Atmospheric Research, New Zealand; Clive Howard-Williams, National Institute of Water and Atmospheric Research, New Zealand; Jianxin Hu, Center for Environmental Sciences, Peking University, China; Min Hu, Center for Environmental Sciences, Peking University, China; Michael Huber, Australia; A.Z.M. Iftikhar Hussain, Ministry of Health and Family Welfare, Bangladesh; Toshiaki Ichinose, Center for Global Environmental Research, National Institute for Environmental Studies, Environment Agency of Japan, Japan; Gen Inoue, Center for Global Environmental Research, National Institute for Environmental Studies, Environment Agency of Japan, Japan; Mylvakanam Iyngararasan, Environment Assessment Program for Asia-Pacific, Asian Institute of Technology, Thailand; Sitanon Jesdapipat, Thailand Environment Institute, Thailand; S. John Joseph, M.S. Swaminathan Foundation, India; Ananda Raj Joshi, South Asia Co-operative Environment Programme, Sri Lanka; Mikiko Kainuma, Global Environment Division, National Institute for Environmental Studies, Environment Agency of Japan, Japan; Abdul Khaleque, Ministry of Environment and Forest, Bangladesh; Jitt Kongsangchai, Forest Research Office, Royal Forest Department, Thailand; Surachai Koomsin, Center for Integrated Plan of Operation, Thailand; Nisakorn Kositratna, Hazardous Substance and Waste Management Division, Pollution Control Department, Ministry of Science Technology and Environment, Thailand; Pradyumna Kumar Kotta, South Asia Co-operative Environment Programme, Sri Lanka; Purushotam Kunwar, Ministry of Population and Environment, Nepal; Daw Yin Yin Lay, Ministry of Foreign Affairs, National Commission for Environmental Affairs, Myanmar; Can Le Thac, Viet Nam Environment and Sustainable Development Center, Viet Nam; Thierry Lefevre, Asian Institute of Technology, Thailand; Cheng Geok Ling, International Environment and Policy Department, Ministry of the Environment, Singapore; Jingyi Liu, Research Center for Eco-Environmental Sciences, Academia Sinica, China; Shengji Luan, Center for Environmental Sciences, Peking University, China; San Lwin, Attorney General's Office, Myanmar; Ahmed Ali Manik, Ministry of Planning, Human Resources and Environment, Republic of Maldives; Mok Mareth, Ministry of Environment, Cambodia; Yasunobu Matoba, Mekong River Commission, Thailand; Orapun Metadilogkul, Occupational and Environmental Medicine Association, Thailand; Choudhury Rudra Charan Mohanty, Asian Institute of Technology, Thailand; Monica Moktan, International Centre for Integrated Mountain Development, Nepal; Hideyuki Mori, Global Environment Department, Environment Agency of Japan; Shunji Murai, Asian Institute of Technology, Thailand; P.G. D.P., and H.J. Mustapha, Ministry of Development, Brunei Darussalam; Samorn Muttamara, Asian Institute of Technology, Thailand; Ma Sein Mya, Environment Unit, Mekong River Commission, Thailand; Vishal Narain, Tata Energy Research Institute, India; Stewart Needham, Department of the Environment, Environment Australia, Australia; Somrudee Nicro, Urbanization and Environment Programme, Thailand Environment Institute, Thailand; Shuzo Nishioka, Global Environment Division, National Institute for Environmental Studies, Environment Agency of Japan, Japan; Ian Noble, Australia National University, Australia; Boo-Ho Noh, International Affairs Division, Ministry of Environment, Republic of Korea; Phonechaleun Nonthaxay, Science, Technology, and Environment Organization, Lao People's Democratic Republic; Azhar Bin Noraine, Ministry of Science, Technology and Environment, Malaysia; Karma C. Nyedrup, National Environment Commission, Bhutan; Akira Ogihara, Environment Department, Pacific Consultants International, Japan; Toshiichi Okita, Obirin University, Japan; Tongroj Onchan, Thailand Environment Institute, Thailand; Yasuyuki Oshima, The Japan Wildlife Research Centre, Japan; Kuninori Otsubo, Water Soil Environment Division, National Institute for Environmental Studies, Japan; Alan Oxley, Australian Asia Pacific Economic Cooperation Study Centre, Australia; R.K. Pachauri, Tata Energy Research Institute, India; Jyoti Parikh, Indira Ghandi Institute of Development Research, India; Kirit Parikh, Indira Ghandi Institute of Development Research, India; Jinxin Peng, State Environmental Protection Administration, China; Yeshey Penjor, National Environment Commission, Bhutan; late Dhira Phantumvanit, Thailand Environment Institute, Thailand; Somphone Phanousith, Science Technology and Environment Organization, Lao People's Democratic Republic; Warasak Phuangcharoen, Environment Policy and Planning Division, Office for Environmental Policy and Planning, Thailand; Ung Phyrun, Ministry of Environment, Cambodia; Suphavit Piamphongsant, Ministry of Science, Technology, and Environment, Thailand; Ken Piddington, Waikato University, New Zealand; Amara Pongsapich, Social Research Institute, Chulalongkorn University, Thailand; Pramod Pradhan, International Centre for Integrated Mountain Development, Nepal; Chalermsak Rabilwongse, Office of the Commission for the Management of Land Transport, Thailand; Atiq Rahman, Bangladesh Centre for Advanced Studies, Bangladesh; R. Rajamani, India; Chatchai Ratanachai, Faculty of Environmental Management, Prince of Songkla University, Thailand; Dev Raj Regmi, Ministry of Water Resources, Nepal; Srivardhana Ruangdej, Kasetsart University, Thailand; Sompop Rungsupa, Sichang Marine Science Research and Training Station, Aquatic Resources Research Institute, Thailand; Sumeet Saksena, Tata Energy Research Institute, India; Wanee Samphantharak, Office of Environmental Policy and Planning, Thailand; Jeffrey Sayer, Center for International Forestry Research, Indonesia; Darrell Sequeira, Mekong River Commission, Thailand; S.B. Sharma, Ministry of Population and Environment, Nepal; Basanta Shrestha, International Centre for Integrated Mountain Development, Nepal; Ram Manohar Shrestha, Asian Institute of Technology, Thailand; Liu Shuqin, National Environmental Protection Agency, China; Nguyen Ngoc Sinh, National Environment Agency, Ministry of Science, Technology and Environment, Viet Nam; Steven M. Smith, Global Resource Information Database, Christchurch, New Zealand; Allan Spessa, Environment Australia, Australia; Kanongnij Sribuaiam, Urbanization and Environment Programme, Thailand Environment Institute, Thailand; Ruangdej Srivardhana, Department of Economics, Faculty of Economics, Kasetsart University, Thailand; Leena Srivastava, Tata Energy

Research Institute, India; Usha Subramaniam, Ministry of Environment and Forests, India; Cicilia Sulastri, Environmental Impact Management Agency, Indonesia; Apichai Sunchindah, Assocation of South-East Asian Nations, Indonesia; C. Surapong, Asian Institute of Technology, Thailand; M. S. Swaminathan, M.S. Swaminathan Research Foundation, India; Monthip Sriratana Tabucanon, Environmental Research and Training Centre, Department of Environmental Quality, Ministry of Science, Technology and Environment, Thailand; Kazuhiko Takemoto, Global Environment Department, Environment Agency of Japan; Xiaoyan Tang, Center for Environmental Sciences, Peking University, China; Supichai Tangjaitrong, Chulalongkorn University, Thailand; Rowan Taylor, c/o Ministry for Environment, New Zealand; Nguyen Cong Thanh, JT-Envi Consultants Ltd, Thailand; Nguyen Thi Tho, International Relations and Planning Department, National Environment Agency, Ministry of Science, Technology, and Environment, Viet Nam; Govinda R. Timilsina, Center for Energy-Environment Research and Development, Asian Institute of Technology, Thailand; Le Anh Tuan, Energy Program, Asian Institute of Technology, Thailand; Nguyen Quang Tuan, National Institute for Science and Technology Policy and Strategy Studies, Viet Nam; Narcisa R. Umali, National Economic and Development Authority, Philippines; Shinsuke Unisuga, Research and Information Office, Global Environment Department, Environment Agency of Japan, Japan; Vicharn Upatising, Environment Division, Department of Mineral Resources, Thailand; Batu Krishna Upreti, Ministry of Population and Environment, Nepal; Mikoto Usui, Shukutoku University, Japan; Dang Ung Van, Viet Nam National University, Viet Nam; Rusong Wang, Chinese Academy of Sciences, China; Zhijia Wang, State Environmental Protection Administration, China; Huixiang Wang, Center for Environmental Sciences, Peking University, China; Chalermsak Wanichsombat, Ministry of Science, Technology and Environment, Thailand; Yingmin Wen, State Environmental Protection Administration, China; Guan Xia, State Environmental Protection Administration, China; Hu Xiulian, Energy Research Institute, State Development and Planning Committee, China; Kazuhito Yamada, Environment Department, Pacific Consultants Co, Japan; Weimin Yang, Foreign Techno-Economic Cooperation Division, Yunnan Provincial Environmental Protection Bureau, China; Yoshifumi Yasuoka, Center for Global Environmental Research, National Institute for Environmental Studies, Japan; Nguyen Hoang Yen, Ministry of Science, Technology and Environment, Viet Nam; Wi Sok Yon, Embassy of Democratic People's Republic of Korea in Thailand, Democratic People's Republic of Korea; Ruisheng Yue, State Environmental Protection Administration, China; Shigang Zhang, State Environmental Protection Administration, China; Shiqiu Zhang, Center for Environmental Sciences, Peking University, China; Yisheng Zheng, Chinese Academy of Social Sciences, China; Shuseng Zhou, Beijing Medical University, China; Jianming Zhou, China Academy of Urban Planning and Design, China.

Europe

Chris Anastasi, British Energy, United Kingdom; Ewa Anzorge, Department of European Integration and International Cooperation, Ministry of Environmental Protection, Natural Resources and Forestry, Poland; Agajan G. Babayev, Turkmenian Academy of Sciences, Turkmenistan; Jan Bakkes, National Institute of Public Health and the Environment, The Netherlands; Petr Ya Baklanov, Far East Branch, Institute of Geography, Russian Academy of Sciences, Russian Federation; Jaroslav Balek, Environmental Engineering Consultancy, Czech Republic; Edward Bellinger, Central European University, Hungary; André Berger, Institut d'Astronomie et de Géophysique, Université Catholique de Louvain, Belgium; Marcel Berk, National Institute of Public Health and the Environment, The Netherlands; Claes Bernes, Swedish Environmental Protection Agency, Sweden; Kornelis Blok, University of Utrecht, The Netherlands; Vladimir P. Bogachev, Ministry for Ecology and Natural Resources, Kazakhstan; Johannes Bollen, National Institute of Public Health and the Environment, The Netherlands; Peter Bosch, European Environment Agency, Denmark; Trevor Bounford, Chapman Bounford and Associates, United Kingdom; Lex Bouwman, National Institute of Public Health and the Environment, The Netherlands; Philippe Bourdeau, Université Libre de Bruxelles, Belgium; Emanuelle Bournay, Global Resource Information Database, Arendal, Norway; Winston H. Bowman,* The Regional Environmental Center for Central and Eastern Europe, Hungary; Joop Brouns, Department of International Affairs, Institute for Forestry and Nature Research and European Centre for Nature Conservation, The Netherlands; Rudolf Bruno, Global Precipitation Climatology Centre, Germany; Budag A. Budagov, Institute of Geography, Azerbaijan Academy of Sciences, Azerbaijan; Françoise Burhenne-Guilmin, Environmental Law Centre, IUCN - The World Conservation Union, Germany; T.D. Button, World Business Council for Sustainable Development, Switzerland; Arcadie Capcelea, Ministry of Environment, Republic of Moldova; Roberto Caponigro, Ministry of the Environment, Italy; M. J. Chadwick, Leadership for Environment and Development - Europe, Switzerland; Mike Cloughley, The Oil Industry International Exploration and Production Forum, United Kingdom; N. Mark Collins, World Conservation Monitoring Centre, United Kingdom; Paul Crutzen, Max-Planck Institute of Chemistry, Germany; Paul Csagoly, Information Exchange Department, The Regional Environmental Center for Central and Eastern Europe, Hungary; András R. Csanady, Strategy Directorate, Ministry for Environment and Regional Policy, Hungary; Tatiana Davydovskaia, Department of International Cooperation and Science, Ministry for Natural Resources and Environment Protection, Belarus; Dick de Bruijn, Ministry of Housing, Spatial Planning and the Environment, The Netherlands; Jos de Bruin, Multimedia and Culture, Free University of Amsterdam, The Netherlands; André de Moor, Environment and Spatial Planning, Ministry of Economic Affairs, Directoraat Generaal Europese Samenwerking, The Netherlands; Michel den Elzen, National Institute of Public Health and the Environment, The Netherlands; V. Demkin, Environmental Policy Division, Ministry for Environmental Protection and Nuclear Safety, Ukraine; R.G. Derwent, Atmospheric Processes Research, Meteorological Office, United Kingdom; Francesco Di Castri, Centre d'Ecologie Fonctionnelle et Evolutive, Centre National de la Recherche Scientifique, France; George Dieca, State University of Moldova, Republic of Moldova; A.M. Dourdiev, National Institute of Desert, Flora and Fauna, Ministry of the Use of Natural Resources and Environmental Protection, Turkmenistan; Nikolai M. Dronin, Faculty of Geography, Moscow State University, Russian Federation; George Duca, State University of Moldova, Republic of Moldova; Hans Eerens, National Institute of Public Health and the Environment, The Netherlands; Bulat K. Esekin, Kazakhstan; Ian A. Fleming, Norwegian Institute for Nature Research, Norway; Karen Fletcher, International Hotels Environment Initiative, United Kingdom; Isabelle Fleuraud, France; Eeva R. Furman, Finnish Environment Institute, Finland; Ainars Gailitis, Environmental Consulting and Monitoring Centre, Ministry of Environmental Protection and Regional Development, Latvia; Gilberto Gallopin, Stockholm Environment Institute, Sweden; Ali S. Gassanov, The State Committee of Ecology and Control of Natural Resources Utilization, Azerbaijan; Pietro Giuliani, Ente Per le Nuove Tecnologie, l'Energia e l'Ambiente - Antartide, Italy; Nikita F. Glazovsky, Institute of Geography, Russian Academy of Sciences, Russian Federation; Genady N. Golubev, Faculty of Geography, Moscow State University, Russian Federation; John Goodall, European Construction Industry Federation, Belgium; Jean Graebling, Permanent Mission of France to Geneva in Switzerland, France; Allan Gromov, Ministry of Environment, Environmental Policy and International Relations, Estonia; Brian Groombridge, IUCN - The World Conservation Union, Switzerland; Cuno Grootscholten, National Center for Sustainable Construction, The Netherlands; Paolo Guglielmi, Mediterranean Programme, WWF International, Italy; Myroula Hadjichristophorou, Ministry of Agriculture, Natural Resources and Environment, Cyprus; David O. Hall, King's College, University of London, United Kingdom; Sigmund Haugsjå, Norwegian State Railways, Norway; William J. Hartnett, Negotiation Internationale Professionnelle, France; Oliver W. Heal, United Kingdom; Irina Herczeg, Central European University, Hungary; Thomas Ietswaart, National Institute of Public Health and the Environment, The Netherlands; David Insull, United Kingdom; Bengt-Owe Jansson, Department of Systems Ecology, Stockholm University, Sweden; Ljubomir Jeftic, Advisory Committee on Protection of the Sea, United Kingdom; Alexander Juras, The Regional Environmental Center for Central and Eastern Europe, Hungary; Ahte Kalle, State Directorate for the Protection of Nature and Environment, Croatia; Albena Karadjova, International Cooperation Department, Ministry of Environment and Waters, Bulgaria; Stefan Karpis, Information Science and Monitoring Department, Ministry of Environment, Slovakia; Andrzej Kassenberg, Institute for Sustainable Development, Poland; Dmitri Kavtaradze, Biological Faculty, Moscow State University, Russian Federation; Stjepan Keckes, Croatia; Nariman Soltangamid Oglu Kerimov, The State Committee of Ecology and Control of Natural Resources Utilization, Azerbaijan; Yann Kermode, Union Bank of Switzerland, Switzerland; Fazlun Khalid, Islamic Foundation for Ecology and Environmental Sciences, United Kingdom; Vitaly Kimstach, Arctic Monitoring and Assessment Programme, Norway; Janosz Kindler, Warsaw University of Technology, Poland; Jósef Kindler, Budapest University of Economics, Hungary; Alexandre Charles Kiss, Centre National de la Recherche Scientifique, France; Kees Klein Goldewijk, National Institute of Public Health and the Environment, The Netherlands; Joost Knoop, National Institute of

Public Health and the Environment, The Netherlands; Margarita Korkhmazyan, Department of International Cooperation, Ministry of Nature Protection, Armenia; Johan Kuylenstierna, Stockholm Environment Institute, United Kingdom; J.W.M. la Riviere, International Institute for Infrastructural, Hydraulic and Environmental Engineering, The Netherlands; Darina Lacikova, Environmental Conceptions and Planning Department, Ministry of Environment, Slovakia; Istvan Lang, Hungarian Academy of Sciences, Hungary; Fred Langeweg, National Institute of Public Health and the Environment, The Netherlands; Rik Leemans, National Institute of Public Health and the Environment, The Netherlands; Mihai Lesnic, Research and Engineering Institute for Environment, Ministry of Waters, Forests and Environmental Protection, Romania; Erich Lippert, Department of Strategy and Environmental Statistics, Ministry of the Environment, Czech Republic; Peter S. Liss, University of East Anglia, United Kingdom; Michael Loevinsohn, International Service for National Agricultural Research, The Netherlands; Vladimir F. Loginov, Institute of Problems of Natural Resources Use and Ecology, National Academy of Sciences, Belarus; Kim Losev, All Russian Institute of Scientific and Technical Information, Russian Federation; Finn Lynge, Indigenous Peoples Secretariat of the Arctic Council, Denmark; Pim Martens, Department of Mathematics, International Centre for Integrative Studies, Maastricht University, The Netherlands; Philippe Martin, European Commission, Italy; Emily E. Matthews, United Kingdom; Malgosia Mazurek, The Regional Environmental Center for Central and Eastern Europe, Hungary; Anthony J. McMichael, London School of Hygiene and Tropical Medicine, United Kingdom; Derek McNally, Department of Physics and Astronomy, University College London, United Kingdom; Jeffrey A. McNeely, IUCN - The World Conservation Union, Switzerland; Dominique van der Mensbrugghe, OECD, France; Ivan Mersich, Hungarian Meteorological Service, Hungary; Ruben Mnatsakanian, Department of Environmental Sciences and Policy, Central European University, Hungary; Bedrich Moldan, Environmental Centre, Charles University, Czech Republic; Michail E. Nikiforov, Institute of Zoology, Byelorussian Academy of Sciences, Belarus; Michael Norton-Griffiths, Centre for Social and Economic Research on the Global Environment, United Kingdom; Tatiana Mikhailovna Novikova, Ministry of Nature Protection, Tajikistan; Karen O'Brien, Center for International Climate and Environmental Research, University of Oslo, Norway; Roel Oldeman, International Soil Reference and Information Centre, The Netherlands; Johannes B. Opschoor, Faculteit de Economische Wetensechappen en Econometrie, Vrije Universiteit Amsterdam, The Netherlands; Nicolae Panin, National Institute for Marine Geology and Geo-ecology, Romania; Jit Peters, International Environmental Policy, Ministry of Environment, The Netherlands; Hanne Petersen, Department of Arctic Environment, National Environmental Research Institute, Denmark; Véronique Plocq-Fichelet, Scientific Committee on the Problems of the Environment, France; Elitsa Polizoova, c/o Earth Council, National Council of Sustainable Development, Bulgaria; Max Posch, National Institute of Public Health and the Environment, The Netherlands; Ferenc Rabar, Department of Economics, Faculty of Law and State Sciences, Pazmany Peter Catholic University, Hungary; Lars-Otto Reiersen, Arctic Monitoring and Assessment Programme, Norway; Polat Reimov, Uzbekistan; Philippe Rekacewicz, Global Resource Information Database, Arendal, Norway; Jean-Pierre Ribaut, Council of Europe, France; Leslie Roberts, World Resources Institute, USA; Henning Rodhe, Department of Meteorology, Stockholm University, Sweden; Melita Rogelij, Central European University, Hungary; Leonid G. Rudenko, Institute of Geography, Ukranian National Academy of Sciences, Ukraine; Leo Saare, Environment Information Centre, Ministry of the Environment, Estonia; Rolf Sagesser, International Federation of Consulting Engineers, Switzerland; Peter H. Sand, University of Munich, Germany; Paul Sands, Earthscan Publications Limited, United Kingdom; Peter Saunders, United Kingdom; Alexandre Sàvàstenko, Secretariat of the Interstate Ecological Council, Belarus; Kai Schlegelmilch, Wuppertal Institute for Climate, Environment and Energy, Germany; Thomas Schmid, Federal Ministry of the Environment, Nature Conservation and Nuclear Safety, Germany; Andrey Semichaevsky,* Central European University, Hungary; Julia Elena Serpa, International Coffee Organization, United Kingdom; N.G. Shadieva, International Relations and Programmes Department, State Committee for Nature Protection, Uzbekistan; Jerome Simpson, The Regional Environmental Center for Central and Eastern Europe, Hungary; Mari Skåre, Norway; Harry Slaper, National Institute of Public Health and the Environment, The Netherlands; Rob Sluyter, National Institute of Public Health and the Environment, The Netherlands; Danièle Smadja, European Commission, Belgium; Valerian A. Snytko, Institute of Geography, Siberian Branch of the Russian Academy of Sciences, Russian Federation; Dmitry I. Soloviev, Permanent Representation Office of the Sakha Republic, Russian Federation; Nicholas Sonntag, Stockholm Environment Institute, Sweden; Menka Spirovska, Division of Environment and Nature Protection, Ministry of Urban Planning, Construction and Environment, Former Yogoslav Republic of Macedonia; David Stanners, European Environment Agency, Denmark; Stephen Stec, The Regional Environmental Center for Central and Eastern Europe, Hungary; Paul Stoop, National Institute of Public Health and the Environment, The Netherlands; Rob Swart, National Institute of Public Health and the Environment, The Netherlands; J.K. Syers, Department of Agricultural and Environmental Science, Faculty of Agriculture and Biological Sciences, University of Newcastle upon Tyne, United Kingdom; Robert L. Sykes, International Council of Tanners, United Kingdom; Olena Sylenok, Ministry for Environmental Protection and Nuclear Safety of Ukraine, Ukraine; László Szendródi, University of Sopron, Hungary; Ben Ten Brink, National Institute of Public Health and the Environment, The Netherlands; Victoria Ter-Nikoghosyan, Armenia; Ketevan Tsereteli, Ministry of Environment, Georgia; Inga Turk, Nature Protection Authority, Ministry of the Environment and Physical Planning, Slovenia; Svein Tveitdal, Global Resource Information Database, Arendal, Norway; Diana Urge-Vorsatz, Central European University, Hungary; Cees van Beers, Department of Economics, Delft University of Technology, The Netherlands; Detlef van Vuuren, National Institute of Public Health and the Environment, The Netherlands; Jaap van Woerden, National Institute of Public Health and the Environment, The Netherlands; György Várallyay, Research Institute for Soil Science and Agricultural Chemistry, The Hungarian Academy of Sciences, Hungary; Evaldas Vebra, International Cooperation Unit, Ministry for Environmental Protection, Lithuania; Gábor Vida, Department of Genetics, Eötvös Loránd University, Hungary; Davor Vidas, Fridtjof Nansen Institute, Norway; Lukas Vischer, University of Bern, Switzerland; Gerrit H. Vonkeman, Institute for European Environmental Policy, Belgium; C. C. Wallen, France; Jacob Werksman, Foundation for International Environmental Law and Development, School of Oriental and African Studies, University of London, United Kingdom; Simon Wilson, Arctic Monitoring and Assessment Programme, The Netherlands; Christa Wolf, Permanent Mission of Germany to Geneva, Switzerland; Alexey V. Yablokov, Center for Russian Environmental Policy, Russian Federation; Oleg N. Yanitsky, Institute of Sociology, Russian Academy of Sciences, Russian Federation; Tony Zamparutti,* Environment Directorate, Organization for Economic Co-operation and Development, France.

Latin America and the Caribbean

Freddy Abarca, Observatorio del Desarrollo, Universidad de Costa Rica, Costa Rica; Ximena Abogabir Scott, Casa de La Paz, Chile; Celeste Acevedo, Subsecretaría de Estado de Recursos Naturales y Medio Ambiente, Ministério de Agricultura y Ganadería, Paraguay; Yosu Rodríguez Z. Aldabe, Secretaría de Medio Ambiente, Recursos Naturales y Pesca, México; Dimas Isaac Arcia González, Autoridad Nacional del Ambiente, Panamá; Paulo Artaxo, Institute of Physics, University of São Paulo, Brazil; Luis Mario Batallés Rivas, Ecosistemas Costeros y Marinos, Ministério Medio Ambiente, Uruguay; Raúl Brañes, Asociación Latinoamericana Derecho Ambiental, México; Francisco Brzovic Parilo, Chile; Gerardo Budowski, Earth Council, Costa Rica; Federico Burone Magariños, Secretariado Manejo Medio Ambiente, International Development Research Council, Oficina Regional America Latina y el Caribe, Uruguay; Melina Carla D'Auria, Misión Rescate Argentina, Argentina; Axel Dourojeanni, Environment and Development Division, Economic Commission for Latin America and the Caribbean, Chile; João Batista Drummond Câmara, Instituto Brasileiro do Meio Ambiente e dos Recursos Naturais Renováveis, Brazil; Juan Escudero Ortúzar, Comisión Nacional del Medio Ambiente, Chile; Exequiel Ezcurra, Centro de Ecologia, UNAM, Mexico; Zilda Faria, Instituto Brasileiro do Meio Ambiente e dos Recursos Naturais Renováveis, Brazil; José Francisco C. Fracchia, Departamento Ordenamiento Territorial, Subsecretaría de Estado de Recursos Naturales y Medio Ambiente, Ministério de Agricultura y Ganadería, Paraguay; Patricia Frenz Yonechi, Comisión Nacional del Medio Ambiente, Chile; Maria Soledad Frías Tapia, Misión Rescate Chile, Chile; Maria Teresa García Aguilar, Secretaria de Medio Ambiente, Recursos Naturales y Pesca (SEMARNAP), Mexico; Guillermo M. García Cornejo, Comisión Nacional del Medio Ambiente, Chile; Randall García Víquez, Ministério de Ambiente y Energía, Costa Rica; José L. Gómez Reintsch, Ministério de Desarrollo Sostenible y Planificación, Bolivia; Adriana Gonçalves Moreira, Instituto Brasileiro do Meio Ambiente e dos Recursos Naturais Renováveis, Brazil; Edgar E. Gutiérrez-Espeleta, Observatorio del Desarrollo, Universidad de Costa Rica, Costa Rica; Ryan Hanson, Observatorio del

Desarrollo, Universidad de Costa Rica, Costa Rica; Vladimir R. Hermosilla, Departamento Desarrollo Rural, Facultad de Ciencias Agrarias y Forestales, Universidad de Chile, Chile; Fabian M. Jaksic, Departamento de Ecología, Catholic University of Chile, Chile; Tom Jolly, Misión Rescate Planeta Tierra Perú, Perú; Maximo T. Kalaw, Jr., Earth Council, Costa Rica; Ivonne Emma Lacombe Lemaitre, Misión Rescate Chile, Chile; Stefan Larenas Riobo, Consumers International, Chile; José Leal, Unidad Economia Ambiental, Comisión Nacional del Medio Ambiente, Chile; Pablo Leyva, Instituto de Hidrología, Meteorología y Estudios Ambientales, Colombia; Ernesto López-Zepeda, Ministério de Medio Ambiente y Recursos Naturales, El Salvador; Manuel Magalhães, Instituto Brasileiro do Meio Ambiente e dos Recursos Naturais Renováveis, Brazil; Pedro Maldonado Grunwald, Programa de Investigaciones en Energía, Universidad de Chile, Chile; Marina Mansilla Hermann, Misión Rescate Argentina, Misión Rescate Planeta Tierra, Argentina; Victor H. Marín, Universidad de Chile, Chile; Marilia Marreco Cerqueira, Ministério do Meio Ambiente, dos Recursos Hídricos e da Amazônia Legal, Brazil; Cristina Martín, Acción y Desarrollo Ecológico, México; Claudia Martínez, Corporación Andina de Fomento, Venezuela; Rodrigo L. Mellado Espinoza, Comité Nacional Pro-Defensa de la Fauna y Flora, Chile; Claudia Maria Mello Rosa, Ministério do Meio Ambiente, dos Recursos Hídricos e da Amazônia Legal, Brazil; Francisca Menezes, International Affairs Advisory of the Ministry of Environment, Brazil; Roberto Messias Franco, Instituto Brasileiro do Meio Ambiente e dos Recursos Naturais Renováveis, Brazil; José Domingos González Míguez, Brazilian Ministry of Science and Technology, Brazil; Jeffrey Orozco, Observatorio del Desarrollo, Universidad de Costa Rica, Costa Rica; Vicente Paeile Maranbio, Departamento Recursos Naturales, Comisión Nacional del Medio Ambiente, Chile; Carlos Piña Riquelme, Comisión Nacional del Medio Ambiente, Chile; Maritza Reechinti, Ministério del Ambiente y de los Recursos Naturales Renovables, Venezuela; Iglando Rey, Ministry of Science, Technology and Environment, Cuba; Carla Roberto, Ministry of Housing, Land-Use Management and Environment, Uruguay; Marisabel Romaggi Chiesa, Programa de Desarrollo Sustentable, Centro de Análisis de Políticas Públicas, Universidad de Chile, Chile; Hugo Romero, Escuela de Geografía, Universidad de Chile, Chile; Marisa Rotenberg, Instituto Brasileiro do Meio Ambiente e dos Recursos Naturais Renováveis, Brazil; Roxana Salazar, Fundación AMBIO, Costa Rica; Hugo Henry Saldivar Canales, Comisión Nacional del Medio Ambiente, Chile; Aida C. Sánchez González, Ministério de Ciencia, Tecnología y Medio Ambiente, Cuba; Mariano Castro Sánchez-Moreno, Gestión Transectorial y Territorial, Consejo Nacional del Ambiente, Perú; Paul Sánchez-Navarro Russell, Pronatura Asociación Civil, México; Hernán Sandoval Orellana, Corporación Chile Ambiente, Chile; Fernando Santibáñez, Facultad de Ciencias Agrarias, Universidad de Chile, Chile; Carmen E. Scholtfeldt Leighton, Instituto de Estudios Urbanos, Pontificia Universidad Católica de Chile, Chile; Javier A. Simonetti, Departamento de Ciencias Ecológicas, Facultad de Ciencias, Universidad de Chile, Chile; Carmiña Soto, Subsecretaría de Estado de Recursos Naturales y Medio Ambiente, Ministerio de Agricultura y Ganadería, Paraguay; Sinfronio Sousa Silva, Instituto Brasileiro do Meio Ambiente e dos Recursos Naturais Renováveis, Brazil; Nella Stewart, University of West Indies, Jamaica; Osvaldo Sunkel, Programa de Desarrollo Sustentable, Centro de Análisis de Politicas Públicas, Universidad de Chile, Chile; Vanessa Tavares, Instituto Brasileiro do Meio Ambiente e dos Recursos Naturais Renováveis, Brazil; Fabián Valdivieso Eguiguren, Comisión Permanente del Pacífico Sur, Ecuador; Jorge Valenzuela, Chilean Ministry of Foreign Affairs, Chile; Raúl Antonio Velásquez Ramos, Comisión Nacional del Medio Ambiente, Guatemala.

North America

Mohammad A. Ansari, The American Institute for Pollution Prevention, United States; Richard Ballhorn, Environment Division, Department of Foreign Affairs and International Trade, Canada; Marcus Ballinger, International Affairs Branch, Environment Canada, Canada; Steve Barg, International Institute for Sustainable Development, Canada; Jane Barr, Commission for Environmental Cooperation of the North American Agreement on Environmental Cooperation, Canada; Diane D. Beal, Office of Prevention, Pesticides and Toxic Substances, Environmental Protection Agency, United States; Pierre Belan, International Joint Commission, Secretariat of the Great Lakes Water Quality Board, Canada; Leonard Berry, Florida Center for Environmental Studies, Florida Atlantic University, United States; J. Michael Bewers, Bedford Institute of Oceanography, Canada; Steve Blight, Environmental Conservation Service, Environment Canada, Canada; Greg Block, Commission for Environmental Cooperation of the North American Agreement on Environmental Cooperation, Quebec, Canada; Philip Bogdonoff, Millennium Institute, United States; Thomas J. Brennan, Office of Environmental Policy, Bureau of Oceans and International Environmental and Scientific Affairs, United States Department of State, United States; Alan Brewster,* World Resources Institute, United States; Don Brown, US Environmental Protection Agency, USA; James P. Bruce, International Institute for Sustainable Development, Canada; Tom Brydges,* Environment Canada, Canada; Brigitte Bryld, Division for Sustainable Development, Department for Policy Coordination and Sustainable Development, United States; Jennifer Castleden, International Institute for Sustainable Development, Canada; Ann Dale, Sustainable Development Research Institute, University of British Columbia, Canada; Anne Dufresne, Canadian Wildlife Services, Environment Canada, Canada; Jill Engel-Cox, United States Environmental Protection Agency, United States; Robert M. Engler, United States Army Engineer Waterways Experiment Station, United States; David Fisher, Geological Survey of Canada, Canada; Mark Fisher, Environmental Conservation Service, Environment Canada, Canada; Richard A. Fleming, Canadian Forest Service, Canada; Liseanne Forand, Fisheries and Oceans, Canada; Amy Fraenkel, Office of International Activities, United States Environmental Protection Agency, United States; Bill Freedman, Dalhousie University, Canada; William S. Fyfe, Department of Earth Sciences, The University of Western Ontario, Canada; Kim Girtel, Trade and Environment, Climate Change and Energy Division, Department of Foreign Affairs and International Trade, Canada; Michael Glantz, National Center for Atmospheric Research, University Corporation for Atmospheric Research, United States; Bill Glanville, International Institute for Sustainable Development, Canada; Jerome C. Glenn, American Council for the United Nations University, United States; Glynn Gomes, University of Toronto, Canada; William Gregg, National Center Biological Resources Division, United States Geological Survey, United States; Paul Griss, New Directions Group, Canada; Venna Halliwell, International Affairs, Policy and Communications, Environment Canada, Canada; Andrew Hamilton, Science Division, Commission for Environmental Cooperation of the North American Agreement on Environmental Cooperation, Canada; Allen Hammond, World Resources Institute, United States; Kevin S. Hanna, Department of Geography, University of Toronto, Canada; Peter Hardi, International Institute for Sustainable Development, Canada; Tom Harner, University of Toronto, Canada; David Henry, Global Resource Information Database, Ottawa, Canada; R. Anthony Hodge, Canada; Christine Hogan, International Affairs Branch, Environment Canada, Canada; Gary Ironside, Indicators and Assessment Office, Environment Canada, Canada; Yvan Jobin, Environmental Relations Division, Department of Foreign Affairs and International Trade, Canada; Eileen Johnson, Toxics Pollution Prevention, Environment Canada, Canada; James Martin Jones, World Wide Fund for Nature, United States; Janet Jones, Commissioner for the Environment and Sustainable Development, Office of the Auditor General, Canada; Sarah Kalhok, Environmental Adaptation Research Group, Institute for Environmental Studies, University of Toronto, Canada; Michael Keating, Canada; Anne, Kerr, Indicators and Assessment Office, Environment Canada, Canada; Jeremy Kerr, York University, Canada; R. Koop, International Joint Commission, Secretariat of the Great Lakes, Canada; Thomas L. Laughlin, Office of International Affairs, National Oceanic and Atmospheric Administration, United States Department of State, United States; Kristin Lauhn-Jensen, Canadian Environmental Defence Fund, Canada; Luis Leigh, Environment Canada; Canada; Michael C. MacCracken, United States Global Change Research Program, United States; Thomas F. Malone, Sigma Xi, The Scientific Research Society, United States; Pamela Matson, Department of Geological and Environmental Sciences, Stanford University, United States; Gordon McBean, Atmospheric Environment Service, Canada; Jim McCallum, National Marine Fisheries Service, United States; Richard A. Meganck, Unit for Sustainable Development and Environment, Organization of American States, United States; Michael Metelits, Office of Environmental Policy, Bureau of Oceans and International Environmental and Scientific Affairs, United States Department of State, United States; Doug Miller, Environics International Ltd, Canada; Elizabeth Mundell, Strategic Priorities, Fisheries and Oceans Canada, Canada; Ted Munn, Institute of Environmental Studies, University of Toronto, Canada; Janine Murray, Northern Science and Contaminants Research, Indian and Northern Affairs, Canada; Vicki Norberg-Bohm, Massachusetts Institute of Technology, United States; John C. O'Connor, OconECO, United States; Darlene Pearson, Office of the Sustainable Development Strategy, Canada; Polly A. Penhale, United States National Science Foundation, United States; László Pinter, International Institute for Sustainable Development, Canada; Jonathan Plaut, Joint Public Advisory Board, North American Free Trade

Agreements, United States; Paul Raskin, Stockholm Environment Institute, Boston, United States; Kal Raustiala, Harvard Law School, United States; Steve Rayner, Pacific Northwest National Laboratory, United States; John Reid, Northern Division, Environmental Conservation Service, Environment Canada, Canada; Paul G. Risser, Oregon State University, United States; Kirk P. Rodgers,* Organization of American States, United States; Jon Rogers, Policy Development and Analysis, Fisheries and Oceans Canada, Canada; Jody Rosen-Berger, Accelerated Reduction Elimination of Toxics Secretariat, Environment Canada, Canada; Marlene Roy, International Institute for Sustainable Development, Canada; Marc Safley, Ecological Sciences Division, Natural Resources Conservation Service, United States; Renée Sauve, Environmental Relations Division, International Environmental Affairs Bureau, Department of Foreign Affairs and International Trade, Canada; Jacob Scherr, International Programs, Natural Resources Defence Council, United States; Nola-Kate Seymoar, International Institute for Sustainable Development, Canada; Dana Silk,* Canadian Environmental Network, Canada; Slobodan P. Simonovic, Department of Civil and Geological Engineering, Natural Resources Institute, University of Manitoba, Canada; Carol Smith-Wright, Environmental Relations Division, International Environmental Affairs Bureau, Department of Foreign Affairs and International Trade, Canada; William T. Sommers, Vegetation Management and Protection Research, United States Forest Service, United States; Janet Stephenson, Natural Resources Canada, Canada; John W.B. Stewart, University of Saskatchewan, Canada; David Stone, Northern Science and Contaminants Research, Indian and Northern Affairs, Canada; Larry Tieszen, International Program, United States Geological Survey, United States; Peter Timmerman, Institute of Environmental Studies, University of Toronto, Canada; David Van der Zwaag, Dalhousie University, Canada; Michael Vechsel, International Joint Commission, Secretariat of the Great Lakes Water Quality Board, Canada; David G. Victor, Council on Foreign Relations, United States; Konrad von Moltke, Environmental Studies Program, Dartmouth College, United States; Diana Freckman Wall, Natural Resource Ecology Laboratory, Colorado State University, United States; John Waugh, IUCN – The World Conservation Union, United States; Gilbert F. White, Institute of Behavioral Science, Natural Hazards Research Center, University of Colorado, United States; Robin White, World Resources Institute, United States; Rodney R. White, Institute for Environmental Studies, University of Toronto, United States; Anne Whyte, Mestor Associates, Canada; G. R. Williams, University of Toronto, Canada; Larry Williams,* Sierra Club, United States; G. M. Woodwell, The Woods Hole Research Center, United States.

West Asia

Asmaa Ali Aba Hussein, School of Graduate Studies, Arabian Gulf University, Bahrain; Jameel Abbas, University of Bahrain, Bahrain; Jilani Abd Al-Jawad, Soil Science Division, Arab Centre for the Studies of Arid Zones and Drylands, Syrian Arab Republic; Anwar Sheikheldeen Abdu, School of Graduate Studies, Arabian Gulf University, Bahrain; Youssef Abdullatif, Marine Research Institute, Tishreen University, Syrian Arab Republic; Mohammad S. Abido, Arab Centre for the Studies of Arid Zones and Drylands, Syrian Arab Republic; Zieyad Hamzah Abu Ghararah, Meteorology and Environment Protection Administration, Saudi Arabia; Mohammed Akbar, Environment Department, Ministry of Municipal Affairs and Agriculture, Qatar; Nazar Al Baharna, University of Bahrain, Bahrain; Saleh Al Share, General Coporation for the Environment Protection, Jordan; Dhari Al-Ajmi, Kuwait Institute for Scientific Research, Kuwait; Khalid Ghanim Al-Ali, Environment Department, Ministry of Municipal Affairs and Agriculture, Qatar; Salem Al-Dhaheri, Federal Environmental Agency, United Arab Emirates; Saif M. Al-Ghais, Environmental Research and Wildlife Development Agency, United Arab Emirates; Hussein Alwai Al-Gunied, Environment Protection Council, Yemen; Wafa Al-Khamees, Environmental Planning and Impact Assessment, Environment Public Authority, Kuwait; Hisham Al-Khatib, Jordan; Zahwa Al-Kuwari, Environmental Affairs, Ministry of Housing, Municipalities and Environment, Bahrain; Muhammad Hassan Al-Malack, Center for Environment and Water Research Institute, King Fahd University of Petroleum and Minerals, Saudi Arabia; Rabih Al-Merestani, Arab Centre for the Studies of Arid Zones and Drylands, Syrian Arab Republic; Mohammed A. Al-Muharrami, Environmental Research and Studies, Ministry of Regional Municipalities and Environment, Oman; Saad Al-Namairy, Federal Environmental Agency, United Arab Emirates; Abdulhadi S. Al-Otaibi, Kuwait Institute for Scientific Research, Kuwait; Baker H. Al-Qudah, Ministry of Agriculture, Jordan; Saud Al-Rasheed, Air Pollution Department, Environment Public Authority, Kuwait; Mohammad A. Al-Sarawi, Environment Public Authority, Kuwait; Abdul Rahman S. Al-Sharhan, College of Science, United Arab Emirates University, Al-Ain, United Arab Emirates; Mouaffak Al-Sheikh, Arab Centre for the Studies of Arid Zones and Drylands, Syrian Arab Republic; Mahmood Mohammed Al-Zakwani, Ministry of Regional Municipalities and Environment, Oman; Mohamed Nabil Alaa El-Din, School of Graduate Studies, Arabian Gulf University, Bahrain; Mohamed Suleiman Alabri, Division of Environmental Studies, Ministry of Regional Municipalities and Environment, Oman; Ibrahim A. Alam, Saudi Environmental Society, Saudi Arabia; Mahmoud Kamel Ali, Agriculture College, Tishreen University, Syrian Arab Republic; Khawla M.A. Alobeidan, International Affairs Section, Environment Public Authority, Kuwait; Adel R. Awad, Department of Environmental Engineering, Tishreen University, Syrian Arab Republic; Dored Awad, Arab Centre for the Studies of Arid Zones and Drylands, Syrian Arab Republic; Yahia Awaidah, Ministry of State for Environment, Syrian Arab Republic; Ali Awadh Banoubi, Federal Environmental Agency, United Arab Emirates; Murad Jabay Bino, Inter-Islamic Network on Water Resources Development and Management, Jordan; Abdulwahab Dakkak, Natural Resources, Meteorology and Environment Protection Administration, Saudi Arabia; Eddy De Pauw, International Centre for Agricultural Research in Dry Areas, Syrian Arab Republic; Ismail El-Bagouri,* School of Graduate Studies, Arabian Gulf University, Bahrain; Khalid Fakhro, Environmental Affairs, Ministry of Housing, Municipalities and Environment, Bahrain; Abousamra Fouad, Syrian Arab Republic; Moustafa M. Fouda, Department of Fisheries, Science and Technology, College of Agriculture, Sultan Qaboos University, Oman; Adnan Ghata, Al-Baath University, Syrian Arab Republic; Adel Gouda, Plant Studies Division, Arab Centre for the Studies of Arid Zones and Drylands, Syrian Arab Republic; Hassan Habib, Arab Centre for the Studies of Arid Zones and Drylands, Syrian Arab Republic; Ibrahim Nabil Hassan, Arab Centre for the Studies of Arid Zones and Drylands, Syrian Arab Republic; Mohamed Ali Hassan, Environmental Affairs, Ministry of Housing, Municipalities and Environment, Bahrain; Youssef Johar, Al-Baath University, Syrian Arab Republic; Zuheir Joue'jati, State Planning Commission, Syrian Arab Republic; Abdelmajid Khabour, Land Protection Division, General Corporation for Environmental Protection, Jordan; Ahmed Khattab, Water Protection and Marine Environment Division, General Corporation for the Environment Protection, Jordan; H.H. Kouyoumjian, The Lebanese National Council for Scientific Research, Lebanon; Abdel Raheem Loulou, Arab Centre for the Studies of Arid Zones and Drylands, Syrian Arab Republic; Ibrahim Jassim Louri, School of Graduate Studies, Arabian Gulf University, Bahrain; Sawsan Mahdi, Ministry of Environment, Lebanon; Tania Mansour, Ministry of Public Works Management, Lebanon; Saeed A. Mohammed, School of Graduate Studies, Arabian Gulf University, Bahrain; Rofail Nabil, Arab Centre for the Studies of Arid Zones and Drylands, Syrian Arab Republic; Ibrahim Nahal, Faculty of Agriculture, Aleppo University, Syrian Arab Republic; Seif Noureddin, Marine Research Institute, Tishreen University, Syrian Arab Republic; Ahmed Obeidat, Jordan Environment Society, Jordan; Khidhir Elias Putres, Directorate of Environmental Protection and Improvement, Ministry of Health, Iraq; Fadi Riachi, Foundation for Human Environment, Lebanon; Najib Saab, Lebanon; Ryad Saad El-Deen, Arab Centre for the Studies of Arid Zones and Drylands, Syrian Arab Republic; Muhammad Sadiq, King Fahd University of Petroleum and Minerals, Saudi Arabia; Ibrahim Saker, Arab Centre for the Studies of Arid Zones and Drylands, Syrian Arab Republic; Solieman Salhab, Arab Centre for the Studies of Arid Zones and Drylands, Syrian Arab Republic; Hassan Seoud, Arab Centre for the Studies of Arid Zones and Drylands, Syrian Arab Republic; Hussein Shafa'amri, Ministry of Planning, Jordan; Hussein Shahin, Air Protection Division, General Corporation for Environmental Pollution, Jordan; Muhammad R. Shatanawi, University of Jordan, Jordan; Raja Shafeck Shoughari, Directorate of International Relations, Environment Public Authority, Kuwait; Nizar Ibrahim Tawfiq, Meteorology and Environmental Protection Administration, Saudi Arabia; Saeed Wahba, School of Graduate Studies, Arabian Gulf University, Bahrain; Waleed Zubari, School of Graduate Studies, Arabian Gulf University, Bahrain.

United Nations Environment Programme

Mahmood Yousef Abdulraheem; Yinka Adebayo; Johannes Akiwumi; Jacqueline Aloisi de Larderel; Adnan Z. Amin; Salvatore Arrico, (Secretariat for the Convention on Biological Diversity); Gertrud Attar; Ali Ayoub;* Alicia Barcena;* Berna Bayinder;* Maria Angélica Beas Millas, TIERRAMERICA, c/o

UNEP-LAC; Françoise Belmont; Hassane Bendahmane; Nancy Bennett; Mark Berman; Eric Blencowe (Secretariat for the Convention on Migratory Species); Cristina Boelcke; Tore Brevik; Amedeo Buonajuti; Ulf Carlsson; Marion Cheatle; Dan Claasen; Uttam G. Dabholkar; Arthur Lyon Dahl; Maria de Amorim;* Matilde Díaz-Almazán; Salif Diop; Ahmed Djoghlaf; Garth Edward;* late K. Anthony Edwards; Sheila Edwards; Omar E. El-Arini (Secretariat of the Multilateral Fund for the Implementation of the Montreal Protocol); Habib El-Habr; Norberto Fernández; Joanne Fox-Przeworski;* Gabriel Gabrielides (Coordinating Unit for the Mediterranean Action Plan); Eduardo Ganem (Secretariat of the Multilateral Fund for the Implementation of the Montreal Protocol); Makram Gerges;* Leonel González; Hiremagular N. B. Gopalan; Tony Gross (Secretariat of the Convention on Biological Diversity); Barry Henricksen;* Ivonne Higuero; Taka Hiraishi;* Arab Hoballah (Coordinating Unit for the Mediterranean Action Plan); Jorge E. Illueca; Manjit Iqbal; Sipi Jaakkola; Sam Johnston (Secretariat of the Convention on Biological Diversity); Shafqat Kakakhel; Kagumaho Kakuyo; James Kamara; Fouad Kanbour;* Donald Kaniaru; Bakary Kante; Lal Kurukulasuriya; Christian Lambrechts; Isabel Martínez Vilardell; Timo Maukonen; Terttu Melvasalo;* Laura Meszáros; Salem Milad; Danielle Mitchell; Elizabeth Mrema; Arnulf W. Müller-Helmbrecht (Secretariat of the Convention on Migratory Species); Agneta Nilsson; D. Bondi Ogolla; Andréane Perrier de la Bathie; Pierre Portas (Secretariat of the Basel Convention on the Control of the Transboundary Movements of Hazardous Wastes and Their Disposal); Naomi Poulton; Iwona Rummel-Bulska (Secretariat of the Basel Convention on the Control of the Transboundary Movements of Hazardous Wastes and Their Disposal); Arsenio Rodríguez;* Nelson Sabogal (Secretariat of the Vienna Convention and the Montreal Protocol); Vijay Samnotra; Madhava K. Sarma (Secretariat of the Vienna Convention and the Montreal Protocol); Frits Schlingemann; Gerhart Schneider;* Miriam Schomaker;* Megumi Seki; Ravi Sharma; Surendra Shrestha; Ashbindu Singh; Jim Sniffen; Cheikh Omar Sow; Anna Stabrawa; Janet Stevens; Bai Mass-Max Taal; Alexander Timoshenko; Klaus Töpfer; Izgrev Topkov* (Secretariat of the Convention on International Trade of Endangered Species); Peter E.O. Usher;* Suvit Yodmani;* Marceil Yeater; Veerle Vandeweerd;* James B. Willis; Peigi Wilson; Ronald G. Witt; Kaveh Zahedi.

Other United Nations bodies

Iyad Abumoghli, United Nations Development Programme; K. Acheampong, Institute for Natural Resources in Africa, United Nations University; Khaled Alloush, United Nations Development Programme; Juan Antonio Escudero, Division for Ocean Affairs and the Law of the Sea, Office of Legal Affairs, United Nations; J. Baidu-Forson, Institute for Natural Resources in Africa, United Nations University; Hussam Bechnak, United Nations Development Programme; Burton Bennett, United Nations Scientific Committee on the Effects of Atomic Radiation; Patricio A. Bernal, Intergovernmental Oceanographic Commission, United Nations Educational, Scientific and Cultural Organization; Patricia Bliss-Guest, Global Environment Facility; Jean-Yves Bouchardy, United Nations High Commission for Refugees; James B.L. Breslin, World Meteorological Organization; William Chambers, Institute of Advanced Studies, United Nations University, Japan; H.S. Cherif, International Atomic Energy Agency; Eleanor Cody, United Nations Centre for Human Settlements; Carlos Corvalan, World Health Organization; Grégoire de Kalbermatten, Secretariat of the United Nations Convention to Combat Desertification; Annick de Marffy, Division of Ocean Affairs and the Law of the Sea, Office of Legal Affairs, United Nations; Liliana de Pauli,United Nations Development Programme; Julian Dumanski, Agricultural and Forestry Systems, The World Bank; Amin El-Sharkawi, Cairo Office, United Nations Development Programme; Ute Enderlein, World Health Organization; Christopher English, Conference Services, United Nations Office in Nairobi; Lowell Flanders, United Nations Department of Economic and Social Affairs; Mohamad Gabr, Agriculture Section, United Nations Economic and Social Commission for West Asia; Peter Gilruth, United Nations Development Programme Office to Combat Desertification and Drought; Gisbert Glaser, United Nations Educational, Scientific and Cultural Organization; Nicolo E. Gligo Viel, Comisión Económica de las Naciones Unidas para América Latina y el Caribe; Stephen Gold, Training Programme for the Climate Change Convention, United Nations Institute for Training and Research; Robert Goodland, World Bank; N. Ishwaran, World Heritage Centre, United Nations Educational, Scientific and Cultural Organization; Terry Jeggle, Department of Humanitarian Affairs, United Nations Secretariat for the International Decade for Natural Disaster Reduction; Muhammad Khan, Centre for Environmental Health Activities, World Health Organization; Richard Kinley, Secretariat of the United Nations Framework Convention on Climate Change; Mikhael G. Kokine, Environment and Human Settlements Division, United Nations Economic Commision for Europe; Sarim Kol, Economic Commission for Africa of the United Nations; Michel Laverdière, Food and Agriculture Organization; Terence Lee, Environment and Development Division, Economic Commission for Latin America and the Caribbean; Lennart Ljungman, Food and Agriculture Organization; Fu-chen Lo, Institute of Advanced Studies, United Nations University, Japan; L. Ludvigsen, United Nations Centre for Human Settlements, Europe; George Martine, Country Support Team, Office for Latin America and the Caribbean, United Nations Fund for Population Activities; Joseph Maseland, United Nations Centre for Human Settlements; J.Z.Z. Matowanyika, Food and Agriculture Organization; Iouri Moiseev, United Nations Centre for Human Settlements; Jay Moor, United Nations Centre for Human Settlements; Christopher Nuttall, Training Programme in Integrated Environmental Information Systems, United Nations Institute for Training and Research; Merle Opelz, International Atomic Energy Agency; Elina Palm, Department of Humanitarian Affairs, Secretariat for the United Nations International Decade for Natural Disaster Reduction; János Pásztor, Secretariat of the United Nations Framework Convention on Climate Change; Bill Phillips, Bureau of the Convention on Wetlands; Vivien Ponniah, Technical and Policy Division, United Nations Fund for Population Activities; Michael Ramos, Secretariat of the United Nations Convention to Combat Desertification; Thomas Reich Ball, United Nations Development Programme; Samir Riad, United Nations Educational, Scientific and Cultural Organization, Regional Office for Science and Technology for Africa; Vladimir Sakharov, Joint United Nations Environment Programme and Office for the Coordination of Humanitarian Affairs; Abdin M.A. Salih, Cairo Office, United Nations Educational, Scientific and Cultural Organization; Colin Summerhayes, Global Ocean Observing System Project Office, Intergovernmental Oceanographic Commission, United Nations Educational, Scientific and Cultural Organization; Jacob Swager, Secretariat of the United Nations Framework Convention on Climate Change; Peter Swan,* United Nations Centre for Human Settlements; Hiko Tamashiro, World Health Organization; Ludivine Tamiotti, Environment Unit, United Nations High Commission for Refugees; Ricardo Tarifa, World Bank; Archalus Tchekanvorian-Asenbauer, United Nations Development Programme; Kyran Thelen, Oficina Regional de la FAO, América Latina y el Caribe; Jeff Tschirley, Food and Agriculture Organization; Alvaro Ugalde, United Nations Development Programme, Costa Rica; Jerry Velasquez, The United Nations University; Galileo Violini, Regional Office for Latin America and the Caribbean, United Nations Educational, Scientific and Cultural Organization; Y. von Schirnding, World Health Organization; Wolfgang Wagner, United Nations Office for the Coordination of Humanitarian Affairs; Fareed Yasseen, Secretariat of the United Nations Framework Convention on Climate Change; Ryutaro Yatsu, Institute of Advanced Studies, United Nations University, Japan.

Note: * since moved or retired

Index

Page references in *italics* refer to tables and other displays.